North-Central Section
of the
Geological Society of America

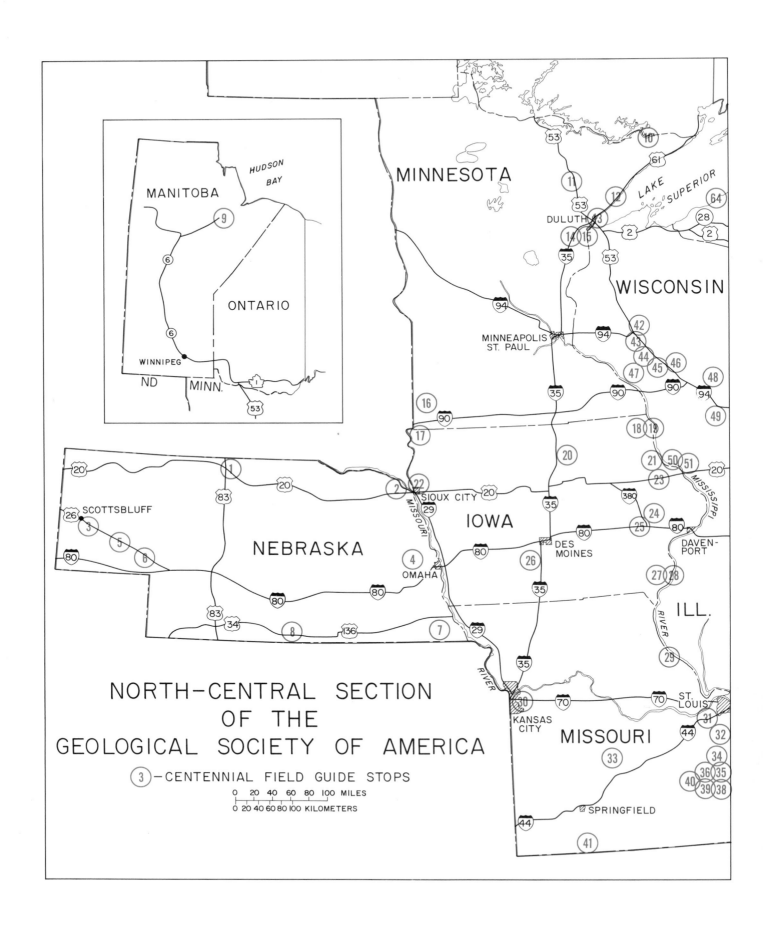

NORTH–CENTRAL SECTION
OF THE
GEOLOGICAL SOCIETY OF AMERICA

③ –CENTENNIAL FIELD GUIDE STOPS

0 20 40 60 80 100 MILES
0 20 40 60 80 100 KILOMETERS

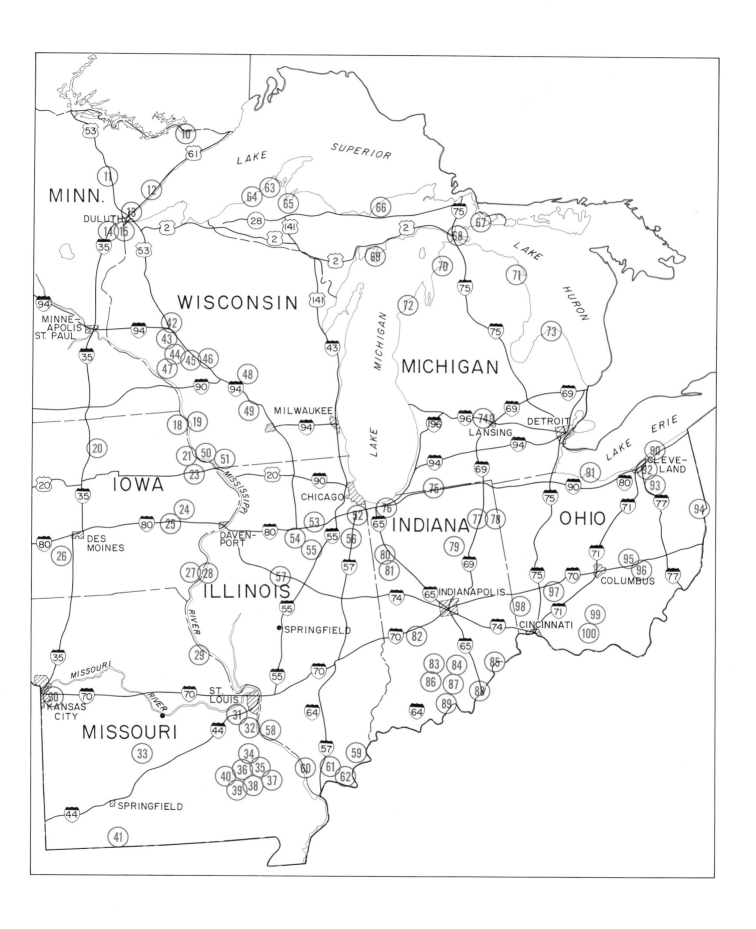

Centennial Field Guide Volume 3

North-Central Section
of the
Geological Society of America

Edited by

Donald L. Biggs
531 Hayward Avenue
Ames, Iowa 50010

1987

Acknowledgment

Publication of this volume, one of the Centennial Field Guide Volumes of *The Decade of North American Geology Project* series, has been made possible by members and friends of the Geological Society of America, corporations, and government agencies through contributions to the Decade of North American Geology fund of the Geological Society of America Foundation.

Following is a list of individuals, corporations, and government agencies giving and/or pledging more than $50,000 in support of the DNAG Project:

ARCO Exploration Company
Chevron Corporation
Conoco, Inc.
Diamond Shamrock Exploration
  Corporation
Exxon Production Research Company
Getty Oil Company
Gulf Oil Exploration and Production
  Company
Paul V. Hoovler
Kennecott Minerals Company
Kerr McGee Corporation
Marathon Oil Company
McMoRan Oil and Gas Company
Mobil Oil Corporation
Pennzoil Exploration and Production
  Company

Phillips Petroleum Company
Shell Oil Company
Caswell Silver
Sohio Petroleum Corporation
Standard Oil Company of Indiana
Sun Exploration and Production Company
Superior Oil Company
Tenneco Oil Company
Texaco, Inc.
Union Oil Company of California
Union Pacific Corporation and
  its operating companies:
    Union Pacific Resources Company
    Union Pacific Railroad Company
    Upland Industries Corporation
U.S. Department of Energy

Published by the Geological Society of America, Inc.
3300 Penrose Place, P.O. Box 9140, Boulder, Colorado 80301

Printed in U.S.A.

Front Cover: Mississippi River bluffs, Clayton County, Iowa, showing limestone of the Middle Ordovician Galena Group. Photo by Gary Hightshoe, Department of Landscape Architecture, Iowa State University, Ames, Iowa.

Library of Congress Cataloging-in-Publication Data
(Revised for volume 3)

Centennial field guide.

"Prepared under the auspices of the regional Sections of the Geological Society of America as part of the Decade of North American Geology (DNAG) Project"—V. 6, pref.
  Vol.   : maps on lining papers.
  Includes bibliographies and index.
  Contents: —v. 3. North-central Section of the Geological Society of America / edited by Donald L. Biggs—   —v. 6. Southeastern Section of the Geological Society of America / edited by Thornton L. Neathery.

  1. Geology—United States—Guide-books. 2. Geology—Canada—Guide-books. 3. United States—Description and travel—1981-   —Guide-books. 4. Canada—Description and travel—1981-   —Guide-books. 5. Decade of North American Geology Project. I. Geological Society of America.
QE77.C46   557.3 s [557.3]   86-11986
ISBN 0-8137-5406-2 (v. 6)

Contents

GEOLOGICAL/GEOGRAPHIC CROSS-REFERENCE CHART

		Nebraska	Manitoba	Minnesota	Iowa	Missouri	Wisconsin	Illinois	Michigan	Indiana	Ohio
Geomorphology		1, 4, 5, 8	9			40, 34		54, 61, 62	72	76, 85, 88	91
	Karst					33				87	
Stratigraphy	Cenozoic	1, 2, 3, 5, 6	9					53, 57	66	77/78, 82	92, 98, 100
	Mesozoic	8			22			59			
	Paleozoic	7			18/19, 20, 21, 23, 24, 25, 26, 27/28	29, 30, 31, 32, 34, 41	42, 43, 44, 45, 47, 50, 51	52, 54, 55, 56, 58, 59,	66, 68, 69, 70, 71, 73, 74	77/78, 82, 83, 84, 85, 86, 88, 89	90, 93, 94, 95, 96, 97, 99, 100
	Precambrian			10, 14/15 16	17	40, 39	42, 46, 48, 49		64, 65, 66		
Cryptoexplosive									63, 75		
Structure		7				30, 31		58, 59, 60	64	80	100
Economic Geology										83	
Glacial Geology			9					53, 57	72	81	98
Paleoecology									67, 71		
Shorelines									68, 73		90, 91
Volcanic Rocks				11, 12		35, 36, 40			64		
Intrusives				13		35, 36, 37					
Metamorphics				11			46		64		
Weathering						36					
Mineralization						38					
Unconformities						37	42	56	65	77/78, 86	
Reefs								52		79	

xiv

Preface

This volume is one of a six-volume set of Centennial Field Guides prepared under the auspices of the regional Sections of the Society as a part of the Decade of North American Geology (DNAG) Project. The intent of this volume is to highlight, for the geologic traveler and for students and professional geologists interested in major geologic features of regional significance, 100 of the best and most accessible geologic localities in the area of the North-Central Section. The leadership provided by the editor, Donald L. Biggs, and the support provided to him by the North-Central Section of the Geological Society of America and the Department of Earth Sciences of Iowa State University of Science and Technology are greatly appreciated.

Drafting services were offered by the DNAG Project to those authors of field guide texts who did not have access to drafting facilities. Particular thanks are given here to Ms. Karen Canfield of Louisville, Colorado, who prepared final drafted copy of many figures from copy provided by the authors.

In addition to Centennial Field Guides, the DNAG Project includes a 29 volume set of syntheses that constitute *The Geology of North America*, and 8 wall maps at a scale of 1:5,000,000 that summarize the geology, tectonics, magnetic, and gravity anomaly patterns, regional stress fields, thermal aspects, seismicity, and neotectonics of North America and its surroundings. Together, the synthesis volumes and maps are the first coordinated effort to integrate all available knowledge about the geology and geophysics of a crustal plate on a regional scale. They are supplemented, as a part of the DNAG project, by 23 Continent-Ocean Transects providing strip maps and both geologic and tectonic cross-sections strategically sited around the margins of the continent, and by several related topical volumes.

The products of the DNAG Project have been prepared as a part of the celebration of the Centennial of the Geological Society of America. They present the state of knowledge of the geology and geophysics of North America in the 1980s, and they point the way toward work to be done in the decades ahead.

Allison R. Palmer
Centennial Science Program Coordinator

Foreword

The purpose of this volume is to provide the professional geologist, curious layman or student with a pre-selected number of 100 localities offering instruction in the geology of the area of the North-Central Section of the Geological Society of America. It was originally thought to make this the 100 most significant localities in the region. More mature reflection, however, indicated that significance is a sometime thing, having a definition that undergoes marked metamorphosis from one specialist to another. In short, a marvelously instructive example of river piracy or a greenstone might be considered of slight use to a paleontologist or carbonate petrologist. As a result of this sort of thinking, a more or less conscious effort was made to give as good a distribution of subject matter as possible consistent with the exposures contained in the region and the expertise of those willing to write of them.

Despite the general lack of alpine structure and, in some areas, the difficulty of finding good exposures of a variety of rocks or a complete stratigraphic section, the North-Central Region of the North American continent exposes many evidences of a long geologic history. Some of the oldest rocks present on the continent are exposed, as are some of the best developed glacial deposits of the latest Pleistocene glaciation. That this region has not been the scene of great tectonism since the beginning of the Phanerozoic in no way militates against it as a repository of earth history. Exposures ranging in age from the Archean to the Pleistocene, with the exception of Triassic and Jurassic deposits, occur in great variety. The upper Mississippi Valley, type locality for the Mississippian System, is particularly well endowed with exposures of Mississippian carbonate rocks, and development of great Late Paleozoic deltas from streams flowing westward from the Appalachian area is recorded in much of the area.

When asked to coordinate the collection of chapters that forms this volume, I considered the task one of great difficulty. Though I had spent a professional lifetime working in midwestern geology, my active efforts had been restricted to Missouri, Illinois, and Iowa. Though I knew, from field trip experience, correspondence, journal reading, and conventioneering, many geologists who work or have worked in the North-Central Section, I was not at all certain that I knew the most important exposures in the area, or the best persons to write of them.

My first thought when faced with the difficulty of getting 100 articles that would best introduce a newly arrived professional geologist to the geology of the area was to take a look at the several geologic maps available to me and try to find the requisite number of exposures that would most profitably offer the desired instruction. Finding the first 100 exposures proved easy indeed. The greatest difficulty was that most of them were in the three states mentioned above. This represented an unacceptable degree of parochiality. Accordingly I decided that the persons best qualified to recommend both sites and authors were to be found in the universities and geological surveys of the several states. Accordingly I sought the advice of a person I considered knowledgable to both the geology and geologists in each state.

The respondents to my request were as follows: Nebraska, James Swinehart; Minnesota, Glen Morey; Iowa, Brian Glenister; Missouri, Thomas Thompson; Wisconsin, Bruce Brown; Illinois, James Baxter; Michigan, Randall Milstein; Indiana, Robert Shaver; and Ohio, Richard Heimlich. These men accepted the challenge of recommending the most instructive sites in their areas, recruiting authors to write the descriptions, collecting the manuscripts, and forwarding them to me. Without their most effective efforts, I should not be writing these words, nor should you be reading them. It is to the nine men mentioned above that much of the credit of assembling this volume belongs.

Behind the state coordinators, of course, stand the authors of the papers herein included. This project offered no cash awards per page of print; many of the people who spent their valuable time developing the chapters that appear here were busy at other research or service activities, and this effort represented simply another commitment to be sandwiched in between existing commitments for no reward other than service to the profession.

To the coordinators and authors, I offer my sincere thanks. To all those whom I have irritated by nagging about deadlines, dangling modifiers, misused verb tenses, and parallel constructions, I apologize and hope that the use of this volume will, in some degree, assuage the pain.

Thanks are also due to the Management Board of the North-Central Section of the Geological Society of America and to the Department of Earth Sciences of Iowa State University for their support in this venture. They provided that indespensible, unobstructive background against which the project was accomplished.

Donald L. Biggs
Editor
August, 1987

Late Cenozoic stratigraphy and geomorphology, Fort Niobrara, Nebraska

M. R. Voorhies, State Museum and Department of Geology, University of Nebraska-Lincoln, Lincoln, Nebraska 68588-0514

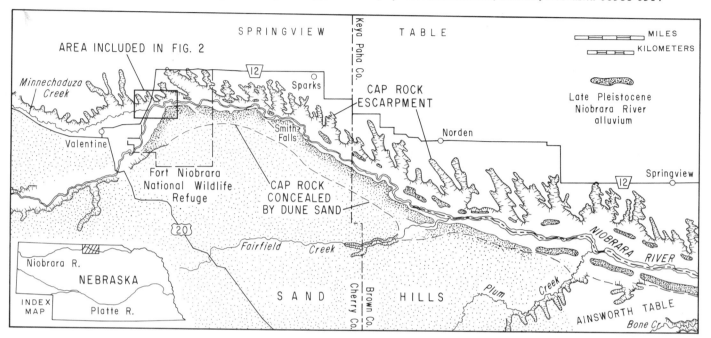

Figure 1. Map of central Niobrara River valley in north-central Nebraska showing major physiographic features and location of Fort Niobrara National Wildlife Refuge, a portion of which is shown in greater detail as Figure 2. The northern rim of the Niobrara Valley is formed by the Cap Rock Escarpment, the erosional edge of a well-indurated tuffaceous sandstone, the Cap Rock Member of the Ash Hollow Formation (Ogallala Group, Miocene). South of the river, the Cap Rock is largely mantled by Holocene sand dunes (Sand Hills) except where breached by major tributaries (Fairfield Creek, Plum Creek). The dune-free area north of the Niobrara River is known as the Springview Table which is mantled, in Keya Paha County only, by fluvial sand and gravel of the Long Pine Formation of late Pliocene (Blancan) age. The northwestern portion of a similar gravel-mantled plain, the Ainsworth Table, is shown in the southeastern corner of the map. The Niobrara River established its present course in geologically very recent time, as indicated by the presence of Rancholabrean-age vertebrate fossils in the topographically highest terrace fill flanking the river, mapped as "Late Pleistocene Niobrara River Alluvium" (heavy stipple).

LOCATION AND SIGNIFICANCE

The central Niobrara River valley in northern Nebraska (Fig. 1) has played an important role in the conceptual development of late Tertiary stratigraphy and vertebrate paleontology in the Great Plains. Much of our current understanding of the evolution of such mammals as horses, camels, and proboscideans is based on fossils collected from the Valentine and Ash Hollow Formations exposed in the deep (by Nebraska standards!) canyons of the Niobrara River and its tributaries in Brown, Cherry, and Keya Paha Counties. The Valentine–Ash Hollow sequence in this area represents the northernmost expression of a complex of alluvial deposits collectively known as the Ogallala (variously termed a group or formation) which blankets much of the Great Plains from southern South Dakota to western Texas and eastern New Mexico.

The only paved road to traverse reasonably good exposures of the classic Ogallala sequence in the Niobrara valley is Nebraska 12 about 3 mi (4.8 km) northeast of Valentine. Two adjacent road cuts at the northwestern margin of Fort Niobrara National Wildlife Refuge (sec.22,T.34N.,R.27W., Cornell Dam Quadrangle, Fig. 2) expose nearly the entire thickness of the Valentine Formation and the Cap Rock Member of the Ash Hollow Formation. Examination of the road cuts, combined with a tour of the Refuge, will provide the visiting geologist with an overview of the later Tertiary mantle of the Great Plains near the boundary between the latter physiographic province and the Central Lowlands Province to the east. Geomorphological relation-

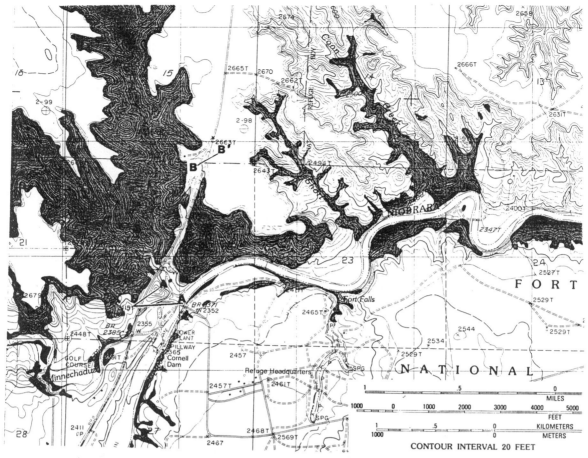

Figure 2. Map of a portion of Fort Niobrara National Wildlife Refuge, showing location of stratigraphic sections A-A' and B-B' (Fig. 3). From Cornell Dam, Nebr.-S. Dak. 7½-minute Quadrangle, Provisional Edition, 1983. (Dark tone is vegetation).

ships are also exceptionally clear along this segment of the Niobrara River; the very late Pleistocene and Holocene age of the present river valley is demonstrated by terrace fills on the Refuge.

Nebraska 12 is an all-weather road providing ready access to the designated exposures. Points of geological interest (e.g., Fort Falls) on the opposite (south) side of the Niobrara River can be reached by way of graveled roads. Field-trip groups wishing to study outcrops besides those normally open to visitors should request permission at the Refuge headquarters. (Large and potentially dangerous animals—bison, wapiti, long-horned cattle—roam free in the area.)

GENERAL INFORMATION

Three regionally flat-lying, nonmarine, stratigraphic units of Tertiary age are exposed in superposition in the road cuts. A measured section (Fig. 3) shows, in ascending order, a well-indurated, pinkish-tan, highly tuffaceous siltstone (the Rosebud Formation) unconformably overlain by a much more friable sandy unit (the Valentine Formation), which is in turn overlain by a ledge-forming tuffaceous sandstone (the Cap Rock Member of the Ash Hollow Formation). The latter forms the well-defined north rim of the Niobrara Valley on the Refuge (Fig. 4).

The Rosebud Formation is a relatively unfossiliferous unit and therefore has received little attention from paleontologists and stratigraphers. On some older reports and maps, in fact, the unit has been confused with the much older and lithologically distinct (more bentonitic) Brule Formation (White River Group), well known in the Big Badlands of South Dakota and the Little Badlands of northwestern Nebraska. (Recent research by Swinehart and others (1985) indicates that considerable thicknesses of previously unrecognized upper White River Group rocks very similar to the Rosebud Formation are present in the subsurface in western Nebraska.)

The Rosebud erodes into vertical bluffs at Fort Niobrara and intermittently for more than 50 mi (80 km) downstream along the Niobrara River. Its uniform horizontal bedding and fine grain size suggest that it was deposited by low-energy fluvial currents on a wide, flat floodplain. Desiccation cracks are com-

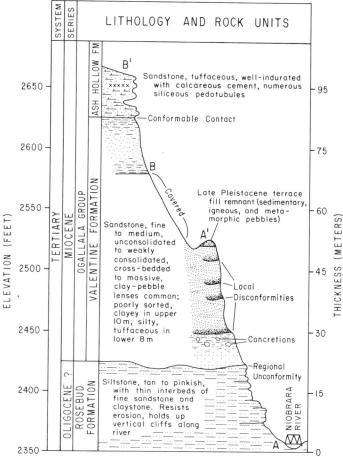

Figure 3. Stratigraphic section of Tertiary sediments exposed in road cuts on Nebraska 12, Fort Niobrara National Wildlife Refuge, Cherry County, Nebraska. See Figure 2 for location and routes of sections A-A′ and B-B′.

Figure 4. Photograph of north wall of Niobrara River valley directly north of headquarters, Fort Niobrara National Wildlife Refuge (Figure 2) with lithologic units labelled. Man pointing out Valentine/Rosebud contact is noted vertebrate paleontologist Henry Fairfield Osborn. Photo taken by E. H. Barbour in 1916.

Figure 5. Looking downstream (east) along north bank of Niobrara River from bridge, point A, Figure 2. Note spring seeps at unconformity between unconsolidated sands of Valentine Formation (above) and compact siltstone of the Rosebud Formation (below). Flat, undissected top of late Pleistocene terrace alluvium is visible on south (right) bank of Niobrara River in the background. Photo by E. H. Barbour, 1916.

mon in the formation but are obvious only on rare horizontal exposures. Sparse collections of mammalian fossils made from the Rosebud at its type locality in South Dakota and in the lower Niobrara Valley in northeastern Nebraska suggest an early Arikareean (late Oligocene) age for the unit in terms of the sequence of North American land mammal ages. The base of the unit is not exposed on the Refuge, but elsewhere it is observed to overlie the Pierre Shale, a black, marine shale of Cretaceous age. Low permeability of the Rosebud compared with that of the loose sands at the base of the overlying Valentine Formation gives rise to numerous springs and seeps at the unconformity between the two formations (Fig. 5).

The Valentine Formation, consisting predominantly of unconsolidated fine to medium sand and semiconsolidated sandstone, overlies the Rosebud. The marked disconformity between the units, clearly exposed in the roadcut (A-A′, Fig. 2), is an erosional surface of regional extent. Primary sedimentary structures indicating fluvial deposition are especially common in the lower 98 ft (30 m) of the formation. Lenses of cross-bedded sand and clay-pebble conglomerate indicate relatively high-

energy currents, while finer-grained, laminated units alternating with the coarser clastic intervals suggest slack-water deposition. An especially prominent and well-exposed, fine-grained unit, a very silty and clayey fine sandstone, occurs at the base of the Valentine Formation in the road cut. This unit is characterized by large (up to 12 in [30 cm]) pedogenic calcareous concretions, white externally but with dark-hued centers exhibiting numerous dendrites of manganese dioxide. Such fine-grained, often concretionary, lithologies are laterally discontinuous and regionally can occur at any level within the Valentine but are most common in the lower half of the formation.

Another lithology that occurs sporadically in the Valentine Formation is hard, dense, feldspathic sandstone cemented with opaline silica. Irregular masses of this material, sometimes measuring 10 or more ft (several m) in thickness and up to 0.6 mi (1 km) in lateral extent, can be found throughout the outcrop belt of the Valentine Formation in the Niobrara River valley and as far northeast as the Bijou Hills in south-central South Dakota, which are capped with this material. The origin of the opaline cement, which usually occurs in lenses of relatively coarse, well-sorted sand, is poorly understood. No silica-cemented sandstone was observed *in situ* in section A–A', but blocks and ovoid "pods" of the material are present on the slope immediately west of A' on the opposite side of the highway.

Near the top of the Valentine Formation, as exposed in section B-B', the cross-bedded, unconsolidated sands characteristic of the lower part of the formation pass upward into more massive, siltier, clay-rich sands. The weathered surface of these sands show desiccation cracks reflecting their high clay content. These upper beds, formally designated the Devils Gulch Member by Skinner and others (1968) have a greenish-yellow cast in contrast with the predominantly gray hues typical of the lower part of the formation (the Crookston Bridge Member of Skinner and others, 1968). The uppermost formal subdivision of the Valentine Formation, the Burge Member, is not present in the road cut, although it has been identified beneath the Cap Rock about 1.9 mi (3 km) farther east on the Refuge. Characteristic Burge lithologies—coarse sand and gravel, including a few igneous and metamorphic clasts—have a very localized distribution, being confined to narrow channel cuts.

The 196 ft (60 m) of Valentine sands measured in the road cut represent about an average thickness for the formation. Like most fluvial units, the Valentine locally varies considerably in thickness. The thickest sections known to the writer occur south of Norden in Brown County (Fig. 1), where more than 290 ft (90 m) of Valentine occupy at least two paleovalleys excavated deeply into the underlying Rosebud Formation. The definitely established distribution of the Valentine Formation is confined to the drainage basin of the Niobrara River in north-central and northeastern Nebraska (as far east as Cedar County) and in south-central South Dakota. Isolated outcrops east of the Missouri River in South Dakota suggest that its eastern extent was originally much greater. Basal Ogallala strata equivalent in age to the Valentine Formation occur in the Republican River basin of southern Nebraska and northern Kansas and have sometimes been included in the Valentine Formation, but this practice is unsound because the beds in question are lithologically very different from the Valentine and were almost certainly derived from different source areas. Pebble-sized clasts in the Valentine are virtually all locally derived, sedimentary lithologies, whereas igneous and metamorphic pebbles are exceedingly common in basal Ogallala channel deposits in southern and southwestern Nebraska.

Paleontological studies of the Valentine Formation began nearly 130 years ago when F. V. Hayden collected vertebrate fossils from the "sands of the Niobrara" while attached to a military expedition under the command of Lt. G. K. Warren in 1857. Hayden's fossils were described in a classic monograph by Leidy (1869), which laid the groundwork for all future studies of North American Tertiary vertebrates. The "Niobrara River" mammalian fossils were of critical importance because they pro-

vided the first evidence of connecting links between the archaic (Oligocene) mammals already known from the Dakota Badlands and the comparatively modernized Quaternary forms familiar in the eastern states. Leidy named 24 species of late Tertiary mammals in Hayden's Niobrara collection, 13 of them representing new genera or subgenera; remarkably, nearly all are still regarded as valid taxa today. Although the exact source of Hayden's collections will never be known, the types of most (20) of his "sands of the Niobrara" species have been duplicated by in situ collections from the Valentine Formation along the route followed by the Warren expedition.

The first explicitly paleontological expedition to the Niobrara was organized by Yale's O. C. Marsh in 1873. His heavily armed packtrain amassed a considerable collection near what later became Fort Niobrara, and this area became a focal point for a good deal of later collecting activity. Our own century has seen intensive prospecting and excavating of Valentine vertebrate fossils by many museums, especially the American Museum of Natural History (Frick Laboratory), University of California Museum of Paleontology, and the University of Nebraska State Museum. Several dozen major quarry sites and numerous surface finds have been discovered.

The mammalian fauna of the Valentine include an impressive variety of large herbivores and their attendant carnivores. By far the largest are gomphotheres and mastodons, extinct relatives of the elephant, both of which make their earliest North American appearance in basal Valentine deposits. As many as three species of rhinoceros, ten of horses, and five of camels have been collected from single quarry sites in the formation along with comparable numbers of other large- and medium-sized herbivores: tapirs, chalicotheres, peccaries, oreodonts, and horned ruminants. Carnivores include large and small members of both the dog and cat families, the totally extinct "bear dogs" or amphicyonids, and mustelids ranging in size from weasels to wolverines. Application of screenwashing techniques to the Valentine has resulted in a large increase in the recovery of small mammals such as rodents and insectivores, and lower vertebrates such as fishes, reptiles, and amphibians. As many as 144 species of extinct vertebrates (exclusive of birds) have been excavated from single Valentine quarries.

Fossil plant remains, although not so intensively studied as vertebrates, are also present in the Valentine Formation. Most common are the hulls, seeds, and stems of siliceous grasses, forbs, and borages. Silicified wood, occurring both as upright stumps and as water-rolled chunks, is also locally abundant. A large stump is on display at the Refuge headquarters. By far the most informative paleobotanical collection from the Valentine Formation, however, was made about 25 mi (40 km) west of Fort Niobrara. Here the Kilgore Flora includes well-preserved leaves, seeds, and pollen representing more than 50 species. The flora include subtropical trees, which imply a considerably milder, moister climate than that existing in northern Nebraska today. The vegetation of north-central Nebraska during the deposition of the Valentine Formation as reconstructed consisted of hardwood forests on the low, broad floodplains and an open, grassy woodland or savanna on the slightly higher interfluves. This reconstruction harmonizes well with the evidence provided by the vertebrates, which include numerous browsing as well as grazing species. The presence of giant land tortoises, alligators, and other cold-sensitive lower vertebrates also supports the paleobotanical evidence for an essentially frost-free climate.

The age of the Valentine Formation, based on fossil vertebrates and preliminary fission-track dating, ranges from medial Barstovian (13 to 14 Ma) at the base to early Clarendonian (approximately 11 Ma) at the top (Burge Member). These land-mammal ages span the transition from medial to late Miocene.

Overlying the relatively unconsolidated Valentine Formation is a well-lithified unit referred to as the Cap Rock Member of the Ash Hollow Formation (section B-B′, Fig. 3). In contrast with the Valentine, which contains relatively little volcaniclastic material, the Ash Hollow Formation contains abundant tuffaceous sediment both as discrete beds and as disseminated glass shards. An impure tuff bed about 2 in (5 cm) thick is present near the top of the road cut. The Cap Rock Member is also characterized, both in this exposure and throughout its known extent, by the presence of numerous small pedotubules, including both calcareous and siliceous examples.

The Cap Rock maintains a rather uniform thickness of approximately 36 to 50 ft (10 to 15 m) throughout most of its area of relatively continuous outcrop in the central Niobrara Valley. Its basal contact shows remarkably little relief and slopes eastward at approximately 8 to 10 ft/mi (1.5 to 2 m/km).

The Cap Rock frequently contains fragmentary fossil bones, but only a few quarriable concentrations of more complete remains have been discovered. Large tortoises, rhinos and horses with tall-crowned teeth, and camels dominate the vertebrate faunas recovered from the Cap Rock. Browsing mammals (those with short-crowned teeth) make up a notably lower percentage of Cap Rock assemblages, both in terms of species and of individuals, than they do in assemblages from the underlying Valentine Formation. This fact, combined with paleobotanical evidence (increased diversity and abundance of fossil grasses, virtual absence of fossil wood), suggests that grassland had spread at the expense of forests in the central plains at the time the Cap Rock deposits were accumulating. The inferred vegetational change may have been triggered by an increase in aridity. The prevalence of caliche or calcrete in the Cap Rock, giving the unit its most obvious lithological character, lends credence to this notion.

Mammals from the Cap Rock Member are regarded as medial Clarendonian in age, being slightly more advanced morphologically than their counterparts (probable ancestors in many cases) in the underlying Burge. Fission-track dates of 10.5 and 9.5 Ma have been determined for volcanic ashes bracketing the Cap Rock Member.

No post–Cap Rock Tertiary deposits have been identified on the Refuge. Farther east in the Niobrara Valley, a succession of relatively narrow channels was incised into the Cap Rock and filled with unconsolidated sand during the late Clarendonian. Still later, during the Hemphillian land mammal age, similar but deeper channels were eroded into the Valentine/Ash Hollow sequence. The youngest of the channel fills, dated at 5 Ma, is exposed east of the mouth of the Niobrara River in Knox County. During the succeeding Blancan land mammal age (4.5 to 2 Ma = Pliocene), a sheetlike accumulation of more than 100 ft (30 m) of sand and western-source gravel was deposited above the Cap Rock in what is now the eastern half of the Niobrara River valley. Large road cuts on U.S. 20 at the town of Long Pine provide an excellent cross section of these strata. The Ainsworth Table and eastern portion of the Springview Table (Fig. 1) are mantled with these deposits.

The youngest fluvial stratigraphic units on the Refuge are terrace fills paralleling the Niobrara River and obviously recording early stages in the formation of its present valley. The oldest (highest) identifiable terrace fill is exposed immediately east of Fort Falls (Figs. 2, 5). Pebble- to boulder-sized clasts, mostly of locally derived Tertiary lithologies, dominate the lenses of coarse gravel which characterize this terrace fill. The distribution of this unit is shown (as "Late Pleistocene Niobrara River alluvium") in Figure 1. At many locations, notably at Smith Falls, the high terrace fill contains vertebrate fossils of late Pleistocene (Rancholabrean) age, including bison, mammoth, horse, camel, wolf, wolverine, and a great diversity of rodents. Nearly all of the high terrace rodents belong to living species, but many of them (e.g., collared lemmings, yellow-cheeked voles, heather voles, red squirrels) no longer inhabit the Great Plains but are confined to boreal life zones. For this reason, the high terrace vertebrate assemblage is assigned to a full-glacial, probably Wisconsinan, interval.

Lower terraces, such as that upon which the Refuge headquarters building rests, are still younger and have yielded remains of exclusively Holocene taxa, such as the modern species of bison, and are clearly less than 10,000 years old. Still younger are the dunes of aeolian sand which overlie the entire terrace sequence south of the river.

To summarize, then, the primary importance of this site is the glimpse it provides of the Ogallala Miocene sequence of the central Niobrara Valley. The Valentine and Ash Hollow units span an approximately 4-m.y. interval in which at least 5 successional vertebrate faunal assemblages can be collected in direct superposition. The sediments contain a wealth of additional information regarding the composition and dynamics of terrestrial ecosystems in mid-latitude North America which is only beginning to be exploited. Radiometric dating of the intercalated ash beds is still in its infancy, and paleomagnetic studies have not yet begun.

A stratigraphic section measured at For Niobrara was first published by Johnson (1936). Other pertinent references that help place the local geology in a regional perspective are also listed below.

SELECTED REFERENCES

Johnson, F. W., 1936, The status of the name "Valentine" in Tertiary geology and paleontology: American Journal of Science (5th ser.) 31, p. 467–475.

Leidy, J., 1869, The extinct mammalian fauna of Dakota and Nebraska: Journal of the Academy of Natural Sciences of Philadelphia (2nd ser.) 7, p. 1–472.

Lugn, A. L., 1939, Classification of the Tertiary System in Nebraska: Geological Society of America Bulletin 50: 1245–1276.

MacGinitie, H. D., 1962, The Kilgore flora: A late Miocene flora from northern Nebraska: University of California Publications in Geological Science 35, p. 67–158.

Skinner, M. F., and Johnson, F. W., 1984, Tertiary stratigraphy and the Frick Collection of fossil vertebrates from north-central Nebraska: American Museum of Natural History Bulletin, v. 178, art. 3, p. 215–368.

Skinner, M. F., Hibbard, C. W., and others, 1972, Early Pleistocene pre-glacial and glacial rocks and faunas of north-central Nebraska: American Museum of Natural History Bulletin 148, p. 1–148.

Skinner, M. F., and Taylor, B. E., 1967, A revision of the geology and paleontology of the Bijou Hills, South Dakota: American Museum Novitates 2300, p. 1–53.

Skinner, M. F., Skinner, S. M., and Gooris, R. J., 1968, Cenozoic rocks and faunas of Turtle Butte, south-central South Dakota: American Museum of Natural History Bulletin 138, p. 379–436.

Swinehart, J. B., and others, 1985, Cenozoic paleogeography of Western Nebraska, *in* Cenozoic paleogeography of the West Central United States, R. M. Flores and S. S. Kaplan, editors, Rocky Mountain Section S.E.P.M., Denver, Colorado, p. 209–229.

Voorhies, M. R., 1985, Vertebrate paleontology of the proposed Norden Reservoir area, Brown, Cherry, and Keya Paha Counties, Nebraska: University of Nebraska Division of Archeological Research Technical Report, 82-09.

Webb, S. D., 1969, The Burge and Minnechaduza Clarendonian mammalian faunas of north-central Nebraska: University of California Publication in Geological Science 78, p. 1–191.

Wood, H. E., and others, 1941, Nomenclature and correlation of the North American continental Tertiary: Geological Society of America Bulletin 52, p. 1–48.

Woodburne, M. O., MacFadden, B. J., and Skinner, M. F., 1981, The North American "Hipparion" datum and implications for the Neogene of the Old World: Geobios 14, p. 493–524.

Late Cretaceous strata exposed at Ponca State Park, Dixon County, Nebraska

Roger K. Pabian and Dennis R. Lawton, *Conservation and Survey Division, IANR, University of Nebraska-Lincoln, Lincoln, Nebraska 68588-0517*

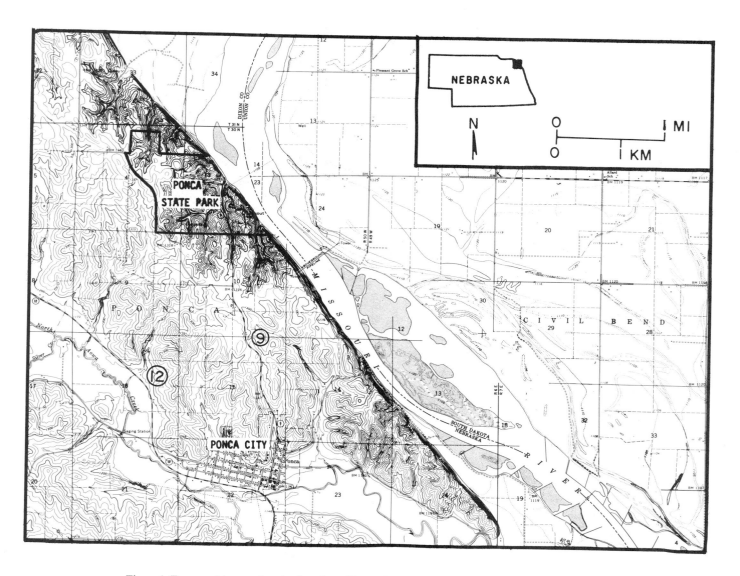

Figure 1. Topographic map showing location of Ponca State Park. Inset marks boundaries. For detailed inset, see Figure 2.

OUTCROPS AND SIGNIFICANCE

The outcrops of the Dakota Group and the Graneros Shale–Greenhorn Limestone interval of Late Cretaceous age at Ponca State Park provide an excellent sedimentary record of the last great marine invasion of the North American midcontinent. The southeastern end of the outcrop area consists of the older continental sandstones. As the viewer examines younger rocks toward the northwest, the original environments of deposition change, and these deposits give way to lagoonal, estuarine, or tidal and mudflat deposits, which are finally overwhelmed by marine shales and limestones.

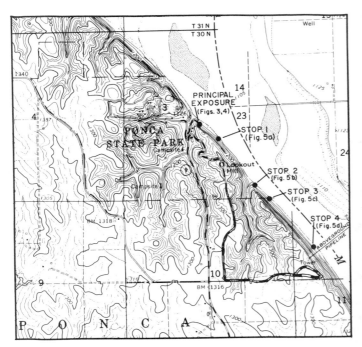

Figure 2. Topographic map of Ponca State Park showing locations of outcrops.

ACCESSIBILITY OF OUTCROPS

All of the outcrops described here are in Ponca State Park. The Nebraska Game and Parks Commission requires that all vehicles on state parks have current entry permits. Annual or daily permits may be purchased from the park ranger.

EXPOSED SECTIONS

Ponca State Park is situated in extreme northeastern Nebraska in the bluffs on the south side of the Missouri River overlooking the broad valley shared by both the Missouri and Big Sioux Rivers (Fig. 1).

Excellent outcrops of Graneros Shale, Greenhorn Limestone, and shales and sandstones of the Dakota Group can be seen at the locations shown in Figure 2. Measured sections at these outcrops are shown in Figures 4 and 5.

At Ponca State Park, the Greenhorn Limestone conformably overlies Graneros Shale, but it is unconformably overlain by the Quaternary Peoria Loess. About 12.5 mi (20 km) to the west of the park, the Greenhorn Limestone is overlain by the Carlile Shale, but the nature of the contact in this area cannot be ascertained in surface outcrops.

The principle geologic section at the boat landing is situated in the NE¼NE¼SW¼,Sec.3,T.30N.,R.6E. (Figs. 2-4). Additional outcrops can be seen along the Missouri River in the CSL,NW¼SE¼Sec.3,T.30N.,R.6E. (Stop 1); the SW¼SE¼SE¼Sec.3,T.30N.,R.6E. (Stop 2); the SE corner, SW¼SE¼Sec.3,T.30N.,R.6E. (Stop 3); and the NW¼SW¼NW¼Sec.11,T.30N.,R.6E. (Stop 4), all in Dixon County, Figure 2.

Figure 3. Principal outcrop of Graneros Shale and Greenhorn Limestone at traffic loop near boat landing.

At the principal exposure, the section consists of about 69 ft (21 m) of Graneros Shale, 8 feet (2.44 m) of Greenhorn Limestone, and up to 75 feet (23 m) of Peoria Loess (Figs. 3 and 4).

Witzke and others (1983) stated that the Graneros Shale conformably overlies the Dakota (Formation) in the eastern margin of the western interior. They further indicated that the contact between the two units is gradational and is commonly placed at the top of the highest iron-cemented sandstone of the Dakota. DeGraw and Pabian (1980) stated that the Graneros-Dakota contact was based on lithology and was somewhat subjective. They further indicated that shales in the lower part of the Graneros in the west interfingered with Upper Dakota sandstones to the east. The Dakota-Graneros contact is not exposed in Ponca State Park, but a driller's log for the water supply well for the swimming pool, about 500 ft (152 m) south of the principal outcrop, places it at about 1100 ft (335 m) above sea level, or about 5 ft (1.5 m) below the normal Missouri River level. Hattin and Cobban (1977) commented on the complexity of the Dakota-Graneros transition and indicated that in Nebraska, as well as in Kansas, the transition suggested a stratigraphically upward replacement of nonmarine and marginal marine environments by uniform marine environments.

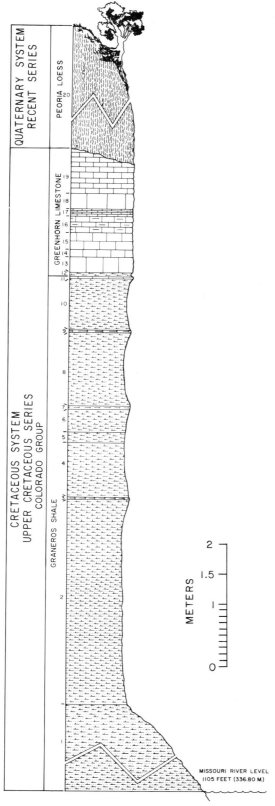

Figure 4. Geologic section of Graneros Shale and Greenhorn Limestone as shown in Figure 3 above.

The lower Graneros Shale at Ponca State Park consists mainly of medium to dark gray, silty, largely noncalcareous shales with occasional lenses of siltstones and sandstones. Upper Graneros shale contains medium to dark gray calcareous shales. Three beds of bentonite are exposed in the Graneros Shale, and these may be of considerable importance in the regional correlation of Graneros units in the western interior.

Kirumakki (1976) performed a detailed petrographic study of Greenhorn units which showed that Greenhorn sediments had generally undergone considerable bioturbation, with burrows and burrowed matrix being common; thin sections showed both skeletal and foraminiferal biomicrites, as well as pelmicrites.

PRINCIPAL ROCK EXPOSURE AT BOAT LOADING RAMP IN NE¼NE¼SW¼SEC.3,T.30N.,R6E., DIXON COUNTY (FIGS. 3 AND 4)

Quaternary System—Pleistocene Series

Peoria Loess
Unit 20.   Light yellowish brown, wind-deposited silt; contains fossil land snail shells. Apparent maximum thickness about 95 ft (29 m); actual thickness about 75 ft (23 m).

Cretaceous System—Upper Cretaceous Series

Colorado Group—Greenhorn Limestone
Unit 19.   Limestone, weathered to light red-brown, slabby to thinly bedded, medium-grained; contains numerous *Mytiloides* (= *Inoceramus*) *labiatus* fragments. Thickness, 2.2 ft (0.67 m).
Unit 18.   Limestone, light gray, medium-grained; contains some *M. labiatus* fragments. Thickness, 1 ft (0.31 m).
Unit 17.   Limestone weathered to light red-brown, slabby, medium-grained; contains some *M. labiatus* fragments. Thickness, 6.5 ft (0.16 m).
Unit 16.   Limestone, gray, medium-grained, shaly; contains some *M. labiatus* fragments. Less dense than unit 17. Thickness, 1.12 ft (0.34 m).
Unit 15.   Limestone, light red-gray to brown, dense, coarsely crystalline; numerous calcite crystals on fracture surfaces; contains numerous *M. labiatus*. Thickness, 1.12 ft (0.34 m).
Unit 14.   Limestone, medium to dark gray, dense, medium-grained, clayey. No calcite crystals as in unit 15. Thickness, 0.5 ft (0.16 m).
Unit 13.   Limestone, light red-brown, medium-grained, dense; calcite crystals on fracture surfaces; contains numerous *M. labiatus*. Thickness, 1.1 ft (0.30 m).
Unit 12.   Siltstone, medium dark gray, thinly laminated, calcareous; contains fish scales, *M. labiatus,* and other clams. Thickness, 0.2 ft (0.09 m).

Colorado Group—Graneros Shale
Unit 11.   Limestone, light red-brown, medium-grained, dense; layer of aragonite crystals, 1 cm thick at top of bed. Thickness, 1.2 in (3 cm).
Unit 10.   Siltstone, dark gray, medium-grained, calcareous; contains a few clam shells and fish scales throughout. Thickness, 3.48 ft (1.06 m).
Unit 9.   Clay, dark yellow-brown, bentonitic, deeply weathered. Thickness, 1.2 in (3 cm).
Unit 8.   Siltstone, medium dark gray, medium-grained, calcareous; contains a few *M. labiatus*. Thickness, 5 ft (1.53 m).

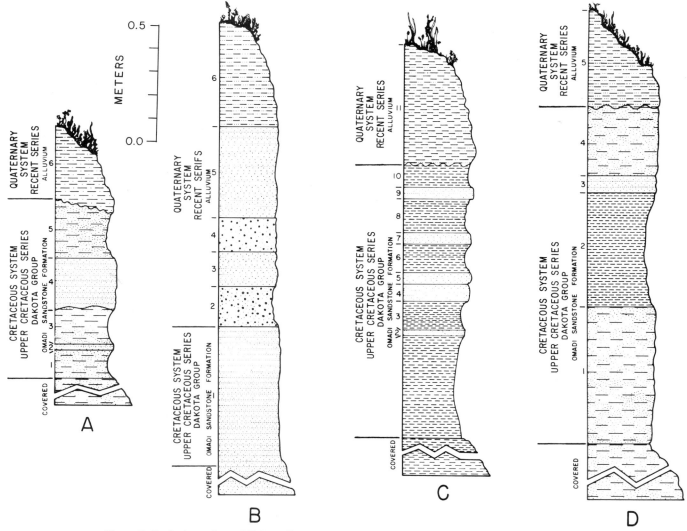

Figure 5. Geologic sections of Dakota Group as seen at stops 1, 2, 3, and 4 (Fig. 2) along the Missouri River walking trail.

Unit 7.   Clay, dark yellow-brown, bentonitic, deeply weathered. Thickness, 2.4 in (6 cm).

Unit 6.   Siltstone, light medium-gray, medium-grained, calcareous; contains a few fish scales, numerous powdery melanterite and halotrichite crystals at contact with unit 5 below. Thickness, 1.5 ft (0.46 m).

Unit 5.   Siltstone, light medium-gray, medium-grained, calcareous; contains a few fish scales; numerous powdery melanterite crystals. Thickness, 0.7 ft (0.21 m).

Unit 4.   Siltstone, dark gray, finely laminated, calcareous; not fossiliferous. Thickness, 3.7 ft (1.13 m).

Unit 3.   Clay, dark yellow-brown, bentonitic, deeply weathered. Thickness, 1.2 in (3 cm).

Unit 2.   Siltstone, medium dark gray, fine- to medium-grained, thinly laminated; not fossiliferous. Thickness, 13.61 ft (4.15 m).

Unit 1.   Colluvium, or slump from above units. Thickness (measured from river level, 43 ft (13.12 m).

The Omadi Sandstone Formation (Condra and Reed, 1959) of the Dakota Group in northeastern Nebraska is about 150 ft (45.8 m) thick in the type area in nearby Dakota County. More recently, Witzke and others (1983) indicated that the Omadi Sandstone and the upper Fuson Shale are equivalent to the Woodbury Member of the Dakota Formation, and Bowe (1972) and Karl (1971) used the Kansas terminology of Plummer and Romary (1942) to designate this interval.

In this area, the upper Dakota units consist largely of mudstone and claystones with some interbedded siltstone and occasional lignites. There are numerous very fine- to fine-grained sandstone bodies, many containing trough or tabular cross beds. These deposits appear to have been laid down in a variety of environments, including fluvial nonmarine, nearshore marine, estuaries, lagoons, and tidal flats. The sandstones are commonly bioturbated, and ichnofossils of marine or marginal marine origin have been recognized in the upper parts of these units (Siemers, 1970).

The source area for Dakota sediments appears to have been

a low-lying land mass of Precambrian-age plutonic and metamorphic rocks in western Minnesota and Iowa and the eastern part of the Sioux Ridge.

The first exposure of Omadi Sandstone (Fig. 5a) to be seen is about 623 ft (190 m) downstream from the light post in the center of the traffic loop at the principal exposure. Here you will see sandstones and shales. The sandstones are massive, and bedding planes may have been destroyed by tidal action. Here you will see rapid lateral and vertical lithologic changes characteristic of tidal flat deposits. The section seen is as follows:

ROCK EXPOSURE AT RIVER TRAIL STOP 1 ON C, SL, NW¼NW¼SE¼SEC.3,T.30N.,R.6E., DIXON COUNTY (FIG. 5A)

Quaternary System—Recent Series

Alluvium
Unit 6. Silty sands and sandy silts, light gray to light brownish-gray. Up to 4 ft (1.22 m).

Cretaceous System—Upper Cretaceous Series

Dakota Group—Omadi Sandstone Formation
Unit 5. Shale, medium gray, slabby, silty, with sandy stringers, 9.4 in (24 cm).
Unit 4. Sandstone, dark orange-brown, massive, medium-grained, with irregular shale stringers, 10.63 in (27 cm).
Unit 3. Shale, medium gray, slabby, silty, with sandy stringers, wavy top, about 5.9 in (15 cm).
Unit 2. Sandstone, dark orange-brown, medium-grained, massive, 0.8 in (2 cm).
Unit 1. Shale, medium gray, slabby, silty, with sandy stringers, 5.2 in (13 cm) measured.

At 2,655 ft (810 m) downstream from the lamp post in the parking area, a second exposure of Omadi Sandstone can be seen. There are about 2 ft (0.6 m) of dark orange-brown, evenly bedded sandstone with sand layers separated by evenly bedded, dark gray, silty shale stringers. The units appear to have undergone no significant bioturbation. The measured section is as follows:

ROCK EXPOSURE AT RIVER TRAIL STOP 2 SW¼ SE¼SE¼SEC.3,T.30N.,R.6E., DIXON COUNTY (FIG. 5B)

Quaternary System—Recent Series

Alluvium
Unit 6. Silt, medium gray, thinly laminated to friable, up to 18 in (46 cm).
Unit 5. Sand, medium gray to dark orange-brown, thinly laminated; contains beetle burrows, 2 to 6 in (5 to 15 cm).
Unit 4. Gravel made of medium gray, silty shale fragments, 2 to 6 in (5 to 15 cm).
Unit 3. Sand, medium gray to dark orange-brown, thinly laminated, 2 to 6 in (5 to 15 cm).
Unit 2. Gravel, medium gray, silty shale fragments, about 11.8 in (30 cm).

Cretaceous System—Upper Cretaceous Series

Dakota Group—Omadi Sandstone Formation
Unit 1. Sandstone, dark orange-brown with dark gray, silty shale stringers and powdery halotrichite coating. About 2 ft (0.61 m) exposed.

At about 3,000 ft (980 m) from the lamp post at the principal outcrop, the following section can be seen:

ROCKS EXPOSED AT RIVER TRAIL STOP 3 SE COR., SW¼SE¼SEC.3,T.30N.,R.6E., DIXON COUNTY (FIG. 5C)

Quaternary System—Recent Series

Alluvium
Unit 11. Silt and sand, light to medium gray, about 24 in (60 cm) exposed.

Cretaceous System—Upper Cretaceous Series

Dakota Group—Omadi Sandstone Formation
Unit 10. Shale, medium gray, fissile to sandy, 4 in (10 cm).
Unit 9. Sandstone, dark orange-brown, medium-grained, massive, 2 in (5 cm).
Unit 8. Shale, medium gray, fissile to sandy, about 6 in (15 cm).
Unit 7. Sandstone, dark orange-brown, medium-grained, massive, about 2 in (5 cm).
Unit 6. Shale, medium gray, fissile to sandy, about 5.1 in (13 cm).
Unit 5. Sandstone, dark orange-brown, medium-grained, massive, about 2 in (5 cm).
Unit 4. Sandstone, dark gray to dark orange-brown, thinly laminated, with shale stringers, about 3.1 in (8 cm).
Unit 3. Shale, dark gray to black, fissile; yellow, powdery halotrichite coating, about 3.1 in (8 cm).
Unit 2. Sandstone, light gray, thinly laminated to cross bedded, about 1.2 in (3 cm).
Unit 1. Shale, dark gray to black fissile; yellow powdery halotrichite coating, about 1.5 ft (46 cm).

The sandstones here are very fine grained and cross bedded and probably show the channel of a low-velocity stream. Sands are separated by thinly bedded shales.

At stop 4, (Fig. 5d) about 3,600 yards (1,105 m) from the lamp post, excellent examples of cross-bedded sandstones are to be seen. The sands are very fine- to fine-grained and are more massively bedded than in the previous exposures. The section at stop 4 is as follows:

ROCKS EXPOSED AT RIVER TRAIL STOP 4 NW¼SW¼NW¼SEC.11,T.30N.,R.6E., DIXON COUNTY (FIG. 5D)

Quaternary System—Recent Series
Alluvium
Unit 5. Mixed sand and silt, not measured.
Cretaceous System—Upper Cretaceous Series
Unit 4. Sandstone, dark orange-brown, fine- to medium-grained, thinly laminated with dark gray, fissile shale stringers, 1 ft (0.30 m).
Unit 3. Sandstone, dark orange-brown, medium-grained, massive, about 2.75 in (7 cm).
Unit 2. Shale, dark gray to black, thinly laminated, with sandy stringers in top 6 in (15 cm), 20 in (51 cm) measured.
Unit 1. Sandstone, light gray, thinly bedded, silty to very finely sandy with numerous micro- to macro-cross beds and dark gray, thinly laminated, silty to clayey shale stringers and halotrichite coating. 2 ft (0.61 m) measured. About 2.95 ft (0.91 m) covered.

The above section shows, in reverse order, a part of the sequence of deposits that took place during the last marine transgression across the western interior. By retracing your steps, the events can be placed in their actual order.

PALEONTOLOGY

Fossils are sparse in most of the Dakota group, although they may be locally abundant. No fossils have been found in the Dakota Group within the boundaries of the park. In several areas in northwestern Iowa, southeastern South Dakota, and northeastern Nebraska, plant remains including angiosperm leaves (Lesquereux, 1874; Retallack and Dilcher, 1981) and freshwater, brackish water, and marine molluscan faunas, have been observed in the upper Dakota Group (Meek, 1876; Stanton, 1922; Tester, 1931; Veatch, 1969; Hattin and Cobban, 1977). Witzke (1981) reported crocodile and turtle remains, teleost fish remains, including teeth of large, predatory ichthyodectids. Agglutinated foraminifers have been observed in the upper Dakota Group by Tester (1952), and the only dinosaur remains from Nebraska were recorded by Barbour (1931).

In the Graneros Shale, teleost fish debris is sometimes common; it is often concentrated in calcareous concretions near the top of the formation. Rich (1975) indicated the presence of many agglutinated foraminifers in the Graneros Shale in northwestern Iowa. Inoceramid bivalves occur sparsely throughout the Graneros section at the park, and horizontal burrows are fairly common.

Beginning with Greenhorn deposition, the bivalve *Mytiloides* (= *Inoceramus*) *labiatus* (Schlotheim) becomes abundant. Crosbie (1941) has recorded 17 genera of foraminifers in the Greenhorn, including *Hedbergella*, *Rotalipora*, and *Heterohelix*. Kirumakki (1976) reported coccoliths and algal calcispheres in the Greenhorn and stated that fossil pellets in depositionally significant amounts were probably the products of annelid worms and decapod crustaceans. Witzke and others (1983) reported rare ammonids from this interval, and Witzke (1981) recorded 8 shark species.

Pleistocene loesses here have yielded some fossil land snail material, but no fossils have been observed in the alluvium.

REFERENCES

Barbour, E. H., 1931, Evidence of dinosaurs in Nebraska: Nebraska State Museum Bulletin, v. 1, no. 21, p. 187–190.

Bowe, R. J., 1972, Depositional history of the Dakota Formation in Eastern Nebraska [M.S. thesis]: University of Nebraska, 87 p.

Condra, G. E., and Reed, E. C., 1959, The geological section of Nebraska: Nebraska Geological Survey Bulletin, 14A, 82 p.

Crosbie, J. H., 1941, Some foraminifera from the Greenhorn Formation and adjacent base in south central Nebraska [M.S. thesis]: University of Nebraska, 57 p.

DeGraw, H. M., and Pabian, R. K., 1980, Collecting in Nebraska's Cretaceous strata: Rocks and Minerals, v. 55, no. 2, p. 97–102.

Georgeson, N. C., 1931, The stratigraphy of the Colorado Group in northeastern Nebraska [M.S. thesis]: University of Iowa, 143 p.

Hattin, D. E., 1967, Stratigraphic and paleoecologic significance of macroinvertebrate fossils in the Dakota Formation (Upper Cretaceous) of Kansas, *in* Essays in paleontology and stratigraphy, R. C. Moore Commemorative Volume, Kansas University, Department of Geology Special Publication 2, p. 570–584.

—— , and W. A. Cobban, 1977, Upper Cretaceous stratigraphy, paleontology, and paleogeology of western Kansas: Mountain Geologist 14, p. 175–218.

Karl, H. A., 1971, Depositional history of Dakota Formation (Upper Cretaceous) sandstones, southeastern Nebraska [M.S. thesis]: University of Nebraska, 87 p.

Kirumakki, N. S., 1976, Stratigraphy, petrography, and depositional environment of Greenhorn Formation of eastern Nebraska [M.S. thesis]: University of Nebraska, 162 p.

Lesquereux, L., 1874, On the fossil plants of the Cretaceous Dakota Group of the United States: U.S. Geologic Survey of the Territories Vol. 6: 136 p.

Meek, F. B., 1876, A report on the invertebrate Cretaceous and Tertiary fossils of the upper Mississippi country: U.S. Geologic Survey of the Territories Vol. 9: 629 p.

Plummer, N., and Romary, J. F., 1942, Stratigraphy of the pre-Greenhorn Cretaceous beds of Kansas: Kansas Geological Survey Bulletin 41, p. 313–348.

Retallack, G., and Dilcher, D. L., 1981, A coastal hypothesis for the dispersal and the rise to dominance of flowering plants, *in* Niklas, K. J., editor, Paleobotany, paleoecology, and evolution: New York, Praeger, p. 27–77.

Rich, M. A., 1975, Foraminifera of the Graneros and Greenhorn Formation, (Upper Cretaceous) from one exposure near Sioux City, Iowa [M.S. thesis]: Ames, Iowa State University, 114 p.

Siemers, C. T., 1970, Facies distribution of trace fossils in a deltaic environmental complex: Upper part of the Dakota Formation (Upper Cretaceous), central Kansas [abs.]: Geological Society of America Abstracts with Programs, v. 2, p. 683–684.

Stanton, J. W., 1922, Some problems connected with the Dakota Sandstone: Geological Society of America Bulletin, v. 33, no. 2 p. 255–272.

Tester, A. C., 1931, The Dakota stage of the type locality: Iowa Geological Survey Annual Report 35: p. 197–332.

—— 1952, Additional facts concerning the age and origin of the type section of the Dakota Stage: Geological Society of America Bulletin, v. 63, no. 12, pt. 2, p. 1386.

Veatch, M. D., 1969, Groundwater occurrence movement-hydrochemistry within a complex stratigraphic framework, Jefferson County, Nebraska [Ph.D. thesis]: Palo Alto, Stanford University, 202 p.

Witzke, B., 1981, Cretaceous vertebrate fossils of Iowa and nearby areas of Nebraska, South Dakota, and Minnesota: Iowa Geological Survey Guidebook 4, p. 105–122.

—— , and others, 1983, Cretaceous paleogeography along the eastern margin of the Western Interior Seaway, Iowa, southern Minnesota, and eastern Nebraska and South Dakota. Rocky Mountain Section, S.E.P.M., Mesozoic Paleogeography of West Central United States. Reynolds, W. W., and Dolly, E. D., editors, Denver, Colorado.

Late Cenozoic geology along the summit to museum hiking trail, Scotts Bluff National Monument, western Nebraska

James B. Swinehart, Conservation and Survey Division, University of Nebraska-Lincoln, Lincoln, Nebraska 68588-0517
David B. Loope, Department of Geology, University of Nebraska-Lincoln, Lincoln, Nebraska 68588-0340

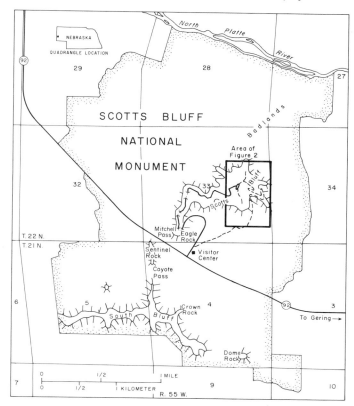

Figure 1. Location of Scotts Bluff National Monument. Base is from Scottsbluff South 7½-minute Quadrangle.

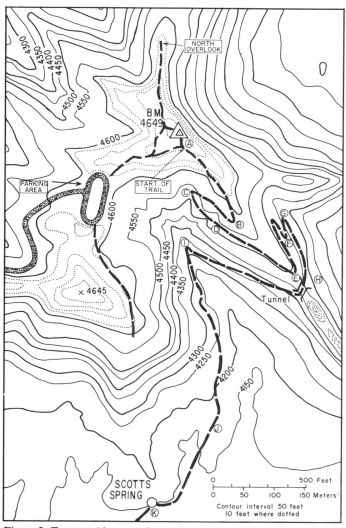

Figure 2. Topographic map of summit to museum hiking trail. Letters refer to stops along trail.

LOCATION AND ACCESSIBILITY

Scotts Bluff National Monument is best reached by starting from the junction of Nebraska 71 and 92 in downtown Gering, Nebraska, and proceeding westward on Nebraska 92 for 2.4 mi (3.4 km) to the visitor center (Fig. 1). The monument is open from 8 A.M. to 8 P.M. Memorial Day to Labor Day, 8 A.M. to 6 P.M. Labor Day to October 1, and 8 A.M. to 5 P.M. October 1 to Memorial Day. Educational groups are not charged the nominal usage fee for admittance to the summit road and parking (Fig. 2). Please stay on the trails and note that collecting rocks or other items is prohibited without a permit from the National Park Service.

SIGNIFICANCE OF LOCALITY

The exposures of the Arikaree and White River Groups (Figs. 3 and 4) along the summit to museum trail are certainly the most accessible of any in western Nebraska. A wide variety of sedimentary structures, diagenetic features, trace fossils, and volcaniclastic sediments is well displayed along the trail. In addition, an excellent panoramic view of the North Platte River valley is provided from the north overlook (Fig. 2).

SITE INFORMATION

The rocks exposed within the monument are Tertiary in age and are of nonmarine origin. They belong to four stratigraphic units (Orella; Whitney; Gering; Monroe Creek–Harrison) that

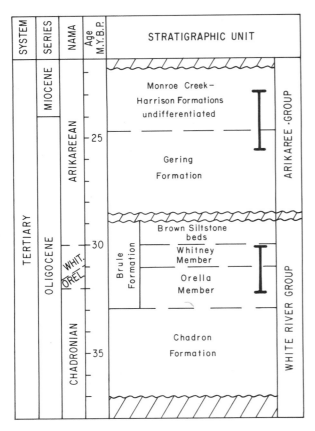

Figure 3. Time stratigraphic chart of a portion of the Cenozoic. Vertical bars indicate approximate stratigraphic intervals exposed within Scotts Bluff National Monument. NAMA refers to North American Mammal Ages (Whit. = Whitneyan; Orel. = Orellan).

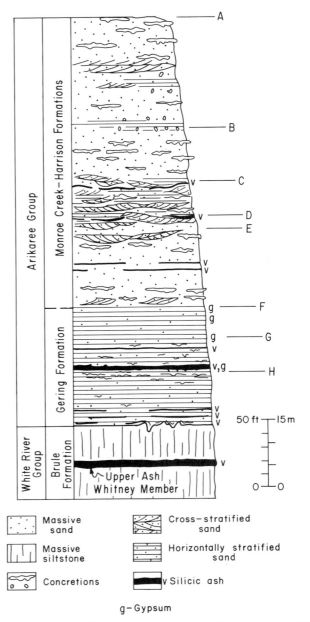

Figure 4. Diagrammatic geologic section along the upper part of the hiking trail. Letters refer to stops along the trail.

can be placed into two groups—the White River and overlying Arikaree (Figs. 3 and 4).

The Orella Member of the Brule Formation (not exposed on the trail) consists of siltstones and mudstones with interbedded thin lenticular sandstones and forms the small badlands area in the northeast part of the monument (Fig. 1). Subsurface information indicates that the base of the White River Group (underlain by Late Cretaceous Pierre Shale) is about 250 ft (75 m) below the lowest Orella exposures. The Whitney Member of the Brule Formation is a massive, pinkish-brown, volcaniclastic siltstone; it contains two vitric ash beds (Upper and Lower Ash) of regional extent. There is a 20-ft (6-m)-thick sequence of interbedded, fine-grained sandstone and siltstone below the Lower Ash on Sentinel and Eagle Rocks (Fig. 1). White River Group siltstones contain 40 to 70 percent silicic glass shards (Swinehart and others, 1985) and an additional 20 to 30 percent crystal and lithic pyroclastic detritus. The pyroclastic material was derived from western-source rhyolitic and volcanic centers, with those in Colorado the most probable sources (Swinehart and others, 1985).

The Gering Formation is about 88 ft (27 m) thick and consists of thin, horizontally stratified pale brown to gray brown, very fine to fine-grained, volcaniclastic sandstone (Fig. 4). It

also contains a number of ash beds. Placement of an upper contact for the Gering is subject to some debate, as the contact is gradational at many localities. The horizontally stratified sequence is present at other sites in the region and appears to occur above the more typical fluvial cut and fill sequences of the Gering.

The overlying pale brown and light gray, silty, very fine to fine-grained sandstones are shown as a combined Monroe Creek-Harrison unit because the criteria for differentiating the two formations outside their type areas in northwest Nebraska have not proven consistent.

Figure 5. Sketch made from a photograph of eolian cross-stratified and horizontally stratified sandstone of the Arikaree Group at Stop D. Person is 6 ft (1.8 m) tall.

North Overlook on Summit Trail

The Orella badlands are visible from the North Overlook (Fig. 2). A large number of vertebrate fossils were collected from this area prior to 1910 when the area was incorporated into the monument and fossil collecting was prohibited.

The overlook is about 800 ft (244 m) above the North Platte River and provides a panoramic view of the fertile cropland in the North Platte Valley, which is about 6 mi (9.7 km) wide at this location. Upstream from this point, the North Platte River drains an area of 24,330 mi^2 (63,000 km^2) heading in the Rocky Mountains. Because reservoirs were built to retain spring snowmelt for irrigation during the summer, the flow of the river formerly was much more variable, and its channel was considerably wider than it is now. Regulation of the river's flow by means of reservoir releases, together with irrigation-seepage returns to the river, results in flow during all seasons. River discharge ranged from 449,000 to 1,700,000 acre-feet (0.55 km^3/yr to 2.10 km^3/yr), averaging 822,000 acre-feet (1.01 km^3/yr) during the 10-year period from 1967 to 1976. Most of the water used for irrigation within the North Platte Valley is obtained from canals that divert water from the North Platte River in Wyoming and Nebraska. A relatively small acreage is irrigated with groundwater.

Summit to Museum Trail

The trail begins between Summit Trail markers 12 and 13 and ends at the museum 1.6 mi (2.6 km) to the south. There are no permanent trail markers, so geologic points of interest are keyed to easily located sites on the trail (Figs. 2 and 4).

Stop A (start of trail). The morphology of the calcite-cemented "pipy" concretions typical of the Arikaree Group is well displayed just north of the start of the trail. The concretions maintain a consistent northeast-southwest orientation over much of the southern Nebraska panhandle. Clasts of these concretions occur in a number of intraformational gully fills in western Nebraska and indicate that the concretions formed shortly after deposition of the host sand. Note the small diameter vertical tubules preserved in many of the concretions.

About 197 ft (60 m) down the trail is a 26-ft (8-m)-thick sequence of low-angle (7° to 15°), large-scale, cross-stratified sand (Fig. 4). This sequence is also well exposed at the south side of the summit parking lot (Fig. 2). A section of horizontally stratified sandstone with numerous small-diameter burrows is exposed another 98 ft (30 m) down the trail.

Both the massive and stratified sandstones of the Monroe Creek–Harrison contain 25 to 50 percent silicic glass shards and an additional 25 to 40 percent crystal and lithic pyroclastic detritus.

Stop B (first switchback). Note the knobby "potato" concretions and the crudely stratified sandstone in this area.

Stop C (second switchback). About 45 ft (14 m) down trail from the museum signpost is a thin, pinkish ash lentil with a meter or more of local relief. It can also be seen in the cliff face to the south. Note both the cross- and horizontally stratified sandstone above and below the ash (Fig. 4).

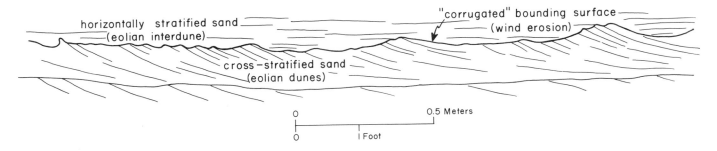

Figure 6. Sketch made from a photograph of a "corrugated" bounding surface exposed about 98 ft (30 m) downtrail from Stop D.

Stop D (at concrete steps on trail, Fig. 5). The exposures between this stop and at Stop E (about 490 ft [150 m] southeast) are worth examining in some detail. The cross-stratification in these very fine to fine-grained sandstones is compound; large-scale wedge planar sets (Fig. 5) up to 5 ft (1.7 m) thick are themselves cross-laminated. These cross-laminated sets are laterally extensive, even and distinct. They contain small-scale foresets (up to a centimeter in length) with shallow dips that are typically nearly perpendicular to the dip of the large-scale foresets. The majority of the laminae are inversely graded. The relatively higher percentage of dark, heavy minerals in the coarse silt versus the very fine sand fraction allows relatively easy recognition of the inverse grading and cross-lamination, especially on surfaces oblique to the laminations. We interpret these structures as products of wind ripple migration (Hunter, 1977) on eolian dunes.

Horizontally stratified sandstones are also present in this sequence (Figs. 5 and 6); laminae in these rocks are typically inversely graded. This stratification was also produced by migration of wind ripples, but across flat interdune surfaces. In modern interdunes, a distinctive ridge and swale topography is produced by differential wind erosion of cohesive or lightly cemented, cross-stratified sand (Ahlbrandt and Fryberger, 1981, Figure 4c and 4d). When buried and viewed in vertical section, such erosional surfaces appear as irregular or "corrugated" bounding surfaces (Simpson and Loope, 1985). Such surfaces occur at several horizons visible from the trail (Fig. 5) between Stops D and E and seem best interpreted as features produced by wind erosion.

A lag deposit composed of evenly spaced, coarse sand to pebble-sized material is located at the base of the cross-stratified sandstones above the trail and can be traced laterally for hundreds of meters. A similar lag deposit in eolian sediments is illustrated by Ahlbrandt and Fryberger (1981, Figure 4a).

Further evidence of an eolian origin for the cross-stratified sandstones comes from flume and field studies, cited by Driese and Dott (1984, p. 583). For sands with grain sizes below 0.11 mm, subaqueous dunes and sand waves are not stable. The median grain size for the cross-stratified sandstone in the Monroe Creek-Harrison exposed here is less than 0.09 mm (Bart, 1974). This suggests that the cross-strata formed by the migration of eolian rather than fluvial bed forms.

About 75 ft (23 m) down the trail from the concrete steps (Fig. 5) is an exposure of a laminated volcanic ash lens up to 20 in (50 cm) thick. Excellent samples of a number of different types of invertebrate burrows are present in the ash and in adjacent sandstones. The burrows are cylindrical, smooth walled, massive or meniscate, and generally nonbranching. They were interpreted by Stanley and Fagerstrom (1974) to represent shelter burrows, deposit-feeding burrows, and vertical passageways made by insects, possibly beetles. Stanley and Fagerstrom (1974) interpreted the cross-stratified sands to have been deposited by migrating sandbars in a braided river system.

Stop E (major bend in trail with museum signpost). Note the volcanic ash lens about 13 ft (4 m) above the trail. Compare the cross-stratified eolian sandstone with the massive sandstone. The mineralogy and grain size of these two units is very similar, and we suggest that the lack of depositional sedimentary structures is due to (1) a much slower sedimentation rate and intense bioturbation, and/or (2) trapping of sediment by vegetation so that laminations were never present.

Stop F (second of three switchbacks along northeast-facing bluff). This stop is at the top of the horizontally stratified sandstone of the Gering Formation (Fig. 4). Here, "sand crystals" up to 1 in (2.5 cm) in diameter are developed at several horizons. The discoidal shape of these sand crystals clearly indicates that they were formed by the growth of gypsum which has since been replaced by calcite.

Stop G (third switchback on northeast-facing bluff). There is at least one horizon of "sand crystals" here also. Note the uniform nature of the stratification in the Gering. Many individual strata can be traced for tens of meters. Small-scale cross-stratification occurs locally. About 200 ft (60 m) down the trail and 11 ft (3.3 m) above the white ash at Stop H, a 0.8-in (2-cm)-thick, gray volcanic ash bed is present and can be traced continuously for a minimum of 650 ft (200 m). A similar ash occurs in the same stratigraphic position above a white ash just north of the highest tunnel on the road approximately 0.6 mi (1 km) west of here.

Stop H (at entrance to tunnel). The ash above the tunnel contains abundant calcite pseudomorphs after lenticular gypsum crystals. The evidence for precipitation of evaporites suggests an arid climatic setting. We interpret these horizontally stratified and

Figure 7. Photograph of concave-up deformation features interpreted to be fossil vertebrate tracks. Two tracks are easily visible near the center, and two others are indicated by arrows. The hammer is 11 in (28 cm) long.

small-scale, cross-laminated, very fine-grained sands as ephemeral stream deposits.

Beginning at the level of the thin gray ash described above, concave-up deformation structures are common (Figs. 4 and 7). Bart (1975) interpreted these and several other types of deformation at this locality as inorganically induced deformation structures. Due to their resemblance to features described by Laporte and Behrensmeyer (1980), we believe that many of these structures are tracks of vertebrates. The scale and bilobed nature of some structures indicate that some of the track-makers were large ungulates—probably entelodonts (Loope, 1986). Other potential track-makers known to have lived during this time in-

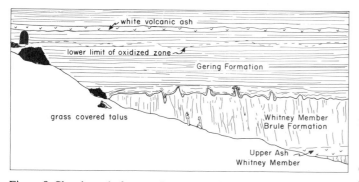

Figure 8. Sketch made from a photograph of the Gering-Whitney contact east of the trail tunnel (visible in the upper left corner). The two people standing on the talus slope are each 6 ft (1.8 m) tall.

clude a variety of oreodonts, hyracodontid rhinos, tapirs, small horses, camels, and a variety of carnivores.

Stop I (at head-of-canyon switchback). Volcaniclastic siltstones of the Whitney Member of the Brule Formation are exposed here. The mineralogy, grain size, texture, and regional mantling nature of the Whitney all suggest that it represents slow accumulation of airfall pyroclastic material. The Upper Ash of the Whitney crops out about 30 ft (9 m) above the trail at this stop. Ash beds in the White River Group are much more continuous than those in the Arikaree Group and can be traced throughout much of the Nebraska panhandle (Swinehart and others, 1985).

Stop J (700 ft [215 m] down the trail from Stop I). The highly irregular nature of the Whitney-Gering contact can be seen in the cliff face to the northeast (Fig. 8). At other exposures within the monument, this contact is flat and has very little relief. Bart (1974) and Stanley and Fagerstrom (1974) suggested that the deformation seen here is evidence against a major hiatus at the White River–Gering contact. However, stratigraphic evidence indicates that this contact represents a hiatus of up to 4 m.y. at the monument (Fig. 3).

The remains of a large rockfall that occurred in October 1974 can be seen on the talus slope below the end of the cliff face described above. Much of the erosion at Scotts Bluff occurs through such rock falls.

Stop K (Scotts Spring). This spring issues from fractures in Whitney Member siltstones.

REFERENCES CITED

Ahlbrandt, T. S., and Fryberger, S. G., 1981, Sedimentary features and significance of interdune deposits, *in* Ethridge, F. G. and Flores, R. M., editors, Recent and ancient nonmarine depositional environments: Models for exploration: Society of Economic Paleontologists and Mineralogists, Special Publication No. 31, p. 293–314.

Bart, H. A., 1974, A sedimentologic and petrographic study of the Miocene Arikaree Group of southeastern Wyoming and west-central Nebraska [Ph.D. thesis]:, Lincoln, University of Nebraska, 106 p.

——1975, Downward injection structures in Miocene sediments, Arikaree Group, Nebraska: Journal of Sedimentary Petrology, v. 45, p. 944–950.

Driese, S. G., and Dott, R. H., Jr., 1984, Model for sandstone-carbonate "cyclothems" based on upper member of Morgan Formation (Middle Pennsylvanian) of northern Utah and Colorado: American Association of Petroleum Geologists Bulletin, v. 68, p. 574–597.

Hunter, R. E., 1977, Basic types of stratification in small eolian dunes: Sedimentology, v. 24, p. 362–387.

Laporte, L. F., and Behrensmeyer, A. K., 1980, Tracks and substrate reworking by terrestrial vertebrates in Quaternary sediments in Kenya: Journal of Sedimentary Petrology, v. 50, p. 1337–1346.

Loope, D. B., 1986, Recognizing and utilizing vertebrate tracks in cross section; Cenozoic hoofprints from Nebraska: Palaios, v. 1, p. 141–151.

Simpson, E. L., and Loope, D. B., 1985, Amalgamated interdune deposits, White Sands, New Mexico: Journal of Sedimentary Petrology, v. 55, p. 361–365.

Stanley, K. O., and Fagerstrom, J. A., 1974, Miocene invertebrate trace fossils from a braided river environment, western Nebraska, U.S.A.: Palaeogeography, Palaeoclimatology, and Palaeoecology, v. 5, p. 63–82.

Swinehart, J. B., Souders, V. L., DeGraw, H. M., and Diffendal, Jr., R. F., 1985 Cenozoic paleogeography of western Nebraska, *in* Flores, R. M., and Kaplan, S., editors, Cenozoic paleogeography of the West-Central United States: Rocky Mountain section—Society of Economic Paleontologists and Mineralogists, p. 209–229.

4

The Platte River and Todd Valley, near Fremont, Nebraska

William J. Wayne, Department of Geology, University of Nebraska-Lincoln, Lincoln, Nebraska 68508

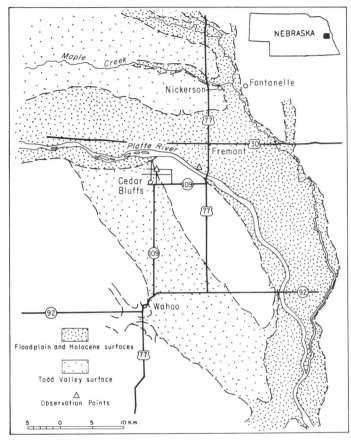

Figure 1. Map of Platte and Todd Valley area, showing locations of overview points, important highways, and landmarks.

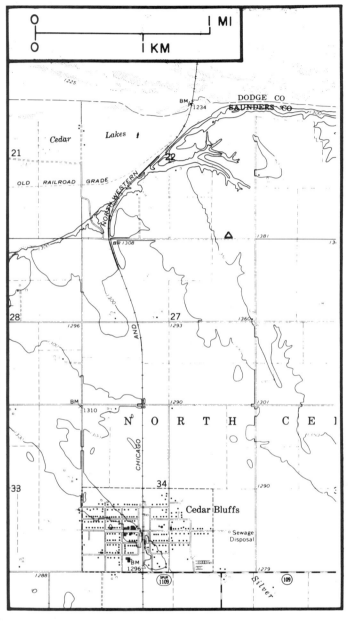

Figure 2. Detailed location map of Todd Valley overlook site (from Fremont West 7½-minute Quadrangle).

LOCATION

From a high point on the south side of the Platte Valley, 4.7 mi (7.5 km) west of Fremont (Fig. 1) and 1.8 mi (2.9 km) north of Cedar Bluffs (SE¼SE¼Sec.22,T.17N.,R.7E.; Fremont West 7½-minute Quadrangle), one can view the present Platte River and valley as well as look across an abandoned valley segment through which at least part of the Platte River once flowed (Fig. 2). In addition, the south bluffs of the Platte, which have remained cleanly swept exposures of a large part of the Pleistocene record of eastern Nebraska since before 1900, can be seen from across the river in Hormel Park (N edge SW¼Sec.34,T.17N.,R.8E.) (Fig. 3).

To reach the overview point, follow Nebraska 109 north from Wahoo to Cedar Bluffs, go east 0.25 mi (400 m), then north 2 mi (3.2 km) to County Road 2. Turn west 0.1 mi (0.2 km) and stop at the top of the slope (Fig. 1). To reach Hormel Park in Fremont, return to Nebraska 109 and turn east 5 mi (8 km). At U.S. 77, turn north for 1.5 mi (2.5 km), then west 0.4 mi (0.6

km). Follow the gravel road to the southwest about 0.2 mi (0.3 km) to Hormel Park. Drive about 0.2 mi (0.3 km) farther, park, and walk to the edge of the Platte River. Field glasses are useful to examine the bluffs, which are across the river, about 600 ft (200 m) away.

Both the overview point and Hormel Park are public sites

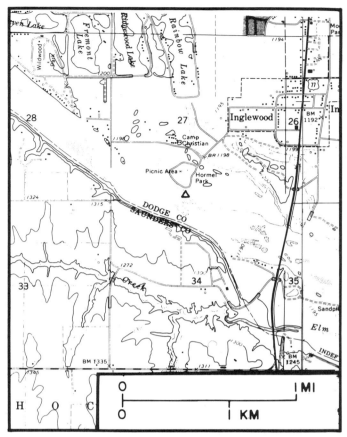

Figure 3. Detailed location map of Fremont bluffs section view site (from Fremont West 7½-minute Quadrangle).

and no permission is required. To reach the bluffs, though, one must either walk through private property or approach them from the river. Neither is recommended.

SIGNIFICANCE OF SITES

The Platte River, much studied by sedimentologists, is a master drainage line from west to east across the high plains and through Nebraska. It flows over a long alluvial plain that slopes eastward from the Rocky Mountains. Although directed by the alluvial slope, in some places its present course has become adjusted to structure. Near its eastern terminus, its course shows ice-marginal control. It is a multichannel river with low sinuousity that carries a heavy bedload of sand. These bedforms and their internal structures have been studied by many for comparison with ancient fluviatile sediments. At one time, the Platte drained southeastward through the broad, high-level trough known as Todd Valley, although today it continues eastward, then south-

ward through a broad valley that it shares with the Elkhorn River. The almost continuous exposures along the south side of the Platte from Todd Valley to Fremont have been considered important sections in the study of the Pleistocene of Nebraska since one of them was described by Todd in 1899 (p. 79).

SITE INFORMATION

That the Platte River had once flowed through this high-level abandoned channel was noted by Todd (1899); later, Condra (1903) named it "Todd Valley." Most of the sediments that underlie it are known largely from well records and samples, because exposures are few. The surface is covered with Peoria Loess, some of which can be seen in a cut along the road to the north just below the top of the hill. Beneath the loess is a thick sand unit that becomes coarser and gravelly toward its base. Well data (Souders, 1967) show it to range from 100 ft (30 m) to more than 150 ft (45 m) in thickness. A till is present at its base in a few places. These Pleistocene sediments overlie the Cretaceous Dakota Sandstone. Lugn (1935) named the sand that underlies the Peoria Loess the "Todd Valley Sand."

The Peoria Loess also caps the adjacent upland outside Todd Valley, but the sediments beneath it differ greatly from those beneath the abandoned valley. The sections exposed in the bluffs south of Fremont have been described by several authors (Todd, 1899, p. 79; Lugn, 1935, p. 41–42; Condra and others, 1947; Lueninghoener, 1947, p. 19–20; Frankel, 1956, p. 165–174; Reed and Dreeszen, 1965, p. 52–55), all of whom considered these to be key sections, but no thorough investigation of the several units exposed there has been undertaken. All authors, however, agree that the section includes Peoria Loess, locally fossiliferous, a strongly developed paleosol, and two or more tills. Reed and Dreeszen (1965) named the uppermost one the Cedar Bluffs till. A different till underlies the Cedar Bluffs, separated from it in some places by a thin layer of noncalcareous material that Lugn (1935) referred to as "gumbotil" and Reed and Dreeszen (1965, p. 54, 55) called the "Fontanelle soil." This second till was considered to extend to below river level by Reed and Dreeszen, who identified it as the Nickerson till of Kansan age. Lugn (1935) and Condra and others (1947) called the lowermost till in the section "Nebraskan" till. Current investigation shows that three tills are present in the eastern end of the long section (Table 1, Fig. 2), the basal one of which is distinctive dark gray in color and corresponds well with Shimek's (1909) original description of the till near Omaha to which he applied the name "Nebraskan."

Studies published during the 1970s showed that the "marker bed" on which many of the Pleistocene units of Nebraska, Kansas, Iowa, and South Dakota were identified and correlated—the Pearlette Ash—was in reality at least three and possibly four separate ash beds with ages of 2.0, 1.2, and 0.6 Ma (Naeser and others, 1973; Izett, 1981; Boellstorff, 1976, 1978). Prior to that discovery, if an ash bed was found to overlie a till, the till was considered to have been deposited during the Kansan glaciation;

Figure 4. Sketch of Platte River bluff section south of Fremont as seen from north side of river in Hormel Park. Letters correspond to units described in Table 1.

if the reverse, the till was post-Yarmouth in age. This is no longer true, and the classical names Yarmouthian, Kansan, Aftonian, and Nebraskan must be used with caution.

A manner and time of abandonment of Todd Valley were suggested by Lugn (1935, p. 157) and elaborated by Lueninghoener (1947); they recognized sediments of Wisconsinan age at its surface and so theorized that Todd Valley had been the route of the Platte, separated from the Elkhorn by a peninsular divide, until post-Wisconsinan time, and that a small tributary of the Elkhorn cut headwardly through the divide and diverted the Platte into the Elkhorn.

The great width of the breach at the junction of the Platte and Elkhorn, the gradient of the Platte River terraces both through Todd Valley and upstream from Fremont, and the drainage patterns of tributaries that enter the Platte valley from the north caused Wayne (1985, p. 116) to question an early Holocene abandonment of the Todd Valley strath and to suggest an alternate explanation. Field data, though still incomplete, indicate that the James Lowland lobe of the Late Wisconsinan glacier blocked the Missouri River at least briefly, and that the overflow followed an overland route to the North Fork of the Elkhorn River and thence southeastward to the Platte. Where the glacially diverted overflow entered the Platte valley at Fremont, it deposited much of its sediment load as a fan in the wide part of the Platte-Elkhorn valley, raising base level sufficiently to cause the Platte to aggrade and reoccupy a former channel in Todd Valley. Thus the original abandonment of the Todd Valley route by the Platte River probably took place at some time prior to the Wisconsinan Glaciation.

TABLE 1. FREMONT CLIFFS SECTION (Measured June 1984 at east end of long series of exposures, in NE¼ NW¼Sec.34,T.17N.,R.8E.)

[F] Peoria Loess

13.	Silt, light gray (10YR 7/2 dry), locally contains snail shells, stands in vertical cliffs; thickness varies, maximum observed	19.7 ft (6.0 m)

[E] Gilman Canyon Loess Bed

12.	Silt, grayish brown (10YR 5/2), massive,	1.6 ft
	contains disseminated organic matter, base horizontal	(0.5 m)

[D] Yarmouth-Sangamon Paleosol

11.	Clay, light yellowish-brown (10YR 6/4 dry), blocky	0.6 ft (0.2 m)
10.	Clay, yellowish-brown (10YR 5/4 dry), blocky, angular	0.6 ft (0.2 m)
9.	Clay, strong brown (7.5YR 5/6 dry), blocky, angular peds 0.12 inch (3 mm) in size	2.1 ft (0.65 m)
8.	Clay, pale olive (5Y 6/3 dry), blocky to prismatic	1.2 ft (0.4 m)

[C] Cedar Bluffs Till

7.	Till, clayey, yellowish brown (10YR 5/4), calcareous, $CaCO_3$ nodules along strongly oxidized joints in upper part, top somewhat uneven	15.0 ft (4.4 m)
6.	Till, clayey, grayish-brown (10YR 5/2 to 2.5Y 5/3), calcareous, elongate cobbles with striae oriented S55°W	9.0 ft (2.8 m)

[B] Nickerson Till

5.	Till, silty, yellowish-brown (10YR 5/4), grading to pale brown (10YR 6/3) below top 20 inches (50 cm), calcareous, joints oxidized to light yellowish-brown (10YR 6/4) 2.4 to 4 inches (6 to 10 cm) along each side of thin (0.12 inches [3 mm] wide) $CaCO_3$-filled core.	15.0 ft (4.5 m)
4.	Till, silty, pale brown (10YR 6/3), calcareous	5.25 ft (1.6 m)

[A] "Nebraskan" Till of Shimek, Lugn

3.	Till, clayey, olive (5Y 5/3), grading upward to light yellowish-brown (2.5Y 6/4), calcareous; wide oxidized zone along fractures	6.5 ft (2.0 m)
2.	Till, clayey, dark olive gray (5Y 3/2), massive, very compact, base not exposed	10.0 ft (3.0 m)
1.	Covered to river level	6.5 ft (2.0 m)
	Total	91.84 ft (28.00 m)

REFERENCES CITED

Boellstorff, John, 1976, The succession of late Cenozoic volcanic ashes in the Great Plains: A progress report:. Kansas Geological Survey Guidebook Series 1, p. 37–71.

——1978a, North American Pleistocene stages reconsidered in light of probable Pliocene-Pleistocene continental glaciation: Science, v. 202, p. 305–307.

——1978b, Chronology of some late Cenozoic deposits from the central United States and the ice ages: Nebraska Academy of Sciences, Transactions, v. 6, p. 35–49.

Condra, G. E., 1903, An old Platte Channel: American Geologist, v. 31, p. 361–369.

Condra, G. E., Reed, E. C., and Gordon, E. D., 1947, Correlation of the Pleistocene deposits of Nebraska: Nebraska Geological Survey Bulletin 15, 73 p.

Frankel, Larry, 1956, Pleistocene geology and paleoecology of parts of Nebraska and adjacent areas [Ph.D. thesis]: University of Nebraska, 297 p.

Izett, G. A., 1981, Volcanic ash beds: Recorders of Upper Cenozoic silicic pyroclastic volcanism in the western United States: Journal of Geophysical Research, v. 86, no. B11, p. 10200–10222.

Lueninghoener, G. C., 1947, The post-Kansan geologic history of the lower Platte Valley area: University of Nebraska Studies, new series no. 2, 82 p.

Lugn, A. L., 1935, The Pleistocene geology of Nebraska: Nebraska Geological Survey, Bulletin No. 10, 223 p.

Naeser, C. W., Izett, G. A., and Wilcox, R. E., 1973, Zircon fission-track ages of Pearlette family ash beds in Meade County, Kansas: Geology, v. 1, p. 93–95.

Reed, E. C., and Dreeszen, V. H., 1965, Revision of the classification of the Pleistocene deposits of Nebraska: Nebraska Geological Survey, Bulletin 23, 65 p.

Souders, V. L., 1967, Availability of water in eastern Saunders County, Nebraska: U.S. Geological Survey, Hydrologic Investigations Atlas, HA-266.

Todd, J. E., 1899, The moraines of southeastern South Dakota and their attendant deposits: U.S. Geological Survey, Bulletin 158, 171 p.

Wayne, W. J., 1985, Drainage patterns and glaciations in eastern Nebraska: TER-QUA Symposium Series, V. 1, 111–117.

Duer Ranch, Morrill County, Nebraska: Contrast between Cenozoic fluvial and eolian deposition

James B. Swinehart and R. F. Diffendal, Jr., Conservation and Survey Division, IANR, University of Nebraska-Lincoln, Lincoln, Nebraska 68588-0517

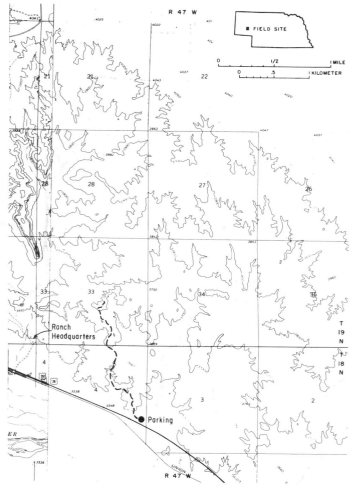

Figure 1. Map showing location of Duer Ranch locality. Base is from Tar Valley SW (contour interval of 20 ft) and Broadwater (contour interval of 10 ft) 7½-minute Quadrangles; scale 1:24,000. Dashed line is a suggested initial travel route that traverses many of the interesting features of this locality.

LOCATION AND ACCESSIBILITY

The Duer Ranch locality is situated on either side of U.S. 26 in southeastern Morrill County, Nebraska, halfway between the villages of Lisco and Broadwater and about 54 mi (90 km) southeast of Scottsbluff, Nebraska. The Rush Creek Land and Livestock Company currently owns all of sections 28, 29, 33, 34, and 35 of T.19N.,R.47W., as well as parts of sections 2, 3, 4, 10, and 11 of T.18N.,R.47W. (Broadwater and Tar Valley SW, 7½-minute Quadrangles; Fig. 1). The owners have allowed

geologists to study the exposures on ranch property provided that visitors stop at the ranch headquarters (Fig. 1) and obtain permission from the ranch foreman. Vehicles should be parked inside the gate on the north side of U.S. 26 in the SW¼Sec.3,T.18N., R47.W. (Fig. 1), and all study of the site should be done on foot. Be certain to close any gates that you open, and do not smoke on the property. Rattlesnakes are found on the ranch. If you have a small group and wish to examine the exposures in the southeast corner of the locality, two vehicles can park safely along the north side of the road cut on U.S. 26 in the NE¼ NW¼ sec. 10, T.18N.,R.47W.

SIGNIFICANCE OF THE LOCALITY

The Duer Ranch locality contains some of the finest and most easily accessible examples of different styles of alluvial cuts and fills in the Cenozoic rocks of Nebraska. It offers a unique area in which to examine the geometries and alluvial fills of several Miocene and Pliocene age paleovalleys and paleo-gullies. Good exposures of eolian volcaniclastic siltstones and a regionally important volcanic ash of the Oligocene age Brule Formation are also present at the Duer Ranch locality. In addition, Quaternary ephemeral stream development and deposits can be studied.

LOCALITY INFORMATION

Deposits of Oligocene through Quaternary age are exposed along the deeply incised, intermittent stream valleys on the Duer Ranch (Fig. 2 and measured sections I-IV). The oldest of these units is the Whitney Member of the Brule Formation, White River Group. The Whitney is usually considered the uppermost of the two Brule members recognized in western Nebraska (the other being the Orella). However, recently Souders and others (1980) and Swinehart and others (1985) have presented evidence that another unit, informally named the Brown Siltstone beds, can be recognized above the Whitney in many places in western Nebraska.

The Whitney Member is a massive to crudely bedded volcaniclastic siltstone containing abundant smectite and carbonate cement. There are at least two prominent vitric ash beds (tuffs) exposed in the Whitney (Fig. 2 and measured section III). The lowest of these ashes, easily visible from U.S. 26 in the NE¼ Sec.10,T.18N.,R.47W., is the Upper Ash of the Whitney, a regional marker bed for much of the Nebraska panhandle. The Lower Ash of the Whitney (Figs. 2 and 3), an even more extensive marker bed, was identified in test hole 24-A-53.

Siltstones of the Whitney Member are very well sorted and contain an average of 50 percent relatively unaltered rhyolitic

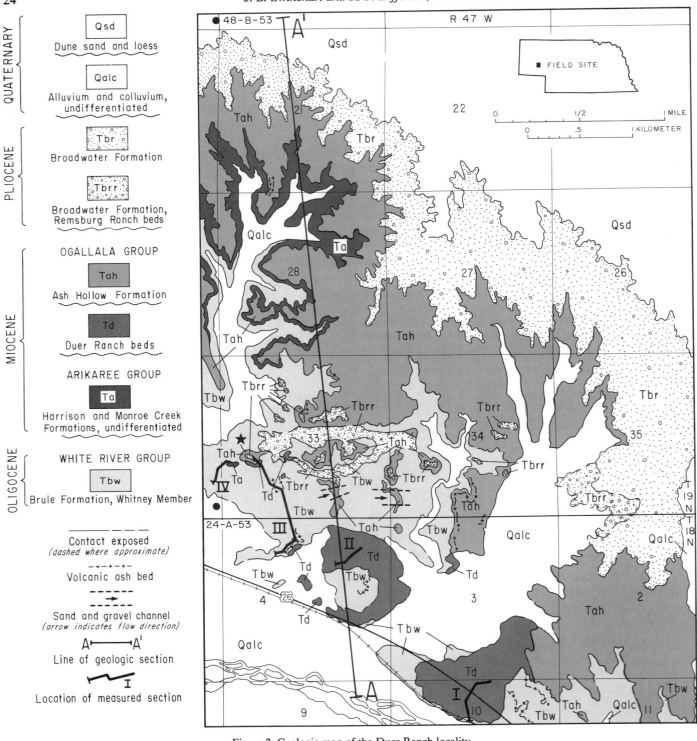

Figure 2. Geologic map of the Duer Ranch locality.

glass shards (Fig. 4) and an estimated 30 percent volcanically derived plagioclase and rock fragments in the coarse silt and very fine sand fractions. These characteristics, combined with the general lack of stratification and scarcity of fluvial sequences, suggest that most of the Whitney was deposited by the wind and

accumulated on upland surfaces where organisms and pedogenic processes destroyed most stratification. The pyroclastic material was derived from volcanic vents in the western United States with the San Juan volcanic field of southwestern Colorado one of the most probable sources. Regional studies of the Whitney

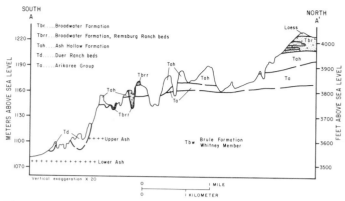

Figure 3A. Geologic section along line A-A'. Lower Ash of Whitney projected from Conservation and Survey Division test hole 24-A-53, 0.75 mi (1.2 km) west of section.

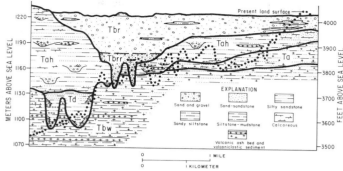

Figure 3B. Restored geologic section.

(Souders and others, 1980; and Swinehart and others, 1985) indicate that it is a blanket deposit with the Upper and Lower ash beds present over thousands of square kilometers of western Nebraska. These studies add additional strength to the concept of a primarily loessic upland origin for the Whitney.

The Arikaree Group is represented by the undifferentiated Harrison and Monroe Creek Formations that disconformably overlie the Brule Formation. These yellow brown to yellow gray, massive, very well-sorted volcaniclastic silty sands with calcareous "pipy" concretions occur only in the northwestern part of the locality (Fig. 2). Poorly defined horizontal bedding and locally abundant vertical tubules (burrows?) are the only common structures present in the Arikaree. The mineralogy of the Arikaree, like that of the Brule, is dominated by volcanically derived grains (Fig. 4). The Arikaree at the Duer Ranch locality is interpreted to be primarily an eolian deposit.

Unconformably overlying both the Arikaree and Brule is a fluvial sequence informally called the Duer Ranch beds (Figs. 2 and 3) and included within the Ogallala Group. The Duer Ranch beds consist of rock types ranging from claystones to coarse conglomerates. Brown, silty, very fine to medium-grained sands are the most common lithology (refer to measured sections I-III). These materials fill several paleogullies cut a minimum of 50 ft (15 m) below the Upper Ash of the Whitney Member and 165 ft (50 m) below the base of the Ash Hollow Formation. In the NE¼Sec.4,T.18N.,R.47W. (Fig. 2), the Duer Ranch paleogullies trend easterly, but less than 0.6 mi (1 km) east they trend southerly.

Coarse sediments, mostly colluvial deposits, are more common in the lower 50 ft (15 m) of the Duer Ranch and also near the edges of the gullies (Fig. 3B), where clasts of Brule siltstone and Arikaree sandstone up to 2 ft (0.6 m) in diameter are locally abundant. Pebbly lenses are generally less than 1 ft (30 cm) thick and are very limited in extent. Indistinct horizontal stratification is the most common sedimentary structure in the Duer Ranch, although some crossbed sets up to 1.5 ft (0.5 m) thick are present. Crossbed orientations essentially agree with the

gully trends defined by gully boundaries. At some sites near these boundaries, Duer Ranch strata have primary dips of up to 7°, and the gully sides dip as steeply as 20°. Gradients up to 100 ft per mi (20 m/km) occur along some of the paleogullies (measured section III) tributary to the main complex.

The fine-grained sediments in the Duer Ranch beds are similar in color and general appearance to the sandy silts of the Brule Formation, but the two units are very distinct mineralogically (Fig. 4).

The Duer Ranch beds represent a significant change from the eolian-dominated environments of the Brule and Arikaree. The complex probably represents a combination of fluvial and mass-wasting processes. The absence of pedogenic structures (pa-

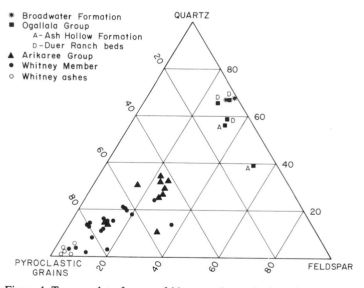

Figure 4. Ternary plot of quartz, feldspar, and pyroclastic grains in the very fine sand fraction of selected samples from the Duer Ranch locality. Pyroclastic grains include glass shards, rhyolitic to andesitic volcanic rock fragments and glass-mantled quartz, plagioclase, and heavy minerals. This class represents the minimum amount of volcanically derived material in any sample. Samples were sieved and treated with 10 percent HCl to remove carbonates and with 3 percent HF to remove authigenic clay (smectite) grain coatings.

leosol horizons) and unconformities within the fills suggests rapid sedimentation, and the Duer Ranch beds are interpreted to have been deposited by ephemeral streams.

Vertebrate fossils collected from the Duer Ranch beds place their deposition during the early part of the Clarendonian Land Mammal Age (about 10 to 12 Ma). These fossils include two horses, *Pseudhipparion* and a primitive *Pliohippus,* a camel, cf. *Protolabis,* and an antilocaprid, *Cosoryx.*

The Ash Hollow Formation rests disconformably on the Duer Ranch beds and is characterized by a hetereogeneous assemblage of sands, sandstones, sandy siltstones, sands and gravels, calcretes, silts and clays, and volcanic ashes. The typical carbonate-cemented sandstones and siltstones ("mortar beds") of the type Ash Hollow (Diffendal, this volume) are common. Ash Hollow sandstone are arkoses to subarkoses, usually containing less than 15 percent volcanically derived grains (Fig. 4). The 16- to 32-mm-size clasts of gravels are characterized by 80 percent granitic (including feldspar and quartz) pebbles and 10 percent rhyolitic to andesitic volcanic pebbles. In this locality, the cuts and fills of the Ash Hollow, in contrast to the Duer Ranch beds, are not deeply incised into older beds and lack steep gradients (Figs. 2 and 3). There is an excellent exposure of a 32-ft (10-m)-thick, sand-and-gravel-filled Ash Hollow channel in the SW¼SE¼Sec.33,T.19N.,R.47W. A light gray volcanic ash bed (vitric tuff) up to 20 ft (6 m) thick occurs in the SE¼SW¼-Sec.34,T.19N.,R.47W. (Fig. 2) and forms the base of a prominent vertical face. The abundance of siliceous rhizolith (= root cast) horizons and calcretes (refer to measured section III) indicates a slower rate of deposition for the Ash Hollow than for the Duer Ranch beds.

Diagnostic vertebrate fossils from the Ash Hollow Formation at this locality include two horses, *Dinohippus* and *Astrohippus,* and a rhinoceros, *Teleoceras.* These fossils allow assignment of an early Hemphillian Land Mammal Age (6 to 9 Ma) to this unit.

The Broadwater Formation of Pliocene age provides an example of a third style of alluvial deposition at the Duer Ranch locality. The formation consists primarily of sands and gravels within a major paleovalley that can be traced along the North Platte River valley for more than 95 mi (150 km).

The Broadwater Formation, as originally defined by Schultz and Stout (1945) from exposures in sections 20 and 21, T.19N.,R.47W., was composed of three members—a basal gravel, a middle finer-grained unit (the Lisco) and an upper gravel. Subsequent fieldwork and test drilling have demonstrated that there is more than one fine-grained unit in this part of the Broadwater (Fig. 3).

In this guide, the Broadwater Formation is divided into the lower Remsburg Ranch beds (a new informal name) and the upper, generally finer-grained alluvial fill originally described by Schultz and Stout (1945). The Remsburg Ranch beds were deposited primarily in the deep, narrow, anastomosing, bedrock-incised inner channels of the Broadwater paleovalley (Figs. 2, 3, 5, and 6). Breyer (1975) first suggested that these deposits were

Figure 5A. Remsburg Ranch beds within partially exhumed Broadwater Formation inner channels cut into Ash Hollow and Brule Formations. View looking ESE from starred location on Figure 2 (NW¼SW¼Sec.33, T19N.,R.47W.).

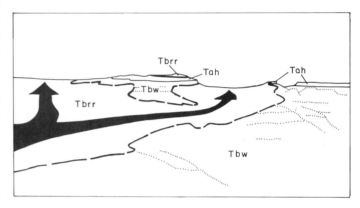

Figure 5B. Sketch of 5A with inner channel boundaries shown by heavy dashed line. Arrows indicate paleoflow direction.

part of the Broadwater Formation but only used indirect evidence to support this idea. Superposition of the Remsburg Ranch beds and the main body of the Broadwater now can be demonstrated in the SW¼SW¼Sec.35,T.19N.,R.47W. The Broadwater increases in thickness from about 60 ft (18 m) in the northwestern corner of the study area to about 264 ft (80 m) in section 35 (Fig. 2). In sections 33 and 34, the Remsburg Ranch beds occur as remnants within inner channel 525 to 1,100 ft (160 to 330 m) wide and up to 150 ft (46 m) deep (Figs. 3, 5, and 6). During the incision of the inner channels, several bedrock islands were formed. The morphology of these deep, narrow cuts is clearly visible from the air (Fig. 6) and resembles closely the inner channels of valleys described by Shepherd and Schumm (1974).

The maximum size of western source clasts (primarily quartzite derived from the Medicine Bow Mountains of southeastern Wyoming) allows differentiation of the two parts of the Broadwater. The intermediate diameters of the ten largest clasts from local sites within the Remsburg beds average 4 to 5.5 in (11 to 14 cm) while the same value for the main body of the Broadwater Formation is 2.7 to 3 in (7 to 8 cm). In addition, the Remsburg Ranch beds locally contain more mafic

Figure 6. Oblique aerial view looking west at anastomosing inner channels filled with sand and gravel of Remsburg Ranch beds. Arrows indicate directions of flow. "I" is bedrock island, "R" is ranch headquarters.

plutonic clasts and fewer sandstone and siltstone clasts than the upper part of the Broadwater. The Broadwater Formation is interpreted to have been deposited by a major braided river system.

The age of the Broadwater Formation based on fossil vertebrates is early to middle Blancan (2.5–3.5 Ma). Fossils from sites along the outcrop belt of the Remsburg Ranch beds indicate that this unit is somewhat older than the original Broadwater Formation.

Quaternary deposits in the Duer Ranch area consist of colluvium, alluvium, thin sandy loess, and dune sand. These deposits are complex and are worth examination by themselves. Some Holocene stream cuts have also exposed sequences of Quaternary buried soils.

MEASURED SECTIONS

Section I: base at center of south line of SE¼NE¼NW¼Sec.10,T.18N., R.47W. (Fig. 2) in a gully 49.5 ft (15 m) north of railroad tracks at elevation of 3,540 ft (1,080 m).
WHITE RIVER GROUP
Brule Formation - Whitney Member
1. Siltstone, slightly clayey, massive; upper 10 ft (3 m) covered .... 16 ft (4.9 m)
OGALLALA GROUP
Duer Ranch beds
2. Sand, very fine to coarse, pebbly, poorly sorted; thick, indistinct stratification, some graded beds and low-angle (10°), large-scale cross-beds; abundant intraformational claystone cobbles and boulders (up to 18 in [45 cm] in diameter); a spring occurs at the Whitney/Duer Ranch contact .... 15 ft (4.6 m)
3. Sand, very fine to very coarse, with some fine gravel lenses; some cross-bedded strata with both trough and planar cross-bedding, southerly paleoflow direction; mostly granitic pebbles with 5 to 10 percent acidic volcanics .... 20 ft (6.1 m)
4. Siltstone, very sandy; pebbly, yellow-brown (10YR 6/4), laminated; thinly interbedded with sandstone, very fine to medium; local thin claystones .... 18 ft (5.5 m)
5. Siltstone to silt, very sandy, pebbly; interbedded with thin sand to sandstone, very fine to fine; beds usually less than 6 in (15 cm) thick .... 8 ft (2.4 m)
6. Sand, very fine to fine, very silty; interbedded with poorly sorted,

pebbly, very fine to fine friable sandstone; local small fills of fine gravel; upper 24 ft (7.6 m) contains a few irregularly cemented sandstone ledges .... 52.5 ft (16 m)
7. Sand, very fine to medium, very silty, slightly clayey, locally pebbly; indistinct horizontal bedding; upper 13 ft (4 m) to hilltop are covered .... 24 ft (7.3 m)

SECTION II: base at NE¼NE¼SE¼NE¼Sec.4,T.18N.,R.47W., (Fig. 2) about 1,980 ft (600 m) north of U.S. 26 at elevation of 3,575 ft (1,090 m).
OGALLALA GROUP
Duer Ranch beds
1. Sand, very fine to fine, moderately silty; moderate, yellowish-brown (10YR 5/4); thick bedded, massive; interbedded with sandy silt; local pebbly sand lesnes; some medium-scale cross-bedding; most stratification is indistinct .... 35 ft (10.7 m)
2. Gravel and sand, gravel clasts up to cobble size and mostly composed of reworked Arikaree and Whitney concretions; granitic pebbles also present; fossil bone fragments common; unit is approximately 200 ft (60 m) wide and pinches out into sandy silts ..... 4 ft (1.2 m)
3. Sand, very fine to fine, very silty, poorly sorted; moderate yellowish-brown (10YR 5/4); indistinct, thick horizontal bedding, locally calcareous .... 16 ft (4.9 m)
4. Sand, very fine to fine, moderately silty, pebbly with local thin granitic gravel lenses; local cut and fill with a maximum of 12 inches (0.3 m) of relief .... 14 ft (4.3 m)
5. Sandstone, very fine to fine, moderately silty, calcareous, thin-bedded and cross-laminated; interbedded with thin clayey siltstones and claystones, yellowish-brown (10YR 6/2); some siltstones contain many small vertical tubules (burrows?) .... 35 ft (10.6 m)

SECTION III: base at center of south line SE¼NE¼NW¼Sec.4, T.18N.,R.47W. at elevation of 3,580 ft (1,092 m).
WHITE RIVER GROUP
Brule Formation - Whitney Member
1. Siltstone, moderately clayey; grayish orange (10YR 7/4) to moderate yellowish-brown (10YR 5/4); massive; compact; several thin 6- to 12-in (15- to 30-cm) mottled zones with irregular clumps of darker colored siltstone ("microbreccias" = reworked soil horizons?) containing small tubules and "vesicles;" small pods of barite crystals common in upper 10 ft (3 m) .... 32 ft (9.7 m)
2. Upper Ash bed, silt to siltstone, moderately sandy; very pale orange (10YR 8/2); 81 percent glass shards in very fine sand fraction; burrows common; gradational upper contact .... 4 ft (1.2 m)
3. Siltstone, slightly clayey; very slightly sandy; massive; several thin, lightly mottled zones or "microbreccias," as below ..... 46 ft (14 m)
4. Volcanic ash bed, siltstone to silt, moderately sandy; 88 percent glass shards in very fine sand fraction; unit pinches out to the east .... 4 ft (1.2 m)
5. Siltstone, slightly clayey; massive; small, irregular calcareous, cemented concretions in upper part; 3.3 ft (1 m) below top is a very silty, brecciated, white (N9) claystone; 3 to 7 in (8 to 17 cm) thick containing very small (<1 mm diameter) tubules ...... 25 ft (7.6 m)
6. Siltstone to silt, slightly sandy; yellow olive grey (5Y 6/2); massive; small, irregular, calcareous nodules; unit becomes sandier about 1,650 yards (500 m) east and contains carbonate-cemented, vertically oriented, and interconnected concretions; this unit may be equivalent to the base of the "Brown Siltstone beds" of Swinehart and others (1985) .... 5 ft (1.5 m)
7. Silt, very sandy; 71 percent glass shards in very fine sand fraction; fining upward; small tubules (burrows?) common; local oxidized zones; small, irregular concretions; gradational upper contact .... 8 ft (2.4 m)
8. Siltstone, very sandy, 68 percent glass shards in very fine

sand fraction; yellowish-gray (5Y 7/2); 1 cm thick indurated ledge at top .... 6 ft (1.8 m)
9. Siltstone, slightly sandy; moderate yellowish-brown (10YR 5/4); massive; a paleogully of the Duer Ranch beds (units 15-18) cuts almost to the base of this siltstone .... 21 ft (6.4 m)
10. Siltstone, slightly sandy; mottled yellowish-gray (5Y 7/2) and moderate yellowish-brown (10YR 5/4) "microbreccias" in upper 3.3 ft (1 m); small (1-mm) diameter vertically oriented tubules (burrows?) common .... 9 ft (2.7 m)
11. Siltstone, moderately sandy; yellowish-gray (5Y 7/2); massive .... 4 ft (1.2 m)
12. Siltstone, slightly to moderately sandy; white (N9) to pale greenish-yellow (10YR 8/2); many small burrows(?), as below ... 3.3 ft (1 m)
13. Siltstone, moderately sandy, fining upward, massive; small vertical burrows(?) common; 3 in (8 cm) white, calcareous siltstone at top .... 11 ft (3.4 m)
14. Siltstone, very sandy; 47 percent glass shards in very fine sand fraction; poorly defined horizontal laminations; abundant small (2-mm) diameter vertical tubules (burrows?) up to 24 in (60 cm) long, some siliceous rhizoliths in upper 24 in (60 cm); carbonate-cemented, vertically oriented, and interconnected concretions with indistinct boundaries .... 9 ft (2.7 m)

OGALLALA GROUP
Duer Ranch beds
15. Conglomerate, clasts up to 10 in (25 cm) in diameter, larger clasts composed of reworked Arikaree and Whitney calcareous concretions; this unit caps a small knob just south of the saddle where the main fill of a Duer Ranch paleogully (unit 16) is exposed. This conglomerate is not present in the fill and probably is a colluvial deposit at the gully edge .... 5 ft (1.5 m)
16. Silt to siltstone, very sandy with interbedded silty sand and pebbly sand lenses; moderate yellowish-brown (10YR 5/4); poorly defined horizontal bedding; this unit fills a narrow (330 ft [100 m]-wide) gully eroded into the Brule .... 69 ft (21 m)
Ash Hollow Formation
17. Sandstone, very fine to medium, slightly silty; calcareous; abundant siliceous rhizoliths .... 4 ft (1.2 m)
18. Sandstone, fine to coarse, locally pebbly, arkosic; interbedded with sandstone, very fine to fine, very silty; calcareous "mortar bed" ledges; several siliceous rhizolith horizons .... 35 ft (10.6 m)
Broadwater Formation—Remsburg Ranch beds
19. Gravel, sandy, larger clasts (maximum axis about 7.5 in [19 cm]) predominantly quartzite; mafic plutonic pebbles common, less than 1 percent anorthosite .... 8 ft (2.5 m)

SECTION IV: base of section in NE¼NE¼SE¼SE¼Sec.32,T.19N., R.47W. at elevation of 3,630 ft (1,106.4 m)
WHITE RIVER GROUP
Brule Formation—Whitney Member
1. Siltstone, moderately clayey, slightly sandy; pale yellowish-brown (10YR 6/4) to pale olive (10Y 6/2 in upper 8 in (20 cm); massive .... 8 ft (2.4 m)
2. Siltstone, moderately sandy; 74 percent glass shards in very fine sand fraction; mottled light grey (N7) to pale olive (10Y 6/2) "microbreccia;" abundant small tubules (burrows?) ..... 2 ft (0.6 m)

3. Siltstone, slightly to moderately sandy; grayish-orange (10YR 7/4) to yellowish-brown (10YR 5/4); massive; several thin, discontinuous "microbreccia" zones; 3.75-in (9.3-cm)-thick light gray (N7) to pale olive (10Y 6/2) siltstone occurs about 40 ft (12 m) above base of unit; mostly vertical exposures above this unit .... 81 ft (24.7 m)
4. Silt, very sandy; well sorted; 64 percent glass shards in very fine sand fraction, grayish-orange (10YR 7/4); massive, iron stain in upper 2 ft (60 cm); upper and lower contacts are indistinct; the base of this unit may be equivalent to the base of the Brown Siltstone beds of Swinehart and others (1985) .... 5 ft (1.5 m)
5. Siltstone, slightly sandy; pale olive (10Y 6/2) in upper 2-ft (60-cm) width; local staining; some small vertical tubules (burrows?) .... 17 ft (5.2 m)
6. Claystone; white (N9); brecciated, possible mud cracks .... 1 ft (0.3 m)
7. Siltstone, slightly sandy; massive; angular claystone clasts occur in basal 2 ft (60 cm); indistinct calcareous concretions .... 14 ft (4.3 m)
8. Siltstone, sandy; contains claystone fragments; 3-ft (90-cm) thick; a 6-inch (15-cm)-thick claystone; white (N9) at base and top of siltstone; 330 ft (100 m) east are 3 thin claystone beds in this unit; easily recognizable marker bed .... 4 ft (1.2 m)
9. Siltstone, very sandy; moderate yellowish-brown (10YR 5/4); the unit becomes slightly finer grained upward; small diameter burrows(?) common; a few thin calcareous ledges ..... 7 to 14 ft (2.1 to 4.3 m)

ARIKAREE GROUP (?)
Harrison and Monroe Creek formations, undifferentiated
10. Siltstone, very sandy at base grading up into very silty, very fine sandstone, 35 percent in very fine sand fraction; massive; locally abundant small vertical burrows(?), some siliceous rhizoliths; calcareous, vertically oriented concretions with indistinct boundaries; this unit could be interpreted to be part of the Brule Formation below .... 11 to 14 feet (3.4 to 4.3 m)

SELECTED REFERENCES

Breyer, J., 1975, The classification of Ogallala sediments in western Nebraska: University of Michigan Papers on Paleontology, v. 3, no. 12, p. 1–8.

Diffendal, R. F., Jr., 1982, Regional implications of the geology of the Ogallala Group (upper Tertiary) of southwestern Morrill County, Nebraska, and adjacent areas: Geological Society of American Bulletin, v. 93, p. 964–976.

Schultz, C. B., and Stout, T. M., 1945, Pleistocene loess deposits of Nebraska: American Journal of Science, v. 243, no. 5, p. 231–244.

——, 1948, Pleistocene mammals and terraces in the Great Plains: Geological Society of America Bulletin, v. 59, no. 6, p. 553–588.

Shepherd, R. G., and Schumm, S. A., 1974, Experimental study of river incision: Geological Society of America Bulletin, v. 85, p. 257–268.

Souders, V. L., Smith, F. A., and Swinehart, J. B., 1980, Geology and ground-water supplies of Box Butte County, Nebraska: Conservation and Survey Division, University of Nebraska, Nebraska Water Survey Paper No. 47, 205 p.

Swinehart, J. B., Souders, V. L., DeGraw, H. M., Diffendal, R. F., Jr., 1985, Cenozoic paleogeography of western Nebraska, *in* Cenozoic paleogeography of West-Central United States, R. M. Flores and S. S. Kaplan, editors, Rocky Mountain section S.E.P.M., Denver, Colorado, p. 209–229.

Ash Hollow State Historical Park: Type area for the Ash Hollow Formation (Miocene), western Nebraska

R. F. Diffendal, Jr., Conservation and Survey Division, IANR, University of Nebraska-Lincoln, Lincoln, Nebraska 68588-0517

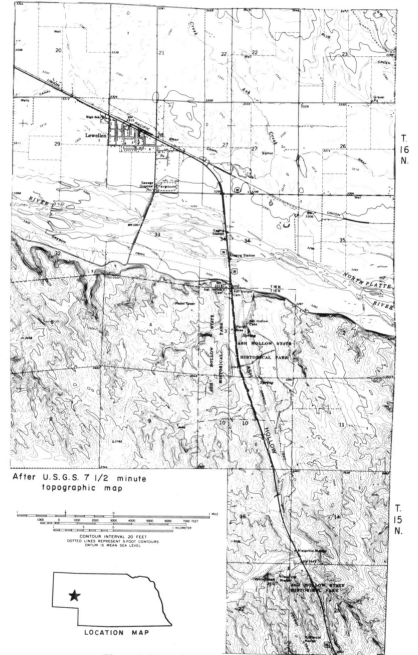

Figure 1. Map of the Ash Hollow area.

LOCATION AND ACCESSIBILITY

Ash Hollow State Historical Park, administered by the Nebraska Game and Parks Commission, is located primarily in Sec.3,N½,Sec.10, and the E½NE¼Sec.22,T.15N.,R.42W. (Fig. 1), Garden County, Nebraska. The park and adjacent lands are shown on the Lewellen and Ruthton 7½-minute Quadrangles; scale 1:24,000. U.S. 26 passes through the park. This highway and other paved roads provide access from I-80. A camping area

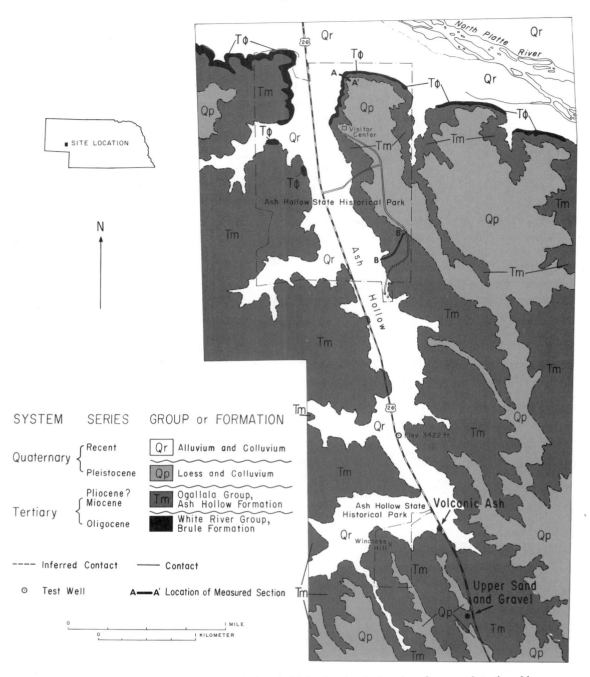

Figure 2. Geologic map of Ash Hollow Park and vicinity showing the location of measured stratigraphic sections.

and a museum with excellent interpretive displays are located in the park.

Rock exposures are easily accessible by foot in the park and at road cuts along U.S. 26 and nearby county roads. No collecting of specimens is permitted in the park, but samples may be collected from road cuts. Fossils found by people in and near the park have been added to the museum collections. Poison ivy can be found in abundance in some parts of the area. Rattlesnakes may live anywhere in this part of Nebraska.

SIGNIFICANCE

Ash Hollow is the type area of the Ash Hollow Formation, Ogallala Group. The formation is a major part of a regional aquifer extending from South Dakota to Texas and covering parts of eight states. Beds within the Ash Hollow contain the best record of late Cenozoic fossil grasses and forbs in the world. Major faunas of fossil vertebrates occur in the formation. The geology of the park area demonstrates that the formational geometry of continental deposits often must be worked out both from outcrop and well information. Furthermore, the human record at the park goes back at least 9,000 years.

GENERAL INFORMATION

Formations exposed in the park area are of Oligocene, Miocene, and Quaternary age (Figs. 2 and 3). The oldest rock unit, the upper part of the Whitney Member of the Brule Formation (White River Group), crops out at and near the mouth of the valley of Ash Hollow Creek. At these locations, the Whitney is a volcaniclastic siltstone exposed on steep slopes (Fig. 4A). The visible part of the member is readily divisible into three siltstone units (Figs. 2, 3, and 4B, and section AA'). The upper unit, which thickens to the west, contains vertically elongate, calcium-carbonate-cemented concretions. Paleosols occur in the middle unit.

The Ash Hollow Formation of the Ogallala Group rests disconformably above the Whitney. The Ash Hollow was probably first described by Stansbury (1852) and later by Engelmann (1876), who is credited with naming the formation. Lugn (1938, 1939) added to the description of the formation.

The basal part of the Ash Hollow Formation is a conglomerate whose grains larger than sand size are primarily composed of concretions derived from erosion by streams of the upper part of the Whitney. This conglomerate and similar beds above it are interstratified with sandstone (Figs. 3, 4B, C, and section AA').

Quaternary stream erosion has produced a peeled-back stratigraphic sequence. Because of this erosion, it is not possible to measure and describe the complete Ash Hollow sequence at any one section. Section BB' (Figs. 2 and 3) contains many of the lithologies typical of the Ash Hollow. These include siltstones, sandstones, sand and gravel bodies, conglomerates, pebbly sandstone mortar ledges, and caliche horizons (Figs. 2, 5A, and section BB').

Air-fall volcanic ash occurs in the Ash Hollow at many sites in the southern Nebraska panhandle. A typical example of a partly indurated light-gray ash crops out along the west side of U.S. 26 and can be sampled (Fig. 2).

Also outside of the park area but on public land is an exposure of sand and gravel of Pliocene(?) age above the Ash Hollow Formation (Fig. 5B). This unit contains distantly derived, stream-transported clasts larger than any found in the underlying Ash Hollow in the park area.

Given only information obtained from a study of the exposed, nearly horizontal Whitney-Ash Hollow contact and the nearly horizontal inclination of the beds in the Ash Hollow, it is easy to assume that the disconformity between the two rock units also is almost horizontal beneath the whole area. Test hole 2-B-72 (Fig. 2) was drilled by the Conservation and Survey Division of the University of Nebraska-Lincoln to aid in constructing a map of the configuration of the base of the Ogallala in this part of western Nebraska. The record of rocks encountered during the drilling (Fig. 3 and description of test hole) shows clearly how

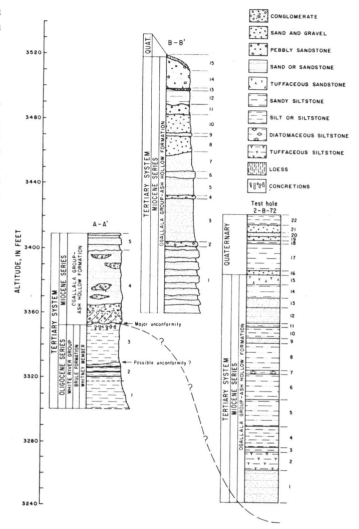

Figure 3. Measured stratigraphic sections. Locations shown on Figure 2.

deceiving judgments about formation geometry made solely on the basis of surficial studies can be. The Brule-Ogallala contact is not horizontal. It changes altitude by at least 100 ft (30 m) in 1 mi (1.6 km).

Thick Quaternary loess sequences occur to the south and southeast of the park. These wind-transported sediments are over 200 ft (60 m) thick in some areas and contain paleosols (Fig. 5C).

One early-man site has been found in the park area and another (Fig. 5D) nearby on private land. These sites have been found in alluvium beneath older stream terraces (Fig. 5A) or in other Holocene deposits. The Ash Hollow Creek area has been a haven for humans for at least 9,000 years because of its shelter and clear spring of water.

The two measured sections and one test hole log given below typify the complexity and rapid facies changes that occur in the Cenozoic deposits in this part of western Nebraska. A hike

A   B   C

Figure 4. Exposures in Ash Hollow: A. Brule Ash Hollow contact (solid line) on the east side of Ash Hollow Creek north of Ash Hollow Cave. Beds are numbered as in Section AA′ and Figures 3. P is a paleosol. Dashed line is edge of erosion surface, probably within the Whitney Member of the Brule Formation. B. Intermittent springs emerging along an erosion surface, probably within the Whitney Member of the Brule Formation. C. Ash Hollow basal conglomerate, bed 4, composed of concretions (c) eroded from bed 3.

through the park between the two section locations will reinforce these ideas of complexity and rapid lateral change.

Section AA′, east valley side at the mouth of Ash Hollow Creek, NW¼NE¼Sec.3,T.15N.,R.42W. (Fig. 3).

Tertiary System: Miocene Series
   Ogallala Group; Ash Hollow Formation
      Unit 5   Sandstone and sand, yellowish gray with pedotubules, fossil seeds, floating pebbles, 10 ft (3 m).
      Unit 4. Lithic conglomerate, cross-bedded and sandstone, massive, with calcrete at top; gravel clasts reworked from concretions; base and top irregular; seed fossils; 45 to 46 ft (13.5 to 13.8 m).
Tertiary System: Oligocene Series
   White River Group; Brule Formation, Whitney Member
      Unit 3. Siltstone, massive, fractured, grayish orange pink, with lime-cemented, vertically elongated concretions in the upper 4 to 5 ft (1.2 to 1.5 m), base and top irregular, 24 ft (7.2 m).
      Unit 2. Siltstone, massive, grayish orange pink, with up to three dark, weathered horizons (paleosols); top irregular, base nearly horizontal, 11.7 ft (3.6 m).
      Unit 1. Siltstone, massive, fractured, grayish orange pink, 17.5 ft (5.4 m).
Section BB′, in Ash Hollow State Park, on the east side of the valley, about 1,000 ft (300 m) north of parking lot at the

schoolhouse, SE¼NE¼ and NE¼NE¼Sec.10,T.15N.,R.42W. (Fig. 2).

Tertiary System: Miocene Series

   Ogallala Group; Ash Hollow Formation
      Unit 15. Covered interval, 7.2 ft (2.2 m).
      Unit 14. Sand and sandstone, light brown, pebbly, 17.5 ft (5.4 m).
      Unit 13. Lithic pebble conglomerate, light brown, 0.7 ft (0.2 m).
      Unit 12. Silt, light brownish gray, sandy, 9.0 ft (2.7 m).
      Unit 11. Sandstone, light brown, pebbly, 6.0 ft (1.8 m).
      Unit 10. Sand and gravel, pink granitic, with some bone fragments, 12.0 ft (3.69 m).
      Unit 9. Sandstone, light brown, 1.0 ft (0.3 m).
      Unit 8. Sand and gravel, pink granitic, partially cemented, with bone fragments, 10.2 ft (3.13 m).
      Unit 7. Sand and sandstone, pale yellowish gray, with pedotubules and seed fossils, 11.7 ft (3.6 m).
      Unit 6. Sandstone, pale yellowish gray, with pedotubules and seed fossils, 4.0 ft (1.2 m).
      Unit 5. Sand, very fine, pale yellowish gray, with pedotubules, 11.0 ft (3.3 m).
      Unit 4. Sandstone, massive, light gray, pebbly, 1.5 ft (0.46 m).
      Unit 3. Sand, very fine, light yellowish gray, 16.7 ft (5 m).

Figure 5. Exposures in Ash Hollow: A. Ash Hollow outcrops and Quaternary terrace deposits (1) and (2), east of Windlass Hill turnoff. B. Pliocene (?) gravel (hammer is 14 in [35.5 cm] long). C. Quaternary loess with buried soil (P). D. University of Nebraska State Museum field party excavating a bison kill site in Ash Hollow.

Unit 2. Sandstone, light brown, pebbly, 3.0 ft (0.9 m).
Unit 1. Sand and sandstone, pale yellowish gray, with pedotubules, seed fossils, and honeycomb caliche in some ledges, 42.0 ft (12.8 m).
Conservation and Survey Division, University of Nebraska-Lincoln, test well 2-B-72, NW¼NE¼NE¼SE¼,Sec.15,T.15N., R.42.W., altitude 3,422 ft (1027 m), Garden County, Nebraska.

Depth in ft (m)

Quaternary System: Pleistocene Series
Unit 22. Silt, yellowish brown, very fine sand — 0–6 (0–18)
Unit 21. Sand and gravel, sandstone pebbles, moderate yellowish brown, with granitic gravel — 6–11 (1.8–3.4)
Unit 20. Very fine sand, dark yellowish brown — 11–13.5 (3.4–4.1)
Unit 19. Gravel, sandstone pebbles, dark — 13.5–16

yellowish brown with granitic gravel — (4.1–4.9)
Unit 18. Silt, dusky yellowish brown (soil?) — 16–18 (4.9–5.5)
Unit 17. Silt and very fine sand, dark yellowish brown, some granitic and lithic gravel, finer grained at 27 ft (8.3 m) — 18–34.5 (5.5–10.5)
Unit 16. Sand and gravel gravel, pale to dark yellowish brown, siltstone clasts with granite — 34.5–37 (10.5–11.3)

Tertiary System: Miocene Series
Ogallala Group; Ash Hollow Formation
Unit 15. Sandstone, fine to very fine, yellowish gray, lime-cemented, very ashy in some fragments — 37–43 ft (11.3–13.1)
Unit 14. Sandstone, very fine to siltstone, — 43–51 ft

yellowish gray to light olive gray, calcite-cemented, some ash shards throughout, finer grained toward base   (13.1–15.5)

Unit 13. Sandstone, very fine, yellowish brown, much calcite cement decreasing downward, probably caliche   51–57 (15.5–17.4)

Unit 12. Sandstone, very fine, light olive gray, hackberry (*Celtis*) endocarps at top, thin caliche occasionally present (limier layers), pedotubules at 64 to 65 ft (19 to 20 m)   57–67 (17.4–20.4)

Unit 11. Sandstone, very fine to siltstone, yellowish brown, much calcite cement (caliche)   67–69.5 (20.4–21.2)

Unit 10. Siltstone with some very fine sand, dark yellowish brown with some pedotubules   69.5–76.5 (21.2–23.3)

Unit 9. Sand and sandstone, very fine to medium, moderate yellowish brown, variable amounts of cement, finer grained downward   76.5–80 (23.3–24.3)

Unit 8. Siltstone to very fine sandstone, dark yellowish brown, poorly cemented, with pedotubules, calcite cement increasing toward base   80–96.5 (24.3–29.4)

Unit 7. Diatomaceous silt, light olive gray   96.5–98.5 (29.4–30.0)

Unit 6. Siltstone to very fine sandstone, dark yellowish brown with pedotubules, calcite cement increasing downward   98.5–115 (30.0–35.0)

Unit 5. Siltstone to very fine sandstone, yellowish gray, much calcite cement (caliche), rodent incisor at 125 feet (38.1 m), coarser downward with fine sand at 125 ft (38.1 m), many ash shards at 125 ft (38.1 m) and below   115–132 (35.0–40.2 m)

Unit 4. Siltstone, very pale orange to pale yellowish brown, much calcite cement   132–145 (40.2–44.2)

Unit 3. Silt, light olive brown   145–148 (44.2–45.1)

Unit 2. Tuffaceous siltstone, dark yellowish brown, with biotite and muscovite   148–159.5 (45.1–48.6)

Unit 1. Sand and sandstone, fine to medium grained, granitic, loose, increasing calcite and silica cement below 165 ft (50.3 m), grains increasingly greenish below 165 ft (50.3 m), some finer grained beds at 172 to 173 ft (52.4 to 52.7 m), very well cemented at 173 ft (52.7 m) and at 179 ft (54.6 m)   159.5–179.2 (48.6–54.6)

REFERENCES CITED

Champe, J. L., 1946, Ash Hollow Cave: A study of stratigraphic sequence in the central Great Plains: University of Nebraska Studies, new series, v. 1, 130 p.

Darton, N. H., 1903, Preliminary report on the geology and water resources of Nebraska west of the one hundred and third meridian: U.S. Geological Survey Professional Paper 17, 69 p.

Diffendal, R. F., Jr., Pabian, R. K., and Thomasson, J. R., 1982, Geologic history of Ash Hollow Park, Nebraska: Conservation and Survey Division, University of Nebraska-Lincoln, Educational Circular 5, 33 p.

Engelmann, H., 1876, Report of the geology of the country between Fort Leavenworth, K. T., and the Sierra Nevada near Carson Valley, *in* Simpson, J. H., Report of explorations across the Great Basin of the territory of Utah for a direct wagon-route from Camp Floyd to Genoa, in Carson Valley, in 1859: Washington, U.S. Government Printing Office, p. 243–336.

Lugn, A. L., 1938, The Nebraska State Geological Survey and the Valentine problem: American Journal of Science, 5th Series, v. 36, no. 213, p. 220–228.

——1939, Classification of the Tertiary System in Nebraska: Geological Society of America Bulletin, v. 50, p. 1245–1276.

Stansbury, H., 1852, Exploration and survey of the valley of the Great Salt Lake of Utah, including a reconnaissance of a new route through the Rocky Mountains: Philadelphia, Lippincott, Grambo and Company, 487 p.

Stout, T. M., and others, 1971, Guidebook to the late Pliocene and early Pleistocene of Nebraska: Conservation and Survey Division, University of Nebraska-Lincoln, 109 p.

Thomasson, J. R., 1979, Late Cenozoic grasses and other angiosperms from Kansas, Nebraska, and Colorado: Biostratigraphy and relationships to living taxa: State Geological Survey of Kansas Bulletin 218, 68 p.

Pennsylvanian and Permian rocks associated with the Humboldt Fault Zone in southeastern Nebraska

Raymond R. Burchett, *Conservation and Survey Division, Nebraska Geological Survey, University of Nebraska-Lincoln, Lincoln, Nebraska 68588*

LOCATION AND SIGNIFICANCE

This outcrop of Pennsylvanian and Permian rocks is along the east side of Nebraska 105 about 3 mi (5 km) south of Humboldt in Richardson County, Nebraska. It is located in the SW¼Sec.22 and NW¼Sec.27,T.2N.,R.13E., as shown in Fig. 1, which reproduces a small part of the Humboldt SW 7½-minute Quadrangle. The rocks are exposed on public right-of-way in the road ditch east of the highway and are easily accessible by foot.

Here is a good exposure of the contact between the Pennsylvanian and Permian rocks and also one of the best localities in Nebraska to observe steeply dipping rocks associated with the Humboldt Fault Zone. The exposure here is approximately 170 ft (51.8 m) thick. The dip is 17°E.

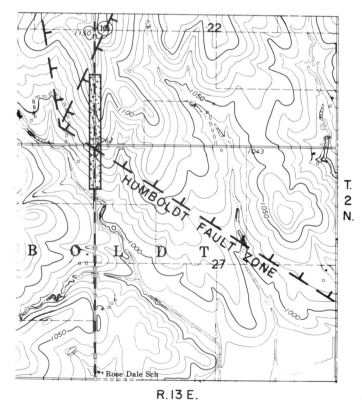

R.13 E.

Figure 1. Map showing location of field guide site (stippled) approximately 3 mi (5 km) south of Humboldt, Nebraska. Dashed lines show trace of Humboldt Fault Zone with tic marks on the downthrown side. For regional context, see Figure 3.

GEOLOGIC INFORMATION

Rocks of Pennsylvanian and Permian age crop out in Nebraska only in the eastern part of the state (Burchett, 1982a, p. 5). These surface exposures have been measured, described, and correlated by several geologists in the last hundred years (Burchett, 1979, p. 1, 4). Pennsylvanian rocks in Nebraska rest unconformably on rocks ranging in age from Precambrian to Mississippian. Generally, the unconformity is marked by a basal sand or a detrital zone of angular quartz sand and weathered chert.

Pennsylvanian and Permian strata are overlain by rocks ranging in age from Jurassic to Quaternary (Burchett, 1969). Where the Pennsylvanian-Permian contact is visible in southeastern Nebraska, a period of erosion without significant diastrophism is indicated; at least one channel sand of Permian age cuts approximately 100 ft (30.5 m) into Pennsylvanian rocks. Placement of the Pennsylvanian-Permian boundary has been summarized by Moore (1940, p. 298–305; 1949, p. 19–22) and by Mudge and Yochelson (1962, p. 116–127).

The thickness of Pennsylvanian rocks in Nebraska ranges from a featheredge in the northeastern part of the state to slightly more than 2,100 ft (6.40 m) in the extreme southeastern part (Burchett, 1982b, p. 5). Permian rocks in Nebraska are absent in the northeastern part of the state and have a maximum thickness of more than 1,500 ft (460 m) in the south-central part (Burchett, 1982b, p. 11). Outcropping Pennsylvanian and Permian rocks have a combined thickness of about 810 ft (250 m) in Richardson County. A composite section (Fig. 2) shows the generalized lithology of Pennsylvanian and Permian rocks in Nebraska. The Pennsylvanian sequence in Nebraska is characterized by a repetition of cycles of marine shale and limestone alternating with nonmarine sandstones and shales. The Permian is composed primarily of shale, limestone, and sandstone in the outcrop area of southeastern Nebraska.

Nebraska can be subdivided into several major geologic and tectonic provinces (Fig. 3). Two major structural features in eastern Nebraska are: the Nemaha Uplift, a north-south feature bounded on the east by the Humboldt Fault Zone, and a northeast-southwest feature, the Eastern Nebraska Uplift, bounded on the south by the Union Fault Zone. Formation of the Nemaha Uplift in post-Mississippian, pre–Des Moines (Middle Pennsylvanian) time accounts for the unconformable relation of Pennsylvanian strata to the underlying older rocks in eastern Nebraska. The area included in this uplift extends from the vicinity of Omaha southward across Kansas and into Oklahoma. In much of Johnson and Pawnee Counties, Nebraska, and in the

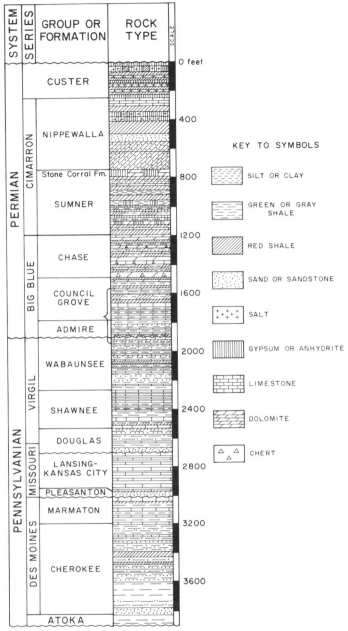

Figure 2. Composite section of Pennsylvanian and Permian rocks in Nebraska. Bracket left of rock type column identifies stratigraphic units exposed in Richardson County road ditch. (Modified from Burchett, 1982b).

adjoining part of Kansas, uplift was so great that all strata from Middle Ordovician thru Mississippian age were removed by concurrent and subsequent erosion, resulting in exposure of Precambrian rocks. Although deposition of Pennsylvanian strata in eastern Nebraska began during Atoka time, complete burial of the uplift did not occur until early Missouri time. Crustal unrest continued through the deposition of the Pennsylvanian rocks and into deposition of the Permian rocks. Contemporary subsidence

of the areas on either side of the arch resulted in deposition of thicker sequences of Pennsylvanian strata in the downwarped basins. Because the area immediately east of the arch subsided at a faster rate, the thickness of Pennsylvanian strata accumulating there was significantly greater than that on the west side. This area of subsidence—known as the Forest City Basin—includes parts of Nebraska, Iowa, Missouri, and Kansas.

The eastern margin of the Nemaha Uplift is sharply defined by the Humboldt Fault Zone. This zone of faulting and/or steep dips extends from Richardson County northward into Nemaha, Otoe, Cass, and Sarpy Counties and continues into Iowa. Rocks along this fault are upthrown on the west side.

An excellent example of faulting and/or steep dip is provided by this outcrop and a report on western Richardson County (Burchett and Arrigo, 1978). A composite section measured at this locality (Fig. 4) is described as follows:

Location. Richardson County, SW¼Sec.22 and along west line NW¼,Sec.27,T.2N.,R.13E., in east road ditch of Nebraska 105, approximately 3 mi (5 km) south of Humboldt, Nebraska.

Elevation. Top of Pennsylvanian (Brownville Limestone 1026 ft above mean sea level.

Remarks. Beds dip to northeast.

	Thickness	
	feet	*(meters)*
PERMIAN SYSTEM - Big Blue Series - Council Grove Group:	10.0	(3.04)
Beattie Formation:		
Cottonwood Member:		
Limestone, cream to light gray, finely to irregularly crystalline; contains abundant fusulinids	10.0	(3.04)
Eskridge Formation:		
Shale (covered interval)	50.0	(15.24)
Grenola Formation:		
Neva Member:		
Limestone, light gray, finely crystalline, massive, soft, limonitic staining, vuggy; contains fossil fragments; pitted; weathers dark brown	3.5	(1.06)
Shale, pale olive; contains thin limestone seams. Limestone, light gray, finely crystalline, thick-bedded, soft; contains brachiopod spines and crinoids interbedded with dark-olive shale	6.8	(2.00)
Salem Point Member:		
Shale, dark greenish-gray	5.0	(1.52)
Burr Member:		
Limestone, light gray, finely crystalline, thick-bedded; interbedded with greenish-		

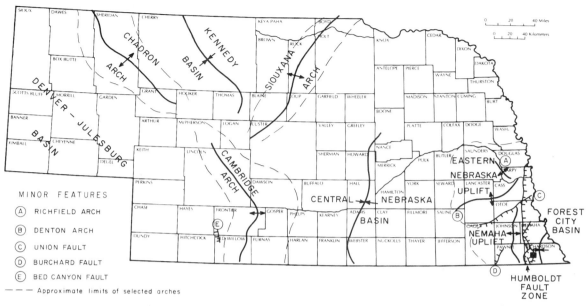

Figure 3. Map showing principal structural features of Nebraska and location of field guide site. (Modified from Burchett, 1983).

gray shale 12.0 (3.65)

Legion Shale Member:
Shale, light greenish-gray 2.0 (0.6)

Sallyards Limestone Member:
Limestone, medium gray, finely
crystalline 1.5 (0.46)

Roca Formation:
Shale, light greenish-gray 6.0 (1.83)
Shale, reddish-gray mottled with greenish-
gray 2.0 (0.6)
Shale, greenish-gray 7.0 (2.13)

Red Eagle Formation:
Howe Member:
Limestone, light to brown, irregularly crystalline, vuggy and porous; weathers dark
brown 3.0 (0.91)

Bennett Member:
Shale, dark gray 1.0 (0.3)
Shale, black, carbonaceous 4.0 (1.21)

Glenrock Member:
Limestone, brownish-gray, finely crystalline;
contains abundant fusulinids and some brachiopods 1.5 (0.46)

Johnson Formation:
Shale, light greenish-gray 5.0 (1.52)

Foraker Formation:
Long Creek Member:

Limestone, tannish-gray, very finely crystalline, thick-bedded, vuggy and porous 4.0 (1.21)

Hughes Creek Member:
Covered interval, normally shale interbedded with thin beds of limestone 0.5 (1.52)
Limestone, bluish-gray, very finely crystalline, one bed; contains abundant crinoids
and some brachiopods 0.6 (0.18)

Faulted Interval: Approximately 80 ft (24 m) of section is missing between the Hughes Creek Member and the West Branch Formation (this includes the lower part of the Hughes Creek Member, the Americus Member, the Hamlin Formation, the Five Point Formation, and the upper part of the West Branch Formation).

Admire Group:

West Branch Formation:
Covered interval (shale) 13.0 (3.96)
Shale, olive with interbedded olive claystone
lenses 6.0 (1.83)

Falls City Formation:
Lehmer Member
Limestone, yellowish-brown, pseudo-oolitic,
porous; one bed; contains *Osagia,* pelecypods, and gastropods; weathers pale
yellow 2.0 (0.6)

Reserve Member:
Shale, light greenish-gray 4.0 (1.21)

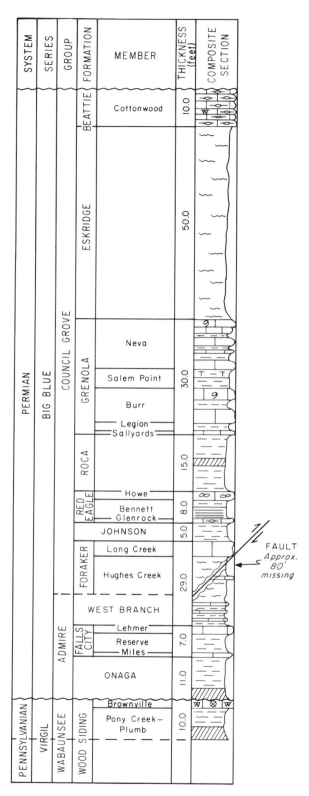

Figure 4. Composite section measured along east road ditch of Nebraska 105 in the SW¼Sec.22 and NW¼Sec.27,T.2N.,R.13E., Richardson County, Nebraska.

Miles Member:
Limestone, light gray, very thin-bedded, shaly; contains abundant brachiopods and crinoids; weathers medium gray mottled with yellowish-brown 0.9   (0.27)

Onaga Formation:
Shale, light greenish-gray 8.0   (2.43)
Shale, reddish-gray 2.0   (0.6)

PENNSYLVANIAN SYSTEM - Virgil Series - Wabaunsee Group:
Wood Siding Formation:
Brownville Member:
Limestone, pale yellow, very finely crystalline, very thin- and irregularly bedded; contains abundant brachiopods and crinoids 0.1   (0.03)
Limestone, pale yellow, very finely crystalline; one bed; contains abundant brachiopods and crinoids 0.8   (0.24)
Pony Creek - Plumb Members:
Shale, light greenish-gray 5.0   (1.52)
Shale, reddish-gray 3.0   (0.9)

Start of measured section.

On both sides of the fault the directions and angles of dip exhibited in these Pennsylvanian and Permian rocks differ markedly within very short distances, thus indicating that the structural pattern along the Humboldt Fault Zone is highly complex.

REFERENCES

Burchett, R. R., (compiler) 1969, Geologic bedrock map of Nebraska: University of Nebraska-Lincoln, Conservation and Survey Division, Nebraska Geological Survey Map. Scale 1:1,000,000.

—— 1979, The Mississippian and Pennsylvanian (Carboniferous) Systems in the United States—Nebraska: U.S. Geological Survey Professional Paper 1110-P., 15 p.

—— 1982a, Tectonics and seismicity of eastern Nebraska: University of Nebraska-Lincoln, Conservation and Survey Division, Nebraska Geological Survey Report of Investigation 6, 63 p.

—— 1982b, Thickness and structure maps of the Pennsylvanian and Permian rocks in Nebraska. University of Nebraska-Lincoln, Conservation and Survey Division, Nebraska Geological Survey Report of Investigation 7, 15 p.

Burchett, R. R., and Arrigo, J. L., 1978, Structure of the Tarkio Limestone along the Humboldt Fault Zone in southeastern Nebraska: University of Nebraska-Lincoln, Conservation and Survey Division Nebraska Geological Survey Report of Investigation 4, 112 p.

Moore, R. C., 1940, Carboniferous-Permian boundary: American Association of Petroleum Geologists Bulletin, v. 24, no. 2, pp. 282–336.

—— 1949, Divisions of the Pennsylvanian System in Kansas: Kansas Geological Survey Bulletin 83, 203 p.

Mudge, M. R., and Yochelson, E. L., 1962, Stratigraphy and paleontology of the uppermost Pennsylvanian and lowermost Permian rocks in Kansas: U.S. Geological Survey Professional Paper 323, 213 p.

The Late Cretaceous Niobrara Formation in south central Nebraska

Roger K. Pabian, Conservation and Survey Division, IANR, University of Nebraska-Lincoln, Lincoln, Nebraska 68588-0517

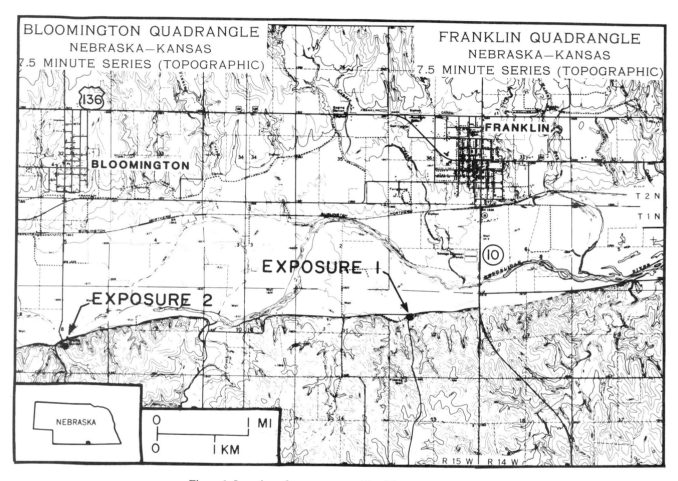

Figure 1. Location of exposures near Franklin and Bloomington.

LOCATION AND ACCESSIBILITY

Excellent outcrops of Niobrara and Pleistocene strata can be seen near Franklin and Bloomington (Exposures 1 and 2, Fig. 1; Figs. 2-5) in Franklin County. The Niobrara Formation generally dips westward at about 15 ft/mi (3 m/km), and at the western end of the outcrop area in Harlan County, the Niobrara is conformably overlain by the Pierre Shale.

All of the outcrops are road cuts and are within public access. The outcrop near Franklin has a very steep north slope and a fairly gentle east slope. The section at the Bloomington bridge is very steep but negotiable by persons without physical limitations. The outcrops near Red Cloud are fairly steep but are also negotiable by persons without physical limitations.

SIGNIFICANCE

The exposures of Late Cretaceous strata in Webster and Franklin Counties, Nebraska (Figs. 1, 6) present a classic example of shallow marine sedimentary rocks that were deposited during the last marine invasion of the North American midcontinent. The erosional surface at the top of the Cretaceous section, which underlies both Pliocene and Pleistocene deposits in this area, also shows a disconformity of regional significance.

EXPOSED SECTIONS

The section south and west of Franklin is situated in the SW¼NW¼Sec.12,T.1N.,R.15W., in Franklin County. It con-

Figure 2. Outcrop of Niobrara Formation (Exposure 1) near Franklin.

sists of about 32 ft (10 m) of chalk with bentonitic seams and is unconformably overlain by Pleistocene alluvium and the Peoria and Bignell Loesses. Although cut through for a road, the Pleistocene here shows reversal of topography. This section is shown in Figures 2 and 3, and the measured units are as follows:

QUATERNARY SYSTEM—PLEISTOCENE SERIES

Bignell Loess
    Unit 7. Loess, light gray, silty, apparent thickness about 2 ft (0.61 m) at top of highest bed of Niobrara Formation.
Peoria Loess
    Unit 6. Loess, light yellow-brown to light gray, silty with humic or soil horizon on top 1.0 ft (0.3 m). Apparent thickness 3.3 ft (1.0 m). Thickens eastward. Soil horizon pinches out eastward.
Alluvium
    Unit 5. Stream deposits, light gray silty sand and sandy silt. Sandier near base and siltier near top. Contains fossil snails, bivalves, small vertebrate fossils. Up to 15 ft (4.5 m).

CRETACEOUS SYSTEM—LATE CRETACEOUS SERIES

Niobrara Formation
  Smoky Hill Chalk Member
    Unit 4. Limestone, light orange-brown to orange-brown, silty, concretionary to knobby. Weathers with box-work-like structures. Some fossils. 15 ft (4.6 m) exposed.

Unit 3. Limestone, massive, light gray to light orange-brown. Beds thinner in lower 3.3 ft (1 m) and slabby in upper 2.6 ft (0.8 m). Thickness, 7.9 ft (2.4 m).
Unit 2. Limestone, light bluish-gray, silty, medium crystalline. Weathers into thin beds. Thickness 2.6 ft (0.9 m).
Unit 1. Covered interval. 6 ft (1.83 m).

A very important site is the classic Bloomington bridge section (Exposure 2, Fig. 1, Figs. 4, 5) located in the NW¼SW¼ Sec.8,T.1N.,R.15W., Franklin County. This section was first described by Miller and others (1964), although it had been a stop for numerous geology field trips in previous years and had served as an important section in Loetterle's (1937) study of Niobrara microfossils. The section here (Figs. 4, 5) consists of about 82 ft (25 m) of chalk with about 25 interbedded seams of bentonite ranging up to 4 in (0.1 m) thick. The Smoky Hill Chalk here is unconformably overlain by Pleistocene Peoria and Bignell Loesses, which show an excellent example of reversal of topography (Fig. 4).

Miller and others (1964, p. 73, 74) published an extremely detailed section of the Bloomington bridge site. Many of their minor units appear to be locally discontinuous, and the following section (Fig. 5) that delineates only the major lithological units is presented here.

QUATERNARY SYSTEM—PLEISTOCENE SERIES

Loess, Peoria and Bignell Formations, Undifferentiated.
    Unit 11. Wind-deposited silt. Grass-covered, badly slumped. Not measured.

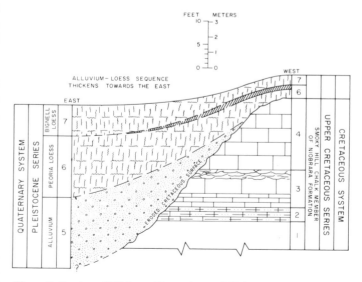

Figure 3. Section of Niobrara Formation and Pleistocene units at Exposure 1 near Franklin.

Figure 4. Outcrop of Niobrara Formation (Exposure 2) near Bloomington. Note reversal of topography.

CRETACEOUS SYSTEM—LATE CRETACEOUS SERIES

Niobrara Formation

Smoky Hill Chalk Member

Unit 10. Limestone, mottled, pale yellow-orange to grayish-orange, silty. Thin bentonite parting about 11.5 ft (3.5 m) above base, 23.5 ft (7.16 m).

Unit 9. Limestone, light bluish-gray, silty, massive to hackly, weathers into thin beds. Contains numerous bentonite seams. 16.8 ft (5.12 m).

Unit 8. Bentonite, white, mottled, stained dark yellow-orange. 3.5 in (0.09 m).

Unit 7. Limestone, light bluish-gray, silty, massive to hackly. Weathers into thin beds. 1.4 ft (0.43 m).

Unit 6. Bentonite, white, mottled, stained dark yellow-orange. 3.5 in (0.09 m).

Unit 5. Limestone, light bluish-gray, silty, massive to hackly. Weathers into thin beds. Contains numerous bentonite seams. 8.6 ft (2.62 m).

Unit 4. Bentonite, white, mottled, stained dark yellow-orange. 2.4 in (0.06 m).

Unit 3. Limestone, light bluish-gray, silty, massive to hackly. 12.5 ft (3.81 m).

Unit 2. Covered interval 16 ft (5.0 m).

Unit 1. Limestone exposure in river bottom. 3.3 ft (1.0 m) measured in river bottom above water level.

The older Fort Hays Member of the Niobrara Formation is restricted to the eastern extreme of the outcrop area in eastern Webster County. There are some exposures of the Niobrara

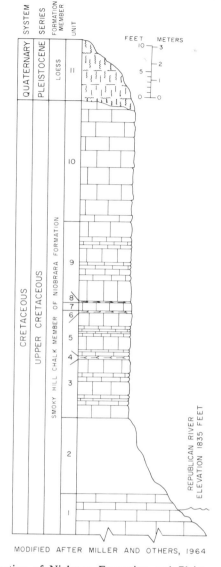

MODIFIED AFTER MILLER AND OTHERS, 1964

Figure 5. Section of Niobrara Formation and Pleistocene units at Exposure 2 near Bloomington.

Formation south of Guide Rock in Webster County; however, these are on the opposite side of a deep irrigation ditch and generally are not easily or safely accessible.

The Cretaceous and Pleistocene strata near Red Cloud in Webster County (Exposures 3 and 4) have been described by Pabian (1980); he recorded about 79 ft (24 m) of the Smoky Hill Chalk Member of the Niobrara Formation in the SW¼ SE¼SE¼Sec.11,T.1N.,R.11W., there. This is overlain unconformably by up to 30 ft (9 m) of Quaternary-age Peoria Loess and 5 ft (1.5 m) of Bignell Loess. In the NE¼NE¼NE¼,Sec. 14, T.1N.,R.11W., the Niobrara Formation has been deeply incised by a stream and is unconformably overlain by over 20 ft (6 m) of alluvium and up to 30 ft (9 m) of loess.

PALEONTOLOGY

The Niobrara Formation in south central Nebraska is best known for the very large inoceramid bivalve shells that are completely covered by the small commensal oyster, *Ostrea congesta* Conrad. Loetterle (1937) reported a diverse microfauna of ostracods and foraminifers from the Niobrara and indicated that the foraminifers, *Bolivina crenulata* Loetterle, *Loxostoma applinae* (Plummer) and *Schackoina trituberculata* (Morrow) were common in the upper 50 ft (15 m) of the Smoky Hill Chalk Member and served as good index fossils for this stratigraphic interval in Nebraska, Kansas, and South Dakota. Loeblich (in Miller and others, 1964, p. 12) identified 16 species of foraminifers from the Niobrara Formation in Franklin County.

Specimens of *Baculites* and two species of the ammonoid *Clioscaphites* have been recovered from near Red Cloud. *C. saxitonianus* (McLearn) and *C. choteauensis* Cobban have been reported from the Pueblo, Colorado, area by Scott and Cobban (1964), suggesting a middle Santonian age for the strata here. Several undetermined genera of bivalves and sparse gastropods are also known from the Red Cloud area.

Teeth from the ray, *Ptychodus,* as well as teeth from isurid sharks and the scales and vertebra of other fishes, are known from the Franklin area.

Pleistocene invertebrate fossils consist mainly of land and aquatic snails, as reported by Leonard (1947, 1950, 1952).

Schultz and Tanner (1957) reported a diverse assemblage of medial Pleistocene vertebrate fossils from this area. The varieties include rabbits, gophers, prairie dogs, beavers, mice, coyotes, elephants, horses, peccaries, camels, and deer.

SELECTED REFERENCES

Leonard, A. B., 1947, Yarmouthian molluscan fauna of the Great Plains and Missouri Valley Pleistocene [abs.]: Geological Society of America Bulletin, v. 58, no. 12, p. 1202.

——1950, A Yarmouthian molluscan fauna in the mid-continent region of the United States: Kansas University Paleontologic Contribution 8, Mollusca, article 3, 48 p.

——1952, Illinoinan and Wisconsin molluscan faunas in Kansas. Kansas University Paleontologic Contribution 9, Mollusca, article 4, 38 p.

Loetterle, G. J., 1937, The micropaleontology of the Niobrara Formation in Kansas, Nebraska, and South Dakota: Conservation and Survey Division, University of Nebraska, Bulletin 12, 2nd Ser., 67 p.

Miller, R. D., Van Horn, R., Dobrovolny, E., and Buck, L. P., 1964, Geology of Franklin, Webster, and Nuckolls Counties, Nebraska: U.S. Geological Survey Bulletin 1165, 91 p.

Pabian, R. K., 1980, Geology along the Republican River near Red Cloud, Nebraska: Conservation and Survey Division, IANR, University of Nebraska, Lincoln, Field Guide 9, 25 p.

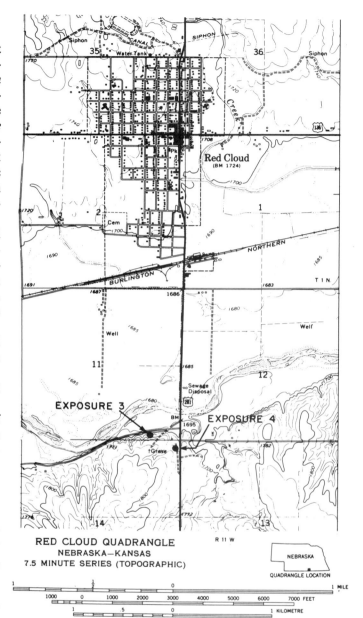

Figure 6. Locations of exposures near Red Cloud.

Schultz, C. B., and Tanner, L. G., 1957, Medial Pleistocene fossil vertebrate localities in Nebraska: Nebraska State Museum Bulletin, v. 4, no. 9, p. 59–81.

Scott, G. R., and Cobban, W. A., 1964, Stratigraphy of the Niobrara Formation at Pueblo, Colorado: U.S. Geological Survey Professional Paper 454L, 30 p.

Glacial and interglacial stratigraphy, Hudson Bay Lowlands, Manitoba

L. A. Dredge, Geological Survey of Canada, 601 Booth Street, Ottawa, Ontario K1A 0E8, Canada
E. Nielsen, Manitoba Department of Mines, 535-330 Graham Avenue, Winnipeg, Manitoba R3C 4E3, Canada

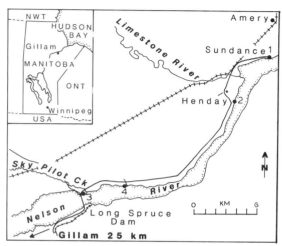

Figure 1. Location map.

Unit	Age
Tyrrell Sea sediments	Holocene
Lake Agassiz clay	
Turbidity flow sediments	
Sky Pilot till	Wisconsinan
Long Spruce till	
Nelson River sediments	Sangamonian
Amery till	"Illinoian"
Paleosol	Oxygen Stage 7?
Sundance till	Oxygen Stage 8? 12?

Figure 2. Stratigraphic column.

LOCATION

Extensive sections along the Nelson River in northern Manitoba lie within NTS sheet 54 D (1:250,000). The nearest town, Gillam, can be reached by air or road from Winnipeg, Manitoba. From Gillam, follow the road eastward across Long Spruce Dam and proceed toward Sundance. Access to each site (Fig. 1) is obtained by parking the vehicle along the roadside and walking a very short distance to the bluff.

SIGNIFICANCE

The sections lie in the heart of the area covered by the Laurentide ice sheet, and are the best exposed and most readily accessible in the Hudson Bay Lowlands. Nine stratigraphic units (Fig. 2) record three glaciations and intervening nonglacial intervals, shifting centers of glacial outflow, and postglacial inundations by glacial lakes and seas. These units, and correlative deposits on adjacent rivers (Nielsen and Dredge, 1982; Dredge and Nielsen, 1985; Dredge and Cowan, 1987), provide a regional stratigraphic framework for the lowlands surrounding Hudson Bay. The age and rank of nonglacial marker beds in the Hudson Bay region has been a subject of considerable controversy for the past 20 years (Terasmae and Hughes, 1960; McDonald, 1968; Skinner, 1973; Shilts, 1982; Andrews and others, 1983). The various arguments have been summarized by Dredge and Cowan (1987). The record for western Hudson Bay, described here, suggests continuous glacial cover during the Wisconsinan Stage and a Sangamonian age for the major subtill nonglacial deposits. The sites described below record the following sequence of events:

1. The earliest recorded ice flow was southward from a center in Keewatin. Glaciers deposited a sandy till (Sundance till) containing granitic clasts. This till is presumed to be older than oxygen isotope stage 8.

2. The bleached and oxidized upper part of the Sundance till is interpreted as a paleosol. The pollen assemblage suggests that it formed under subarctic conditions, possibly during oxygen isotope stage 7.

3. During the subsequent glaciation, ice flowed westward across the region from a center in central Quebec. It deposited a silty calcareous till (Amery till) containing foraminifera and shell fragments, as well as distinctive Proterozoic erratics derived from eastern Hudson Bay. The till is probably Illinoian in age, but could possibly be mid-Sangamonian if the underlying paleosol formed during oxygen isotope stage 5e instead of stage 7.

4. A major interglacial sequence of sand, peat, and thinly bedded silt (Nelson River sediments) was deposited over the Amery till during a period when Hudson Bay was ice free. On the basis of their stratigraphic position, climatic indications, and correlation with the Missinaibi Formation in northern Ontario, these units are assigned to a cool phase of the Sangamon Interglaciation.

5. The Wisconsin glaciation is represented by a thick, calcareous, silty till sheet, which is grey at the base (Long Spruce till) but becomes brown near the top (Sky Pilot till). The till was emplaced by ice flowing out of, or across Hudson Bay. The brown color in the upper part results from the inclusion of Devon-

ian reddish shale and siltstone, which underlies the south-central part of Hudson Bay; it relates to a shift in the Hudson ice flow trajectory during the late phases of glaciation.

6. Because of the natural slope of the land and isostatic depression created by the Laurentide Ice Sheet, Glacial Lake Agassiz expanded northward as the glacier retreated. This vast lake extended in various stages from the Dakotas to northern Manitoba. Loose brown tilly diamictons, sandy turbidity grain flow deposits, and varved clays are associated with deposition from the ice margin into Glacial Lake Agassiz.

7. Upon disintegration of the ice sheet occupying Hudson Bay about 8,000 years ago, the Gillam area was inundated by the Tyrrell Sea. This water body began as a high-level sea, at an elevation of 425 ft (130 m) ASL, and gradually regressed to the present level of Hudson Bay.

STOP 1. SUNDANCE SECTION

Access to the cliffs is through the town of Sundance. The 100-ft-high (30 m) bluff can be descended near the southeastern corner of the townsite by following a 165 ft (50 m) cut line linking the roadway and the river. Boulders at the base of the cliff illustrate the variety of transported erratics derived from the tills. These include limestone, iron formation, concretionary gray-wacke, and pink arkose of eastern provenance; and granites, pebble conglomerates, and mauve volcanics from northern sources. Precambrian basement rocks form the rapids, while Paleozoic dolomitic limestone forms the base of the section (Fig. 3). The limestone is striated 240° to 260° and 170° to 190°. The former record a very early glaciation; the latter were produced by the ice flow that created the Sundance till.

A dense, compact gray till (Sundance till; Nielsen and others, 1986) that tends to crumble into small fragments directly overlies the bedrock. This till is sandier than other tills in the area, being approximately 10% gravel, 47% sand, 32% silt, and 12% clay. About half the clasts are carbonates derived from the underlying bedrock, and half are undifferentiated "granites" transported from the northwest. The <63um fraction of the Sundance till comprises 16% carbonate. The sandy texture, granitic lithology, the NNW–SSE pebble fabric of the till, and the striations on the adjacent bedrock indicate that the Sundance till was deposited by ice flowing southward from a Keewatin source area.

The upper 6.6 ft (2 m) of the Sundance till contains a bleached and oxidized zone, considered to be the base of a paleosol which is traceable for 500 ft (150 m) along the bluff. Its position corresponds to a boulder pavement. Microscopic examination shows an abundance of charcoal fragments and a moderate amount of pollen. The abundant nonarboreal pollen and minor birch assemblage is indicative of tundra conditions. The terrestrial nature of the unit demands an unglaciated Hudson Bay.

The Amery till (Klassen, 1986) truncates and directly overlies the paleosol. This till is silty, calcareous, and contains transported foraminifera and mollusc fragments. The till is blocky and light olive-gray, with brown oxidation rinds along joint surfaces.

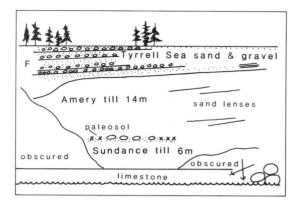

Figure 3. Sundance section, Stop 1.

The till typically consists of 8% gravel, 27% sand, 40% silt, and 25% clay. The pebble fraction is 80% carbonates, 15% "granite," and 5% graywacke, including pebbles from the Omaralluk Formation on the east side of Hudson Bay. The silt-clay fraction contains 25% carbonate. The till contains discontinuous sand seams varying from 0.4 in (1 cm) to more than 3 ft (1 m) in thickness. These intratill units suggest a basal melt-out origin for the till. The fine texture, erratics of eastern provenance, pebble fabric trending northeast to southwest, and striations oriented 225° on the faceted tops of boulders situated along the contact with the underlying Sundance till indicate that the ice center responsible for the Amery till was in central Quebec. The Amery till is probably Illinoian in age. The abundant Quaternary fossils in the till were derived from marine sediments laid down in Hudson Bay.

About 6.5 to 13 ft (2 to 4 m) of sandy marine gravel cap the Amery till and overlie it unconformably. *Hiatella arctica* molluscs, aged 7,180 ± 70 BP (GSC-3326), date the regression of the postglacial isostatic sea from this site.

STOP 2. HENDAY SECTION

Approximately 0.9 mi (1.5 km) west of the Henday switching station at a major bend in the road, the section (Fig. 4) is reached easily by walking 500 ft (150 m) through the bush and descending the 100-ft-high (30 m) cliff. The bottom 40 ft (12 m) of the section consists of the compact, gray, oxidized Amery till. The steep weathering base formed by this till distinguishes it from other tills exposed higher in the section. The petrologic and textural character, and pebble fabric of this till, is similar to the Amery till at the Sundance section, with which it is correlated. Shell fragments in the till gave an aspartic D/L ratio of 0.423, which is also comparable with those in the Amery till at Sundance.

The top of the Amery till rapidly grades to a 10-ft-thick (3 m) nonglacial unit comprising compressed peat, including spruce twigs, and interbedded fine-textured sediment. This unit, called the Nelson River sediments, can be observed at numerous sites along the Nelson River between Long Spruce dam and Sun-

Figure 4. Henday Section, Stop 2. Amery till (A), overlain by Nelson River interglacial sediments (B), overlain by Long Spruce and Sky Pilot tills (C), and Tyrrell Sea deposits (D).

Figure 5. Long Spruce dam, Stop 3. Turbidity flow deposits (A), Lake Agassiz glaciolacustrine varves (B), and Tyrrell Sea marine silt (C). GSC Photo 204314-X.

dance. Pollen profiles from these sediments are dominated by spruce, with pine and some birch. Grasses and sedges are abundant. This assemblage is similar to modern pollen spectra from forest-tundra ecotones. The pollen profiles are similar to, and are correlated with, those of the Missinaibi Formation in northern Ontario (Terasmae and Hughes, 1960; Skinner, 1973). Aspartic and D/L ratios on wood are also similar (0.18 to 0.22). Beetle assemblages from the peat, particularly the species *Blethisa catenaria,* indicate pond environments at or above treeline, such as the present-day tundra ponds at Churchill 100 mi (150 km) to the north. The paleoecological studies suggest that the Nelson River sediments at this site represent a cool phase of an interglaciation, when temperatures were 3° to 4°C cooler than those at Gillam today. Warmer indicators have been found at other sites. The Nelson River sediments are assigned a Sangamonian age, based on stratigraphic position, paleoecologic interpretations, and similarity with the Missinaibi Formation.

Interglacial deposits are sharply truncated by the light olive-gray Long Spruce till, which has very thin oxidation rinds around joints.The pebbles from this till are 24% "granites", 72% carbonates, and 4% graywackes. The silt plus clay fraction averages 24% carbonate. It is texturally similar to the Amery till, but is less compact. Shear planes are common at this site. Marine mollusc fragments and foraminifera are also common, indicating in part a marine interglacial source deposit for this till. The composition of the till suggests an eastern source. Pebble fabrics are variable but generally suggest flow in a southwesterly direction. Long Spruce till is assigned a Wisconsinan age, and may span from the early to the late Wisconsinan.

The Long Spruce till grades upward into a brown till (Sky Pilot till), which forms the surface till throughout most of the region. At Henday this till is compact, fissile, and slightly oxidized. It is considered to be a subglacial basal melt-out deposit. At other sites, however, its upper part is relatively soft, unoxidized, and quasistratified: this facies may have been deposited by an ice sheet retreated against a glacial lake. The brown till has a pebble lithology similar to that of the gray tills, but has more clay (46%). The pebble fabric suggests westerly ice flow. The brown color is due to the incorporation of red-brown siltstones derived from Devonian formations in south-central Hudson Bay. The till is considered to be of late Wisconsinan age. The tills of the Henday section therefore suggest that the area was continuously ice covered during the Wisconsinan Stage but that there were gradual shifts in ice-flow trajectory during the glaciation.

The upper 13 ft (4 m) of the section consist of 20 in (50 cm) of lacustrine varved clay, overlain by 11 ft (3.5 m) of marine sand and gravel.

STOP 3. LONG SPRUCE DAM SECTION

A gravel access road on the north front side of the dam leads to a small section that clearly illustrates the upper part of the glaciolacustrine sequence and the overlying marine deposits. The lowermost unit (Fig. 5, unit A) is a sandy diamicton, with sand inclusions and pebbles suspended in a silty sand matrix. Flow folds and streamlined fluid forms indicate deposition under turbid conditions. Faint horizontal lineations near the top of the unit suggest a change to laminar flow, and a less dense sediment-water mix. This unit is an ice-proximal glaciolacustrine sediment, deposited where sediment laden meltwater debouched into deep water from conduits at the base of the ice. At Long Spruce, the contact between this unit and the overlying rhythmites is abrupt (erosion interval?), and is marked in some places by a pebble lag deposit.

The second major unit is a glaciolacustrine varve sequence composed of alternating beds of brown and gray silty clay (unit B). The brown clay layer contains brecciated rip-ups of the gray unit, and is thus considered to have been deposited under higher energy conditions. Eighty couplets were counted at this site. Both parts of the couplet thin upward from a maximum of about 0.4 in (1 cm) at the base, reflecting increasing distance from the source sediments. The upper 20 in (50 cm) is a "massive" or microlaminated red-brown clay. The varves were deposited into Glacial Lake Agassiz.

The red-brown clay is overlain by gray-brown marine clay (C); the fact that there is so little disruption between the upper part of the lacustrine sediment and the base of the marine unit suggests that the two water bodies might have been at about the same level at the time of the change over. The marine clay grades upward into a poorly bedded buff sandy silt containing marine molluscs. The uppermost part of the section is a brown wavy-laminated silt, which may be a tidal deposit. Paired valves of *Macoma calcarea* from a thick marine section nearby (Dredge and Nielsen, 1987) were dated at 7780 ± 80 BP (GSC-3916) and 8000 ± 200 BP (BGS-812), and are the oldest dates from uncontaminated molluscs in the Hudson Bay Lowlands pertaining to the Tyrrell Sea inundation. This glacioisostatic sea reached an elevation of about 425 ft (130 m) ASL in the Gillam area, and eventually regressed to the present level of Hudson Bay.

STOP 4. DEBRIS FLOW (GLACIOLACUSTRINE TURBIDITE)

Debris flow deposits are reached by parking at the small creek that crosses the road 2.5 mi (4 km) east of Sky Pilot Creek, and walking through the bush to the cliff. Long Spruce till is overlain by a thick section of cross-bedded sand and gravel. The current direction is westerly. This glaciofluvial unit was deposited toward the end of glaciation near the margin of the easterly retreating ice front. The glaciofluvial unit is overlain by subaquatic turbidity flows in a generally fining upward sequence (Fig. 6). The lowest facies (A) consists of highly deformed sand and gravel. The flow folds and streamlined structures in this unit indicate turbulent flow caused by bed irregularities at the bottom of the debris flow. The turbulent zone is overlain by a relatively massive facies representing grain flow (B). This subunit is in places separated from the underlying turbulent facies by a glide plane, but elsewhere the contact is gradational. There are relatively few clasts in this facies, and though it is described as massive, there are zones of well-sorted fine sand and silt, and weakly developed stratification in places. The upper facies (C) consists of undeformed laminated pebbly sand and silt. The contact with the underlying massive facies is gradational. This facies was deposited by laminar flow possibly resulting from increased liquifaction and shear at the water interface on top of the plug or massive facies.

Figure 6. Debris flow, Stop 4. Turbulent flow (A), massive grain flow (B), and laminar flow (C) facies, GSC Photo 204037-T.

The subaquatic debris flows are overlain by fine-textured gray and brown rhythmites. The thinning-upward succession of both parts of the couplets in all sections examined suggests these rhythmites are varves, and they represent approximately 70 years of glaciolacustrine deposition in Lake Agassiz.

REFERENCES CITED

Andrews, J. T., Shilts, W. W., and Miller, G. H., 1983, Multiple deglaciations of the Hudson Bay Lowlands, Canada: Quaternary Research, v. 19, p. 18–37.

Dredge, L. A. and Cowan, W. R., 1987, Quaternary geology of the Southwestern Shield, *in* Fulton, R. J., Heginbottom, J. A., and Funder, S., eds., Quaternary geology of Canada and Greenland: Geological Survey of Canada, no. 1 (also Geological Society of America, The Geology of North America, v. K-1 [in press]).

Dredge, L. A. and Nielsen, E., 1985, Glacial and interglacial deposits in the Hudson Bay Lowlands; A summary of sites in Manitoba: Geological Survey of Canada Paper 85-1A, p. 247–257.

—— , 1987, Guide to the Gillam and Churchill areas: International Union for Quaternary Research, Excursion Guide Book C-13, Ottawa, 45 p.

Klassen, R. W., 1986, Surficial geology of north-central Manitoba: Geological Survey of Canada Memoir 419, 57 p.

McDonald, B. C., 1968, Glacial and interglacial stratigraphy, Hudson Bay Lowlands: Geological Survey of Canada Paper 68-53, p. 78–99.

Nielsen, E., and Dredge, L. A., 1982, Quaternary stratigraphy and geomorphology of a part of the lower Nelson River: Geological Association of Canada Guidebook for Trip 5, Winnipeg, 56 p.

Nielsen, E., Morgan, A. V., Morgan, A., Mott, R. J., Rutter, N. W., and Causse, C., 1986, Stratigraphy, paleoecology, and glacial history of the Gillam area, Manitoba: Canadian Journal of Earth Sciences, v. 23, p. 1641–1661.

Shilts, W. W., 1982, Quaternary evolution of the Hudson/James Bay region: Naturaliste Canadien, v. 109, p. 309–332.

Skinner, R. G., 1973, Quaternary stratigraphy of the Moose River Basin, Ontario: Geological Survey of Canada Bulletin 225, 73 p.

Terasmae, J., and Hughes, O. L., 1960, A palynological and geological study of Pleistocene deposits in the James Bay Lowlands, Ontario: Geological Survey of Canada Bulletin 62, 14 p.

Seagull Lake–Gunflint Lake area: A classical Precambrian stratigraphic sequence in northeastern Minnesota

J. D. Miller, Jr., G. B. Morey, and P. W. Weiblen, Minnesota Geological Survey, St. Paul, Minnesota 55114

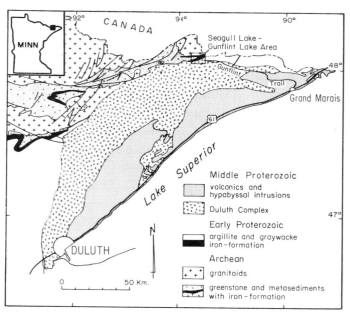

Figure 1. Generalized geologic map of northeastern Minnesota showing the location of the Seagull Lake–Gunflint Lake area.

LOCATION

The Seagull Lake–Gunflint Lake area along the Gunflint Trail (Cook County Highway 1) in T.65N.,R.3 and R.4W. in the northwestern part of Cook County, Minnesota (Fig. 1) is north of Grand Marais, the county seat and starting point for the Gunflint Trail. Although accessible by automobile or bus, much of the site is rugged and heavily wooded and requires considerable off-road travel. Except for obvious resorts and summer homes, the site is on public lands managed by the U.S. Forest Service.

There is no single place where the entire geologic succession is laid out for easy study, so one must visit several localities to appreciate the significance of the area. This guide suggests 12 stops (Figure 2) that can be easily visited in one day and that collectively illustrate the major rock types and their stratigraphic and structural attributes. However, exposure is of such quantity and quality that independent exploration is heartily encouraged.

SIGNIFICANCE OF SITE

An exceptionally complete record of Precambrian history is recorded in the rocks of the Seagull Lake–Gunflint Lake area (Fig. 1). Typical exposures of the Archean greenstone-granite terrane of the Canadian Shield are represented by a succession of metavolcanic rocks (greenstone), intruded by the Saganaga Tonalite (ca 2700 Ma). The Archean rocks are unconformably overlain by the Early Proterozoic sediments of the Animikie Group (ca 2000 Ma), which include classic banded iron-formation. A record of Middle Proterozoic rifting is preserved in hypabyssal dikes and sills of the Logan intrusions (ca 1200 Ma) and several plutonic phases of the Duluth Complex (ca 1100 Ma) that intrude and truncate the Archean and Early Proterozoic rocks.

GEOLOGIC SETTING

Archean

The oldest rocks in the site comprise the easternmost extension of the Vermilion district, a typical Archean greenstone-granite terrane (Southwick, this volume). The Vermilion district is divided into a number of fault-bounded segments (Fig. 1), each with distinct geologic characteristics that cannot be easily correlated from place to place. The Seagull Lake–Gunflint Lake area is part of the eastern end of the Gabimichigami segment of Gruner (1941). The supracrustal component of this segment is dominantly basalt and diabase (greenstone) but also contains andesite and volcanogenic graywacke and argillite. Volcanoclastics become more abundant to the west, which is stratigraphically upsection. The intrusion of the Saganaga Tonalite has deformed and metamorphosed the supracrustal rocks along the northern edge of the segment. The intrusion of the Middle Proterozoic Duluth Complex, which forms the southern boundary of the segment, had a less extensive thermal metamorphic effect on the Archean rocks.

Pillows are commonly developed in the basalts of the area (locality 4, Fig. 2) where they are as much as 3 ft (1 m) in diameter, although many have been deformed so that they are now 2 or 3 times as long as they are wide. Chilled rinds, a centimeter or so thick, are well developed and are typically lighter in color than the dark-green or dark-greenish-gray pillow interiors. Interpillow material consists of tuffaceous material, chert, or pillow-rind fragments. Several tabular bodies of diabase, typically less than 200 ft (60 m) thick, intrude the metavolcanic rocks.

The basaltic and diabasic rocks show intense alteration indicative of greenschist-facies metamorphism. Recognizable minerals in the metabasalts include relict augite and calcic plagioclase, and secondary sodic plagioclase, actinolite, chlorite, epidote, calcite, quartz, leucoxene, and opaques. The tabular bodies of diabase have a relict poikilitic texture (actinolite pseudomorphs after augite) and a mineralogy very similar to that of the metabasalt.

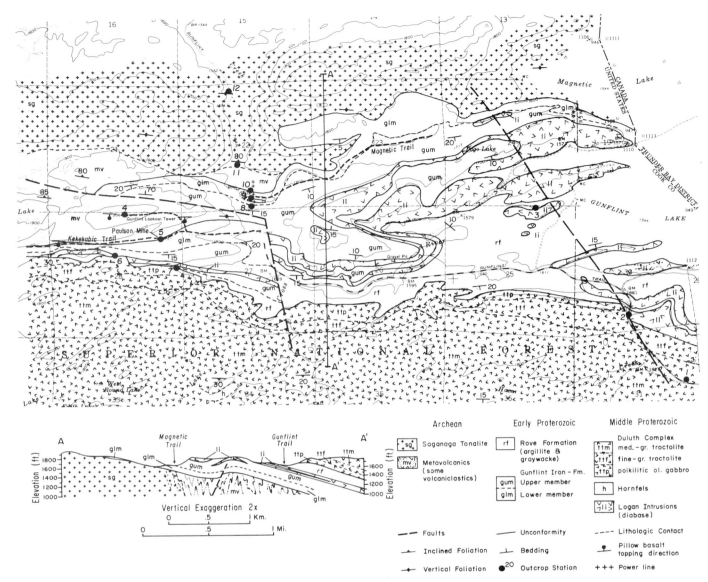

Figure 2. Geologic map and cross section of the Seagull Lake-Gunflint Lake area (after Morey and others, 1981). Recommended field stop localities numbered 1 through 12.

Thin beds of texturally and mineralogically immature volcanoclastic rocks are locally intercalated with the basalts but are rare in the Seagull Lake–Gunflint Lake area. Several beds have textures indicative of a pyroclastic origin, but most contain pebble- to silt-size clasts of locally derived metadiabase in a finer-grained matrix of chert, plagioclase, hornblende, chlorite, and sericite.

The metamorphic effects of the Saganaga Tonalite on the supracrustal sequence become apparent only within several tens of meters of the contact. Near the contact, basalt has been recrystallized to amphibolite consisting dominantly of hornblende and calcic plagioclase (locality 10 at powerline north of Magnetic Trail and south of locality 11 on Gunflint Trail, Fig. 2). The intrusion of the tonalite also imparted a strong schistosity to the

amphibolite and a weaker foliation to the greenstone farther from the contact. Evidence for this is given by the parallelism between the steep, intrusive contact, the metavolcanic foliation, and the well-developed foliation within the tonalite (Fig. 2). Based on pillow topping directions and foliation, the metavolcanic succession in the Seagull Lake–Gunflint Lake area defines a south-dipping homoclinal structure that forms the northern limb of a southeast-plunging synclinorium (Gruner, 1941).

The Saganaga Tonalite (Hanson, 1972) is a syn- to late-kinematic batholithic intrusion, which in the field area can be divided into two major phases—a mafic border phase and a medium- to coarse-grained tonalite.

The mafic border phase (locality 11, Fig. 2) is a foliated, medium- to coarse-grained, quartz-bearing hornblende diorite that is gradational over several tens of meters with the more felsic tonalite. The border phase crops out extensively in the Seagull Lake–Gunflint Lake area, where it is as much as 1,000 ft (300 m) wide. The generally mafic nature of the border phase relative to the tonalite has been explained by Hanson (1972) as partial assimilation of amphibolitic material from the country rock, and by Grout (1929), in his classic study of these rocks, as strictly igneous processes that involved crystal settling of mafic phases.

The tonalitic phase, which comprises about 85 percent of the outcrop area, is a gray, medium- to coarse-grained granitic rock with plagioclase in much greater abundance than microcline. Ubiquitous to the tonalite are large quartz aggregates or "eyes," about 1 cm in diameter, that resemble phenocrysts (locality 12 and numerous roadside outcrops in area, Fig. 2). The quartz "eyes" contain many grains, 1 to 2 mm in diameter, that have different optical orientations. Quartz also occurs as an interstitial mineral to subhedral plagioclase (An_{20-28}). Sparse amounts (<10%) of microcline occur as antiperthitic exsolution in plagioclase, as discrete rims on plagioclase, and as small interstitial grains. Euhedral to subhedral grains of typically unaltered hornblende occur as the dominant ferromagnesian mineral. Other minor phases in the tonalite include augite, biotite, epidote, and chlorite.

Early Proterozoic

Sedimentary rocks of the Early Proterozoic Animikie Group crop out in an east–northeast-trending belt that extends from Thunder Bay on Lake Superior to a point 12 mi (19 km) west of Gunflint Lake where the belt is truncated by the Middle Proterozoic Duluth Complex. The Animikian section is composed of the basal Gunflint Iron Formation and the overlying slates of the Rove Formation. This sequence is correlative with similar Early Proterozoic lithologies throughout the Lake Superior region (Morey, 1972). The rocks form a homocline that dips 10–15° SE except where Middle Proterozoic intrusions, folding, and faulting have distorted or disturbed the beds. There is no evidence in this area for a major orogenic event at the end of Early Proterozoic time as there is elsewhere in Minnesota.

Just across the border in Canada, the Gunflint Iron Formation is essentially unmetamorphosed and consists of quartz, hematite, iron carbonates, greenalite, and trace amounts of magnetite. In Minnesota however, the original mineralogic character of the iron-formation is obscured by a thermal metamorphic overprint produced by the contact effects of Middle Proterozoic igneous rocks as described in the next section. Consequently, many of the fine-scale textural and mineralogical attributes used to subdivide the iron-formation in Canada (Goodwin, 1956) have been destroyed. Therefore, a four-fold nomenclatural scheme, originally proposed for the correlative Biwabik Iron Formation of the Mesabi range, which emphasizes various bedding aspects, has been used to describe the Gunflint Iron Formation since the early work of Broderick (1920). This scheme

defines four informal members—the Lower cherty, Lower slaty, Upper cherty, and Upper slaty—whose names emphasize the relative proportions of thick-bedded, granular (cherty), and thin-bedded non-granular (slaty) iron-formation. Only the upper and lower members are distinguished on the geologic map (Fig. 2). Although the boundaries of these members do not coincide with those recognized in Canada by Goodwin (1956), the two schemes can be equated with only slight difficulty (Morey, 1972).

The Lower cherty member ranges in thickness from 15 to 45 ft (4.5 to 13.5 m). A conglomerate of Archean granite and greenstone pebbles set in a matrix of feldspasthic quartzite or an algal-rich chert (locality 10, Fig. 2) intermittently occurs at the base. Throughout most of the area, however, the basal conglomerate or chert is absent. Instead, a persistent, very thick-bedded to massive, magnetite-rich, silicate-bearing unit, 5 to 15 ft (1.5 to 4.5 m) thick, composes the base of the Lower cherty member and unconformably overlies Archean rocks (locality 5 at Paulson Mine and other localities along the Kekekabic Trail, Fig. 2). The magnetite-rich unit is in turn overlain by a thick-bedded, chert-rich, magnetite-poor unit about 15 ft (4.5 m) thick.

The overlying Lower slaty member is 80 to 95 ft (24 to 29 m) thick. The lowermost 10 ft (3 m) is a black, thinly bedded, nearly magnetite-free argillite composed predominantly of volcanically derived material called the Intermediate slate. Massive and cherty beds that resemble the upper part of the Lower cherty member occur above the Intermediate slate, but they pass abruptly upward into a sequence of thick, chert- and silicate-rich beds with sparse magnetite intercalated with intervals of thinly laminated silicate-rich beds. The remaining 50 ft (15 m) consists of thinly bedded to laminated strata that contain various silicates and from 20 to 35 percent magnetite (locality 9, Fig. 2).

The Upper cherty member is approximately 50 ft (15 m) thick. The gradational contact between the Lower slaty and the Upper cherty members is marked by the first appearance of irregularly bedded to lenticular chert-rich layers and by thin irregular layers consisting almost entirely of magnetite (many exposures along Magnetic Trail, Fig. 2). The upper part of the Upper cherty member is characterized by several thick units of granular chert that contain algal structures, conglomerate fragments, and abundant disseminated magnetite (locality 8, powerline south of Magnetic Trail, Fig. 2).

The Upper slaty member is approximately 150 ft (45 m) thick. Thick lenticular beds of chert containing disseminated magnetite occur in the lower few meters, but most of the member consists of thin-bedded to laminated, silicate-rich units that are intercalated with intervals of thinly laminated graphitic argillite and centimeter-thick beds of relatively pure chert. The upper 10 ft (3 m) are nearly magnetite-free and consist of limestone and chert interbedded with argillite.

The Rove Formation gradationally overlies the Gunflint Iron Formation and is at least 3,200 ft (970 m) thick. It can be divided into a lower argillaceous unit, a transition sequence, and an upper thin-bedded graywacke-rich unit (Morey, 1969). The lower argillite consists of intercalated argillaceous siltstone, silty argillite, and carbonaceous argillite (locality 3, Fig. 2). The transition sequence separates dominantly argillaceous rocks below from an interval containing intercalated beds of coarse- and fine-grained graywacke. The sandy beds contain many primary sedimentary structures indicative of deposition by turbidity currents that flowed dominantly from the north.

The contact metamorphic effects of the Duluth Complex on the Gunflint Iron Formation have been studied in detail by Floran and Papike (1975, 1978), and the metamorphism of the Rove Formation has been described by Labotka and others (1981). The Logan intrusions have also metamorphosed both formations. The sizes of the metamorphic aureoles, up to 33 ft (10 m) wide, and their mineral assemblages (up to pyroxene-hornfels facies) are directly related to the thicknesses of the sills.

In general, three metamorphic zones have been distinguished within the Gunflint Iron Formation adjacent to the Duluth Complex:

— An outer zone of slightly metamorphosed iron-formation consisting of quartz, iron carbonates, minnesotaite, stilpnomelane, and hematite partially replaced by magnetite. This part of the Gunflint Iron Formation is much like that of "unmetamorphosed" iron-formation in many other parts of the Lake Superior region.
— A 1.2-mi-wide (2-km-wide) intermediate zone of moderately metamorphosed iron-formation containing grunerite-cummingtonite, hornblende, and actinolite, as well as quartz and fine-grained magnetite (e.g., locality 8, Fig. 2).
— A proximal zone of highly metamorphosed iron-formation occurring within 0.3 mi (0.5 km) of the contact with the Duluth Complex composed chiefly of quartz, magnetite, iron-rich pyroxenes, and fayalite. Very commonly, euhedral or subhedral grains of partially recrystallized magnetite are poikilitically enclosed within large silicate grains (e.g., locality 6, Fig. 2). Actinolite is common in the magnetite-rich layers, and both prograde and retrograde grunerite are abundantly present.

In the Rove Formation, in contrast to the iron-formation, evidence for recrystallization and the development of metamorphic mineral assemblages is barely discernable in the field even adjacent to the Duluth Complex (locality 7, Fig. 2). However, a complex mixture of rock types suggestive of partial melting, as well as mineral and textural variations due to original inhomogeneities and degree of metamorphism, occur within a meter or so of the contact with the Duluth Complex. Within this contact zone the argillaceous rocks retain large-scale layering but have a vague granoblastic texture. Individual layers may contain cordierite and hypersthene with minor biotite and ilmenite; hypersthene, plagioclase, biotite, and ilmenite; augite, plagioclase ± minor olivine, biotite, and ilmenite; or hypersthene, plagioclase, K-feldspar, and biotite. At varying distances from the contact but generally no more than 100 ft (30 m) away, biotite is the only well-developed metamorphic mineral in the pelitic rocks. Calcareous beds near the contact contain a grossular garnet.

Middle Proterozoic

Gabbroic rocks of Middle Proterozoic age comprise the remaining exposures in the Seagull Lake–Gunflint Lake area. Diabasic dikes and sills intrusive into the Animikian strata are collectively referred to as the Logan intrusions (Weiblen and others, 1972). Medium- to coarse-grained gabbros and troctolites exposed in the southern quarter of the field area are part of the Tuscarora intrusion of the Duluth Complex (Morey and others, 1981). The complex underlies more than 4,960 mi^2 (8,000 km^2) in northeastern Minnesota and composes the plutonic component of the 1,100 Ma Midcontinent rift system.

Logan intrusions crop out along a series of east-trending ridges formed by the differential erosion of the diabase sills and sedimentary rocks, particularly the Rove Formation (cross section, Fig. 2). Individual sills, as much as 1,100 ft (330 m) thick, can be traced along strike for several kilometers. However, branching and merging of individual sills is common, and many individual sills thicken and thin down-dip. Some sills terminate against joints and possibly against fault planes. In places within the Rove Formation, open fractures are occupied by thin dikes, which, together with the sills, give a box-work configuration to the hypabyssal rocks.

Rock types within the sills include aphyric basalt; fine- to medium-grained diabase with ophitic clinopyroxene enclosing plagioclase; plagioclase cumulates; and granophyre (Jones, 1984). Chilled margins form sharp contacts with country rocks (localities 9 [south of trail] and 3, Fig. 2). Diabase grades in grain size from fine to medium toward the center of sills. Clinopyroxene is ophitic throughout. Plagioclase crystals enclosed in clinopyroxene oikocrysts rarely exceed 5 mm in length, although phenocrysts may be as much as 4 in (10 cm) long in some of the thicker sills. Diabasic rocks with plagioclase phenocrysts grade into accumulations of essentially coarse-grained plagioclase in the upper parts of some of the thicker sills. Similarly, granophyric intergrowths consisting of quartz, sodic plagioclase, and orthoclase are common in the upper part of the thicker sills where they impart a pink mottling that can be recognized in hand specimens.

The Middle Proterozoic Duluth Complex is a sequence of generally discordant plutonic rocks consisting of many separate intrusions (Weiblen and Morey, 1980). Rocks informally referred to as the Tuscarora intrusion crop out in the Seagull Lake-Gunflint Lake area.

The basal part of the Tuscarora intrusion consists of a fine-grained, augite-poikilitic, olivine gabbro (localities 2 and 7, unit ttp, Fig. 2). Contacts with overlying fine-grained troctolite (unit ttf, Fig. 2) are not exposed, and it is not clear whether both units formed in the same intrusive event (Morey and others, 1981). Within 0.3 mi (0.5 km) of the basal contact, fine-grained troctolite coarsens to medium grained (unit ttm, locality 1, Fig. 2).

The troctolite units consist of 65 to 70 percent plagioclase, and 10 to 15 percent olivine. Relative amounts of poikilitic augite and iron-titanium oxides vary locally. Orthopyroxene mantles olivine and occurs in symplectic intergrowth with plagioclase.

Biotite is associated with the iron-titanium oxides. Planar orientation of plagioclase and modal layering are locally well developed and mutually concordant in the ttm unit and gently dip to the south.

The ttp and ttf units commonly contain chalcopyrite, pyrrhotite, and minor pentlandite interstitial to plagioclase and olivine (e.g., locality 2, Fig. 2). Although the sulfide concentrations are not economic, they are sufficiently large to form discontinuous areas of gossan and visible sulfides that can be mapped at the quadrangle scale (Morey and others, 1981).

South of the Seagull Lake–Gunflint Lake area, the troctolite becomes more rich in augite and plagioclase. Mafic hornfels, anorthosite inclusions, and granophyric bodies are common throughout the Tuscarora intrusion (Morey and others, 1981).

Most of the structural features in the Seagull Lake–Gunflint Lake area are probably Middle Proterozoic in age, although many may be older but became reactivated in Middle Proterozoic time. Faulting in the area clearly offsets Archean and Early Proterozoic geology, but appears to have had a minor affect on the Duluth Complex (Fig. 2). The apparent folding and fault-termination of Logan intrusions as sills in Early Proterozoic strata may indicate that deformation is post-Logan and syn- or pre-Tuscarora. Field, paleomagnetic, and radiometric data indicate that the Logan intrusions are older than the Tuscarora (Jones, 1984). It is also possible, however, that the Logan sills were intruded into previously deformed Early Proterozoic strata.

REFERENCES CITED

Broderick, T. M., 1920, Economic geology and stratigraphy in the Gunflint iron district, Minnesota: Economic Geology, v. 15, p. 422–452.

Floran, R. J., and Papike, J. J., 1975, Petrology of the low-grade rocks of the Gunflint iron-formation, Ontario-Minnesota: Geological Society of America Bulletin, v. 86, p. 1169–1190.

—— 1978, Mineralogy and petrology of the Gunflint Iron Formation, Minnesota-Ontario: Journal of Petrology, v. 19, p. 215–288.

Goodwin, A. M., 1956, Facies relations in the Gunflint Iron Formation: Economic Geology, v. 51, p. 565–595.

Grout, F. F., 1929, The Saganaga granite of Minnesota-Ontario: Journal of Geology, v. 37, p. 562–576.

Gruner, J. W., 1941, Structural geology of the Knife Lake area of northeastern Minnesota: Geological Society of America Bulletin, v. 52, p. 1577–1642.

Hanson, G. N., 1972, Saganaga batholith, *in* Sims, P. K., and Morey, G. B., eds., Geology of Minnesota: A centennial volume: Minnesota Geological Survey, p. 102–107.

Jones, N. W., 1984, Petrology of some Logan sills, Cook County, Minnesota: Minnesota Geological Survey Report of Investigations 29, 40 p.

Labotka, T. C., Papike, J. J., Vaniman, P. T., and Morey, G. B., 1981, Petrology of contact metamorphosed argillite from the Rove Formation, Gunflint Trail, Minnesota: American Mineralogist, v. 66, p. 70–86.

Morey, G. B., 1969, The geology of the Middle Precambrian Rove Formation in northeastern Minnesota: Minnesota Geological Survey Special Publication 7, 62 p.

—— 1972, Gunflint range, *in* Sims, P. K., and Morey, G. B., eds., Geology of Minnesota: A centennial volume: Minnesota Geological Survey, p. 218–226.

Morey, G. B., Weiblen, P. W., Papike, J. J., and Anderson, D. H., 1981, Geology of the Long Island Lake quadrangle, Cook County, Minnesota: Minnesota Geological Survey Miscellaneous Map Series M-46, scale 1:24,000.

Southwick, D. L., 1986, Geologic highlights of an Archean greenstone belt, western Vermilion district, northeastern Minnesota (this volume).

Weiblen, P. W., and Morey, G. B., 1980, A summary of the stratigraphy, petrology, and structure of the Duluth Complex: American Journal of Science, v. 280-A, pt. 1, p. 88–133.

Weiblen, P. W., Mathez, E. A., and Morey, G. B., 1972, Logan intrusions, *in* Sims, P. K., and Morey, G. B., eds., Geology of Minnesota: A centennial volume: Minnesota Geological Survey, p. 394–406.

Geologic highlights of an Archean greenstone belt, western Vermilion district, northeastern Minnesota

D. L. Southwick, Minnesota Geological Survey, St. Paul, Minnesota 55114

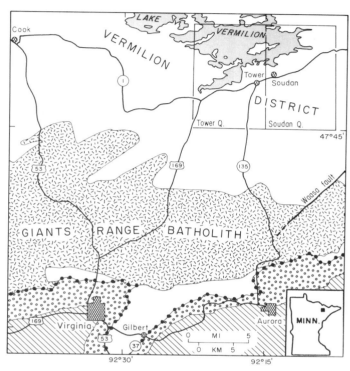

Figure 1. Map showing location of the western Vermilion district relative to major geologic boundaries and cultural features, and the Tower and Sudan quadrangles where described stops are located. Heavy dotted line is a gently south-dipping unconformity at base of the Proterozoic Animikie Group; Archean rocks crop out north of this line. Virginia, Gilbert, and Aurora are important mining towns on the Mesabi iron range, where the mines exploit the Biwabik Iron Formation of Proterozoic age (open circles on map). Geology adapted from Sims and others (1970).

LOCATION

The western Vermilion district is located in northeastern Minnesota, about 18 mi (30 km) north of the towns of Virginia and Aurora on the Mesabi iron range (Fig. 1). The major towns in the western Vermilion district are Cook, Tower, Soudan, and Ely. The area to be described here lies within the Tower and Soudan 7½-minute Quadrangles (Fig. 1).

SIGNIFICANCE

The Vermilion district is the best example in the United States of an Archean greenstone belt. Although the district is small and the rocks are not very well exposed in comparison to many greenstone belts in Canada, it contains the characteristic lithologic, stratigraphic, and structural attributes of larger green-

stone belts within a relatively accessible small area. Because there is no single place where the entire picture is laid out for easy viewing, one must visit several localities and integrate them conceptually. This guide suggests seven stops, all easily reached from U.S. 169 and subsidiary roads, that collectively illustrate the major lithologies and structures of an Archean greenstone belt.

DESCRIPTION

General Background

Archean greenstone belts throughout the world share the following characteristics:

1. The rocks are dominantly supracrustal, consisting of volcanic flows, pyroclastic deposits, and assorted sedimentary rocks derived chiefly from volcanic sources.

2. The stratigraphic succession includes rocks of one or more volcanic cycles. Typically, the earliest rocks of a cycle are komatiitic to tholeiitic basalt; these pass upward into calc-alkaline basalt and andesite, which in turn give way to late-stage calc-alkaline dacite, and, more rarely, to rhyolite. Most of the basalt is in pillowed flows that were erupted under water; repeated eruptions are interpreted to have built up one or more volcanic piles on the sea floor, which, in many instances, emerged above sea level at about the time that dacite became the principal volcanic product. The dacitic rocks commonly are pyroclastic and pass laterally into sedimentary aprons of reworked tuff and associated volcanogenic graywacke.

3. Laminated, cherty iron-formation may occur almost anywhere in the stratigraphic succession in minor amounts, but typically it is most abundant in the upper parts of a volcanic cycle. The iron-formation apparently precipitated as a chemical sediment during periods of reduced volcanic activity and relative tectonic stability, probably in close proximity to submarine hot springs and fumaroles.

4. The volcanic and sedimentary rocks of greenstone belts invariably are tightly folded and extensively faulted. Most commonly they have been metamorphosed to mineral assemblages typical of the greenschist facies. Cleavage, schistosity, and lineation are widely developed. Multiple deformation, indicated by superimposed fold structures, is very common and is becoming more widely recognized as more detailed mapping is done. Large longitudinal strike-slip faults that record displacements of 12 mi (20 km) or more are common late-stage features of most greenstone belts, and seem to be especially characteristic of those in the Superior Province of the Canadian Shield.

5. The supracrustal rocks of greenstone belts have been in-

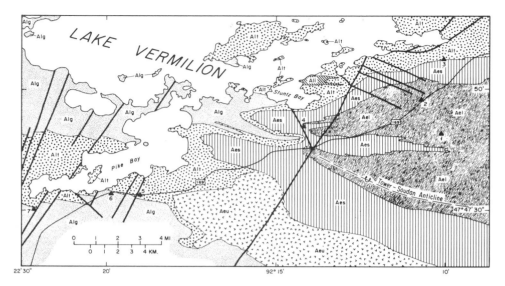

Figure 2. Simplified geologic map of the Tower–Soudan area showing locations of field trip stops. Adapted from Ojakangas and others (1978), and Sims and Southwick (1980). See Figure 3 for explanation of rock units.

vaded by granitoid plutons of diverse size and shape. The granitoid intrusions tend to be diapiric in style and to range in composition from tonalite to granodiorite; monzogranite is relatively rare. The wall rocks commonly are metamorphosed to amphibolite grade near large granitoid intrusions.

6. The basement on which the volcanic and sedimentary rocks of greenstone belts were deposited is rarely preserved, and its attributes generally are not well defined. Continental crust has been identified beneath a few greenstone belts; oceanic crust is postulated beneath others. The basement beneath the supracrustal rocks of the Vermilion district is not exposed, and its characteristics therefore are unknown.

The field trip stops described below illustrate the stratigraphy of the upper part of the lowermost volcanic and sedimentary cycle in the Vermilion district, as exposed on the steeply dipping, north and southwest limbs of the Tower–Soudan anticline (Fig. 2). The applicable stratigraphic nomenclature is summarized in Figure 3.

Descriptions of Selected Stops

Stop 1. Cuts along abandoned railroad about 0.4 mi (0.65 km) ENE of Murray logging road, SW¼SE¼Sec.30,T.62N., R.14W., Soudan 7½-minute Quadrangle (Fig. 4). This locality displays typical pillowed and massive metabasalt belonging to the lower member of the Ely Greenstone (Unit Ael). The upper part of a thick flow is exposed at the northwest end of the north cut; it consists of irregular, bulbous pillows, about 2.6 ft (0.8 m) in largest dimension, that have thin, glassy rinds. The pillows have been flattened tectonically in the plane of foliation that strikes about 75° and dips 72° NW. The actual flow top is not exposed. Stratigraphically downward (eastward along railroad grade) the

flow becomes massive as the pillows lose their definition and gradually disappear. The interior of this thick flow is essentially massive and homogeneous metabasalt with a texture almost coarse enough to be called metadiabase. Indeed, if it were not for its pillowed upper part and the widely scattered small amygdules of nearly black chlorite, the flow might be interpreted as a hypabyssal rock. Although the metabasalt of these outcrops has undergone some plastic deformation, as indicated by the flattening of pillows, it is obvious that much structural accommodation

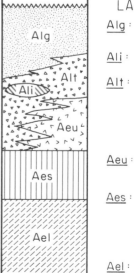

LAKE VERMILION FORMATION

Alg: Volcanogenic graywacke – slate. (stops 6, 7)

Ali: Iron – formation

Alt: Dacite tuff, tuff – breccia, and allied rocks; variably reworked. (stops 3, 5)

ELY GREENSTONE

Aeu: Upper member; chiefly pillowed tholeiitic basalt.

Aes: Soudan Iron-formation Member. Cherty iron-formation interbedded with felsic to mafic volcanic rocks. (stops 3, 4)

Ael: Lower member; chiefly pillowed calc-alkaline basalt. (stops 1, 2)

Figure 3. Simplified stratigraphic column for the map area of Figure 2. Nomenclature modified from Morey and others, (1970).

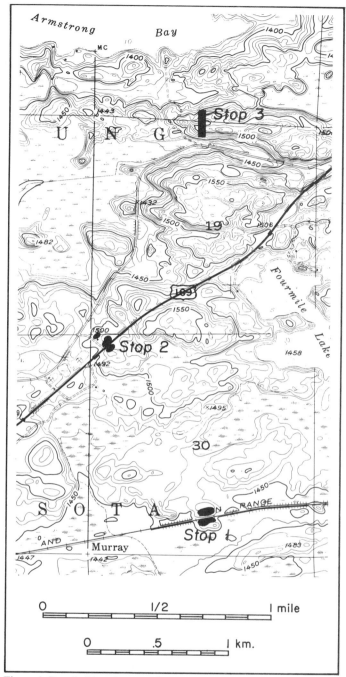

Figure 4. Detailed location map for stops 1, 2, and 3. Topographic base from Soudan 7½-minute Quadrangle, 1956 edition (photorevised 1969).

has occurred by brittle fracturing. The rock is cut by countless small shear surfaces having fine-scale slickensides on chlorite films, and also by several mullion-like fluted surfaces on which the flute crests are 4 to 6 in (10 to 15 cm) apart. Multiple shear motions are indicated by successive generations of slickensides on some surfaces.

Stop 2. Outcrops on both sides of U.S. 169 and in adjacent

Figure 5. Breccia, possibly of debris-flow origin, containing clasts of dacite (light color) and iron-formation (layered) in a matrix of smaller volcanic fragments; Stop 3.

clearing, NW¼NW¼Sec.30,T.62N.,R.14W., Soudan 7½-minute Quadrangle (Figure 4). The highway exposures at this locality are metabasalt belonging to the lower member of the Ely Greenstone, and are similar to the rocks at the previous stop. The important differences between this stop and Stop 1 are to be seen in outcrops at the east edge and northwest corner of the clearing, where thin-bedded, schistose, felsic tuff-breccia and associated beds of ferruginous chert are intercalated with metabasalt flows, and under the power line at the north edge of the clearing, where a coarse-grained diabasic sill crops out. The upper part of the lower Ely characteristically contains thin units of felsic and intermediate tuff, indicating a temporal transition toward more silicic volcanism. Silicic volcanic rocks increase gradually in abundance upward (northward) and become about equal to mafic rocks within the Soudan Iron-formation Member. The diabase sill at Stop 2 is an interflow intrusion that is compositionally identical to the flows it intrudes. Sills of this general type are widespread throughout the lower member of the Ely, where they are easily confused with thick, massive flows of the sort exposed at Stop 1.

Stop 3. Outcrops along power line swath extending northward from abandoned highway in the northern part of Sec.19, T.62N.,R.14W., and continuing into southern part of Sec.18, T.62N.,R.14W., Soudan 7½-minute Quadrangle (Figure 4). This series of outcrops illustrates the complex stratigraphy within and

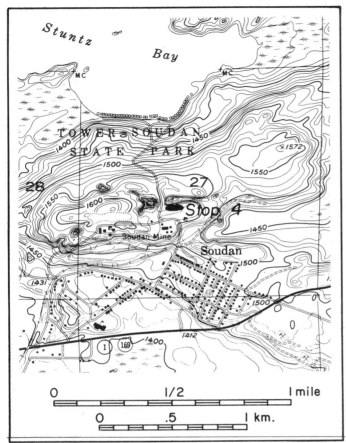

Figure 6. Detailed location map for Stop 4. Topographic base from Soudan 7½-minute Quadrangle, 1956 edition (photorevised 1969).

Figure 7. Multiply folded iron-formation, Soudan Hill (Stop 4). Note "eye" and "mushroom" interference folds, upper left part of photograph. Pocket knife is 7.5 cm long.

Stop 4. Soudan Hill; glacially polished outcrop near Stuntz Bay road in Tower-Soudan State Park, NE¼SW¼Sec.27,T.62N., R.15W., Soudan 7½-minute Quadrangle (Fig. 6). This classical exposure of the Soudan Iron-formation displays two generations of close folding (Fig. 7) in delicate laminations of chert (creamy white), chert-hematite (red), and magnetite-chert (black). The second-generation folds are of tectonic origin; their subvertical axial surfaces trend east-west, their axes plunge steeply east, and they are dominantly of Z-symmetry. The first-generation folds have been sharply refolded by the second generation, resulting in fold interference patterns of several types. Some geologists (e.g. Hooper and Ojakangas, 1971) have interpreted the first folds to be of tectonic origin, and others (e.g. Hudleston, 1976) have suggested that they may have been formed by soft-sediment processes. Small faults and kink bands of brittle origin cut across the folds.

The deep open pits a few meters north of the outcrop are early workings of the Soudan iron mine, the first iron mine in Minnesota. The mine produced about 16 million tons of high-grade hematite lump ore between 1884 and 1962, when high mining costs and changes in steelmaking technology forced it to close. The early open-cut method of mining was replaced by underground operations in about 1900; most of the historic production has come from underground mining. The mine was deeded to the state of Minnesota in 1962 and is now operated as a tourist facility featuring guided underground tours.

Stop 5. Highway cuts along U.S. 169 about 1.7 mi (2.7 km) west of Tower, NW¼NE¼Sec.1,T.61N.,R.16W., Tower 7½-minute Quadrangle (Fig. 8). An important lithofacies of the Lake Vermilion Formation is the nearly massive, thick-bedded dacitic wacke in unit Alt, interpreted to be a sedimentary rock derived more or less directly from dacite tuff, tuff-breccia,

just above the Soudan Iron-formation Member of the Ely Greenstone (unit Aes and immediately overlying part of unit Alt). The southernmost exposures are laminated magnetite-chert iron-formation intruded by quasi-concordant, sill-like bodies of dacite porphyry. The porphyry contains prominent phenocrysts of quartz. A continuous section about 66 ft (20 m) thick of vertically dipping iron-formation holds up the sharp ridge crest. To the north, the iron-formation is overlain in stratigraphic order by 20 ft (6 m) of pillowed andesitic greenstone (the top 2 ft [0.6 m] is fragmental), 6.5 ft (2 m) of laminated iron-formation, 72 ft (22 m) of interbedded dacitic tuff-breccia and dacite porphyry sills, and 3 ft (1 m) of laminated iron-formation with centimeter-scale interbeds of dacite tuff. About 160 ft (50 m) farther north, beyond a covered interval, is a low but spectacular outcrop of andesite-dacite tuff-breccia and block breccia that contains many angular clasts of iron-formation as long as 12 or 15 in (30 or 40 cm) in addition to volcanic fragments (Fig. 5). This rock is interpreted to be some sort of debris flow, but the exact manner of its deposition is open to debate. Outcrops on the steep hill to the north, across a narrow swamp-filled gulch, are mainly dacitic tuff breccia and thick dacite porphyry sills.

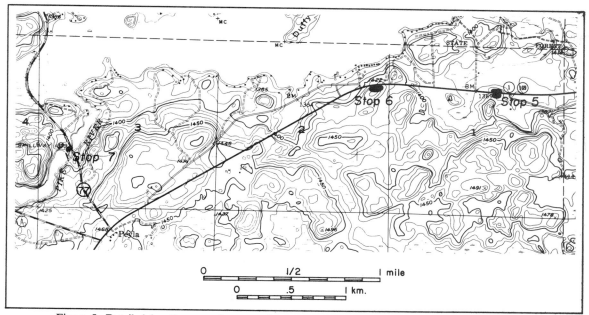

Figure 8. Detailed location map for stops 5, 6, and 7. Topographic base from Tower 7½-minute Quadrangle, 1956 edition (photorevised 1969).

and allied volcanic rocks. The rock is uncommonly well displayed in these cuts, where its subtle bedding, textural grading, and folding can be observed. This lithology is associated regionally with finer, more typically turbidite-like volcanic wacke (Unit Alg of Fig. 2; Stops 6 and 7), which presumably was a more distal facies in the basin of deposition, and with coarser dacite-dominated conglomerate and breccia (also mapped in unit Alt), which presumably were proximal facies or channel deposits within submarine fans (Ojakangas, 1985).

Stop 6. Highway cuts along U.S. 169 about 2.5 mi (40 km)

Figure 9. Multiply folded graywacke and slate of the Lake Vermilion Formation at Stop 6. Note the strong cleavage related to the second folds (dashed lines), and the bulbous, buckle-like morphology of the folded light gray graywacke beds. Film box is about 6 cm long.

Figure 10. Small-scale fault and fault breccia at high angle to bedding and cleavage, Lake Vermilion Formation (Stop 7). Brunton compass gives scale.

west of Tower, NE¼NE¼Sec.2,T.61N.,R.16W., Tower 7½-minute Quadrangle (Fig. 8). The rock at this stop is well graded, thin-to-medium-bedded, intricately folded metagraywacke and slate of the Lake Vermilion Formation (unit Alg). Here, as at Soudan hill (Stop 4), the rocks have undergone two generations of folding—a second phase, clearly tectonic, associated with an axial plane cleavage in which sedimentary clasts are visibly flattened, and an earlier phase of debated soft-sediment or tectonic origin. Interference phenomena are widely developed, especially "eye" and "mushroon" forms. The exceptional abundance of topping information from graded beds allows one to identify small-scale antiformal synclines and synformal anticlines, both indicative of stratigraphic inversion prior to the second folding. All of the measurable finite strain in the rocks of this outcrop can

be ascribed to the second phase of folding (Hudleston, 1976). This, together with the nearly concentric, bulbous morphology of the early folds (Fig. 9), strongly suggests that the first folds were the product of soft-sediment deformation.

Stop 7. Glacially scoured outcrop west of St. Louis County 77, immediately north of the Pike River bridge, NW¼SW¼ Sec.3,T.61N.,R.16W., Tower 7½-minute Quadrangle (Fig. 8). This outcrop of graded, thin-bedded, south-topping metagraywacke and slate contains a good cleavage (left of bedding), but no small folds. Apart from the "textbook" refraction of cleavage as it crosses layers of contrasting competence, the main features of interest in this outcrop are the late kink bands and minor faults that cross bedding at high angles (Fig. 10). These are related to regional northeast-trending faulting (Fig. 2) that was the latest tectonic event to affect this part of the Vermilion district.

REFERENCES CITED

Hooper, P. R., and Ojakangas, R. W., 1971, Multiple deformation in Archean rocks of the Vermilion District, Minnesota: Canadian Journal of Earth Sciences, v. 8, p. 423–434.

Hudleston, P. J., 1976, Early deformational history of Archean rocks in the Vermilion district, northeastern Minnesota: Canadian Journal of Earth Sciences, v. 13, p. 579–592.

Morey, G. B., Green, J. C., Ojakangas, R. W., and Sims, P. K., 1970, Stratigraphy of the Lower Precambrian rocks in the Vermilion district, northeastern Minnesota: Minnesota Geological Survey, Report of Investigations 14, 33 p.

Ojakangas, R. W., 1985, Review of Archean clastic sedimentation, Canadian Shield: Major felsic volcanic contributions to turbidite and alluvial fan—

fluvial facies associations, *in* Ayres, L. D., ed., Evolution of Archean supracrustal sequences: Geological Association of Canada Special Paper 28, p. 23–47.

Ojakangas, R. W., Sims, P. K., and Hooper, P. R., 1978, Geologic map of the Tower quadrangle, St. Louis County, Minnesota: U.S. Geological Survey Geologic Quadrangle Map GQ-1457, scale 1:24,000.

Sims, P. K., and Southwick, D. L., 1980, Geologic map of the Soudan quadrangle, St. Louis County, Minnesota: U.S. Geological Survey Geologic Quadrangle Map GQ-1540, scale 1:24,000.

Sims, P. K., Morey, G. B., Ojakangas, R. W., and Viswanathan, S., 1970, Geologic map of Minnesota, Hibbing sheet: Minnesota Geological Survey, scale 1:250,000.

Plateau basalts of the Keweenawan North Shore Volcanic Group

John C. Green, Department of Geology, University of Minnesota, Duluth, Minnesota 55812

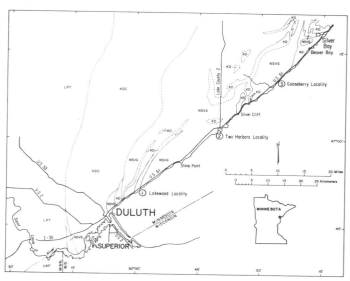

Figure 1. Sketch map of the Duluth, Minnesota region showing the southwestern part of the North Shore Volcanic Group (NSVG) and field trip localities described in text. LPT—Lower Proterozoic Thomson Formation; KDC—Keweenawan Duluth Complex; KD—Keweenawan diabase intrusion; UKF—Upper Keweenawan Fond du Lac sandstone.

LOCATION

Three specific outcrop areas are included: (1) Lakewood, 2.1 mi (3.5 km) northeast of the Lester River (in Duluth, Minnesota) on St. Louis County 61 (S½Sec.34,T.51N.,R.13W.; Lakewood 7½-minute Quadrangle), in Congdon (city) Park; (2) Two Harbors, 0.5 mi (0.8 km) south of U.S. 61 on First Street (NW¼Sec.6,T.52N.,R.10W.; Two Harbors 7½-minute Quadrangle), in Burlington Bay (city) Tourist Park, and (3) Gooseberry Falls, 13.5 mi (23 km) northeast of Two Harbors on U.S. 61 (SW¼Sec.22,T.54N.,R.9W., Split Rock Point 7½-minute Quadrangle), in Gooseberry Falls State Park. Note: State Park rules prohibit defacing or collecting rocks. Removal of bedrock specimens showing the special structures in all areas is strongly discouraged for the benefit of future observers.

SIGNIFICANCE

These three locations illustrate a variety of physical-volcanological features associated with the Proterozoic flood basalts of Keweenawan age (1100 Ma) that erupted along the Midcontinent rift. The lack of contemporaneous weathering or erosion, along with the absence of internal deformation, has preserved original textures and structures of these lava flows. Post-Pleistocene erosion by lake waves and streams has enhanced the features by providing both cross sections and surface exposures of

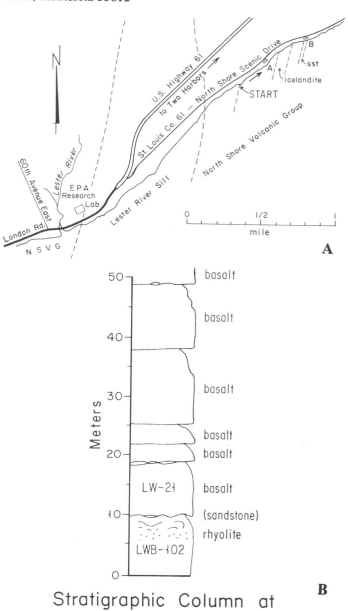

Stratigraphic Column at Lakewood Locality A

Figure 2. A) Sketch map of Locality 1, Lakewood area, Duluth. Park at A, walk back to "Start," then northeast along shore, following text description. Second portion of stop is at B. B) Stratigraphic column at Lakewood locality A.

many individual flows. The only other large flood basalt province in North America is the Columbia River Plateau, and its physical features are generally less well exposed and more limited in variety.

TABLE 1. CHEMICAL ANALYSES OF LAVAS AT FIELD
TRIP LOCALITIES

	A	B	C	D	E
SiO_2	49.54	72.32	50.19	47.38	46.91
TiO_2	2.14	0.39	1.51	0.99	0.89
Al_2O_3	15.05	12.22	15.15	17.32	17.65
Fe_2O_3	10.05	1.36	5.51	7.78	6.43
FeO	3.23	3.14	5.82	2.29	3.38
MnO	0.17	0.05	0.15	0.12	0.16
MgO	5.05	0.05	5.91	7.65	8.04
CaO	8.38	0.62	9.13	11.50	11.46
Na_2O	2.67	2.93	2.71	2.09	2.22
K_2O	1.29	6.27	0.62	0.13	0.17
P_2O_5	0.29	0.01	0.17	0.10	0.04
H_2O	2.53	0.39	3.68	3.33	2.98
CO_2	n.d.	n.d.	0.13	n.d.	n.d.
Others	0.03*	0.15*	–	0.04*	–
Total	100.43	99.90	100.68	100.72	100.37

*A: Cr_2O_3; B: 0.01 S and 0.14 ZrO_2); D: Cr_2O_3.
A. Fe-tholeiite, sample LW-21, from 30-ft- (9-m-) thick
basalt overlying rhyolite at Lakewood locality. From
Green, 1979.
B. Rhyolite, sample LWB-102, from 260 ft (78 m) flow
beneath basalt at Lakewood locality. From Brannon, 1984.
C. Quartz tholeiite, sample TH-2, from main 95 ft (29 m)
flow at Two Harbors locality. From Green, 1972.
D. Olivine tholeiite, sample SR-9, from colonnade of
main flow at Gooseberry Falls locality. D. Blanchard,
NASA-Houston, analyst.
E. Olivine tholeiite, sample SR-10, from lower falls
below highway at Gooseberry Falls locality. D.
Blanchard, NASA-Houston, analyst.

GEOLOGIC SETTING

The North Shore Volcanic Group (NSVG) (Green, 1972, 1982) comprises a thickness of about 5.4 mi (8.7 km) as measured in continuous sequence along the lakeshore. It has the form of half of a large dish tilted to the southeast—near Duluth (Fig. 1) the lavas strike roughly north and dip gently east; at the northeast end (Grand Portage) they strike west and dip gently south. The three localities described here occur at roughly 7,000 ft (2,100 m), 19,000 ft (5,800 m) and 21,000 ft (6,500 m), respectively, above the base of the section exposed near Duluth. The NSVG consists predominantly of subaerial tholeiitic flood basalt and basaltic andesite, and includes several small to large rhyolite units and smaller units of andesite and icelandite. The volcanic features indicate large-volume eruptions in general, and high eruptive rates and very fluid lava for the basalts. Among Cenozoic analogs, these features are most similar to the Tertiary plateau basalts of Iceland. The rocks have been chemically altered to varying degrees (some very little, others strongly) by burial metamorphism.

1. Lakewood Locality

In the eastern part of Duluth, U.S. 61 (London Road) crosses the mouth of the Lester River and soon becomes a four-lane expressway. Keep right on County 61 ("Scenic North Shore Drive") (Fig. 2). At about 1.3 mi (2.1 km) from this junction (just before the first significant right bend), park in a

A

B

Figure 3. A) Photo of ropy structures on pahoehoe surface of basalt flow at Lakewood locality. B) Photo of gently tilted, lumpy surface of basalt flow at Lakewood, area B (Fig. 2).

paved area on the lake (right) side of the road and walk back (southwest) about 1,000 ft (300 m) to a wide grassy area. From the widest part of the grassy area, descend the brushy bluff on an unofficial trail to the lakeshore. The bluff is made of red clay deposited in Glacial Lake Duluth, which occupied the southwest end of the Lake Superior basin as the late Wisconsinan ice sheet melted back about 10,000 years ago.

At the shore, the base of a 29-ft-thick (9-m-thick) Fe-tholeiite or transitional basalt flow overlies the lumpy, irregular, flow-banded top of a thick, red rhyolite flow (Fig. 2B). For chemical analyses see Table 1, samples A and B. Thin lenses and patches of interflow sandstone occur in places along the contact, and calcite and some fluorite fill vesicles and fractures. The base of the basalt has a thin amygdaloidal zone that passes upward as one walks northeastward along the shore into the massive phaneritic interior. The interior contains vesicle (amygdule, cylinders and, toward the top, the flow becomes patchily, then generally amygdaloidal. The top of this flow is made of several thin flow units that range from a few centimeters to a couple of meters thick, and the uppermost surfaces show well-preserved ropy pa-

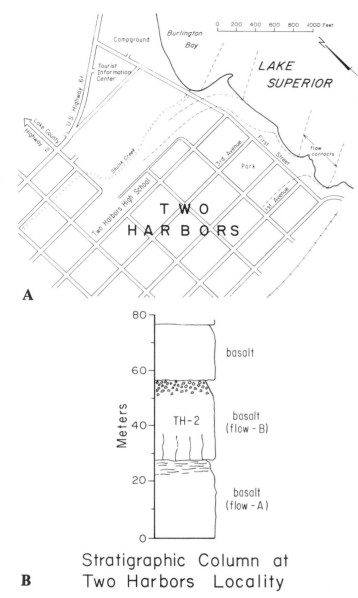

Figure 5. Photo of base of quartz tholeiite flow forming massive layer in cliff and underlying slabby pahoehoe top of amygdaloidal basalt. Locality 2, Two Harbors.

Stratigraphic Column at
B Two Harbors Locality

Figure 4. A) Sketch map of Locality 2, Two Harbors. B) Stratigraphic column at Two Harbors locality.

hoehoe forms (Fig. 3A). (Please do not damage them.) The tight, unbrecciated base, the massive interior, vesicle cylinders, subophitic texture, and pahoehoe top indicate that this flow advanced rapidly over the surface in a highly fluid condition and ponded before crystallization had begun to interfere with flow.

This flow is overlain by two thinner basalt units that are somewhat less well exposed. The units have pipe amygdules in the base with a very flat, vesicular crust, and must have been highly fluid at the time of deposition.

The succeeding flow is larger and has especially coarse grain size and large vesicle cylinders several centimeters across. Joints and fractures showing various stages of alteration and mineralization cut the basalt. Another similar but thinner flow follows. At the next flow contact (at a small creek) climb back to the parking area above.

Around the right bend just ahead on County 61 is another parking area just above a cross-bedded, fluvial, interflow sandstone unit overlain by another basalt flow, both capped by red glacial lake clay. This flow has large, lumpy-topped, vesicular flow units in its base and a well-preserved lumpy pahoehoe top which has been exposed by wave erosion (Fig. 3B).

2. Two Harbors Locality

Continue northeast on U.S. 61 (accessible by the next paved road heading inland) to Two Harbors. Drive through town (Fig. 4A) past the third traffic light at County 2. Turn right just beyond the Tourist Information Center onto 1st Street. Park between 3rd and 2nd Avenues and walk down to the lakeshore. Here is a sequence of thick but fine-grained quartz tholeiite or Fe-tholeiite flows, dipping gently southeast (Fig. 4B). The flow that is directly accessible (A) shows a vuggy upper-middle zone that grades rapidly upward into a highly amygdaloidal top zone, 6.5 to 10 ft (2 to 3 m) thick, of slabby pahoehoe. Quartz lines the vugs and laumontite and calcite fill the amygdules. Patches and lenses of red volcanic sandstone occur between flow A and the overlying 95-ft-thick (29-m-thick) flow (B) in the wave-cut cliff (Fig. 5). Danger: unstable rock! A large rock fall occurred here in 1981, and more are possible. Examine the fallen blocks, not the overhang. Note the fluid-appearing, non-rubbly base of the overlying flow (B), which contains a few small amygdules. The massive portion just above, with crude columnar jointing, also shows millimeter-scale oxidation laminations, subparallel to the base, which are common in this general composition of flood basalt (SiO$_2$ 50-54%). There are also a few lensoid amygdules filled with black chlorite or gray agate. For rock analysis see Table 1, sample C.

By following a trail along the top of the cliff to the south,

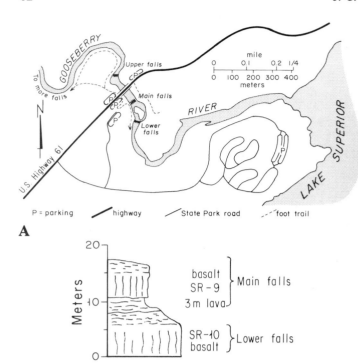

A

B

Schematic Stratigraphic Column
of Basalt Flows
Gooseberry Falls Locality

Figure 6. A) Sketch map of Gooseberry Falls area, Locality 3. There are many more foot trails (not shown here) in the park. B) Stratigraphic column at Gooseberry Falls locality.

one soon comes to more broad wave-washed ledges of the upper-middle part of flow B, which shows hackly fractures and local vugs, and passes up into a thick rubbly top zone. Laumontite and calcite occur between the autobrecciated fragments, and within amygdules. This flow evidently advanced much too fast to develop rubble at the base (as typical Hawaiian aa flows do), but eventually produced a thick aa top. Return to 1st Street and to U.S. 61.

3. Gooseberry Falls Locality

Stop in parking area along the highway southwest of the bridge over the Gooseberry River (Fig. 6A). The stone steps, a WPA project, were built of highly iron-enriched cumulate ferro-

Figure 7. Photo of main waterfall over colonnade of an olivine tholeiite flood basalt, Gooseberry River locality. The lower half of the falls is in a thinner, 10-ft-thick (3-m-thick) flow that has a billowed surface.

gabbro from a small quarry in the Beaver Bay Complex a few miles to the northeast (see Green, 1972). They lead down to a series of falls over a series of olivine tholeiite basalt flows (Fig. 6B). Looking upriver here beneath the bridge, one can see the upper waterfall which is made by the same flow colonnade as the main waterfall and which is accessible by another trail. One can approach the top of the main falls (Fig. 7) over the weathered, largely eroded amygdaloidal top of the main flow. Depending on the water level, one can see more or less well the outlines of the joint columns that control the main waterfall. Take care with your footing! A flow contact occurs about two-thirds of the way down this waterfall, the upper part being formed by the colonnade of the main basalt and the rest being in a thinner, 10-ft (3-m) thick lava. The tops of this lava and the underlying flow exposed in the river bed below are of the billowy pahoehoe type, without ropy structures on the broad convex forms, generally about 6.5 ft (2 m) across, that characterize the amygdaloidal crust of this type of flood basalt. These surfaces are like many developed on large Quaternary flood basalts in Iceland. Some small potholes and a small natural arch are found nearby. The lower falls is made of the colonnade of the lower flow. For chemical analyses, see Table 1, samples D and E.

Other field trip localities for the NSVG, including flows of a wide variety of compositions, are described in Green (1979).

REFERENCES CITED

Brannon, J. C., 1984, Geochemistry of successive lava flows of the Keweenawan North Shore Volcanic Group [Ph.D. dissertation]: Washington University, St. Louis, 328 p.

Green, J. C., 1972, North Shore Volcanic Group, in Sims, P. K., and Morey, G. B., eds., Geology of Minnesota: A centennial volume: Minnesota Geological Survey, p. 294–332.

——— 1979, Field Trip Guidebook for the Keweenawan (Upper Precambrian)

North Shore Volcanic Group, Minnesota: Minnesota Geological Survey Guidebook Series 11, 22 p.

——— 1982, Geology of Keweenawan extrusive rocks, in Wold, R. J., and Hinze, W. J., eds., Geology and tectonics of the Lake Superior Basin: Geological Society of America Memoir 156, p.47–55.

——— 1983, Geologic and geochemical evidence for the nature and development of the Middle Proterozoic (Keweenawan) Midcontinent Rift of North America: Tectonophysics, v. 94, p. 413–437.

The Middle Proterozoic Duluth Complex at Duluth, Minnesota

John C. Green, Department of Geology, University of Minnesota, Duluth, Minnesota 55812

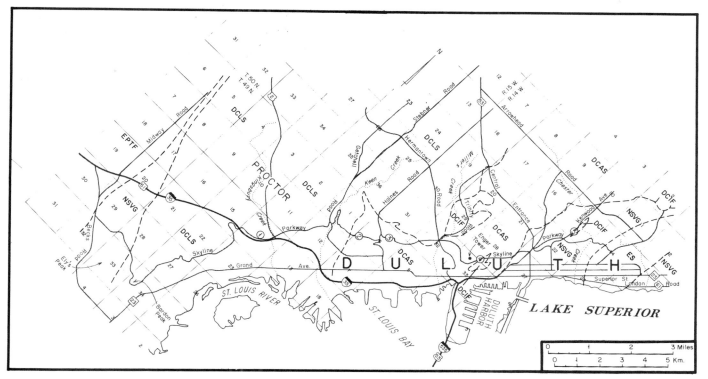

Figure 1. Geologic sketch map of Duluth showing locations of field trip stops. ES—Endion sill; DCIF—Duluth Complex intermediate and felsic intrusions; DCLS—Duluth Complex layered series; DCAS—Duluth Complex anorthositic series; NSVG—North Shore Volcanic Group; EPTF—Early Proterozoic Thomson Formation; dashed lines—geologic contact.

LOCATION

The Duluth Complex underlies an area of approximately 2,265 mi^2 (6500 km^2), but over much of its extent it is covered by glacial deposits. Some of the best and most representative exposures are in the city of Duluth. Several outcrops are described here along the Skyline Parkway, a well-marked scenic road near the top of the slope on which Duluth is built, facing the Lake Superior Basin. They are in the West Duluth, Duluth Heights, and Duluth 7½-minute Quadrangles. The first stop is adjacent to I-35 just southwest of Duluth (Fig. 1).

SIGNIFICANCE

The Duluth Complex, one of the world's largest mafic intrusive complexes (Sims and Morey, 1972; Weiblen and Morey, 1980), was emplaced in late Middle Proterozoic (Keweenawan) time under a cover of plateau lavas. The complex is one of the major magmatic products of the Midcontinent rift of North America (Wold and Hinze, 1982). Although the term "lopolith" was coined by Grout (1918) to describe the form of this complex,

we now know that the "Duluth lopolith" was never a single body of magma. Dated at 1.1 Ma (Van Schmus and others, 1982), it consists of dozens of small to large, separate but adjacent intrusions that range from troctolite through gabbro, gabbroic anorthosite, syenodiorite, and ferrogranodiorite to granite. Chilled margins and estimates of bulk compositions of the mafic intrusive units show that the magmas were not as primitive as those that formed some of the world's other major mafic intrusions such as the Bushveld, Stillwater, and Muskox, although some of the comagmatic Keweenawan plateau basalts are more primitive (Green, 1982). The origin of this immense volume of evolved magma is not well understood.

Mapping throughout the exposed part of the Duluth Complex has generally supported the finding by Taylor (1964) in the Duluth area that most intrusive units can be assigned to either a "troctolitic series" or an "anorthositic series." Intermediate and felsic portions are relatively minor, except in Cook County, Minnesota, near the northeastern end of the complex. The troctolitic series intrusions (troctolite, anorthositic troctolite, olivine gabbro) tend to be layered cumulates, and intrude members of the anorthositic series (gabbroic and troctolitic anorthosites, anorthositic

Figure 2. Photo of layered cumulate olivine gabbro at Stop 1. Notebook is 12 × 19 cm.

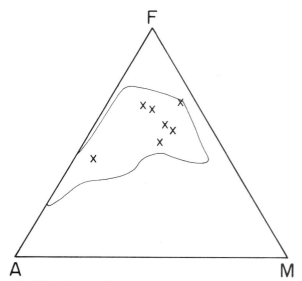

Figure 3. AFM diagram of Duluth Complex rocks of field trip area. A = $Na_2O + K_2O$; F = $FeO + 0.9Fe_2O_3$; M = Mgo. Circled area shows distribution of 27 analyses of Duluth Complex rocks.

gabbro) that are typically not layered but flow-foliated (intruded as a crystal mush) and are somewhat more altered than the troctolites. Only minute amounts of ultramafic rocks are known; these are generally dunites and peridotites associated with the troctolitic series and pyroxenites and peridotites associated with anorthositic series rocks. Inclusions of anorthositic series rocks are found in troctolitic units and within other anorthositic intrusions. Some anorthositic troctolites and troctolitic anorthosites are not readily assignable to either series, and represent transitional compositions.

Large but currently subeconomic concentrations of copper and nickel sulfides have been found near the base of the complex to the north of Duluth but not in the Duluth area. Potential for Pt group, Cr, and Ti deposits is also being investigated.

DESCRIPTION

In Duluth, Taylor (1964) found the bulk of the complex to consist of a layered series of troctolitic and olivine-gabbroic cumulate rocks (Fig. 1). The layering (banding), which occurs on a scale ranging from millimeters to tens of centimeters or more, dips toward the east at about 25–35°, in general conformity to the attitude of the Keweenawan plateau basalt above and below it. Approximately 17,000 ft (5200 m) of thickness is thus indicated for the layered series here, but Taylor did not find significant compositional changes in the cumulate minerals (cryptic layering) in this sequence. At its top the layered series intrudes rocks of the anorthositic series, which in turn have lavas of the North Shore Volcanic Group for their roof. Small bodies of ferrogranodiorite and granite occur near the top of the layered series and within the anorthositic series, and also penetrate the overlying flows. This trip includes stops at several of these units and rock types, which are fairly typical of the Duluth Complex as a whole. All of the intrusive rocks in Duluth, as well as the overlying lavas, show normal magnetic polarity. Plateau basalts of the North Shore Volcanic Group are described in another field trip of this series.

Stop 1

Thomson Hill Information Center, on the frontage road beside I-35 just west of the junction with U.S. 2 in the NW¼ Sec.14,T.49N.,R.15W. Layered rocks of the troctolitic series are exposed in cuts at the back of the parking area (Fig. 2) and the northeast access road. Some of the glaciated surface of the bedrock is visible beneath a till cover formed by the Superior lobe of the Wisconsinan ice sheet flowing to the southwest out of the lake basin. The layering here represents differences in modal amounts of cumulus plagioclase (An63), olivine, augite, and oxides (ilmenite and magnetite) and their post cumulus overgrowths. A chemical analysis is given in Table 1. (See also Figure 3.) The rocks show a foliation resulting from the accumulation of tabular plagioclase and ilmenite on the crystal-magma interface. According to the trend and dip of the layering, these rocks are roughly 7,389 ft (2250 m) above the base of the complex. Very olivine-rich, picritic to dunitic layers can be seen nearer the base below Bardon Peak (Fig. 1), about 4.5 mile (7.6 km) to the southwest along and below the Skyline Parkway, past the Spirit Mountain Recreation Area.

To reach the next stop, continue northeast on the frontage road, cross U.S. 2 and continue on Skyline Parkway. The last outcrops of obviously layered rocks of the troctolitic series are seen in cuts as the road makes a long hairpin excursion to cross Keene Creek (Figs. 1, 4). These gabbros are very iron-rich (D-105, Table 1). About 0.4 mi (0.6 km) beyond Oneota Cemetery (on the right) is a large road cut with a white rock in the center (Figs. 4, 5).

Stop 2

SW¼Sec.6,T.49N.,R.14W. Pale-gray xenoliths of pure or

TABLE 1. CHEMICAL ANALYSES OF ROCKS OF THE DULUTH COMPLEX
AT FIELD TRIP STOPS

	GKS-2	D-105	D-106	M-3762	D-108	D-30	DG-66
SiO_2	46.4	38.0	47.5	49.65	48.2	50.40	62.6
TiO_2	5.50	12.3	1.28	3.93	3.97	2.92	1.02
Al_2O_3	16.0	11.2	23.1	13.22	13.6	15.01	15.3
Fe_2O_3	2.2	3.5	1.7	1.58	5.0	6.33	3.8
FeO	9.2	16.3	6.5	11.76	11.1	6.92	3.6
MnO	0.155	0.25	0.104	0.21	0.28	0.20	0.114
MgO	6.15	7.74	4.52	5.44	4.50	4.11	1.18
CaO	11.6	9.58	10.4	8.98	7.86	4.20	2.46
Na_2O	2.65	1.71	3.03	2.71	3.13	4.36	4.69
K_2O	0.30	0.18	0.58	0.97	1.32	2.08	3.57
P_2O_5	0.05	0.08	0.104	0.53	0.66	0.51	0.20
H_2O+	–	–	–	0.59	–	2.77	–
CO_2	–	–	–	0.04	–	0.15	–
S	–	–	–	0.08	–	–	–
Total	100.2	100.9	98.8	99.69	99.6	99.96	98.5

GSK-2: ilmenite-olivine gabbro, cut behind Thomson Hill Information Center (Stop 1). R. Knoche, analyst.
D-105: Fe-rich gabbro, Skyline Parkway east of Keene Creek. R. Knoche, analyst.
D-106: gabbroic anorthosite, northeast of Haines Road on Skyline Parkway (Stop 2). R. Knoche, analyst.
M-3762: microgabbro dike, Skyline Parkway northeast of Haines Road (Stop 3). (Taylor, 1964).
D-108: Fe, Ti-rich basaltic dike, Skyline Parkway just east of Twin Ponds (Stop 4). R. Knoche, analyst.
D-30: porphyritic basalt, First Street at 5th Ave. E (on strike with Kenwood Ave. Quarry, Stop 5). K. Ramlal, analyst.
DG-66: red granite, Kenwood Avenue quarry (Stop 5). R. Knoche, analyst.

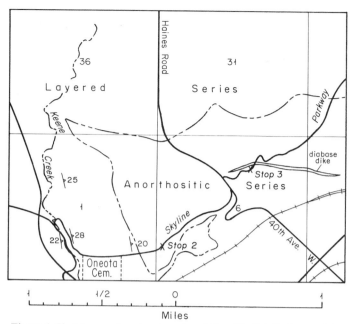

Figure 4. Sketch map showing stops 2 and 3 on Skyline Parkway. After Taylor (1964).

nearly pure anorthosite are found in a few places in the Duluth Complex and very commonly in the olivine diabases of the Beaver Bay Complex, which intrudes the North Shore Volcanic Group in Lake County, Minnesota, to the northeast of Duluth. This example is fairly typical, though more sodic than most. It is included in ophitic ilmenite gabbro assigned by Taylor to the anorthositic series, and has granoblastic and tabular igneous texture and a composition of An64. These anorthosite xenoliths, which have a variety of textures including igneous, cataclastic, and annealed, are unrelated to the Keweenawan magmatism and were evidently broken off and rafted up from the lower crust or uppermost mantle by the Keweenawan magmas (Morrison and others, 1983).

Continue northeast on Skyline Parkway, crossing Haines Road by a jog to the left.

Stop 3

Approximately 0.2 mi (0.3 km) northeast of Haines Road; NE¼Sec.6,T.49N.,R.14W. Here the anorthositic series is well exposed in the form of flow-foliated, coarse-grained gabbroic anorthosite and anorthositic gabbro. The foliation is nearly vertical here but varies considerably from place to place. No layering or banding is present. Cumulate plagioclase crystals (An70) are surrounded by interstitial poikilitic olivine, augite, and opaques (Table 1). The rock appears to have been intruded as a crystal mush dominated by plagioclase. As with much of the anorthositic series, the rock is considerably altered, perhaps by a pulse of water and heat above the intruding layered series. Around the left

bend a few tens of meters ahead is a large diabase (microgabbro) dike with chilled contacts (Table 1) that is intruded into the anorthositic series. It is an ophitic olivine tholeiite with some later, interstitial orthoclase, quartz, hornblende, and biotite.

Continue northeast on Skyline Parkway; at 0.6 mi (1.0 km) an overlook gives a fine view of the U.S. Steel iron ore docks and, across the bay in Superior, Wisconsin, a shipping facility for western low-sulfur coal. Cross Miller Creek at the six-way intersection and continue on Skyline Parkway through and past a large gravel deposit. This was a large delta deposited by Miller Creek in Glacial Lake Duluth, whose shoreline is generally followed by the Skyline Parkway. At 0.85 mi (1.4 km) past the junction, keep right to drive around the outside of the hill on which Enger Tower is situated. Here is another overlook with a good view of the Duluth-Superior harbor with its grain elevators, and Minnesota Point, a 3.5 mi-long (6.0 km) baymouth bar. The St. Louis River estuary visible below was formed by isostatic uplift of the outlet of Lake Superior at Sault Ste. Marie, which raised the water level over the past five thousand years, drowning the valleys that had been eroded in the red clays and silts deposited in the bottom of the former Glacial Lake Duluth.

At the north side of the hill the road crosses between Twin Lakes. Stop at the next left bend.

Stop 4

Skyline Parkway just northeast of Twin Lakes. SW¼SE¼ Sec.28,T.50N.,R.14W. Here a road cut and surface outcrop display a brick-red, strongly altered augite-quartz monzodiorite, probably a differentiate of one of the gabbro series, and a cross-

Figure 5. Photo of anorthosite xenolith at Stop 2.

Figure 6. Photo of intrusive relations between red granite and basalt in Kenwood Avenue Quarry, Stop 5.

cutting dike of high FeTi basalt (Table 1). The dike shows excellent chilled contacts and xenoliths of the country rock. Because the dike intrudes the quartz monzodiorite, it is evidently one of the youngest rocks in the Keweenawan succession in the Duluth area. The composition of this dike is similar to that of many other Keweenawan dikes and plateau lavas in Minnesota, and probably resulted from extensive Fe-enrichment during fractionation in deep crustal chambers during rifting.

Continue northeast on Skyline Parkway; in about 0.8 mi (1.3 km) the roof of the Duluth Complex is reached. Jog left on Mesaba Avenue to cross Central Entrance, and keep to the right above First Methodist Church, where the first outcrops of contact-metamorphosed, porphyritic basalt lava of the North Shore Volcanic Group are seen. Continue on Skyline Parkway to Kenwood Avenue; here turn uphill to the left. At 0.4 mi (0.7 km) pull off the road to the right in unpaved area at the start of a broad left bend.

Stop 5

Kenwood Avenue Quarry (abandoned), SE¼Sec.15,T.50N., R.14W. Across the road from the parking area is gray porphyritic basalt of the North Shore Volcanic Group, intruded by red granitic rock associated with the underlying Duluth Complex. The basalt (Table 1) is somewhat altered and recrystallized from the effects of the intrusion. Scattered amygdules are filled with quartz, sodic plagioclase, augite, hornblende, epidote, biotite, and chlorite in place of original zeolites and lower temperature chlorite. The granitic rock (Table 1) is a rather altered hornblende-quartz monzonite that contains primary magnetite, sphene, apatite, and zircon and secondary epidote, chlorite, and calcite. Excellent intrusive relations are exhibited in the back face of the quarry (Fig. 6) where complex veins and dikes of the quartz monzonite surround angular and fretted xenoliths of basalt.

Many more outcrops of the Duluth Complex can conveniently be located by consulting the map of Taylor (1964).

REFERENCES CITED

Green, J. C., 1982, Geology of Keweenawan extrusive rocks, *in* Wold, R. J. and Hinze, W. J., eds., Geology and Tectonics of the Lake Superior Basin: Geological Society of America Memoir 156, p. 47–55.

Grout, F. F., 1918, The lopolith; an igneous form exemplified by the Duluth gabbro: American Journal of Science, 4th series, v. 46, p. 516–522.

Morrison, D. A., Ashwal, L. D., Phinney, W. C., Shih, Chi-Yu, and Wooden, J. L., 1983, Pre-Keweenawan anorthosite inclusions in the Keweenawan Beaver Bay and Duluth Complexes, northeastern Minnesota: Geological Society of America Bulletin, v. 94, p. 206–221.

Sims, P. K., and Morey, G. B., 1972, Geology of Minnesota: A centennial volume: Minnesota Geological Survey, St. Paul, 632 p.

Taylor, R. B., 1964, Geology of the Duluth Gabbro Complex near Duluth,

Minnesota: Minnesota Geological Survey Bulletin 44, 63 p.

Van Schmus, W. R., Green, J. C., and Halls, H. C., 1982, Geochronology of Keweenawan rocks in the Lake Superior region: A summary, *in* Wold, R. J., and Hinze, W. J., eds., Geology and Tectonics of the Lake Superior Basin: Geological Society of America Memoir 156, p. 165–171.

Weiblen, P. W., 1982, Keweenawan intrusive igneous rocks, *in* Wold, R. J., and Hinze, W. J., eds., Geology and Tectonics of the Lake Superior Basin: Geological Society of America Memoir 156, p. 165–171.

Weiblen, P. W., and Morey, G. B., 1980, A summary of the stratigraphy, petrology, and structure of the Duluth Complex: American Journal of Science, v. 280-A, p. 88–133.

Wold, R. J., and Hinze, W. J., 1982, Geology and Tectonics of the Lake Superior Basin: Geological Society of America Memoir 156, 280 p.

Jay Cooke State Park and Grandview areas: Evidence for a major Early Proterozoic-Middle Proterozoic unconformity in Minnesota

M. A. Jirsa and G. B. Morey, Minnesota Geological Survey, St. Paul, Minnesota 55114

SITE 14- JAY COOKE STATE PARK

Location

Jay Cooke State Park occupies nearly 15 mi² (38 km²) in the northeastern corner of Carlton County in T.48N.,R.15 and 16W. (Fig. 1). It is about midway between the village of Thomson and Fond du Lac, the westernmost suburb of Duluth. The park may be reached by following Minnesota 210 from the Carlton-Cromwell interchange with I-35 eastward for approximately 3.5 mi (6 km) through the town of Carlton to the St. Louis River. Immediately after crossing the river at Thomson, the highway turns sharply to the south and 0.1 mi (0.2 km) later enters Jay Cooke State Park (Fig. 2). The Park Headquarters, a picnic area, and an information building are located approximately 2 mi (3.4 km) into the park. As with all Minnesota state parks, a daily fee or a yearly permit is required to use the park facilities. Minnesota 210 continues through the park along the north side of the St. Louis River for an additional 5.5 mi (9.3 km) where it joins Minnesota 23 in Fond du Lac.

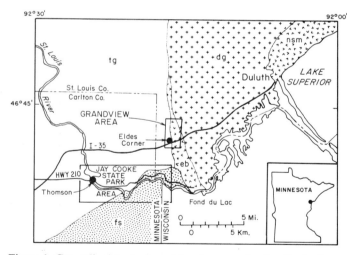

Figure 1. Generalized geologic and location map for Jay Cooke State Park and Grandview areas (insets). Geologic units: tg - Thomson Formation; eb - Ely's Peak basalts; nsm - North Shore Volcanic Group; dg - Duluth (gabbro) Complex; fs - Fond du Lac Formation.

Significance of the Site

Excellent exposures of the Early Proterozoic Thomson Formation and the Middle Proterozoic Fond du Lac Formation occur in Jay Cooke State Park. The Thomson Formation is a typical flysch sequence consisting of intercalated graywacke and slate. The Fond du Lac Formation is one of several sedimentary units in the Keweenawan Supergroup and is a classic redbed sequence deposited by fluvial-alluvial processes. An angular unconformity between the two is well exposed at several places in the park.

Geochronometric data indicate that the Thomson Formation was deformed and metamorphosed during the Penokean orogeny (ca 1850 Ma), whereas the Fond du Lac Formation probably was deposited about 1000–900 Ma. Thus the unconformity is an erosional surface representing about 900 million years of unrecorded geologic history, an interval about 1.5 times longer than all of the Phanerozoic Eon. Interestingly the unconformity between the Fond du Lac Formation and the overlying Pleistocene glacial deposits also represents a hiatus of nearly 1000 m.y., from 1000–900 Ma to possibly less than 20,000 years ago.

Geology of the Park

Pleistocene glacial deposits in the park include a sequence of red clay, silt, and sand of lacustrine origin about 200 ft (60 m) thick (Wright and others, 1970). The ancestral St. Louis River, swollen with glacial meltwater, cut deeply into this easily eroded material to expose a long (8 mi; 13 km) narrow window of resistant graywacke and slate that now make up a series of gorges, falls, and rapids. Although a dam near the east edge of the park has considerably modified the original morphology of the valley, it is still evident that the valley widens and flattens where the river has cut into softer sandstone and shale of the Fond du Lac Formation (Fig. 2).

The Thomson Formation underlies much of east-central Minnesota (Morey, 1978). The total thickness of the formation is unknown inasmuch as neither the top nor bottom is exposed. Furthermore the lack of continuous exposures and suitable marker beds and the deformed nature of the strata make it difficult to estimate the total thickness. Wright and others (1970) have shown that at least 2000 ft (600 m) are exposed along the St. Louis River in the Cloquet-Carlton-Thomson area.

The type locality of the Thomson Formation occurs along the St. Louis River west of the village of Thomson (Fig. 2, Locality 1). There, approximately 650 ft (200 m) of strata are exposed between the dam north of Highway 210 and the railroad bridge south of it. These exposures contain approximately equal proportions of graywacke, siltstone, and slate, which, except on a small scale, vary independently of stratigraphic position.

The graywacke units have variable texture and consist of 7 to 73 percent very fine to medium framework grains, 20 to 85 percent matrix material, and 1 to 18 percent cement. The compositions of selected samples are shown in Figure 3. Mineralogi-

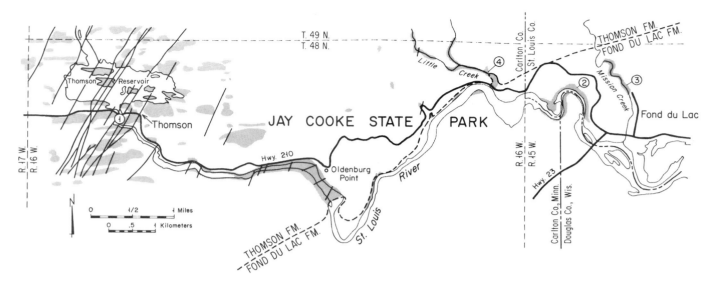

Figure 2. Geology and reference locations (numbered) of the Jay Cooke State Park area. Shaded areas are exposures; northeast-trending solid lines are basaltic dikes (after Kilburg and Morey, 1977; Wright and others, 1970).

cally, the siltstone and slate units are the fine-grained and very fine grained counterparts of the graywacke. Major framework grains in the graywacke include quartz, sodic plagioclase (albite-oligoclase), and, in the coarser grained beds, granitic and quartzose sedimentary rock fragments; trace amounts of microcline and orthoclase also occur. Matrix material consists of chlorite and muscovite as grains finer than 0.03 mm. Calcite is the major cement. The shape and size of both the framework grains and the matrix particles have been considerably modified by recrystallization during metamorphism to the greenschist facies. However, the grain size and mineralogy are closely related, inasmuch as the coarse-grained rocks have a smaller detrital component than their finer grained counterparts. Abundant carbonate-rich concretions characterize much of the formation and are useful in delineating master bedding in otherwise massive beds (Morey and Ojakangas, 1970).

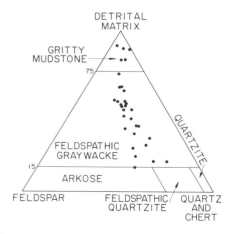

Figure 3. Summary of the mineralogic composition of the Thomson Formation (from Morey and Ojakangas, 1970).

Sedimentological aspects of the Thomson Formation have been described by Morey and Ojakangas (1970). The slate units are very fine grained, laminated to very thinly bedded, and probably accumulated under quiet water conditions. Intercalated graywacke units range in thickness from 1 in (2 cm) to 14 ft (4 m), and commonly display a wide variety of sedimentary structures considered indicative of turbidite deposition. Graded bedding is well developed throughout; cross-bedding is common in the uppermost parts of the many beds. Laminations, sole marks, flute casts, flame structures, ball structures, and other related features also are present. The dip azimuth directions of cross-bedding show consistent current flow to the south (Fig. 4). In contrast, current directions inferred from flute and groove casts are more scattered with poorly defined trends to the east and west. Considered by themselves, the sole marks have a bimodal distribution with both westward and southeastward orientations (not shown on Fig. 4). Consequently, Morey and Ojakangas (1970) inferred that the cross-bedding probably was caused by currents flowing down the paleoslope, perpendicular to a shoreline, whereas the diverse sole mark directions resulted from flowage over and around an irregular submarine topography. Alternatively, the sole marks could reflect a fanning-out of the currents over gently sloping paleoslope in the axial part of the basin.

Numerous structural features related to the Penokean orogeny are found in the Thomson Formation (Holst, 1984). At and near the type locality, there are gentle to open folds on a scale of centimeters to kilometers (Wright and others, 1970). The fold axes trend to the east, have vertical and steep, southward dips, and plunge gently east and west. A well-developed axial-planar cleavage, vertical or dipping steeply to the south, is present in the slaty beds. Kink bands, which deform this axial cleavage, are common in exposures north of the bridge, as are concretions and

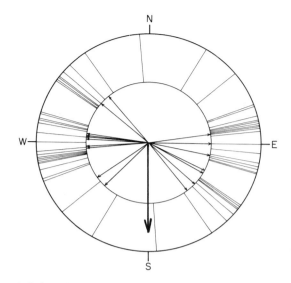

Figure 4. Orientation of 17 flute casts (inner circles) and 26 groove casts (outer circle), and the mean of 201 cross-beds (large arrow) in the Thomson Formation (after Morey and Ojakangas, 1970).

mud chips flattened in the plane of the cleavage. Joints that may be associated with the folding, as well as some younger joint sets, have been described by Wright and others (1970).

Dikes of ophitic microgabbro ranging in width from a few centimeters to about 200 ft (65 m) form a dike swarm that occupies a northeast-trending joint set in the Thomson Formation (Fig. 2). Some dikes are offset by a younger, northwest-trending joint set. The larger dikes are medium-grained microgabbro with chilled margins of fine-grained basalt. The thin dikes are fine-grained basalt throughout. Flow banding is common near the edges of many dikes, regardless of thickness. Where the dikes are thick and closely spaced, the country rocks were metamorphosed to the albite-epidote-hornfels facies.

The chronometric age of the dike swarm has not been determined. However, dikes of the same orientation and composition intrude rocks of the North Shore Volcanic Group but are truncated by plutonic rocks of the Duluth Complex (both of the Keweenawan Supergroup) approximately 4.5 mi (7 km) to the northeast and east of Jay Cooke State Park (Kilburg and Morey, 1977).

Quartz veins ranging in width from several centimeters to 10 ft (3 m) are common in the Thomson Formation. The veins are predominantly crystalline milky quartz with minor amounts of pyrite, chalcopyrite, and other sulfides. The largest exposed quartz vein occurs just north of Minnesota 210 and west of the bridge over the St. Louis River at Locality 1 (Fig. 2). This vein occupies an extensional fracture near the crest of an anticline. Smaller veins, generally less than 6 in (15 cm) wide, occur along joint and bedding surfaces. Some of these have been folded with the adjacent rock and now resemble ptygmatic folds.

Downstream from the Thomson Formation exposures, the St. Louis River has cut down into the Fond du Lac Formation.

This redbed sequence is exposed along both river banks, but most outcrops are relatively inaccessible. More easily visited exposures occur on the north side of the river just upstream from the Fond du Lac picnic area at the Minnesota 23 bridge over the river (Fig. 2, Locality 2). The formation also is exposed in several old quarries on the south side of the river directly across from the picnic area. The rocks that crop out along Mission Creek about 0.5 mi (1 km) to the northeast can be reached easily by automobile or bus (Fig. 2, Locality 3).

The approximately 300 ft (90 m) of redbeds exposed along the St. Louis River constitute the type locality of the Fond du Lac Formation (Morey, 1967). This is only a small fraction of the estimated maximum thicknesses in this area (800 ft; 240 m) and to the south in east-central Minnesota (3000 ft; 910 m).

The Fond du Lac Formation consists of a basal quartz-pebble conglomerate overlain by lenticular beds of sandstone, siltstone, and shale. Together these rocks form a simple homoclinal structure with beds generally striking about N. 60° E. and dipping 5° to 12° SE. Therefore, a pronounced angular unconformity exists between it and the strongly deformed Thomson Formation. Map relationships also imply that the redbeds unconformably overlie volcanic rocks of the North Shore Volcanic Group, but the actual contact has neither been observed nor penetrated by drilling.

The quartz-pebble conglomerate at the base of the formation is approximately 60 ft (18 m) thick. The contact between it and the underlying Thomson Formation crops out at several places, including the valley bottom downstream from Oldenburg Point. The conglomerate is best seen along Little Creek about 0.4 mi (0.6 km) upstream from Minnesota 210 (Fig. 2, Locality 4). Although the actual unconformity is buried at this locality, steeply dipping beds of the Thomson Formation crop out in the river bottom some 200 ft (60 m) north of the lowermost exposed conglomerate.

Pebbles and cobbles in the basal conglomerate are as much as 6 in (15 cm) in diameter and consist predominantly of vein quartz, although minor amounts of chert, quartzite, graywacke, and slate can be found. The matrix is mostly a red-colored coarse grit of angular quartz and feldspar, with some clay-size matrix material and dolomite cement. Pyrite and marcasite are common and occur as concretions or individual grains in the matrix; locally the sulfides had been altered to limonite. The upper part of the conglomerate grades upward through a stratigraphic interval of approximately 3 ft (1 m) into an arkosic sandstone identical to that which composes the bulk of the formation.

A basalt-pebble conglomerate crops out along Little Creek, midway between the quartz-pebble conglomerate and the highway. This conglomerate also has a reddish-brown, arkosic matrix, but the pebble-size clasts consist of highly altered basalt and basalt porphyry. Basalt-pebble conglomerate units also occur higher in the section, where they make up a small but significant part of the formation.

Sandstone units are generally arkosic or subarkosic in composition (Fig. 5). They consist of 36 to 68 percent quartz, 5 to 29

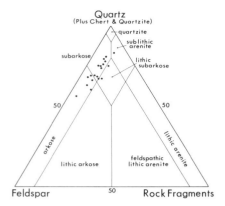

Figure 5. Summary of mineralogic composition of sandstones of the Fond du Lac Formation (from Morey, 1967).

percent feldspar (microcline and albite-oligoclase), 1 to 10 percent rock fragments, 1 to 15 percent matrix material composed of quartz, illite, and rare kaolinite, and 1 to 20 percent cement of hematite, calcite, quartz, and dolomite. Siltstone lenses are mineralogically equivalent to the sandstone units enclosing them. Shale units occur as thin lenses and as uniformly thick, laterally persistent units; both consist predominantly of illite with minor amounts of kaolinite, montmorillonite-illite, quartz, and feldspar.

Medium- to large-scale cross-bedding is the most abundant primary sedimentary structure in the Fond du Lac Formation. It is predominantly of the trough type and occurs in units that fill channels cut into previously deposited sediments. The azimuths of the cross-bedding, as well as other directional sedimentary features such as channel axes, grain lineations, ripple marks, and flutes, are consistent with an eastward paleoslope (Fig. 6).

These sedimentary structures, as well as the presence of mud cracks and rain imprints, indicate that the Fond du Lac Formation was deposited by fluvial processes. The sandstone beds were probably deposited by a series of streams that meandered across a broad alluvial plain. The intercalated shale may represent deposi-

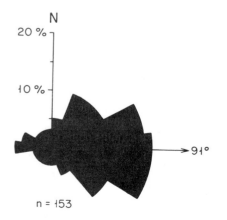

Figure 6. Orientation of 20 trough axes, 20 grain lineations, 10 flute and ripple marks and 103 cross-beds from the Fond du Lac Formation (from Morey, 1967).

tion on floodplains or in lakes. The presence of associated mud cracks, ripple marks, and rain imprints indicates that the mudstones were subjected to alternate periods of wetting and drying. The Fond du Lac Formation is considered to be a clastic wedge deposited in a shallow, oxidizing, alluvial environment by a system of streams emerging from a western highland and dispersing material to the east and southeast. Regional studies have shown that the depositional basin is a half-graben structure that formed late in the history of the Midcontinent rift system. Thus the Fond du Lac Formation represents the kinds of rocks that mark the waning development of this major tectonic feature, which extends from the Lake Superior region southward to Kansas.

SITE 15—GRANDVIEW AREA

Location

The Grandview area is located in sections 17 and 20, T.49N.,R.15W. in the southwestern corner of St. Louis County, a short distance north of Eldes Corner off I-35 (Fig. 1). The interchange to Eldes Corner (Midway Road) is 4 mi (6.7 km) west of the interchange of I-35 with U.S. 2, and 4 mi (6.7 km) east of the interchange to the village of Esko. After leaving the freeway, proceed north on Midway Road through Eldes Corner for 0.95 mi (1.6 km) and park along the narrow gravel road, which extends to the east (Fig. 7). This road provides access to the locality.

Significance of the Site

The Grandview area is well known in geologic literature because it contains exposures showing the unconformable relationship between the Early Proterozoic Thomson Formation and the overlying Middle Proterozoic sedimentary and volcanic rocks of the Keweenawan Supergroup. Additionally, several small outcrops of the plutonic Duluth Complex, which is coeval with and intrusive into the North Shore Volcanic Group, also occur in the northeastern part of section 17 (Fig. 7), but they are not easily accessible from the west.

Geologic Setting

The Thomson Formation crops out sporadically between Midway Road and the base of the steep, west-facing hill 0.4 mi (0.6 km) to the east (Fig. 7). The unit consists of folded and metamorphosed feldspathic graywacke, siltstone, and mudstone or slate. Although poorly exposed, these rocks are structurally and compositionally identical to those described at the Jay Cooke State Park locality.

In contrast to the deformed rocks of the Thomson Formation, the Keweenawan rocks have been little affected by tectonism since deposition, as shown by their gentle dip to the east. The Keweenawan stratigraphic section in this area consists of a basal sedimentary unit—the Nopeming Formation—which

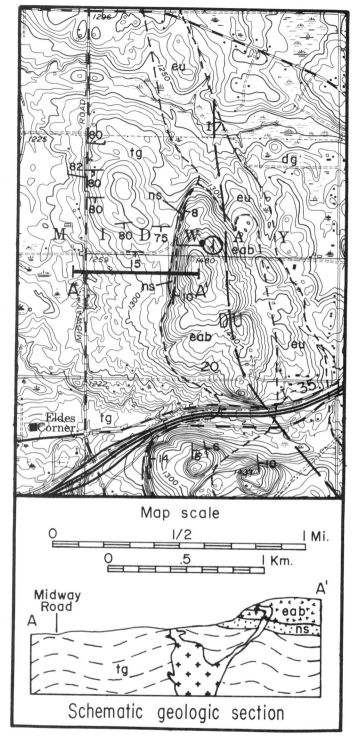

Figure 7. Geology of the Grandview area, T. 49 N., R. 15 W., St. Louis County, Minnesota. Geologic map from Kilburg and Morey (1977). Units: tg - Thomson Formation; ns - Nopeming Formation; eu, eab - Ely's Peak basalts; dg - Duluth Complex. Schematic geologic section below map is not to scale.

is conformably overlain by the Ely's Peak basalts (Figs. 1 and 7). The latter are mafic volcanic rocks that comprise the oldest flows of the North Shore Volcanic Group in this area of the Midcontinent rift and were emplaced over a relatively short interval of time at about 1200 to 1100 Ma (Green, 1972).

The contact between the Nopeming Formation and the underlying Thomson Formation is not exposed, but appears to be an angular unconformity (Fig. 7). In the closest exposures of the two formations, separated by a covered area 350 ft (110 m) wide, the Thomson strikes N. 85° E. and dips 84° S., whereas the Nopeming strikes N. 10° W. and dips 20° E.

The Nopeming Formation is well exposed in section 20 (Fig. 7, Locality 1) at the base of the west-facing slope. In its best exposure beneath the basal volcanic flow, it consists of approximately 30 ft (10 m) of interbedded conglomerate and quartz arenite, with minor siltstone beds in the uppermost part of the unit. The conglomerate contains well-rounded, pebble-size clasts of quartz and to a lesser extent quartzite, set in a sandy matrix. This sandy matrix is compositionally and texturally identical to the intercalated beds of sandstone. Both are a medium- to coarse-grained, well-sorted, well-rounded, silica-cemented quartz arenite. Both contain on the average 88 percent quartz, 1 percent stable rock fragments, and 10 percent epitaxial quartz cement. Some samples contain small amounts (<5 percent) of very fine grained matrix material—mostly sericite, epidote, and chlorite (Mattis, 1972). Both rock types are thick to very thick bedded and contain sedimentary structures indicative of deposition in very shallow water by paleocurrents flowing to the north-northwest.

The thin beds of siltstone intercalated with sandstone in the upper meter of the section are very fine grained and consist on the average of 38 percent quartz, 4 percent orthoclase, 1 percent plagioclase, and 53 percent sericite-epidote matrix. Other constituents, such as fragments of altered felsite, basalt, granite, and slate, and grains of zircon, apatite, and actinolite, make up about 4 percent of a typical mode.

The contact between the Nopeming sedimentary rocks and the overlying Ely's Peak basalts is marked by small-scale load casts, clastic dikes, and other sedimentary structures indicative of soft-sediment deformation. Furthermore the basal part of the lowermost lava flow contains well-developed pillow structures. This combination of sedimentary and volcanic structures implies that relatively low energy clastic sedimentation was abruptly replaced by subaqueous volcanism, most likely in the same body of water.

In the most complete section, the Ely's Peak basalts consist of at least 20 individual flows that have a total thickness of 1200 ft (365 m). However, only four flows with an aggregate thickness of approximately 70 to 75 ft (21 to 23 m) are exposed at the Grandview locality. The flows range in thickness from 10 to 30 ft (3 to 10 m) and are generally fine- to medium-grained tholeiitic basalts characterized by single or glomeroporphyritic phenocrysts of augite and altered olivine (Kilburg, 1972). Individual flow units may be recognized by the presence of rubbly amygdaloidal flow tops, by basal chilled margins, and by abrupt changes in composition and grain size in rocks of adjacent outcrops.

A tabular body of diabasic gabbro inferred to be about 20 ft (6 m) thick crops out between several exposures of flows in section 17. Unfortunately no contacts are exposed and this intrusion may be either a dike or a sill.

Mineral assemblages that contain epidote, sericite, actinolite, and chlorite in the matrix material of the sedimentary rocks and in veins and amygdules in the volcanic rocks are evidence of metamorphism. However the metamorphic history is complex and probably involves both contact metamorphism by the subjacent lavas and intrusions and a regional metamorphic event prior to emplacement of the Duluth Complex (Kilburg, 1972).

The unique stratigraphic and compositional attributes of the rocks at this locality are apparently typical of the basal part of the Keweenawan Supergroup. Similar relationships are observed some 150 mi (240 km) northeast of the Grandview area near Grand Portage. Thus, these rocks are representative examples of those that define the initiation of the Midcontinent rift system.

The Grandview locality should be visited in conjunction with the Jay Cooke State Park locality. The combination of geologic relationships at both places provides insight into the early and late stages of development of the Keweenawan Supergroup (Fig. 8).

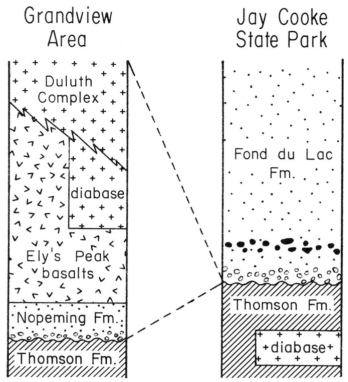

Figure 8. Schematic geologic correlation chart for the Grandview and Jay Cooke State Park areas (not to scale).

REFERENCES

Green, J. C., 1972, North Shore Volcanic Group, *in* Sims, P. K., and Morey, G. B., eds., Geology of Minnesota: A centennial volume: Minnesota Geological Survey, p. 294–332.

Holst, T. B., 1984, Evidence for nappe development during the Early Proterozoic Penokean orogeny in Minnesota: Geology, v. 12, p. 135–138.

Kilburg, J. A., 1972, Petrology, structure, and correlation of the Upper Precambrian Ely's Peak basalts (northeast Minnesota and Ontario, Canada) [M.S. thesis]: University of Minnesota, Duluth, 97 p.

Kilburg, J. A., and Morey, G. B., 1977, Reconnaissance geologic map of the Esko quadrangle, St. Louis and Carlton Counties, Minnesota: Minnesota Geological Survey Miscellaneous Map Series M-25, scale 1:24,000.

Mattis, A. F., 1972, Puckwunge Formation of northeastern Minnesota, *in* Sims, P. K., and Morey, G. B., eds., Geology of Minnesota: A centennial volume: Minnesota Geological Survey, p. 412–415.

Morey, G. B., 1967, Stratigraphy and petrology of the type Fond du Lac Formation, Duluth, Minnesota: Minnesota Geological Survey Report of Investigations 7, 35 p.

——1978, Lower and Middle Precambrian stratigraphic nomenclature for east-central Minnesota: Minnesota Geological Survey Report of Investigations 21, 52 p.

Morey, G. B., and Ojakangas, R. W., 1970, Sedimentology of the Middle Precambrian Thomson Formation, east-central Minnesota: Minnesota Geological Survey Report of Investigations 13, 32 p.

Wright, H. E., Jr., Mattson, L. A., and Thomas, J. A., 1970, Geology of the Cloquet quadrangle, Carlton County, Minnesota: Minnesota Geological Survey Geologic Map Series GM-3 [with text 30 p.], scale 1:24,000.

Pipestone National Monument: The Sioux Quartzite—An Early Proterozoic braided stream deposit, southwestern Minnesota

G. B. Morey and Dale R. Setterholm, *Minnesota Geological Survey, St. Paul, Minnesota 55114*

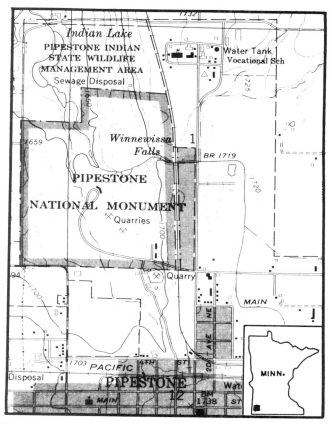

Figure 1. Location map for Pipestone National Monument (modified from Pipestone North 7½-minute Quadrangle, 1967).

LOCATION

Pipestone National Monument is located in parts of sections 1 and 2, T.106N.,R.46W., north of the City of Pipestone, Pipestone County, southwestern Minnesota (Fig. 1). The monument is located 0.9 mi (1.5 km) northwest of the intersection of U.S. 75 and Minnesota 23. Proceed north on U.S. 75 for 0.4 mi (0.7 km); turn west and follow monument signs 0.5 mi (0.8 km) to the entrance. As with all national parks and monuments, no samples may be collected without a permit from the Department of Interior, U.S. National Park Service.

SIGNIFICANCE OF THE SITE

The Sioux Quartzite is a thick unit of generally hard, vitreous, maroon to gray quartzite of Early Proterozoic age that forms an east-trending belt in southwestern Minnesota and adjacent parts of South Dakota and Iowa. Although the Sioux is distributed widely in the subsurface, exposures of relatively thick stratigraphic sections occur in only a few places. One such place is Pipestone National Monument where approximately 100 ft (30 m) of strata are exposed. Although these readily accessible exposures represent only a very small part of the formation, they nonetheless provide a unique opportunity to examine many of the sedimentologic attributes associated with a braided stream system. Additionally, the Sioux Quartzite is a representative example of several redbed sequences in the Lake Superior region that were deposited under stable conditions in an epicratonic setting after the end of the Penokean orogeny at approximately 1850 Ma and prior to the onset of Keweenawan rifting at 1200–1100 Ma.

Pipestone National Monument also has considerable historical and archeological significance. The quartzite contains intercalated beds of red-colored claystone that have been quarried since at least A.D. 1200 (Corbett, 1980). Because of its peculiar mineralogical properties, the claystone is easily carved and has been used by American Indian tribes of the Northern Great Plains to make ceremonial pipes and other items of religious significance. The importance of this site to the Indians was recognized by the white populace after a visit in 1836 by the artist George Catlin. Among other things, he collected several samples of the claystone for subsequent geologic study. The mineralogist C. J. Jackson established that the samples had a unique composition, and, because they were believed to occur only where Catlin found them, the claystone was given the name "catlinite" in his honor. This term is still used today for this particular kind of rock.

GEOLOGY OF THE MONUMENT

The geology of the monument (Fig. 2) has been mapped by Morey (1984). Pleistocene glacial deposits, possibly of Kansan age, mantle the Sioux Quartzite. The Sioux consists predominantly of orthoquartzite, but fine-grained rocks, including quartz-rich siltstone, silty mudstone, and catlinite are also present. In general, the quartzitic rocks are highly resistant to erosion and weathering and form prominent outcrops like the north-trending ridge that presents a west-facing escarpment, 25 to 30 ft (7.5 to 10 m) high along the eastern edge of the monument (Fig. 2). This ridge is part of a major bedrock high that extends for 3 mi (3.5 km) to the north of the monument. Smaller outcrops also occur either as north-trending ridges that rise a meter or so above the general land surface, or in the bottoms of shallow valleys dissected through the drift by small streams. The finer grained rocks, particularly the catlinite beds, rarely form outcrops. They are best seen in the string of small quarries that extend through the center of the monument.

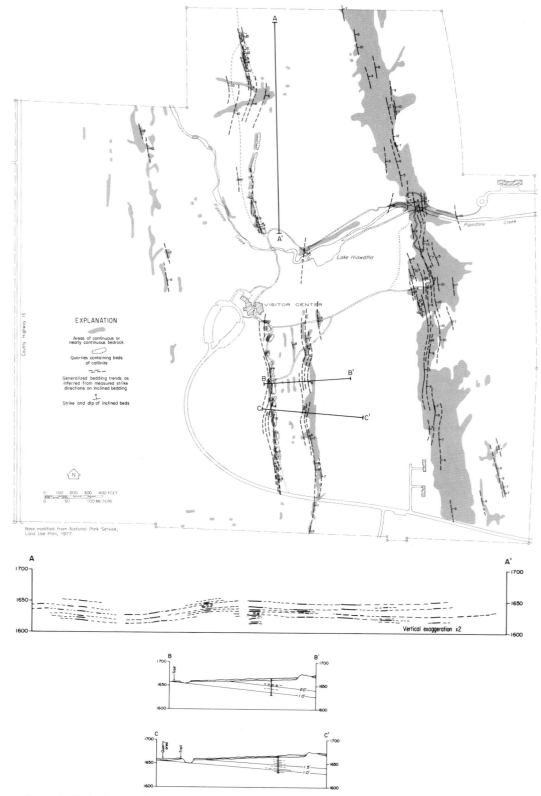

Figure 2. Geologic map of the Sioux Quartzite in Pipestone National Monument (modified from Morey, 1984).

The Sioux Quartzite is characterized by extreme compositional and textural maturity. The quartzite is typically medium pink in color, but beds vary from light pink to deep red. The lighter hues result from finely disseminated hematite around the edges of quartz grains, whereas the deep reds result from a hematite-stained matrix of admixed sericite and very fine grained quartz. This kind of matrix material, together with an epitaxial quartz cement, forms only 2 to 9 percent of a typical sample. The remainder consists of well-sorted, medium to fine sand-size framework grains. Quartz grains predominate, but minor amounts of polycrystalline chert, iron-formation, and metamorphic quartzite are also present. Many of the quartz grains have abraded authigenic quartz overgrowths indicative of several cycles of deposition and erosion. Heavy minerals, including zircon, tourmaline, magnetite, and rutile, are present. In a few samples, small grains of diaspore, kaolinite, and pyrophyllite are intergrown with or replace the matrix material.

Beds of catlinite contrast markedly with the quartz-rich rocks. The catlinite is typically very fine grained and deep red to pale orange in color. Beds are generally structureless except for a few shaly parting planes parallel to bedding. True catlinite typically lacks quartz, and this attribute makes it soft and easy to carve. Most catlinite beds consist of various proportions of very fine grained sericite, hematite, and diaspore, together with smaller amounts of pyrite, kaolinite, chlorite, and possibly rutile. The catlinite units exposed in the monument also contain small grains of pyrophyllite, a mineral found at only one other place in Minnesota.

Some recrystallization is indicated by the ubiquitous presence of sericite in the matrix of the quartz-rich rocks and pyrophyllite in the catlinite beds. Textural evidence implies that late in the history of the rock the matrix recrystallized from some clay-size protolith whose original composition has not yet been established. The pyrophyllite may have formed by a metamorphic reaction involving sericite, quartz, and water with the concurrent release of some potassium, or alternatively by metamorphic reactions involving silica and either kaolinite or diaspore, or both, with neither the release nor the consumption of potassium. Regardless of the specific reactions involved, the presence of pyrophyllite implies that the rocks now exposed in the monument were subjected to temperatures and/or pressures above those normally associated with diagenesis.

SEDIMENTOLOGIC FRAMEWORK

A variety of sedimentary structures including trough and planar cross-bedded sands, horizontally laminated and cross-laminated sands, scour and fill deposits, ripple marks, mud cracks, and mud chips are well developed in the monument. However trough cross-bedding, especially of the festoon type, is by far the most common sedimentary structure and its ubiquitous presence complicates the recognition of master bedding surfaces and any structural interpretation of these rocks.

The orientations of presumed master bedding surfaces (Fig. 2) were derived from thick catlinite units or from thin siltstone or mudstone beds whose upper surfaces are ripple marked or mud cracked. Although the master bedding strikes consistently to the north and dips to the east (section A–A′ on Fig. 2), the surfaces display gentle rolls or warps of diverse size and orientation, as illustrated in section B–B′ of Figure 2.

The rolled or warped surfaces could be of tectonic origin, but Morey (1984) suggested from subsurface data that they are the local manifestations of curvilinear master bedding surfaces that separate large lenticular bodies of strata. These bodies of strata have strike lengths of 200 ft to more than 800 ft (60 to 240 m), widths of as much as 500 ft (150 m), and appear to be elongated in a south-southeastward direction.

As noted previously, trough cross-bedding is the most prevalent sedimentary structure in the monument. It is well exposed on the quartzite ridge, where stacked units form stratigraphic intervals as much as 30 ft (10 m) thick. Most of the trough cross-bedded strata bear little evidence of an erosional relationship to underlying strata. However some trough cross-bedded units do occupy scoured depressions that have a channel-like geometry. These scour and fill deposits are characterized by a basal, intraclast-strewn, structureless unit, which in places contains scattered granule-size quartz grains or disc-shaped mud chips. These basal deposits are always thin and inconspicuous and everywhere give way upward rather abruptly to the trough cross-bedded strata.

Planar cross-bedded strata occur in about 10 percent of the studied outcrops of the Sioux Quartzite, but are not particularly abundant in the monument. These strata can occur as single entities, but more commonly they form superimposed multi-story accumulations as much as 3 ft (1 m) thick that lack internal reactivation surfaces.

Horizontally laminated beds of very fine grained sandstone or siltstone a few inches (2 to 5 cm) thick occur at several places, but laminated beds that are ripple marked, either as solitary trains or as climbing ripples, are more common. Nearly planar beds or laminae of mudstone and fine-grained silty mudstone occur as patchy sedimentation units a few centimeters thick. These muddy units form drape deposits over underlying irregularities, and many of them have ripple-marked upper surfaces. The ripples are mostly of the symmetrical type and have amplitudes of only a few millimeters. The thick catlinite units quarried by the Indians are generally structureless to vaguely laminated. Their upper surfaces are commonly, but not always, marked by sets of asymmetrical and symmetrical ripple marks, or by mud cracks.

The azimuths of current directions associated with the trough cross-bedding indicate a paleocurrent direction predominantly to the south-southeast (Fig. 3). The azimuths of other current-direction indicators are diversely oriented. The planar cross-beds show a generally southward current flow, but lack a strong central tendency. Asymmetrical ripple marks yield even more diverse orientations and some indicate flow directions nearly 90° to those measured from immediately subjacent cross-beds.

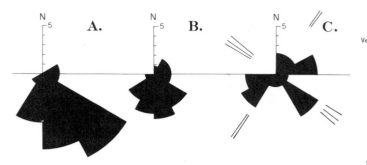

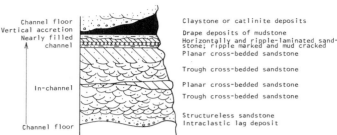

Figure 4. Inferred composite vertical profile of sedimentary structures.

Figure 3. Summary of paleocurrent data; A—trough cross-beds; B—planar cross-beds; C—asymmetrical ripple marks.

Although there is no definitive evidence of cyclicity, many of the sedimentary structures appear in recurring combinations (Morey, 1984) that form an upward-fining vertical profile (Fig. 4). The presence of these repetitive fining-upward sequences, the general lack of silt- and clay-size detritus, the extensive development of trough cross-bedding, and the lenticular nature of the quartz-rich strata imply that the rocks in the monument were deposited by fluvial processes associated with a braided-stream system.

In general modern braided-stream systems consist of a number of branching and coalescing sand- and silt-filled channels separated by clayey accumulations deposited in abandoned channels or during periods of high water when the river spilled beyond the channels. Thus, four major lithotopes can be recognized in modern braided-stream systems: channel-floor lag deposits, in-channel deposits, nearly filled channel deposits, and vertical accretion deposits (Miall, 1977).

These same lithotopes can be recognized in the Sioux Quartzite. The coarser grained scour and fill deposits in the basal parts of several exposed channels appear to have accumulated as channel-floor lag deposits that formed as through-flowing currents removed finer materials. However, the abundance of trough cross-bedded material implies that most of the sand-size detritus was transported in the channels as linguoid dunes or sand bars that migrated in a downstream direction. This form of sediment transport gives rise to cross-bedding with azimuths that are subparallel to the dominant paleoslope direction. In this regime, finer sediments are deposited in shallow water toward the edges of the channels, in channels nearly filled with sediments, or in inter-channel areas that receive deposits during periods of high water. Because currents during high-water periods are not entirely constrained by the original channel geometry, these deposits are characterized by current ripple marks and small-scale planar cross-beds that exhibit widely divergent paleocurrent transport directions.

Sharply bounded channel deposits in the Sioux Quartzite are rare, mainly because the river flowed over previously formed sandy accumulations in a constantly branching and rejoining pattern. Thus the original distribution of any particular channel deposit has been obscured by dissection and redeposition during subsequent changes in the courses of individual channels, and the deposits have coalesced to form extensively distributed sandy bodies having a generally lenticular shape.

The thick catlinite units, which formed as vertical accretion deposits, are uncommon features in braided-stream environments. Most fine material, such as silt and clay, is transported through individual channels without significant accumulation and vertical accretion deposits form only when the river spills from its main channel system onto the surrounding floodplain during major floods. When the river reverts back to its main channels, it leaves the ripple-marked fine-grained detritus to become desiccated and mud cracked. However, these fine-grained vertical accretion deposits tend to be quickly eroded by relatively rapidly migrating channels. Consequently, the presence of clayey units in a braided-stream system, such as those preserved at the monument, is somewhat unique. Therefore the monument is an excellent place to examine sand-clay relationships in a braided-stream system.

REFERENCES CITED

Corbett, W. P., 1980, Pipestone—the origin and development of a national monument: Minnesota Historical Society, Minnesota History, v. 47, p. 83–92.

Miall, A. D., 1977, A review of the braided-river depositional environment: Earth Science Reviews, v. 13, p. 1–62.

Morey, G. B., 1984, Sedimentology of the Sioux Quartzite in the Fulda basin, Pipestone County, southwestern Minnesota: Minnesota Geological Survey Report of Investigations 32, p. 59–74.

Precambrian Sioux Quartzite at Gitchie Manitou State Preserve, Iowa

Raymond R. Anderson, Iowa Geological Survey, Iowa City, Iowa 52242

LOCATION

Gitchie Manitou State Preserve is located in the northwesternmost corner of Iowa, in the W½,Sec.11,T.100N.,R.49W., about 9 mi (14.5 km) northwest of Larchwood in Lyon County, Iowa (Fig. 1). The preserve is situated in the valley of the Big Sioux River, which forms the western limits of the preserve and the boundary between Iowa and South Dakota. The highest elevation within the preserve is about 1,286 ft (392 m) above sea level, about 40 ft (12 m) above river level but about 150 ft (45 m) below the elevation of nearby upland areas.

SIGNIFICANCE

The Sioux Quartzite is a thick, Early Proterozoic clastic sequence, dominated by pink to maroon, fine- to medium-grained, quartz cemented, fluvial quartz arenite, but also includes minor intercalated mudstones and conglomerates. A number of workers have correlated the Sioux Quartzite with similar rocks in Wisconsin including the Baraboo, Waterloo, Barron, Flambeau, and Rib Mountain quartzites; lithologically similar units encountered in the subsurface of southeastern Iowa and south-central Nebraska may also be related to the Sioux. These rocks may represent scattered remnants of a vast sedimentary wedge that once blanketed the southern, passive margin of the proto–North American craton (Dott, 1983). This passive margin (Fig. 2) can be traced south into New Mexico and Arizona where the Ortega, Mazatzal, and Deadman quartzites appear to be correlative with the Sioux and other midcontinent quartzites.

There are no reliable determinations of the absolute age of deposition of the Sioux Quartzite; however, the age is constrained by a 2280 ± 110 Ma U-Pb age (Van Schmus and others, 1986) from a rhyolite underlying the quartzite in northwest Iowa, and K-Ar deformation age of 1120 Ma (Goldich and others, 1961) from a sericitic argillite bed in the quartzite at Pipestone in southwest Minnesota. The ages of the correlative units in Wisconsin are much more tightly constrained, overlying a rhyolite dated at 1760 ± 10 Ma (U-Pb by Van Schmus, 1979) and intruded by a granite that yielded a date of 1500 ± 30 Ma (U-Pb by Van Schmus, 1973). The minimum age probably can be better constrained by a 1640 ± 40 Ma (Van Schmus and others, 1975) regional deformational event, which is probably responsible for a fracture cleavage that developed in some of the Wisconsin quartzites (Greenburg and Brown, 1983). The Sioux Quartzite was probably also deposited in this 1760 to 1640 Ma time frame, an interval frequently referred to as the Baraboo interval (Dott, 1983).

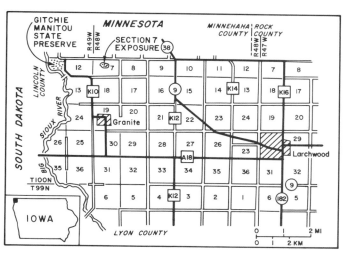

Figure 1. Location of Sioux Quartzite exposures described in this chapter.

About 15,000 mi^2 (38,000 km^2) of Sioux Quartzite is preserved in an east-west–trending belt stretching 330 mi (535 km) from New Ulm, Minnesota, to 45 mi (75 km) west of Pierre, South Dakota (Fig. 3). The belt is about 90 mi (150 km) at its widest with the northernmost subcrops near Brookings, South Dakota, the southernmost near Niobrara, Nebraska. The quartzite is extremely resistant to weathering and erosion, and has stood as a positive feature throughout most of the Phanerozoic. Although the unit was inundated by some of the many Phanerozoic marine transgressions into the Midcontinent, it was submerged only during the maximum stages. All Paleozoic sediments were eroded from the quartzite ridge as the seas receded, and today the only rock units known to overlie the Sioux Quartzite are of Cretaceous age. Because of the flanking relationships of adjacent Paleozoic units, the area of the Sioux Quartzite has been interpreted by many workers as a structural feature and called the "Sioux Uplift." There is, however, *no* structural or stratigraphic evidence suggesting actual uplift. Instead, the available data suggest that the Sioux Quartzite is a stable, positive topographic feature created by differential erosion of less resistant bedrock surrounding the quartzite. The topographic feature should properly be called the Sioux Ridge.

The Sioux Quartzite displays extreme compositional and textural maturity. The dominant quartz-arenites are composed of fine to medium, rounded quartz sand as framework grains, composing about 95 percent of the rock, and most commonly ce-

Figure 2. Distribution of Sioux quartzite and related late Precambrian quartzites in the central United States.

mented by quartz overgrowths. The unit ranges in color from white to deep red and purple, with pink the predominant hue. The pink coloration is caused by thin hematite coatings that formed on the grains prior to overgrowth formation. The deeper hues result from hematite staining of the cement (Morey, 1985). Claystone beds, generally less than 3 ft (1 m) thick, are composed of various proportions of very fine grained sericite, hematite, kaolinite, and chlorite, generally a deeper red color than the associated quartz-arenites. These claystones are often referred to as Catlinite (named after an early explorer of the area, George Catlin) or Pipestone (since various Indian groups have traditionally carved pipe bowls from the stone). The claystone contact with the quartz arenite is gradational, with silt-sized quartz grains increasingly abundant near the contact.

Conglomerate beds are also present in the Sioux Quartzite. A basal conglomerate is exposed near New Ulm, Minnesota, north of Pipestone, Minnesota, and in drill core from southwest Minnesota and northwest Iowa. The most common coarse clasts in the conglomerate are vein quartz, chert, quartzite, iron-formation, and rhyolite, but local basement lithologies are also found. The matrix includes coarse to medium quartz sand and locally abundant white argillaceous material (generally kaolinite), usually cemented by quartz overgrowths. Rare orthoquartzite conglomerate beds occur higher in the Sioux Quartzite section. They are generally composed of 0.5 to 2.5 in (1 to 6 cm) pebbles of vein quartz, quartzite, hematitic chert, and rare banded iron-

formation, all well rounded and well sorted and forming clast-supported units (Ojakangas and Weber, 1985).

The primary depositional environment for the Sioux Quartzite was braided fluvial systems, with possible marine influence in the upper part of the unit (Ojakangas and Weber, 1985). Basal conglomerates, with their coarseness, poor sorting, and fining-upward nature, are probably the product of proximal regions of the braided fluvial systems. The majority of the section, however, displays large-scale trough and planar cross-bedding, ripple marks, horizontal laminations, mudcracks, mud chips, and lag deposits suggestive of the distal regions of a sand-dominated braided fluvial system (Ojakangas and Weber, 1985). The upper third of the formation displays local bimodal-bipolar and herringbone cross-bedding, interpreted by Ojakangas and Weber (1985) as tidal or tidally influenced deposition. The preservation of marine shales and banded iron-formation conformably above the Baraboo (Wisconsin) Quartzite tends to support this interpretation. Paleocurrent studies of the Sioux Quartzite suggest that the braided fluvial depositional systems trended south-southeast with a relatively low gradient. These systems were apparently controlled, at least in their eastern outcrop belt, by a series of low-relief, fault-bounded, south-southeast–striking basins (Morey, 1985).

As it can be seen today, the Sioux Quartzite is a relatively undeformed and unmetamorphosed unit. Bedding dips are generally 5° to 10° to the north (Morey, 1985), although local tectonic effects have produced dips up to 70°. Post-consolidation alteration of the unit seems to be best described as high-grade diagenesis or very low grade, regional metamorphism. Morey (1985, p. 63) reviewed available data and concluded that the presence of pyrophyllite and sericite (in the claystone units) implied pressures and/or temperatures above those normally associated with diagenesis; however, Baldwin (1951) and Weber (1981) concluded that the rocks showed no evidence of metamorphism.

Baldwin (1951) estimated a maximum thickness of 9,800 ft (3,000 m) for the Sioux Quartzite, slightly thicker than Southwick and Mossler's (1985) estimate of about 7,900 ft (2,400 m). This represents an extremely thick sequence of primarily first-generation quartz sandstone.

The maturity of the sediments suggests that they were derived from a deeply weathered granitic terrain. This deep weathering has been observed beneath the Sioux Quartzite at a number of locations, for example a core in northwest Iowa where a 65 ft (20 m) weathered horizon is present, with quartz and kaolinite the dominant residuum. The landscape at the time of Sioux deposition has been interpreted (Southwick and Mossler, 1985) as low to moderate relief, partly mantled by residuum. Paleomagnetic interpretations suggest that the Superior Province moved from near-equatorial latitudes to near-polar and back to near-equatorial between about 1800 and 1600 Ma (Irving, 1979). The thick regolith is consistent with humid weathering conditions; however, actual deposition of the Sioux Quartzite could have occurred under climatic conditions ranging between polar and equatorial (Southwick and Mossler, 1985).

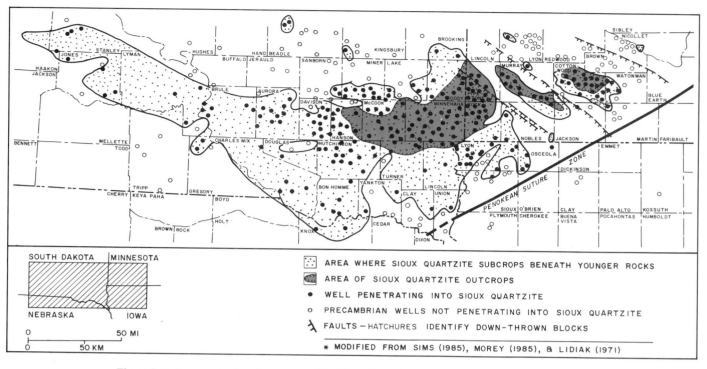

Figure 3. Outcrop and subsurface extent of the Sioux Quartzite in South Dakota and adjacent states.

EXPOSURES AT GITCHIE MANITOU STATE PRESERVE

History

The first descriptions of the area around Gitchie Manitou State Preserve came from eighteenth century French explorers, including Coureurs de Bois who called the upland areas "Coteau de Prairies." Early settlers in the area called the area Gitchie Manitou, the Sioux Indian name for "Great Spirit." The Big Sioux River west of the preserve is shallow and rocky and was a commonly used crossing (Rock Ford) in the 1800s. At that time a post office and land office were constructed within the present area of the preserve. C. A. White, an early Iowa State Geologist, visited the area of the preserve and named the exposed beds of red quartzite the "Sioux Quartzite" (White, 1870, p. 168), making the preserve the type area of the Sioux Quartzite. Shortly thereafter a quarry was opened in the area, probably to provide local building stone. No record of this quarry activity presently exists; however, a photograph of the water-filled quarry was published by Beyer in 1897 (p. 73), entitled "Jasper Pool," a name still applied to the quarry because of its green-colored water. The rock shelter at the preserve was constructed in the 1930s as a part of the federal WPA program. The Gitchie Manitou area was dedicated as a state geological, natural, and archaeological preserve in 1969.

SITE DESCRIPTION

About 30 ft (10 m) of quartzite can be seen in natural exposures and in the quarry at the preserve. The quartzite is fine- to coarse-grained (dominantly medium-grained), quartz-cemented, pink to red, quartz arenite. Individual quartz clasts are rounded, coated by iron oxides, and cemented by quartz overgrowths. Ripple marks and cross-bedding can be observed within the quartzite, but for the most part the Sioux Quartzite is an extremely uniform quartz arenite at the preserve, and no conglomerates or argillites are exposed. Fractures at Gitchie Manitou State Preserve are generally bimodal trending primarily about N50°E and N40°W. Measurements on bedding planes suggests strikes ranging from N50°E to N80°E, with a dip of about 6° to the northwest.

The area of Gitchie Manitou State Preserve was overrun by numerous Pleistocene continental glacial ice advances; however, the grooves and striae that are observed on the quartzite in other areas have not been identified in the preserve. The highly polished quartzite surfaces seen throughout the preserve are the product of constant bombardment by wind-transported dust and sand particles (Koch, 1969).

The only other exposure of the Sioux Quartzite in Iowa lies about 2 mi (3.2 km) due east of Gitchie Manitou State Preserve (Fig. 1). This exposure is located on private property near the center of Sec.7,T.100N.,R.48W., just south of the road that follows the Iowa-Minnesota border (Fig. 3). Called the Section 7 exposure (Anderson, in preparation), about 120 ft (37 m) of Sioux Quartzite is exposed over an area of about 2 acres (0.8 ha). Cross-bed measurements at this exposure suggest a transport direction averaging S40°E. Beds at the Section 7 exposure strike an average of N64°E and dip to the northwest at about 7°. The exposure is cut by a 25-ft-wide (7 m) zone of anastomosing fractures trending about N37°E. A number of additional exposures occur in Minnesota, north of the road that forms its border with Iowa in the area of Gitchie Manitou State Preserve and the Section 7 exposure.

REFERENCES CITED

Baldwin, W. B., 1951, The geology of the Sioux Formation [Ph.D. thesis]: Columbia University, 161 p.

Beyer, S. W., 1897, Sioux Quartzite and certain associated rocks: Iowa Geological Survey Annual Report, v. 6, p. 67–112.

Dott, R. H., Jr., 1983, The Baraboo interval; A tale of red quartzites: Geological Society of America Abstracts with Programs, v. 15, p. 209.

Greenburg, J. K., and Brown, B. A., 1983, Middle Proterozoic to Cambrian rocks in central Wisconsin; Anorogenic sedimentary and igneous activity: Geological Society of America North-Central Section Field Trip Guide Book 8, 50 p.

Goldich, S. S., Nier, A. O., Baagsgaard, H., Hoffman, J. H., and Krueger, H. W., 1961, The Precambrian geology and geochronology of Minnesota: Minnesota Geological Survey Bulletin 41, 193 p.

Irving, E., 1979, Paleopoles and paleolatitudes of North America and speculations about displaced terrains: Canadian Journal of Earth Science, v. 16, no. 3, pt. 2, p. 669–694.

Koch, D. K., 1969, The Sioux Quartzite Formation in Gitchie Manitou State Preserve: Iowa State Advisory Board for Preserves Development Series Report 8, 28 p.

Lidiak, E. G., 1971, Buried Precambrian rocks of South Dakota: Geological Society of America Bulletin, v. 82, p. 1411–1420.

Morey, B. G., 1985, Sedimentology of the Sioux Quartzite in the Fulda Basin, Pipestone County, southwestern Minnesota, *in* Southwick, D. L., ed., Shorter contributions to the geology of the Sioux Quartzite (Early Proterozoic), southwestern Minnesota: Minnesota Geological Survey Report of Investigations 32, p. 59–74.

Ojakangas, R. W., and Weber, R. E., 1985, Petrography and paleocurrents of the Lower Proterozoic Sioux Quartzite, Minnesota and South Dakota, *in* Southwick, D. L., ed., Shorter contributions to the geology of the Sioux Quartzite (Early Proterozoic), southwestern Minnesota: Minnesota Geological Survey Report of Investigations 32, p. 1–15.

Sims, P. K., 1985, Precambrian basement map of the northern midcontinent, U.S.A.: U.S. Geological Survey Open-File Report 85-0604, 16 p.

Southwick, D. L., and Mossler, J. H., 1985, The Sioux Quartzite and subjacent regolith in the Cottonwood County basin, Minnesota, *in* Southwick, D. L., ed., Shorter contributions to the geology of the Sioux Quartzite (Early Proterozoic), southwestern Minnesota: Minnesota Geological Survey Report of Investigations 32, p. 17–44.

Van Schmus, W. R., 1973, Chronology of Precambrian igneous and metamorphic events in eastern Wisconsin and Upper Michigan [abs.]: EOS Transactions of the American Geophysical Union, v. 54, no. 4, p. 495.

—— , 1979, Geochronology of the southern Wisconsin rhyolites and granites: University of Wisconsin Extension, Geoscience Wisconsin, v. 2, p. 9–24.

Van Schmus, W. R., Thurman, M. E., and Petermann, A. E., 1975, Geology and Rb-Sr chronology of middle Precambrian rocks in eastern and central Wisconsin: Geological Society of America Bulletin, v. 86, p. 1255–1265.

Van Schmus, W. R., Bickford, M. E., and Zietz, I., 1986, Early and Middle Proterozoic provinces in the central United States, *in* Kroner, A., ed., Proterozoic Lithospheric Evolution: American Geophysical Union Geodynamics series, v. 17, p. 43–68.

Weber, R. E., 1981, Petrology and sedimentation of the upper Precambrian Sioux Quartzite, Minnesota, South Dakota, and Iowa [M.S. thesis]: Duluth, University of Minnesota, 151 p.

White, C. A., 1870, Report on the geological survey of the State of Iowa: Iowa Geological Survey, v. 1, 391 p.

Cambrian and Ordovician stratigraphy in the Lansing area, northeastern Iowa

Brian J. Witzke and Robert M. McKay, Iowa Geological Survey, Iowa City, Iowa 52242

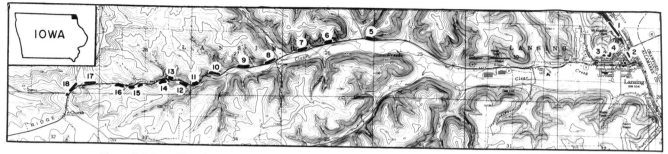

Figure 1. Map of Lansing–Church area showing locations of measured stratigraphic sections. Base map taken from Lansing and Church 7½-minute Quadrangles. Contour interval 20 ft (6 m). Localities 1–4 in T.99N.,R.3W.; localities 5–18 in T.99N.,R.4W..

LOCATION

The oldest Paleozoic rocks exposed in Iowa are found in the northeastern corner of the state in Allamakee County. Cambrian exposures in the county form part of the classic reference area in the Upper Mississippi Valley for the Upper Cambrian Croixan Series. Overlying dolomites of the Prairie du Chien Group are exposed as bold cliffs in the bluffs along the Mississippi River and its tributaries. Cambrian and Lower Ordovician strata collectively encompass the Sauk Sequence, the first large-scale transgressive-regressive cratonic marine cycle of the Phanerozoic (Sloss,1963). A major episode of erosion followed deposition of the Sauk Sequence. A widespread unconformity separates Lower and Middle Ordovician strata in the region, marking the boundary between the Sauk Sequence and the succeeding Tippecanoe Sequence.

The most complete sequence of Cambrian and Ordovician strata in northeast Iowa is exposed between Lansing and Church as one ascends from the Mississippi River up the valley of Clear Creek to the crest of Lansing Ridge along Iowa 9 (Fig. 1). Vertical relief along this profile is 630 ft (192 m), and the composite stratigraphic section measured totals 619 ft (189 m). The base of the section begins along Iowa 26 north of the Mississippi River bridge (loc. 1, Fig. 1), and continues through a series of roadcuts and natural exposures in Mount Hosmer Park (loc. 4, Fig. 1). Strata equivalent to the lower half of the section at locality 1 can be seen behind Knopf's Standard Station, 115 North Second Street, in Lansing (loc. 2, Fig. 1). The section resumes west of Lansing in an overlapping series of roadcuts along Iowa 9 (localities 5–18, Fig. 1). Highway construction during 1978–1979 expanded and improved these roadcuts, but subsequent slumping and plant growth partially obscured the sections at some localities (locs. 5, 6, 15; Fig. 1). The composite stratigraphic section is described in the measured section and illustrated in Figure 2.

SITE 18—CAMBRIAN STRATIGRAPHY

The Cambrian sequence in northeast Iowa directly overlies Precambrian crystalline rocks in the subsurface. The basal Cambrian (Dresbachian) Mt. Simon Sandstone and the overlying Eau Claire Formation are not exposed in Iowa, but are penetrated by water wells in the area. The exposed Cambrian section begins in the Ironton Member of the Wonewoc Formation. The Ironton is not represented in the Lansing section (Fig. 2), but exposures north of Lansing can be visited, including: 1) 4 mi (6.5 km) north of Lansing where Iowa 26 crosses a small creek (NE¼SW¼NE¼-Sec.12,T.99N.,R.4W.) and 2) 2.9 mi (4.7 km) southwest of New Albin (center SE¼NW¼Sec.21,T.100N.,R.4W.) north of the Upper Iowa River on County Road A26 (Anderson and others, 1979). The Ironton is a fine- to medium-grained quartzose sandstone containing abundant inarticulate brachiopods (*Obolus*) and trilobite debris (*Camaraspis convexa*) in some beds (*Elvinia* Zone of the lower Franconian Stage).

The Lone Rock Formation (Franconia Formation of Minnesota classification) overlies the Ironton and is dominated by very fine-grained feldspathic and glauconitic greensands. The formation is about 150 ft (45 m) thick at Lansing, but the basal Tomah and Birkmose members are not exposed in Iowa. The bulk of the Lone Rock is represented by the Reno Member. The glauconitic Reno sandstones are horizontally laminated to ripple cross-laminated with shale and siltstone interlaminae and drapes. Some beds with prominent vertical and horizontal burrows have been termed "wormstone." The Reno has yielded trilobite faunas of the *Ptychaspis-Prosaukia* Zone (Franconian) in Allamakee County. Collections at Lansing (Walter, 1924) include the following trilobites: *Ptychaspis striata, Monocheilus micros, Idahoia wisconsinensis,* and *Pseudagnostus josepha*. A collection from the Reno in the upper Iowa River Valley northwest of Lansing (Schuldt, 1940) includes trilobites from a slightly higher strati-

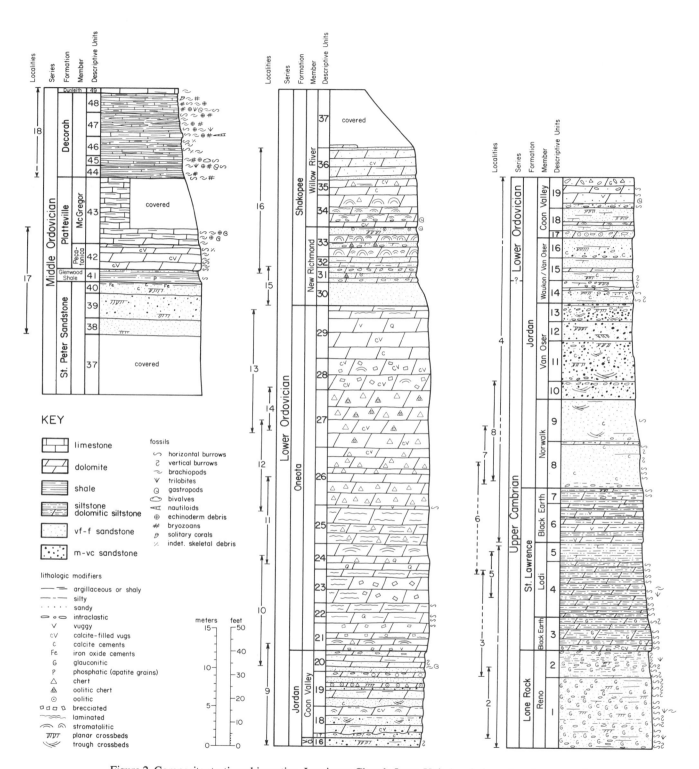

Figure 2. Composite stratigraphic section, Lansing to Church, Iowa. Unit descriptions given in measured section.

graphic position: *I. hamulus, Chariocephalus whitfieldi,* and *P. josepha.* The upper unit of the Reno at locality 1 (unit 2, Fig. 2) is less glauconitic than underlying beds and has yielded trilobites of the basal *Saukia* Zone: *Dikelocephalus postrectus, Saukiella minor,* and *Illaenurus* sp. The appearance of *Saukia* Zone faunas marks the base of the Trempealeauan Stage in the upper part of the Lone Rock Formation.

The overlying St. Lawrence Formation includes two members: medium-bedded silty dolomites and dolomitic silt-stones of the Black Earth Member, and thin-bedded dolomitic siltstones of the Lodi Member. Although fossils are not common in the formation, the Lodi locally has produced *Saukia* Zone faunas at Lansing and other localities in Allamakee County. Spec-imens of *Dikelocephalus minnesotensis* (= *D. gracilis*) are the most common, but species of *Illaenurus, Tellerina,* and *Calvin-ella* have been noted. Rare aglaspid merostome arthropods (*Aglaspis, Glypharthrus*) are known from the Lodi at Lansing (Raasch, 1939). Additional fossils have also been collected in the Lodi of northeast Iowa: inarticulate (*Lingulella, Westonia*) and articulate (*Finkelnburgia*) brachiopods, gastropods, and cono-donts (*Proconodontus*). Access to Lodi strata at locality 1 (Fig. 1) is difficult because of steep and hazardous slopes.

The Jordon Sandstone is the best exposed Cambrian forma-tion in northeast Iowa, forming steep sandstone cliffs in places beneath the Prairie du Chien dolomite escarpment. The Jordan is divided into four members in northeast Iowa (Fig. 2) following Odum and Ostrom (1978). The basal member, the Norwalk, is dominantly a very fine-grained feldspathic sandstone. The overly-ing Van Oser is much coarser grained, typically fine or medium to coarse-grained quartzose sandstone. The Waukon Member, a very fine to fine-grained feldspathic sandstone, wedges into the Van Oser in northeast Iowa. The Waukon type section is in the Upper Iowa River Valley (center NW¼Sec.6,T.99N.,R.5W.) northwest of Lansing, where it is both underlain and overlain by Van Oser strata (Anderson and others, 1979). However, the Waukon at Lansing (loc. 4, Fig. 1) overlies the Van Oser, but incorporates some medium to coarse-grained lithologies of Van Oser aspect, especially in the upper part (unit 16, Fig. 2). The upper member of the Jordan, the Coon Valley, previously was included in the Oneota Formation by some workers. It is charac-terized by interbedded sandstones and sandy dolomites, in part oolitic.

The Jordan Sandstone is a prominently cross-bedded unit displaying trough, low-angle planar, and herringbone cross-bed forms. Intraclastic beds are present in all members, and sandy dolomite beds occur in all but the Van Oser. The Jordan sand-stones typically are friable or poorly cemented, but irregular calcite-cemented zones are common including unusual "grape-shot"-cemented sand clusters. The finer-grained facies of the Jor-dan are bioturbated, in part, by horizontal and vertical burrowers. Aside from some poorly preserved gastropod and brachiopod molds in the Coon Valley, the Jordan is notably unfossiliferous in northeast Iowa. Conodont studies of Jordan strata in northeast Iowa have not been done, but faunas in Wisconsin (Miller and

Melby, 1971) suggest that the Cambrian-Ordovician boundary falls within the Jordan and not above it as previously thought.

SITE 19—ORDOVICIAN STRATIGRAPHY

The Lower Ordovician Prairie du Chien Group is divided into two formations, a lower Oneota Formation and an upper Shakopee Formation. The Oneota is a pure dolomite and cherty dolomite sequence lacking detrital quartz. Diagenetic processes have altered severely the original depositional fabrics of Oneota strata, although laminations, stromatolites, burrows, and oolitic chert are recognized (Fig. 2). The Oneota served as host rock for calcite-sulfide mineralization in the Lansing area. Secondarily brecciated zones, fractures, and void spaces are silicified locally along their margins and commonly are lined with calcite and marcasite (usually oxidized to iron oxides). Such mineralized zones can be seen at locality 13 (Fig. 1). Although outside the main Upper Mississippi Valley Zinc-Lead District, lead ore deposits were mined from a vertical fracture in the Oneota and upper Jordan 5.4 mi (8.7 km) northwest of Lansing in the 1890s, producing more than 250 tons of cerrusite and galena (Leonard, 1897; Heyl and others, 1959).

The Shakopee unconformably overlies the Oneota in Alla-makee County and locally displays erosional truncation beneath the contact (Ostrom and others, 1970). The Shakopee markedly contrasts with the pure dolomites of the Oneota, containing sandy dolomite, sandstone, and some green shale (Davis, 1970). The lower New Richmond Member is an interbedded sequence of very fine to medium-grained sandstone and laminated, intraclas-tic, and stromatolitic dolomite. The overlying Willow River Member is dominated by dolomite and sandy dolomite, in part laminated to stromatolitic, but some thin sandstones are present. Oolitic cherts are common in the Shakopee. The Prairie du Chien Group is only sparingly fossiliferous. Gastropod molds are pres-ent locally, and additional fossils are found occasionally in chert nodules. Conodonts have been recovered from the dolomites and shales.

A major episode of erosion followed deposition of the Prairie du Chien Group. The unconformity surface that devel-oped was buried by pure quartzose sandstones of the St. Peter Sandstone when deposition resumed later in the Middle Ordovi-cian. Karst and channel incision truncated the entire Prairie du Chien sequence in places, and thick St. Peter sections locally overlie Cambrian strata south of Allamakee County (e.g., Pikes Peak State Park near McGregor, Iowa). The St. Peter Sandstone is exposed along Iowa 9 above a covered interval (loc. 17, Fig. 1). The overlying noncalcareous and feldspathic Glenwood Shale is sandy, in part, and yields conodont faunas. The Platteville For-mation is represented by a lower dolomite unit, the Pecatonica Member, and an overlying wavy-bedded fossiliferous limestone interval, the McGregor Member. The upper Platteville is covered between localities 17 and 18 (Fig. 1).

The Decorah Formation is exposed at locality 18 (Fig. 1), but some beds are overgrown. The Decorah is a highly fossiliferous formation comprising interbedded gray-green calcareous shales and limestones, primarily skeletal wackestones and packstones. Brachiopods and bryozoans are especially abundant in some beds. The Decorah undergoes a significant facies change in northeast Iowa. South of Allamakee County, the Decorah includes three members: a basal Spechts Ferry Shale; a middle Guttenberg Limestone; and an upper interbedded limestone and shale, the Ion (see the Guttenberg, Iowa, section described elsewhere in this volume). However, the limestone-dominated Guttenberg Member interfingers with shale-dominated facies to the north. Strata equivalent to the Guttenberg in central Allamakee County (units 45, 46, lower 47; Fig. 2) contain numerous interbeds of shale, and distinguishing features of the member are generally absent in this area. All members of the Decorah are replaced to the north near the Minnesota line by a shale-dominated and undifferentiated Decorah Formation. A zone of hemispherical stony bryozoans (*Prasopora*) extends from limestone facies to the south, across an intermediate facies belt (unit 48, Fig. 2), into shale-dominated facies to the north.

DEPOSITION

The sequence of Upper Cambrian through Middle Ordovician strata in the Upper Mississippi River Valley was deposited in a series of transgressive-regressive sedimentary cycles. Each cycle typically includes medium-grained quartzose sandstones in the lower part with finer-grained feldspathic sandstones or shales, in part glauconitic, above; carbonate, sandy carbonate, or calcareous siltstone caps each sequence (Ostrom, 1978). The upward-fining sequence of lithosomes in each cycle records deposition in progressively less-agitated environments, probably ranging from nearshore littoral settings to offshore carbonate shelf environments. Source terranes for terrigenous clastic detritus were primarily to the northeast, including the Wisconsin Dome and surrounding parts of the Canadian Shield. Feldspars, micas, and clays were derived from Precambrian crystalline and sedimentary source rocks, although many feldspar grains display authigenic overgrowths. The clean quartzose sandstones in each cycle include polycyclic quartz grains derived from older Precambrian sandstones.

The following sedimentation cycles are displayed in the Lansing area: 1) Wonewoc through St. Lawrence formations, 2) Jordan and Oneota formations, 3) New Richmond and Willow River members, and 4) St. Peter through Platteville formations. An erosional unconformity, which expands in magnitude towards the area of the Wisconsin Arch, marks the top of each cycle. The coarsest sandstones exposed at Lansing are found in the Van Oser Member of the Jordan, a facies that records the central part of a regressive subcycle within the larger transgressive cycle. A decrease in terrigenous clastic influx during the Lower Ordovician permitted expansion of Prairie du Chien carbonate facies over the region. Oolite shoals and subtidal to intertidal stromatolitic facies were represented in the shallow carbonate environments. Clastic source areas shifted from the Wisconsin

Dome–Canadian Shield area to the Transcontinental Arch during the Middle Ordovician. The shales of the Decorah Formation were derived from the weathering of Precambrian rocks on the Transcontinental Arch in Minnesota (Witzke, 1980).

MEASURED SECTION

Composite stratigraphic sequence, Lansing to Church, Alamakee County, Iowa (Fig. 2). Lithologic descriptions.

Galena Group (Middle Ordovician)

Dunleith Formation

Unit 49. Poorly exposed. Limestone, slightly argillaceous, fossiliferous; 2 ft (0.6 m).

Decorah Formation (Decorah Subgroup of Illinois classification)

Unit 48. Limestone, thin to medium wavy-bedded, argillaceous skeletal wackestone, including skeletal packstone near base; scattered thin shale interbeds; very fossiliferous in lower portion, abundant *Prasopora* near base, brachiopods include orthids, rhynchonellids, strophomenids; upper portion covered to poorly exposed; 7.5 ft (2.3 m).

Unit 47. Shale-dominated, gray-green, calcareous; interbedded with thin argillaceous limestone, in part nodular, predominantly skeletal wackestone, minor skeletal packstone, part abraded grains, sparse to very fossiliferous, sowerbyellids and resserellids noted; lower 23 in (58 cm) is prominent ledge former, limestone, argillaceous, massive below, thinly bedded above, burrowed, skeletal wackestone; 10 ft (3.1 m).

Unit 46. Shale, gray-green, calcareous, with scattered thin argillaceous limestone interbeds, skeletal wackestone, part abraded grain; 7.6 ft (2.3 m).

Unit 45. Limestone, argillaceous, skeletal wackestone to packstone, part abraded grain, interbedded with thin shales; fossiliferous, brachiopods include sowerbyellids, strophomenids, resserellids, interval partly covered; 4 ft (1.2 m).

Unit 44. Shale, olive-green, calcareous, with thin argillaceous limestone beds, skeletal wackestone to packstone; fossiliferous, brachiopods include orthids, strophomenids, sowerbyellids, trilobites moderately common; interval partly covered; 6.3 ft (1.9 m).

Platteville Formation (Platteville Group of Illinois classification

McGregor Member (Grand Detour and Mifflin formations of Illinois classification)

Unit 43. Limestone, thin wavy-bedded, part slightly argillaceous, very dolomitic at base, becoming less dolomitic above, partly fossiliferous; basal 8 ft (2.4 m) exposed in roadcut, approximately 22 ft (6.7 m) covered (covered interval accessible at quarry west of locality 18, SW¼SW¼Sec.29).

Pecatonica Member (Pecatonica Formation of Illinois classification)

Unit 42. Dolomite, extremely fine to very fine crystalline, thin to medium-bedded, scattered fossil molds, argillaceous burrow mottlings near base; small to large vugs scattered, part calcite filled; becomes calcitic at top; 10 ft (3 m).

Glenwood Shale

Unit 41. Shale, noncalcareous; basal 20 in (49 cm) shale, orange-brown with thin olive laminae; next 7.2 in (18 cm), hard green shale, small apatite clasts, interlaminae of fine sand, scattered horizontal burrows; next 31.6 in (79 cm), green shale, soft to hard and flaky, some orange mottlings; top 4.8 in (12 cm) shale, green to yellow-brown, very sandy to fine sandstone at base, silty to dolomitic at top; total 5.3 ft (1.6 m).

St. Peter Sandstone

Unit 40. Sandstone, very fine to fine, friable, top 7.2 in (18 cm) hematite-cemented, some calcite cement; some west-dipping cross-laminae, in part irregular wavy-bedded, possible hummocky bedding; top 2 in (5 cm) very sandy mudstone; 5.3 ft (1.6 m).
Unit 39. Sandstone, fine to medium, irregular to thick-bedded, faint west-dipping planar cross-laminae; 10.6 ft (3.2 m).
Unit 38. Sandstone, fine, massive, faint cross-laminae; 6.9 ft (2.1 m).
Unit 37. Sandstone, covered; approximately 25 ft (7.5 m).

Prairie du Chien Group (Lower Ordovician)

Shakopee Formation

Willow River Member

Unit 37. Covered; some sandy dolomite float; approximately 25 ft (7.5 m).
Unit 36. Dolomite, very fine to fine crystalline, slightly sandy to very sandy, very fine to fine sand, thin to medium-bedded; thin sandstone units, dolomite cements, at top and bottom, 16 in (40 cm) sandstone at base; 12.5 ft (3.8 m).
Unit 35. Dolomite, very fine to medium crystalline, dense; basal 4.3 ft (1.3 m) with horizontal calcite fills; next 16 in (40 cm) with abundant chert nodules (to 3.3 ft long by 6 in thick; 1 m by 15 cm), thin green shale at top; top 28 in (70 cm) with calcite void fills; 7.9 ft (2.4 m).
Unit 34. Dolomite, very fine to medium crystalline; basal 28 in (70 cm), abundant calcite void and fracture fill; next 3.6 ft

(1.1 m) is cross-bedded with flat-pebble intraclasts to 2 in (5 cm), gastropod mold noted; middle 6.6 ft (2 m) is horizontally laminated 0.04 to 0.8 in (1 mm to 2 cm) to stromatolitic (small stroms 1.2 to 2 in thick by 4 to 10 in wide; 3 to 5 cm by 10 to 25 cm), pale green dolomitic shale at top; top 5 ft (1.5 m) with large domal stromatolites to 24 in thick by 6.6 ft wide (60 cm by 2 m), scattered chert nodules, pale green argillaceous dolomite at top; 17.5 ft (5.3 m).

New Richmond Member

Unit 33. Interbedded stromatolitic dolomite and sandstone, very fine to fine; large coalescing stromatolites to 3.3 ft thick by 6.6 ft wide (1 m by 2 m); middle sandstone, part dolomite-cemented, interbedded with thin laminated green shales, sandy dolomite, and oolitic chert; upper sandstone unit capped by sandy, cross-bedded, oolitic dolomite with common small gastropods; 12.5 ft (3.8 m).
Unit 32. Dolomite, slightly sandy to silty, part thinly laminated, part slightly argillaceous; approximately 4.3 ft (1.3 m).
Unit 31. Sandstone, very fine to medium, and dolomite, part sandy; intraclast conglomerate at top and bottom with clasts of dolomite and green shale, rare mudcracks; middle sandstone with silicified oolites and oolitic chert; poorly exposed; approximately 6 ft (1.8 m).
Unit 30. Sandstone, fine to medium, friable, part slightly dolomitic; trough cross-beds in lower part; poorly exposed; approximately 8.9 ft (2.7 m).

Oneota Formation

Unit 29. Dolomite, dense, very fine to fine crystalline, scattered calcite void and fracture fill, scattered chalcedony void fill near top, elongate vugs 6.6 ft (2 m) down, some pin-point porosity; laminated towards top; part covered; 20.8 ft (6.3 m).
Unit 28. Dolomite, dense, very fine crystalline, scattered voids; extremely altered and mineralized, large irregular calcite void fills to 5 ft (1.5 m) wide surrounded by iron-oxides and silicified areas, some mineralized zones occupy stromatolitic structures; rocks are part brecciated with breccia clasts 8 in (20 cm) or larger; 13.5 ft (4.1 m).
Unit 27. Dolomite, dense, cherty; lower half, very fine to fine crystalline, irregular mottles, calcite void and fracture fill, some chert nodules, part oolitic; top half fine to medium crystalline, abundant irregular chert nodules, oolitic to pisolitic; coarse crystalline dolomite at top; 24.8 ft (7.5 m).
Unit 26. Dolomite, cherty, extra fine to fine crystalline, irregular thin to thick-bedded, some calcite void fills in lower and upper parts; part with irregular mottled texture; cherty to very cherty, scattered to abundant chert nodules, smooth to chalky, white to light gray, rare quartz void linings; 25.4 (7.7 m).
Unit 25. Dolomite, laminated; lower half very fine to fine crystalline, prominently laminated to undulatory (stromatolites), scattered chert nodules, smooth to chalky, nodules to 2 in thick

by 20 in wide (5 by 50 cm); upper half very fine to coarse crystalline, massive, part laminated with slight undulations; top 8 in (20 cm) extra fine to very fine crystalline, white to off-white, good marker bed, trace laminations, irregular mottling, burrows?, mudcracks noted; 16.5 ft (5 m).

Unit 24. Dolomite, very fine to fine crystalline, laminated to stromatolitic (low amplitude); lower half locally with quartz-lined voids, scattered smooth chert nodules; approximately 10 ft (3 m).

Unit 23. Dolomite, very fine to fine crystalline, irregular coarse laminations; locally brecciated, breccia clasts 0.4 to 8 in (1 to 25 cm), angular to rounded; scattered chalcedony-lined voids; 13.9 ft (4.2 m).

Unit 22. Dolomite, very fine to fine crystalline, dense, part laminated, some interlaminae with horizontal to diagonal burrows (to 0.4 in long; 1 cm); some quartz-lined voids near middle; 8.6 ft (2.6 m).

Unit 21. Dolomite, very fine crystalline; top 6.6 ft (2 m) weakly brecciated, clasts 0.08 to 0.8 in (2 mm to 2 cm), prominent horizontal burrows (2.4 to 3.2 in long; 6 to 8 cm) at top; bottom 5.3 ft (1.6 m), thin to medium-bedded; upper half stromatolitic (0.8 to 10 in high by 3.3 to 5 ft long; 2 to 25 cm by 1 to 1.5 m) with abundant chalky chert along stromatolitic layers; very fine to medium crystalline in lower half, slightly brecciated to bioturbated; 12 ft (3.6 m).

Jordan Sandstone Formation

Coon Valley Member

Unit 20. Dolomite, slightly to very sandy (very fine to fine, some medium to coarse), part nonsandy, very fine to fine crystalline; minor sandstone, very fine to fine, scattered medium to coarse, friable to dolomitic; rare pale green dolomitic shale to argillaceous dolomite partings; small shale intraclasts in dolomite locally at top; vertical burrows, brachiopod and gastropod molds locally in lower half; approximately 9.3 ft (2.8 m).

Unit 19. Variable unit; dolomite, slightly to very sandy (very fine to fine), part silty, very fine to medium crystalline, in beds 8 to 40 in (20 to 100 cm) thick, flat-pebble intraclastic to oomoldic near top, cross-bedded dolomitized oolitic grainstone locally in middle part, horizontal burrows and gastropod molds locally present, scattered chert nodules and oolitic chert, slightly glauconitic near top; sandstone, subordinate to dolomite, in beds 12 to 44 in (30 to 110 cm) thick, very fine to fine and fine to coarse, friable to dolomite- or calcite-cemented, low-angle planar to trough cross-beds noted; base locally irregular; approximately 15.5 ft (4.7 m).

Unit 18. Variable unit; dolomite, sandy to very sandy (very fine to fine and fine to coarse), part silty, very fine to coarse crystalline, thin to thick-bedded, part with small stromatolites or algal laminations; sandstone, very fine to fine and very fine to medium, part dolomite- to calcite-cemented, part slightly argillaceous, horizontal laminated to low-angle planar cross-beds,

some small trough cross-beds; irregular base; 9.6 to 11.9 ft (2.9 to 3.6 m).

Unit 17. Sandstone, very fine to fine and fine to medium, part calcite- to dolomite-cemented, part intraclastic with sandy dolomite and/or green shale clasts, locally argillaceous with green shale drapes over rippled sandstone, locally oolitic; some sandy dolomite; irregular upper surface up to 16 in (40 cm) relief; 28 to 36 in (70 to 90 cm).

Waukon and Van Oser Members (interbedded)

Unit 16. Sandstone, very fine to fine, some very fine to medium and fine to coarse, part calcite-cemented, low-angle to planar cross-beds, large ripples near top, slightly argillaceous with vertical burrows in upper part, small intraclasts in lower part; 8.6 ft (2.6 m).

Unit 15. Sandstone, very fine to fine, part calcite- to dolomite-cemented, horizontal laminated to low-angle planar cross-beds, small trough cross-beds and vertical burrows near base; dolomite, very sandy, 20.8 in (52 cm) thick near middle; laminated to low-angle planar cross-beds, some herringbone cross-sets; 10 ft (3 m).

Unit 14. Sandstone, very fine to fine, very fine to medium, and fine to medium in middle part, part dolomite-cemented, some calcite; intraclastic in lower and middle part, clasts silty to sandy dolomite; horizontal laminae to low-angle cross-beds, undulatory truncation surfaces, horizontal to vertical burrows; dolomite to 34 in (85 cm), sandy to silty, locally truncated; 9.3 ft (2.8 m).

Van Oser Member (Upper Cambrian—precise top not known)

Unit 13. Sandstone, medium to coarse and medium to very coarse, minor fine to medium, part calcite-cemented; large trough cross-beds in lower 4.3 ft (1.3 m); upper 3.3 ft (1 m) with low-angle planar cross-beds, part intraclastic, clasts argillaceous to silty, very sandy intraclastic dolomite 8 in (20 cm) thick; 7.6 ft (2.3 m).

Unit 12. Sandstone, fine to coarse, some very coarse, minor calcite cements, bi-directional planar cross-beds, large trough near top; 8.3 ft (2.5 m).

Unit 11. Sandstone, medium to coarse, some very coarse, minor calcite cements, prominent large trough cross-beds, some planar to low-angle cross-beds, possible hummocky cross-stratification in lower part; locally intraclastic near top and bottom; 17.5 ft (5.3 m).

Unit 10. Sandstone, fine to medium, fine to coarse, some very fine to fine and very fine to medium, part calcite-cemented, small to large trough cross-beds, minor low-angle to planar cross-beds; intraclast conglomerate locally in lower part, very fine sandstone and siltstone clasts in fine to very coarse sandstone matrix; 7.3 ft (2.2 m).

Norwalk Member

Unit 9. Sandstone, very fine to fine, locally fine to medium, soft, part calcite-cemented, prominent small to large trough cross-beds, part burrow-mottled, minor intraclasts at base; 18.2 ft (5.5 m).

Unit 8. Sandstone, very fine, locally very fine to fine (trace medium to coarse in some laminae), minor calcite cements, horizontal to vertical burrow mottles abundant, small to large trough cross-beds locally; intraclastic silty to sandy dolomite locally at top; very fine sandstone to siltstone at base with scattered mud clasts; 20 ft (6.1 m).

St. Lawrence Formation

Black Earth Member

Unit 7. Siltstone, dolomitic, with very silty dolomite in lower part, scattered argillaceous laminae, horizontal and subvertical burrow mottles, low-angle planar cross-beds noted, very fine sandy in upper part; intraclastic near top; 5.3 ft (1.6 m).

Unit 6. Interbedded silty dolomite and siltstone; silty dolomite in beds 2 to 5.3 ft (0.6 to 1.6 m), part faintly laminated, some vugs, very fine to medium crystalline; siltstone, soft, part calcite- or dolomite-cemented, some silty very fine sandstone in lower part; 18 ft (5.5 m).

Lodi Member

Unit 5. Siltstone to silty very fine sandstone, slightly argillaceous, trace glauconite, finely laminated, part ripple cross-laminated; difficult accessibility; approximately 6.6 ft (2 m).

Unit 4. Siltstone, dolomitic, trace very fine sand, part finely laminated, slightly argillaceous, scattered ripples and starved ripples, scattered horizontal burrows; rare trilobites (*Dikelocephalus minnesotensis*), aglaspid fragments, lingulids, articulate brachiopods (*Finkelnburgia* sp.); 23 ft (7 m).

Black Earth Member

Unit 3. Ledge former; siltstone, dolomitic, and dolomitic siltstone to very fine sandstone, glauconitic to nonglauconitic (glauconite pellets locally to 1 mm), part slightly argillaceous; some silty dolomite; part finely laminated; ripple cross-laminated, intraclastic, clay drapes, and calcite void fills in basal part; horizontal argillaceous burrows common; 13.2 ft (4.3 m).

Lone Rock Formation (Franconia Formation of Minnesota classification)

Reno Member

Unit 2. Siltstone to very fine sandstone, nonglauconitic to very glauconitic, some greensand, part micaceous, part finely laminated to low-angle cross-beds, some clay drapes, ripple cross-laminae in lower part, scattered intraclastic beds; horizontal and vertical burrows common; scattered trilobite molds (*Dikelocephalus posterectus, Saukiella minor, Illaenurus* sp.); 11.6 ft (3.5 m).

Unit 1. Sandstone, very fine, some very fine to fine in lower one-third, glauconitic to very glauconitic, common greensands, slightly micaceous, part calcite-cemented, horizontally laminated to low-angle planar cross-beds, common ripple cross-laminae, clay and siltstone interlaminae and drapes, scattered mudchip intraclasts in some beds; scattered to common horizontal and vertical burrows, rare trilobite molds (*Ptychaspis sp.*) and inarticulate brachiopod fragments; 29.7 ft (9 m).

REFERENCES CITED

Anderson, R., McKay, R. M., and Witzke, B. J., 1979, Field trip guidebook to the Cambrian stratigraphy of Allamakee County: Geological Society of Iowa, Guidebook 32, 12 p.

Davis, R. A., Jr., 1970, Prairie du Chien Group in the Upper Mississippi Valley, *in* Field Trip Guidebook for Cambrian–Ordovician Geology of Western Wisconsin: Wisconsin Geological and Natural History Survey, Information Circular 11, p. 35–44.

Heyl, A. V., Jr., Agnew, A. F., Erwin, J. L., and Behre, C. H., Jr., 1959, The geology of the Upper Mississippi Valley Zinc–Lead District: U.S. Geological Survey Professional Paper 309, 310 p.

Leonard, A. G., 1897, Lead and zinc deposits of Iowa: Iowa Geological Survey, Annual Report, v. 6, p. 11–66.

Miller, J. F., and Melby, J. H., 1971, Trempealeauan conodonts (and comments), *in* Clark, D. L., ed., Conodonts and biostratigraphy of the Wisconsin Paleozoic: Wisconsin Geological and Natural History Survey, Information Circular 19, p. 4–9, 78–81.

Odum, I. E., and Ostrom, M. E., 1978, Lithostratigraphy, petrology and sedimentology of the Jordan Formation near Madison, Wisconsin, *in* Lithostratigraphy, petrology, and sedimentology of Late Cambrian–Early Ordovician rocks near Madison, Wisconsin: Wisconsin Geological and Natural History Survey, Field Trip Guidebook no. 3, p. 23–45.

Ostrom, M. E., 1978, Stratigraphic relationships of Lower Paleozoic rocks of Wisconsin: Wisconsin Geological and Natural History Survey, Field Trip Guidebook, no. 3, p. 3–22.

Ostrom, M. E., Davis, R. A., Jr., and Cline, L. M., field trip committee, 1970, Field trip guidebook for Cambrian–Ordovician geology of western Wisconsin: Wisconsin Geological and Natural History Survey, Information Circular 19, 131 p.

Raasch, G. O., 1939, Cambrian Merostomata: Geological Society of America Special Paper 19, 146 p.

Schuldt, W. C., 1940, Cambrian strata of northeastern Iowa [Ph.D. thesis]: Iowa City, University of Iowa, 229 p.

Sloss, L. L., 1963, Sequences in the cratonic interior of North America: Geological Society of America Bulletin, v. 74, p. 93–114.

Walter, O. T., 1924, Trilobites of Iowa and some related Paleozoic forms: Iowa Geological Survey, Annual Report, v. 31, p. 167–389.

Witzke, B. J., 1980, Middle and Upper Ordovician paleogeography of the region bordering the Transcontinental Arch, *in* Fouch, T. D., and Magathan, E. R., eds., Paleozoic paleogeography of the west-central United States: Rocky Mountain Section, Society of Economic Paleontologists and Mineralogists, p. 1–18.

The Lime Creek Formation of north-central Iowa

Wayne I. Anderson, *Department of Earth Science, University of Northern Iowa, Cedar Falls, Iowa 50614*
William M. Furnish, *Department of Geology, University of Iowa, Iowa City, Iowa 52242*

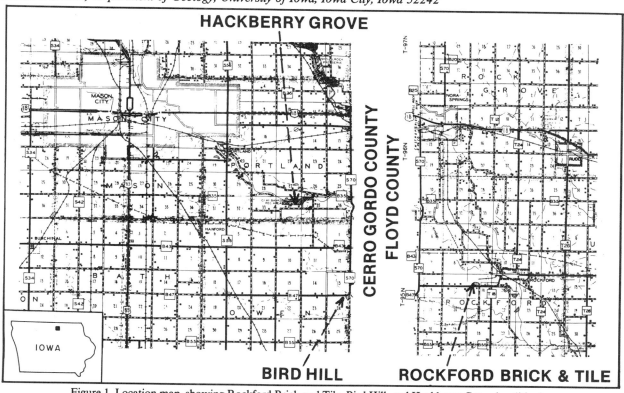

Figure 1. Location map, showing Rockford Brick and Tile, Bird Hill, and Hackberry Grove localities in Floyd and Cerro Gordo counties, Iowa.

LOCATION

There are three principal exposures of the Lime Creek Formation in the area between Mason City and Charles City, northern Iowa. By a considerable margin, the most extensive site is the quarry of the former Rockford Brick and Tile Company, only 1 mi (1.6 km) west of town in Floyd County (NW¼Sec.16, T.95N.,R.18W.). The Juniper Hill State Preserve is 0.5 mi (0.8 km) to the northwest. Bird Hill is located just a few miles farther west at a road cut on an east-west county right-of-way in eastern Cerro Gordo County (NE¼Sec.24,T.95N.,R.19W.). The stratigraphic section known historically as Hackberry Grove, now called Claybanks Forest, lies just south of the Winnebago River (near center NE¼Sec.34,T.96N.,R.19W.). The Winnebago River was originally called Lime Creek (hence the name Lime Creek Formation). A location map (Fig. 1) has been prepared from the general highway and transportation maps of Cerro Gordo and Floyd counties. Topographic sheets for the Mason City Southeast and Rockford 7½-minute Quadrangles cover the area.

The Rockford Brick and Tile property lies adjacent to County Road B47, west of Rockford. The kilns have been inactive for several years, and permission to visit the grounds must be

secured from the present owner, Allied Construction Company, 1211 South Main Street, Charles City, Iowa 50616.

Bird Hill is also situated on County Road B47, with a large parking area on the north side of the road. The site has recently been designated a State Preserve.

At Hackberry Grove, the Claybanks Forest consists of a 56-acre (22-ha) State Preserve, administered by the Cerro Gordo County Conservation Board. The bluff exposing the Lime Creek Formation is a short distance north of a township road, about 2 mi (3.2 km) west of County Road S70 (Fig. 1).

SIGNIFICANCE

During a bleak northern Iowa winter, it might be difficult to imagine that the area contains evidence of an ancient subtropical ocean. Yet, when the snows melt in the spring, the indications may be easily observed. Hoards of superbly preserved fossils, relics of warm shallow seas, weather free from the soft shales and limestones of the local bedrock. Many choice specimens can be found in a few minutes time. For this reason, the Lime Creek Formation of eastern Cerro Gordo and western Floyd counties is known internationally. Nearly all paleontologists are aware that these rocks provide some of the best collecting in the entire Midwest. During field operations of Iowa's first state survey, Edward Hungerford and J. D. Whitney collected spiriferid brach-iopods, such as *Theodossia hungerfordi* and *Cyrtospirifer whitneyi,* from the fossiliferous Lime Creek beds, west of Rock-ford. These species were named and illustrated by the renowned paleontologist-geologist James Hall in 1858 (Hall and Whitney, 1858).

SITE INFORMATION

Lithostratigraphic sections of the Lime Creek Formation at Rockford Brick and Tile, Bird Hill, and Hackberry Grove are shown in Figure 2. The Lime Creek Formation is currently di-vided into three members, in ascending order: Juniper Hill Member, Cerro Gordo Member, and Owen Member. The Cerro Gordo Member is the most fossiliferous of the three members and a variety of fossils can be collected from it at the three localities mentioned above.

Belanski (1931), in an article published two years after his untimely death, divided the Hackberry Formation (Owen and Cerro Gordo members of current usage) into 45 zonules or beds, based on lithology and faunal content. One of the authors of this article (Furnish) verified Belanski's stratigraphic divisions, to the inch, by making trenches at the Rockford Brick and Tile and Bird Hill localities in 1935 and 1936. Everything was as Belanski reported.

Unfortunately, Belanski died at an early age, just as he was starting to make important professional contributions. The bulk of his scientific observations remain unpublished (Anderson and Furnish, 1983).

The Cerro Gordo section at Hackberry Grove is probably complete, but a portion is covered by debris. The lower half of the Cerro Gordo Member can be studied in detail only at the Rockford Brick and Tile locality, where the contact with the underlying Juniper Hill Member can be seen (Fig. 3). The upper Juniper Hill shales and lower Cerro Gordo beds from this locality were used for the manufacture of brick and tile. The contact of the Juniper Hill Shale with the underlying Shell Rock Formation can be seen along the north bank of the Winnebago River, west of Rockford (SE¼SE¼Sec.9,T.95N.,R.18W.) and in the Weaver Quarry, west of Portland (SW¼NE¼Sec.19, T.96N.,R.19W.). The Bird Hill locality exposes the upper portion of the Cerro Gordo Member and its contact with the overlying Owen Lime-stone Member (Fig. 2).

Evolution of terminology for the Upper Devonian strata in northern Iowa has presented some problems. As early as 1883, it was realized that the name Rockford was unavailable for these beds because it had been used for a Mississippian formation in Indiana. A designation of Lime Creek beds by H. S. Williams (1883) can be regarded as only a casual usage, and there were no formal procedures on stratigraphic nomenclature to follow when Williams' article was published. The first valid proposal may, therefore, have been by Clement L. Webster, a school teacher at Charles City. Webster published several articles on the rocks of

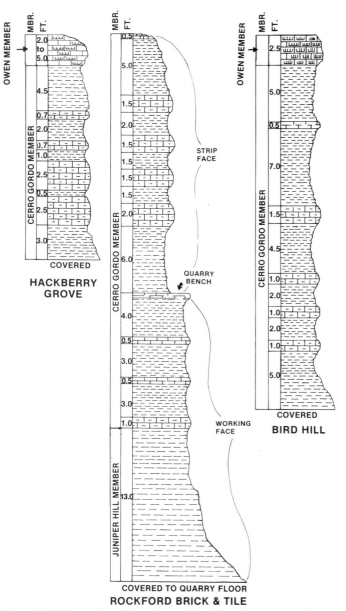

Figure 2. Lithostratigraphic sections of the Lime Creek Formation at Hackberry Grove, Rockford Brick and Tile, and Bird Hill localities, Cerro Gordo and Floyd counties, Iowa.

the Rockford area, referring to the strata as the Hackberry beds (Webster, 1889). Webster's name was derived from Hackberry Grove, then, and still, a primary reference site.

The term Hackberry has been utilized extensively, especially in publications by the Fentons (Fenton, 1919; Fenton and Fen-ton, 1924). However, the Iowa Geological Survey, since the 1890s, has consistently favored use of the name Lime Creek Formation for these beds. Nevertheless, as late as 1942, the Geo-logical Society of America's correlation chart (G. A. Cooper, editor) utilized Hackberry for Upper Devonian strata of northern

CERRO
GORDO

———

JUNIPER
HILL

Figure 3. South-facing exposure of the Lime Creek Formation at the Rockford Brick and Tile locality. Portions of the Cerro Gordo and Juniper Hill members are exposed. Grain elevators at Rockford, Iowa, visible on the right. Photograph taken June, 1984 by Wayne I. Anderson.

Iowa. In 1944, Stainbrook recommended that to avoid confusion the name Hackberry be abandoned in favor of Lime Creek Formation. His recommendation has been followed since.

The portion of Cerro Gordo Member composed of soft calcareous shale encompasses tens of square miles in northern Iowa, with some known outliers. This type of rock, sometimes called marl, provides perfectly preserved fossils as it weathers. While many thousands of these fossil remains have been added to private collections and carried to museums, schools, colleges, and universities, the supply is replenished by natural weathering processes.

Fenton (1920) published a faunal list for the Owen and Cerro Gordo members (known then as the Hackberry Stage). Although taxonomic practices and species concepts have changed since Fenton's time, this faunal list still provides some indication of the diversity of the "Hackberry Fauna." Fenton reported 83 taxa from the Owen Member and 143 taxa from the Cerro Gordo Member. Included in the list were species of corals, stromatoporoids, echinoids, bryozoans, brachiopods, pelecypods, gastropods, and cephalopods.

Fenton and Fenton illustrated the megafauna of the Cerro Gordo and Owen members in 1924. Obviously, some of their taxonomy needs to be revised, but the Fentons' study still represents a thorough and systematic study of the "Hackberry Fauna." It is significant in attempting to deal with all elements of the megafauna. Subsequent studies have been less comprehensive.

Brachiopods from the Lime Creek Formation have received special attention. For example, C. L. Fenton's (1931) *Studies of the Genus Spirifer* used material from the Lime Creek Formation. Mallory (1968) conducted bed-by-bed collecting from the Cerro Gordo Member in a paleoecological study and demonstrated that several of the brachiopod taxa range throughout the member. Comparisons have been made between the Lime Creek brachiopod fauna and similar brachiopod faunal communities in

France (Wallace, 1978). Beus (1978) reported that six species of brachiopods, known from the Cerro Gordo Member, also occur in the Martin Formation of Arizona. Cooper and Dutro (1982) reported Devonian brachiopod taxa from New Mexico that are also present in the Devonian of Iowa. They redescribed 11 species from the Cerro Gordo Member as part of their study.

Brachiopod faunas, along with the ammonoid cephalopod *Manticoceras,* document the Late Devonian (Frasnian) age of the Cerro Gordo. Conodonts described by Anderson (1966) provide additional support for an age assignment of Late Devonian (Frasnian) for the Lime Creek Formation.

Megafossils from the Cerro Gordo Member commonly bear epibionts. Epibionts are particularly common on brachiopods (Anderson and Megivern, 1982). The occurrence and distribution of the epibionts can best be explained by their attachment and subsequent growth on either living or dead brachiopods. Common epibionts include *Spirorbis* sp. (a calcareous worm tube), *Cornulites* sp. (a conical shell of uncertain affinity), auloporid tabulate corals, *Hederella* sp. (a bryozoan), other bryozoans, juvenile horn corals, and *Petrocrania* sp. (an inarticulate brachiopod). In addition, many brachiopods contain circular borings (possibly produced by polychaete worms) and dendritic grooves and channels of the type associated with clionid sponges. Although less common, horn corals, gastropods, and bryozoans also served as hosts for epibionts. Because the Cerro Gordo Member was deposited on a muddy sea floor, attachment sites for small suspension feeders were limited. In this environment, brachiopod shells and other larger invertebrates provided relatively mud-free sites where epibionts could attach, grow, and survive.

Dorheim and others (1970) prepared a faunal list of the common megafossils that can be collected readily in the Rockford area:

Owen Member
Stromatoporoids. *Idiostroma* sp. (now identified as *Amphipora* sp.; see Stock, 1984b).

Cerro Gordo Member
Brachiopods. *Atrypa (Pseudoatrypa) devoniana, Cyrtospirifer whitneyi, Douvillina* sp. *Productella walcotti, Schizophoria iowensis, Spinatrypa rockfordensis, Spinocyrtia (Platyrachella) oweni, Strophonelloides hybrida (Strophonelloides hybridus), Tenticospirifer cyrtiniformis, Theodossia hungerfordi.*

Bryozoans. *Lioclema* sp., *Orthopora* sp., *Petalotrypa* sp.

Coelenterates. *Pachyphyllum woodmani.*

Pelecypods. *Paracyclas sabini.*

Gastropods. *Floydia concentrica, Holopea* (?) *iowensis, Westernia pulchra.*

Juniper Hill Member
Brachiopods. *Lingula fragilis.*

A variety of microfossils have been recovered from residues of the Juniper Hill and Cerro Gordo beds at the Rockford Brick and Tile and Bird Hill localities (Wilson and McNamee, 1984). Microfossils include the following: foraminifera, ostracods, charophytes, tentaculitids, scolecodonts, conodonts, holothurian schlerites, megaspores, juvenile gastropods, juvenile bryozoans, and juvenile brachiopods.

Additional information on the Lime Creek fauna can be obtained from the following sources: Baker and others (1986); Conklin and others (1972), Cushman and Stainbrook (1943); Gibson (1952); Johnson (1974); Kier (1968); Newport and Urban (1974); Peck and Morales (1966); Stock (1984a and 1984b); and Strimple and Levorson (1971). Kier's 1968 report of *Nortonechinus,* a cidarid echinoid, is significant in that paleontologists consider echinoids of this type to be the ancestral stock of all post-Paleozoic echinoids. J. E. Sorauf, Department of Geological Sciences and Environmental Studies, State University of New York, Binghamton is currently preparing a manuscript on the corals of the Lime Creek Formation.

The Cerro Gordo Member was probably deposited in a sublittoral environment of moderately high energy. The underlying Juniper Hill Member may have been deposited in deeper water and is characterized by fewer fossils and lower faunal diversity than the Cerro Gordo Member. The Owen Member was deposited under less turbid conditions than the Cerro Gordo Member. Whereas shales are dominant in the Cerro Gordo Member, carbonates predominate in the Owen Member. The basal beds of the Owen Member are present at the Bird Hill and Hackberry Grove localities, where they constitute a biostrome composed chiefly of the branching stromatoporoid *Amphipora* sp.

Characteristics of the three members of the Lime Creek Formation in other parts of Iowa are discussed by Anderson (1984), Koch (1963), and Witzke and Bunker (1984). In general, dolomite-dominated facies characterize the Lime Creek Formation in the subsurface of central and western Iowa.

REFERENCES CITED

Anderson, W. I., 1966, Upper Devonian conodonts and the Devonian Mississippian boundary of north-central Iowa: Journal of Paleontology, v. 40, p. 395–415.

——— , 1984, General geology of north-central Iowa: Guidebook for the 48th Annual Tri-State Geological Field Conference, 149 p.

Anderson, W. I., and Furnish, W. M., 1983, Iowa's self-trained paleontologists: Proceedings of the Iowa Academy of Science, v. 90, p. 1–12.

Anderson, W. I., and Megivern, K., 1982, Epibionts from the Cerro Gordo Member of the Lime Creek Formation (Upper Devonian), Rockford, Iowa: Proceedings of the Iowa Academy of Science, v. 89, no. 2, p. 71–82.

Baker, C., Glenister, B. F., and Levorson, C. O., 1986, Devonian ammonoid *Manticoceras* from Iowa: Proceedings of the Iowa Academy of Science, v. 93, p. 7–15.

Belanski, C. H., 1931, Introduction; The stratigraphy of the Hackberry Stage, *in* Fenton, C. L., ed., Studies of the evolution of the genus *Spirifer*: Philadelphia, Publication of the Wagner Free Institute of Science, v. II, p. 1–7.

Beus, S. S., 1978, Late Devonian (Frasnian) invertebrate fossils from the Jerome Member of the Martin Formation, Verde Valley, Arizona: Journal of Paleontology, v. 52, p. 40–54.

Conkin, J. E., Sawa, T., Takshi, C., Salman, R. G., and Abdullah, M., 1972, The charophyte genus *Sycidium* in the Upper Devonian of Iowa: Micropaleontology, v. 18, no. 1, p. 74–80.

Cooper, G. A., 1942, Correlation of the Devonian sedimentary formations of North America, chart no. 4: Geological Society of America Bulletin, v. 53, p. 1729–1794.

Cooper, G. A., and Dutro, J. T., Jr., 1982, Devonian brachiopods of New Mexico: Bulletins of American Paleontology, v. 82 and 83, no. 315, 196 p., 39 plates.

Cushman, J. A., and Stainbrook, M. A., 1943, Some foraminifera from the Devonian of Iowa: Cushman Laboratory for Foraminiferal Research Contributions, no. 19, pt. 4.

Dorheim, F. H., Koch, D. L., and Laudon, L. R., 1970, The Mississippian and Devonian of Iowa: Geological Society of America North-Central Section Field Trip no. 3, Guidebook, 36 p.

Fenton, C. L., 1919, The Hackberry Stage of the Upper Devonian of Iowa: American Journal of Science, v. 48, no. 285, p. 355–376.

——— , 1931, Studies of evolution of the genus *Spirifer*: Philadelphia, Publication of the Wagner Free Institute of Science, v. II, 436 p.

Fenton, C. L., and Fenton, M. A., 1924, The stratigraphy and fauna of the Hackberry Stage of the Upper Devonian: Ann Arbor, University of Michigan, Contributions from the Museum of Geology, v. 1, 260 p.

Gibson, L. B., 1952, Upper Devonian ostracoda from the Cerro Gordo Formation of Iowa: Bulletin of American Paleontology, v. 35, no. 154, 38 p.

Hall, J., and Whitney, J. D., 1858, Report on the geological survey of the state of Iowa, embracing the results of investigations made during 1855, 56, and 57: Albany, New York, v. 1, pt. 1 Geology; pt. 2 Paleontology, 724 p., 29 plates.

Johnson, M. E., 1974, Occurrence of a ctenacanthoid shark spine from the Upper Devonian of north-central Iowa: Proceedings, Iowa Academy of Science, v. 81, no. 2, p. 56–60.

Kier, P. M., 1968, *Nortonechinus* and the ancestry of the cidarid echinoids: Journal of Paleontology, v. 42, p. 1163–1170.

Koch, D. L., 1963, The Lime Creek Formation in the area of Garner, Iowa: Proceedings, Iowa Academy of Science, v. 70, p. 245–252.

Mallory, B. F., 1968, Paleoecologic study of a brachiopod fauna from the Cerro Gordo Member of the Lime Creek Formation (Upper Devonian), north-central Iowa [Ph.D. thesis]: Columbia, University of Missouri, 53 p.

Newport, R. L., and Urban, J. B., 1974, Acritarchs of the Lime Creek Formation (Upper Devonian) of Iowa: Dallas, University of Texas, Geoscience Manual 9, 76 p.

Peck, R. E., and Morales, G. A., 1966, The Devonian and Lower Mississippian charophytes of North America: Micropaleontology, v. 12, no. 3, p. 303–324.

Stainbrook, M. A., 1944, The Devonian System in Iowa: Illinois Geological Survey Bulletin 68, p. 182–188.

Stock, C. W., 1984a, Upper Devonian (Frasnian) Stromatoporoidea of north-central Iowa; Redescription of the type specimens of Hall and Whitfield (1873): Journal of Paleontology, v. 58, p. 773–788.

——— , 1984b, The distribution of stromatoporoids in the Upper Devonian of north-central Iowa: Guidebook for the 48th Annual Tri-State Geological Field Conference, p. 125–129.

Strimple, H. L., and Levorson, C. O., 1971, New flexible crinoids from the Upper Devonian of north-central Iowa: Proceedings, Iowa Academy of Science, v. 78, p. 9–11.

Wallace, P., 1978, Homeomorphy between Devonian brachiopod communities in France and Iowa: Lethaia, v. 11, p. 259–272.

Webster, C. L., 1889, A general preliminary description of the Devonian rocks of Iowa: American Naturalist, v. 23, p. 229–243.

Williams, H. S., 1883, On a remarkable fauna at the base of the Chemung Group in New York: American Journal of Science, v. 25, p. 97–104.

Wilson, C., and McNamee, L., 1984, Microfossils from the Juniper Hill and Cerro Grodo members of the Lime Creek Formation: Guidebook for the 48th Annual Tri-State Geological Field Conference, p. 130–141.

Witzke, B. J., and Bunker, B. J., 1984, Devonian stratigraphy of north-central Iowa: Iowa Geological Survey Open-File Report 84-2, pt. II, p. 107–149.

The Ordovician Sequence in the Guttenberg Area, northeast Iowa

Brian J. Witzke and Brian F. Glenister, Iowa Geological Survey and University of Iowa, Department of Geology, Iowa City, Iowa 52242

LOCATION

Guttenberg is a picturesque community nestled in the Mississippi River Valley, where local relief approaches 400 ft (120 m). Lock and Dam No. 10 at Guttenberg is a major conduit for upper Mississippi barge traffic. Visitors can view the lock in operation at close range from an observation platform.

Excellent roadcuts and many additional exposures of Middle Ordovician strata are easily accessible in the Guttenberg area, where virtually every bed is available in continuous fresh exposures over a stratigraphic interval of more than 250 ft (75 m). Three reference sections (Fig. 1B) enable the sequence of Ordovician formations (St. Peter, Glenwood, Platteville, Decorah, Dunleith, Wise Lake) to be composited for general discussion (Fig. 2): 1) Guttenberg South Roadcut, U.S. 52 (SW¼Sec.29,T.92N., R.2W.), a superb continuous section that offers a spectacular view of the Mississippi River (Delgado, 1983); 2) Guttenberg North Roadcut, U.S. 52 (SW¼SW¼Sec.5 and NW¼Sec.8,T.92N., R.2W.), a well-studied but now partly overgrown sequence (Templeton and Willman, 1963; Bakush, 1985); and 3) X56 Roadcut, Great River Road (NW¼NW¼Sec.32 and SW¼SW¼Sec.29, T.93N.,R2W.), a similar sequence to Guttenberg North, but better exposed by recent road construction. Excellent exposures of Lower and Middle Ordovician rocks are also accessible at Clayton (especially private barge terminal road) and the Pikes Peak State Park area (Fig. 1A). Good exposures of Jordan Sandstone (Cambrian) can be seen along the river from McGregor north to Effigy Mounds (Fig. 1A).

Ordovician stratigraphic terminology in the area follows that of the Iowa Geological Survey (Witzke, 1983), but the general Illinois classification is also applicable (Templeton and Willman, 1963; Levorson and Gerk, 1972; Willman and Kolata, 1978). Conodonts are found in all formations in the area except the St. Peter, providing important biostratigraphic control.

Brecciated and fractured Middle Ordovician carbonates in the lower Dunleith Formation of the Guttenberg area are sites of Upper Mississippi Valley-type mineralization. Galena, associated with calcite, marcasite, and barite, was mined along Miner's Creek immediately west of Guttenberg (Fig. 1) from about 1855 into the early 1900s (Heyl and others, 1959).

STRATIGRAPHY

A major unconformity, which divides the Sauk Sequence below from the Tippecanoe Sequence above, separates the St. Peter Sandstone from underlying Lower Ordovician Prairie du Chien Group dolomites in the study area. Prairie du Chien rocks first appear at the surface north of Guttenberg. The upper part of the St. Peter Sandstone and the Glenwood Shale can be seen at

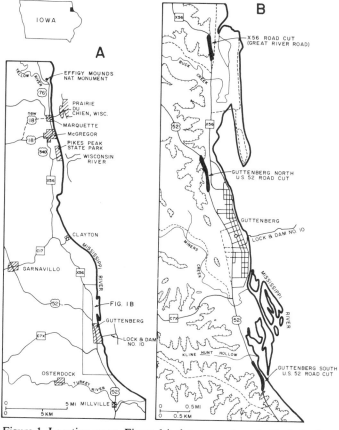

Figure 1. Location maps, Figure 1A shows paved roads (dashed where new U.S. 18 is under construction). Fig. 1B shows location of three main roadcut sections: paved roads solid; secondary gravel roads dashed; hachured line marks 900-foot contour and shows general position of upland areas.

the North U.S. 52 and X56 roadcuts. The St. Peter is an exceptionally pure quartzarenite. The overlying Glenwood Shale, often poorly exposed, is a noncalcareous partly feldspathic greenish shale containing common conodonts and rare other fossils. The St. Peter/Glenwood sequence, or Ancell Group, probably spans parts of the upper Chazyan and Blackriveran stages.

The Platteville Formation (or Group in the Illinois classification) includes two members in Iowa: 1) a lower Pecatonica Member, dolomite with some fossil molds, sandy at the base with a prominent hardground at the top; and 2) the McGregor Member (includes Mifflin and Grand Detour formations of Illinois classification), a wavy-bedded fossiliferous limestone and dolomitic limestone unit with thin argillaceous to shaly partings. The McGregor is dominantly a sparse biomicrite (Mossler, 1985), in part extensively burrowed; some hardgrounds occur. Well-

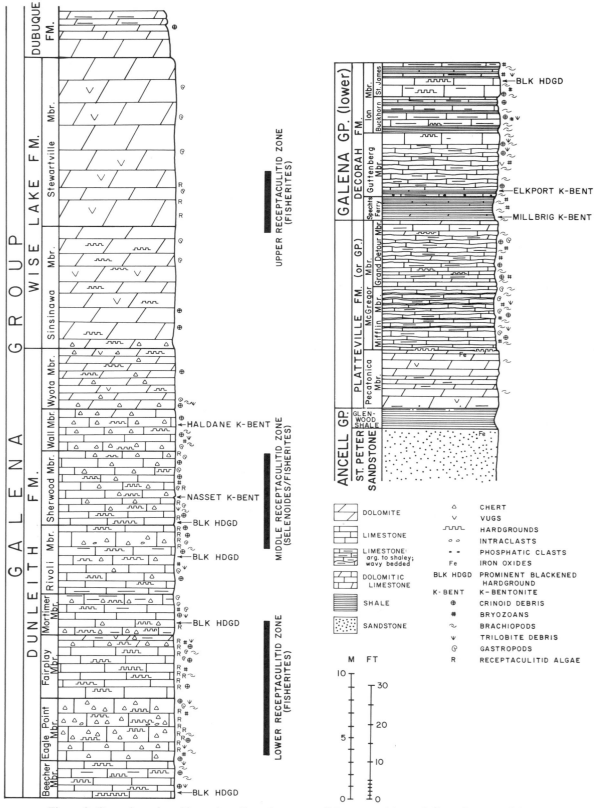

Figure 2. General stratigraphic section, Guttenberg area. Column on right underlies column on left. Adapted from Delgado (1983), Bakush (1985), and Templeton and Willman (1963).

preserved fossils are common on bedding planes and include brachiopods (especially *Oepikina, Strophomena, Hesperorthis, Pionodema, Campylorthis*), trilobites, giant ostracodes *(Eoleperditia)*, echinoderm debris, and some bryozoans. Gastropods, especially bellerophontids, are abundant in some beds. Giant endocerid nautiloids (to 10 ft; 3 m) occur sporadically throughout the area. The Platteville is, at least in part, of Rocklandian age.

The Decorah Formation (or Subgroup of the Illinois classification) overlies the Platteville, possibly with minor disconformity, and is subdivided into three members in northeast Iowa (Fig. 2). It can be seen at all three roadcuts (Fig. 1), although only the upper half is exposed at Guttenberg South. The Decorah marks the basal part of the Galena Group. It is distinguished from adjacent units by the abundance of greenish calcareous shales in the upper and lower parts. The basal Spechts Ferry Member is dominantly a shale unit with thin coquinoid limestone layers containing abundant brachiopods (especially *Pionondema, Doleroides*), trepostome bryozoans, and some other fossils. A prominent K-bentonite, termed the Millbrig (Willman and Kolata, 1978), is seen as a thin orange-colored layer 10 in (25 cm) above its base. The Guttenberg Member, whose type locality is the Guttenberg North section, is characterized by nodular to wavy-bedded limestones with thin red-brown shale partings. Some of the shales, as seen in unoxidized sections and core, are extremely organic. The thin Elkport K-bentonite (Willman and Kolata, 1978) is seen near the base. The Guttenberg is dominated by skeletal calcilutites (sparse biomicrites), but beds and lenses of skeletal calcisiltites and calcarenites (packed biomicrites and biosparites) are present in the middle and upper parts. The Guttenberg contains common fossils, including brachiopods (especially *Sowerbyella, Rafinesquina*), trilobites, crinoid debris, bryozoans, and gastropods.

The Ion Member of the Decorah, which correlates with the Buckhorn and St. James members of the Dunleith Formation to the south, is characterized by interbedded limestone (skeletal calcilutite to calcarenite) and greenish shale. Several hardgrounds are present. The Ion contains a normal marine biota; brachiopods (*Dalmanella, Glyptorthis*, etc.) and bryozoans are common in some beds. The upper bed of the Ion is highly fossiliferous and is characterized by abundant hemispherical stony bryozoans (the "*Prasopora* zonule"). The Decorah is probably Rocklandian to Kirkfieldian in age.

The Dunleith Formation overlies the Decorah conformably and is easily accessible at all three roadcuts (Fig. 1). It differs from the Decorah in lacking prominent shale interbeds. The Dunleith is divided into eight members at Guttenberg (Fig. 2); bedding style, chert content, argillaceous character, and key fossils are useful in recognition of the constituent members (Levorson and Gerk, 1983). The extreme lateral continuity of stratigraphic units within the Dunleith permits tracing of members throughout most of the Upper Mississippi Valley outcrop belt. Unlike the Platteville and Guttenberg, bedding is generally parallel and even. The Dunleith is dominated by bioturbated biomicrites (Korpel, 1983) which are dolomitic to varying degrees. Dolomitization typically

follows large burrow-mottled networks, strikingly similar to those noted in equivalent strata in Manitoba (Kendall, 1977) and elsewhere. The Wyota Member and upper bed of the Fairplay Member are pervasively dolomitized (Fig. 2). Thin biosparite beds (0.25 to 5 in; 1 to 13 cm) are scattered through the Dunleith and overlie scoured surfaces with sharp contact; bioclasts may be imbricated.

Hardgrounds are especially well developed in the Dunleith; they "are one of the most spectacular, as well as important, petrologic features of the Galena Group" (Delgado, 1983). Many hardgrounds are burrowed and bored and show overhanging sculpture; some are prominently blackened with pyrite or iron oxides (rarely colophane); a few are encrusted with epibionts. At least 67 separate hardgrounds are recognized in the Dunleith at the Guttenberg South Roadcut (Delgado, 1983). Nodular cherts are abundant in the Dunleith, especially in the Eagle Point, Rivoli, Sherwood, Wall, and Wyota members. Two K-bentonites, the Haldane and Nasset (Fig. 2), have been recognized in the Dunleith of the Guttenberg area.

The paleontology of the Dunleith remains poorly studied, but petrographic analysis reveals a stenohaline biota throughout (Korpel, 1983; Bakush, 1985), dominantly echinoderms, mollusks, brachiopods, and trilobites. Identified brachiopods include *Sowerbyella, Strophomena* and *Dalmanella*. Bryozoans, ostracodes, and sponge spicules are common in some beds. Nautiloids, bivalves, and corals are noted locally. A fossil association typified by large receptaculitid green algae and thick-shelled snails is conspicuous in two stratigraphic intervals and differs biotically from that found in adjacent strata. Receptaculitid-bearing intervals in the Dunleith are termed: 1) the lower receptaculitid zone (dominantly *Fisherites oweni*) in the Eagle Point and Fairplay members, and 2) the middle receptaculitid zone (dominantly *Selenoides iowensis*) in the upper Rivoli and Sherwood members. Dasyclad algae *(Vermiporella)* commonly are observed in thin sections from the receptaculitid zones (Delgado, 1983). The Dunleith is approximately Kirkfieldian and Shermanian in age.

The upper two formations of the Galena Group, the Wise Lake and Dubuque, form the upland bedrock surface in the Guttenberg area. The lower part of the Wise Lake (lower Sinsinawa Member) can be seen at the three roadcuts, and upper Wise Lake (Stewartville Member) and lower Dubuque strata are exposed at Guttenberg South but are accessed with considerable difficulty. The Wise Lake Formation, dominated by vuggy dolomite, is distinguished from the underlying Dunleith by its thick to massive bedding and absence of chert (except at the base). The Sinsinawa Member contains hardgrounds similar to those in the Dunleith; molds of crinoid debris and gastropods occur. The Stewartville Member contains scattered receptaculitid *(Fisherites)* molds (the upper receptaculitid zone) with prominent large gastropods *(Maclurites)*. Although it has been suggested that the Stewartville biota lived in hypersaline stressed environments, to the northwest (near Decorah) the Wise Lake Formation in limestone facies contains a stenohaline crinoid-brachiopod-mollusk fauna throughout (Bakush, 1985). These skeletal constituents probably have been obscured by dolomitization in the southern sections. The base of the Wise Lake is near the base of the Upper Ordovician Edenian Stage.

The Dubuque Formation, which is pervasively dolomitized in the Guttenberg area, is distinguished from the Wise Lake primarily by the interbedded shales and shaly partings and its more argillaceous character. The Dubuque Formation is best observed at other localities in northwest Iowa (Levorson and others, 1979). A prominent phosphorite and brown shale unit in the lower Maquoketa Formation caps a widespread hardground or corrosion surface at the top of the Dubuque.

DEPOSITION AND DIAGENESIS

Ancell Group sandstones and shales were deposited as the Middle Ordovician Seaway transgressed northward toward the Transcontinental Arch. Further deepening of the sea resulted in deposition of subtidal Platteville and Galena carbonates. An influx of clay from Transcontinental Arch source terranes in Minnesota is recorded in the Decorah Formation (Witzke, 1980). Faunas of the Platteville and Galena indicate that normal-marine benthic environments, primarily below normal wave base, characterized the bulk of deposition. Strata encompassing the three receptaculitid zones clearly were deposited in the photic zone, probably at the shallower end of a depositional spectrum. The absence of algae in other strata is equivocal. Ash falls from Taconic (Appalachian) volcanoes, some 900 mi (1,500 km) distant,

periodically swept across the area, providing convenient stratigraphic datums.

The thin biosparite beds in the Galena Group have been interpreted to represent storm deposits (Delgado, 1983; Korpel, 1983; Bakush, 1985). Normal quiet-water bottom conditions periodically were punctuated by turbulent storm events. The abundant hardgrounds, so spectacularly displayed at the Guttenberg South Roadcut, represent "significant periods of early submarine cementation followed by mechanical erosion and/or chemical dissolution during slow or nondeposition of carbonate sediments" (Bakush, 1985). They are of submarine origin and do not indicate subaerial exposure.

Dolomitization is an important feature of Platteville and Galena strata, but its origins remain unclear. Preferential dolomitization of burrow-mottled networks indicates that the burrows formed preferred pathways for dolomitizing fluids. Dolomitization of such networks may be of submarine origin (Delgado, 1983) or may delineate flow paths for later metasomatic dolomitization. Pervasive regional dolomitization, as seen in the Wise Lake Formation at Guttenberg, may be related to the progression of freshwater and marine phreatic mixing-zone diagenetic environments that accompanied the withdrawal of the seaway near the close of the Ordovician. Nevertheless, the possible role of mixing-zone and brine reflux processes in Galena Group dolomitization remains speculative (Witzke, 1983). Chertification of carbonate was roughly coincident with regional dolomitization; the silica ultimately was derived from biogenic opal, primarily sponge spicules.

REFERENCES CITED

Bakush, S. H., 1985, Carbonate microfacies, depositional environments, and diagenesis of the Galena Group (Middle Ordovician) along the Mississippi River (Iowa, Wisconsin, Illinois, and Missouri), U.S.A. [Ph.D. thesis]: Urbana-Champaign, University of Illinois, 223 p.

Delgado, D. J., 1983, Deposition and diagenesis of the Galena Group in the Upper Mississippi Valley, in Delgado, D. J., ed., Ordovician Galena Group of the Upper Mississippi Valley: Great Lakes Section, Society of Economic Paleontologists and Mineralogists, 13th Annual Field Conference, p. A1–A17, Road log p. R1–R38.

Heyl, A. V., Jr., Agnew, A. F., Erwin, J. L., and Behre, C. H., Jr., 1959, The geology of the Upper Mississippi Valley zinc-lead district: U.S. Geological Survey Professional Paper 309, 310 p.

Kendall, A. C., 1977, Origin of dolomite mottling in Ordovician limestones from Saskatchewan and Manitoba: Bulletin of Canadian Petroleum Geology, v. 25, p. 480–504.

Korpel, J. A., 1983, Depositional and diagenetic history of the Middle/Upper Ordovician Dunleith Formation in northeast Iowa [M.S. thesis]: Iowa City, University of Iowa, 91 p.

Levorson, C. O., and Gerk, A. J., 1972, A preliminary stratigraphic study of the Galena Group of Winneshiek County, Iowa: Proceedings, Iowa Academy of Science, v. 79, p. 111–122.

—— , 1983, Field recognition of stratigraphic position within the Galena Group in northeast Iowa (limestone facies), in Delgado, D. J., ed., Ordovician

Galena Group of the Upper Mississippi Valley: Great Lakes Section, Society of Economic Paleontologists and Mineralogists, 13th Annual Field Conference, p. C1–C11.

Levorson, C. O., Gerk, A. J., and Broadhead, T. W., 1979, Stratigraphy of the Dubuque Formation (Upper Ordovician) in Iowa: Proceedings, Iowa Academy of Science, v. 86, p. 57–65.

Mossler, J. H., 1985, Sedimentology of the Middle Ordovician Platteville Formation, southeastern Minnesota: Minnesota Geological Survey Report of Investigations 33, 27 p.

Templeton, J. S., and Willman, H. B., 1963, Champlainian Series (Middle Ordovician) in Illinois: Illinois State Geological Survey Bulletin 89, 260 p.

Willman, H. B., and Kolata, D. R., 1978, The Platteville and Galena groups in northern Illinois: Illinois State Geological Survey Circular 502, 75 p.

Witzke, B. J., 1980, Middle and Upper Ordovician paleogeography of the region bordering the Transcontinental Arch, in Fouch, T. D., and Magathan, E. R., eds., Paleozoic paleogeography of the west-central United States: Rocky Mountain Section, Society of Economic Paleontologists and Mineralogists, p. 1–18.

—— , 1983, Ordovician Galena Group in Iowa subsurface, in Delgado, D. J., ed., Ordovician Galena Group of the Upper Mississippi Valley: Great Lakes Section, Society of Economic Paleontologists and Mineralogists, 13th Annual Field Conference, p. D1–D26.

Cretaceous exposures, Big Sioux River Valley north of Sioux City, Iowa

Brian J. Witzke and Greg A. Ludvigson, Iowa Geological Survey, Iowa City, Iowa 52242

LOCATION

Cretaceous exposures in the Sioux City area, primarily along the Big Sioux River in Iowa and the Missouri River Valley in northeast Nebraska, were of considerable importance in the early formulation of stratigaphic nomenclature in the Western Interior Province (Tester, 1931). It was from this area that Meek and Hayden (1862) originally named the "Dakota Group," the basal subdivision of Cretaceous rocks as used throughout much of the western U.S. A summary of Cretaceous stratigraphy and paleogeography in the Iowa area is given by Witzke and others (1983).

The east side of the Big Sioux River Valley reveals a series of Cretaceous exposures (Fig. 1) that are among the best in Iowa. Many of the exposures are along the Iowa 12 right-of-way and are publicly accessible (localities 1, 2, 3, 4, 8). Roadside exposures in Stone State Park (localities 5, 6) and old shale pit operations at Kirk Hanson Recreation Complex in Riverside (locality 9) are also publicly accessible, but collecting is discouraged. Other exposures are on private land, and permission must be secured before entering (e.g., shale pit of Siouxland Sand and Gravel, locality 7). Many of the exposures are extremely steep and potentially hazardous; please exercise caution.

A general description of Cretaceous rock units examined and measured at localities 1 through 9 (Fig. 2) is given in the Measured Section. Strata at locality 10 (Fig. 2) are poorly exposed and the illustrated section is adapted from Tester (1931). The rock sequence at locality 2 (Fig. 2) is adapted, in part, from Rich (1975), who recovered foraminifera through much of the section. Dakota and Graneros strata are commonly pyritic in the subsurface. However, weathered exposures typically display iron oxide–cemented zones in the sandstone and siltstone beds and an abundance of gypsum crystals on mudstone and shale slopes. Thin native sulfur rinds occur on some exposed surfaces.

The study area lies within the northern extremity of Iowa's Western Loess Hills. Wisconsinan loess blankets much of the area and locally overlies the eroded Cretaceous surface. The modern drainage as well as Holocene alluvial deposits cut across the loess, pre-Illinoian till and sand units, and Cretaceous strata along the Big Sioux Valley.

DAKOTA FORMATION

The Dakota Formation reaches thicknesses of 490 ft (150 m) in the study area (Munter and others, 1983), but only the upper 88 ft (27 m) are accessible in outcrop. The exposed Dakota section is closely comparable in thickness, lithology, and stratigraphic position to the type Dakota sequence in Dakota County, Nebraska (Brenner and others, 1981), 15 mi (25 km) to the

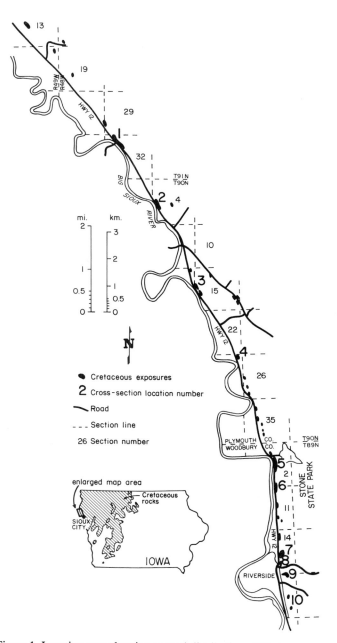

Figure 1. Location map showing general distribution of Cretaceous exposures along Iowa 12 north of Sioux City, Iowa. Location numbers correspond to measured sections shown on cross-section (Fig. 2).

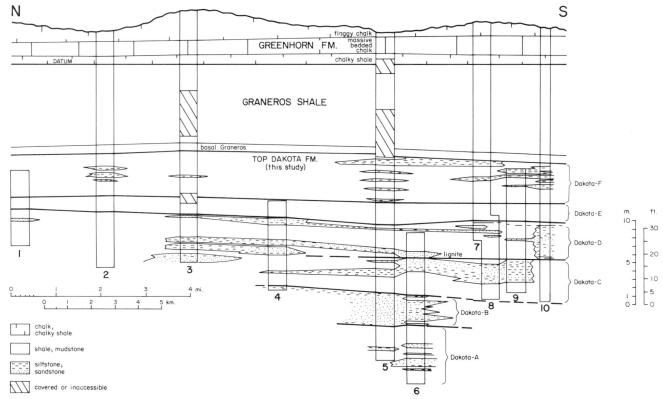

Figure 2. Generalized cross-section of Cretaceous strata north of Sioux City, Iowa (see Fig. 1 for section locations). Description of stratigraphic units given in Measured Section.

south. The Dakota Formation in Iowa is divided into two members, a lower sandstone-dominated Nishnabotna Member (not exposed in northwest Iowa) and an upper mudstone-dominated Woodbury Member. Locality 5 in Stone State Park serves as the type section of the Woodbury.

Kaolinite is the dominant clay mineral in the upper Dakota, with lesser quantities of illite, mixed-layer clays, muscovite, and rare smectites. The sandstones are quartzarenites, dominated by quartz and chert, with minor feldspar, muscovite and lithic grains. Most mudstones are interlaminated with siltstones, commonly displaying ripple and starved-ripple cross-laminae. Sandstone and siltstone beds also are, in part, ripple laminated. Some sandstone beds contain imbricated mud clasts and tabular crossbeds. Most sandstone/siltstone beds are poorly cemented, but some are partially to pervasively cemented with iron oxides (pyrite cements in fresh unoxidized exposures). Siltstones in the uppermost Dakota commonly are calcite-cemented. Lignite and lignitic mudstones are noted locally on the outcrop, including a relatively widespread horizon in the southern sections (Fig. 2).

The exposed portion of the Dakota Formation is informally subdivided into six units (Figs. 2). The upper four units (Fig. 3) are known to contain foraminifera (Rich, 1975), and the top unit (Dakota-F) contains benthic bivalves and scattered to abundant fish debris. Horizontal and vertical burrows are scattered to

common in portions of the exposed Dakota sequence. Indeterminate small carbonaceous plant debris or carbonized wood fragments are recognized on outcrop, primarily in the lower four units (Fig. 2). Plant cuticle and angiosperm leaf fossils are known from mudstones and sandstones in the lower half of the exposed sequence. The prominent lignite at Stone Park has yielded abundant palynomorphs (primarily miospores from fern, bryophyte, and lycopod taxa with rare angiospermous and gymnospermous pollen) indicative of an early Cenomanian age (Ravn, 1981).

GRANEROS SHALE

The Graneros Shale overlies the Dakota Formation, but the contact between the two formations has been placed at various positions in the sequence by different workers. The contact commonly has been drawn at the top of the highest sandstone, iron oxide–cemented silstone, or nodular limonite unit. However, such criteria remain inconsistent and stratigraphically variable. The Graneros-Dakota contact in this report is drawn at the base of the calcareous shale-dominated unit above the noncalcareous to weakly calcareous shales, mudstones, and siltstones (part calcite-cemented) of the Dakota (Figs. 2, 3). Using this definition, the Graneros becomes an easily recognizable rock unit beneath Greenhorn Limestone chalky strata, which ranges between 33 and 40 ft (10 and 12 m) in thickness in the study area (Hattin, 1965, reported 35 ft [10.6 m] in the area). By contrast, the Graneros in Kansas is dominated by noncalcareous silty shale and includes sandstone/siltstone beds to 9 ft (2.7 m) thick (Hattin, 1965). The Graneros Shale is broadly diachronous from west to east; the Graneros of Iowa correlates with lower Greenhorn strata in western Kansas (Witzke and others, 1983). The Graneros clay shales are dominated by kaolinite and illite, derived from eastern source areas.

The basal Graneros Shale in the study area contains siltstone laminae and ripples (starved and climbing). Siltstone laminae are less common in the remainder of the Graneros sequence. White specks (<1 mm) are common in the shales and represent fecal pellets and calcareous foraminifera. Fish scales, coccoliths, and planktonic formanifera occur through much of the unit. Inoceramids and ammonites are noted rarely, primarily in the lower 5 ft (1.5 m) of the Graneros. Plesiosaur bones also are known from the Graneros in the study area (Witzke, 1981). The fauna in the Iowa Graneros indicates a late Cenomanian age (Cobban and Merewether, 1983).

GREENHORN FORMATION

The Greenhorn Formation caps the Cretaceous sequence in most of the study area, and forms resistant ledges or cliffs on the outcrop (Fig. 3). Chalk and shaly chalk, containing vast numbers of coccoliths and calcareous formainifera, characterizes the Greenhorn. Inoceramid bivalves, both broken and whole-shell, are the primary macrofauna. Fish and shark material is present in noteworthy amounts. The fauna indicates an early to middle Turonian age (Cobban and Merewether, 1983).

CARLILE SHALE

The Carlile Shale caps the Greenhorn at the northern edge of the study area (northernmost exposure, Fig. 1) and scattered exposures are present along Iowa 12 to the north. The Carlile is dominated by silty shale that becomes less calcareous upward in the sequence. The Carlile contains planktonic microfossils and fish debris, but macrofauna is sparse in most beds. However, ammonites are common locally (*Collignoniceras, Subpriono-cyclus*), and inoceramid, ostreid, and venerid bivalves also occur. The Carlile in Iowa is of middle Turonian age (Cobban and Merewether, 1983).

DEPOSITIONAL SEQUENCE

The Cretaceous sequence exposed in the Sioux City area was deposited during a portion of the transgressive-regressive Greenhorn Cyclothem. Nonmarine fluvial sediments were progressively overstepped by deltaic, marginal marine, marine shale, and pelagic carbonate facies as the Western Interior Seaway transgressed eastward.

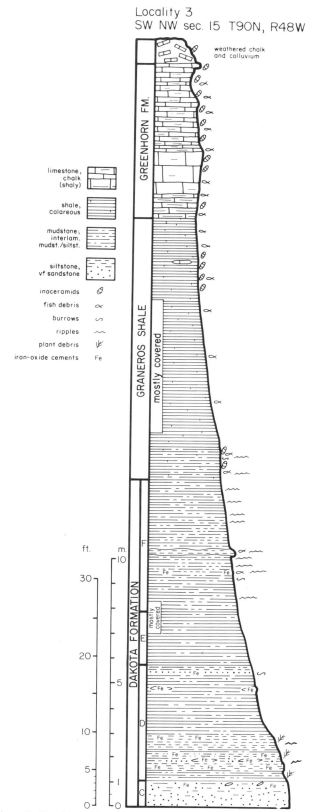

Figure 3. Graphic section of Cretaceous sequence exposed at Locality 3 (Fig. 1).

Braided to meanderbelt fluvial systems, which drained Precambrian-Paleozoic cratonic terrains to the east, deposited the bulk of the Dakota strata (Witzke and others, 1983), but most of the fluvial Dakota sequence occurs in the subsurface in the Sioux City area. A major portion of the Dakota Formation presently exposed in the area, including the type locality in adjacent Nebraska, apparently was deposited in deltaic and marginal marine settings. The presence of foraminifera through much of the exposed Dakota section is inconsistent with fluvial deposition (Rich, 1975). The general absence of paleosols and rooting (except Dakota-A interval) and the abundance of ripple cross-laminae suggest deposition under generally subaqueous conditions, probably in delta-front and prodeltaic settings. Likewise, sandstone and siltstone bodies are not channeled into subjacent strata, suggesting aggradation in subaqueous settings distal to the fluvial facies. However, distributary channel sandstones are noted in equivalent strata to the south (especially the Homer Channel) (Brenner and others, 1981). In general, upper Dakota deposition in the study area occurred as marginal marine environments transgressed across shoreline and deltaic environments. Lignites exposed in the study area apparently accumulated as peats in delta plain and coastal lowland settings. Destructive processes modified delta geometries as the seaway transgressed eastward.

The Graneros Shale in the study area was deposited in prodeltaic and offshore marine environments. Fluvial influx supplied clay and silt to the eastern area of the seaway. Coccoliths and foraminifera supplied calcareous material to the sediments. As the shoreline migrated further eastward, the influx of eastern-derived siliciclastic sediments was reduced in the study area, and pelagic carbonate deposition became dominant (Greenhorn Formation). Eastern-derived siliciclastic sediments prograded westward during Carlile deposition, recording, in part, the regressive phases of the Greenhorn Cyclothem. Benthic conditions during Graneros, Greenhorn, and Carlile deposition was inhospitable for most organisms, probably due to bottom oxygen stresses, and low-diversity inoceramid paleocommunities are characteristic.

MEASURED SECTION

Greenhorn Formation

Where capped by Carlile Shale, approximately 24 to 29 ft (7.5 to 9 m) thick, study area exposures vary 0 to 21 ft (0 to 6.5 m) thick.

Lithologies: chalky, shaly chalk, chalky shale; some hard limestone beds; generally oxidized on outcrop to pale gray, white, or pale orange-gray. Greenhorn can be subdivided into three general units as follows:

Upper flaggy beds: Shaly to argillaceous chalk, weathers into flaggy beds 0.4 to 2 in (1 to 5 cm) thick; generally less resistant than unit below; some resistant beds to 6 in (15 cm) thick; beds are commonly soft and crumbly with some interbedded harder limestone beds (in part, inoceramite lenses); scattered to abundant inoceramid bivalves, broken to whole shell; scattered to common fish debris.

Middle massive beds: Approximately 5 to 7.25 ft (1.7 to 2.2 m) thick; chalk, slightly argillaceous to argillaceous (especially in lower half); more resistant than beds above or below, most resistant in upper half; in massive beds 18 to 36 in (45 to 95 cm); in part faintly laminated; fish debris; upper half with scattered to abundant inoceramid bivalves, including whole-shell horizons.

Lower chalky shale: Approximately 3 to 4.5 ft (0.9 to 1.4 m); chalky shale, chunky, some faint shale and chalk laminae; gradational below with chalky shale lenses in Graneros-like shale at base; part slightly silty; common to abundant fish debris; inoceramid scraps rare to absent.

Fauna: Inoceramid bivalves (Mytiloides mytiloides) are common through much of the sequence; rare ostreid bivalves; telost fish scales and bones common, rare articulated fish; rare shark teeth; extremely abundant planktonic calcareous foraminifera and coccoliths; calcareous benthonic foraminifera and rare benthonic arenaceous foraminifera in basal bed.

Graneros Shale

Approximately 33 to 44 ft (10 to 12 m) thick.

Dominated: Shale, medium to dark gray (unoxidized), fissile to blocky; calcareous with scattered to abundant white specks; part slightly silty.

Secondary: Shale interlaminated with lensoid siltstones (0.04 to 2.5 in (1 mm to 6 cm) thick, forming starved current ripples that generally decrease in amplitude upward in sequence; some siltstones locally display climbing ripples; scattered horizontal burrows; most siltstones are calcite-cemented; siltstone interlaminations well developed only in basal 3 to 6 ft (1 to 1.8 m) of Graneros.

Minor: Concretions, calcareous, argillaceous, locally with spetarian structure, 4 to 40 in (10 cm to 1 m) diameter; septarians best developed in lower 5 ft (1.5 m) of Graneros; scattered small limestone concretions present in upper 10 ft (3 m).

Fauna: Fish scales and bones scattered to common throughout Graneros; plesiosaur bones locally noted; common to abundant planktonic calcareous foraminifera and coccoliths throughout; basal 6.6 ft (2 m) of Graneros with scattered to common inoceramid bivalves (*Inoceramus prefragilis;* specimens in concretions to 15 in [40 cm] diameter) and rare ammonoids (*Metengonoceras dumbli, Dunveganoceras pondi*); upper 13 ft (4 m) of Graneros with scattered inoceramids, mostly fragmentary (*I. ginterensis*).

Dakota Formation

Total thickness in study area (mostly subsurface) 400 to 500 ft (120 to 150 m). Dakota Formation overlies Ordovician or Devonian strata in Sioux City area. Note: All unoxidized Dakota mudstones are light medium to medium dark gray and are micaceous to varying degrees. Carbonaceous shales and mudstones are dark gray to black. Upper Dakota exposures in the study area are informally subdivided into six stratigraphic units, Dakota-A through Dakota-F.

Dakota-F: Approximately 16 to 18 ft (5 to 5.6 m) thick.

Dominated: Mudstone, clayey to very silty, noncalcareous to part weakly calcareous; most mudstones are interlaminated with thin (less than 1.2 in; 3 cm) siltstones that commonly display low-angle current ripples (including starved and climbing ripples); scattered to common silt-filled burrows (horizontal, vertical, rare U-shaped).

Secondary: Siltstone in beds 1 to 40 in (3 cm to 1 m), part argillaceous; calcite cement common; part iron-cemented; abundant low-angle current ripples (some climbing ripples); locally with hummocks and symmetrical megaripples, rare small-scale cross-beds with 0.8 in (2 cm) basal truncation; tool marks and grooves present at the base of some beds; rare fine plant debris; rare very fine sandstone.

Fauna: Molluscan fauna locally present in siltstones (dominated by veneroid bivalves, especially *Aphrodina;* rare nuculoid and arcoid bivalves; rare gastropods); compressed *Aphrodina* locally present in mudstones; top 20 in (50 cm) of unit with occasional *Inoceramus prefragilis* (to 6 in; 15 cm diameter); fish bones and scales locally common to abundant in siltstones, calcareous planktonic foraminifera present in some beds (generally uncommon).

Dakota-E: Approximately 4.25 to 8.5 ft (1.3 to 2.6 m) thick.

Dominated: Clayey mudstone to shale, generally featureless; noncalcareous to weakly calcareous; rare silt laminae, part with low-angle ripple lenses; rare horizontal burrows; rare septarian concretions.

Fauna: Calcareous planktonic foraminifera common in some beds.

Dakota-D: Approximately 12.5 to 15 ft (3.8 to 4.6 m) thick.

Dominated: Mudstone, clayey to silty; interlaminated siltstone common; mudstones part blocky and nonlaminated; ripple cross-laminae common in some beds; scattered horizontal to subvertical burrows; scattered plant debris; siltier beds part iron-cemented.

Secondary: Siltstone to very fine sandstone in beds (0.4 to 36 in (1 to 90 cm) thick; part ripple laminated; part with abundant subvertical burrows; rare mud clasts in some sandstones, scattered plant debris; part iron-cemented.

Minor: Thin carbonaceous to lignitic shale interbedded with mudstones in upper portion of unit in southern sections (locations 7, 10).

Fauna: Some mudstones in this interval contain benthonic arenaceous foraminifera; rare plantonic calcareous foraminifera.

Dakota-C: Approximately 12 to 18 ft (3.7 to 5.5 m) thick.

Dominated: Silty mudstone, part with thin siltstone laminae; some ripple cross-laminae; scattered plant debris; part iron-cemented.

Secondary: Siltstone to very fine sandstone, primarily in upper half of unit; intervals to 8.5 ft (2.6 m) thick; part iron-cemented; part horizontal to ripple laminated (asymmetrical current ripples); scattered plant debris, burrows; part with interlaminated mudstone.

Minor: Lignite and lignitic shale, 2 to 6 in (5 to 15 cm) thick, part cross-cut by very fine sandstone-filled burrows; lignite present only in southern area (locations 5–10); clayey mudstones also present, part with abundant plant debris.

Fauna: Some mudstones in this interval contain benthonic arenaceous foraminifera; rare planktonic calcareous foraminifera.

Dakota-B: Approximately 12 to 13 ft (3.6 to 4 m) thick.

Dominated: Very fine sandstone (rare fine sandstone), part with argillaceous laminae; part iron-cemented; common ripple cross-laminae (asymmetrical current ripples), some tabular cross-beds; scattered to common horizontal to vertical burrows; scattered to common plant debris; sandstone units replaced southward by siltstone and mudstone; rare imbricated mud clasts in some beds.

Secondary: Siltstone, part iron-cemented, and mudstone, silty to very fine sandy; scattered to abundant plant debris; plant cuticle locally preserved.

Dakota-A: Greater than 21 ft (6.6 m) thick.

Dominated: Interlaminated mudstone-siltstone with scattered to abundant ripple cross-laminae (including starved ripples); some plant debris.

Secondary: Laterally discontinuous very fine sandstone and siltstone beds 2 to 31.5 in (5 to 80 cm) thick, part iron-cemented; scattered to abundant plant debris; rare burrows; part rooted.

Minor: Mudstone to shale with abundant plant debris, including plant cuticle and angiosperm leaf fragments.

REFERENCES CITED

Brenner, R. L., and 7 others, 1981, Cretaceous stratigraphy and sedimentation in northwest Iowa, northeast Nebraska, and southeast South Dakota: Iowa Geological Survey Guidebook Series, no. 4, 172 p.

Cobban, W. A., and Merewether, E. A., 1983, Stratigraphy and paleontology of Mid-Cretaceous rocks in Minnesota and contiguous areas: U.S. Geological Survey Professional Paper 1253, 52 p.

Hattin, D. E., 1965, Stratigraphy of the Graneros Shale (Upper Cretaceous) in Kansas: Kansas State Geological Survey Bulletin 178, 83 p.

Meek, F. B., and Hayden, F. V., 1862, Description of new Lower Silurian (Primordial), Jurassic, Cretaceous, and Tertiary fossils, collected in Nebraska by the exploring expedition: Proceedings of the Acadmy of Natural Sciences of Philadelphia, v. 13, p. 415–447.

Munter, J. A., Ludvigson, G. A., and Bunker, B. J., 1983, Hydrogeology and stratigraphy of the Dakota Formation in northwest Iowa: Iowa Geological

Survey Water Supply Bulletin, no. 13, 55 p.

Ravn, R. L., 1981, Preliminary observations on the palynology of upper Dakota Formation lignites in northwest Iowa and northeast Nebraska: Iowa Geological Survey Guidebook Series, no. 4, p. 123–127.

Rich, M. A., 1975, Foraminifera of the Graneros and Greenhorn Formation (Upper Cretaceous) from one exposure near Sioux City, Iowa [M.S. thesis]: Ames, Iowa State University, 114 p.

Tester, A. C., 1931, The Dakota stage of the type locality: Iowa Geological Survey Annual Report, v. 35, p. 197–332.

Witzke, B. J., 1981, Cretaceous vertebrate fossils of Iowa and nearby areas of Nebraska, South Dakota, and Minnesota: Iowa Geological Survey Guidebook Series, no. 4, p. 105–122.

Witzke, B. J., Ludvigson, G. A., Poppe, J. R., and Ravn, R. L., 1983, Cretaceous paleogeography along the eastern margin of the Western Interior Seaway, Iowa, southern Minnesota, and eastern Nebraska and South Dakota, in Reynolds, M. W., and Dolly, E. D., eds., Mesozoic paleogeography of the west-central United States: Denver, Rocky Mountain Section, Society of Economic Paleontologists and Mineralogists, p. 225–252.

Upper Ordovician Maquoketa Formation in the Graf Area, eastern Iowa

Brian J. Witzke and Brian F. Glenister, *Iowa Geological Survey and University of Iowa, Department of Geology, Iowa City, Iowa 52242*

LOCATION

James Hall (1858) was the first to recognize that a shale-dominated interval, the Hudson River Group, separates the Galena Limestone (Middle–Upper Ordovician) from the overlying Niagara Limestone (Silurian) in eastern Iowa. Hall's primary exposure of these rocks was located in the Little Maquoketa River Valley near present-day Graf, Iowa. There he observed "orthoceratite beds," graptolitic shales, and an abundance of "extremely small" fossils. Charles White (1870) named this sequence of bluish and brownish shales the Maquoketa Shales after "typical localities on the Little Maquoketa River about twelve miles westward from Dubuque" (i.e., near Graf). Joseph James (1890) described a railroad cut exposed in 1886 near the Graf station, which he termed the "typical locality of the shales in Iowa." This same section is accessible today, but has been expanded to the northwest by subsequent road construction. Additional descriptions of the Graf section are given by Calvin and Bain (1900), Thomas (1914), and Tasch (1955).

The Graf section (S½NW¼SW¼Sec.29,T.89N.,R.1E., Dubuque County) is recognized as the type locality of the Upper Ordovician Maquoketa Formation, although the bulk of the Maquoketa Shale sequence in the area is represented by covered slopes. Access to this section is most convenient from the south on gravel roads that begin 0.5 mi (0.8 km) west of the intersection of Old U.S. 20 and County Road Y21 (Figs. 1, 2). The Maquoketa Formation throughout the outcrop belt of northeast Iowa is expressed geomorphically as covered slopes to rolling surfaces, commonly row-cropped, between the steeper slopes and cliffs of the Galena Group below and Silurian dolomites above. Silurian rocks form the upland bedrock surface in central Dubuque County at elevations between about 970 and 1150 ft (295 and 350 m) (the Mississippi River at Dubuque is 592 ft [180 m]). The Silurian outcrop edge is marked by a prominent bedrock escarpment, commonly wooded, along the valley walls (Fig. 1).

MAQUOKETA STRATIGRAPHY

Upland well penetrations indicate that the total Maquoketa sequence near Graf (Fig. 3) approximates 235 ft (72 m). The formation in the area is subdivided into three units (Brown and Whitlow, 1960; Bunker and others, 1985) in ascending order: 1) the brown shaly unit, a dolomite and shale facies of the Elgin Member that correlates with most of the Scales Formation of the Maquoketa Group in northwestern Illinois (Kolata and Graese, 1983), and encompasses the entire exposure at Graf; 2) a greenish gray, dolomitic, shale-dominated interval informally termed

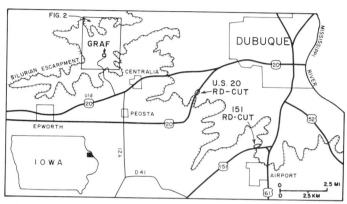

Figure 1. Map of the Dubuque–Graf area showing selected highways and locations of upper Brainard Shale localities discussed in text. Hachured line marks position of Silurian edge.

"Clermont-Brainard undifferentiated"; and 3) the Neda Member, a red to green shale, locally with ironstone ooids.

The Maquoketa overlies, apparently conformably, a prominent hardground or corrosion surface at the top of the Dubuque Formation (Galena Group). The basal 1 to 4 ft (30 to 120 cm) of the Maquoketa in the area is characterized by two or three thin phosphorites interbedded with brown shale or phosphatic dolomite. The phosphorites contain phosphatic nodules and molds of diminutive fossils in a groundmass dominated by small (<0.5 mm), concentrically-laminated, "discoidal," apatite pellets (Brown, 1966, 1974). The silty and argillaceous phosphorites are dolomitic to varying degrees and commonly are cemented by pyrite in unoxidized sections. The basal Maquoketa phosphatic interval has been termed the "depauperate zone" (Ladd, 1929), a misnomer referring to the diminutive nature of the contained fossils. The basal sequence is not exposed at the Graf roadcut, but is seen on nearby private land and at numerous localities in northeast Iowa.

The base of the Graf section begins within a brown shale unit that overlies the "depauperate zone" (units 1–3, Fig. 3). This interval is about 13 ft (4 m) thick at Graf. The medium to dark brown-black, organic shales are in part pyritic and contain scattered to abundant graptolite fragments (*Orthograptus truncatus peosta, Climacograptus typicalis putillus*; see Herr, 1971). The shales are nonlaminated in the lower half of unit 1, but become partly laminated above (units 1–3, Fig. 3). Inarticulate brachiopods are common along some laminae (small *Leptobolus* and larger lingulids to 2 cm). Thin phosphatic beds with diminutive

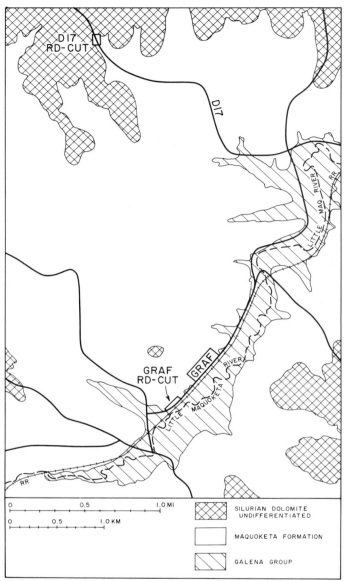

Figure 2. The area around Graf showing location of roadcuts, county roads, and general bedrock geology.

molluscan faunas similar to those in the basal Maquoketa occur in unit 2 (Fig. 3).

The upper part of the exposure at Graf (units 4–19, Fig. 3) is an interbedded sequence of argillaceous phosphatic dolomite and brown to black shale. The first prominent dolomite bed (unit 4, Fig. 3) contains abundant apatite discoids (<1 mm) and small molluscan fossil molds with scattered phosphatic clasts (to 5 mm); the discoidal pellets and fossils are packed in the upper half. This is the most phosphatic bed at the Graf exposure, containing up to 60% apatite by weight (Bromberger, 1965).

Additional beds of argillaceous phosphatic dolomite are present higher in the section (units 6, 10–13, 17–19), but are

notably less phosphatic than unit 4. These lack discoidal apatite pellets (except unit 6), but contain scattered to abundant phosphatic molds of diminutive fossils easily recovered in acid residues. Most fossils are less than 3 mm in size, but high-spired gastropods reach 20 mm in some beds. The diminutive faunas are molluscan-dominated and typically contain an abundance of conical shells referred to the scaphopod *Plagioglypta iowaensis* by Bretsky and Bermingham (1970). A variety of small gastropods are present; following the form taxonomy of Harrison and Harrison (1975) most are referable to *Bucanella, Liospira, Cyrtolites, Loxoplocus, Holopea, Murchisonia, Cyclora,* and others. Additional molluscan elements include common hyolithids, infaunal bivalves (*Nuculites, Palaeoneilo, Palaeoconcha*), polyplacophoran plates (*Septemchiton*), and small nautiloids (mostly *Isorthoceras*). Ostracode molds (*Primitia, Milleratia,* etc.) and small inarticulates (*Leptobolus*) generally are associated. Small bryozoans molds (trepostomes, cryptostomes) and phosphatized pelmatozoan debris (primarily crinoid, some cystoid) are present in most beds. Additional fossils include small articulate brachiopod molds (*Diceromyonia, Zygospira*), trilobites, starfish plates (units 17–19, Fig. 3), *Hindia* sponges (primarily unit 6), rare conularids, three-dimensional graptolites (*Orthograptus*), conodonts, and scolecodonts.

The Graf exposure is renowned for the exceptional accumulation of orthoconic nautiloids, especially in units 7 and 9 (Fig. 3). These latter nautiloid-rich beds are slightly argillaceous to argillaceous, microcrystalline to finely crystalline dolomites that are slightly to highly phosphatic. The nautiloids average about 0.7 in (2 cm) in diameter and occur in packed concentrations variably displaying parallel or random orientations. The shells typically are unbroken, but some individuals are telescoped into other shells penetrating fractured septa (Miller and Youngquist, 1949). The abundant nautiloids are overwhelmingly dominated by a single species, *Isorthoceras sociale*; a few other nautiloid taxa are noted (Miller and Youngquist, 1949). Many specimens of *I. sociale* display a lustrous sheen that resembles pearly aragonite. However, the original shell material, including the nacreous layer, was "secondarily phosphatized at an early diagenetic stage" preserving the "original ultrastructure" (Mutvei, 1983). This remarkable form of preservation adds further significance to the Graf exposure.

Unlike other carbonate beds in the sequence, the nautiloid-rich intervals at Graf lack abundant diminutive fossils (except ostracodes). However, the nautiloid beds contain rare additional macrofaunal elements represented by phosphatic molds 5 to 20 mm in diameter. These include gastropods (*Murchisonia, Liospria*), bivalves (*Nuculites*), and brachiopods (*Diceromyonia*). Normal-sized cranidia and pygidia of the trilobite, *Thelecalymene mammilata,* are not uncommon in the nautiloid beds (Whittington, 1971); the phosphatized exoskeletons are finely granulated and preserve original shell ultrastructure (Mutvei, 1981).

Carbonate beds at Graf are interbedded with brown to black shales. The shales commonly are laminated and contain scattered graptolite debris (three-dimensional *Orthograptus* common in

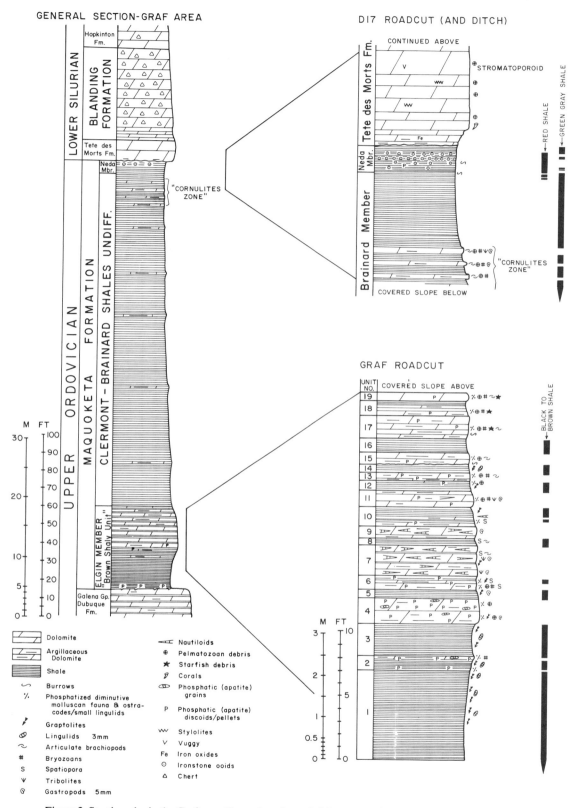

Figure 3. Stratigraphy in the Graf area. General section at left is a composite of outcrop sections and well penetrations in the area. Enlarged sections at right from measured descriptions in the Graf area (1978–1983).

unit 6, Fig. 3). Compressed nautiloids in some shale beds (units 6, 7, 8, 10) display finely reticulated patterns over their entire outer surfaces, which have been interpreted as encrusting bryozoans (*Spatiopora*). The shales contain phosphatized diminutive molluscan faunas in the more dolomitic portions. Large compressed trilobites (*Anataphrus*) have been found in unit 10 (Fig. 3). The shaly beds lose the brown and black hues high in the Graf section (unit 18, Fig. 3).

Skeletal material is phosphatized through most of the Graf sequence. However, dolomitized wackestones and packstones, primarily echinoderm debris and *Plagioglypta,* are seen in some of the upper beds (units 11, 15, 17, 19; Fig. 3). The uppermost bed (unit 19) includes dolomitized echinoderm-bryozoan wackestones and packestones/grainstones, in part cross-bedded. Burrowing is absent through much of the Graf section, but bioturbated facies become noteworthy upward in the sequence (units 14–18, Fig. 3).

The Maquoketa shale sequence above the Graf exposure (Clermont-Brainard undifferentiated) typically forms a covered slope throughout the area. This interval, as seen in core, is dominated by greenish gray dolomitic shale with scattered argillaceous dolomite beds. It is poorly fossiliferous, but some phosphatized diminutive molluscan faunas and trilobites occur. Near the top of this thick sequence the shales are interbedded with highly fossiliferous dolomite beds; this fossiliferous interval, about 10 to 30 ft (3 to 9 m) thick, has been termed the "*Cornulites* zone" of the upper Brainard Shale by Ladd (1929). Dolomitized skeletal wackestone and packstone/grainstone beds, some with abraded grains or graded bedding, contain an abundant fauna. Siderite void fills are common in some beds. Poor exposures of "*Cornulites* zone" strata (and associated float) can be seen in the ditches and slopes along County Road D17 north of Graf (near D17 roadcut, Fig. 2; NW¼NE¼Sec.18,T.89N.,R.1E.). Better exposures are accessed easily along U.S. 20 west of Dubuque (U.S. 20 roadcut, Fig. 1; SW¼NE¼Sec.6,T.88N.,R.2E.) and U.S. 151 south of Dubuque (151 roadcut, Fig. 1; SE¼NE¼Sec.22,T.88N.,R.2E.).

The abundant "*Cornulites* zone" fauna contains a diversity of taxa in Dubuque County. Sowerbyellid brachiopods (*Eoplectodonta* or "*Thaerodonta*") commonly dominate (except at the U.S. 20 roadcut). Other brachiopods include orthids (*Plaesiomys, Austinella, Diceromyonia, Onniella, Platystrophia*), spiriferids (*Zygospira*), strophomenids (*Megamyonia, Strophomena, Holtedahlina, Tetraphalerella*), and rhynchonellids (*Lepidocyclus, Hypsiptycha*). Bryozoans are abundant locally (numerous trepostome genera with rarer cyclostomes and cryptostomes). Other fossils include echinoderm debris, solitary corals (*Helicelasma*), tapering worm-like tubes (*Cornulites*), gastropods, epifaunal bivalves (*Pterinea, Ambonychia*), nautiloids, and trilobites (*Flexicalymene, Ceraurus,* etc.). Phosphatized diminutive molluscan faunas are recovered in acid residues from some beds.

The uppermost unit of the Maquoketa, the Neda Member, can be found at localities in Dubuque County only where the Maquoketa is thickest, as pre-Silurian erosion removed the Neda over most of the area. A good section of the Neda was exposed

beneath the Silurian dolomites during 1981–1982 road construction along County Road D17 (D17 roadcut, Fig. 2; SE¼-SE¼NW¼NE¼Sec.18,T.89N.,R.1E.); the section is still accessible but trenching may be needed. The Neda is about 2 ft (0.6 m) thick at this locality (Fig. 3) and contains scattered to abundant flattened ironstone ooids about 1 mm in diameter in a red-gray claystone to mudstone matrix. The ooids are composed dominantly of goethite, but laminae of apatite (especially outer cortex) and chamosite are present. Phosphatic clasts (1–5 cm) occur with the ooids in the clay-rich matrix. Horizontal and vertical burrows, some filled with ooids, have disrupted the sediment. The Neda contains some thin green-gray shales, and red mottling is noted in the uppermost Brainard. The upper Brainard shales above the "*Cornulites* zone" are unfossiliferous and plastic when wet.

SILURIAN STRATIGRAPHY

Silurian strata unconformably overlie an eroded Maquoketa surface that displays up to 100 ft (30 m) of relief in Dubuque County (Whitlow and Brown, 1963). Although Silurian rocks are not considered in detail here, the general stratigraphic sequence in the area is noteworthy. Silurian strata are accessible at many localities in the area; a good sequence along U.S. 151 progresses westward from the Brainard Shale roadcut noted previously (Fig. 1). The basal Silurian formation, the Mosalem, filled topographic depressions on the Maquoketa surface and varies accordingly from 0 to 100 ft (30 m) in thickness. It is characterized by thinly bedded argillaceous to shaly dolomite. The Mosalem is absent where the Maquoketa is thickest (e.g., the D17 roadcut, Figs. 2, 3). Overlying Silurian units in the area include, in ascending order (Bunker and others, 1985): 1) Tete des Morts Formation, a thick-bedded cliff-forming dolomite with common corals and stromatoporoids that overlies the Maquoketa in places to a maximum thickness of 24 ft (7 m); 2) Blanding Formation, dominated by medium-bedded very cherty dolomite with common corals and stromatoporoids, which averages 55 ft (17 m) in thickness; and 3) Hopkinton Formation, a fossiliferous dolomite and cherty dolomite sequence with abundant *Pentamerus* in the middle part and is about 130 ft (40 m) thick.

MAQUOKETA DEPOSITION

The Graf roadcut exposes a rock sequence that, in many respects, is atypical of the lower Maquoketa regionally. At no other locality is the development of nautiloid beds and phosphatic dolomites so pronounced. Graf occupies a unique facies position within the Maquoketa outcrop belt, separating limestone-dominated strata of the Elgin Member to the northwest and brown shale-dominated strata of the Scales Formation to the southeast (Fig. 4). Trilobite faunas and other normal-sized shelly benthos inhabited environments to the north during Elgin deposition, whereas the shales to the south were deposited under generally low-oxygen to anoxic bottom conditions. The transition between these facies at Graf marks the position where low-

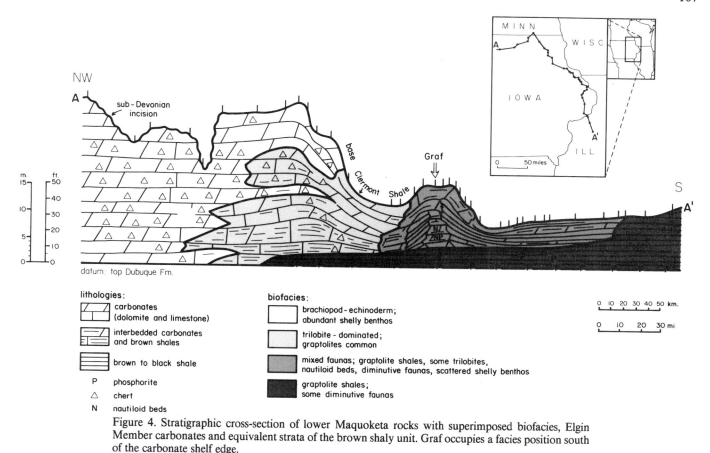

Figure 4. Stratigraphic cross-section of lower Maquoketa rocks with superimposed biofacies, Elgin Member carbonates and equivalent strata of the brown shaly unit. Graf occupies a facies position south of the carbonate shelf edge.

oxygen, nutrient-rich waters may have impinged or upwelled (Brown, 1974) along the carbonate shelf edge. The surface waters in such a setting may have been highly productive, permitting an abundance of nautiloids and graptolites to flourish. Bottom environments at Graf, by contrast, probably were subjected to conditions varying between anoxic (laminated black shales) and oxygenated. The diminutive molluscan faunas, apparently dominated by true paedomorphic individuals (Snyder and Bretsky, 1971), probably were uniquely adapted for life in environments exposed to recurrent oxygen stresses (Witzke, 1980; Kolata and Graese, 1983). "During times of high influx of oxygen-poor, phosphate-rich water, the benthic faunas presumably were killed and their skeletons replaced by phosphate. When oxygenated waters returned, even for short periods, specially adapted opportunistic species were able to establish themselves" (Kolata and Graese, 1983).

Previous workers (Miller and Youngquist, 1949; Tasch, 1955) suggested that the nautiloid beds at Graf were deposited in shallow, nearshore environments by violent wave action. The unabraded character and poor sorting of the contained grains as well as the general facies position of the strata would tend to preclude such an interpretation. Instead, the bulk of the Graf

sequence probably was deposited below wave base in oxygen-stressed environments. Dense concentrations of nautiloids accumulated on the sea floor during deposition of the nautiloid beds, perhaps when episodic mass mortalities occurred in the surface waters. Periodic failures of gas-filled phragmocones of newly-settled *Isorthoceras* shells may have resulted in implosion under depth-dependent hydrodynamic pressure. The violent entry of water into the voids resulting from such failures may have caused chain reactions that nested or telescoped up to six shells, each in close proximity and oriented by gentle bottom currents. Elgin Member strata at Graf record a general shallowing-upward sequence: 1) "basinal" anoxic shales in the lower part, 2) shales and carbonates in the middle part apparently deposited near the interface between anoxic and oxic water masses, and 3) packstones and grainstones in the upper part deposited in more agitated and shallower-water environments (cross-bedded at top).

The largely covered Clermont-Brainard interval lacks shelly benthos through most of the sequence. Much of this interval probably was deposited in relatively deep, low-oxygen environments (Kolata and Graese, 1983). The Clermont-Brainard interval in parts of northern Illinois and northeast Iowa is split in the middle part by a shallowing-upward carbonate sequence, the

Fort Atkinson Limestone Member, but this limestone facies is absent in Dubuque County. Burrowing becomes more pronounced upward in the Maquoketa Shale sequence of the Graf area. The presence of abundant shelly benthos in the "*Cornulites* zone" near the top indicates deposition in fully oxygenated, normal-marine settings. These factors document a general shallowing-upward sequence in the upper Brainard. The overlying Neda represents the final phases of Maquoketa deposition in the area. Flattened ironstone ooids were reworked by burrowers and incorporated into phosphatic clasts during Neda deposition (Witzke and Heathcote, 1983). The ironstone ooids apparently are not the product of later diagenetic replacement of carbonate, and their composition precludes an origin by lateritic weathering.

The Maquoketa shales represent the western extension of a vast clastic wedge derived from distant Taconic (Appalachian) sources (Witzke, 1980). The Maquoketa Formation spans much of the Late Ordovician, ranging in age from mid-Maysvillian through Richmondian. Maquoketa strata were eroded following the withdrawal of the sea near the close of the Ordovician. The eroded Maquoketa surface in Dubuque County was buried by carbonates as Early Silurian seas transgressed westward into Iowa.

REFERENCES

Bretsky, P. W., and Bermingham, J. J., 1970, Ecology of the Paleozoic scaphopod genus *Plagioglypta* with special reference to the Ordovician of eastern Iowa: Journal of Paleontology, v. 44, p. 908–924.

Bromberger, S. H., 1965, Mineralogy and petrology of basal Maquoketa (Ordovician) phosphatic beds [M.S. thesis]: Iowa City, University of Iowa, 159 p.

Brown, C. E., 1966, Phosphate deposits in the basal beds of the Maquoketa Shale near Dubuque, Iowa: U.S. Geological Survey Professional Paper 550-B, p. B152–B158.

—— , 1974, Phosphatic zone in the lower part of the Maquoketa Shale in northeastern Iowa: U.S. Geological Survey Journal of Research, v. 2, p. 219–232.

Brown, C. E., and Whitlow, J. W., 1960, Geology of the Dubuque South Quadrangle, Iowa–Illinois: U.S. Geological Survey Bulletin 1123A, p. 1–93.

Bunker, B. J., Ludvigson, G. A., and Witzke, B. J., 1985, The Plum River Fault Zone and the structural and stratigraphic framework of eastern Iowa: Iowa Geological Survey Technical Information Series, no. 13, 126 p.

Calvin, S., and Bain, H. F., 1900, Geology of Dubuque County: Iowa Geological Survey Annual Report, v. 10, p. 381–622.

Hall, J., 1858, Report of the Geological Survey of the State of Iowa: v. 1, 472 p.

Harrison, W. B., III, and Harrison, L. K., 1975, A Maquoketa-like molluscan community in the Brassfield Formation (Early Silurian) of Adams County, Ohio: Bulletins of American Paleontology, v. 67, p. 193–234.

Herr, S. R., 1971, Biostratigraphy of the graptolite-bearing beds of the Upper Ordovician Maquoketa Formation, Iowa [M.S. thesis]: University of Iowa, 161 p.

James, J. F., 1890, On the Maquoketa shales and their correlation with the Cincinnati Group of southwestern Ohio: American Geologist, v. 5, p. 335–356.

Kolata, D. R., and Graese, A. M., 1983, Lithostratigraphy and depositional environments of the Maquoketa Group (Ordovician) in northern Illinois: Illinois State Geological Survey Circular 528, 49 p.

Ladd, H. S., 1929, The stratigraphy and paleontology of the Maquoketa Shale of Iowa: Iowa Geological Survey Annual Report, v. 34, p. 307–448.

Miller, A. K., and Youngquist, W., 1949, The Maquoketa coquina of cephalopods: Journal of Paleontology, v. 23, p. 199–204.

Mutvei, H., 1981, Exoskeletal structure in the Ordovician trilobite *Flexicalymene:* Lethaia, v. 14, p. 25–234.

Mutvei, H., 1983, Flexible nacre in the nautiloid *Isorthoceras,* with remarks on the evolution of cephalopod nacre: Lethaia, v. 16, p. 233–240.

Snyder, J., and Bretsky, P. W., 1971, Life habits of diminutive bivalve molluscs in the Maquoketa Formation (Upper Ordovician): American Journal of Science, v. 271, p. 227–251.

Tasch, P., 1955, Paleoecologic observations on the orthoceratid coquina of the Maquoketa at Graf, Iowa: Journal of Paleontology, v. 29, p. 510–518.

Thomas, A. O., 1914, A new section of the railway cut near Graf, Iowa: Proceedings, Iowa Academy of Science, v. 21, p. 225–229.

White, C. A., 1870, Report on the Geological Survey of the State of Iowa: Des Moines, v. 1, 381 p.

Whitlow, J. W., and Brown, C. E., 1963, The Ordovician–Silurian contact in Dubuque County, Iowa: U.S. Geological Survey Professional Paper 475C, p. C11–C13.

Whittington, H. B., 1971, A new calymenid trilobite from the Maquoketa Shale, Iowa: Smithsonian Contributions to Paleobiology, no. 3, p. 129–136.

Witzke, B. J., 1980, Middle and Upper Ordovician paleogeography of the region bordering the Transcontinental Arch, *in* Fouch, T. D., and Magathan, E. R., eds., Paleozoic paleogeography of the west-central United States: Rocky Mountain Paleogeography Symposium 1, Rocky Mountain Section, Society of Economic Paleontologists and Mineralogists, p. 1–18.

Witzke, B. J., and Heathcote, R. C., 1983, The distribution, composition, and deposition of the Upper Ordovician Neda Member in Iowa: Geological Society of America Abstracts with Programs, v. 15, p. 222.

Silurian Carbonate Mounds, Palisades-Kepler State Park, Iowa

Brian J. Witzke, Iowa Geological Survey, Iowa City, Iowa 52242

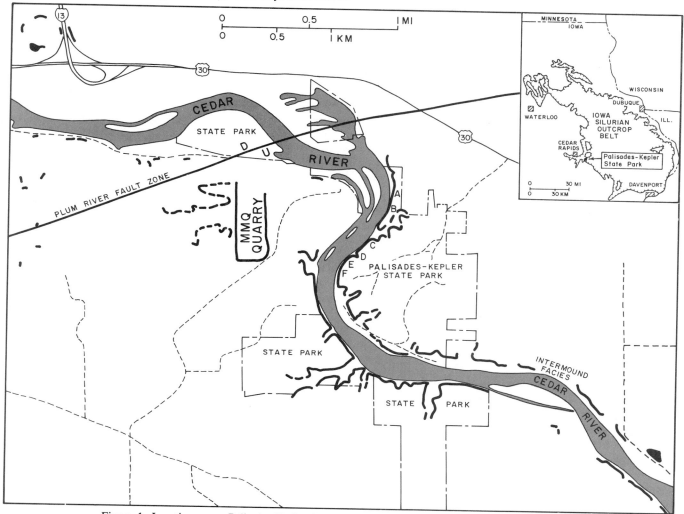

Figure 1. Location map, Palisades-Kepler State Park and surrounding area, Linn County, Iowa. Secondary roads are dashed; bedrock outcrop shown in black; localities A through G correspond to mound crest locations of Philcox (1970).

LOCATION

Silurian strata are exposed at numerous localities in east-central and northeast Iowa (Fig. 1), where the total Silurian sequence approaches 500 ft (150 m) in thickness. The bulk of the sequence is dominated by dolomite and cherty dolomite characteristically lacking a notable siliciclastic component. The Silurian carbonates are pervasively dolomitized, except along portions of the northern and northwestern margins of the outcrop belt, where limestones occur. The Silurian (early Llandoverian–Ludlovian) dolomite sequence is divided into six formations (Bunker and others, 1985), and a variety of fossil associations characterize the Silurian sequence (Witzke, 1983).

The youngest Silurian rocks in Iowa are exposed in the southern region of the outcrop belt (Fig. 1) and are assigned, in ascending order, to the Scotch Grove and Gower formations. Strata encompassing a part of the Scotch Grove–Gower interval are well exposed and publicly accessible in the valley of the Cedar River in and around the area of Palisades-Kepler State Park (Fig. 1), about 10 mi (16 km) southeast of downtown Cedar Rapids and 4 mi (6.5 km) west of Mt. Vernon. The Cedar River has cut through Silurian carbonate mound and intermound facies in the area, displaying a series of exposures along forested slopes and bold cliffs, in part overhanging, that form the scenic palisades. The cliffs reach heights to 65 ft (20 m) in places. Hiking trails in the state park provide access to many of the exposures. The most spectacular and instructive exposures encompass localities A through F (Fig. 1).

The woodland and river-bottom settings in the area are home to a variety of plants and animals. Be forewarned that stinging nettles are common in the valley bottoms. Woodland Indian burial mounds can be seen in the uplands. Camping and picnicking facilities are available in the park. A modest visitor's fee is payable at the park headquarters. Rock hammers and specimen collecting are not permitted in the state park.

The area around Palisades-Kepler State Park is located near the western terminus of the Plum River Fault Zone (Fig. 1), which trends for 108 mi (180 km) across parts of northwestern Illinois and eastern Iowa (Bunker and others, 1985). The fault zone is not well exposed in the Palisades-Kepler area, but limited exposures to the west of the park as well as gravity-magnetic profiling have been used to interpret the structure (Bunker and others, 1985). The north or downthrown side of the fault exposes Middle Devonian strata preserved in a plunging syncline. Devonian carbonates of the Otis Formation can be seen at the Iowa 13-U.S. 30 interchange (Fig. 1). Silurian strata are exposed south of the fault zone, which are upthrown approximately 250 ft (75 m) with respect to the downthrown edge. Mississippi Valley–type sulfide mineralization is noted south of the fault at the Martin-Marietta Quarry (MMQ, Fig. 1).

SCOTCH GROVE FORMATION

The Scotch Grove Formation reaches thicknesses to about 295 ft (90 m) in eastern Iowa. The uppermost part of the formation is exposed in the Palisades-Kepler area, where it is characterized by both carbonate mound and intermound facies, assigned respectively to the Palisades-Kepler Member and Buck Creek Quarry Member. Strata presently assigned to the Palisades-Kepler Member were classified by most previous workers as the LeClaire facies of the Gower Formation. As shown by Philcox (1970, 1972) and Witzke (1981, 1983), crinoidal carbonate mounds at Palisades-Kepler State Park do not correlate with typical laminated dolomites of the Gower Formation, but occur consistently below them. As such, this crinoidal mound facies was removed from the Gower Formation and formally named as a member in the upper Scotch Grove Formation (Bunker and others, 1985). Exposures along the east bank of the Cedar River in the state park (Locations A–F, Fig. 1) form the type locality of the Palisades-Kepler Member.

The Palisades-Kepler Member at the state park encompasses a fascinating series of coalesced carbonate mounds. The mounded complex is over 1.2 mi (2 km) in diameter. Beds dip at varying angles (<45°) from the tops of six or more mound crests (Fig. 2). Bedding approximates horizontal through much of the complex. The member is characterized by very crinoidal dolomite, displaying skeletal wackestone to packstone fabrics (rare grainstone). Although the mounds have been termed "reefs" by many workers, the general absence of skeletal boundstone fabrics precludes their interpretation as organic-framework reefs. Crinoid

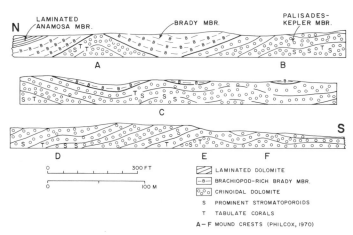

Figure 2. Generalized cross-section of exposures along east side of Cedar River in Palisades-Kepler State Park (adapted from Philcox, 1970). Localities A through F shown on Figure 1. Datum is Cedar River. Slight vertical exaggeration.

debris (either moldic or dolomitized) is abundant throughout, which should be apparent to even the causal observer. Crinoid stems reach diameters to 0.8 in (2 cm) but most are smaller. Articulated cups of crinoids (especially *Eucalyptocrinites*) and cystoids (*Caryocrinites*) are seen occasionally. Additional skeletal material is noted, but is much less common than the crinoid debris. Individual colonies (most less than 10 in; 25 cm) of tabulate corals (primarily *Favosites*) and stromatoporoids, in part overturned, are conspicuous in places, but do not form an organic framework. Solitary rugosans are not uncommon. Scattered brachiopods (e.g., *Atrypa*, rhynchonellids, pentamerids), and bryozoans are common in places. Gastropods, nautiloids, ischaditid algae, and trilobites can also be seen.

Although echinoderm debris forms the major skeletal constituent in the mounds of the Palisades-Kepler Member, the more abundant matrix material apparently represents dolomitized carbonate mud. Relict fibrous submarine cements are observed in thin section. Graded and wedge-shaped beds, apparently debris flows, are observed in the more steeply dipping portions of some mounds (especially mounds A and B, Fig. 2). Dense accumulations of nautiloids and trilobites (especially *Bumastus*) are present within discrete pockets in a few places within the mounds (especially mound A, Fig. 2). Philcox (1970) interpreted these as pene-contemporaneous cave and fissure fills developed within the lithified mounds. The Palisades-Kepler Member is overlain by the Brady Member of the Gower Formation, and the sharp contact is well displayed around mounds A and B (Fig. 2).

As one progresses downstream from the main mounded complex at the state park, the strata become horizontally bedded and progressively less skeletal. The series of nearly continuous exposures displays the transition between mound and intermound facies. Unfortunately, the region labelled "intermound facies" on Figure 1 lies outside the park boundary on private land. The

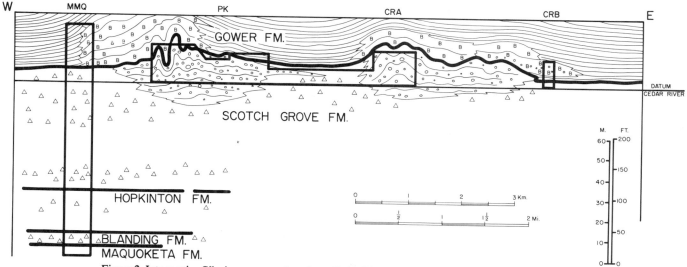

Figure 3. Interpretive Silurian cross-section along Cedar River in vicinity of Palisades-Kepler State Park (from Witzke, 1981). Localities MMQ through CRA shown on Figure 1 (PK = state park). Locality CRB on south bank of Cedar River about 4.5 km from state park. Boxed areas correspond to outcrop or core sections; remainder is reconstructed. Symbols as in Figure 2; triangles represent cherty strata.

intermound facies is characterized by sparsely skeletal-moldic finely crystalline dolomite, in part with common chert nodules. Small echinoderm debris, scattered brachiopods (*Atrypa, Dalejina,* etc.), small cup corals, and bryozoans are noted. Similar rocks can be seen in the lower quarry walls at locality MMQ (Fig. 1). The intermound facies is assigned to the Buck Creek Quarry Member of the Scotch Grove Formation. Further downstream from the area labelled intermound facies (Fig. 1), a second mounded complex of the Palisades-Kepler Member is encountered at locality CRA (Figs. 1, 3). The Buck Creek Quarry Member is overlain by the Gower Formation, primarily the laminated Anamosa Member.

GOWER FORMATION

The Gower is the youngest Silurian formation recognized in Iowa. The formation includes two members in the Palisades-Kepler area: 1) a skeletal-rich mound facies, the Brady Member, and 2) an unfossiliferous laminated carbonate facies in the intermound position, the Anamosa Member. Both of these members contrast markedly with mound and intermound facies in the underlying Scotch Grove Formation. Carbonate mounds of the Brady Member are well exposed at localities A and B (Figs. 1 and 2), which represent a second stage of mound building above the mounds of the Palisades-Kepler Member. Unlike the highly crinoidal mound facies of the Palisades-Kepler Member, the Brady mounds typically lack a notable crinoidal component. However, the Brady Member does contain a phenomenal abundance of skeletal material, primarily molds and dolomitized shells of small brachiopods (about 0.4 in; 1 cm in size). Although extremely abundant, the Brady brachiopod faunas are of relatively

low diversity, dominated by rhynchonellids (*Ancillotoechia*) and athyrids (*Protathyris*). The athyrids commonly display well-preserved dolomitized spiralia internally. Solitary rugosans (*Fletcheria*) are also common in the Brady Member, and small favositids are presnt. The basal portion of the Brady Member contains a slightly more diverse fossil assemblage, still dominated by *Ancillotoechia* and *Protathyris*, but including additional rhynchonellid (uncinulids), spiriferid (*Reticulatrypa*, meristellids, *Spirinella*), and strophomenid (*Fardenia, Leptaena*) brachiopods as well as gastropods, bivalves (*Plethomytilus, Cyrtodonta*), and nautiloids, Beds in the Brady Member dip at varying angles; some are wedge-shaped and apparently represent debris flows off the mounds.

The fossiliferous Brady Member interfingers down-dip with, and its replaced laterally by, unfossiliferous laminated dolomites of the Anamosa Member. These relations are displayed north of mound A as well as at the MMQ quarry (Figs. 1, 2, 3). The laminations are laterally continuous on outcrop, and the member primarily is horizontally bedded. Although the Anamosa Member is characteristically unfossiliferous, enigmatic rod-shaped fossils (0.4 in; 1 cm) occur on some bedding surfaces, and brachiopods and other fossils are seen locally in the vicinity of the Brady Member mounds. The Anamosa Member overlies intermound facies of the upper Scotch Grove Formation (as seen at intermound facies and MMQ localities, Fig. 1).

DEPOSITION

The abundance of fossil organisms in the Scotch Grove Formation that are generally regarded as stenohaline, especially echinoderms, suggests that stable marine salinities were maintained in eastern Iowa throughout deposition of the formation. Once initiated, carbonate mound building in the upper Scotch Grove probably was a self-perpetuating process; the deposition of abundant crinoid, coral, stromatoporoid, and other grains provided suitable attachment sites for the next generation of mound-building skeletal organisms. As skeletal debris and mud accumulated, the mounds became elevated above the surrounding sea floor, apparently providing more hospitable environments for many organisms in the shallower mound settings than were available in the intermound areas. The absence of a boundstone framework generally precludes deposition in highly agitated environments, and vertical growth of mounds in the Palisades-Kepler Member probably was limited by the position of normal fair-weather wave base. Nevertheless, periodic water turbulence, probably generated during storm events, exerted considerable influence on mound deposition. Wedge-shaped debris flows, periodic sediment removal, and varying sedimentation rates in the mounds probably relate to episodic high-energy conditions across the mound crests (Philcox, 1971; Witzke, 1983). The mounds apparently became rigid features on the sea floor primarily through submarine cementation processes. The pervasive submarine cementation required movement of large volumes of water through the mounds. Fissures developed in the rigid mounds, locally serving as traps for trilobite molts and nautiloid shells.

The intermound areas of the Scotch Grove Formation were sites dominated by carbonate mud deposition, but skeletal remains indicate that benthic conditions remained suitable for a variety of stenohaline organisms, primarily smaller or thinner-shelled forms than found in the mounds. However, the onset of Gower deposition was marked by the widespread appearance of unfossiliferous laminated carbonate sediments (Anamosa Member) over much of east-central Iowa. The depositional environment of the laminated Gower is interpreted as one of quiet subtidal conditions of restricted circulation and high salinity (Witzke, 1983). The dramatic change in sedimentation and paleontology marked at the Scotch Grove–Gower contact probably reflects a major change in water circulation patterns in the area during deposition. A drop in sea level during the mid to late Wenlockian apparently left east-central Iowa as a restricted embayment of the Silurian sea, thereby cutting off open circulation with the adjacent areas. The absence of shelly and burrowing

biotas in the Anamosa Member underscores the unsuitable character of intermound bottom environments for benthic organisms.

Although bottom conditions during Anamosa deposition generally excluded benthic organisms, contemporaneous mound facies of the Brady Member, which were deposited in shallower environments, were sites of abundant brachiopod and coral faunas. This suggests that a water stratification was developed in eastern Iowa: 1) an upper oxygenated layer suitable for shelly benthos, and 2) a lower hypersaline water layer without benthos (Witzke, 1983). The Brady Member encompasses a second stage of carbonate mound building above the older mounds of the Palisades-Kepler Member (Fig. 2). The older mounds served as loci for the lateral and vertical expansion of skeletal-rich Brady mound facies. As in the Palisades-Kepler Member, the Brady mounds lack an organic boundstone framework and became rigid features on the sea floor primarily through submarine cementation processes (relict isopachous and botryoidal fibrous submarine cements are abundant). However, the dramatic contrast between the faunas of the crinoidal Palisades-Kepler Member and the overlying brachiopodal Brady Member indicates that the mound environments underwent significant changes between Scotch Grove and Gower deposition. The Brady faunas, while extremely abundant, are generally of low diversity and characteristically lack several normally stenohaline groups, most notably the echinoderms. The waters of the upper surface layer apparently posed stresses that tended to exclude several groups of marine organisms, and slightly elevated salinities are suggested (Witzke, 1983). In general, surface-water salinites probably increased shoreward within the restricted Gower embayment. The sinking of saline waters within the embayment, coupled with a barrier to bottom outflow near the mouth of the embayment, apparently enabled a dense hypersaline bottom layer to form across the area.

The entire Silurian sequence across most of eastern Iowa is pervasively dolomitized. Dolomitization may relate to the descent of brines within the Gower embayment and/or to the lateral and vertical movement of mixing-zone diagenetic environments across the area during offlap of the Silurian seaway. A prolonged period of erosion, spanning some 30 m.y., separated Silurian and Middle Devonian carbonate deposition in the eastern Iowa area. The Silurian sequence is bevelled beneath the Devonian sequence in the area and was erosionally removed across large areas of the Midcontinent.

REFERENCES CITED

Bunker, B. J., Ludvigson, G. A., and Witzke, B. J., 1985, The Plum River Fault Zone and the structural and stratigraphic framework of eastern Iowa: Iowa Geological Survey Technical Information Series, no. 13, 126 p.

Philcox, M. E., 1970, Geometry and evolution of the Palisades reef complex: Journal of Sedimentary Petrology, v. 40, p. 177–183.

Philcox, M. E., 1971, Growth forms and role of colonial coelenterates in reefs of the Gower Formation (Silurian), Iowa: Journal of Paleontology, v. 45, p. 338–346.

Philcox, M. E., 1972, Burial of reefs by shallow-water carbonates, Silurian Gower Formation, Iowa, U.S.A.: Geologische Rundschau, v. 61, p. 686–708.

Witzke, B. J., 1981, Silurian stratigraphy of eastern Linn and western Jones counties, Iowa: Geological Society of Iowa, Guidebook 35, 38 p.

Witzke, B. J., 1983, Silurian benthic invertebrate associations of eastern Iowa and their paleoenvironmental significance: Transactions of Wisconsin Academy of Sciences, Arts, and Letters, v. 71, p. 21–47.

Cedar Valley Formation of the Coralville Lake area, Iowa

Bill J. Bunker and Brian J. Witzke, Iowa Geological Survey, Iowa City, Iowa 52242

INTRODUCTION

Coralville Lake (Fig. 1) was authorized under the Flood Control Act of 1938 as a means of moderating stream flow on the Iowa River. Construction of the dam was begun in 1949, then delayed by the Korean conflict, and consequently not completed until 1958. Coralville Dam regulates runoff from 3,084 mi^2 (8,018 km^2) of land upstream, providing flood protection to 1,703 mi^2 (4,428 km^2) of the Iowa River Valley below the dam. The U.S. Army Corps of Engineers maintains Coralville Lake as a multiple-use project providing primary benefits in flood control and low-flow augmentation, and secondary benefits in recreation, fish and wildlife management, forest management, and water quality improvement. Lake MacBride (Fig. 1) is a subimpoundment of Coralville Lake. The lake and surrounding land comprise Lake MacBride State Park, which is managed by the Iowa Conservation Commission.

Exposures along the valley walls of that portion of the Iowa River which comprises Coralville Lake, afford one the opportunity to examine at numerous points along its extent the uppermost portion of the Wapsipinicon Formation and the entire Cedar Valley Formation. Excellent exposures of the Coralville Member of the Cedar Valley Formation can also be seen in the valley walls of the Iowa River in Iowa City (Witzke, 1984a).

STRATIGRAPHY

Wapsipinicon Formation

The Wapsipinicon Formation was named for exposures along the Wapsipinicon River in northeastern Linn County (Norton, 1895). The Wapsipinicon Formation consists of three members (Bunker and others, 1985), in ascending order, the Kenwood, Spring Grove, and Davenport. Within the Coralville Lake area, only the upper portion of the Davenport Member is exposed (stations 14 and 15, Figs. 1 and 2).

Davenport Member. The Davenport Member consists of lithographic limestone, in part laminated and extremely brecciated in the Coraville Lake area. The brecciation is considered to be the result of dissolution of gypsum-anhydrite in these strata (Bunker and others, 1985) followed by collapse of the overlying beds.

Cedar Valley Formation

The Cedar Valley Formation was originally named for the series of exposures of Devonian carbonate rocks along the valley of the Cedar River (McGee, 1891). As originally defined, it included the Devonian sequence above the Silurian unconformity and below the Hackberry Shale (Lime Creek Formation). Subse-

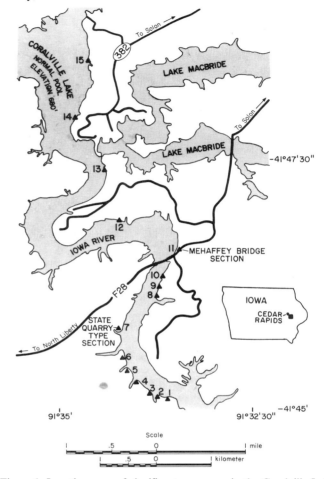

Figure 1. Location map of significant exposures in the Coralville Lake area. Stratigraphic sections exposed at selected stations are depicted in Figure 2.

quent recognition of the Wapsipinicon and Shell Rock formations restricted the Cedar Valley Formation to a position between these two formations. Although no specific type section was ever designated for the Cedar Valley Formation, a surface reference section (Conklin Quarry, Secs. 32,33,T.80N.,R.6W., approximately 2 mi (3.2 km) south of the Coralville Dam) has been established (Bunker and others, 1985) in the Johnson County, Iowa, area. This reference section has been utilized by many authors in the past as representative of the Cedar Valley Formation in east-central Iowa. However, access to this reference section is limited, occurring on private property. Fortunately the entire Cedar Valley Formation as represented in the reference section is exposed in the valley walls of Coralville Lake.

The Cedar Valley Formation, as presently defined (Witzke

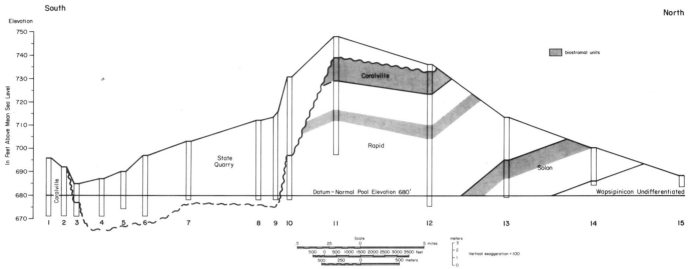

Figure 2. Simplified north-south cross-section of the Cedar Valley Formation in the Coralville Lake area. Station locations are noted on the map in Figure 1. Shaded portions correspond to coralline or biostromal strata.

and Bunker, 1984; Bunker and others, 1985), consists of four members, in ascending order, the Solon, Rapid, Coralville, and State Quarry. The type sections of these four members are located in or near the vicinity of the Iowa River Valley in Johnson County.

Solon Member. The Solon consists predominantly of dense, thick-bedded, fine-grained, skeletal calcarenite, averaging about 20 ft (6.1 m) thick in the Coralville Lake area. The base of the Solon is characterized by thin discontinuous transgressive sandstones and/or arenaceous carbonates, which extend downward into brecciated carbonates of the upper part of the Wapsipinicon Formation (these relationships can be observed in exposures between stations 13 and 14, Figs. 1 and 2). Stainbrook (1941) divided the Solon into two subunits based on faunal content: (1) the lower "*independensis* zone," which derives its name from the common atrypid brachiopod *Desquamatia independensis*; and (2) the upper "*profunda* zone," which derives its name from the colonial rugose coral *Hexagonaria profunda*. Mitchell (1977) recognized a further subdivision of the *profunda* zone, as well as the localized development of bioherms within this biostromal interval, in the Coralville Lake area near Lake MacBride (near station 13, Fig. 1). Lamellar stromatoporoids are abundant to dominant elements of the lower portions of the biostromes and bioherms. These stromatoporoids probably helped to stabilize the sea floor providing a substrate for the massive rugose corals (*Asterobillingsa* and *Hexagonaria*) that dominate the middle and upper portions of the *profunda* zone. A marked decrease in mud content also occurs across this interval. Growth of Solon bioherms may have been initiated on local topographic highs. Relief on the sea floor may have been caused by differential collapse of the lower Solon due to dissolution of interbedded evaporites in the underlying Davenport Member of the Wapsipinicon Forma-

tion by seepage of normal marine waters from the southward-transgressing Solon sea. Further growth of the biostromes and bioherms was terminated by the deposition of carbonate and terrigenous mud that marked the beginning of Rapid sedimentation. Therefore, the boundary between the Solon and Rapid is generally drawn at the base of the first prominent shaley calcilutite above the *profunda* zone calcarenites. Within the Lake MacBride area (station 13) a prominent nautiloid-bearing interval occurs near the top of the Solon.

Rapid Member. The Rapid is dominated by argillaceous, skeletal calcilutites, which are bioturbated to varying degrees. Thin shales or very shaley limestones commonly separate the calcilutites. Within the Coralville Lake area the Rapid averages approximately 50 ft (15.2 m) in thickness. Stainbrook (1941) subdivided the Rapid into three macrofaunal zones, as follows:

(1) The "*bellula* zone," named after the characteristic atrypid *Spinatrypa bellula*, encompasses the lower interval of the Rapid. The top of the *bellula* zone is generally drawn above the highest occurrence of *S. bellula*.

(2) The "*Pentamerella* zone" is named after the characteristic pentamerid brachiopod. This zone is dominated by a brachiopod fauna, which includes *Schizophoria, Pseudoatrypa, Orthospirifer, Tylothyris,* and *Strophodonta.* Two prominent and widely traceable coralline biostromes occur in the upper part of the *Pentamerella* zone (stations 11 and 12, Figs. 1 and 2; units 4 and 6, Fig. 3). The lower biostrome contains abundant hemispherical and encrusting stromatoporoids and common colonial rugose corals (*Hexagonaria*); favositid, alveolitid, and solitary rugose corals are less common (Zawistowski, 1971). The upper biostrome is a skeletal framework of solitary and colonial rugose corals; favositid and alveolitid tabulate corals are also common. Stromatoporoids are sparse (Zawistowski, 1971).

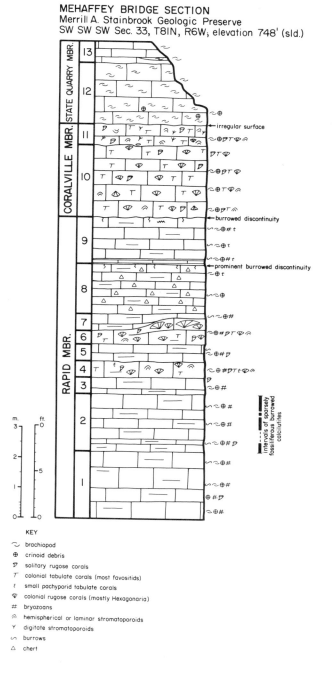

MEHAFFEY BRIDGE SECTION
Merrill A. Stainbrook Geologic Preserve
SW SW SW Sec. 33, T8IN, R6W; elevation 748' (sld.)

KEY

〜 brachiopod
⊕ crinoid debris
𝒫 solitary rugose corals
T colonial tabulate corals (most favositids)
t small pachyporid tabulate corals
♈ colonial rugose corals (mostly Hexagonaria)
bryozoans
⌒ hemispherical or laminar stromatoporoids
Y digitate stromatoporoids
⌒ burrows
△ chert

Figure 3. Generalized lithostratigraphic section exposed at the east end of Mehaffey Bridge (station 11 of Figs. 1 and 2).

(3) The "*waterlooensis* zone" includes the upper Rapid above the biostromes. This interval derives its name from the characteristic large atrypid brachiopod *Desquamatia waterlooensis*. This zone is dominated by argillaceous skeletal calcilutites with calcarenite lenses. It is glauconitic to varying degrees in the lower part with local concentrations of white chert nodules (especially evident at station 11, Fig. 1 and 2; unit 8, Fig. 3). The contact between the Rapid and the overlying Coralville is drawn at a sharp, burrowed discontinuity surface near the top of the *waterlooensis* zone. This surface marks a boundary separating argillaceous calcilutites below from coralline calcarenites above. Calcarenite-filled burrows to 8 in (20 cm) in depth penetrate the discontinuity surface.

Coralville Member. The Coralville in Johnson County has been subdivided into two lithologically distinct rock units. The lower unit consists of brown, massively bedded, coral- or stromatoporoid-rich, skeletal calcarenites, ranging in thickness from 16 to 23 ft (4.9 to 7 m). Stainbrook (1941) divided this interval into two macrofaunal zones. In ascending order, they are: (1) the coral-rich "*Cranaena* zone"; and (2) the "*Idiostroma* zone," a branching stromatoporoid-rich interval that has locally developed into a biostromal unit. Kettenbrink (1973) recognized five distinct lithologic subdivisions in the upper Coralville, in ascending order: (1) gastropod-oncolite calcilutite ("*Straparollus* zone" of Stainbrook, 1941); (2) laminated lithographic limestone; (3) *Amphipora*-rich pelleted calcilutite; (4) intraclastic calcilutite to calcirudite; and (5) birdseye calcilutite. Maximum observed thickness of the Coralville in Johnson County is 45 ft (13.7 m).

State Quarry Member. Based upon litho- and biostratigraphic studies of Devonian rocks in north-central Iowa (Witzke and Bunker, 1984) and extreme east-central Iowa (Witzke and others, 1985), the State Quarry is considered a member of the Cedar Valley Formation (Bunker and others, 1985). The State Quarry beds, however, are geographically restricted within Johnson County, and form a linear northeast-to-southwest-trending pattern within the Coralville Lake area. The State Quarry consists of skeletal calcarenites and calcilutites that represent a channel-filling sequence, occupying erosional channels that cut across Coralville and Rapid strata (Fig. 2). The State Quarry is dominated by skeletal calcarenites; packstones and grainstones contain varying proportions of echinoderm, brachiopod, branching stromatoporoid, and coral grains and intraclasts. Pelleted calcarenites and stromatoporoidal calcirudites also occur. Fish teeth and plates are common in the basal calcarenites ("bone bed") in the lower and marginal parts of the State Quarry channel. The State Quarry reaches thicknesses of about 40 ft (12 m).

The old State Quarry northeast of North Liberty, Johnson County, is designated as the type locality (Figs. 1 and 2, station 7; SW¼SW¼NW¼,Sec.8,T.80N.,R.6W.). Operation of this quarry dates from the late 1830s, when stone from it was utilized in the construction of the Old Capital Building in Iowa City. In the 1870s, it also produced foundation stones for the new State Capitol in Des Moines. In 1969, the old State Quarry was designated as a geological preserve.

Pennsylvanian

Pennsylvanian detrital sediments are preserved locally as scattered outliers within the Coralville Lake area, commonly within depressions on the Devonian bedrock surface. These Pennsylvanian strata are characterized by shaley feldspathic sandstones, siltstones, and mudstones, containing scattered plant debris and thin lenses of carbonaceous or coaly material (Witzke and Kay, 1984; Witzke, 1984b). Repeated attempts to define a more precise biostratigraphic determination (by palynology) of the age of these sediments within this area have failed (Witzke, 1984b). However, based upon lithologic criteria (Ludvigson, 1985; Fitzgerald, 1985) in the Davenport-Rock Island area to the southeast, where these sediments have been dated, the Pennsylvanian strata within the Coralville Lake area are considered to be Middle Pennsylvanian in age. A Pennsylvanian sandstone is exposed in the first minor reentrant to the south of Mehaffey Bridge (station 11, Fig. 1).

MEHAFFEY BRIDGE SECTION

In 1960, Congress authorized construction of Mehaffey Bridge to replace a structure which was built prior to the turn of the century. Johnson County Road F28 (Fig. 1) is the major access route to Mehaffey Bridge. It was during road excavations at the east end of the bridge (station 11, Fig. 1) that a unique combination of geological features was exposed. Immediately above and to the southeast of the roadcut, removal of approximately 50 ft (15.2 m) of drift and loess revealed a glacial pavement surface developed on the State Quarry Member. Here, a smooth knob of the State Quarry has been etched with parallel grooves and striations. These furrows mark the direction of glacial movement over the area, and were gouged by rock fragments carried at the base of the ice mass as it inched across the rock surface. Beneath the glacially inscribed bedrock surface, spectacular exposures of richly fossiliferous limestones of the Cedar Valley Formation occur along the roadcut. In 1969, 33 acres (13.3 ha) of hilly, partly wooded terrain, encompassing these geological features, which are of exceptional scientific and public interest, were dedicated as the Merrill A. Stainbrook Geological Preserve.

Exposed in the roadcut of Stainbrook Preserve are the *waterlooensis* beds of the upper Rapid, the coral-stromatoporoid biostrome of the lower Coralville, and skeletal calcarenites of the State Quarry (units 7–13, Fig. 3). Immediately to the north of the roadcut along the Coralville Lake front additional exposures of the lower Rapid are displayed extending downward into the upper part of the *bellula* zone (units 1–6, Fig. 3). The two prominent biostromes of the Rapid, which are not exposed at the roadcut, can be traced at the top of the cut to the north along the lake front.

Within the *Pentamerella* zone of the lower Rapid, beneath the biostromes, intervals of unfossiliferous to sparsely fossiliferous burrowed calcilutites occur, which Zawistowski (1971) termed the "key bed" (unit 2, Fig. 3). The key bed differs from other argillaceous calcilutites of the Rapid by the paucity of the skeletal grains and sparse faunal content. However, psilophyte plant remains, inarticulate brachiopods, bivalves, branchiopods, crustaceans, conularids, bryozoans, and trilobites have been noted.

At the first reentrant to the north of Mehaffey Bridge an 18-ft (5.5-m) faulted interval occurs, which has displaced State Quarry calcarenites into juxtaposition with the Rapid biostromes. To the south of Mehaffey Bridge the State Quarry channel incision is readily evident (Fig. 2).

MEASURED SECTION

Mehaffey Bridge Section. (Fig. 3) (Measured by B. J. Witzke, B. J. Bunker, and J. E. Day, Iowa Geological Survey, 1985.)

Merrill A. Stainbrook Geological Preserve. (Encompasses units 7–13, as well as the overlying glacial pavement developed on the State Quarry.) SW¼SW¼SW¼,Sec.33,T.81N., R.6W.; elevation of the bedrock surface at the top of the roadcut is approximately 748 ft (244 m) (sea level datum).

Cedar Valley Formation (Middle Devonian)

State Quarry Member

Unit 13. Skeletal calcilutite, scattered to rare brachiopods, lower contact slightly irregular; locally (far northeast exposure) an atrypid packstone is developed 3 to 8 in (8 to 20 cm) above base of unit; *Variatrypa* (*Radiatrypa*) sp. cf. v. (R.) clarkei, Schizophoria; 27.5 in (70 cm).

Unit 12. Skeletal calcarenite, brachiopod-rich, scattered crinoid debris; some small intraclasts; disconformable below (2 in; 5 cm relief along exposure), locally with low-angle cross-beds; *Variatrypa* sp., *Cranaena,* spiriferids, rostroconchs, gastropods; 6.6 ft (2.0 m).

Coralville Member

Unit 11. Biostrome, skeletal calcarenite; abundant stromatoporoids and corals, some overturned; intraclastic near base, irregular basal surface; *Idiostroma* (digitate stromatoporoids), hemispherical stromatoporoids, branching *Favosites,* pachyporid corals, horn corals, *Cranaena*; 28 in (70 cm).

Unit 10. Biostrome, skeletal calcarenite; abundant corals and stromatoporoids, many overturned; scattered brachiopods, *Desquamatia* common near base, *Pseudoatrypa* above, ?*Orthopleura* near top; encrusting and hemispherical stromatoporoids, *Favosites,* pachyporids, horn corals, *Hexagonaria, Conocardium*; irregular surface; 7.5 ft (2.3 m).

Rapid Member

Unit 9. Argillaceous skeletal calcilutite, some calcarenitic lenses and stringers, burrowed; discontinuity surface at top (sharp break from calcilutite below to Coralville calcarenite above), vertical burrows infilled with Coralville lithologies extend downward locally to 8 in (20 cm); basal 4 to 6 in (11 to 15 cm) with two prominent shale partings; abundant crinoid debris (including *Megistrocrinus*), bryozoans (including *Sulcoretopora*), scattered pachyporid corals, large *Desquamatia waterlooensis*, *Orthospirifer*, *Tylothyris*, *Cranaena* (upper part), *Strophodonta*, *Schizophoria*, other brachiopods; 4.9 ft (1.5 m).

Unit 8. Argillaceous skeletal calcilutite to calcarenite, burrowed; prominent discontinuity surface at top with abundant vertical burrows to 4 in (10 cm); scattered to abundant nodular chert (smooth to chalky), nodules 0.4 to 12 in (1 to 30 cm); glauconitic to very glauconitic in upper part; abundant crinoid debris, scattered small to large brachiopods including *Orthospirifer* and ?*Desquamatia*; pachyporid corals near top; 5.4 ft (1.7 m).

Unit 7. Argillaceous skeletal calcilutite with lenses or stringers of calcarenite; oversteps upper surface of underlying biostrome; shaley to very argillaceous at top; small *Favosites* and horn corals near base; crinoid debris (including *Megistrocrinus*), bryozoans, *Orthospirifer*, *Tylothyris*, *Schizophoria*, *Pseudoatrypa*, *Athyris*, *Cranaena*, *Pentamerella*; 10 to 27 in (27 to 70 cm).

Unit 6. Biostrome, slightly argillaceous skeletal calcilutite to calcarenite; shaley parting at top; biostrome varies laterally, where thickest includes massive accumulations of *Hexagonaria*; crinoid debris, fenestellid bryozoans, horn corals, *Favosites*, alveolitids, *Hexagonaria*, encrusting stromatoporoids, *Pseudoa-*

trypa, *Tylothyris*, *Strophodonta*, *Cranaena*, *Gypidula*, *Pentamerella*; 15 to 33 in (40 to 85 cm).

Unit 5. Argillaceous skeletal calcilutite, burrowed, shaley at top; crinoid debris (including *Megistocrinus*), bryozoans (including fenestellids and cystodictyonids), horn corals at top, *Orthospirifer*, *Pseudoatrypa*, *Strophodonta*, *Schizophoria*; 13 to 27 in (35 to 70 cm).

Unit 4. Biostrome, argillaceous skeletal calcilutite to calcarenite; *Favosites*, alveolitids, pachyporids near top, *Hexagonaria*, horn corals, hemispherical stromatoporoids, crinoid debris, *Orthospirifer*, *Pseudoatrypa*; 19 in (50 cm).

Unit 3. Argillaceous skeletal calcilutite to calcarenite at top; crinoid debris, bryozoans (including fenestellids), *Orthospirifer*, *Pseudoatrypa*, *Schizophoria*, horn corals at top; 21 in (53 cm).

Unit 2. Sparsely fossiliferous to unfossiliferous argillaceous calcilutite, contains lenses or stringers or argillaceous skeletal calcilutite to calcarenite; extensively burrow-mottled throughout; fossiliferous horizons include crinoid debris (including *Megistocrinus*), bryozoans (including fenestellids), strophomenids, *Orthospirifer*, *Tylothyris*, *Pseudoatrypa*, *Schizophoria*, *Retichonetes* (upper part); lower part locally includes horn corals and *Hexagonaria* (to 11 in; 30 cm); 6.2 ft (1.9 m).

Unit 1. Argillaceous skeletal calcilutite to calcarenite, burrowed; crinoid debris (including *Megistocrinus*), bryozoans (including fenestellids and cystodictyonids), horn corals in lower part, *Phacops*, *Spinatrypa bellula*, *Pseudoatrypa*, *Orthospirifer*, *Tylothyris*, *Schizophoria*; 7.2 ft (2.2 m).

REFERENCES CITED

Beinert, R. J., 1968, Development report of Coralville Reservoir geological preserve by Mehaffey Bridge, and State Quarry geological preserve: Iowa State Advisory Board for Preserves, Development Series Report 1, 40 p.

Bunker, B. J., Ludvigson, G. A., and Witzke, B. J., 1985, The Plum River Fault Zone and the structural and stratigraphic framework of eastern Iowa: Iowa Geological Survey, Technical Information Series 13, 126 p.

Fitzgerald, D. J., 1985, Pennsylvanian rocks of Muscatine County, Iowa, *in* Hammer, W. R., Anderson, R. C., and Schroeder, D. A., eds., 15th Annual Field Conference Guidebook: Great Lakes Section, Society of Economic Paleontologists and Mineralogists, p. 94–129.

Kettenbrink, E. C., 1973, Depositional and post-depositional history of the Devonian Cedar Valley Formation, east-central Iowa [Ph.D. thesis]: Iowa City, University of Iowa, 191 p.

Ludvigson, G. A., 1985, Observations of the basal Pennsylvanian sandstones of eastern Iowa, with comments on their stratigraphic and tectonic significance, *in* Hammer, W. R., Anderson, R. C., and Schroeder, D. A., eds., 15th Annual Field Conference Guidebook: Great Lakes Section, Society of Economic Paleontologists and Mineralogists, p. 65–93.

McGee, W. J., 1891, The Pleistocene history of northeastern Iowa: U.S. Geological Survey, 11th Annual Report, part 1, p. 199–577.

Mitchell, J. C., 1977, Biostromes and bioherms of the Solon Member of the Cedar Valley Limestone Middle Devonian, eastern Iowa [M.S. thesis]: Iowa City,

University of Iowa, 179 p.

Norton, W. H., 1985, Geology of Linn County: Iowa Geological Survey, Annual Report, v. 4, p. 121–195.

Stainbrook, M. A., 1941, Biotic analysis of Owen's Cedar Valley Limestone: Pan–American Geologist, v. 75, p. 321–327.

Watson, M. G., 1974, Stratigraphy and environment of deposition of the State Quarry Limestone, Johnson County, Iowa [M.S. thesis]: Iowa City, University of Iowa, 140 p.

Witzke, B. J., ed., 1984a, Geology of the University of Iowa campus area, Iowa City: Iowa Geological Survey Guidebook No. 7, 76 p.

—— , 1984b, Pennsylvanian strata at Conklin Quarry: Geological Society of Iowa Guidebook 41, p. 21–23.

Witzke, B. J., and Bunker, B. J., 1984, Devonian stratigraphy of north-central Iowa: Iowa Geological Survey Open-File Report 84-2, p. 107–149.

Witzke, B. J., Bunker, B. J., and Klapper, G., 1985, Devonian stratigraphy in the Quad Cities area, eastern Iowa–northwestern Illinois, *in* Hammer, W. R., Anderson, R. C., and Schroeder, D. A., eds., 15th Annual Field Conference Guidebook: Great Lakes Section, Society of Economic Paleontologists and Mineralogists, p. 19–64.

Zawistowski, S. J., 1971, Biostromes in the Rapid Member of the Cedar Valley Limestone (Devonian) in east-central Iowa [M.S. thesis]: Iowa City, University of Iowa, 120 p.

Pennsylvanian cyclothem near Winterset, Iowa

Philip H. Heckel, Department of Geology, University of Iowa, Iowa City, Iowa 52240

LOCATION

This section is located on the west side of U.S. 169 about 1.7 mi (2.7 km) south of its junction with Iowa 92 on the western outskirts of Winterset (Fig. 1; E line NE¼SW¼Sec.11,T.75N., R.28W., Madison County, Iowa). Because traffic on this highway can be heavy, pull well off onto the shoulder, and please do not cross the highway.

INTRODUCTION

The cyclic alternation of limestone and shale formations that dominate the Middle and Upper Pennsylvanian sequence (Fig. 2) along the Midcontinent outcrop belt has intrigued geologists ever since Moore (1931) first described it, and Wanless and Weller (1932) applied the name "cyclothem" to the component unit of repeating rock types in Illinois. This term was soon adopted in the Midcontinent outcrop area by Moore (1936). Weller (1930) invoked a model of periodic tectonism to explain both the overall alternation and the individual cyclothem. In contrast, Wanless and Shepard (1936) related both these features to periodic eustatic changes in sea level brought about by waxing and waning of Gondwanan ice caps. More recently, autocyclic models of delta shifting have been applied to cyclic sequences in the Appalachians (Ferm, 1970) and Texas (e.g., Galloway and Brown, 1973), and these have been extended to Illinois (Merrill, 1975). In the meantime, Wanless (1964, 1967) suggested that the glacial eustatic model readily accommodates delta-shifting as a mechanism to explain otherwise anomalous clastic wedges in many of the cyclothems. This view recently has been more fully developed by Heckel (1977, 1980, 1984a), who recognized the cyclothems as basically marine transgressive-regressive sequences, centered on the thin, nonsandy, black phosphatic ("core") shales, which represent maximum inundation of the shelf, and with most deltas forming during the succeeding regressive phase.

The term cyclothem has been applied to a number of different, but related and quite specific, repeating lithic successions in the Pennsylvanian (e.g., Moore, 1936, 1950; Weller, 1958; see review in Heckel, 1984b). Current work on the mid-Desmoinesian to mid-Virgilian sequence in the Midcontinent has established the nature of the basic transgressive-regressive ("Kansas") cyclothem (Figs. 3, 4). This cyclothem resulted from a major rise and fall of sea level over the northern Midcontinent shelf, which extended from the Arkoma-Anadarko basinal region of central Oklahoma across the lower parts of the North American craton northward into Iowa and Nebraska. In ascending order, this cyclothem consists of:

1. Transgressive ("middle") limestone, deposited in deepening water. This is typically a thin marine skeletal calcilutite with a diverse open-marine biota, deposited below effective wave

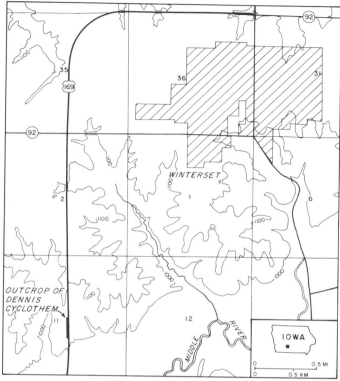

Figure 1. Part of Winterset 7½-minute Quadrangle showing location of Dennis cyclothem exposure near Winterset.

base (e.w.b., Fig. 4) later during transgression. These limestones typically are dense, dark, nonpelleted calcilutites with neomorphosed aragonite grains, and lack evidence of early marine cementation or meteoric leaching and cementation. This is because they remained in the marine phreatic environment of deposition until buried by higher marine strata of the cyclothem, which acted as a barrier to meteoric diagenesis (Heckel, 1983). Thus they underwent slow compaction before cementation in a burial environment of decreasing oxygen concentration in which much of the fine organic matter became preserved in the rock.

2. Offshore ("core") shale formed at maximum transgression. This is typically a thin, nonsandy, gray to black phosphatic shale deposited under conditions of near sediment starvation. In most cyclothems the water became deep enough for a thermocline to develop over much of the northern Midcontinent shelf. The thermocline reduced vertical circulation and bottom-oxygen replenishment to produce the gray dysaerobic facies over shallower areas, with only low-oxygen-tolerant benthic invertebrates, such as certain brachiopods. It eliminated oxygen from bottom water over the deeper areas to produce the black anoxic facies

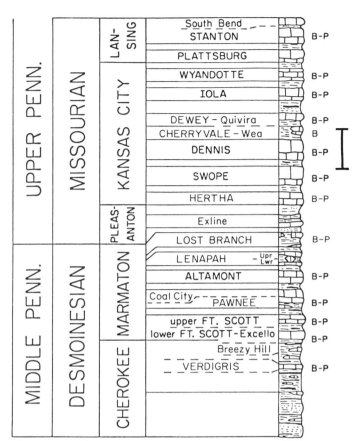

Figure 2. Part of Middle and Upper Pennsylvanian sequence along Mid-continent outcrop belt, showing stratigraphic position of Winterset exposure (bar at right). Horizontal names are formations and members (mainly limestones) that are developed to various degrees as cyclothems. B-P denotes presence of black phosphatic shales that characterize classic cyclothems.

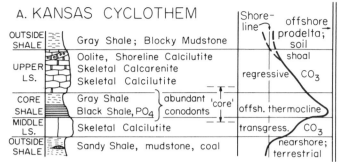

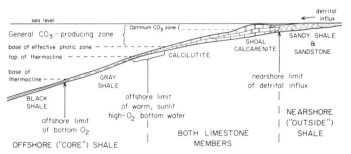

Figure 3. A. Basic "Kansas" cyclothem characterizing with minor modification all major and many intermediate marine cycles of deposition across northern Midcontinent Shelf (Iowa to southern Kansas). Dark line on right is sea-level curve. B. General model for limits of carbonate deposition on gently sloping tropical shelf, showing lateral distribution of rock types that became superposed by transgression and regression to form classic "Kansas" cyclothem. (From Heckel, 1986.)

with only pelagic fossil remains such as certain conodonts, fish debris, and, if preserved in nodules, radiolarians and occasional ammonoids. Conodonts, in particular, are abundant in both gray and black facies of these offshore shales.

Sedimentation was so slow at this time in the northern Midcontinent that in the gray facies, aragonite fossils apparently were dissolved and calcite fossils were locally corroded (Malinky, 1984). Further evidence for this lies in the appearance of great numbers of originally aragonitic molluscs (snails, clams, ammonoids; Boardman and others, 1984), now preserved as siderite, pyrite, or phosphorite in thicker developments of these gray shales in southern Kansas and Oklahoma. Ammonoids also are preserved locally in early diagenetic carbonate nodules ("bullion") in the black facies in Kansas and Missouri. In the first case, early rapid burial and mineralization, and in the second case early matrix mineralization, prevented sea floor dissolution of the fossils in the colder undersaturated waters below the top of the thermocline.

Quasi-estuarine circulation and upwelling associated with the thermocline in this deeper phase of the shelf sea caused deposition, in both the gray and black shale facies, of nonskeletal phosphorite as granules, laminae, and nodules. These are analogous to modern phosphorite nodules, forming under similar conditions of periodic upwelling associated with a thermocline, in low-oxygen sediment on the offshore shelf along the coast of Peru (Kidder, 1985).

3. Regressive ("upper") limestone deposited in shallowing water. This is typically a thick marine skeletal calcilutite with a diverse biota deposited below effective wave base, and grading upward into skeletal calcarenite with algae, and abraded grains and cross-bedding, evidence for traction transport in shallow water. The tops of many regressive limestones, particularly northward (Fig. 4), display sparsely fossiliferous, laminated to birdseye-bearing lagoonal to peritidal carbonates that represent passage of the shoreline during later regression.

Even without development of these shoreline facies, passage of the shoreline is recorded in subaerial exposure surfaces on the tops of many regressive limestones. These exposure surfaces resulted from the effects of meteoric weathering and include cracking, in-place brecciation, pitting, and formation of "solution-

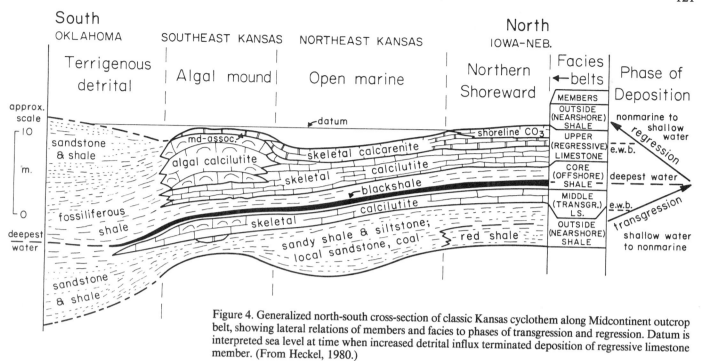

Figure 4. Generalized north-south cross-section of classic Kansas cyclothem along Midcontinent outcrop belt, showing lateral relations of members and facies to phases of transgression and regression. Datum is interpreted sea level at time when increased detrital influx terminated deposition of regressive limestone member. (From Heckel, 1980.)

tubes" from plant rooting and infiltration of fresh water and terrestrial organic matter. This often led to a mottled, rubbly, rusty, or punky appearance when various amounts of overlying shale and iron oxides were carried down into the open spaces in the limestone. The recognition and lateral tracing of these widespread exposure surfaces on regressive limestones, which cannot be explained by a depositional model involving delta shifting alone, has more firmly corroborated the eustatic model for cyclothem formation in the northern Midcontinent (Watney, 1984).

Subaerial exposure and infiltration of oxygenating, undersaturated meteoric water down into the regressive limestone before much grain-to-grain compaction took place also oxidized much of the original organic matter in the sediment, leached most of the aragonitic grains and eventually, as it became saturated, precipitated blocky calcite in both the intergranular and moldic voids. This preserved original peloidal fabric, original depositional packing of grains, and, in the case of incomplete cementation, much porosity. Thus the lighter-colored, more porous and more conspicuously sparry, upper regressive limestones often stand in contrast with the darker, denser, overcompacted transgressive limestones and the lower, more offshore facies of regressive limestones (Heckel, 1983).

4. Nearshore ("outside") shales (which lie outside the limestone formations comprising the three previously described marine members of the cyclothem) encompass a great variety of nearshore marine and terrestrial deposits on the shelf, all deposited at lower stands of sea level. They include thick, sparsely fossiliferous prodeltaic shales that prograded out over regressive

limestones in areas where deltas were active during later stages of sea-level fall. In places these grade upward into delta-front and delta-plain sandstones and coals.

Outside shales also include thinner blocky mudstones, particularly in Iowa, which range from gray to red in color, from inches to several feet in thickness, and typically overlie exposure surfaces on regressive limestones. Those that have been studied in detail show upward decrease in crystallinity of illite and upward increase in mixed-layer and kaolinite proportions relative to illite, which are characteristics of clay mineralogy expected in a soil profile. The blocky fabric of the paleosol mudstone is a result of disturbance of the normally flat-lying clay minerals (which account for the fissility of marine shales) by plant rooting, animal burrowing, and rain-water illuviation of clay minerals into holes formed by the organic agents and into cracks formed by periodic desiccation. This blocky fabric provides a strong suggestion of the origin of these mudstones as paleosols in cases where the clay minerals have not been analyzed. In addition, these mudstones often contain irregular carbonate nodules, some of which show internal clotted, mottled, and cracked fabric characteristic of formation in a soil (Watney, 1980).

Because many marine units in the Pennsylvanian sequence represent lesser inundations of the shelf that did not develop all members of a classic cyclothem, I have classified the irregular spectrum of transgressive-regressive cycles into three categories for the purpose of further analysis, based partly on the extent of transgression on the shelf: 1) Major cycles are those inundations far and deep enough onto the shelf to form a conodont-rich shale

P. H. Heckel

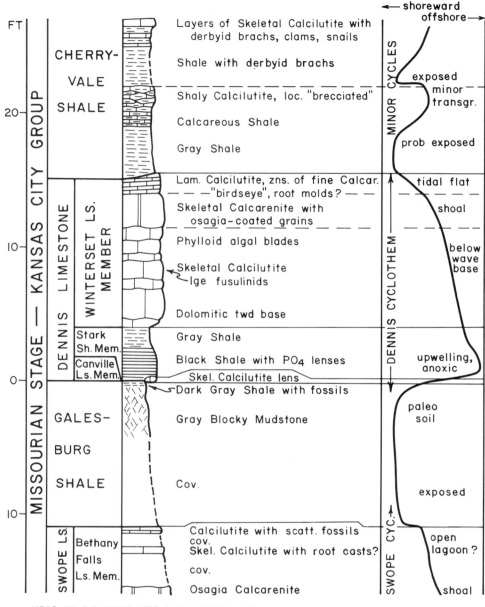

Figure 5. Rock strata exposed near Winterset with interpreted changes in depositional environments related to sea-level curve on right.

to the northern limit of outcrop in Nebraska and Iowa, and to develop enough of the other facies to be recognized as classic cyclothems over much of their extent. 2) Intermediate cycles extend as marine horizons into Iowa and Nebraska, but carry conodont-rich horizons only on the lower shelf, and most have been recognized only in some places as cyclothems on parts of them. 3) Minor cycles typically extend as marine horizons only a short distance from the basinal region of Oklahoma into Kansas or Missouri, or represent a minor reversal within a larger cycle,

and have not generally been recognized as cyclothems nor in many cases been named as separate units. The recognition of minor eustatic cycles requires correlation of the horizon over a reasonable geographic area, because in any one section or small geographic area, autocyclic processes such as delta shifting can produce minor cycles of deposition.

Analysis of the probable lengths of time during which each of these types of marine cycles of transgression and regression took place, based on sets of assumptions explained elsewhere

(Heckel, 1986), has estimated a range of lengths of 235,000 to about 400,000 years for the major cyclothems, 120,000 to 220,000 years for the intermediate cycles, and 44,000 to 120,000 years for the minor cycles. The estimated ranges for all cycles fall within the range of periods of the earth's orbital cycles that constitute the Milankovitch insolation theory for control of the Pleistocene ice ages. These cyclic orbital parameters are: eccentricity, with two dominant periods, one about 413,000 years, and the other ranging from 95,000 to 136,000 years and averaging about 100,000 years; obliquity, with a dominant period near 41,000 years; and precession, with two dominant periods averaging 19,000 and 23,000 years (Imbrie and Imbrie, 1980). Because glacial deposits of an age spanning the entire Pennsylvanian are known on the southern continent of Gondwanaland (Crowell, 1979), these ranges of periodicities for the Pennsylvanian cycles give further firm support for the glacial-eustatic control over formation of Pennsylvanian cyclothems.

THE DENNIS CYCLOTHEM

The field guide site described here shows the entire Dennis cyclothem (Fig. 5), which has only a lenticular transgressive limestone, but a relatively thick black shale and well-developed regressive limestone. It also includes the top of the underlying Swope cyclothem at the base, and parts of minor cycles of deposition at the top.

Several of the upper beds of the Bethany Falls Limestone, the regressive limestone of the Swope cyclothem, are exposed southward down the road ditch. These include calcarenite and skeletal calcilutite with sparse fossils, which represent shallow-water carbonates deposited late during regression. One of the beds contains brownish stained vertical structures that may represent root casts.

The Galesburg Shale is a light gray blocky mudstone, which has clay mineralogical characteristics of a paleosol at this locality, e.g., low crystallinity of illite and a high mixed-layer to illite ratio (Schutter, 1983) as well as the typical blocky fabric. This ancient Galesburg soil can be traced from Nebraska to here and southward 200 mi (320 km) to east central Kansas, south of which it contains fluvial deposits that grade to deltaic deposits around Tulsa, Oklahoma, 200 mi (320 km) farther south. In Decatur County, Iowa, about 30 to 50 mi (48 to 80 km) south of here, the Galesburg has a thin coal at the top, which probably originated in swamps that formed ahead of the transgression that was responsible for the overlying Dennis marine cycle. The dark fossiliferous shale at the top of the Galesburg here represents the early shallow-water part of this transgression.

The lenticular Canville Limestone was found only on the east side of the highway just after construction, and before extensive slumping covered it. It is a skeletal calcilutite laid down in patches below effective wave base during a rapid transgression that did not keep the sea bottom in strongly sunlit water long enough to produce enough algal-derived carbonate mud to form a laterally extensive limestone as in most other cycles.

The black, fissile, phosphatic Stark Shale was deposited below a thermocline during maximum marine transgression of the Dennis marine incursion, when bottom conditions became anoxic because of loss of vertical circulation, and when periodic upwelling caused deposition of phosphate in thin lenses, as occurs today on the offshore shelf along the coast of Peru. The Stark Shale is traced along outcrop throughout the neighboring states and has been identified around Tulsa, Oklahoma, 400 mi (640 km) to the south, attesting to the extremely widespread nature of the Dennis transgression above the Galesburg Shale, which is terrestrial throughout the same entire region. The gray shale at the top of the Stark formed as the sea shallowed enough to destroy the thermocline, and vertical circulation resumed to begin supplying oxygen to the bottom again.

The Winterset Limestone, the regressive limestone of the Dennis cyclothem, is named from this area and is traced readily throughout the outcrop belt in neighboring states along with the underlying Stark and Galesburg Shales. The lower two thirds of the Winterset here are skeletal calcilutite with a diverse biota, which was deposited below effective wave base when the sea shallowed to the point that enough sunlight again reached the bottom to promote growth of carbonate-mud-producing algae. Above this is an *"Osagia"*-grain calcarenite, which represents a shoal developed above effective wave base later during regression when minute blue-green algae and encrusting foraminifers (together termed *"Osagia"*) coated sand-sized skeletal grains in an agitated environment where mud was winnowed out. The top of this bed and also the overlying capping bed of the Winterset, exposed northward along the roadcut, consist of barren laminated calcilutite, with thin layers of fine-grained calcarenite. These layers probably represent storm-washed debris on a tidal flat that formed late during regression when the ancient shoreline was migrating southward across this area toward the basin in Oklahoma. The small sparry spots in this calcilutite, known as "birds-eye," probably represent small gas bubbles trapped beneath algal mats on the tidal flat as well as vertical root molds from plants that established themselves on the emergent surface after regression.

The overlying Cherryvale Shale, which is typically poorly exposed, contains a gray shale at the base followed upward by a rubbly, shaly calcilutite that locally shows a brecciated fabric. This probably represents the nearshore end of a more minor cycle of marine inundation, which became incorporated into a soil profile that formed after regression took the shoreline back southward toward the basin. It may be one of the several minor cycles of marine inundation and retreat, illustrated by Heckel and Watney (1985), in the thicker development of Winterset Limestone in east-central Kansas. The overlying shales and thin limestones that carry brachiopods and other fossils represent another, probably intermediate, cycle of marine inundation. In this cycle enough sediment was deposited that the succeeding regression did not totally leach the rock and destroy the fossils, as apparently happened in the minor cycle below.

REFERENCES CITED

Boardman, D. R., II, Mapes, R. H., Yancey, T. E., and Malinky, J. M., 1984, A new model for the depth-related allogenic community succession within North American Pennsylvanian cyclothems and implications on the black shale problem, *in* Hyne, N. J., ed., Limestones of the Mid-Continent: Tulsa Geological Society Special Publication 2, p. 141–182.

Crowell, J. C., 1978, Gondwanan glaciation, cyclothems, continental positioning, and climate change: American Journal of Science, v. 278, p. 1345–1372.

Ferm, J. C., 1970, Allegheny deltaic deposits: Society of Economic Paleontologists and Mineralogists Special Publication 15, p. 246–255.

Galloway, W. E., and Brown, L. F., Jr., 1973, Depositional systems and shelf-slope relations on cratonic basin margin, uppermost Pennsylvanian of north-central Texas: American Association of Petroleum Geologists Bulletin, v. 57, p. 1185–1218.

Heckel, P. H., 1977, Origin of phosphatic black shale facies in Pennsylvanian cyclothems of Midcontinent North America: American Association of Petroleum Geologists Bulletin, v. 61, p. 1045–1068.

—— , 1980, Paleogeography of eustatic model for deposition of Midcontinent Upper Pennsylvanian cyclothems, *in* Fouch, T. D., and Magathan, E. R., eds., Paleozoic paleogeography of west-central United States: Rocky Mountain Section, Society of Economic Paleontologists and Mineralogists Paleogeography Symposium I, p. 197–215.

—— , 1983, Diagenetic model for carbonate rocks in Midcontinent Pennsylvanian eustatic cyclothems: Journal of Sedimentary Petrology, v. 53, p. 733–759.

—— , 1984a, Factors in Mid-Continent Pennsylvanian limestone deposition, *in* Hyne, N. J., ed., Limestones of the Mid-Continent: Tulsa Geological Society Special Publication 2, p. 25–50.

—— , 1984b, Changing concepts of Midcontinent Pennsylvanian cyclothems, North America, *in* International Congress on Carboniferous Stratigraphy and Geology, 1979: Compte Rendu, Carbondale, Southern Illinois University Press, v. 3, v. 535–553.

—— , 1986, Sea-level curve for Pennsylvanian eustatic marine transgressive-regressive depositional cycles along Midcontinent outcrop belt: Geology, v. 14, p. 330–334.

Heckel, P. H., and Watney, W. L., 1985, Stop 5A, US-69 roadcut by Jingo; Swope and Dennis cycles, *in* Watney, W. L., and others, eds., Recent interpretations of Late Paleozoic cyclothems: Proceedings 3rd Annual Meeting and Field Conference, Midcontinent Section, Society of Economic Paleontologists and Mineralogists, p. 44–51.

Imbrie, J., and Imbrie, J. Z., 1980, Modeling the climatic response to orbital variations: Science, v. 207, p. 943–953.

Kidder, D. L., 1985, Petrology and origin of phosphate nodules from the Midcontinent Pennsylvanian epicontinental sea: Journal of Sedimentary Petrology, v. 55, p. 809–816.

Malinky, J. M., 1984, Paleontology and paleoenvironment of "core" shales (Middle and Upper Pennsylvanian) Midcontinent North America [Ph.D. thesis]: Iowa City, University of Iowa, 327 p.

Merrill, G. K., 1975, Pennsylvanian conodont biostratigraphy and paleoecology of northwestern Illinois: Geological Society of America Microform Publication 3.

Moore, R. C., 1931, Pennsylvanian cycles in the northern Mid-Continent region: Illinois Geological Survey Bulletin 60, p. 247–257.

—— , 1936, Stratigraphic classification of the Pennsylvanian rocks of Kansas: Kansas Geological Survey Bulletin 22, 256 p.

—— , 1950, Late Paleozoic cyclic sedimentation in central United States: 18th International Geological Congress, Great Britain 1948, pt. 4, p. 5–16.

Schutter, S. R., 1983, Petrology, clay mineralogy, paleontology, and depositional environments of four Missourian (Upper Pennsylvanian) shales of Mid-Continent and Illinois Basin [Ph.D. thesis]: Iowa City, University of Iowa, 1208 p.

Wanless, H. R., 1964, Local and regional factors in Pennsylvanian cyclic sedimentation: Kansas Geological Survey Bulletin 169, p. 593–606.

—— , 1967, Eustatic shifts in sea level during the deposition of late Paleozoic sediments in the central United States: West Texas Geological Society Publication 69-56, p. 41–54.

Wanless, H. R., and Shepard, F. P., 1936, Sea level and climatic changes related to late Paleozoic cycles: Geological Society of America Bulletin, v. 47, p. 1177–1206.

Wanless, H. R., and Weller, J. M., 1932, Correlation and extent of Pennsylvanian cyclothems: Geological Society of America Bulletin, v. 43, p. 1003–1016.

Watney, W. L., 1980, Cyclic sedimentation of the Lansing-Kansas City Groups in northwestern Kansas and southwestern Nebraska: Kansas Geological Survey Bulletin 220, 72 p.

—— , 1984, Recognition of favorable reservoir trends in Upper Pennsylvanian cyclic carbonates in western Kansas, *in* Hyne, N. J., ed., Limestones of the Mid-Continent: Tulsa Geological Society Special Publication 2, p. 201–245.

Weller, J. M., 1930, Cyclic sedimentation of the Pennsylvanian period and its significance: Journal of Geology, v. 38, p. 97–135.

—— , 1958, Cyclothems and larger sedimentary cycles of the Pennsylvanian: Journal of Geology, v. 66, p. 195–207.

Starrs Cave Park, Burlington area, Des Moines County, southeastern Iowa

Brian F. Glenister, *Department of Geology, University of Iowa, Iowa City, Iowa 52242*
Alan C. Kendall, *Department of Geology, University of Toronto, Toronto, Ontario M5S 1A1, Canada*
Jennifer A. Person (Collins), *2011 Goldsmith Street, Houston, Texas 77030*
Alan B. Shaw, *1315 Kamira Drive, Kerrville, Texas 78028*

LOCATION AND ACCESSIBILITY

Starrs Cave Park and Preserve is a facility situated on Flint Creek (Fig. 1) in the northern outskirts of Burlington, Des Moines County, Iowa (W½Sec.19,T.70N.,R.2W.). It is administered by the Des Moines County Conservation Board, Courthouse, Burlington, Iowa 52601. Continuous exposures extending from the Late Devonian English River Siltstone through the Early Mississippian (Kinderhookian) carbonates of the North Hill Group and into the Medial Mississippian (Osagean) Burlington Limestone extend for 0.5 mi (0.8 km) along the east bank of Flint Creek. At low creek levels, the exposures are accessible directly from the Starrs Cave parking lot, on the west side of Flint Creek off Irish Ridge Road. A steep foot trail down the cliffs to the east provides access from the northern extension of Irish Ridge Road during periods of high water.

SITE 27—UPPER MISSISSIPPI VALLEY REFERENCE SECTION FOR UPPER DEVONIAN AND MISSISSIPPIAN

Significance of Site. Exposures in the vicinity of Burlington form part of the body stratotype for the Mississippian System (Williams, 1891), and the Flint Creek section comprises one of the fuller developments of the Kinderhookian Series in the Upper Mississippi Valley. Graphic correlation (Shaw, 1964; Miller, 1977) demonstrates striking stratigraphic diachronism, lithic units being consistently older in the Le Grand area of central Iowa than at Burlington (Fig. 2).

The sequence of facies that accumulated following the Mississippian transgression is logically predictable—both primary lithofacies and biofacies become less restricted upward. Selective dolomitization near the bottom of the Mississippian can be demonstrated to have occurred prior to full lithification (Site 28 herein).

Diverse Mississippian faunas of the Burlington area have been investigated for well over a century. Crinoids in particular are renowned for their abundance, diversity, and excellence of preservation. However, collecting is not permitted within the park boundaries.

SITE INFORMATION

Lithologies of the Devonian and Mississippian strata exposed along Flint Creek and elsewhere in the Burlington area are so distinctive that they have been differentiated consistently and

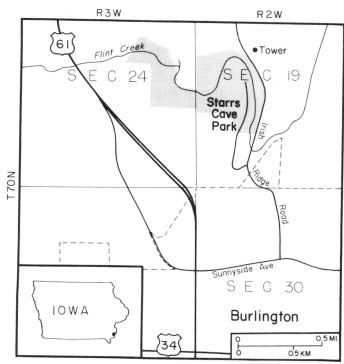

Figure 1. Location map, Starrs Cave Park, Burlington, Des Moines County, Iowa.

are generally recognizable in publications dating as far back as Owen (1852). Additionally, the faunas are sufficiently spectacular to have commanded repeated attention (e.g., Hall and Whitney, 1858; Wachsmuth and Springer, 1897; Weller, 1900a, 1901, 1914; Laudon, 1931). The bottom 15 ft (4.5 m) of the exposure comprises fossiliferous Late Devonian siliciclastics referred to the English River Siltstone (Fig. 3). Unconformably overlying the Devonian is the 29.5 ft (9 m) dominantly carbonate succession of the Early Mississippian (Kinderhookian) North Hill Group; in ascending order are the McCraney Limestone, Prospect Hill Siltstone, Starrs Cave Oolite, and Wassonville Dolomite. A second major unconformity separates the Wassonville from the Medial Mississippian (Osagean) Burlington Limestone; here the Burlington is a 60 ft (18 m) sequence of cherty fossiliferous grainstones whose base may overhang the recessive, underlying sections by as much as 25 ft (7.5 m). A poorly exposed, thin till veneer caps the glaciated surface of the bedrock sequence.

Joint systems serve as conduits for groundwater and commonly are marked at the surface by lush growths of moss that may be partly fossilized as calcareous tufa. One such joint has been enlarged in the Wassonville and basal Burlington to form the small cave for which the park is named.

STRATIGRAPHY

English River Siltstone. Fifteen ft (4.5 m) of massively bedded gray siltstone forms the bedrock slopes, beneath the limestones, at the base of the Flint Creek section. Exposures along Iowa 99 at North Hill, 2.5 mi (4 km) southeast of Starrs Cave, reveal that this same interval is gradational with the underlying Maple Mill Shale, the contact being placed approximately 25 ft (7.5 m) below the overlying carbonates. A comparable sequence along the English River in northern Washington County, 60 mi (95 km) northwest of Burlington, forms the type localities for both the Maple Mill Shale and the English River Siltstone.

Fossils occur in the English River Siltstone at Starrs Cave as well-preserved molds; size, diversity, and abundance increase upward, and the fauna is especially rich in the 6-ft (2-m) interval beneath the Mississippian limestones. Weller (1900a) described a fauna of 81 species from this interval, dominated by bivalves (32 species), gastropods (21 species) and brachiopods (20 species), but including cephalopods (6 species) as well as conulariids, scaphopods, and bryozoans.

Abundance of the productellid brachiopod *Chonopectus fischeri* (Norwood and Pratten, 1855) motivated Weller to name the interval the "Chonopectus sandstone," a name that still persists.

Originally assigned as a portion of the Mississippian body stratotype, the age of the "Chonopectus sandstone" remained controversial until the recognition that specimens referred to *Goniatites opimus* (White and Whitfield, 1862) are properly assignable to the exclusively Late Devonian (Famennian) cylmeniid genus *Cyrtocylmenia* (*C. strigata* House, 1962). Presumed age differences between the "Chonopectus sandstone" and the type English River Siltstone have led to the suggestion (Straka, 1968) that the Starrs Cave sequence be referred to the Maple Mill Shale. However, in view of the continuing uncertainty as to the precise age of the type English River, and the probability of diachronism in such silt units, the term English River Siltstone is still appropriate for the sequence at Burlington.

Interpretation. Virtual absence of benthic biota, high carbonaceous content, and fine grain-size throughout most of the Late Devonian allochthonous sequences in the Upper Mississippi Valley attest to low-energy conditions, either below wave base or beyond wave reach. Progressively greater proportions of coarser clastics in the transition to the English River Siltstone may be interpreted as resulting from eustatic fall that exposed Early Paleozoic clastic sources on the Transcontinental Arch. Alternately, these changes may have accompanied deepening in an extensive shallow seaway with restricted circulation. Although still equivocal, regional relationships tend to favor the shoaling hypothesis, which is also readily compatible with emergence across the sys-

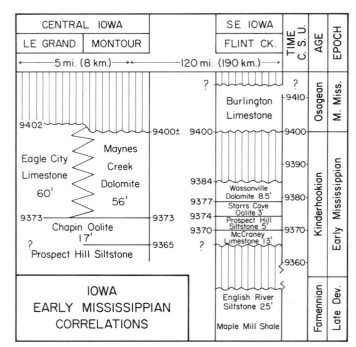

Figure 2. Correlation of Early Mississippian sections, based on graphic methods (Shaw, 1964; Miller, 1977) that utilize all known taxa. Time is expressed as Composite Standard Units (C.S.U.). These C.S.U. result from an equal subdivision of the Mississippian into 1,400 parts, accepting the Elsevier (1975) estimate of 35 m.y. for the period. The value 9,400 was chosen for the base of the Burlington Limestone at Flint Creek and has been projected elsewhere by graphic correlation. Flint Creek is the Starrs Cave Park and Preserve section in the Burlington area of southeastern Iowa. Approximately 120 mi (190 km) distant in central Iowa, Montour is the B. L. Anderson Quarry, Sec.9,T.83N.,R.16W., Tama County; and Le Grand is the Cessford Construction Company Quarry, Sec.1,T.83N.,R.17W., Marshall County.

temic boundary. Benthic environments became strikingly more favorable upward in the clastic sequence, presumably due to progressively higher energy and better circulation. Bioturbation is common first near the top of the Maple Mill Shale, and benthic fossils are abundant sporadically first at that level. However, it was not until the time of the upper few meters of the English River Siltstone at Starrs Cave that circulation improved sufficiently to sustain a diverse and abundant benthic fauna, much of which achieved large size. Molluscan dominance may reflect instability of the substrate or some lingering environmental hostility. However, excellent preservation of many delicate structures, such as spines, and common articulation of both bivalves and brachiopods attest to limited turbulence. Final emergence is indicated by the unconformity between the English River Siltstone and the McCraney Limestone.

Precise duration of the interval missing at the contact of the English River Siltstone and the McCraney Limestone is still uncertain. Ammonoids have not been collected from the Flint Creek section, but the clymeniid cephalopod fauna that indicates

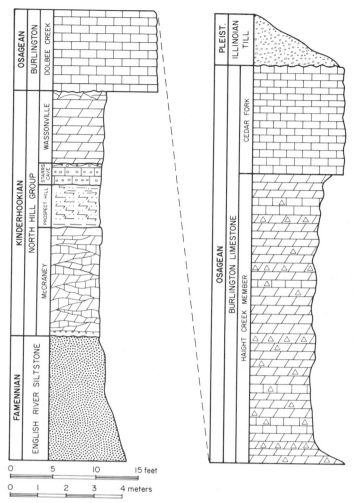

Figure 3. Stratigraphic column, Starrs Cave Park, Burlington, Des Moines County, Iowa.

youngest Devonian (late Famennian) *Clymenia* or *Wocklumeria* zones (House, 1962) is now known to be present in the upper 6.5 ft (2 m) of the English River Siltstone at Sullivan Slough, 7.5 mi (12 km) south of Starrs Cave. It is thus improbable that the English River Siltstone at Burlington extends into the Mississippian. However, graphic correlation (Fig. 2) suggests that much of the early Kinderhookian is missing at the unconformity.

McCraney Limestone. The 13-ft (4-m) McCraney Limestone succeeds the English River Siltstone with abrupt unconformable contact. Its strikingly banded appearance (Site YY herein) results from an alternation of buff-colored sublithographic limestone with thinner lenses and irregular beds of dark-colored laminated microcrystalline dolomite. The basal 2–4 in (5–10 cm) of the formation at Starrs Cave, as in all other Des Moines County exposures, is a distinctive coquina composed almost exclusively of a single minute species of chonetid brachiopod set in a spar matrix with later-infilling lime mud. The massive uppermost 2 ft (0.6 m) of the formation is also distinctive; it has an orange color and commonly lacks the conspicuous banding that characterizes the remainder of the unit. X-ray diffraction mineral percentages are available for sampled intervals within this and other Mississippian formations. They are omitted here due to space limitations.

The nature of the contacts between the buff limestone bands and the brown dolomite deserves detailed description (Site YY herein). Limestone bands commonly grade imperceptibly, both vertically and laterally, into the dolomite, with the transition occupying several millimeters to as much as 1 cm. Limestones display varying degrees of fracturing and brecciation, and fractures and areas between breccia fragments are generally filled with dolomite. All fracture contacts are abrupt, exhibiting no gradation in dolomite content.

The McCraney is the least fossiliferous formation of the North Hill Group, characterized by low diversity, relatively small individual size, and low abundance except for the chonetid coquina at the base of the formation. The fauna of the basal chonetid coquina (Weller, 1901) comprises four species of brachiopods, but specimens other than those of *Retichonetes loganensis* (*R. loganensis* [Hall and Whitfield, 1877] = *Chonetes gregarius* [see previous note] Weller, 1901) are rare. Greater diversity (brachiopods, 4 species; bivalves, 3 species; gastropods, 3 species; cephalopods, 1 species) characterizes a superjacent thin oolite that occurs sporadically in the Burlington area, but the main body of the McCraney is virtually devoid of fossils. The McCraney fauna is distinctive: only *Chonopectus fischeri* and *Syringothyris extenuatus* (Hall *in* Hall and Whitney, 1858) are now known to extend up from the English River Siltstone, and only *Retichonetes logani* (Norwood and Pratten, 1855) has been traced beyond the McCraney, into the Burlington Limestone.

Conodonts are extremely rare in the McCraney Limestone, but a sparse fauna from elsewhere in the Burlington area (Scott and Collinson, 1961) includes *Siphonodella cooperi* Hass. Although *S. cooperi* ranges as high as the Wassonville (Straka, 1968, Fig. 2), its first occurrence in the McCraney is significant. In the *Siphonodella* zonation of the Early Mississippian (Sandberg and others, 1978), the lowest occurrence of *S. cooperi* is in the Upper *duplicata* Zone. This evidence indicates that the McCraney is not earliest Mississippian.

Our graphic correlation (A.B.S.) of all faunal elements in the Starrs Cave section with those of two sites in the Le Grand area of central Iowa (Fig. 2) indicates that the McCraney at Flint Creek is contemporaneous with the lower part of the Chapin Oolite at Le Grand.

Interpretation. Both lithofacies and biofacies indicate that a more open environment existed at the time the base of the McCraney was formed than occurred higher in the formation. Small size and low diversity of the biota in the chonetid coquina suggest stressful conditions, perhaps associated with shallow early stages of the transgression, but paucity of primary mud matrix (that which occurs in the coquina was introduced after deposition) attests to circulation sufficient to achieve effective washing.

Sporadic occurrence of ooids in the basal McCraney is compatible with this interpretation. Mud removal was ineffectual throughout deposition of most of the formation above the basal coquina, presumably reflecting even shallower water depths insufficient to sustain wave motion and currents. Paucity and low diversity of fossils indicate extreme stress possibly resulting from environmental fluctuation in shallow water.

Prospect Hill Siltstone. The Prospect Hill Siltstone ("yellow sandstone" of authors) at Flint Creek is a highly fossiliferous, gray to yellowish brown, dolomitic quartz siltstone 5 ft (1.5 m) thick. Grains range in diameter from 0.03 to 0.07 mm. Thin green shale beds and partings occur sporadically throughout the unit, especially in the bottom 3 in (7.5 cm) and the top 1 ft (0.3 m). Weathered surfaces and polished slabs display extensive bioturbation, vertical burrows commonly exceeding 6 in (15 cm) in depth. Body fossils are sporadically abundant, occurring as molds that in some cases are filled with white kaolinite. Contact with the underlying McCraney is abrupt, across an undulating surface with up to 18 in (0.5 m) of relief. Lag concentrates of phosphatic pebbles and fish teeth and spines are common in depressions within the basal Prospect Hill.

The Prospect Hill maintains its identity and is recognizable in outcrop as far west as central Iowa. As with other Kinderhookian formations, it is older in that area, predating the McCraney at Burlington (Fig. 2).

A diverse Prospect Hill fauna was described by Weller (1901), and additional elements were listed by Moore (1928) to provide the following totals for species: bryozoans, 1; brachiopods, 31; bivalves, 24; gastropods, 10; scaphopods, 1; cephalopods, 3; and fish, 16. With the addition of numerous conodonts (Straka, 1968), the total Prospect Hill fauna exceeds 100 taxa.

As noted earlier, few species of megafossil are common to the McCraney and the Prospect Hill. However, approximately 20 percent of the taxa from the latter extend up into the Starrs Cave Oolite.

Conodont faunas from the Prospect Hill (Straka, 1968, Fig. 4) include *Siphonodella cooperi, S. obsoleta,* and *S. quadruplicata* and indicate a range from the *sandbergeri* Zone to the *isosticha*–Upper *crenulata* Zone (Sandberg and others, 1978).

Interpretation. Presence of more than 100 taxa of marine body fossils in the Prospect Hill Siltstone, some of large size, indicates open marine conditions that should have been favorable for carbonate production. High energy is attested by scour and by cross-bedding, presumably produced under shallow-water conditions. Quartz content within the formation reaches 78 percent at the base and averages 67 percent; no other part of the North Hill Group has a quartz content as high as 10 percent. Consequently, we conclude that the Prospect Hill represents a pulse of allochthonous sediment that was introduced into a shallow sea at a rate that diluted the rapid carbonate production. The original carbonate was removed subsequently by dissolution, further increasing the clastic percentage seen today.

Starrs Cave Oolite. The Starrs Cave Oolite is a thick-bedded, fossiliferous, buff to gray, oolitic limestone whose type locality lies within the preserve. At Flint Creek, it is only 2.5 ft (0.7 m) thick and is conformable with both the underlying Prospect Hill and the superjacent Wassonville Dolomite. Skeletal grainstones and oolites occupying comparable stratigraphic positions are recognizable in almost all Iowa Mississippian sections, and reach a thickness exceeding 60 ft (18 m) in south-central Iowa (Lawler, 1981). Like other Mississippian formations, these grainstones are older in central Iowa, where the equivalent Chapin Oolite wholly predates the Starrs Cave of southeastern Iowa (Fig. 2).

Bedding is characteristically tabular, although thinner sets of low-angle cross-beds are observed rarely. Ooids and associated bioclasts range in diameter from 0.1 mm to 1 mm, and are closely packed in sparry matrix. Where discernible, nuclei are echinoderm grains. Diagenetic alteration has been minimal, but includes some recrystallization and limited late dolomitization in the form of clear euhedral rhombs that cut across both grains and matrix.

More than 30 species of megascopic invertebrates have been recorded from the Starrs Cave Oolite (Weller, 1901; Moore, 1928), approximately one-half extending up from the Prospect Hill and one-quarter ranging into the Wassonville Dolomite. Included are the oldest species of coral and the one rostroconch (*Conocardium pulchellum* White and Whitfield, 1862) known from the Kinderhookian part of the section. Brachiopods predominate, with 20 species. Abundance and large size of the strophomenid brachiopod *Schellwienella planumbona* Weller (1914) served as a basis for definition of the *Schellwienella* zone (Laudon, 1931); two other species of *Schellwienella* are present, *S. crenullicostata* Weller (1914) extending into the Starrs Cave from the underlying Prospect Hill, and *S. inflata* (White and Whitfield, 1862) ranging from the Starrs Cave into the Burlington Limestone.

Interpretation. Environment of deposition for the Starrs Cave Oolite would have been basically similar to that envisioned for the Prospect Hill, except that provenance of grains was in oolite shoals rather than allochthonous siliciclastic sources. Paucity of cross-bedding and other evidence of high energy, as well as the brachiopod-dominated biofacies suggest that the Starrs Cave Park area was not itself a site where ooids were generated. No such "factory" is known for southeastern Iowa, although several have been recognized in the vicinity of Le Grand, central Iowa (Ressmeyer, 1983).

Wassonville Dolomite. At Flint Creek the Wassonville comprises 8.5 ft (2.6 m) of thick-bedded, tan to gold, iron-stained dolomite. The lower half of the unit is less dolomitized, coarser grained, and contains more recognizable fossils. A deeply weathered, rust-colored interval up to 2 ft (0.6 m) thick forms the top of the formation directly beneath the overhanging Burlington ledge. Silicification is confined to individual grains at Flint Creek, but highly fossiliferous chert lenses characterize the Wassonville at its type locality on the English River in Washington County. Unlike other formations of the North Hill Group, the Wassonville thickens to the northwest, where it merges with the Maynes Creek Dolomite of the Hampton Formation (Hager, 1981).

However, this thickening is primarily an expression of the unconformity beneath the Burlington Limestone (Fig. 2).

Relative paucity of identifiable fossils in the Wassonville is partly due to extensive dolomitization, and well-preserved fossils are abundant at the type locality where they are protected by early silicification (e.g., Rollins, 1975). A meagre fauna that includes large specimens was recorded from the Burlington area by Weller (1900b); additional listings for the Wassonville (e.g., Moore, 1928; Straka, 1968; Rollins, 1975) increase the total number of species to more than 50. Graphic correlation indicates that the Wassonville of the Burlington area represents a brief interval of late Kinderhookian time, and that a major hiatus separates this unit from the Middle Mississippian (Osagean) Burlington Limestone (Fig. 2).

Interpretation. Pervasive dolomitization precludes confident interpretation of environments of deposition for the Wassonville Dolomite. However, moderately large size and high diversity of fossils as well as preponderance of bioclasts in the lower one-half of the formation indicate open marine environments with sufficient water depth to permit winnowing of mud. Paucity of whole fossils or even bioclastic grains in the upper part of the formation suggests more restricted conditions.

Burlington Limestone. All three members of the Burlington Limestone are present in the Starrs Cave section, although access to the precipitous faces at the top and bottom of the unit is difficult. Type sections for the members (Harris and Parker, 1964) are within 12 mi (19 km) of the site. The cliff of massive limestone at the base of the formation corresponds approximately to the Dolbee Creek Member, the steep talus slope in the middle is the Haight Creek Member, and the cliff at the top is the Cedar Fork Member (Fig. 3). The Dolbee Creek comprises approximately 10 ft (3 m) of coarse crinoidal grainstone, sparsely dolomitized and with only rare chert nodules. It overlies the Wassonville with an erosional unconformity, and is succeeded conformably by the recessive Haight Creek Member. The latter consists of thin interbeds of brown dolomite and lighter-colored limestone, approximately 35 ft (11 m) thick. Irregular chert beds and nodules are characteristic, and glauconite occurs profusely throughout the unit. The Cedar Fork Member comprises 13 ft (4 m) of massive coarse crinoidal limestone, with glauconite disseminated throughout and nodules and lenses of chert appearing in the upper portion. It overlies the Haight Creek conformably, and is succeeded by Pleistocene (Pre-Illinoian) diamictite (till). Thickness of full sections of the Cedar Fork Member in the Burlington area averages 25 ft (7.6 m), so it can be assumed that pre-Pleistocene or glacial erosion has removed approximately 12 ft (3.7 m) of the uppermost Burlington at this site.

The Burlington Limestone is noted for the large size, high diversity, great abundance and excellent preservation of its fossils. Crinoid faunas are classic (e.g., Hall and Whitney, 1858; Wachsmuth and Springer, 1897) and serve for zonation and correlation (Laudon, 1937). Van Tuyl (1925) recorded over 100 species of invertebrates from the type Burlington: corals, 8; bryozoans, 9; brachiopods, 35; bivalves, 1; gastropods, 10; trilobites,

1; crinoids, 36; and blastoids, 3. Significant additions have followed, and conodonts are especially notable (Collinson and others, 1971; Lane, 1978; Chauff, 1981). The oldest conodont zone of the Middle Mississippian Osagean Series (*Polygnathus communis carina*) is not represented in the type Burlington, but succeeding zones (*Pseudopolygnathus multistriatus* through *"Polygnathus" mehli*) are present (Chauff, 1981).

Interpretation. Preponderance of grainstones and large size and high diversity of the Burlington biota all attest to open marine conditions and water depths sufficient to sustain wave motion and currents. Shallow-water correlatives of the Burlington Limestone are encountered in the upper Gilmore City Limestone ("Humboldt Oolite") of northcentral Iowa (Glenister and Sixt, 1982; Baxter and Brenckle, 1982; Sixt, 1983).

Pleistocene. The bluffs along the east bank of Flint Creek are capped by a thin veneer of Pleistocene glacial deposits. Exposures are inadequate for detailed observation, but some information is available in erosion gullies associated with the foot trail and from rare boulders scattered on the surface. General petrologic analysis indicates that the sequence comprises Pre-Illinoian tills assignable to the Hickory Hills Till Member of the Wolf Creek Formation (Hallberg, 1980).

SITE 28—DOLOMITIZATION IN THE McCRANEY LIMESTONE (MISSISSIPPIAN)

The McCraney Limestone is the basal unit of the Early Mississippian (Kinderhookian) North Hill Group, which is exposed widely beneath the Burlington Limestone in road, quarry, and stream sections in Des Moines County, southeastern Iowa. The most extensive accessible exposures are within Starrs Cave Park and Preserve (Site 27 herein), along Flint Creek in the northern outskirts of Burlington (W½Sec.19,T.70N.,R.2W., Des Moines County, Iowa). There and elsewhere the McCraney exhibits a strikingly banded appearance (Fig. 4A, 4B), resulting from alternation of buff-colored sublithographic limestone with thinner lenses and irregular beds of dark-colored laminated microcrystalline dolomite. Limestone beds commonly grade imperceptibly, both vertically and laterally, into the dolomite (Fig. 4B), with the transition occupying several millimeters to as much as 1 cm. The limestones display varying degrees of fracturing and brecciation (Fig. 4C–E), and fractures and areas between breccia fragments are generally filled with dolomite. All fracture contacts are abrupt, exhibiting no gradation in dolomite content. These relationships have been considered enigmatic, and it is the purpose of the present essay to explain their origin.

Differences between the brown and buff layers in the McCraney are largely of diagenetic origin. Irregular dark laminations in the brown dolomitic layers (Fig. 4E) are unlikely to be a modified primary feature because the dolomitic layers commonly grade laterally into the buff limestone, and in rare cases individual laminations may extend from the dolomite into the periphery of the limestone but are absent within the limestone bands themselves (Fig. 4B–E).

Figure 4. Exposure of Devonian–Mississippian section at Starrs Cave Park, southeastern Iowa, and details of basal Mississippian McCraney Limestone. *A.* Section on east bank of Flint Creek with figure for scale: ER = Late Devonian English River Siltstone, McC = Early Mississippian (Kinderhookian) McCraney Limestone, PH = Prospect Hill Siltstone, SC = Starrs Cave Oolite, W = Wassonville Dolomite, B = overhanging ledge of Burlington Limestone. *B-E.* Details of banding and brecciation in McCraney Limestone at Starrs Cave; light beds are limestone, dark areas are dolomite; *B* is natural exposure, with hammer for scale; *C-E* are polished slabs with 5 cm scale. Note gradational contacts between limestone and dolomite, except at fracture boundaries where contacts are abrupt. Light-colored circular area near top of *E* is secondary calcite.

We hypothesize that banding in the McCraney originated through coalescence of layers of carbonate nodules that formed repeatedly close to the sediment-water interface within a more-or-less homogeneous sediment. Carbonate was possibly supplied to the nodules by dissolution of aragonite near the top of the sediment column. Nodule formation was initiated close to the sediment-water interface (Fig. 5A) where diffusion paths were short and carbonate transfer rapid. Lateral coalescence of nodules produced irregular tabular sheets as sedimentation continued and diffusion paths lengthened (Fig. 5B). Eventually diffusion to individual tabular sheets became so slow that Ca^{2+} and HCO_3^- were no longer utilized as fast as they were supplied by surface dissolution: $CaCO_3$ concentration then initiated formation of a new nodule layer nearer to the sediment-water interface (Fig. 5C). This part of the interpretation is identical with that proposed by Jenkyns (1974) to explain nodular limestones from the Mediterranean Jurassic.

The matrix between the tabular sheets remained permeable and relatively uncompacted, and therefore susceptible to preferential dolomitization, perhaps by fluids derived from the highly restricted McCraney environment. Because the boundaries of the cemented tabular sheets were gradational, dolomitization of the uncemented matrix and the transitions between sheets and matrix produced the gradational dolomitized zones at the tabular sheet margins (Fig. 5D, E). Only isolated dolomite rhombs formed within the limestone sheets themselves. Finally, differential physical and chemical compaction of the dolomitized but unlithified matrix produced the dark laminations, which represent pressure solutional surfaces, as well as tensional fracturing and brecciation of the lithified tabular sheets. Dolomite matrix was injected into the voids, with coincident distortion of the dark laminations (Fig. 5F).

The significant corollary of these relationships is that here dolomitization was probably early diagenetic and occurred prior to complete lithification.

The processes involved in nodule formation were not carried to completion in the upper 2 ft (0.6 m) of the McCraney. Permeability thus remained relatively uniform throughout that interval, thereby allowing more-regular dolomitization and precluding development of distinct banding. Silt from the overlying Prospect Hill Siltstone was also able to penetrate into the interval.

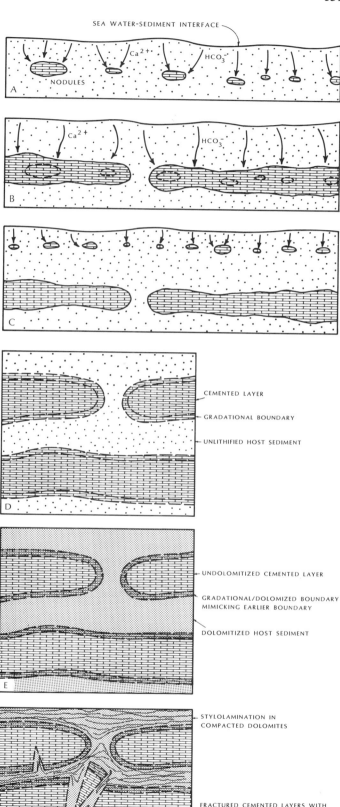

Figure 5. Carbonate nodule formation and selective penecontemporaneous dolomitization of McCraney Limestone. *A–C* represent repeated growth of nodule layers close to the sea floor and lateral coalescence of individual nodules to form lithified tabular sheets. Diffusion paths are marked by arrows; nodule outlines of previous stage are represented by dashed lines. *D,E* show dolomitization of unlithified matrix and gradational alteration around margins of limestone sheets. Density of stippling reflects dolomite concentration. *F* shows differential compaction and dissolution of dolomite matrix, producing tensional fracturing and brecciation of nodular sheets. Note that in *D–F,* upper few feet of sediment are omitted.

REFERENCES CITED

Baxter, J. W., and Brenckle, P. L., 1982, Preliminary statement on Mississippian calcareous foraminiferal successions of the Midcontinent (U.S.A.) and their correlation to western Europe: Newsletter on Stratigraphy, v. 11, p. 136–153.

Chauff, K. M., 1981, Multielement conodont species from the Osagean (Lower Carboniferous) in Midcontinent North America and Texas: Palaeontographica, Abteilung A, Band 175, p. 140–169.

Collinson, C., Rexroad, C. B., and Thompson, T. L., 1971, Conodont zonation of the North American Mississippian, *in* Sweet, W. C., and Bergström, S. M., ed., Conodont biostratigraphy: Geological Society of America Memoir 127, p. 353–394.

Glenister, B. F., and Sixt, S. C., 1982, Mississippian biofacies-lithofacies trends, northcentral Iowa: Geological Society of Iowa Guide, 21 p.

Hager, R. C., 1981, Petrology and depositional environments of the Hampton Formation (Kinderhookian) in central Iowa [M.S. thesis]: Iowa City, University of Iowa, 201 p.

Hall, J., and Whitfield, R. P., 1877, Palaeontology: United States geological exploration of the fortieth parallel, v. 4, pt. 2, p. 198–302.

Hall, J., and Whitney, J. D., 1858, Report of the geological survey of the State of Iowa, v. 1: Charles van Benthuysen, 724 p.

Hallberg, G. R. (ed.), 1980, Illinoian and Pre–Illinoian stratigraphy of southeast Iowa and adjacent Illinois: Iowa Geological Survey Technical Information Series 11, 206 p.

Harris, S. E., Jr., and Parker, M. C., 1964, Stratigraphy of the Osage Series in southeastern Iowa: Iowa Geological Survey Report of Investigations 1, 52 p.

House, M. R., 1962, Observations on the ammonoid succession of the North American Devonian: Journal of Paleontology, v. 36, p. 247–284.

Jenkyns, H. C., 1974, Origin of red nodular limestones (Ammonitico Rosso, Knollenkalk) in the Mediterranean Jurassic; A diagenetic model: International Association of Sedimentologists Special Publication 1, p. 249–271.

Lane, H. R., 1978, The Burlington Shelf (Mississippian, north-central United States): Geologica et Palaeontologica, v. 12, p. 165–176.

Lawler, S. K., 1981, Stratigraphy and petrology of the Mississippian (Kinderhookian) Chapin Limestone of Iowa [M.S. thesis]: Iowa City, University of Iowa, 118 p.

Laudon, L. R., 1931, The stratigraphy of the Kinderhook series of Iowa: Iowa Geological Survey Annual Report, 1929, v. 35, p. 333–451.

—— , 1937, Stratigraphy of northern extension of Burlington Limestone in Missouri and Iowa: American Association of Petroleum Geologists Bulletin, v. 21, p. 1158–1167.

Miller, F. X., 1977, The graphic correlation method in biostratigraphy, *in* Kauffman, E. C., and Hazel, J. E., ed., Concepts and methods of biostratigraphy: Stroudsburg, Pennsylvania, Dowden, Hutchinson and Ross, Inc., p. 165–186.

Moore, R. C., 1928, Early Mississippian formations in Missouri: Missouri Bureau of Geology and Mines, 2nd. ser., v. 21, 283 p.

Norwood, J. G., and Pratten, H., 1855, Notice of the genus Chonetes as found in the western states and territories, with descriptions of eleven new species: Academy of Natural Sciences of Philadelphia Journal, v. 3, p. 23–32.

Owen, D. D., 1852, Report of a geological survey of Wisconsin, Iowa and Minnesota: Philadelphia, Lippincott, Grambo and Co., 638 p.

Ressmeyer, P. F., 1983, Biostratigraphy and depositional environments of the Hampton Formation, Lower Mississippian (Kinderhookian) of central Iowa [M.S. thesis]: Iowa City, University of Iowa, 171 p.

Rollins, H. B., 1975, Gastropods from the Lower Mississippian Wassonville Limestone in Southeastern Iowa: American Museum of Natural History Novitates, no. 2579, 35 p.

Sandberg, C. A., Ziegler, W., Leuteritz, K., and Brill, S. M., 1978, Phylogeny, speciation, and zonation of Siphonodella (Conodonta, Upper Devonian and Lower Carboniferous): Newsletter on Stratigraphy, v. 7, p. 102–120.

Scott, A. J., and Collinson, C., 1961, Conodont faunas from the Louisiana and McCraney formations of Illinois, Iowa, and Missouri: Twenty-sixth Annual Field Conference of the Kansas Geological Society Guidebook, p. 110–141.

Shaw, A. B., 1964, Time in Stratigraphy: New York, McGraw-Hill, 365 p.

Sixt, S. C. (Smith), 1983, Depositional environments, diagenesis, and stratigraphy of the Gilmore City Formation (Mississippian) near Humboldt, north-central Iowa [M.S. thesis]: Iowa City, University of Iowa, 164 p.

Straka, J. J., 1968, Conodont zonation of the Kinderhookian Series, Washington County, Iowa: University of Iowa Studies in Natural History, v. 21, no. 2, 71 p.

Van Tuyl, F. M., 1925, The stratigraphy of the Mississippian formations of Iowa: Iowa Geological Survey Annual Reports, 1921 and 1922, v. 30, p. 33–374.

Wachsmuth, C., and Springer, F., 1897, The North American Crinoidea Camerata: Museum of Comparative Zoology at Harvard College Memoirs, v. 21, v. 22, 897 p.

Weller, S., 1900a, Kinderhook faunal studies; II. The fauna of the Chonpectus sandstone at Burlington, Iowa: Academy of Science of St. Louis Transactions, v. 10, p. 57–129.

—— , 1900b, The succession of fossil faunas in the Kinderhook beds at Burlington, Iowa: Iowa Geological Survey Annual Report, 1899, v. 10, p. 43–79.

—— , 1901, Kinderhook faunal studies; III. The faunas of beds No. 3 to No. 7 at Burlington, Iowa: Academy of Science of St. Louis Transactions, v. 11, p. 146–214.

—— , 1914, The Mississippian Brachiopoda of the Mississippi Valley Basin: Illinois Geological Survey Monograph 1; pt. 1, 508 p.; pt. 2, 187 p.

White, C. A., and Whitfield, R. P., 1862, Observations upon the rocks of the Mississippi Valley which have been referred to the Chemung Group of New York, together with descriptions of new species of fossils from the same horizon at Burlington, Iowa: Boston Society of Natural History Proceedings, v. 8, p. 289–306.

Williams, H. S., 1891, The lower Carboniferous or Mississippian series; The development of the nomenclature, and classification of the lower Carboniferous formations of the Mississippian province: U.S. Geological Survey Bulletin, no. 80, p. 135–172.

ACKNOWLEDGMENTS

Many of the interpretations presented in the text were developed during our tenure as joint leaders of the Amoco Production Company Seminar on Stratigraphic Principles. We thank Amoco for release of proprietary data, and acknowledge the stimulation of discussions with numerous seminar participants. Reviews of portions of the manuscript by Gilbert Klapper (Univ. Iowa); and Bill J. Bunker, George R. Hallberg, and Brian J. Witzke (Iowa Geological Survey) resulted in improvements.

Lover's Leap, type section Hannibal Shale, Missouri

Thomas L. Thompson, *Department of Natural Resources, Missouri Division of Geology and Land Survey, Box 250, Rolla, Missouri 39216*

LOCATION

This is a major Mississippi River bluff exposure above a road next to the railroad yard on the southeast edge of Hannibal, Marion County, Missouri, just west of center of SE¼ Sec. 28, T.57N.,R.4W., Hannibal East 7½-minute Quadrangle (Fig. 1).

SIGNIFICANCE

Located in the stratotype region of the Mississippian System, this section is in the type region of the Kinderhookian Series and Hannibal Shale, as proposed by Keyes (1892), and has been considered the type locality of the Hannibal since the description by Koenig and others (1961).

The excellent exposure of the contact of the Hannibal and the underlying Louisiana Limestone was used by Scott and Collinson (1961) in their definition of the Devonian-Mississippian boundary for the Mississippian stratotype region. Their study clearly placed the Devonian-Mississippian boundary between the Upper Devonian Louisiana Limestone and the overlying Lower Mississippian (Kinderhookian) Hannibal Shale. Thus, at this section are exposed both the youngest Devonian and the oldest Mississippian formations in the upper Mississippi Valley region.

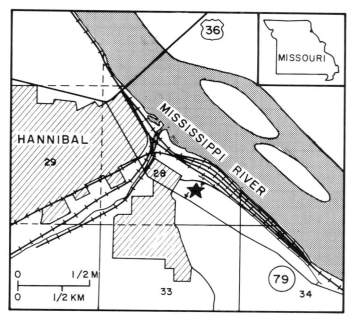

Figure 1. Portion of Hannibal East 7½-minute Quadrangle, Marion County, northeastern Missouri, showing location of bluff called "Lover's Leap," the type section of the Hannibal Shale.

DESCRIPTION

This section comprises three well-exposed formations: the Burlington Limestone, and underlying Hannibal Shale and Louisiana Limestone (Fig. 2). Because the upper Hannibal and Burlington strata are difficult to reach physically on the bluff section, the roadcut on Missouri 79, less than 0.25 mi (0.4 km) southwest of the bluff section, can be used to view close-up the upper part of the Hannibal and the Burlington Limestone (SW¼SE¼ Sec. 28,T.57N.,R.4W.). The description of the type section of the Hannibal Shale is as follows (from Koenig and others, 1961; and Collinson and others, 1979):

MISSISSIPPIAN SYSTEM

Osagean Series

Burlington Limestone

8. Limestone, gray to cream colored, coarsely crystalline, beds 3 to 4 ft (0.9 to 1.2 m) thick, containing beds and concretions of chert, some beds of brown sandy limestone, one 18-in (45 cm) bed of brown sandstone near the top. Exposed as vertical cliffs, and not accessible for detailed description . . . . 80-90 ft (24-27 m)
7. Limestone, brown, coarsely crystalline, highly crinoidal . . . . 10-12 ft (3-3.6 m)
6. Sandstone, brown, in one bed . . . . 1 ft (0.3 m)
5. Limestone, brown, finely crystalline, highly crinoidal . . . . 6-8 ft (1.8-2.4 m)

Kinderhookian Series

Hannibal Shale

4. Siltstone, gray, very argillaceous, massive, slightly calcareous, vermicular in upper part; shale, gray, very silty, calcareous, pyritic, interbedded with siltstone; all unfossiliferous . . . . 13-14 ft (4-4.25 m)
3. Siltstone, gray, argillaceous, slightly calcareous, massive, vermicular, discontinuous shaly streaks; shale, gray, very silty in lower part . . . . 12-14 ft (3.6-4.25 m)
2. Shale, gray, very silty, vermicular in lower part; siltstone, very argillaceous, interbedded with shale; accumulations of eroded material on lower slopes give appearance of more shale than is present. Some gradation at contact with underlying carbonates, with some brown coloration of lowermost beds . . . . 35-40 ft (10.6-12 m)

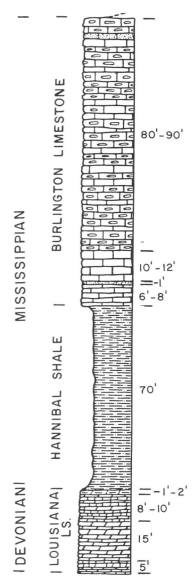

DEVONIAN SYSTEM

Upper Devonian Series

Louisiana Limestone

1. Limestone, light gray to gray, lithographic, pure, irregularly interbedded with dolomitic clay partings and buff to brown sucrosic dolomite. Beds thin in uppermost 10 ft (3 m), increasing in thickness downward. Beds of dolomite most numerous in upper part, decreasing downward. Limestone weathers very light gray and breaks with conchoidal fracture. Formation capped by thick dolomite bed whose uneven upper surface apparently represents an eroded surface . . . . 30-35 ft (9-10 m)

Road level, base of section.

REFERENCES CITED

Collinson, C., Norby, R. D., Thompson, T. L., and Baxter, J. W., 1979, Stratigraphy of the Mississippian Stratotype—Upper Mississippi Valley, U.S.A.: Field trip #8, 9th International Congress of Carboniferous Stratigraphy and Geology, Illinois Geological Survey, 108 p.

Keys, C. R., 1892, The principal Mississippian section: Geological Society of America Bulletin, v. 3, p. 283–300.

Koenig, J. W., Martin, J. A., and Collinson, C. W., 1961, Guidebook, 26th Regional Field Conference, Kansas Geological Society, Northeastern Missouri and west-central Illinois: Missouri Geological Survey and Water Resources, Report of Investigation 27, p. 1–74.

Scott, A. J., and Collinson, C., 1961, Conodont faunas from the Louisiana and McCraney Formations of Illinois, Iowa, and Missouri, *in* Guidebook 26th Regional Field Conference, Kansas Geological Society, Northeastern Missouri and west-central Illinois: Missouri Geological Survey and Water Resources, Report of Investigations 27, p. 110–141.

Figure 2. Columnar section of exposure at Lover's Leap, type section of Hannibal Shale, Marion County, northeastern Missouri. Section taken from Koenig and others (1961).

Geology of the Circumferential Highway System at Kansas City, Missouri

Richard J. Gentile, Department of Geosciences, University of Missouri-Kansas City, Kansas City, Missouri 64110

LOCATION

This section is a roadcut at the I-470 exit at Raytown Road (SW¼NW¼Sec.33,T.48N.,R.32W., Jackson County, Missouri, Lees Summit 7½-minute Quadrangle (Figs. 1 and 2). Park along Raytown Road north of the bridge over I-470 and walk east along the access road and onto the grassy strip of right-of-way about 200 ft (60 m) wide on the north side (westbound lane) of I-470. The section, with faults, is exposed about 1000 ft (300 m) east of Raytown Road.

SIGNIFICANCE

A "textbook example" of normal faulting is well exposed in the thick limestone members of the Pennsylvanian lower Kansas City Group. The section includes the Bethany Falls Limestone Member of the Swope Formation. The Bethany Falls is a ledge of limestone exploited commercially in quarry and mining operations throughout the metropolitan Kansas City area.

DESCRIPTION

The bedrock underlying metropolitan Kansas City and ad-

jacent areas is mostly a thick sequence of limestone and shale beds that alternate in regular fashion throughout the section. A total thickness of about 350 ft (105 m) of strata is exposed in bluffs, along streams and rivers, and in artificial exposures such as quarries and highway excavations. A complete section of Pennsylvanian strata can be constructed from the numerous exposures in road cuts along I-470 and I-435 of the Circumferential Highway System of Kansas City. Figure 3 is a columnar section of the lower part of the exposed Pennsylvanian strata in the Kansas City, Missouri area.

The thickness and rock characteristics of individual beds of limestone, shale, and sandy shale are persistent throughout the metropolitan Kansas City area and beyond. Some beds are a few inches (several cm) thick, whereas others are over 50 ft (15 m). Many of the units have been traced along the outcrop band that extends southwest-northeast from near the Kansas-Oklahoma border into Iowa.

These rocks consist of sediments that were deposited in a variety of environments including shallow seas, estuaries, lagoons, tidal flats, alluvial plains, and swamps. Thin coal beds in the rock sequence indicate that swamp conditions existed for short periods. Remains of nonmarine organisms, particularly plants found in some of the shale and sandy shale beds, indicate emergence of

Figure 1. Map of highway system in the Kansas City, Missouri region, northwestern Missouri, with location of intersection of Raytown Road and I-470 indicated by black dot.

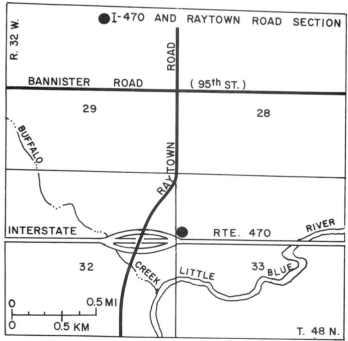

Figure 2. Detailed map of intersection of Raytown Road and I-470 showing location of section described. (From U.S. Geological Survey, Lees Summit 7½-minute Quadrangle).

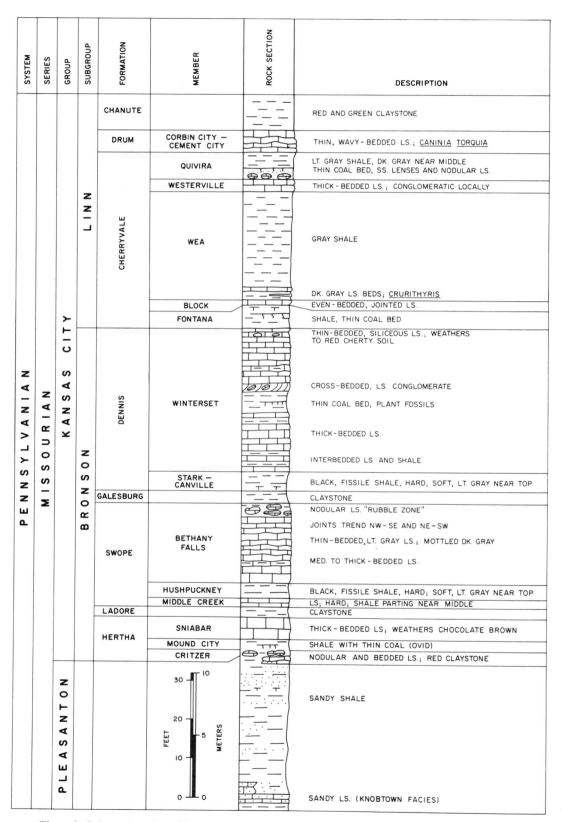

Figure 3. Columnar section of Pennsylvanian strata exposed in the vicinity of I-470 and Raytown road.

Figure 4. Photograph of exposure at intersection of Raytown Road and I-470, Kansas City, Missouri, showing two small normal faults (below arrows in upper part of photo) revealed during construction of I-470. Photograph by R. J. Gentile.

the land, with sediment deposition under riverine conditions. However, the limestone and shale beds that constitute most of the section were deposited in open marine or brackish water environments; the beds contain the remains of abundant shallow marine organisms. Representatives of almost all major invertebrate phyla that occur as fossils have been found, including algae, protozoans, sponges, corals, brachiopods, bryozoans, molluscs, arthropods, and echinoderms; in addition, fossils of unknown zoological affinities, such as conodonts and conularids, have been recovered. The remains of vertebrates, particularly shark teeth, are found occasionally. The types of fossils and the cyclical sedimentary succession indicate that shallow seas transgressed and regressed repeatedly across the flat, low lying interior of the North American Continent during the Late Pennsylvanian.

The Pennsylvanian bedrock in the region is covered by surficial deposits of glacial till and outwash, loess, and soil. The accumulation of till and outwash of early Pleistocene (Kansan) age is over 30 ft (9 m) thick and caps many of the higher hills north of the Missouri and Kansas Rivers, but it has been removed by erosion along the valleys of the major tributary streams and in most places south of the Kansas and Missouri River valleys. However, till and outwash gravels uncovered in excavations in

localized areas south of the major rivers, including downtown Kansas City, are evidence that glacial lobes pushed south of the modern Missouri and Kansas River channels.

The Missouri and Kansas River flood plains are underlain in places by over 100 ft (30 m) of Late Pleistocene and Holocene alluvial sands and gravels. A loess cover overlies the glacial till and outwash gravels except in areas where drift is absent; at these places, loess rests on Pennsylvanian bedrock.

Loess along the Missouri River bluffs is over 75 ft (22 m) thick and is exposed in the faces of high road cuts at the junction of I-435 and Missouri 210, and from 3½ miles (5.6 km) west of I-435, along Missouri 210, to near the municipality of North Kansas City. The loess is Late Pleistocene windblown silt consisting mostly of quartz, calcite, feldspar, and various accessory minerals and rock fragments, and is easily recognized by its brown color and its property of standing in vertical fracture faces in excavations. The silt is believed to have been blown from the floodplain of the Missouri River by westerly winds and deposited on the river bluffs. Loess thickness diminishes with distance from the major rivers; it is less than 15 ft (5 m) in most of the area included in the Circumferential Highway System.

At the I-470 location the rocks have been fractured by

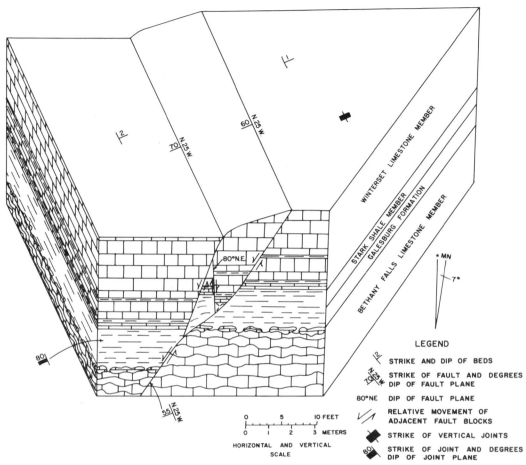

Figure 5. Block diagram showing interpretation of fault exposed at intersection of Raytown Road and I-470 in Kansas City, Missouri.

faulting. The section of Pennsylvanian strata exposed along I-470 is shown in Figure 3. The strata along the route are faulted at several locations. Faults are particularly conspicuous in the thick limestone members of the lower Kansas City Group and are textbook examples of normal faults. Figure 4 shows two of them shortly after they were exposed in a highway excavation. Fault scarps have been removed and the road bed graded. The faulting does not extent into the thick covering of unconsolidated loess and soil. The faults shown in Figure 4 form the sides of a northwest-trending structural block about 200 ft (60 m) wide. Figure 5 is a detailed sketch of the fault on the northeastern side of the block.

The *Bethany Falls Limestone Member of the Swope Formation,* exposed at road level, is a light gray, finely crystalline limestone mottled with dark-gray spots or blotches. The upper 2 or 3 ft (½–1 m) consists of limestone nodules in a greenish-gray matrix.

The Bethany Falls is the most extensively quarried limestone in the metropolitan Kansas City area and probably in western Missouri and eastern Kansas. The exploitation of the Bethany

Falls Limestone is closely related to the industrial development of Kansas City. Large-scale mining operations began before the turn of the century. The Bethany Falls is 20–25 ft (6–7½ m) thick throughout the Kansas City area and commonly forms bluffs along valley walls. Underground mining is particularly common where thickness of overburden makes surface mining uneconomical. Such underground mining operations have left several square miles of mined-out space, much of which is currently used for business operations and commercial storage.

The Bethany Falls is overlain by the *Galesburg Formation,* a medium gray claystone about 3 ft (1 m) thick. The Galesburg forms a seal that prevents water seepage into mined-out areas. Above the Galesburg, the *Dennis Formation* includes, in ascending order, the *Canville Member,* represented by a thin bed of fossiliferous shale about an inch (2 cm) thick; the *Stark Shale Member,* about 4 ft (1½ m) thick, the lower half black fissile shale with abundant conodonts; and the *Winterset Limestone Member,* 30–40 ft (9–12 m) thick, the lower part interbedded with shale. At this stop the upper half of the Winterset is deeply weathered. Overlying the Winterset, about 20 ft (6 m) of soil and loess form

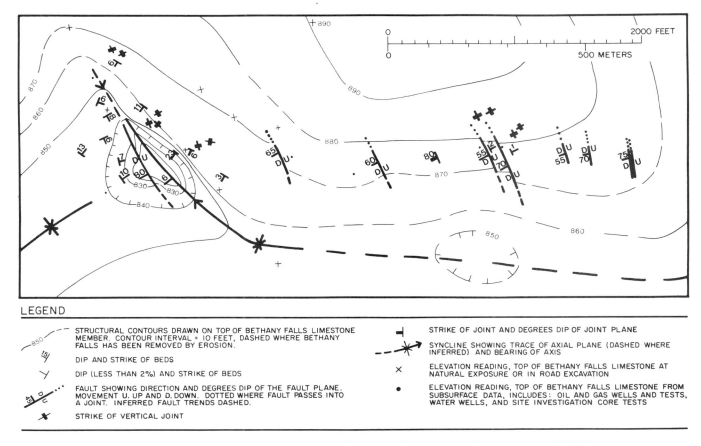

LEGEND

STRUCTURAL CONTOURS DRAWN ON TOP OF BETHANY FALLS LIMESTONE MEMBER. CONTOUR INTERVAL = 10 FEET, DASHED WHERE BETHANY FALLS HAS BEEN REMOVED BY EROSION.

DIP AND STRIKE OF BEDS

DIP (LESS THAN 2%) AND STRIKE OF BEDS

FAULT SHOWING DIRECTION AND DEGREES DIP OF THE FAULT PLANE. MOVEMENT U. UP AND D. DOWN. DOTTED WHERE FAULT PASSES INTO A JOINT. INFERRED FAULT TRENDS DASHED.

STRIKE OF VERTICAL JOINT

STRIKE OF JOINT AND DEGREES DIP OF JOINT PLANE

SYNCLINE SHOWING TRACE OF AXIAL PLANE (DASHED WHERE INFERRED) AND BEARING OF AXIS

ELEVATION READING, TOP OF BETHANY FALLS LIMESTONE AT NATURAL EXPOSURE OR IN ROAD EXCAVATION

ELEVATION READING, TOP OF BETHANY FALLS LIMESTONE FROM SUBSURFACE DATA, INCLUDES: OIL AND GAS WELLS AND TESTS, WATER WELLS, AND SITE INVESTIGATION CORE TESTS

Figure 6. Detailed structure map of complex faulting exposed in excavations for construction of I-470 in fall and winter of 1976.

a grass covered slope.

The part of the section below the Bethany Falls Limestone Member shown in Figure 3 is exposed along Buffalo Creek west of Raytown Road. Beds overlying the Winterset were exposed in the roadbed excavation through a low hill west of Buffalo Creek, but the exposure is mostly overgrown by vegetation.

ORIGIN OF STRUCTURE

The faulting was exposed in construction excavations for I-470 in the fall and winter of 1976, including a 4000-ft (1200-m) long segment of the route from just west of Buffalo Creek to about 3000 ft (910 m) east of the Raytown Bridge (Fig. 6). The faulting comprises a series of ten northwest striking parallel faults, which could be traced along the strike, perpendicular to the roadway, for about 400 ft (120 m) across the area under construction until they became concealed by regolith. All are high-angle normal faults, the fault planes dipping southwest toward a deep structural depression along Buffalo Creek. They are step faults, the downthrown blocks on the southwest sides of several parallel faults; the strata between having moved downward stepwise in relation to the adjacent faults to the northeast.

The steeply dipping strata exposed in the bed of Buffalo Creek can be projected many feet below the level of the valley floor, as determined by the records of core-drill test borings. Moreover, in most places the hill slopes are normal to the fault strikes, therefore, the faulting is not the result of downhill creep by slump blocks toward the lower elevation of the valley.

Several small faulted areas similar to the one at this stop have been recognized in western Missouri and northeastern Kansas (Gentile, 1984a, b); they contrast sharply with the relatively undisturbed nature of the bedrock surrounding them. The following paragraph discusses their probable origin.

It is reasonable to assume that the type of faulting mapped along I-470 is caused by removal of support from below. Gentile (1984a, b) has proposed that these small faulted areas formed when the Pennsylvanian strata collapsed into caverns enlarged by dissolution along fracture zones in thick Mississippian limestones and dolomites underlying the region at depths of 600–800 ft (180–240 m). An extensive system of filled solution cavities is known to exist in Mississippian limestones throughout the Midcontinent. Moreover, these limestones are thick enough to account for the displacement recorded at the surface, using key marker beds in Pennsylvanian strata as datums. In comparison, the Pennsylvanian limestones are not thick enough, if removed by cavern development, to allow this much displacement. These structures did not form in modern times, because the faulting does not extend into the thick covering of unconsolidated Late Pleistocene and Holocene surficial materials.

Deep drill tests and detailed geophysical investigations, especially seismic reflector surveys, are needed to determine subsurface structure.

REFERENCES CITED

Gentile, R. J., 1984a, Paleocollapse structures: Longview region, Kansas City, Missouri: Association of Engineering Geologist Bulletin, v. 21, n. 2, p. 229–247.

—— , 1984b, The Geology of the Belton 7½-minute Quadrangle Missouri-Kansas with special emphasis on the Belton Ring-fault complex: Missouri Department of Natural Resources Division of Geology Report of Investigations no. 69, 110 p.

Howe, W. B., 1961, The stratigraphic succession of Missouri: Missouri Geological Survey and Water Resources, 2nd Series, v. 40, 185 p.

Parizek, E. J., 1968, Geology of the Lees Summit Quadrangle, Jackson County, Missouri: Missouri Geological Survey and Water Resources, Geologic Quadrangle Map Series, No. 1, one sheet, scale 1:24,000.

Wanless, H. R., and Wright, C., 1978, Paleoenvironmental maps of Pennsylvanian rocks, Illinois basin and northern Midcontinent region: Geological Society of America MC-23, 32 p., 164 figures.

Zeller, D. E., 1968, The stratigraphic succession in Kansas: Kansas Geological Survey, Bulletin 189, 81 p.

ACKNOWLEDGMENTS

Dan Chappell, Dept. Geosciences, University of Missouri Kansas City, drafted the illustrations.

House Springs, Missouri, roadcut and fault

Thomas L. Thompson, *Department of Natural Resources, Missouri Division of Geology and Land Survey, Box 250, Rolla, Missouri 39216*

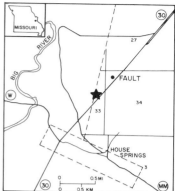

Figure 1. Portion of House Springs 7½-minute Quadrangle map, northern Jefferson County, east-central Missouri, showing locations of roadcut described and exposure of fault crossing Missouri 30.

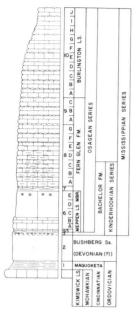

Figure 2. Columnar section of strata exposed in roadcut on Missouri 30 near House Springs, east-central Missouri. Section taken from Collinson and others (1979).

LOCATION

This section is a roadcut on Missouri 30 approximately 0.5 mi (0.8 km) northeast of the junction of Missouri 30 with Jefferson County road MM, the turnoff to the town of House Springs; SE¼SE¼NE¼ Sec. 33,T.43N.,R.4E., Jefferson County, Missouri, House Springs 7½-section Quadrangle Map (Fig. 1). An excellent exposure of a northwest-trending fault is in the NW¼NW¼ Sec. 34,T.43N.,R.4E.

SIGNIFICANCE

Roadcuts at this site expose a sequence of strata from the Middle Ordovician Plattin Limestone through the late Early Mississippian Burlington Limestone (Figs. 2, 3). The Bushberg Sandstone and the overlying Bachelor Sandstone illustrate the unusual relationship of two faunally distinct Kinderhookian sandstones in contact. The Bachelor is middle Kinderhookian, whereas the Bushberg is very early Kinderhookian (some believe very Late Devonian). This section is in a region of southeastern Missouri where there are no strata equivalent to the Chouteau (Upper Kinderhookian); the Middle Kinderhookian Bachelor is overlain by Osagean strata, the Meppen Limestone Member of the Fern Glen Formation (Thompson, 1975). There is no evidence of erosional removal of post-Bachelor Kinderhookian strata; a period of nondeposition, or "still-stand," during the Late Kinderhookian is suggested.

Roadcuts 1 mi (1.6 km) northeast of this first exposure also expose a northwest-southeast–trending normal fault that appears strongly asymmetrical (Figs. 4, 5). Evidence of apparent vertical displacement and resulting drag are on the northeastern upthrown side of the fault. The downthrown southwestern side, however, shows little evidence of vertical displacement, but it is distorted (Fig. 6), giving the appearance of compression instead of vertical drag. Vertical displacement appears to be small, and

the fresh nature of the gouge area makes it difficult to determine the width of the fault zone. This exposure is along a northwest-southeast trending structure, the Eureka–House Springs Anticline, which appears to plunge northwest and southeast. The exposures are readily accessible roadcuts along a two-lane paved highway.

DESCRIPTION

The following description (Fig. 3) is from the guidebook for the Mississippian Stratotype field trip of the 9th International Carboniferous Congress held in Urbana, Illinois, during the summer of 1979 (Collinson and others, 1979).

MISSISSIPPIAN SYSTEM

Osagean Series

Burlington Limestone

Limestone, light gray, medium crystalline to crinoidal; chert, white, 50 percent in lower 10 ft (3 m), 20 percent in upper 15 ft (4.5 m) .... 30 ft (9.1 m)

Fern Glen Formation

Limestone, grayish-brown, finely crystalline, fossiliferous zones; chert, bluish-gray to white, 40 to 50 ...... 10 ft (3 m)

Limestone, grayish-green lower part, very argillaceous, change to gray-brown 10 ft (3 m) above base; lower 3 ft (0.9 m) chert-free; upper part with bluish-gray and pink nodules; very fossiliferous in upper part .... 20 ft (4.5 m)

Limestone, brown, finely crystalline to sublithographic;

Figure 3. Photograph of roadcut on Missouri 30 near House Springs (SE¼SE¼NE¼ Sec. 33, T.43N., R.4E.; Jefferson County, Missouri) exposing Bushberg and Bachelor Sandstones (unit 1), Fern Glen Formation (units 2 and 3; unit 2 is Meppin Limestone Member), and Burlington Limestone (unit 4). Photograph by T. L. Thompson.

shale, green, siliceous nodules .... 2 ft (0.6 m)

Meppen Limestone Member

Limestone, brownish-gray to brown, very fine at bottom becoming coarser toward top; scattered fossil debris; large calcite-filled vugs in upper part; lowermost bed arenaceous with elongate calcite vugs .... 15 ft (4.5 m)

Kinderhookian Series

Bachelor Sandstone

Sandstone, light green, calcareous; phosphatic nodules in lower part .... 1 ft, 6 in (0.45 m)

Bushberg Sandstone

Sandstone, yellowish-brown, calcareous, fine to coarse, massive, porous .... 8 ft (2.4 m)

ORDOVICIAN SYSTEM

Cincinnatian Series

Maquoketa Shale

Shale, blue-gray to dark gray, silty, hard fossiliferous .... 3 ft (0.9 m)

Mohawkian Series (Champlainian Series)

Kimmswick Limestone

Limestone, poorly exposed, mostly removed by pre-Maquoketa erosion .... 0-3 ft (0-0.9 m)

Decorah Formation

Limestone and calcareous shale, forms reentrant at top of Plattin exposure; very fossiliferous .... 5 ft (1.5 m)

Plattin Limestone

Limestone, dark gray, finely crystalline to sublithographic, burrowed .... 25+ ft (7.6+ m)

REFERENCES CITED

Collinson, C., Norby, R. D., Thompson, T. L., and Baxter, J. W., 1979, Stratigraphy of the Mississippian Stratotype—Upper Mississippi Valley, U.S.A.: Field trip #8, 9th International Congress of Carboniferous Stratigraphy and

Figure 4. Photograph of roadcut on Missouri 30 about 1/3 mile northeast of roadcut shown in Figure 3 (NW¼NW¼ Sec. 34.T.43N, R.4E., Jefferson County, Missouri) exposing northwest-trending normal fault. Arrows locate approximate plane of fault. Photograph by T. L. Thompson.

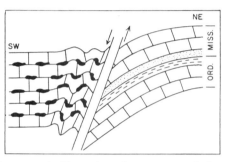

Figure 5. Diagrammatic illustration of strata across fault exposed on Missouri 30 near House Springs. Illustration taken from Collinson and others (1979).

Figure 6. Photograph of quarry face adjacent to roadcut shown in Figure 5, illustrating distortion of Mississippian chert and limestone beds on the apparent downthrown side of the fault. Photograph by T. L. Thompson.

Geology, Illinois Geological Survey, 108 p.

Thompson, T. L., 1975, Redescription and correlation of the Fern Glen Formation of Missouri: Missouri Department of Natural Resources, Division of Geological Survey, Report of Investigations 57, pt. 5, p. 141–172.

I-55 roadcuts, Herculaneum-Barnhart, Missouri

Thomas L. Thompson, Department of Natural Resources, Missouri Division of Geology and Land Survey, Box 250, Rolla, Missouri 39216

LOCATION

This is a sequence of roadcuts on I-55, extending from the NW¼ Sec.19,T.41N.,R.6E., to the SE¼ Sec.19,T.42N.,R.6E. (Fig. 1). These sections are accessible from the frontage road on the west side and from the Interstate highway. All exposures are in the Herculaneum 7½-minute Quadrangle.

SIGNIFICANCE

These are excellent exposures of nearly all the Middle and Upper Ordovician formations of eastern Missouri; they are also representative of Lower Mississippian rocks of the region. They are within a few mi (km) of the type sections of the Joachim, Plattin, and Kimmswick (Middle Ordovician), and Glen Park and Bushberg (Devonian and Mississippian-?) formations.

DESCRIPTION

Middle (Whiterockian and Mohawkian) and Upper (Cincinatian) Ordovician formations exposed in eastern Missouri consist of the following formations:

Cincinnatian	Maquoketa Shale (Maquoketa Group)
	Cape Limestone
Mohawkian	Kimmswick Limestone
	Decorah Formation
	Plattin Limestone
	Joachim Dolomite
	St. Peter Sandstone
Whiterockian	Everton Formation

Of these, all but the Everton are represented in this sequence of roadcuts; the Everton is exposed approximately 2 mi (3.2 km) south of the I-55 bridge over Joachim Creek at the south end of the sequence of exposures (Fig. 1).

Upper Ordovician (Cincinnatian) rocks are represented by the Cape Limestone and overlying Maquoketa Shale. The contact of the Cape Limestone with the underlying Kimmswick Limestone is a "welded contact," not visibly apparent. This contact, however, represents a disconformity between Late Middle Ordovician and Middle Late Ordovician (Sweet and others, 1975). Kimmswick strata were partially removed before Upper Ordovician deposition. Post-Ordovician pre–Late Devonian erosion removed most Upper, and in some cases Middle, Ordovician rocks before Upper Devonian deposition. The lenticular limestone overlying the Maquoketa Shale has been identified as the Upper

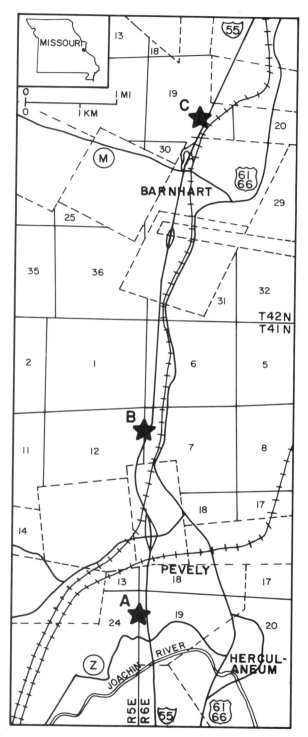

Figure 1. Portion of Herculaneum 7½-minute Quadrangle, eastern Jefferson County, east-central Missouri, showing locations of sections described from roadcuts on I-55.

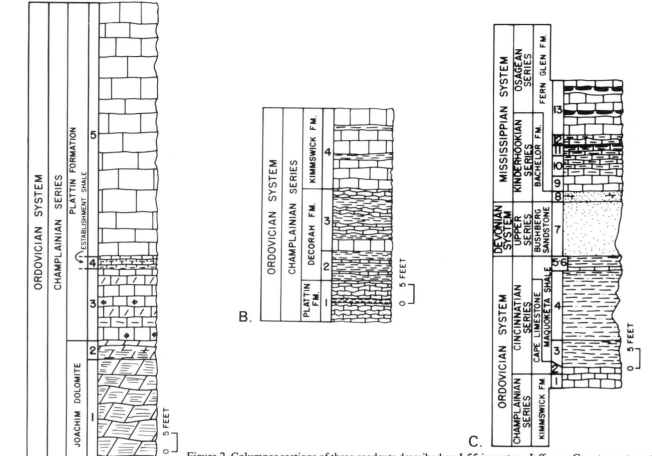

Figure 2. Columnar sections of three roadcuts described on I-55 in eastern Jefferson County, east-central Missouri. Sections taken from Thacker and Satterfield (1977).

Devonian Glen Park Limestone, which is overlain by the Lower Mississippian (Kinderhookian) Bushberg and Bachelor Sandstones. The Osagean Series (Late Early Mississippian) is represented by the Fern Glen Formation, no Kinderhookian equivalents of the Chouteau or Hannibal formations being present in the region (Thompson, 1975).

The following section descriptions (Fig. 2) by Thacker and Satterfield (1977), are exemplary of the strata exposed in this sequence of roadcuts. All exposures are in the Herculaneam 7½-minute quadrangle.

A. Exposure on east side of north-bound lane of I-55, approximately 0.5 mi (0.8 km) north of bridge over Joachim Creek, in the W½SW¼NW¼ Sec.19,T.41N.,R.6E., Jefferson County, Missouri, on the southwestern edge of the town of Pevely (Figs. 2A, 3).

ORDOVICIAN SYSTEM

Champlainian Series (Mohawkian Series)
Plattin Limestone
5. Limestone; lithographic, massively bedded, chocolate brown, weathers to buff; fucoidal; a 3-in (7.6-cm) green shale bed occurs 19 ft (5.8 m) above the base; upper half of unit weathers to a nodular surface; thin stringers of highly fossiliferous limestone occur throughout unit; an 8-in-(20-cm) thick bed of coarse-grained limestone cal carenite occurs approximately 2 ft (60 cm) above bench in roadcut; uppermost 5 ft (1.5 m) of unit is highly fos siliferous with brachiopods and bryozoan debris standing out in relief on the weathered . . . . 62–65 ft (18-20 m)
4. Interbedded thin limestone and green shales; lime stone beds are 1 to 3 in (2.5 to 7.6 cm) thick, and shale beds are 2 to 8 in (5 to 20 cm) thick. (Establishment Shale Member of Plattin Formation . . . . 3 ft (0.9 m)
3. Alternating lithographic limestone and finely crystal line calcareous dolomite beds with very thin beds of green shale; limestone beds are chocolate brown and en tire unit weathers to buff; thin to massively bedded and fucoidal; oolitic limestone beds occur at base of unit and at midsection: 2 ft (0.6 m) above base is a 2-ft (0.6 m) greenish, argillaceous limestone bed; base of unit is high ly undulating . . . . 19 ft (5.8 m)

Figure 3. Photograph of roadcut on I-55 immediately north of Pevely (Section A, Fig. 2; (W½SW¼NW¼ Sec. 19, T.41N., R.6E., Jefferson County, Missouri), exposing Joachim Dolomite (units 1, 2) and lower Plattin Limestone (units 3 and 5). Units 1 and 2 are Joachim Dolomite, units 3 and 5 are Plattin Limestone (4 is Establishment Shale Member). Photograph by J. L. Thacker.

Figure 4. Photograph of roadcut on I-55 approximately 1½ mi north of Pevely (Section B, Fig. 2; SW¼NW¼NW¼ Sec. 7, T.41N, R.6E, Jefferson County, Missouri) exposing uppermost Plattin Limestone (unit 1), Decorah Formation (unit 2), and basal beds of the Kimmswick Limestone (units 3 and 4). Photograph by J. L. Thacker.

Joachim Dolomite

2. Dolomite, finely crystalline; massively bedded (beds 8 to 18 in [20 to 45 cm] thick); light brown, weathers to buff; upper 18 in (45 cm) and lowest 18 in (45 cm) of unit are fucoidal; entire unit appears stromatolitic; 2- to 4-in (5- to 10-cm) green shale beds occur at base and midsection .... 4 ft (1.2 m)

1. Dolomite, finely crystalline, massively bedded, each massive bed contains abundant planar stromatolites, giving the bed a thinly laminated appearance; chocolate brown, weathers to buff; massive, white calcite nodules occur in a horizon approximately at midsection; fucoidal at top of unit .... 28 ft, 4 in (8.6 m)

B. Exposure on west side of south-bound lane of I-55, approximately 1.25 mi (2 km) north of the intersection with Jefferson County Highway Z, SW¼NW¼NW¼ Sec. 7,T.41N. R.6E., Jefferson County, Missouri Herculaneum 7½-minute Quadrangle (Figs. 2B, 4).

ORDOVICIAN SYSTEM

Champlainian Series (Mohawkian Series)

Kimmswick Limestone

4. Limestone, coarsely crystalline; massively bedded with beds of up to 5 ft (1.5 m) thick interbedded with thin shale beds; entire unit is highly fossiliferous and burrowed; upper 11 ft (3.3 m) appears cross-bedded on weathered surfaces; basal 3 ft (0.9 m) contains white chert nodules; light gray to white .... 19 ft, 6 in (6 m)

3. Limestone, fine to coarsely crystalline; thin, nodular

bedded and interbedded with thin green shales—lowest 2 ft (0.6 m) in one massive bed; appears to be an alternation of highly fossiliferous, coarser crystalline beds with beds of more lithographic, nonfossiliferous limestone; light to dark gray, weathers to buff-gray .... 15 ft (4.5 m)

Decorah Formation

2. Shale, green, interbedded with thin beds of limestone; shale is dominant lithology; extremely fossiliferous with bryozoans and brachiopods; a 1-ft-(0.3-m) thick montmorillonite bed occurs 1 ft (0.3 m) from base .... 7 ft, 1 in (2.14 m)

Plattin Limestone

1. Limestone, lithographic; thin to massively bedded; most beds are thin and wavy producing a nodular appearance; a 5-in (13-cm) shale bed occurs in center of unit; chocolate brown, weathering to buff-gray; locally portions of unit appear calcarenitic with much fossil debris; calcite fillings noted throughout .... 19 ft, 5 in (5.9 m)

C. Exposure on west side of south-bound lane of I-55 approximately 0.5 mi (0.8 km) north of the junction with Jefferson County Highway M, just north of the town of Barnhart, Jefferson County, Missouri; SW¼NW¼SE¼ Sec. 19,T.42N,R.6E., Herculaneum 7½-minute quadrangle (Figs. 2C, 5).

MISSISSIPPIAN SYSTEM

Osagean Series

Fern Glen Formation

13. Limestone, fine to coarsely crystalline; light brown

to light gray—weathers to buff-gray; thin to massively
bedded; very fossiliferous; alternating beds of limestone
and chert .... 12 ft (3.6 m)
12. Limestone, very shaly; reddish in color with purplish
cast; and limestone, finely crystalline, red, thin and highly
irregularly bedded; fossiliferous .... 2 ft (0.6 m)
11. Limestone, argillaceous; finely crystalline; light gray;
highly fossiliferous; 4- to 6-in (10- to 15-cm) chert bed
in upper portion .... 2 ft (0.6 m)
10. Limestone, very shaly; finely crystalline; red; thin
and highly irregularly bedded; fossiliferous ...... 5 ft (1.5 m)
9. Limestone, finely crystalline; light brown-tan; massively bedded; abundant fossils stand out in relief on
weathered surface; weathers to greenish-gray (Meppen
Ls. Member) .... 3 ft (0.9 m)

Kinderhookian Series

Bachelor Sandstone

8. Nodular sandstone; nodules are slightly calcareous;
sand grains are rounded and frosted and medium in size;
light green .... 2-3 in (5-8 cm)

MISSISSIPPIAN SYSTEM(?)

Bushberg Sandstone

7. Sandstone; medium grained, grains well sorted; rounded
and frosted; contains abundant iron as entire section is
stained a yellow-brown; very friable; one single
bed .... 13 ft, 6 in (4.1 m)

DEVONIAN SYSTEM

Upper Devonian Series

Glen Park Limestone (of Sulphur Springs Group)

7a. Limestone, lenticular bed beneath Bushberg Sand
stone, usually covered by talus .... 1 ft (0.3 m)

ORDOVICIAN SYSTEM

Cincinnatian Series

Maquoketa Shale

6. Shale, medium gray; very fissile; abundant
graptolites .... 3 ft, 3 in (1 m)
5. Limestone, very sandy; finely crystalline; dark gray;
contains phosphatic material .... 3 in (8 cm)
4. Shale, gray to greenish-gray; appears to be silty;
fissile; abundant graptolites; phosphatic
debris .... 17 ft, 4 in (5.2 m)
3. Shale, gray; appears to be mud shale;
thin bedded .... 7 ft, 6 in (2.3 m)

Figure 5. Photograph of roadcut on I-55 immediately north of Barnhart
exit at Jefferson County Road M (Section C, Fig. 2; SW¼NW¼SE¼
Sec. 19, T.41N., R.6E., Jefferson County, Missouri) exposing the upper
part of the Kimmswick Limestone (unit 1), Cape Limestone (unit 2),
Maquoketa Shale (units 3-6), Bushberg and Bachelor Sandstones (unit
7), and Fern Glen Formation (units 8-13). Photograph by J. L. Thacker.

Cape Limestone

2. Limestone, calcarenitic; coarse grained; medium gray;
limonite coatings on top of bed; brachiopods and other
fossils weathering out on upper surface; single bed
welded to top of underlying Kimmswick
Formation .... 6-10 in (15-25 cm)

Champlainian Series (Mohawkian Series)

Kimmswick Limestone

1. Limestone; coarsely crystalline; pinkish-tan; massively
bedded with 1- to 1.5-ft-(0.3- to 0.5-m) thick beds;
fossiliferous; masses of FeS_2 altered to limonite common.
Old quarry of Kimmswick immediately south of
this exposure, north of Jefferson County
Highway M .... 3 ft, 6 in (1.06 m)

REFERENCES CITED

Sweet, W. C., Thompson, T. L., and Satterfield, I. R., 1975, Conodont stratig-
raphy of the Cape Limestone (Maysvillian) of eastern Missouri: Missouri
Department of Natural Resources, Division of Geological Survey, Report of
Investigations 57, pt. 1, p. 1–60.
Thacker, J. L., and Satterfield, I. R., 1977, Guidebook to the geology along
Interstate-55 in Missouri: Missouri Department of Natural Resources, Divi-
sion of Geology and Land Survey, Report of Investigations 62, 132 p.
Thompson, T. L., 1975, Redescription and correlation of the Fern Glen Forma-
tion of Missouri: Missouri Department of Natural Resources, Division of
Geological Survey, Report of Investigations 57, pt. 5, p. 141–172.

33

Hahatonka karst landform complex, Camden County, Missouri

Arthur W. Hebrank, Department of Natural Resources, Missouri Geological Survey, Rolla, Missouri 65401

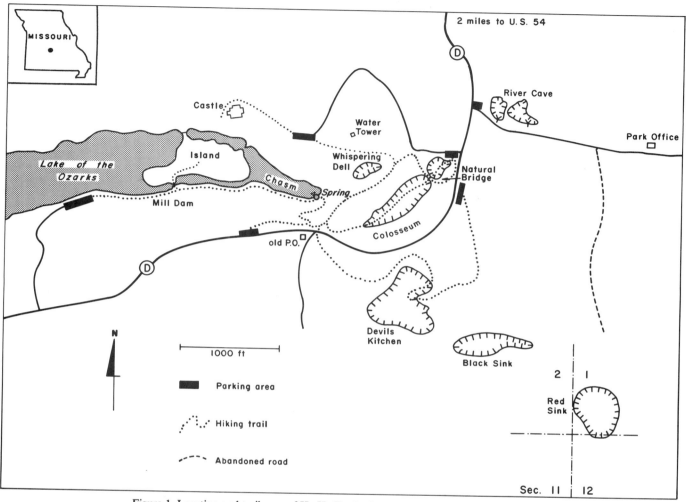

Figure 1. Location and trail map of Ha Ha Tonka State Park, Camden County, Missouri.

LOCATION

Ha Ha Tonka State Park, Sec. 2 and W½SW¼Sec.1, T.37N., R.17W., Camden County, Missouri; Hahatonka 7½-minute Quadrangle. Entrance to Ha Ha Tonka State Park is on State Road D (Fig. 1), about 2.3 mi (3.7 km) southeast of its junction with Missouri 54, west of Camdenton, Missouri. Well-designed and well-maintained trails within the park provide access to all major features of interest (Fig. 1). The site is on public land. Note: State park rules prohibit defacing or collecting rocks.

SIGNIFICANCE

Hahatonka is the premier example in Missouri of a karst landform complex. Interrelated solution features dramatically exhibited here include an awesome cavern-collapse chasm, a large karst spring, several very impressive sinkholes, a well-formed

natural bridge, a sizeable cave with sinkhole and swallow-hole entrances, and several small caves. The major karst features of the complex appear to be related to a single large cave system in various stages of collapse. The significance of the site is that so many spectacularly exhibited, interrelated karst features can be seen in a small area. The central karst complex at Ha Ha Tonka State Park was designated a State of Missouri Geologic Natural Area in 1981.

DESCRIPTION

Ha Ha Tonka State Park is in an intricately dissected portion of the central Ozarks, in the drainage basin of the Niangua River (now an arm of Lake of the Ozarks). All karst landforms exhibited in the park are developed in Late Cambrian Eminence

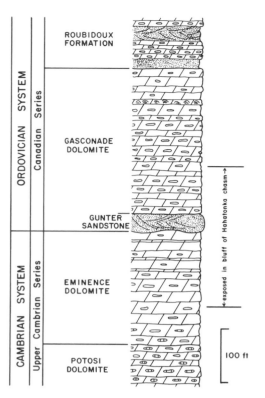

Figure 2. Generalized stratigraphic column of Ha Ha Tonka State Park area, Camden County, Missouri.

Dolomite, and Early Ordovician Gasconade Dolomite and Gunter Sandstone (Fig. 2). The Eminence and Gasconade formations are predominantly coarsely crystalline dolomite, but locally contain abundant chert; they are the most important cave-host stratigraphic units in the central Ozarks. The chasm at Hahatonka Spring (formerly called Gunter Spring) is the type locality of the cross-bedded, quartzose Gunter Sandstone, the basal member of the Gasconade Dolomite. The general geology of the area was described in detail by Hendriks (1942) and Schmitz (1982).

The featured scenic and geologic attractions of Ha Ha Tonka State Park are the dramatically exhibited karst landforms. These were described in superlative detail by Bretz (1956) and more briefly by Beveridge (1978).

The central geologic feature of the site, the ½-mile-long Hahatonka chasm, or gorge, is one of the most spectacular canyons in Missouri (Fig. 3). The north side is bounded by precipitous slopes, including a sheer cliff nearly 250 ft (75 m) high and 800 ft (240 m) long; the south side, by a combination of cliffs and very steep rocky hillslopes. The canyon walls expose the most complete geologic section in the immediate area: approximately 135 ft (41 m) of Eminence Dolomite, 35 ft (11 m) of Gunter Sandstone, and 85 ft (41 m) of Gasconade Dolomite.

The chasm, near its west end, consists of two branches separated by a large island of Eminence Dolomite. At the east end of the island is the entrance to Island Cave, a 300-ft-long (90 m)

joint-determined cave characterized by high, narrow, slot-like passages. Near the northwest corner of the island is Island Pit Cave, a narrow, near-vertical pit, filled with water to within 12 ft (3.6 m) of its rim.

Hahatonka chasm is a classic example of a cavern-collapse canyon. The trunk of a major cave system formerly occupied a position near the floor of the present canyon. Strata above the cavern were thinned by erosion, roof failure occurred, and a major segment of the large cave passage collapsed. Subsequent erosion removed much of the rubble and further modified the canyon.

Several easily observed features attest to the chasm's subterranean heritage. Most obvious is the configuration of the canyon itself; linear, with precipitous walls and head, not typical of a headwater valley formed by simple surface erosion; but from all appearances, predominantly of catastrophic collapse.

Giant, fallen and tilted, dolomite blocks litter the canyon floor north of the island and are scattered within the main part of the chasm. Many of them, as well as some outcrops on the island, are extensively solution pitted, exhibiting a type of sponge-work which characteristically forms inside caves and along solution-enlarged joints. These blocks are the wreckage—the broken and fallen wall rock and roof rock—of a collapsed cave (Bretz, 1956).

Two other evidences allude to the chasm's cavern origin. Extensively weathered masses and fragments of travertine (cave onyx) occur throughout the canyon and on the island. Finally, Island Cave and Island Pit Cave are interpreted as surviving upper-level portions of the once extensive, now ruined cave system.

At the head (east end) of the Hahatonka chasm, rising from beneath a bluff of Eminence Dolomite, is Hahatonka Spring, the twelfth-largest spring in Missouri (Vineyard and Feder, 1982). Hahatonka Spring drains a sizable upland area to the south and east, and has an average flow of almost 50 million gallons per day; the maximum recorded daily flow was 123 million gallons on November 9, 1925. Regular clouding of the spring after severe rainy periods indicates a rather open recharge system, but water quality is good, about typical of large Ozark springs. Chemical analyses were published by Vineyard and Feder (1982).

In addition to local, probable recharge points (Whispering Dell sinkhole, Colosseum sinkhole, and River Cave water trap), two distant recharge sites have been identified by dye tracing: a small sinkhole about 11 mi (17 m) southeast, and a losing reach of Dry Auglaize Creek, about 18 mi (29 km) south-southeast of Hahatonka. Unfortunately, both sites are potential pollution sources. The sinkhole is a popular local dump site; the losing stream swallows effluent, greatly diluted, from the City of Lebanon sewage treatment plant (Skelton and Miller, 1979). Water quality at Hahatonka Spring, however, has not yet been notably affected.

Divers have penetrated the spring to a depth of about 38 ft (12 m) below pool level, and describe the upper spring conduit as a narrow, steeply inclined chamber with a gravel floor and irregular, solution-pitted bedrock ceiling (Ward and Ward, 1977). Flow of water from Hahatonka Spring is impeded by two dams,

Figure 3. Hahatonka chasm, a 0.5-mi-long (0.8 km), 250-ft-deep (76 m) cavern-collapse canyon. The pooled waters of Hahatonka Spring flood the chasm floor, and the ruins of a 30-plus-room, turn-of-the-century mansion rise above the sheer cliff which exposes Eminence Dolomite, Gunter Sandstone, and Gasconade Dolomite. Photo by James E. Vandike.

one on each side of the island at the west end of the chasm, which form a sizable artificial lake, the Trout Glen pool. Hahatonka Spring, as we know it today, is certainly not the groundwater mechanism responsible for ancient cavern development in the Hahatonka area, but it uses some of the old subterranean channels and is the successor of the hydraulic circulation that formed the now collapsed cave system.

About 500 ft (150 m) northeast of the spring orifice is Whispering Dell, a spectacular, 150-ft-deep (48 m), precipitously walled sinkhole. Gasconade Dolomite is exposed on its upper slopes, but most of the structure is developed in Eminence Dolomite and Gunter Sandstone.

Entrances of three small caves are in the walls of the sinkhole. The largest, Bear Cave #1, at the northeast end of the sink, just below the Gunter Sandstone, is about 140 ft (42 m) long. Robbers Cave and Counterfeiters Cave, in the south-wall cliff, nearer the bottom of the sink, are little more than entrances. All three are developed completely in Eminence Dolomite.

The relationship of these features to those already described is apparent. The deep sinkhole was formed by collapse into the same cave system responsible for the Hahatonka chasm; the small caves are surviving upper-level remnants of that system.

About 1,000 ft (300 m) east of Hahatonka Spring and 500 ft (150 m) southeast of Whispering Dell is another large sinkhole, usually called the Colosseum. Elongate in a northeast-southwest direction, and measuring 1,000 ft (300 m) long by 300 ft (90 m)

wide by 180 ft (55 m) deep (maximum), the Colosseum is walled by a sheer cliff on the northwest side, and precipitous rocky slopes on the southeast. Eminence Dolomite, Gunter Sandstone, and Gasconade Dolomite are all exposed in the sinkhole walls.

A superb natural bridge spans the narrow northeast end of the Colosseum sinkhole (Fig. 4). The bridge has a span of 60 ft (18 m), a ceiling height of 13 ft (4 m), and a width of about 70 ft (21 m) on the underside. Its roof is approximately 45 ft (14 m) thick; the upper 15 ft is Gunter Sandstone, the remainder Eminence Dolomite. A small, intermittent surface stream flows southwestward under the bridge and into the Colosseum. The Hahatonka natural bridge was formed by partial cavern collapse: the bridge is an uncollapsed cave-roof remnant; the cave has completely collapsed on either side.

About 1,000 ft (305 m) northeast of the natural bridge is River Cave, also called Mystic River Cave, the largest cave in the Hahatonka karst complex. River Cave's large, vaulted main entrance is at the base of the nearly vertical south wall of a 40-ft-deep (12 m), 150-ft-diameter (46 m) sinkhole developed in Eminence Dolomite and Gunter Sandstone. Immediately east is a similar sinkhole housing a second (upper) cave entrance nearly choked with surface debris.

This second sinkhole is a wet-weather swallow hole. The entire flow of Dry Hollow, an intermittent stream, plunges over a small waterfall and into the upper sinkhole entrance of River Cave. Waters of the pirated surface stream next appear at the

Figure 4. Natural bridge spanning the narrow northeast end of the Colosseum sinkhole. The bridge, with a 60 ft (18 m) span, is developed in Eminence Dolomite and Gunter Sandstone. Photo by James E. Vandike.

main cave entrance, then cascade through large underground chambers to a water trap 700 ft (213 m) southwest of that entrance. The water trap at the lower end of River Cave is undoubtedly a recharge point for the half-mile-distant Hahatonka Spring (Bretz, 1956).

River Cave is entirely in Eminence Dolomite. The narrow slot-like passage between the two entrances, the spacious 30-ft-high (9 m) chambers of the main passage, and the sinuous water-crawl sidepassage have a total combined length of about 1,500 ft (460 m). Geologic features of the cave include a large natural bridge, a rimstone cascade, and an 18-ft-high (6 m) stalagmitic column, the *Christmas Tree,* described by Bretz (1956) as "one of the outstanding stalagmitic columnar domes of the Ozark country." River Cave is the largest enterable uncollapsed remnant of the mostly ruined, ancestral "Hahatonka Cave" system.

South and southeast of the Colosseum and River Cave sinks are three additional sinkholes: Devils Kitchen sinkhole, Black Sink, and Red Sink. The first is named for a 30-ft-wide (9 m) by 20-ft-high (6 m) by 30-ft-long (9 m) cave with a solution-enlarged-joint skylight—suggestive of an old cookstove or fireplace—developed in Eminence Dolomite, near the eastern margin of the sinkhole. Red Sink, the most impressive of the three, is about 400 ft (120 m) in diameter. Its precipitous south wall is nearly 200 ft (60 m) high; the breached north wall rises only 60 ft (18 m) above the sinkhole floor. Somewhat higher in elevation than the other Hahatonka landforms, Red Sink is de-

veloped in Gasconade Dolomite. Reddish rimrock blocks on the high south wall are Roubidoux sandstone; the sink takes its name from red, iron-rich residuum derived from the Roubidoux Formation.

The Hahatonka karst landform complex comprises a collection of cavern-collapse features unparalleled in Missouri. A major cave system once existed at this locality. The prominent karst landforms of today—a collapse chasm, a spring, sinkholes, and a natural bridge—are all products of its destruction, the relicts and wreckage of a collapsed cave.

The exceptional diversity and concentration of interrelated karst landforms at this site are outstanding for their educational value in geologic interpretation. Bretz (1956) summarized it perfectly: "Should we attempt to appraise cavern localities in Missouri, none could take precedence over Hahatonka. This is not because its four [now known to be 8] caves are superlatively showy, not because of the presence of the great spring, nor because of the striking local scenery. Hahatonka and its vicinity stand out in our appreciation because the caves, the spring, the sinks, and the great cliffs fit perfectly into a coherent and understandable sequence of events, and almost nothing is lacking to make a complete picture of the physiographic history of the region."

In addition to its importance as a geologic site, Hahatonka is botanically and historically interesting. The State Park includes at least seven natural terrestrial community types, featuring plant species inhabiting various forest, sinkhole, bluff, talus, savanna, and glade environments. Historically, Hahatonka was an Indian gathering place, pioneer gristmill site, outlaw hideout, and wealthy businessman's estate. Perched atop the 250 ft (75 m) bluff of Hahatonka chasm—vying with the natural features for attention—are the stark, dramatic, stone ruins of a 30-plus-room, turn-of-the-century, Scottish-style castle.

REFERENCES CITED

Beveridge, T. R., 1978, Geologic wonders and curiosities of Missouri: Missouri Department of Natural Resources, Division of Geology and Land Survey, Educational Series no. 4, p. 330, 345–348.

Bretz, J H., 1956, Caves of Missouri: Missouri Geological Survey and Water Resources, v. 39, 2nd Series, p. 123–132.

Hendriks, H. E., 1942, Geology of the Macks Creek quadrangle, Missouri [M.S. thesis]: Iowa City, University of Iowa, 122 p.

Schmitz, L. J., 1982, The general geology of Ha Ha Tonka State Park and surrounding area near Camdenton, Missouri [Masters thesis]: Columbia, University of Missouri, 70 p.

Skelton, J. and Miller, D. E., 1979, Tracing subterranean flow of sewage-plant effluent in Lower Ordovician dolomite in the Lebanon area, Missouri: Ground Water, National Well Association, v. 17, p. 476–486.

Vineyard, J. D., and Feder, G. L., 1982, Springs of Missouri (2nd ed.): Missouri Department of Natural Resources, Division of Geology and Land Survey, Water Resources Report 29, p. 148–149.

Ward, H., and Ward, G., 1977, Ha Ha Tonka Spring: Unpublished report on file at Missouri Department of Natural Resources, Division of Geology and Land Survey, Rolla, Missouri.

Potosi-Highway 8 Cambrian exposure, southeastern Missouri

Thomas L. Thompson and James Palmer, Missouri Geological Survey, Rolla, Missouri 64501

LOCATION

This composite section consists of a sequence of roadcuts on Missouri 8, between the town of Flat River and the bridge over Big River, St. Francois County, Missouri (Fig. 1). It comprises strata from upper Bonneterre Formation to basal Derby-Doerun Dolomite; the entire Davis Formation is represented. This exposure is within 3 mi (4.8 km) of the type section of the Davis Formation (SW¼Sec.7,T.36N.,R.5E., to NE¼Sec.13,T.36N., R.4E.), where approximately 170 ft (52 m) of Davis strata have been described (Buckley, 1908).

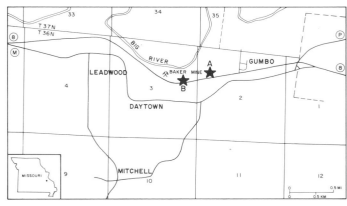

Figure 1. Copy of portion of Flat River 7½-minute Quadrangle, St. Francois County Missouri, showing locations of sections described. Town of Flat River is approximately 2 mi (3.2 km) east of small town of Gumbo.

SIGNIFICANCE

This sequence of roadcuts, representing nearly the entire Davis Formation as known from its type section, exhibits unusual lithic features characteristic of the Davis: "edgewise," "mud-chip," or "lithoclast" conglomerates; "marble boulder beds"; and the clastic shale sequence unique to this formation in the Cambrian outcrop region of the Ozark uplift. Origin of this unit is controversial, there being two nearly opposite viewpoints; deep water, offshore versus shallow water, nearshore.

DESCRIPTION

This is a composite of three sections. The first is a short section of uppermost Bonneterre Formation with the basal shale and dolomite beds of the Davis Formation (NW¼NW¼NW¼ Sec.3,T.36N.,R.4E.); the other two consist of exposures of Davis and Davis and Derby-Doerun strata (sections A, and B respectively; Fig. 2). The Bonneterre-Davis exposure clearly exhibits the disconformable relationship of Bonneterre to Davis, the contact being a distinct, abrupt lithic break.

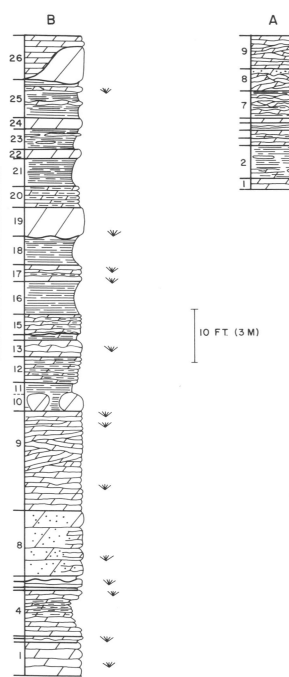

Figure 2. Columnar sections of two exposures of Upper Cambrian Davis Formation in roadcuts on Missouri 8, St. Francois County, Missouri. Fan-like symbols indicate beds containing "edge-wise" conglomerate.

The lower part of the Davis sequence is repeated in a roadcut immediately east of the Missouri 8 bridge over Big River, where the "marble boulder beds" are also exposed.

A. Road cut on Missouri 8 immediately west of sign reading "Gumbo," in NE¼NW¼Sec.2,T.36N.,R.4E., St. Francois County, Missouri (Flat River 7½-minute Quadrangle; Missouri Geological Survey section 1360-76; Fig. 2A).

Cambrian System

Upper (Croixian) Series

Davis Formation

9. Dolomite, mudstone or fine grainstone, in shaly, nodular beds, with zones of very even beds 1 to 6 in (2.5 to 15 cm) thick, to top of exposure; "edgewise conglomerate" 1 ft (30 cm) above base; laminated dolomite with thin shale partings, and with cross-laminated and ripple-marked? shale laminae in dolomite .... 6 ft (1.8 m)

8. Dolomite, sandy, glauconitic, medium bedded, capped by 8 to 18 in (20 to 45 cm) sandy, cross-bedded dolomite grainstone bed; much of unit cross-bedded; very irregularly bedded .... 4 ft (1.2 m)

7. Dolomite and interbedded shale, nodular, wavy beds; dolomite very finely crystalline mudstone or fine grainstone .... 5 ft (1.5 m)

6. Dolomite, single bed; dark bluish-black mudstone or fine grainstone .... 4 to 6 in (10 to 15 cm)

5. Shale and dolomite; shale, lower 2 in (5 cm) and upper 6 in (15 cm); dolomite mudstone or fine grainstone bed 3 to 4 in (7.6 to 10 cm) in lower third of unit .... 1 ft (0.3 m)

4. Dolomite, silty, horizontally burrowed, wavy bedded; upper 4 to 6 in (10 to 15 cm) dense .... 1 ft (0.3 m)

3. Dolomite; "edgewise conglomerate;" calcite filled vugs; two beds 8 in (20 cm) and 4 in (10 cm) thick .... 1 ft (0.3 m)

2. Shale; dolomitic in lower part, upper 4 ft (1.2 m) fissile, with 1 to 2 in (2.5 to 5 cm) cross-laminated mudstone or fine grainstone dolomite beds and thin lenses throughout. Shale, green-brown, clayey in lower 2 ft (60 cm), more resistant, dolomitic, and silty in upper 4 ft (1.2 m) .... 6 ft (1.8 m)

1. Dolomite, tan; even 6 in (13 cm) beds; with incipient wavy ½ to 1 in (1.2 to 2.5 cm) bedding. Small ¼ to ½ in (0.63 to 1.27 cm) calcite-filled vugs upper 3 in (7.62 cm); very finely crystalline, sparsely skeletal (mudstone) .... 2 ft (0.6 m)

Base of Exposure

B. Roadcut on Missouri 8, approximately 1 mi (1.6 km) east of Leadwood, St. Francois County, Missouri, consisting of two cuts, the eastern in the SW¼NW¼Sec.2, the western in the Center S½NE¼Sec.3,T.36N.,R.4E, Flat River 7½-minute Quadrangle (Baker Mine exposure; Missouri Geological Survey section 1360-77; Fig. 2B).

Cambrian System

Upper (Croixian) Series

Derby-Doerun Dolomite

26. Dolomite, dense, brown, very irregular base, several "bioherm" cores with depressed bases, similar to the algal stromatolite reefs described from the Bonneterre Formation of the "old lead belt"; bioherms are thrombolitic mud boundstone, flank beds of mudstone, shale, indeterminate and lithoclast packstone .... 6 to 8 ft (1.8 to 2.4 m)

Davis Formation

25. Shale, gray-green, clayey, fissile to blocky; thin even beds and nodular beds of blue-black mudstone dolomite. Very irregular discontinuous bed of "edgewise conglomerate" 1 ft (0.3 m) below top, ranging from 0 to 1 ft (0.3 m) in thickness .... 6 to 8 ft (1.8 to 2.4 m)

24. Dolomite, dense, dark blue-gray; single bed; thrombolitic mud boundstone biostrome .... 1 to 2 ft (0.3 to 0.6 m)

23. Shale, gray-green, clayey, fissile to blocky, discontinuous 1 to 2 in (2.5 to 5 cm) nodular beds of finely crystalline dolomite; mudstone, dolomite approximately 25% of volume ..... 3 to 4 ft (0.9 to 1.2 m)

22. Dolomite, single bed, skeletal lithoclast wackestone to packstone .... 18 in (0.45 m)

21. Shale, gray-green, clayey, fissile to blocky, like #23 .... 5 ft (1.5 m)

20. Dolomite mudstone with interbedded shale, weathers into thin slabby beds; more argillaceous than #19 .... 4 ft (1.2 m)

19. Dolomite, dense, finely crystalline; essentially one bed; base irregular, "edgewise conglomerate" in base, vuggy middle zone of echinoderm-trilobite packstone .... 5 ft (1.5 m)

18. Shale, gray-green, clayey, fissile .... 5 ft (1.5 m)

17. Dolomite; "edgewise conglomerate"; 6 in (15 cm) shale in middle; fossils, brachiopod, and echinoderm packstone .... 3 ft (0.9 m)

16. Shale, gray-green, clayey, fissile .... 6 ft (1.8 m)

15. Dolomite, in dense 6 to 12 in (15 to 30 cm) beds; shale partings, "edgewise conglomerate"; very irregular beds, "oncolitic packstone" upper 1 ft (30 cm) .... 4 ft (1.2 m)

14. Shale, gray-green, clayey, fissile .... 1 ft (0.3 m)

13. Dolomite, "edgewise conglomerate," 3 beds separated by shale; top bed with very irregular base .... 4 ft (1.2 m)

12. Dolomite, shaly, thin, slabby, rippled and cross-laminated; interlaminated shale and carbonate mudstone .... 5 to 6 ft (1.5 to 1.8 m)

11. Shale, dark gray-green, clayey, fissile to blocky, between and above "marbles" of unit #10 .... 2 to 5 ft (0.6 to 1.5 m)

10. "Marble boulder bed"; subspherical dolomite "biohermal" units; very finely crystalline dolomite; bases rest on thin "edgewise conglomerate" dolomite bed; thrombolitic algal-echinoderm-mud boundstone .... 0 to 3 ft (0 to 0.9 m)

9. Dolomite; mudstone or fine grainstone; cross-laminated and rippled; interlaminated with thin shale; middle 5 ft (1.5 m) mudstone dolomite; "edgewise conglomerate" lense 5 ft (1.5 m) above base and 3 ft (0.9 m) below top; unit capped by "edgewise conglomerate" ... 18 ft (5.5 m)

8. Dolomite, gray, sandy, with horizontal burrows, and green shale blebs; conglomeratic, even beds; thick bedded when fresh; weathers to thin and slabby to medium 4 to 6 in (10 to 15 cm) beds; upper 2 to 3 ft (0.6 to 0.9 m) dense; algal or laminated; thin zones of "edgewise conglomerate" .... 12 ft (3.6 m)

7. Shale, green-gray, clayey, fissile .... 4 to 6 in (10 to 15 cm)

6. Dolomite, lower half very thin beds, slabby, upper half "edgewise conglomerate," skeletal and lithoclast wackstone-packstone; detrital, very irregular .... 6 to 18 in (15 to 46 cm)

5. Shale, gray-green, fissile, clayey; very prominent reentrant .... 8 in (20 cm)

4. Dolomite, mudstone or fine grainstone, interbedded with shale, regular nodular thin beds; distinct reentrant in middle shaly zone; top and bottom more dolomitic; "edgewise conglomerate" top 1 ft (0.3 m) .... 8 ft (2.4 m)

3. Dolomite, sandy; "edgewise conglomerate," very lenticular, irregular beds; one bed .... 6 in (15 cm)

2. Shale and dolomite, green-gray, fissile, lower 4 in (10 cm) very dolomitic; upper 4 in (10 cm) fissile, brown .... 8 in (20 cm)

1. Dolomite, silty, thin to medium even beds, finely crystalline; weathers to shaly slabby beds in some parts, "edgewise conglomerate" 3 ft (0.9 m) above base, and at top .... 6 ft (1.8 m)

Interval difference between unit #9 of section A and unit #1 of Section B unknown.

REFERENCE CITED

Buckley, R. R., 1908, Geology of the disseminated lead deposits of St. Francois and Washington Counties: Missouri Bureau of Geology and Mines, 2nd series, v. 9, 259 p.

Knob Lick Mountain section: Intrusive contact of Precambrian granite with rhyolite in the St. Francois Mountains, Missouri

Eva B. Kisvarsanyi and Arthur W. Hebrank, Missouri Geological Survey, Rolla, Missouri 65401

LOCATION

Knob Lick Mountain, in the eastern part of the St. Francois Mountains, is about 9.5 mi (15.2 km) south of Farmington, in St. Francois County, Missouri (Wachita Mountain 7½-minute Quadrangle). Rising to a gentle elevation of 1,333 ft (400 m) above sea level, Knob Lick Mountain may be reached by following Knob Lick Tower Road for 1.1 mi (1.75 km) west of its junction with U.S. 67, just south of the village of Knob Lick (Fig. 1a), to the parking lot by the lookout tower atop Knob Lick Mountain (NE¼NE¼SE¼,Sec.8,T.34N.,R.6E.). A panoramic view of the St. Francois Mountains is afforded from the bald, 100 ft (30 m) south of the lookout tower. The intrusive contact of granite with rhyolite is exposed in an abandoned granite quarry, marked on Figure 1a (SW¼NE¼NE¼,Sec.8,T.34N.,R.6E.), about 2,000 ft (600 m) northwest from the lookout tower. Access

to the quarry is provided by the old ridge road; walk 2,000 ft (600 m) due northwest from the lookout tower on old ridge road, then 200 ft (60 m) due east through the woods to the quarry. Visit of this site involves about a 1.5-hour hike. The site is on public land.

SIGNIFICANCE

The St. Francois Mountains constitute the exposed portion of an extensive Precambrian terrane of anorogenic, granitic ring complexes and associated rhyolites that underlie most of southeastern Missouri (Kisvarsanyi, 1981). This igneous terrane is of regional interest not only because it is a splendid example of an unmetamorphosed, Proterozoic granite-rhyolite terrane, but also because it forms the only extensive outcrops of rocks whose ages range from 1.48 to 1.45 Ga in the southern mid-continent region

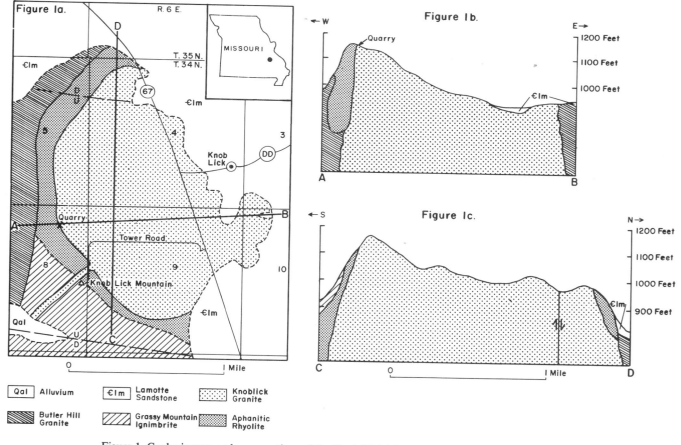

Figure 1. Geologic map and cross sections of the Knob Lick Mountain area. Map 1a is modified from Davis (1969), and Bickford and Sides (1983). East-west (1b) and north-south (1c) cross sections are from Davis (1969).

TABLE 1.   PRECAMBRIAN ROCK UNITS IN THE ST. FRANCOIS MOUNTAINS

St. Francois Mountains Volcanic Supergroup	St. Francois Mountains Intrusive Suite	
	Hypabyssal Rocks[3]	Plutonic Rocks[3]
Taum Sauk Group[1]		
Johnson Shut-ins Rhyolite		Graniteville Granite
Taum Sauk Rhyolite		
Royal Gorge Rhyolite		
Bell Mountain Rhyolite	Buford Granite Porphyry	
Wildcat Mountain Rhyolite	Munger Granite Porphyry	
Russell Mountain Rhyolite	Carver Creek Granite Porphyry	
Lindsey Mountain Rhyolite		
Ironton Rhyolite		
Buck Mountain Shut-ins Formation		
Pond Ridge Rhyolite		
Cedar Bluff Rhyolite		
Shepherd Mountain Rhyolite		
Butler Hill Group[2]		
Pilot Knob Felsite		Silvermine-Knoblick Granites
Grassy Mountain Ignimbrite	Brown Mountain Rhyolite Porphyry	Slabtown-Stono Granites
Lake Killarney Formation		Butler Hill-Breadtray Granites

[1]Volcanic units defined by Berry (1976).
[2]Volcanic units defined by Sides (1976).
[3]Formal names from Tolman and Robertson (1969).

(Bickford and others, 1981). The Precambrian terrane has been deeply eroded and dissected, resulting in a rugged topography and the unroofing of granite. Upper Cambrian marine sedimentary rocks are in nonconformable contact with the underlying igneous rocks. Near the crest of the Ozark dome, the dominant regional structure, the Precambrian outcrops of the St. Francois Mountains represent a structural and topographic high. The granitic ring complexes correspond to the deeply eroded root region of a formerly more extensive volcanic terrane comprising several calderas, cauldron subsidence structures, ring intrusions, and resurgent cauldrons with central plutons.

In the entire Precambrian outcrop area of the St. Francois Mountains, the Knob Lick Mountain quarry affords the best exposure illustrating the intrusive relationship of granite into rhyolite. The granite, mapped as Knoblick Granite (Tolman and Robertson, 1969; Pratt and others, 1979), is one of three principal types recognized in the Precambrian terrane, and is representative of the ring intrusions in the St. Francois Mountains. The view from the lookout tower illustrates the strikingly different topographic expressions of granite and volcanic-rock terrains in the St. Francois Mountains.

DESCRIPTION

Rhyolitic ash-flow tuffs are exposed at the top of Knob Lick Mountain, along its southern slope, and in a narrow belt for about 10 mi (16 km) southward. This area of rhyolite, mapped as Grassy Mountain Ignimbrite (Bickford and Sides, 1983), is a roof

pendant in Butler Hill-Breadtray Granites (Table 1), which are exposed to the west of it (Fig. 1a). East of the rhyolite outcrop, the Knoblick and Slabtown Granites are exposed.

The Knoblick Granite, exposed on the northeastern part of Knob Lick Mountain (Fig. 1a), is a small pluton emplaced along the eastern boundary of the Butler Hill caldera (Sides and others, 1981), and is part of a multiple ring intrusion comprising the Silvermine, Slabtown, and Knoblick Granites. In the granite quarry, the intrusive contact of Knoblick Granite with aphanitic rhyolite is exposed for about 20 ft (6 m) along the west quarry face; granite forms the lower 6 ft (2 m) of the quarry face, rhyolite the upper few feet (Fig. 2). The contact is sharp and gently undulating; granite apophyses project into the rhyolite, and rhyolite xenoliths are included in the granite. Thin seams of epidote are common along the contact. The rhyolite above the contact is recrystallized to a fine hornfelsic aggregate of quartz and alkali feldspar; only occasional relict pumice fragments indicate the ash-flow origin of this rock. The near-vertical orientation of these flattened fragments suggests the steep dip of the rhyolite above the granite contact. Volcanic rocks exposed southwest of the quarry have steep southwesterly dips, the result of structural deformation caused by the forceful intrusion of the Knoblick pluton (Davis, 1969). The east-west section across the quarry (Fig. 1b) shows the volcanic roof pendant "wedged" between the Knoblick pluton and the Butler Hill Granite.

The Knoblick Granite is a medium-grained amphibole-biotite adamellite containing an average of 30 percent orthoclase-microperthite, 35 percent plagioclase, 23 percent quartz, and 10

Figure 2. Distant and close-up views of intrusive contact between Knoblick Granite (below) and rhyolite (above). Note prominent rhyolite xenolith near point of hammer in distant view. Photos by Art Hebrank.

percent mafic minerals. Chemically, it is among the intermediate-to low-silica granites, characteristic of the ring intrusions of the St. Francois terrane (Kisvarsanyi, 1972; Pratt and others, 1979). The early-crystallized, euhedral, zoned plagioclase in Knoblick Granite tends to impart a porphyritic aspect to the rock, especially on weathered surfaces. Another conspicuous characteristic of Knoblick Granite is the presence of many mafic clots of variable size; some are partially assimilated basaltic xenoliths and some are basic segregations in the granite. Xenoliths of mica schist, possibly brought up from the metamorphic basement by intrusion of the pluton, have been reported (Davis, 1969).

The dense, aphanitic rhyolite in the quarry is overlain by a porphyritic unit, the Grassy Mountain Ignimbrite, forming most of the prominent outcrops on the southern slope of Knob Lick Mountain (Fig. 1a). The somewhat bleached and recrystallized ignimbrite suggests that the intrusive contact of Knoblick Granite may not be far below. A north-south section across Knob Lick Mountain shows the relationship of these volcanic units and the granite (Fig. 1c). Both aphanitic and porphyritic rhyolites are intruded by a 10- to 20-ft-(3- to 6-m) wide dike of porphyritic Knoblick Granite (Fig. 1a), well exposed in small prospect pits on the barren southern slope of the mountain, just below the lookout tower. The dike strikes N 40° E and has been mapped for 3,000 ft (900 m) down the mountain slope by Bickford and Sides (1983).

The different weathering characteristics of granite and rhyolite, resulting in strikingly different topographic expressions, are displayed in a panoramic view from the upper southern slope of Knob Lick Mountain, immediately south of the lookout tower and parking area (Fig. 3). The large area of relatively low topographic relief to the southwest, called The Flatwoods, is underlain by Butler Hill and Breadtray Granites; hills and knobs in the distance are formed by rhyolite. Granite areas in the St. Francois Mountains tend to be gently rolling, whereas erosion-resistant volcanic rocks are commonly expressed as knobs or areas of dramatic high relief. The highest point in the State of Missouri, 1,772-ft (532-m) Taum Sauk Mountain, is within the most extensive outcrop and thickest section of volcanic rocks in the St. Francois Mountains (Kisvarsanyi, Hebrank, and Ryan, 1981).

Figure 3. View of "The Flatwoods," an area of relatively low topographic relief, looking southwest from the top of Knob Lick Mountain. The Flatwoods is underlain by Butler Hill and Breadtray Granites; hills along the distant horizon are volcanic rocks more resistant to weathering. Outcrops in the foreground are Grassy Mountain Ignimbrite. Photo by Jerry Vineyard.

REFERENCES

Berry, A. W., 1976, Proposed stratigraphic column for Precambrian volcanic rocks, western St. Francois Mountains, Missouri, *in* Kisvarsanyi, E. B., ed., Studies in Precambrian Geology of Missouri (Contribution to Precambrian Geology No. 6): Missouri Department of Natural Resources, Geological Survey, Report of Investigations 61, p. 81–90.

Bickford, M. E., Harrower, K. L., Hoppe, W. J., Nelson, B. K., Nusbaum, R. L., and Thomas, J. J., 1981, Rb-Sr and U-Pb geochronology and distribution of rock types in the Precambrian basement of Missouri and Kansas: Geological Society of America Bulletin, Part I, v. 92, p. 323–341.

Bickford, M. E., and Sides, J. R., 1983, Geologic map of exposed Precambrian rocks in the Wachita Mountain (Fredericktown NW 1/4) quadrangle, Missouri (Contribution to Precambrian Geology No. 11): Missouri Department of Natural Resources, Geological Survey Open-File Map OFM-83-161-MR, scale 1:24,000.

Davis, J. W., 1969, Petrogenesis and structure of the Knoblick Granite [Masters thesis]: St. Louis, St. Louis University.

Kisvarsanyi, E. B., 1972, Petrochemistry of a Precambrian igneous province, St. Francois Mountains, Missouri (Contribution to Precambrian Geology No. 4): Missouri Geological Survey and Water Resources, Report of Investigations 51, 103 p.

—— 1981, Geology of the Precambrian St. Francois terrane, southeastern Mis-

souri (Contribution to Precambrian Geology No. 8): Missouri Department of Natural Resources, Geological Survey, Report of Investigations 64, 60 p.

Kisvarsanyi, E. B., Hebrank, A. W., and Ryan, R. F., 1981, Guidebook to the geology and ore deposits of the St. Francois Mountains, Missouri (Contribution to Precambrian Geology No. 9): Missouri Department of Natural Resources, Geological Survey, Report of Investigations 67, 119 p.

Pratt, W. P., Anderson, R. E., Berry, A. W., Jr., Bickford, M. E., Kisvarsanyi, E. B., and Sides, J. R., 1979, Geologic map of exposed Precambrian rocks, Rolla 1°×2° quadrangle, Missouri: U.S. Geological Survey, Miscellaneous Investigations Series, I-1161, scale 1:125,000.

Sides, J. R., 1976, Stratigraphy of volcanic rocks in the Lake Killarney quadrangle, Iron and Madison Counties, Missouri, *in* Kisvarsanyi, E. B., ed., Studies in Precambrian Geology of Missouri (Contribution to Precambrian Geology No. 6): Missouri Department of Natural Resources, Geological Survey, Report of Investigations 61, p. 105–113.

Sides, J. R., Bickford, M. E., Shuster, R. D., and Nusbaum, R. L., 1981, Calderas in the Precambrian terrane of the St. Francois Mountains, southeastern Missouri: Journal of Geophysical Research, v. 86, p. 10349–10364.

Tolman, C., and Robertson, F., 1969, Exposed Precambrian rocks in southeast Missouri (Contribution to Precambrian Geology No. 1): Missouri Geological Survey and Water Resources, Report of Investigations 44, 68 p.

Elephant Rocks: A granite tor in Precambrian Graniteville Granite, the St. Francois Mountains, Missouri

Eva B. Kisvarsanyi and Arthur W. Hebrank, Missouri Geological Survey, Rolla, Missouri 64501

LOCATION

SE¼SE¼NE¼Sec.15,T.34N.,R.3E., Elephant Rocks State Park, Iron County, Missouri; Graniteville 7½-minute Quadrangle. Entrance to the park is from Missouri RA, 0.7 mi (1.2 km) west of its junction with Missouri 21 (Fig. 1). From the parking lot, follow marked trail to the lookout point near the top of the hill. The site is on public land.

SIGNIFICANCE

This site is a classic example of a granite tor. It displays giant boulders, or core-stones, that resemble a herd of sitting elephants (Fig. 2). The boulders are of Graniteville Granite, representative of tin-granite central plutons, and one of three principal granite types recognized in the Precambrian St. Francois terrane (Kisvarsanyi, 1981). The granite is exposed on the margin of one of the resurgent cauldron subsidence structures in the region. The site was designated a State of Missouri Geologic Natural Area in 1978.

DESCRIPTION

The prime scenic and geologic attraction of the park are the giant, picturesque residual boulders ("elephant rocks") of Graniteville Granite. The boulders are the result of a two-stage proc-

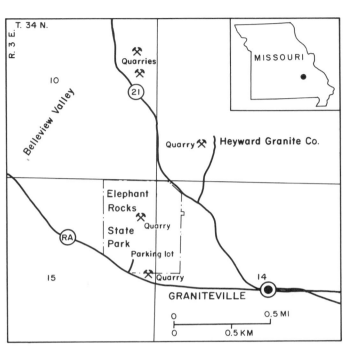

Figure 1. Location map of Elephant Rocks State Park, Iron County, Missouri.

Figure 2. "Elephant Rocks": spheroidally weathered giant boulders of Graniteville Granite in Elephant Rocks State Park. In this view to the northeast, the largest "elephant" atop the granite bald is 25 ft (7.5 m) high. Photo by Art Hebrank.

ess: spheroidal weathering of block-jointed granite by circulating groundwater, followed by erosional stripping of the weathered fines.

The outcrop of Graniteville Granite is restricted to three small areas along the eastern and southeastern boundaries of a sediment-filled depression, the Belleview Valley (Tolman and Robertson, 1969). Morphologically the valley is square, bounded by straight topographic escarpments, which are especially pronounced along its northwestern and northeastern sides. Graves (1938) suggested that the valley is a fault-bounded, down-dropped Precambrian structural block and that the topographic escarpments resulted from erosion along nearly vertical faults. The square-shaped Belleview Valley is a prominent feature on satellite imagery of the region (Kisvarsanyi and Kisvarsanyi, 1976).

Looking northwest from the lookout point in the park, Belleview Valley appears to be a gently rolling, bowl-shaped depression surrounded on the horizon by higher hills of rhyolite. Drillholes indicate that Graniteville Granite underlies the sedimentary rocks in Belleview Valley (Kisvarsanyi, 1981). Aeromagnetic maps show the pluton as an oval magnetic low coincident with Belleview Valley (Cordell, 1979).

The Graniteville Granite is a medium- to coarse-grained, muscovite-biotite alkali granite averaging 55 percent alkali feldspar, 40 percent quartz, and less than 5 percent mafic minerals. In this rock, the common alkali feldspar is typically microcline-microperthite, but albite and orthoclase-microperthite are also present. Both primary and secondary muscovite (sericite) occur. The rock contains a varied suite of accessory minerals, including abundant fluorite, zircon, and magnetite; cassiterite and molybdenite are less common. Locally, the granite contains complex pegmatites with topaz, beryl, muscovite, fluorite, rutile, cassiterite, and sulfide minerals (Tolman and Goldich, 1935). Anomalously high levels of Sn, Be, Y, Nb, F, and U in Graniteville Granite led to its identification as a tin-granite central pluton, the only one exposed, in the St. Francois terrane (Kisvarsanyi, 1981).

Graniteville Granite, known commercially as "Missouri Red," has been quarried in the area since 1869. By the turn of the century, building, paving, and monumental stone were being produced from several quarries. Blocks of Graniteville Granite were used as paving and curbing stone in St. Louis city streets. Buildings and monuments from San Francisco, California to Pittsfield, Massachusetts bear witness to the popularity and widespread use of this beautiful rock as a construction and monumental stone. Many older homes and commercial buildings in the area are also constructed from Graniteville Granite. Today only one quarry, 0.5 mi (0.8 km) northeast of Elephant Rocks State Park (Fig. 1) survives as an intermittent producer of monumental stone. The quarry may be visited by permission to collect samples of the granite.

REFERENCES CITED

Cordell, L., 1979, Gravity and aeromagnetic anomalies over basement structure in the Rolla quadrangle and the southeast Missouri lead district: Economic Geology, v. 74, p. 1383–1394.

Graves, H. B., 1938, The Pre-Cambrian structure of Missouri: St. Louis, Transactions of the Academy of Science, v. 29, n. 5, p. 111–164.

Kisvarsanyi, E. B., 1981, Geology of the Precambrian St. Francois terrane, southeastern Missouri (Contribution to Precambrian Geology No. 8): Missouri Department of Natural Resources, Geological Survey, Report of Investigations 64, 60 p.

Kisvarsanyi, G., and Kisvarsanyi, E. B., 1976, Ortho-polygonal tectonic patterns in the exposed and buried Precambrian basement of southeast Missouri, *in* Hodgson, R. A., Gay, S. P., and Benjamins, J. Y., eds., Proceedings of the First International Conference on the New Basement Tectonics: Salt Lake City, Utah Geological Association Publication n. 5, p. 169–182.

Tolman, C., and Goldich, S. S., 1935, The granite, pegmatite, and replacement veins in the Sheahan quarry, Graniteville, Missouri: American Mineralogist, v. 20, p. 229–239.

Tolman, C., and Robertson, F., 1969, Exposed Precambrian rocks in southeast Missouri (Contribution to Precambrian Geology No. 1): Missouri Geological Survey and Water Resources, Report of Investigations 44, 68 p.

Roadcuts in the St. Francois Mountains, Missouri: Basalt-dike swarm in granite, Precambrian-Paleozoic nonconformity, and a Lamotte channel-fill deposit

Eva B. Kisvarsanyi and Arthur W. Hebrank, Missouri Geological Survey, Rolla, Missouri 64501

LOCATION

This composite site features roadcuts along U.S. 67 and Missouri 72 in the eastern St. Francois Mountains, in Madison County, Missouri (Fig. 1).

STOP 1 (NW¼SW¼NE¼Sec.35,T.34N.,R.6E.), along U.S. 67, is 14.8 mi (23.7 km) south of the junction, west of Farmington, of U.S. 67 and Missouri W. It is located on the Fredericktown 7½-minute Quadrangle. Roadcuts at this stop display a basalt-dike swarm in the Precambrian Slabtown Granite.

STOP 2 (NE¼SE¼NW¼Sec.11,T.33N.,R.6E.) is along Missouri 72; from STOP 1, proceed 2.7 mi (4.3 km) south on U.S. 67, then 1.8 mi (2.9 km) west on Missouri 72 (Fig. 1). Roadcuts at this locality expose the nonconformable contact between Precambrian and Paleozoic rocks in a spectacular manner.

STOP 3 (NE¼NE¼NE¼Sec.9,T.33N.,R.6E.) is 1.7 mi (2.7 km) west of STOP 2, along Missouri 72 (Fig. 1). It is located on the Rhodes Mountain 7½-minute Quadrangle. This site displays horizontally bedded, channel-fill sediments within basal boulder conglomerate mantling the Precambrian surface.

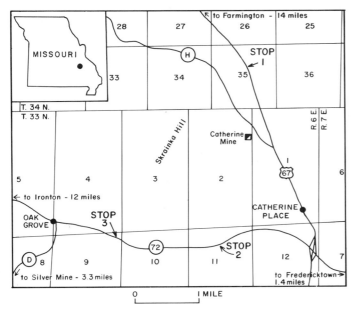

Figure 1. Route map along U.S. 67 and Missouri 72. STOP 1: Basalt-dike swarm in Slabtown Granite; STOP 2: Precambrian-Paleozoic nonconformity; STOP 3: Lamotte channel fill.

SIGNIFICANCE

The Precambrian rocks exposed in the St. Francois Mountains are part of an extensive, buried Proterozoic terrane of epizonal granite and rhyolite that forms the crystalline basement in the southern mid-continent region (Denison and others, 1984). This granite-rhyolite terrane extends in an arcuate, southwest-northeast-trending belt along the southern margin of the North American craton and represents significant addition of sialic material to the continental crust. Its age ranges from 1.48 to 1.45 Ga in the eastern part of the belt, including most of the rocks in the St. Francois Mountains, and from 1.4 to 1.35 Ga in the western part of the belt with some overlap of the younger plutons in the St. Francois Mountains (Thomas and others, 1984). Outcrops of this terrane in the St. Francois Mountains, near the crest of the Ozark dome, have an unsurpassed scenic beauty, and are geologically unique and significant because they represent an easily accessible, exposed "sample" of Proterozoic crust.

The route along U.S. 67 traverses the eastern margin of the Butler Hill caldera, one of several granitic ring complexes mapped in the region (Pratt and others, 1979; Kisvarsanyi, 1980, 1981; Sides and others, 1981). The roadcuts at STOP 1 provide excellent exposures of the Slabtown Granite, one of several ring intrusions recognized in the St. Francois terrane, and display the best example of a basalt-dike swarm, comprising more than 30 dikes, in outcrop. The roadcuts at STOP 2 (Fig. 1) display a textbook example of fundamental age relationships: the erosional surface of the Precambrian Grassy Mountain Ignimbrite, the major ash flow produced by the collapse of the Butler Hill caldera, is nonconformably overlain by basal boulder conglomerate overlapped by Paleozoic strata; a vertical diabase dike in the ignimbrite is truncated by the boulder bed. At STOP 3 (Fig. 1), an excellent example of horizontally bedded, channel-fill sediments is exposed within the basal boulder conglomerate mantling the Precambrian surface. This is one of the type sections (Oak Grove section) illustrating depositional environments in the Lamotte Sandstone in southeastern Missouri (Houseknecht and Ethridge, 1978).

DESCRIPTION

At STOP 1, Slabtown Granite is exposed in the large cuts on both sides of U.S. 67 (Fig. 1). The Slabtown Granite forms numerous small outcrops overlapped by Lamotte Sandstone, in the southeastern part of the igneous outcrop area (Pratt and others, 1979). Drillholes between the isolated outcrops en-

Figure 2. Part of a basalt-dike swarm intruded along joints and fractures in Slabtown Granite. East side of U.S. 67, STOP 1. Photo by Art Hebrank.

Hill, the type locality of the Skrainka Diabase (Tolman and Robertson, 1969), is just 1.5 mi (2.4 km) southwest of this road-cut; the dike swarm is believed to be an offshoot of the large gabbro sill exposed there.

Enroute to STOP 2, U.S. 67 passes through the western part of the historic Mine La Motte-Fredericktown subdistrict of the Southeast Missouri Lead district. The subdistrict, located mostly in T. 33 and 34 N., R. 6 and 7 E. (Fig. 1), includes some of the oldest lead-mining areas in Missouri, and was responsible for the district's only important cobalt-nickel production. Surface lead was discovered at Mine La Motte in 1720 and first mined in 1723. The Catherine Mines, located just west of the junction of U.S. 67 and Missouri H (Fig. 1), opened in the late 1860s and operated intermittently for nearly 90 years. This tract produced an estimated 55,000 tons of lead from the sandy dolomites in the lower part of the Bonneterre Formation.

At STOP 2, along U.S. 72 (Fig. 1), the Grassy Mountain Ignimbrite, a massive Precambrian rhyolite porphyry with well-developed joint sets, is exposed on both sides of the road. A 4- to 5-ft- (1.2- to 1.5-m)-wide diabase dike has intruded the rhyolite along one of the prominent northeast-trending (N 15° E) joints and is exposed on both sides of the road. On the south side the dike is terminated at the Precambrian surface and is truncated by the overlying basal boulder conglomerate (Fig. 3); on the north side the boulder bed is absent and the dike is exposed at the surface, on top of the roadcut (Fig. 4). The dike is deeply weathered in the southern cut and in the upper portion of the northern cut, but is relatively fresh near the road level in the northern cut. The dike contacts are sharp, but fractured and sheared, with calcite and quartz filling narrow fractures along the sheared contact with the rhyolite. Near the west end of the cut, on the south side, a similar but much smaller dike, about 1 ft (0.3 m) wide, has intruded the rhyolite.

On the south side of the road, the basal Paleozoic strata lap onto the Precambrian erosional surface from the east (Fig. 3). This cut is one of many displaying the Precambrian-Paleozoic

countered Slabtown Granite at depths of less than 100 ft (30 m) to more than 300 ft (90 m), suggesting moderate topographic relief on its erosional surface. The Slabtown Granite is part of a multiple ring intrusion, formed by the Knoblick, Slabtown, and Silvermine Granites, along the eastern boundary of the Butler Hill caldera (Pratt and others, 1979; Kisvarsanyi, 1981; Sides and others, 1981). The roadcuts at this locality expose typical Slabtown Granite: fine-grained amphibole granite consisting of about 55 percent orthoclase-microperthite, 12 percent albite-oligoclase, 20 percent quartz, 10 percent fibrous, blue-green amphibole mostly altered to chlorite, and 3 percent magnetite. Slabtown Granite commonly exhibits granophyric texture. In these cuts, Slabtown Granite is intruded by small mafic dikes. Near the north end of the cut, a dike swarm is exposed on both sides of the road. Thirty or more nearly vertical basalt dikes, most of them less than 3 in (8 cm) wide, have intruded joints and fractures of a 25-ft- (7.5-m)-wide, sheared interval of granite (Fig. 2). Skrainka

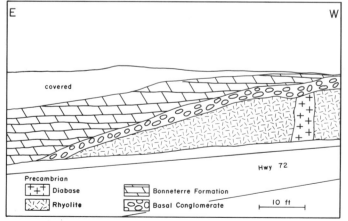

Figure 3. Mafic dike in Grassy Mountain Ignimbrite truncated by overlying basal boulder conglomerate. The conglomerate is overlapped from the east by sandy Bonneterre dolomite. South side of Missouri 72, STOP 2.

Figure 4. Mafic dike in Grassy Mountain Ignimbrite, north side of Missouri 72, STOP 2. This is the same dike shown in Figure 3, but here the overlying boulder conglomerate has been removed by erosion, and the dike is exposed at the top of the cut. Photo by Art Hebrank.

nonconformity in the St. Francois Mountains region. The uneven, weathered surface of the rhyolite is overlain by a 6 ft (1.8 m) section of coarse boulder conglomerate; most of the boulders are weathered Precambrian rhyolite porphyry. Coarse sandy dolomite and dolomite of the Bonneterre Formation overlie the boulder bed.

About 0.3 mi (0.5 km) east of STOP 3 (Fig. 1), the large cuts on both sides of Missouri 72 are in massive Grassy Mountain Ignimbrite, the major ash flow produced by the collapse of the Butler Hill caldera (Sides and others, 1981). Relicts of collapsed pumice indicate an ash-flow origin for the massive rhyolite, although devitrification and recrystallization of the groundmass all but completely obliterated textural features characteristic of welded ash-flow tuffs. The massive ignimbrite displays spectacular joint sets and shear planes. Two prominent shear planes

on the north side of the road are parallel, with a northeasterly striking, nearly vertical joint set; some displacement along these planes is indicated. Near the west end of the roadcut, on the north side, two nearly vertical, weathered diabase dikes parallel the major joint set, while a third intrusive body of diabase is in nearly horizontal contact with the rhyolite above. The lower contact of the third intrusive is obscured, but the diabase may be part of a sill or stock. This roadcut is directly south of Skrainka Hill, the type locality of the Skrainka Diabase (Tolman and Robertson, 1969).

At STOP 3 (Fig. 1), the roadcuts expose 45 ft (14 m) of coarse boulder conglomerate mantling a Precambrian diabase knob. The diabase exposed at the east end of the cut, on the south side of the road, is deeply weathered; in situ masses of exfoliated diabase, up to 10 ft (3 m) in diameter, are surrounded by disintegrated diabase. The basal boulder conglomerate is exposed in the western part of the cut on both sides of the road. Many boulders are weathered red granite; some are exfoliated. The center of the cut, on the south side of the road, is dominated by a large angular block 20 ft (6 m) wide and 5 ft (1.5 m) high of pale pink, partly recrystallized, porphyritic amphibole granite. This block and similar, smaller blocks exposed on the north side of the road appear to be wall-rock remnants of the country rock intruded by the diabase.

On the north side of the highway, 10 to 12 ft (3 to 3.5 m) of coarse boulder conglomerate are exposed in the upper face of the graded roadcut. Near the western end of the lower cut, part of the boulder bed is underlain by a water-deposited sedimentary sequence, 4 ft (1.2 m) thick in maximum exposure, consisting of very thin-bedded siltstone and silty shale (Fig. 5); the irregular contacts of this sequence with the boulder bed suggest that this material is a channel fill. Similar lithologies occur in the Lamotte Sandstone exposed in the drainage ditch immediately northwest of and below this roadcut, and in better exposed Lamotte sequences to the west (Oak Grove area). Houseknecht and Ethridge

Figure 5. Thin-bedded siltstone and silty shale filling a channel in coarse boulder conglomerate (STOP 3). Photo by Art Hebrank.

(1978) attribute the deposition of the conglomerate to a series of high-viscosity debris flows that formed an alluvial fan in a paleo-topographic valley. The lenses of horizontally bedded, fine-grained sediments within a conglomerate (Fig. 5) represent channel-fill or overbank sediment deposited by fluvial systems on the surface of the fan.

REFERENCES

Denison, R. E., Lidiak, E. G., Bickford, M. E., and Kisvarsanyi, E. B., 1984, Geology and geochronology of Precambrian rocks in the central interior region of the United States: U.S. Geological Survey Professional Paper 1241-C, 20 p.

Houseknecht, D. W., and Ethridge, F. G., 1978, Depositional history of the Lamotte Sandstone of southeastern Missouri: Journal of Sedimentary Petrology, v. 48, p. 575–586.

Kisvarsanyi, E. B., 1980, Granitic ring complexes and Precambrian hot-spot activity in the St. Francois terrane, Midcontinent region, United States: Geology, v. 8, p. 43–47.

—— 1981, Geology of the Precambrian St. Francois terrane, southeastern Missouri (Contribution to Precambrian Geology No. 8): Missouri Department of Natural Resources, Geological Survey, Report of Investigations 64, 60 p.

Pratt, W. P., Anderson, R. E., Berry, A. W., Jr., Bickford, M. E., Kisvarsanyi, E. B., and Sides, J. R., 1979, Geologic map of exposed Precambrian rocks, Rolla 1° × 2° quadrangle, Missouri: U.S. Geological Survey, Miscellaneous Investigation Series I-1161, 1:125,000 scale.

Sides, J. R., Bickford, M. E., Shuster, R. D., and Nusbaum, R. L., 1981, Calderas in the Precambrian terrane of the St. Francois Mountains, southeastern Missouri: Journal of Geophysical Research, v. 86, p. 10349–10364.

Thomas, J. J., Shuster, R. D., and Bickford, M. E., 1984, A terrane of 1,350- to 1,400-m.y.-old silicic volcanic and plutonic rocks in the buried Proterozoic of the mid-continent and in the Wet Mountains, Colorado: Geological Society of America Bulletin, v. 95, p. 1150–1157.

Tolman, C., and Robertson, F., 1969, Exposed Precambrian rocks in southeast Missouri (Contribution to Precambrian Geology No. 1): Missouri Geological Survey and Water Resources, Report of Investigations 44, 68 p.

Silver Mine district: Precambrian Ag-W-Pb mineralization and granite shut-ins, the St. Francois Mountains, Missouri

Eva B. Kisvarsanyi and Arthur W. Hebrank, Missouri Geological Survey, Rolla, Missouri 64501

LOCATION

NW¼SW¼SE¼Sec.12,T.33N.,R.5E., Silver Mine Dam on St. Francis River, Madison County, Missouri; Rhodes Mountain 7½-minute quadrangle. From the junction of Missouri 72 and Missouri D at Oak Grove, follow Missouri D south and southwest for 3.3 mi (5.2 km) to the parking lot adjacent to the picnic area on the east side of the road, about 200 ft (60 m) south of the bridge across the St. Francis River (Fig. 1). Cross the road and walk uphill west then north on the prominently marked foot trail leading to the historic Einstein Silver Mine and the Silver Mine Dam. The trail roughly parallels the river for about 2,000 ft (600 m) to the mine. About 1,500 ft (450 m) Missouri D, the trail joins an old wagon road carved into bedrock; bear right and follow the old wagon road downhill to the mine and dam. To return to the parking lot retrace the trail to the mine or, if conditions permit, cross the St. Francis River at the dam and follow the trail along the eastern bluffs of the river. The site is on public land.

SIGNIFICANCE

The vein deposits in the Silver Mine area represent the only known pneumatolytic ore-mineral association in mid-continent Precambrian rocks. The historic mining district is inactive now, but representative samples of the interesting mineral assemblage, unique in the region, may be collected on the old dumps. The bluffs along the St. Francis River expose typical Silvermine Granite, one of the ring intrusions of the Precambrian terrane. An excellent outcrop of a diabase dike in granite is near the dam. The site is one of the best known granite shut-ins of the region.

DESCRIPTION

Upstream from the one-lane bridge across the St. Francis River for a distance of about 0.5 mi (0.8 km), river erosion has developed a scenic, canyon-like gorge, called a "shut-in" or "narrows," in Silvermine Granite; it is one of the best granite shut-ins

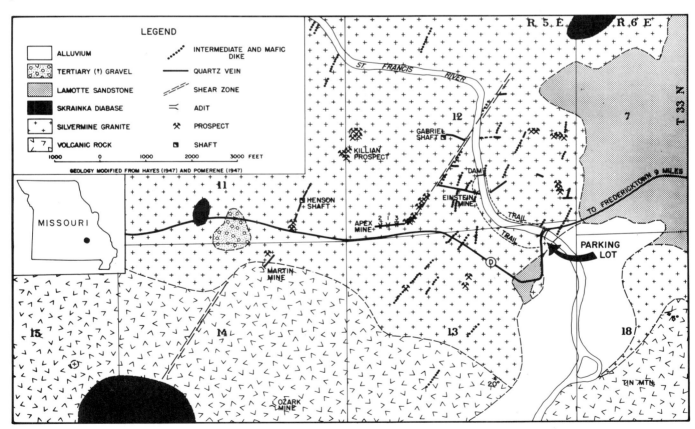

Figure 1. Geologic map of the Silver Mine area.

Figure 2. Outcrop of the "Big Dike," on the east bank of the St. Francis River, just downstream from the Silver Mine dam. This near-vertical, 4-ft- (1.2 m-) wide basalt dike intrudes Silvermine Granite. Photo by Art Hebrank.

in the region. Shut-ins are regionally unique physiographic features in the St. Francois Mountains but are more commonly developed where streams flow over volcanic rock.

Silvermine Granite, exposed in the bluffs along the St. Francis River (Tolman and Robertson, 1969), is part of a multiple ring intrusion formed by the Knoblick, Slabtown, and Silvermine Granites emplaced along the margins of the Butler Hill caldera (Sides and others, 1981). The rock is a medium-grained amphibole-biotite granite averaging 40 percent orthoclase-microperthite, 30 percent sodic oligoclase, 20 percent quartz, and 10 percent mafic minerals. Chemically, Silvermine Granite is intermediate between less silicic Knoblick Granite and more silicic Slabtown Granite (Kisvarsanyi, 1972).

The Einstein Mine (Fig. 1) is but one of several mines and prospects that began operations in the 1870s to produce silver from argentiferous galena in quartz veins cutting Silvermine

Granite. The deposits have been described in detail by Tolman (1933) and Lowell (1975). There were two distinct periods of mining in the Silver Mine area: 1877 to 1894 and 1916 to 1946. During the earlier period, an estimated 50 tons (45 t) of lead and 3000 oz (84 kg) of silver were produced; during the later, an estimated 120 short tons (108 t) of tungsten concentrates were produced, largely by high-grading the old dumps and from shallow surface diggings. The remains of the dam, constructed in 1879, can be seen upstream from the mine dumps. Foundation remnants on the west hillslope south of the dam are all that remain of the large mill constructed during the silver- and tungsten-mining periods.

Of the several quartz veins mined and prospected in the area, vein no. 1, the Einstein, was the most productive; it accounted for the bulk of the early silver and lead production. It was entered by the River Tunnel, the entrance to which is about 50 ft (15 m) above the river. (*Caution: Entry through this old adit is dangerous and should not be attempted!*) The vein strikes N 80° E and dips 35° S; it pinches and swells, having a maximum width of not over 7 ft (2 m) and an ore zone as much as 2 ft (0.6 m) wide (Tolman, 1933). A pinched outcrop of the vein, where it is less than 2 in (5 cm) wide, is visible above a small mine opening in the hillside, about 50 ft (15 m) uphill from the River Tunnel. Near the contacts the intruded granite is intensely greisenized.

The high-temperature, pneumatolytic mineral assemblage at Silver Mine includes argentiferous galena, wolframite, arsenopyrite, sphalerite, cassiterite, chalcopyrite, covellite, hematite, stolzite, and scheelite. Quartz, topaz, sericite, fluorite, zinnwaldite, chlorite, and garnet are among the gangue minerals. Persistent search on the dump downhill from the mine may turn up good specimens.

Numerous intermediate to mafic dikes older than the quartz veins have been mapped in the Silver Mine area (Fig. 1). One of these is well exposed on the east side of the St. Francis River, just below the dam (Fig. 2). The dike is about 4 ft (1.2 m) wide, strikes N 65° E, and is nearly vertical. Its borders against Silvermine Granite are chilled, but its central part is coarser grained. The rock contains a few small plagioclase phenocrysts in a groundmass of andesine and augite with intergranular texture. Euhedral magnetite and pyrite are abundantly disseminated through the groundmass, and there is a small amount of interstitial quartz.

REFERENCES

Hayes, W. C., 1947, Geology of the Ozark-Martin mine area, Madison County, Missouri [Masters thesis]: Rolla, Missouri School of Mines and Metallurgy.

Kisvarsanyi, E. B., 1972, Petrochemistry of a Precambrian igneous province, St. Francois Mountains, Missouri (Contribution to Precambrian Geology No. 4): Missouri Geological Survey and Water Resources, Report of Investigations 51, 103 p.

Lowell, G. R., 1975, Precambrian geology and ore deposits of the Silver Mine area, southeast Missouri: A review, *in* Lowell, G. R., ed., A fieldguide to the Precambrian geology of the St. Francois Mountains, Missouri: Cape Girardeau, Southeast Missouri State University, p. 81–88.

Pomerene, J. B., 1947, Geology of the Einstein-Apex tungsten mine area [Masters thesis]: Rolla, Missouri School of Mines and Metallurgy, 39 p.

Sides, J. R., Bickford, M. E., Shuster, R. D., and Nusbaum, R. L., 1981, Calderas in the Precambrian terrane of the St. Francois Mountains, southeastern Missouri: Journal of Geophysical Research, v. 86, p. 10349–10364.

Tolman, C., 1933, The geology of the Silver Mine area, Madison County, Missouri: Missouri Bureau of Geology and Mines, 57th Biennial Report of the State Geologist, 1931-1932, Appendix 1, 39 p.

Tolman, C., and Robertson, F., 1969, Exposed Precambrian rocks in southeast Missouri (Contribution to Precambrian Geology No. 1): Missouri Geological Survey and Water Resources, Report of Investigations 44, 68 p.

Taum Sauk Power Plant section: Buried and exhumed hills of Precambrian rhyolite, the St. Francois Mountains, Missouri

Eva B. Kisvarsanyi and Arthur W. Hebrank, Missouri Geological Survey, Rolla, Missouri 64501

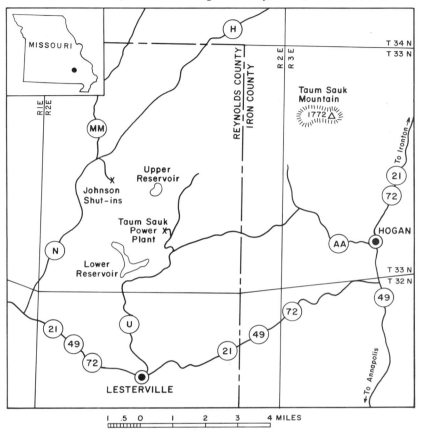

Figure 1. Location map of the Taum Sauk Hydroelectric Power Plant, Reynolds County, Missouri.

LOCATION

NE¼SE¼SW¼Sec.21,T.33N.,R.2E., Taum Sauk Power Plant, Reynolds County, Missouri; Johnson Shut-ins 7½-minute Quadrangle. The site is 10.8 mi (17.3 km) west of the village of Hogan, in Iron County, and can be reached by Missouri AA or, from Lesterville, in Reynolds County, by Missouri U (Fig. 1). The site is owned by the Union Electric Company and is open daily to visitors. A small museum is in the nearby Visitor Center.

SIGNIFICANCE

One of the most spectacular geologic sites in the state, this three-dimensional cut around the power plant displays the erosional unconformity between Precambrian rhyolite and overlying Upper Cambrian sedimentary rocks. Both a buried rhyolite knob and a partially exhumed rhyolite knob can be observed in the 100-ft-high (30-m-high) cut. The site is within the thickest section of volcanic rocks in the St. Francois Mountains, and is part of the Taum Sauk caldera (Berry and Bickford, 1972).

DESCRIPTION

The U-shaped cut at the power station exposes massive Precambrian ash-flow tuff, the Taum Sauk Rhyolite (Berry, 1976), overlain by Upper Cambrian sedimentary rocks. The cut stands some 100 ft (30 m) above the tailrace and reveals a spectacular three-dimensional cross-section of the Precambrian-Paleozoic erosional unconformity. Weathering has produced a few tens of feet of relief on the rhyolite surface. In the north face of the cut, the rhyolite knob is exposed at the ground surface, the overlying sediments having been removed by erosion (Fig. 2). In the east and south faces of the cut, the knob is still buried by sediments.

The rhyolite is overlain by beds of shaly and arkosic

Figure 2. Power station of the Taum Sauk Hydroelectric Plant at the base of Proffit Mountain. The U-shaped cut exposes a Precambrian knob of Taum Sauk Rhyolite; the rhyolite knob is exhumed in the north face of the cut (left), but is still buried beneath sedimentary strata in the east face (right). Note the prominent columnar jointing exhibited by the rhyolite. Photo by Art Hebrank.

dolomite, a sequence of alternating stromatolitic and burrowed carbonate muds assigned by Howe (1968) to the upper Davis Formation and the Derby-Doerun Dolomite. The dolomite laps onto the Precambrian surface from the west and has a maximum dip of 25 degrees. The steep dips are attributed to differential compaction of unconsolidated sediments deposited over the uneven rhyolite surface. The combined effects of carbonate solution, dolomitization, and compaction of argillaceous layers caused a loss of volume in the sedimentary beds; some relative movement of the sediments with respect to the rhyolite knob may have occurred.

REFERENCES CITED

Berry, A. W., 1976, Proposed stratigraphic column for Precambrian volcanic rocks, western St. Francois Mountains, Missouri, in Kisvarsanyi, E. B., ed., Studies in Precambrian Geology of Missouri (Contribution to Precambrian Geology No. 6): Missouri Department of Natural Resources, Geological Survey, Report of Investigations 61, p. 81–90.

Berry, A. W., and Bickford, M. E., 1972, Precambrian volcanics associated with the Taum Sauk caldera, St. Francois Mountains, Missouri, U.S.A.: Bulletin Volcanologique, v. 36-2, p. 303–318.

Howe, W. B., 1968, Planar stromatolite and burrowed carbonate mud facies in Cambrian strata of the St. Francois Mountain area: Missouri Geological Survey and Water Resources, Report of Investigations 41, 113 p.

Johnson Shut-ins: A shut-in canyon exposing a sequence of Precambrian ash-flow tuffs, the St. Francois Mountains, Missouri

Arthur W. Hebrank and Eva B. Kisvarsanyi, Missouri Geological Survey, Rolla, Missouri 64501

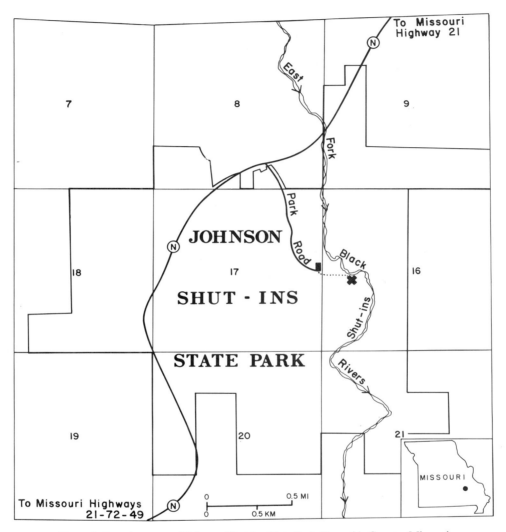

Figure 1. Location map of Johnson Shut-ins State Park, Reynolds County, Missouri.

LOCATION

Johnson Shut-ins State Park, W½Sec.16 and NW¼Sec.21, T.33N.,R.2E., Reynolds County, Missouri; Johnson Shut-ins 7½-minute Quadrangle. Entrance to the park is from Missouri N (Fig. 1), 13.2 mi (21 km) southwest of its junction with Missouri 21, north of Ironton, Missouri, and 6.2 mi (10 km) northeast of its junction with Missouri 21-49-72, west of Lesterville, Missouri. From the parking lot, walk east on well-marked (paved) trail about 1800 ft (540 m) to the shut-ins overlook (Fig. 2). The site is on public land. Note: State park rules prohibit defacing or collecting rocks.

SIGNIFICANCE

Johnson Shut-ins, one of the most picturesque volcanic-rock gorges in Missouri's St. Francois Mountains, is an excellent example of the regionally unique geomorphic features called shut-ins, and dramatically displays a variety of stream-erosion features, including cascades, potholes, plunge pools, and joint-determined channelways and chutes. A 2100 ft (650 m) thick sequence of Precambrian rhyolitic ignimbrites and intercalated volcaniclastic sediments is well exposed in the shut-ins; the rocks display well-preserved volcanic-rock textures and constitute the best easily accessible and well-described section of such rocks in the St.

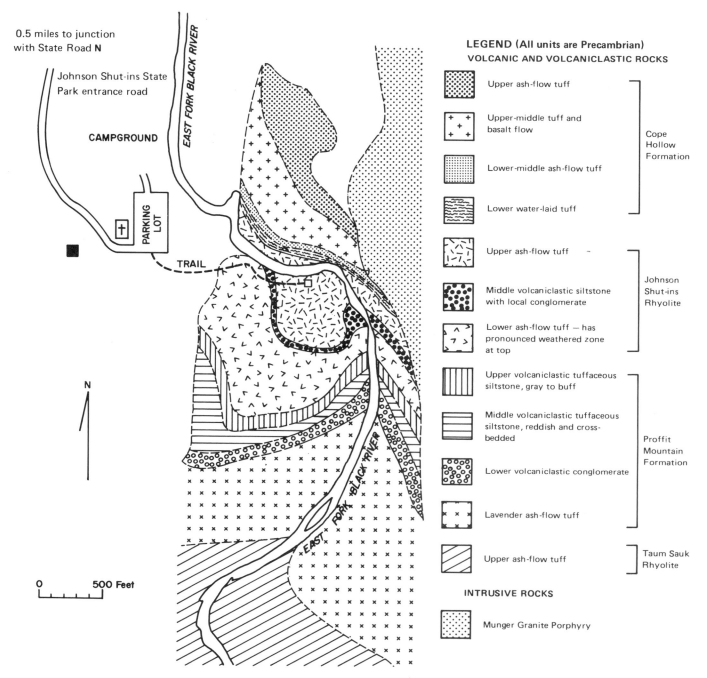

Figure 2. Geologic map of Johnson Shut-ins area. Modified from Blades and Bickford (1976); formal nomenclature after Berry (1976).

Francois Mountains. Johnson Shut-ins was designated a State of Missouri Geologic Natural Area in 1978.

DESCRIPTION

The featured scenic and geologic attractions of this popular state park are the shut-ins. As locally defined, a shut-in is a narrow, constricted stream-valley segment where a stream has cut through resistant igneous rocks. Upstream and downstream from a shut-in, a relatively wide or open valley is developed on less-resistant sedimentary rock. The most popular interpretation of origin contends that a shut-in is formed where resistant igneous rocks are encountered by an antecedent stream originally developed on sedimentary strata that completely or partially buried an old igneous-rock surface.

Johnson Shut-ins is a classic example of volcanic-rock

Figure 3. Johnson Shut-ins. View east from the head of the upper shut-in. Resistant, rounded, potholed rock in the foreground is Johnson Shut-Ins Rhyolite. Photo by Art Hebrank.

shut-ins. Confined within a narrow, steep-walled canyon, a mile-long segment of the East Fork of the Black River cuts through a thick sequence of erosion-resistant volcanic rocks (Fig. 3). The same river has formed an alluvial flood plain more than 0.25 mi (0.4 km) wide in sedimentary rocks upstream and downstream from the shut-ins.

Within the shut-ins, stream erosion is controlled by vertical joints; Beveridge (1978) describes three important joint sets. The major set trends northeast, at right angles to the valley; the valley drains to the southeast parallel to a secondary set; and a third set, trending due east, has been enlarged by erosion, making the channelways all the more complex.

While technically a single geomorphic feature, Johnson Shut-ins is commonly considered to be two tandem shut-ins, an upper and lower cascade, each with its own distinctive character. The upper shut-in is characterized by a maze of potholes, plunge pools, and tortuous, narrow channelways; the lower is dominated by a single, long, deep chute developed along a joint paralleling the flow direction of the river.

The Precambrian geology of the Johnson Shut-ins area has been mapped by Anderson (1970) and by Blades and Bickford (1976), the most detailed geologic description is by Zeller [nee Blades] (1980). The mile-long shut-ins expose a 2100 ft (650 m) thick sequence of ignimbrites (mostly rhyolite) and intercalated volcaniclastic sedimentary rocks (Fig. 4) that dip about 15 degrees to the northeast.

The upper (potholed) cascade, immediately east of the over-look platform, is developed in ash-flow tuffs of the Johnson Shut-ins Rhyolite (Fig. 2), a unit described in detail by Blades and Bickford (1976). It is predominantly a series of ash-flow tuffs, dark gray to red in color, which exhibit well-preserved textures and features indicative of pyroclastic origin. Fiamme (flame-shaped, compacted pumice fragments), lithic fragments, and piso-lites are readily observable. In thin section, the pisolites are seen

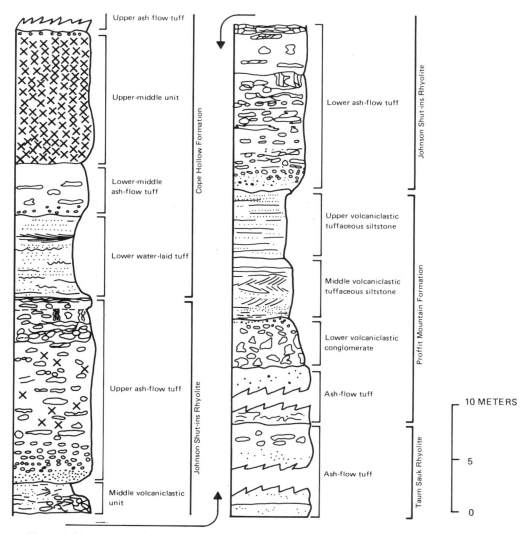

Figure 4. Measured stratigraphic section of Precambrian volcanic-rock units exposed in the Johnson Shut-ins area. Modified from Blades and Bickford (1976); formal nomenclature after Berry (1976).

to be accretionary lapilli composed of fine ash and shards. Lithophysal units are also present in the sequence; the products of vapor-phase crystallization within the individual lithophysae are anhedral quartz, feldspar, and muscovite.

Interbedded with the ash-flow tuffs are several volcaniclastic sedimentary rock units. Easily examined at the head of the upper shut-in, directly across the river from the observation platform, is a prominently exposed bed of water-laid tuff (Fig. 2), a unit described by Blades and Bickford (1976) as a uniform, gray, fine-grained, water-laid tuff with ripple marks, cross-bedding, and finely graded bedding.

It is suggested that interested groups with sufficient time examine the entire sequence of zoned ash-flow tuffs and volcaniclastic sediments exposed at the shut-ins. Possibly the most meaningful procedure would be to walk up through the exposed section, starting at the lower end of the constriction. Plan this adventure for a nice warm day and expect to get wet!

REFERENCES CITED

Anderson, R. E., 1970, Ash-flow tuffs of Precambrian age in southeast Missouri (Contribution to Precambrian Geology No. 2): Missouri Geological Survey and Water Resources, Report of Investigations 46, 50 p.

Berry, A. W., Jr., 1976, Proposed stratigraphic column for Precambrian volcanic rocks, western St. Francois Mountains, Missouri, *in* Kisvarsanyi, E. B., ed., Studies in Precambrian geology of Missouri (Contribution to Precambrian Geology No. 6): Missouri Division of Geology and Land Survey, Report of Investigations 61, p. 81–90.

Beveridge, T. R., 1978, Geologic wonders and curiosities of Missouri: Missouri Department of Natural Resources, Division of Geology and Land Survey, Educational Series no. 4, p. 43–48.

Blades, E. L., and Bickford, M. E., 1976, Rhyolitic ash-flow tuffs and intercalated volcaniclastic tuffaceous sedimentary rocks at Johnson Shut-ins, Reynolds County, Missouri, *in* Kisvarsanyi, E. B., ed., Studies in Precambrian geology of Missouri (Contribution to Precambrian Geology No. 6): Missouri Division of Geology and Land Survey, Report of Investigations 61, p. 91–104.

Zeller, E.L.B., 1980, Rhyolitic ash-flow tuff and intercalated volcaniclastic sedimentary rocks at Johnson Shut-ins, Reynolds County, Missouri [Masters thesis]: Lawrence, University of Kansas, 45 p.

Baird Mountain Quarry, southwestern Missouri

Thomas L. Thompson, Department of Natural Resources, Missouri Division of Geology and Land Survey, Box 250, Rolla, Missouri 39216

LOCATION

This section is in a quarry on the west side of Baird Mountain, 1.3 mi (2 km) southeast of Table Rock Dam; NW¼SW¼ NW¼Sec.26,T.22N.,R.22W., Taney County, Missouri, Table Rock Dam 7½-minute Quadrangle (Fig. 1). It has been designated the type section of the Baird Mountain Limestone Member of the Northview Formation (Thompson and Fellows, 1970, p. 149).

SIGNIFICANCE

The section in the Baird Mountain quarry exposes more than 180 ft (55 m) of Lower and Middle Mississippian shelf carbonates (Fig. 2). It is the most complete and accessible section for this sequence in the state and is an excellent example of the limestones and cherty limestones that typify the broad, shallow carbonate shelf conditions prevalent during Early and Middle Mississippian time (Manger and Thompson, 1982).

This quarry is the type section for the Baird Mountain Limestone Member of the Lower Mississippian (Kinderhookian) Northview Formation; the member is truncated and absent to the north, but is present to the south in Arkansas. This unit is younger than Kinderhookian strata elsewhere in the region and fills a gap left at the top of the Kinderhookian section in the stratotype region of the Mississippi River Valley. The Baird Mountain Limestone Member of the Northview Formation (Fig. 3) has been identified in the St. Joe Limestone in Arkansas and Oklahoma.

The Baird Mountain quarry, located high above the scenic southwestern Missouri Ozark region, is easily accessible by car, and it is only a short walk to the lower portion of the quarry. In the early 1950s this was one of the major sources for rock in the construction of Table Rock Dam.

DESCRIPTION

The Baird Mountain Quarry exposes a section (Fig. 2) from the Lower Ordovician (Ibexian or Canadian) dolomites to the middle Osagean Elsey Formation. Cross-stratified carbonates are exposed in the main quarry face (Fig. 4). Red and black chert and red and green limestone constitute the lower part of the Pierson Formation. The following description is from Thompson and Fellows (1970, p. 150–155).

MISSISSIPPIAN SYSTEM

Osagean Series

Elsey Formation (5 ft [1.5 m] exposed)
44. Chert (75 percent), white with gray to brown mottles,

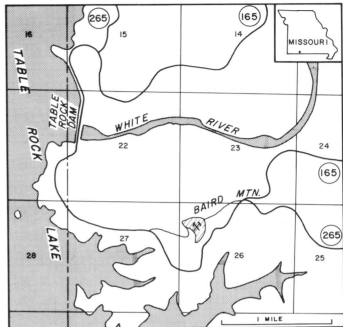

Figure 1. Portion of Table Rock Dam 7½-minute Quadrangle map, western Taney County, southwestern Missouri, showing location of Baird Mountain Quarry.

as large nodules; limestone, dark-gray to gray-brown, finely crystalline, cross-laminated, in thin nodular beds between chert nodules; large float-block at top of quarry .... 5 ft, 6 in. (1.6 m)

Reeds Spring Formation (108 ft; 33 m)
43. Covered interval, or inaccessible .... 33 ft (10 m)
42. Limestone, dark gray, dense, very irregularly bedded; chert, light blue-gray, mottled ...... 4 ft, 2 in (1.3 m)
41. Limestone (50 percent), dark gray-brown, finely crystalline, nodular; chert, tan to white and gray .... 1 ft (0.3 m)
40. Covered interval .... 16 ft, 6 in (5 m)
39. Limestone, gray to gray-green, medium crystalline, medium bedded, nodular; chert, brown, vuggy, weathers white .... 10 ft (3 m)
38. Limestone, gray-green, semicrinoidal to crinoidal; chert, blue-black, thinly bedded, nodular; one bed .... 3 ft, 5 in (1.1 m)
37. Limestone (75 percent), gray to tan, fine to medium crystalline with scattered crinoid debris, thinly bedded in gray shale partings, nodular; chert, mottled gray and brown, in thin beds and scattered nodules .... 1 ft, 11 in (0.6 m)

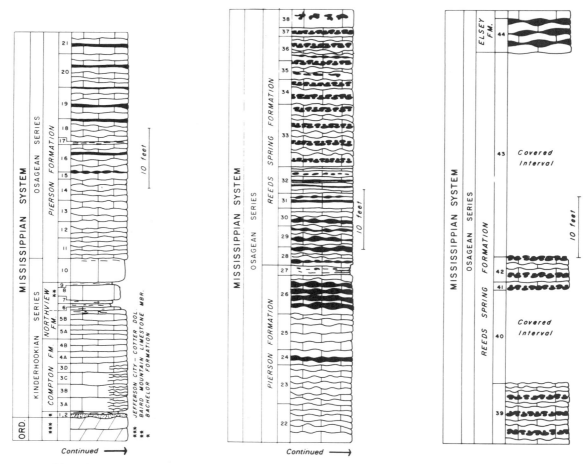

Figure 2. Columnar section of rocks exposed in Baird Mountain Quarry, Taney County, southwestern Missouri. Section taken from Thompson and Fellows (1970).

36. Limestone, gray-green, crinoidal, cross-laminated; chert, brown, crinoidal, very irregularly bedded .... 4 ft, 2 in (1.3 m)

35. Limestone (50 percent), same as unit 37; chert, opaque, brown .... 3 ft (0.9 m)

34. Limestone and chert, same as unit 37 .... 4 ft (1.2 m)

33. Limestone and chert, same as unit 37 ... 10 ft, 6 in (3.2 m)

32. Limestone (60 percent), gray-tan, finely crystalline, argillaceous, in thin, slabby beds; chert, blue-black, as thin beds and nodules .... 4 ft, 4 in (1.3 m)

31. Limestone, gray to gray-green, with streaks of red, medium crystalline, argillaceous; chert, dark blue, nodular upper part; as green opaque and blue translucent nodules .... 2 ft, 6 in (0.8 m)

30. Limestone (60 percent), gray to gray-green, finely crystalline, with scattered crinoid debris, thinly bedded, nodular; chert, blue-gray, tan, and green, with brown rinds; as scattered .... 3 ft, 1 in (0.9 m)

29. Limestone and chert, same as unit 30; thin to medium bedded .... 3 ft, 7 in (1.1 m)

28. Limestone and chert, same as unit 30; thin to medium bedded, nodular .... 3 ft (0.9 m)

Pierson Formation (70 ft; 21 m)

27. Shale, green and gray-green, calcareous, fossiliferous; contains dark-gray chert nodules ..... 1 ft, 6 in (0.5 m)

26. Chert (60 percent), brown and blue-gray, as smooth irregular nodules and beds, with brown rinds; limestone, gray-green to brown, finely crystalline, thin to medium bedded, nodular .... 6 ft, 10 in (2 m)

25. Limestone, green-gray, finely crystalline, argillaceous; chert, gray-black, in beds and small nodules .... 6 ft (1.8 m)

24. Limestone, green-gray, fine to medium crystalline, medium bedded, nodular; chert, dark gray and brown, with green rinds, as nodules and beds, contains inclusions of limestone .... 2 ft, 5 in (0.7 m)

23. Limestone, green-gray, fine to medium crystalline, with green shale partings, nodular .... 6 ft, 6 in (2.0 m)

22. Limestone, red, crinoidal, in thin to medium beds, nodular .... 5 ft, 5 in (1.6 m)

21. Limestone, brick-red, argillaceous, crinoidal, in medium beds; chert, red, with brown rinds, crinoidal .... 3 ft, 9 in (1.1 m)

Figure 3. Photograph of part of face in lower crusher pit of Baird Mountain Quarry (NW¼SW¼NW¼Sec.26,T.22N.,R.22W., Taney County, Missouri) illustrating the relationship of Baird Mountain Limestone Member of the Northview Formation (unit 2) to the underlying Northview (unit 1) and overlying Pierson (unit 3) limestones. Photograph by T. L. Thompson.

20. Limestone, brick-red and green, semicrinoidal; chert, red and green mottled . . . . 5 ft, 5 in (1.6 m)
19. Limestone and chert, same as unit 20 . . . 5 ft, 3 in (1.6 m)
18. Limestone and chert, same as unit 20; chert contains white crinoid debris, with green-white rinds . . . . 4 ft (1.2 m)
17. Shale, red and green, calcareous, crinoidal, platy to blocky . . . . 1 in (2.5 cm)
16. Limestone, light gray, medium crystalline to semicrinoidal, medium beds; chert, red with white rinds at base, dark gray to blue-gray at top, contains crinoid debris . . . . 4 ft, 8 in (1.4 m)
15. Limestone, gray-green, fine to medium crystalline, in medium beds . . . . 1 ft (0.3 m)
14. Limestone, red to gray, fine to medium crystal line, with red shale partings, nodular . . . . 3 ft, 6 in (1.1 m)
13. Limestone, red, fine to medium crystalline, with scattered crinoid debris, medium bedded, with red shaled partings . . . . 3 ft (0.9 m)
12. Limestone, red, finely crystalline at base, semicrinoidal toward top; in single bed with red shale partings; top of unit at floor of main quarry . . . . 2 ft, 10 in (0.9 m)
11. Limestone, green, crinoidal, in slabby nodular beds with green and red shale partings . . . . 3 ft, 8 in (1.1 m)

Figure 4. Photograph of upper quarry face of Baird Mountain Quarry, Taney County, Missouri, showing Pierson (unit 1) and Reeds Spring Formations (unit 2). Photograph by T. L. Thompson.

10. Limestone, red, medium crystalline with scattered crinoid debris, argillaceous at top, with green shale partings, one bed . . . . 4 ft (1.2 m)

Kinderhookian Series
Northview Formation (8 ft 4 in; 6.5 m)
    Baird Mountain Limestone Member (4 ft; 1.2 m)
9. Shale, brick-red, calcareous blocky . . . . 6 in (15.2 cm)
8. Limestone, red and green, crinoidal, shaly . . . . 2 ft, 6 in (0.8 m)
7. Limestone, brick-red with green mottles, medium crystalline, shaly at top . . . . 5 in (12.7 cm)
6. Shale, dark brick-red, calcareous, blocky . . . . 6 in (15.2 cm)
    Unnamed Member
5. Limestone, green, argillaceous, crinoidal, dense at base, shaly toward top, medium bedded . . . . 4 ft, 4 in (1.3 m)
Compton Formation (12 ft; 3.6 m)
4. Limestone, gray-tan, finely crystalline, medium bedded . . . . 4 ft, 3 in (1.3 m)
3. Limestone, gray-green, fine to medium crystalline, wavy bedded, with green shale partings on weathered surface, massive on fresh surface . . . . 8 ft (2.4 m)
Bachelor Formation (1 ft; 0.3 m)
2. Shale, green, sandy and clayey, blocky to platy . . . . 3 in (7.5 cm)
1. Sandstone, green, quartzose, well-cemented with "glint" ("poikilitic") calcite; contains phosphatic nodules and abraded chert pebbles . . . . 3–6 in (7.5–15 cm)

ORDOVICIAN SYSTEM

Ibexian (Canadian) Series
Jefferson City–Cotter Dolomites
undifferentiated . . . . 4 ft (1.3 m)

REFERENCES CITED

Manger, W. L., and Thompson, T. L., 1982, Regional depositional setting of Lower Mississippian Waulsortian Mound facies, southern Midcontinent, Arkansas, Missouri, and Oklahoma, *in* Symposium on the paleoenvironmental setting and distribution of the Waulsortian facies: El Paso Geological Society and University of Texas at El Paso, p. 43–50.

Thompson, T. L., and Fellows, L. D., 1970, Stratigraphy and conodont biostratigraphy of Kinderhookian and Osagean rocks of southwestern Missouri and adjacent areas: Missouri Geological Survey and Water Resources, Report of Investigations 45, 263 p.

Precambrian/Paleozoic unconformity at Chippewa Falls, Wisconsin

M. E. Ostrom, Wisconsin Geological and Natural History Survey, 3817 Mineral Point Road, Madison, Wisconsin 53705

LOCATION

Stream cut in east bank of Duncan Creek just north of first bridge south of Glen Lock Dam in Irvine Park near north city limits of Chippewa Falls (Fig. 1) in the NW¼,SW¼,NE¼, Sec.31,T.29N.,R.8W., Chippewa County (Chippewa Falls 7½-minute Quadrangle, 1972). Exposure can be reached by foot path from northeast side of bridge northward for about 300 ft (90 m).

SIGNIFICANCE

This exposure illustrates the Precambrian/Paleozoic unconformity in Wisconsin and the character of the initial Paleozoic deposits.

DESCRIPTION

The Mount Simon Sandstone (Fig. 2), presumably the oldest Cambrian formation in Wisconsin, rests unconformably on weathered Precambrian gneissic granite. This relationship with the Precambrian persists throughout the state where Paleozoic rocks are exposed, except that lithology and degree of alteration of the Precambrian rocks vary.

The Mount Simon Sandstone is believed to have been deposited in a shallow marine near-shore environment by an advancing sea that migrated from southeast to northwest over a weathered and eroded Precambrian rock surface (Ostrom, 1964).

The areal extent of such environments in the Gulf of Mexico today is limited to the length of the shoreline and a maximum width of about 20 mi (32 km). The Cambrian Ordovician sandstones are believed to be principally a result of spreading out shallow near-shore marine deposits as blankets during transgression (Ostrom, 1964). For example, Calvert (1962) shows that the Mount Simon or its equivalent, the Erwin Sandstone, overlaps to the northwest from Tennessee to Wisconsin, that it was deposited during a period of transgression, and that its age is Early Cambrian in Tennessee and Late Cambrian in Wisconsin.

The Mount Simon has been described as a predominantly quartz sandstone with minor shale, siltstone, and fine conglomerate, with feldspar as a minor constituent (Crowley and Thiel, 1940; Potter and Pryor, 1960). However, Asthana (1969) reported that the feldspar content of the Mount Simon Sandstone averages 18 percent, of which 81 percent is potash feldspar, 10 percent is plagioclase feldspar, and 9 percent is microcline. The range in feldspar content reported is from 2.85 to 40.07 percent. Asthana reported that all of the microcline and plagioclase grains are detrital, as are a part of the potash-feldspar grains. Authigenic orthoclase is very common but occurs as rhombic overgrowths on detrital grains. On this basis it appears that the Mount Simon

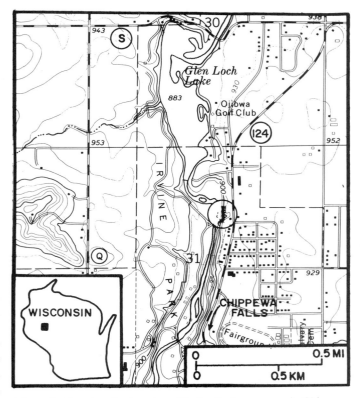

Figure 1. Map showing location of Irvine Park exposures in Chippewa Falls.

contains far more feldspar than has been previously noted, which helps to distinguish it from other Upper Cambrian and Ordovician sandstones in Wisconsin.

The source of the Cambrian sands has long been an enigma. Various hypotheses have been proposed that involve weathering and long transport of eroded Precambrian rocks to produce a relatively clean quartz sand. It should be pointed out that there is a ready source available in the local quartzites of Precambrian age. Distribution of the Baraboo Quartzite today is probably the result of a combination of factors including local and regional variations in intensity of metamorphism, erosion, and disintegration. It has been noted (Ostrom, 1966) that the Baraboo Quartzite disaggregates to yield already rounded monocrystalline quartz sand grains, and that this and similar quartzites may have been major sources of sand found in Cambrian and Ordovician rocks of the region. In a quarry located near North Freedom, Wisconsin, steeply tilted beds of quartzite, weathered in the upper few

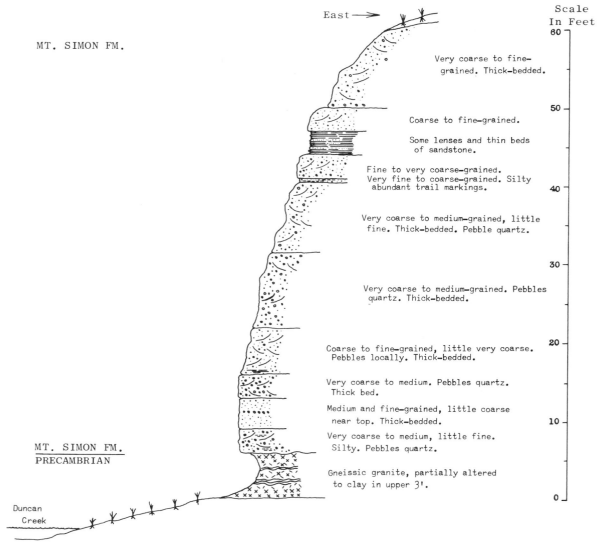

Figure 2. Stratigraphy of Irvine Park exposure.

feet, are overlain by flat-lying beds of the Galesville Sandstone. Here, weathering of the quartzite released rounded quartz grains that went to make up the Galesville Sandstone; an example of a Cambrian beach deposit. Other quartzites exhibit similar disintegration, namely the Rib Hill and Barron in Wisconsin and the Sioux in Minnesota (Austin, 1970). The fact that these quartzites are extensive and thick, and weather to yield already rounded quartz grains, suggests that they may have been a major source of sand supplied to Cambrian and Ordovician seas.

REFERENCES CITED

Asthana, V., 1969, The Mount Simon Formation (Dresbachian Stage) of Wisconsin [Ph.D. thesis]: Madison, University of Wisconsin, 159 p.

Austin, G. S., 1970, Weathering of the Sioux Quartzite near New Ulm, Minnesota, as related to Cretaceous climate: Journal of Sedimentary Petrology, v. 40, no. 1, p. 184–193.

Calvert, W. L., 1962, Sub-Trenton rocks from Lee County, Virginia, to Fayette County, Ohio: Ohio Geological Survey Report of Investigation 45, 57 p.

Crowley, A. J., and Thiel, G. A., 1940, Precambrian and Cambrian relations in east-central Minnesota: American Association of Petroleum Geologists Bulletin, v. 24, p. 744–749.

Ostrom, M. E., 1964, Pre-Cincinnatian Paleozoic cyclic sediments in the upper Mississippi valley; A discussion: Kansas Geological Survey Bulletin 169, v. 2, p. 381–398.

—— , 1966, Cambrian stratigraphy in western Wisconsin, Guidebook to the Michigan Basin Geological Society 1966 annual field conference: Wisconsin Geological and Natural History Survey Information Circular 7, 79 p.

Potter, P. E., and Pryor, W. A., 1960, Dispersal centers of Paleozoic and later clastics of the upper Mississippi valley and adjacent areas: Geological Society of America Bulletin, v. 72, p. 1195–1250.

The Mount Simon Formation at Eau Claire, Wisconsin

M. E. Ostrom, Wisconsin Geological and Natural History Survey, 3817 Mineral Point Road, Madison, Wisconsin 53705

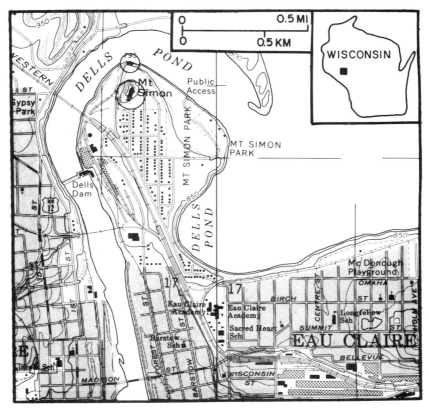

Figure 1. Map showing location of exposures discussed in text.

LOCATION

Exposure in bluff of Chippewa River and in hill called Mount Simon in City of Eau Claire (Fig. 1), in the SW¼SW¼Sec.8,T.27N.,R.9W., Eau Claire County (Elk Mound 7½-minute Quadrangle). Section includes all rock exposed from top of hill called Mount Simon northwest to base of river bluff.

SIGNIFICANCE

This is the type section of the Mount Simon Formation. The lithologic character of the Mount Simon Formation and its stratigraphic boundaries are illustrated in Figure 2.

DESCRIPTION

The Mount Simon Formation here grades upward from well-sorted, thick-bedded, coarse-grained sandstone in the lower part to finer grained, thinner bedded, transitional beds at the top. Although the section contains brachiopod shells in the upper few

feet, these beds are assigned to the Mount Simon rather than the Eau Claire Formation on the basis of lithologic similarity. The Mount Simon is assigned a Dresbachian age because it is transitional with the overlying Eau Claire Formation, which has a Dresbachian fauna (the trilobites *Cedaria* and *Crepicephalus*).

Previous mineralogical analyses of the Mount Simon at this site indicate a range in feldspar content of from 2.06 to 5.0 percent (Stauffer and Thiel, 1941; Crowley and Thiel, 1940; Potter and Pryor, 1961). However, a study by Asthana (1969) shows that the range in feldspar content of samples collected at regular 5-ft (1.5 m) intervals from the exposure is from 1.4 to 40.0 percent with an average of 17.5 percent. Combined plagioclase-microcline percentages range from 0.64 to 12.7 percent.

Predominant heavy minerals in the Mount Simon Sandstone are ilmenite, leucoxene, zircon, tourmaline, and garnet (Tyler, 1936).

The overlying Eau Claire Formation, where sampled near its base at Mount Washington (type section of the Eau Claire Formation), has a minimum feldspar content of 42 percent and a combined plagioclase/microcline content of 12 percent.

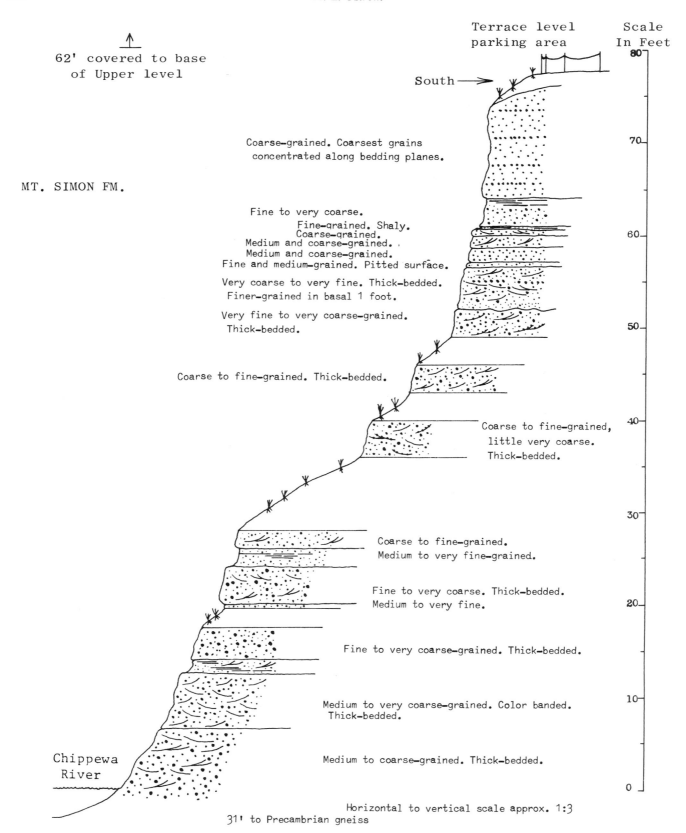

Figure 2. Stratigraphy of Mount Simon Formation in its type section.

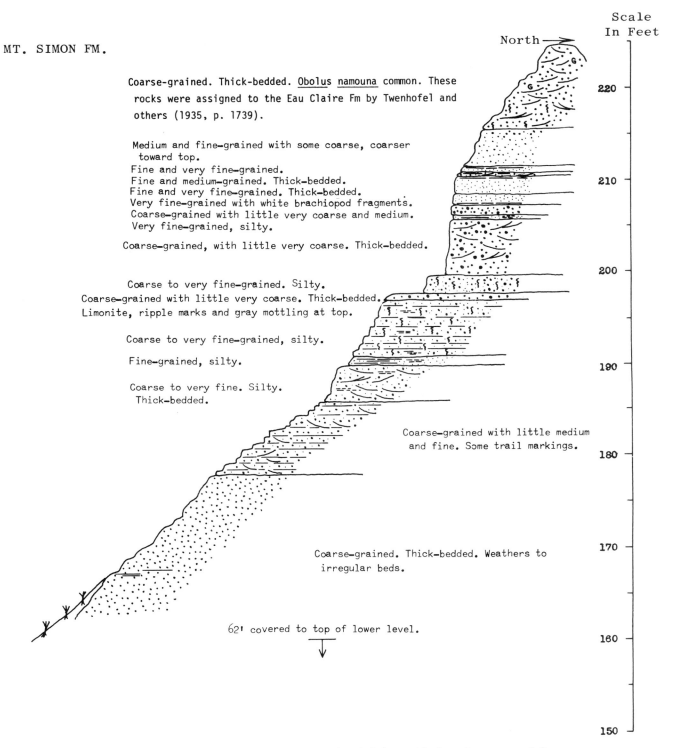

North →

MT. SIMON FM.

Coarse-grained. Thick-bedded. <u>Obolus</u> <u>namouna</u> common. These
rocks were assigned to the Eau Claire Fm by Twenhofel and
others (1935, p. 1739).

Medium and fine-grained with some coarse, coarser
toward top.
Fine and very fine-grained.
Fine and medium-grained. Thick-bedded.
Fine and very fine-grained. Thick-bedded.
Very fine-grained with white brachiopod fragments.
Coarse-grained with little very coarse and medium.
Very fine-grained, silty.

Coarse-grained, with little very coarse. Thick-bedded.

Coarse to very fine-grained. Silty.
Coarse-grained with little very coarse. Thick-bedded.
Limonite, ripple marks and gray mottling at top.

Coarse to very fine-grained, silty.

Fine-grained, silty.

Coarse to very fine. Silty.
Thick-bedded.

Coarse-grained with little medium
and fine. Some trail markings.

Coarse-grained. Thick-bedded. Weathers to
irregular beds.

62' covered to top of lower level.
↓

Horizontal to vertical scale approx. 1:3

The only other mineralogical information available on the Eau Claire Formation is an analysis by Potter and Pryor (1961) that indicates 12.5 percent feldspar in outcrops near Merrillan in northwestern Jackson County. Other analyses from scattered outcrops of the Eau Claire show variable amounts of tourmaline, zircon, ilmenite, magnetite, and garnet.

Of particular interest at this exposure are the transitional beds, which are also well exposed at the Rest Haven Gardens Town Road exposure south of the Eau Claire city limits (Ostrom, 1966, 1970). These beds have been recognized at many outcrops in this vicinity, but have not been traced to other areas due to lack of outcrops of this part of the section.

The transition beds are believed to have formed in a near-shore marine environment located seaward of the beach. The transition beds are characterized by a wide range in grain size from clay to very coarse sand and granules, well-defined bedding, different lithology from bed to bed, uniform lithology of individual beds, and by vertical burrows that are confined to certain beds.

REFERENCES CITED

Asthana, V., 1969, The Mount Simon Formation (Dresbachian Stage) of Wisconsin [Ph.D. thesis]: Madison, University of Wisconsin, 159 p.

Crowley, A. J., and Thiel, G. A., 1940, Precambrian and Cambrian relations in east-central Minnesota: American Association of Petroleum Geologists Bulletin, v. 24, p. 744–749.

Ostrom, M. E., 1966, Cambrian stratigraphy in western Wisconsin, Guidebook to the Michigan Basin Geological Society 1966 annual field conference: Wisconsin Geological and Natural History Survey Information Circular 7, 79 p.

—— , 1970, Field trip guidebook for Cambrian–Ordovician geology of western Wisconsin: Wisconsin Geological and Natural History Survey Information Circular 11, 131 p.

Potter, P. E., and Pryor, W. A., 1961, Dispersal centers of Paleozoic and later clastics of the upper Mississippi valley and adjacent areas: Geological Society of America Bulletin, v. 72, p. 1195–1250.

Stauffer, G. R., and Thiel, G. A., 1941, The Paleozoic and related rocks of southeastern Minnesota: Minnesota Geological Survey Bulletin 29, 261 p.

Twenhofel, W. H., Raasch, G. O., and Thwaites, F. R., 1935, Cambrian strata of Wisconsin: Geological Society of America Bulletin, v. 46, p. 1687–1743.

Tyler, S. A., 1936, Heavy minerals of the St. Peter Sandstone in Wisconsin: Journal of Sedimentary Petrology, v. 6, p. 77–79.

Late Cambrian Eau Claire Formation at Strum, Wisconsin

M. E. Ostrom, Wisconsin Geological and Natural History Survey, 3817 Mineral Point Road, Madison, Wisconsin 53705

LOCATION

Abandoned quarry at east side of County Trunk Highway D 0.2 mi (0.3 km) north of its junction with U.S. 10 at the north edge of the village of Strum in the NW¼,NE¼,SE¼,Sec.18, T.24N.,R.8W., Trempealeau County, Wisconsin, Strum 7½-minute Quadrangle (Fig. 1).

SIGNIFICANCE

This is one of the better, and certainly one of the most accessible, exposures of the Eau Claire Formation in Wisconsin (Ostrom, 1978). The Eau Claire Formation was named by Wooster (1882, p. 109) in referring to "Eau Claire Trilobite Beds" exposed in Eau Claire, Wisconsin. It is assigned to the Upper Ironton Dresbachian Stage on the basis of its trilobite fauna. In ascending order the trilobite assemblage zones in the Dresbachian are the *Cedaria, Crepicephalus, Aphelaspis,* and *Dunderbergia.* The Eau Claire strata in the region have been characterized faunally by the presence of the Upper Cambrian trilobite genera *Cedaria* and *Crepicephalus* (Ulrich, 1914, p. 354). The *Aphelaspis* Zone is known only from outcrops in the city of Hudson at the west edge of Wisconsin in the east bluff of the St. Croix River. The *Dunderbergia* Zone is not present in Wisconsin. In addition to being faunally zoned, the Eau Claire Formation has been subdivided on the basis of lithology. Morrison (1968, p. 21) subdivided the formation into five distinct lithologic units principally on the basis of bedding character and clay content. The lithologic units are:

E. Upper Massive Beds (±20 ft; 6 m). Sandstone; fine and very fine grained, light yellowish gray to brownish and greenish gray, massive and submassive, glauconitic; exposed near Strum and near Whitehall.

D. Upper Thin Beds (±15 ft; 4.5 m). Sandstone; fine and very fine grained, greenish gray; thin bedded; very glauconitic; usually missing.

C. Lower Massive Beds (±25 ft; 8 m). Sandstone, fine and very fine grained, light greenish gray to light brownish gray, thick to massive bedded; may be very glauconitic; few very argillaceous irregularly bedded units separating the more characteristic massive beds.

B. Lower Thin Beds (±20 ft; 6 m). Sandstone, fine and very fine grained, light greenish gray; thin and thick bedded; glauconitic and micaceous.

A. Shaley Beds (±15 ft; 4.5 m). Sandstone shale; dark gray to greenish gray, sandstone very fine and fine grained; very thin bedded; beds less than 3 in (8 cm) thick; very argillaceous.

The Eau Claire Formation is comformable with the underlying Mount Simon Formation. It is believed to represent a transgressive shallow marine shelf depositional environment deposit

Figure 1. Location of exposure at Upper Cambrian Eau Claire Formation in abandoned quarry north of Strum, Wisconsin.

(Ostrom, 1966; 1970). The formation is unconformable with the overlying Galesville Member of the Wonewoc Formation (Ostrom, 1970). The Eau Claire thins from a thickness of about 90 ft (27 m) in the subsurface of western Wisconsin to zero in outcrops at Sheep Pasture Bluff in Juneau County and at Friendship Mound in Adams County. Thinning is attributed to postdepositional erosion, which produced the unconformity.

At the Strum North exposure the section (Fig. 2) extends from the upper part of lithologic Unit A to near the top of Unit C. In addition, both the *Cedaria* and *Crepicephalus* faunal zones are present and readily distinguishable. The faunal break between the two zones occurs near the base of Unit C. One can speculate that in Wisconsin the reason for the almost complete absence of the overlying *Aphelaspis* Zone, and the complete absence of the succeeding *Dunderbergia* Zone, is that they were removed partially or completely by post–Eau Claire and pre-Galesville erosion. The Eau Claire Formation is succeeded by the Galesville Sandstone Member of the Wonewoc Formation. The Galesville Sandstone is interpreted to have formed in a broad and shallow beach/nearshore environment that transgressed over the eroded Eau Claire surface. The unconformable relationship is clear in an exposure in

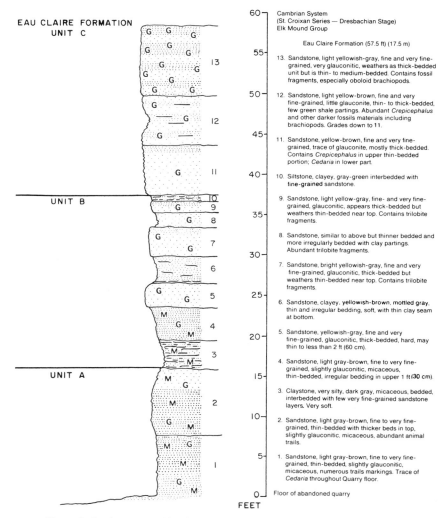

Figure 2. Section exposed in abandoned quarry north of Strum, Wisconsin.

the face of an abandoned quarry in the east bluff of Bruce valley located at the east side of County Highway D about 5 mi (8 km) south-southeast of Strum (NE¼SE¼NW¼Sec.9,T.23N.,R.8W., Trempealeau County, Wisconsin) and in the type section of the Galesville Sandstone, which is in a high and precipitous cliff cut by Beaver Creek where it passes through the city of Galesville about 32 mi (51 km) south of Strum (NE¼NW¼Sec.33, T.19N.,R.8W., Trempealeau County). The Galesville Sandstone is essentially unfossiliferous and is conformable with the overlying Ironton Sandstone, which contains the *Elvinia* fauna of the Franconian Stage. The lack of distinctive fossils in the Galesville Sandstone, its unconformable relationship with the underlying Eau Claire Formation, the eastward thinning and disappearance of the Eau Claire Formation and the conformable relationship of the Ironton Sandstone, which contains the Franconian *Elvinia* fauna, suggest that the Galesville Sandstone should be assigned to the Franconian rather than to the Dresbachian. Viewed from another perspective, it should be noted that no Dresbachian fossils are known to occur above the base of the Galesville Sand-

stone, which is transitional with strata containing a Franconian Fauna. From this perspective, one can reasonably assign the Galesville to the Franconian rather than to the Dresbachian as has been done traditionally.

REFERENCES CITED

Morrison, B. C., 1968, Stratigraphy of the Eau Claire Formation of west-central Wisconsin [M.S. thesis]: Madison, University of Wisconsin, 41 p.

Ostrom, M. E., 1966, Cambrian stratigraphy in western Wisconsin, Guidebook to the Michigan Basin Geological Society 1966 annual field conference: Wisconsin Geological and Natural History Survey Information Circular 7, 79 p.

—— , 1970, Field trip guidebook for Cambrian-Ordovician geology of western Wisconsin: Wisconsin Geological and Natural History Survey Information Circular 11, 131 p.

—— , 1978, Strum north exposure, *in* Geology of Wisconsin; Outcrop descriptions: Wisconsin Geological and Natural History Survey, TR-18/24N/8W, 3 p.

Ulrich, E. O., 1914, Cambrian geology and paleontology: Smithsonian Miscellaneous Collection, v. 57, p. 354.

Wooster, L. C., 1882, Geology of the lower St. Croix district, *in* Geology of Wisconsin: Wisconsin Geological Survey, v. 4, p. 99–159.

Cambrian stratigraphy at Whitehall, Wisconsin

M. E. Ostrom, Wisconsin Geological and Natural History Survey, 3817 Mineral Point Road, Madison, Wisconsin 53705

LOCATION

Road cuts north of Whitehall on Trempealeau County Highway "D" and 1.7 mi (2.7 km) north of its juncture with Wisconsin 53 in the SW¼,SW¼,Sec.12,T.22N.,R.8W., Trempealeau County, on the Pleasantville 7½-minute Topographic Quadrangle, 1973) (Fig. 1).

SIGNIFICANCE

This stop illustrates the contact relationship of the first and second sedimentary cycles in the Cambro-Ordovician rocks of Wisconsin (Ostrom, 1964, 1970). Also, it affords an opportunity to examine and compare three "lithotopes" of the second cycle equivalent to lithotopes of the first cycle, which can be seen at Irvine Park, Mount Simon, and Strum.

DESCRIPTION

This is an excellent exposure to show the interrelationships of the various Cambrian lithostratigraphic units, beginning with the Eau Claire Formation at the base and extending upward into the Lone Rock Formation (Fig. 2). The section is complete except for about 15 ft (4.5 m) of covered interval midway in the Ironton Member.

Beginning with the sharp and unconformable contact of Galesville on Eau Claire near the base, one can proceed upward through the remainder of the section without evidence of major erosional break.

Regionally, the Eau Claire Formation thins to the east until, in the vicinity of Wisconsin Dells, it is not recognized. North and northwest of the Dells, what is believed to be thin Eau Claire can be seen at Friendship Mound, north of Friendship, and at Sheep Pasture bluff located south of Mauston.

At Friendship Mound there is a 1-ft (0.3 m) bed of fine-grained, silty, iron-oxide–cemented sandstone that separates two thick-bedded, medium-grained, well-sorted sandstone units. The upper of these two units is positively identified as the Galesville Sandstone. At Sheep Pasture Bluff the situation is similar except that the thickness assignable to the separating unit is 7 ft (2 m) and it contains only minor iron oxide. Also at Sheep Pasture Bluff, sandstone clasts occur in the base of the Galesville. A possible interpretation is that the separating unit is the Eau Claire Formation, thinned by pre-Galesville erosion, and that the lower sandstone is the Mount Simon.

A study by Asthana (1969) indicates that the feldspar content of the Mount Simon Sandstone ranges from 3 to 40 percent and averages 18 percent. On the other hand, it is known from numerous analyses of the Galesville Sandstone that it seldom contains more than 1 percent feldspar. Asthana determined that the feldspar content of the sandstone unit below the Eau Claire is higher than that above by a factor of 2 at Sheep Pasture Bluff and

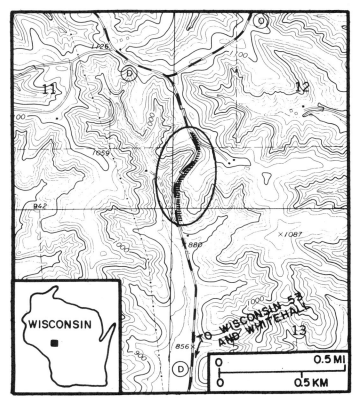

Figure 1. Map showing location of exposures discussed in text.

of 9 at Friendship Mound. This sharp decrease in feldspar content corresponds to a similar difference between the feldspar content of the Mount Simon and Galesville sandstones elsewhere and is interpreted to indicate that at these exposures the Eau Claire is much reduced in thickness, probably due to post–Eau Claire erosion. Thus, the Eau Claire apparently thins eastward due to pre-Galesville erosion.

The Eau Claire–Galesville contact marks the end of one transgressive/regressive sequence and the beginning of the transgressive phase of the subsequent sequence (Ostrom, 1964). The Galesville Sandstone formed during transgression as the result of a process of coalescing of shallow marine near-shore deposits. Rather persistent high-energy conditions are indicated by a noticeable lack of clay, silt, and very fine sand, and the paucity of fossils.

Overlying and transitional with the Galesville is the Ironton Member. The Ironton is interpreted to have formed in an environment located seaward of the beach, where high- and low-energy conditions alternated. Whereas the Galesville is thick bedded, the Ironton is medium bedded and even bedded. Silt and other fine particles are abundant in certain beds. Burrows are common and fossils are present locally. Also, there is commonly

WHITEHALL ROADCUT
SW¼,SW¼,Sec.12,T.22N.,R.8W.

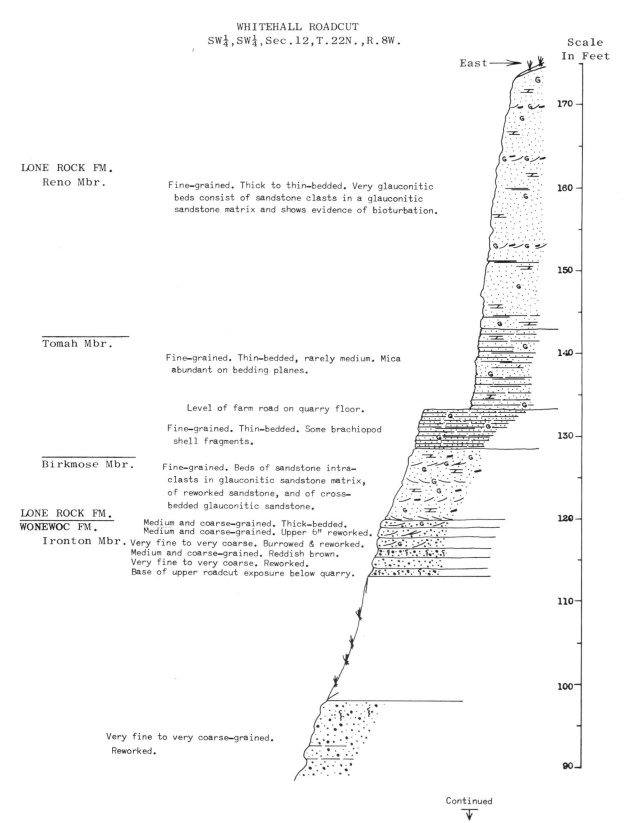

Scale
In Feet

East ➤

LONE ROCK FM.
Reno Mbr.
Fine-grained. Thick to thin-bedded. Very glauconitic
beds consist of sandstone clasts in a glauconitic
sandstone matrix and shows evidence of bioturbation.

Tomah Mbr.
Fine-grained. Thin-bedded, rarely medium. Mica
abundant on bedding planes.

Level of farm road on quarry floor.

Fine-grained. Thin-bedded. Some brachiopod
shell fragments.

Birkmose Mbr.
Fine-grained. Beds of sandstone intraclasts in glauconitic sandstone matrix,
of reworked sandstone, and of cross-
bedded glauconitic sandstone.

LONE ROCK FM.
WONEWOC FM.
Ironton Mbr.
Medium and coarse-grained. Thick-bedded.
Medium and coarse-grained. Upper 6" reworked.
Very fine to very coarse. Burrowed & reworked.
Medium and coarse-grained. Reddish brown.
Very fine to very coarse. Reworked.
Base of upper roadcut exposure below quarry.

Very fine to very coarse-grained.
Reworked.

Continued
▽

Figure 2. Stratigraphy of Cambrian rocks exposed in SW¼,Sec.12T.22N.,R.8W., Pleasantville 7½-
minute Quadrangle (Fig. 1).

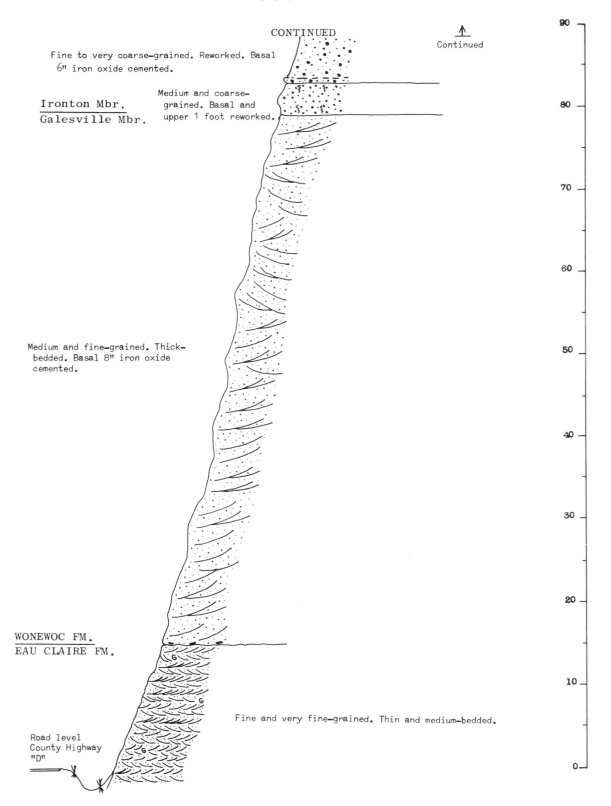

CONTINUED

Continued

Fine to very coarse-grained. Reworked. Basal 6" iron oxide cemented.

Ironton Mbr.
Galesville Mbr.

Medium and coarse-grained. Basal and upper 1 foot reworked.

Medium and fine-grained. Thick-bedded. Basal 8" iron oxide cemented.

WONEWOC FM.
EAU CLAIRE FM.

Fine and very fine-grained. Thin and medium-bedded.

Road level
County Highway
"D"

90

80

70

60

50

40

30

20

10

0

carbonate cement and glauconite in the upper few feet of the unit. Beds alternate between well-sorted, clean, medium- and coarse-grained, cross-bedded quartzarenite and poorly sorted, reworked and burrowed quartzarenite. Emrich (1966) traced certain of the burrowed beds for as much as 100 mi (160 km) across western Wisconsin, which suggests a broad and flat shelf bottom on which the effects of storm or quiet were widely impressed.

The Ironton Member thins eastward toward the Wisconsin Arch. At Whitehall, the Ironton is about 40 ft (12 m) thick. Traced east and south it thins to disappearance in the bluffs of the Wisconsin River south of Lone Rock. In northeastern Illinois, the Ironton thickens again to a maximum thickness of 150 ft (46 m) (Buschbach, 1964; Emrich, 1966). The Ironton also thickens westward into Minnesota. It is assigned a Franconian age on the basis of fossils.

At Whitehall the Ironton is in sharp contact with the overlying fine-grained glauconitic, shaly and thin-bedded Lone Rock Formation of the Tunnel City Group. The Tunnel City Group consists of two distinct facies in the upper Mississippi valley area, namely a glauconitic facies, the Lone Rock Formation, and a nonglauconite facies, the Mazomanie Formation (Trowbridge and Atwater, 1934; Wanenmacher and others, 1934; Twenhofel and others, 1935; Ericson, 1951; Berg, 1954; Ostrom, 1966,

1970). The Lone Rock facies intertongues with and is laterally and vertically transitional with the Mazomanie facies in the direction of the Wisconsin Arch (Ostrom, 1966).

Abundant burrows and trails in the Lone Rock indicate prolific animal life. Thin bedding and fine particles suggest persistent low-energy conditions. Occasional beds, up to 2 ft (0.6 m) thick and rarely up to 8 ft (2.4 m) thick, of sandstone clasts in a greensand matrix suggest occasional episodes of high energy such as storms. The environment of Lone Rock deposition is interpreted to have been located seaward of that of the Ironton in an area of deeper water and lower overall available energy, as attested to by thin beds, fine sediment, abundant fossils, and lateral persistence of beds.

The similarity of the lower part of the Lone Rock Formation at this site to the lower part of the Eau Claire is believed to be significant. In both cases the upward change is from transitional beds characterized by persistent beds of medium and coarse-grained quartzarenite to fine-grained, shaly glauconitic sandstone with abundant trail markings on bedding surfaces. The two units are interpreted as the manifestation of a single environment repeated by two episodes of transgression separated by a minor regression, which is marked by the Galesville Sandstone and the erosion surface at its base (Ostrom, 1964).

REFERENCES CITED

Asthana, V., 1969, The Mount Simon Formation (Dresbachian Stage) of Wisconsin [Ph.D. thesis]: Madison, University of Wisconsin, 159 p.

Berg, R. R., 1954, Franconia Formation of Minnesota and Wisconsin: Geological Society of America Bulletin, v. 65, p. 857–882.

Buschbach, T. C., 1964, Cambrian and Ordovician strata of northeastern Illinois: Illinois Geological Survey Report of Investigation 218, 90 p.

Emrich, G. H., 1966, Ironton and Galesville (Cambrian) sandstones in Illinois and adjacent areas: Illinois Geological Survey Circular 403, 55 p.

Erickson, D. M., 1951, The detailed physical stratigraphy of the Franconia Formation in southwest Wisconsin [M.S. thesis]: Madison, University of Wisconsin, 28 p.

Ostrom, M. E., 1964, Pre-Cincinnatian Paleozoic cyclic sediments in the upper Mississippi valley; A discussion: Kansas Geological Survey Bulletin 169, v. 2, p. 381–398.

——— , 1966, Cambrian stratigraphy in western Wisconsin, Guidebook to the Michigan Basin Geological Society 1966 annual field conference: Wisconsin Geological and Natural History Survey Information Circular 7, 79 p.

——— , 1970, Field trip guidebook for Cambrian-Ordovician geology of western Wisconsin: Wisconsin Geological and Natural History Survey Information Circular 11, 131 p.

Trowbridge, A. C., and Atwater, G. I., 1934, Stratigraphic problems in the upper Mississippi valley: Geological Society of America Bulletin, v. 45, p. 21–80.

Twenhofel, W. H., Raasch, G. O., and Thwaites, F. R., 1935, Cambrian strata of Wisconsin: Geological Society of America Bulletin, v. 46, p. 1687–1743.

Wanenmacher, J. M., Twenhofel, W. H., and Raasch, G. O., 1934, The Paleozoic strata of the Baraboo area, Wisconsin: American Journal of Science, v. 228, p. 1–30.

Archean gneiss at Lake Arbutus Dam, Jackson County, Wisconsin

R. S. Maass and B. A. Brown, Wisconsin Geological and Natural History Survey, 3817 Mineral Point Road, Madison, Wisconsin 53705

LOCATION

SE¼,Sec.3,T.22N.,R.3W., Hatfield 7½-minute Topographic Quadrangle. Outcrop along the Black River below the eastern half of Arbutus Dam (Fig. 1). Approach is on a 0.2-mi-long (0.3 km) gravel road that intersects Clay School Road 0.1 mi (0.2 km) west of the Green Bay and Western Railroad tracks. Additional outcrop occurs for 0.6 mi (1 km) downstream from dam.

SIGNIFICANCE

The main outcrop area immediately below the dam is one of the largest, if not the largest, outcrop of Archean rocks in Wisconsin (Brown and others, 1983). The principal unit is the Hatfield Gneiss, an interlayered sequence of quartzo-feldspathic gneisses and minor amphibolite. The rocks are interpreted as a metavolcanic sequence that was formed about 2,815 Ma and deformed at least twice, with the latest deformation and metamorphism occurring during the Penokean orogeny, about 1,850 Ma. Postdeformational cross-cutting mafic dikes are also present at this locality.

DESCRIPTION

The principal unit exposed (Fig. 2) is the Hatfield Gneiss, an interlayered sequence of granitic to tonalitic gneiss with concordant layers of amphibolite. Over much of the outcrop the layers are 0.1 to 3 cm thick, pink to gray, quartzo-feldspathic gneiss. In some parts the layers are thicker, approaching several meters of massive gneiss. Folding and foliation are best displayed in the more thinly banded portions. The quartzo-feldspathic gneiss has a granoblastic texture and consists of equal amounts of quartz, plagioclase, and microcline. Mafic minerals represent less than 10 percent in most instances. Normative abundances based on bulk chemical analyses show that the amount of quartz is approximately constant and that plagioclase/orthoclase ratios vary from about 1:1 (adamellite) to primarily plagioclase (tonalite).

The amphibolite is interlayered with the quartzo-feldspathic gneiss and consists primarily of hornblende with about 20 percent epidote. The amphibolite has been deformed along with the rest of the gneiss and is interpreted as originally concordant. The entire assemblage is interpreted as having formed from an interlayered sequence of volcanic flows, pyroclastic rocks, or sills (DuBois and Van Schmus, 1978). The major metamorphism currently recorded by the rocks is amphibolite facies. Relict pyroxene has been found in some of the quartzo-feldspathic gneiss samples, suggesting either primary volcanic pyroxene or an earlier, higher grade period of metamorphism.

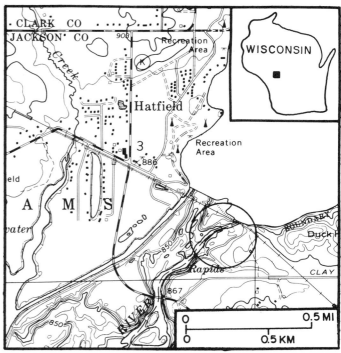

Figure 1. Map showing location of site at Arbutus Dam (circle).

The Hatfield Gneiss has been subjected to an isoclinal folding event (F_1), which produced an axial planar foliation parallel to the layering, except in fold hinges where the foliation transects the layering. The foliation was then tightly to openly folded during F_2 deformation; the axial planes of these folds are at high angles to the foliation. F_1 folds are rarely visible, but F_2 folds are conspicuous wherever the banding is readily apparent.

Lineations (fold axes, crenulations, mineral lineations) and foliations were measured in the gneiss along a 0.4-mi-long (0.6 km) stretch of the river. Poles to foliation define a β axis, which is essentially identical to the orientation of the main grouping of the linear structural elements (Fig. 3). Fold axes, when plotted separately, fall into the two groups in the southeast quadrant of the stereonet, with the vast majority plotting in the main group.

A group of F_1 fold axes in the core of a large, tight F_2 fold were plotted separately from the rest of the linear structural elements. The folding in this vicinity is highly complex, resulting in numerous and diverse interference patterns from the folding of F_1 axes. The axes of these F_1 folds plot in all four quadrants of the stereonet with plunges ranging from horizontal to vertical. Girdles that would indicate a later simple folding pattern of the F_1 axes do not exist, and the interference patterns are therefore believed

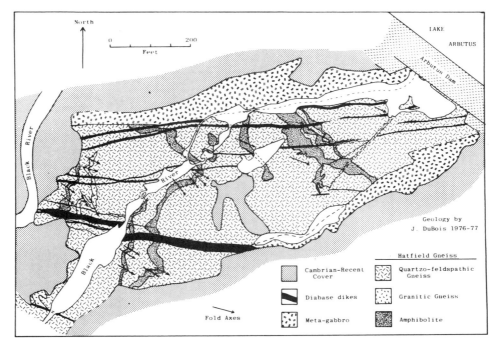

Figure 2. Geologic map of Archean bedrock exposed south of Arbutus Dam.

to be the result of inhomogeneous deformation in the core of the F₂ fold.

Although F_2 deformation is inhomogeneous in this relatively small area of the outcrop, the outcrop as a whole exhibits homogeneous deformation, as demonstrated by the tight distribution of 94.5 percent of the linear structural elements. F_1 fold axes are never exposed in three dimensions (except in the anomalous area just discussed); thus their trend and plunge are unknown. The age of F_1 folding is unclear, but F_2 folding is probably Penokean.

Zircon has been separated from a tonalitic layer of the gneiss on the west bank of the Black River, about 0.6 mi (1 km) downstream from the dam. The zircon crystals are brown, euhedral with normal igneous zoning and no signs of significant overgrowths or relict cores. U-Pb analyses on several fractions show that this unit is essentially the same age (2,815 ± 20 Ma) as other Archean gneisses in central Wisconsin. This age is interpreted as the time of crystallization (volcanism) of the protolith of the Hatfield Gneiss. Rb-Sr analyses from several samples collected in the area of this stop and further downstream do not plot coherently on an isochron diagram, indicating partial resetting during subsequent metamorphism.

REFERENCES CITED

Brown, B. A., Clayton, L., Madison, F. W., and Evans, T. J., 1983, Three billion years of geology; A field trip through the Archean, Proterozoic, Paleozoic, and Pleistocene geology of the Black River Falls area of Wisconsin; Guidebook for the 47th Tri-State Geological Field Conference: Wisconsin Geolog-

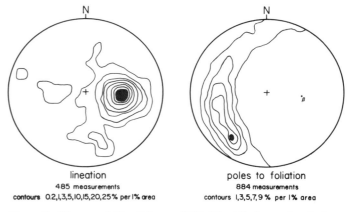

Figure 3. (Maass and Van Schmus, 1980, Fig. 11). Lower hemisphere stereographic projections of structures in the Hatfield Gneiss. (left) Lineations defined by fold axes, crenulations, and mineral elongations. The mean orientation of the lineations is S.84°E. with a plunge of 51°ESE. (right) Plot of poles to foliation yields a mean for β trending S.84°E. with a plunge of 52°ESE, virtually identical to the mean orientation for the lineations.

ical and Natural History Survey Field Trip Guidebook 9, 51 p.

DuBois, V. F., and Van Schmus, W. R., 1978, Petrology and geochronology of Archean gneiss in the Lake Arbutus area, west-central Wisconsin [abs.]: Proceedings of the 24th Institute on Lake Superior Geology, Milwaukee, p. 11.

Maass, R. S., and Van Schmus, W. R., 1980, Precambrian tectonic history of the Black River valley: 26th Institute on Lake Superior Geology, Guidebook 2, 43 p.

St. Lawrence and Jordan formations (Upper Cambrian) south of Arcadia, Wisconsin

M. E. Ostrom, Wisconsin Geological and Natural History Survey, 3817 Mineral Point Road, Madison, Wisconsin 53705

LOCATION

Composite section from outcrops and quarries located along Wisconsin 93 (Fig. 1) and extending from 1.8 mi (2.9 km) to 3.5 mi (5.6 km) south of its intersection in Arcadia with Wisconsin 95, in the W½,NW¼,Sec.9,T.20N.,R.9W., Trempeauleau County (Tamarack 7½-minute Quadrangle, 1973).

SIGNIFICANCE

This stop is an excellent exposure of the St. Lawrence and Jordan formations (Fig. 2). It also illustrates one of the problems in the Upper Cambrian stratigraphy in Wisconsin, namely determining the limits and relationships of the Black Earth Dolomite and Lodi Siltstone members of the St. Lawrence Formation. This stop illustrates the contact relationship of the Lone Rock Formation with the St. Lawrence Formation and of the second and third cycles of sedimentation in the Cambro-Ordovician of Wisconsin. It also illustrates a major economic use of Wisconsin's dolomite formations.

DESCRIPTION

Nelson (1956) studied these units in the upper Mississippi valley area. He defined the Black Earth as " . . . sandy dolomite and interbedded dolomitic siltstone and fine-grained sandstone. In the vicinity of Black Earth and Madison, and at localities along the Mississippi Valley, it generally is massive, brown to buff, slightly glauconitic . . . (with) . . . algal structures locally" (p. 173). The Lodi Member consists of " . . . siltstone, generally dolomitic, and dolomitic sandstone" (p. 173).

The fact that his definitions indicate that both the Black Earth and the Lodi can consist of dolomitic siltstone and fine-grained sandstone is the reason why it is commonly very difficult to distinguish the two members. Here Nelson assigned the lower 17 ft (5 m) of the St. Lawrence to the Lodi, the middle 12-ft (3.7 m) portion to the Black Earth, and an overlying 15-ft (4.5 m) section to the Lodi for a total thickness of about 44 ft (13 m). Close examination of the outcrop shows that if a Black Earth Dolomite occurs here, it is probably the 7 ft (2.1 m) of very silty dolomite in the interval from 19 ft (5.8 m) to 26 ft (7.9 m) above the base of the exposure. However, there does not appear to be any marked difference in lithology to suggest the presence of Black Earth rather than Lodi. The Wisconsin Geological and Natural History Survey recognizes the Black Earth as a medium to thick-bedded, medium to coarsely crystalline dolomite that is locally silty, sandy, and glauconitic, with fossil algae. With the possible exception of several thin beds, all of the St. Lawrence Formation at this exposure is assigned to the Lodi Member.

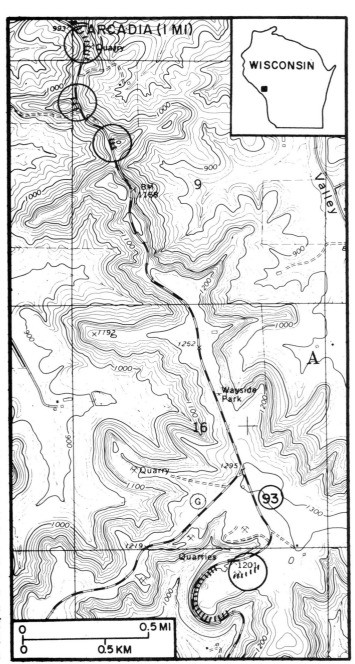

Figure 1. Map showing location of exposures discussed in text.

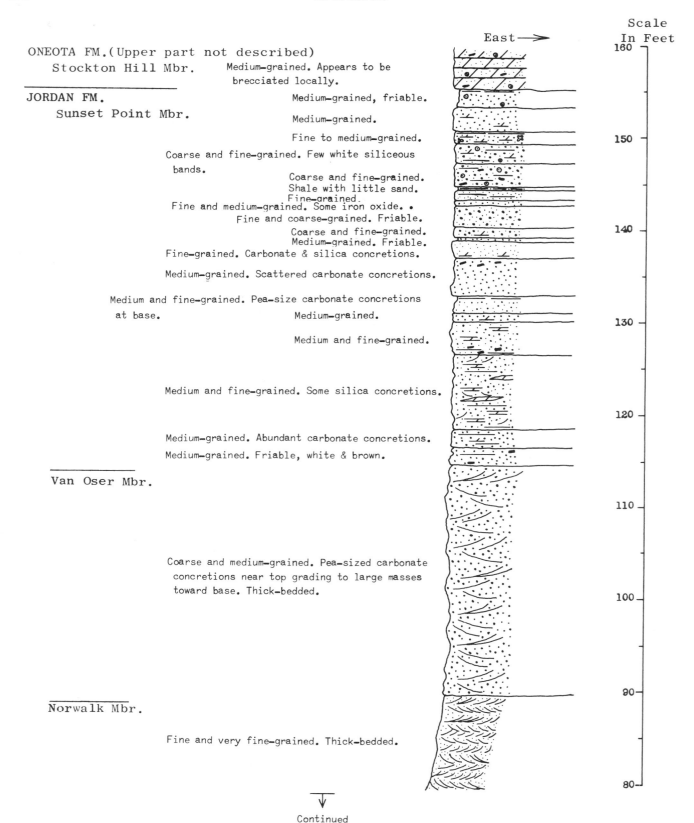

ONEOTA FM.(Upper part not described)
    Stockton Hill Mbr.   Medium-grained. Appears to be
                                brecciated locally.

JORDAN FM.   Medium-grained, friable.
    Sunset Point Mbr.   Medium-grained.
                                   Fine to medium-grained.
            Coarse and fine-grained. Few white siliceous
                bands.
                                Coarse and fine-grained.
                                Shale with little sand.
                            Fine-grained.
            Fine and medium-grained. Some iron oxide.
                Fine and coarse-grained. Friable.
                            Coarse and fine-grained.
                            Medium-grained. Friable.
            Fine-grained. Carbonate & silica concretions.

            Medium-grained. Scattered carbonate concretions.

        Medium and fine-grained. Pea-size carbonate concretions
            at base.   Medium-grained.

                                Medium and fine-grained.

        Medium and fine-grained. Some silica concretions.

        Medium-grained. Abundant carbonate concretions.
        Medium-grained. Friable, white & brown.

    Van Oser Mbr.

        Coarse and medium-grained. Pea-sized carbonate
        concretions near top grading to large masses
        toward base. Thick-bedded.

    Norwalk Mbr.

        Fine and very fine-grained. Thick-bedded.

East —→

Scale
In Feet
160
150
140
130
120
110
100
90
80

Continued

Figure 2. Stratigraphy of Cambrian rocks exposed in NW¼,Sec.9,T.20N.,R.9W., Tamarack 7½-minute
Quadrangle (Fig. 1).

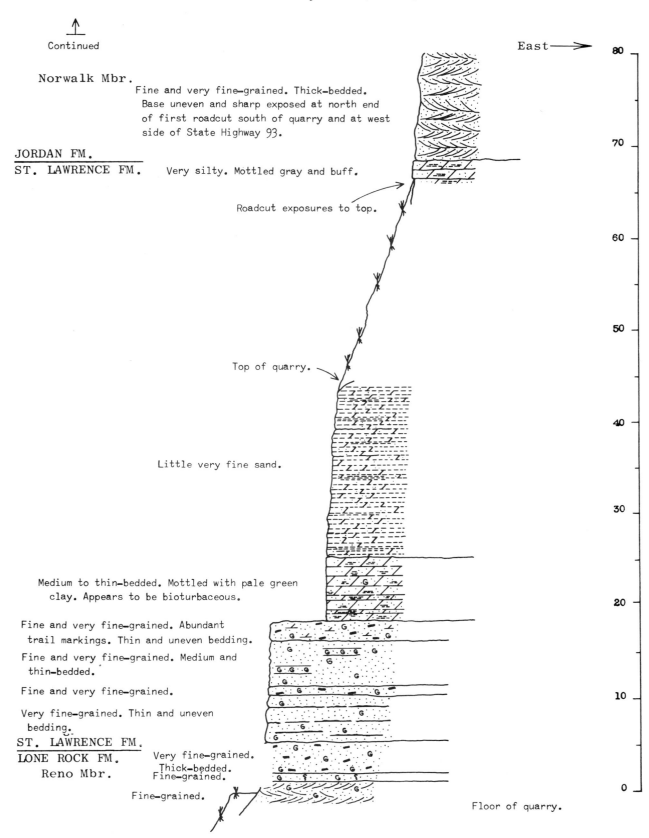

↑
Continued

East ——→   80

Norwalk Mbr.

Fine and very fine-grained. Thick-bedded.
Base uneven and sharp exposed at north end
of first roadcut south of quarry and at west
side of State Highway 93.

70

JORDAN FM.
ST. LAWRENCE FM.   Very silty. Mottled gray and buff.

Roadcut exposures to top.

60

50

Top of quarry.

Little very fine sand.   40

30

Medium to thin-bedded. Mottled with pale green
clay. Appears to be bioturbaceous.   20

Fine and very fine-grained. Abundant
trail markings. Thin and uneven bedding.
Fine and very fine-grained. Medium and
thin-bedded.
Fine and very fine-grained.   10

Very fine-grained. Thin and uneven
bedding.
ST. LAWRENCE FM.
LONE ROCK FM.   Very fine-grained.
Reno Mbr.   Thick-bedded.
Fine-grained.

Fine-grained.   0

Floor of quarry.

The Norwalk Member of the Jordan Formation consists of very fine and fine-grained nonsilty sandstone, which is thick bedded to thin bedded. At this exposure it is separated from the underlying silty and dolomitic Lodi by a sharp and uneven surface interpreted to indicate post-Lodi erosion. At the majority of outcrops of this interval in Wisconsin, the contact appears to be completely gradational. The Norwalk and Van Oser members constitute a thick body of sandstone similar in character to the Galesville Formation(?) and other Cambrian and Ordovician sandstones of this region. This suggests that the Jordan probably had a similar origin, namely that it formed on an erosion surface by a process of coalescing of beach deposits in a transgressing sea. In at least this area, erosion was a minor factor.

The Van Oser Member of the Jordan Sandstone is characterized by medium-grained sandstone, with some coarse and a little fine. The contact of the Van Oser with the Norwalk is commonly, though not always, sharp. At this exposure the contact is slightly uneven. Contact relations of overlying beds are best examined elsewhere.

Field study by McGannon (1960) led him to propose the name Stockton Hill Formation for those strata between the Lone Rock Formation below and Jordan Formation above. The top of the Lone Rock is marked by 3 ft (1 m) of flat-pebble conglomerate overlain by 0.7 ft (0.2 m) of "wormstone," a burrowed, glauconitic, calcareous fine-grained sandstone. McGannon's Stockton Hill Formation extends 36.7 ft (11.2 m) upward to 6.4 ft (2 m) below the top of the quarried section. The lower 24.1 ft (7.3 m) are assigned to the Lodi Member; the upper 12.6 ft (3.8 m) to what he has named the Red Wing Member. The upper 6.4 ft (2 m), which he assigned to the Jordan Formation, contains from 28 to 50 percent carbonate and from 35 to 50 percent silt

and finer particles, although this does not conform to other descriptions of the Jordan. The contact with fine-grained Jordan Sandstone containing only minor carbonate and silt can be observed in the first roadcut above the quarry and west of Wisconsin 93. It is believed that this is the actual contact. The Wisconsin Geological and Natural History Survey retains the name St. Lawrence Formation for what McGannon proposes to call the Stockton Hill Formation and assigns all of the St. Lawrence at this exposure to the Lodi Member.

The Jordan Sandstone is divided on the basis of composition, texture, and bedding characteristics into three members: the lower Norwalk, the middle Van Oser, and the upper Sunset Point. These can be traced throughout southwestern Wisconsin and into eastern Minnesota and Iowa.

The Jordan Sandstone is overlain by the Oneota Dolomite Formation of the Prairie du Chien Group.

Quarries in the Oneota Formation are located along Wisconsin 93 south of the Jordan Sandstone outcrops and at the crest of the ridge that forms the Skyline Drive. The Oneota is a primary source of crushed stone used for construction throughout much of southern Wisconsin. This is a "portable operation;" the crushing and processing equipment is portable, as opposed to stationary.

REFERENCES CITED

McGannon, D. E., Jr., 1960, A study of the St. Lawrence Formation in the upper Mississippi valley [Ph.D. thesis]: Minneapolis, University of Minnesota, 355 p.

Nelson, C. A., 1956, Upper Croixan stratigraphy: Geological Society of America Bulletin, v. 67, p. 165–184.

Proterozoic quartzite at Hamilton Mound, central Wisconsin

B. A. Brown and J. K. Greenberg, *Wisconsin Geological and Natural History Survey, 3817 Mineral Point Road, Madison, Wisconsin 53705*

LOCATION

The quarry at Hamilton Mound is in the NE¼,Sec.36,T.-20N.,R.6E., Coloma NW 7½-minute Quadrangle. It can be reached by turning east from Wisconsin 13 on Archer Drive, just north of Dorro Couche Lake, and proceeding about 4 mi (6 km) to a turnoff leading south into the quarry in the middle of Hamilton Mound. The turnoff from Wisconsin 13 is about 15 mi (24 km) south of Wisconsin Rapids (Fig. 1).

SIGNIFICANCE

Hamilton Mound is an inlier of folded Proterozoic quartzite similar to the Baraboo Syncline and the Waterloo area exposures (Brown, 1986). The quartzite is exposed on a series of low hills; Upper Cambrian sandstone of the Elk Mound Group overlaps the quartzite and is exposed on the slopes. Hamilton Mound is a prominent feature on the flat sand plains of central Wisconsin. Sand dunes are scattered over the plain, which was a Quaternary lake bed. A quarry developed in the quartzite exposes a granite intrusive into the quartzite, and an unusual zone of contact metasomatism and alteration within the quartzite.

DESCRIPTION

The quartzite was originally a fine- to medium-grained quartz sand. Sericite and clays constitute from 1 or 2 percent to 25 percent of the rock, suggesting that the sandstone varied in content of clay (or feldspar?), which is now represented by mica or has been realtered to kaolin. Typical samples contain 5 to 10 percent sericite, 90 percent recrystallized quartz grains, and traces of hematite, chlorite, zircon, and other detrital minerals. Small feldspar grains (less than 1 mm) are common near the granite contact, and chlorite, zircon, sericite, and clay minerals are concentrated near the intrusion.

Primary sedimentary structures include bedding, cross-bedding, and, less commonly, ripple marks. Fine laminated units commonly are slumped and faulted, possibly due to tectonic as well as sedimentary deformation.

STRUCTURAL FEATURES

The macroscopic structure of the Hamilton Mound exposures was mapped by Ostrander (1931) who identified four major folds trending N75°W (Fig. 2). The roughly east-west axial trend is similar to that of the Baraboo Syncline and the Waterloo area. Other structures, including distortion of bedding, several sets of fracture cleavage, foliation, shear zones, and zones of brecciation are well developed in the area of the granitic intrusion (Fig. 3) and increase in intensity as the intrusive contact is approached. The intensity of deformation is evident in thin section (Fig. 4) where quartz grains become highly strained.

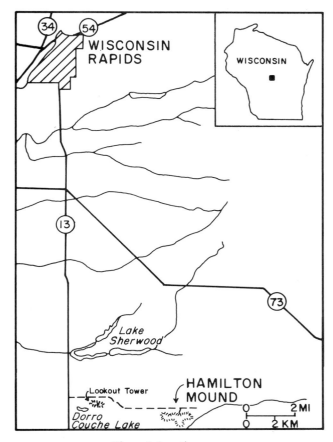

Figure 1. Location map.

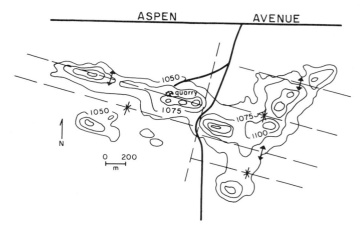

Figure 2. Major structures at Hamilton Mound (after Ostrander, 1931).

Figure 3. Excavation face at Hamilton Mound quarry showing steeply dipping beds in quartzite cut by nearly horizontal fracture cleavage.

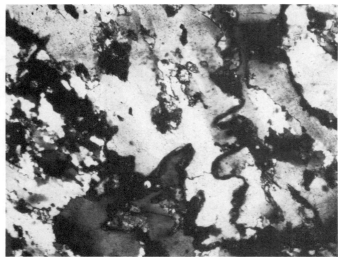

Figure 4. Micrograph of intensely deformed quartzite from near the intrusive contact. Note microstylolite developed between quartz grains. Long dimension is about 6 mm.

Important structural features are zones of quartzite breccia cemented by white vein quartz that extend upward from near the intrusive contact in the quarry. Similar brecciated zones are common in other areas where Baraboo interval quartzites are intruded by granitic rocks (Greenberg, 1986). Taylor and Montgomery (1986) observed porphyritic granitic fragments in the breccia zones, suggesting that they are late hydrothermal phenomena.

THE INTRUSIVE ROCKS

From the present extent of exposure, there is no certain way of knowing the original igneous character of the granitic intrusion at Hamilton Mound. Contaminated igneous material is of two types. The more original-appearing rock is exposed near the pit entrance, and contains bright red-orange phenocrysts (to 0.8 in; 2 cm in length) of potassium feldspar and plagioclase, colored by hematite inclusions. Some larger quartz grains also occur as clasts in a matrix of highly strained quartz (to 50 percent of total), chlorite, opaque minerals, and sericite (Fig. 5). Much of the sericite may have been derived from altered feldspars. Zircon is common. Larger inclusions in the granitic rock are composed of quartz, biotite, chlorite, and sericite. These inclusions are unlike the overlying quartzite and may be remnants of digested basement rocks. Chemical analyses of samples of the porphyritic granite are consistent with a granitic intrusion contaminated by mafic and aluminous material (Taylor and Montgomery, 1986). Initial U-Pb zircon data from the porphyritic granitic rock suggest 1760 Ma (W. R. Van Schmus, unpublished data) as a possible age. This age would further establish a link between Baraboo-interval sedimentation and 1760 Ma magmatism. Rb-Sr analyses (Taylor and Montgomery, 1986) indicate that whatever the original age, the granite at Hamilton Mound was isotopically reset at 1585 ±

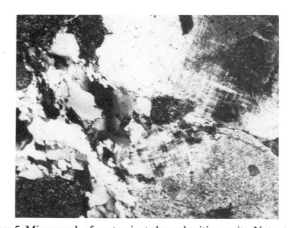

Figure 5. Micrograph of contaminated porphyritic granite. Note strained quartz and chlorite surrounding feldspar phenocrysts. Long dimension is about 8 mm.

30 Ma, an age overlapping the uncertainties of both the 1630 Ma regional disturbance and the 1500 Ma (Wolf River) episode of anorogenic magmatism.

At the west end of the quarry, quartzite and intrusive rocks appear to be very complexly mixed. The gray foliated rock exposed here ranges from a highly deformed micaceous quartzite into a very quartz-rich banded rock containing large amounts of fresh fine-grained feldspar (microcline and plagioclase), biotite, and less common hornblende near the granite. In this zone of transition or mixing, fine banding with the appearance of sedimentary laminations, becomes contorted and indistinguishable from tectonic foliation (Fig. 6). Enigmatic round inclusions (xe-

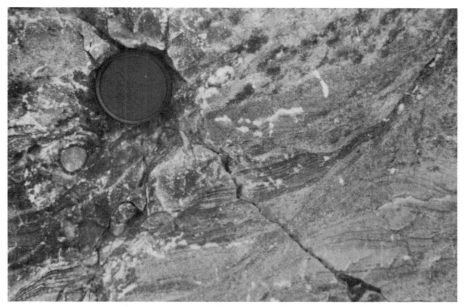

Figure 6. Distorted laminations and inclusions with reaction rims, from the quartzite-intrusion mixing zone at Hamilton Mound. Lens cap is 2 in (5 cm) in diameter.

Figure 7. Excavation face at Hamilton Mound quarry showing quartzite blocks in Cambrian sandstone overlying quartzite. Sandstone beds become more regular and flaggy to the right of the photo. Quartzite bluff is to left. Horizontal dimension is about 33 ft (10 m).

noliths?) of mafic material with reaction rims occur in the mixed zone. U-Pb analyses by W. R. Van Schmus (unpublished data) determined an age for the zircon crystals from this mixed zone as 2500 Ma. One interpretation is that these zircons and inclusions in the magma represent basement assimilated and brought up from below. Another possibility is that the Hamilton Mound quartzite contains detrital zircons derived from eroded Archean basement.

All thin sections of quartzite and intrusive rock collected from within or near the mixed zone have the high-strain deformational fabric associated with the quartzite-intrusion contact. Unusually strain-free grains, feldspar and biotite in particular, appear to be late magmatic (metasomatic?) phases that grew during or after deformation. Rare dikelets of granitic rock containing tourmaline are also known to postdate deformation (Taylor and Montgomery, 1986; Greenberg, 1986). These observations, along with the extensive brecciation, suggest both a forceful intrusion and a substantial chemical interaction between magma and overlying quartzite.

A definite influence of granitic intrusion on the quartzite is color alteration. Although Hamilton Mound quartzite away from the intrusion is characteristically pink-red (as seen on the ridge southeast of the quarry ridge), quartzite in proximity to the granitic rock is distinctly greenish. The color change is probably explained by the reduction of iron in hematite during heating. Similar color variations can be seen at Necedah and in the contact zone of the Baxter Hollow Granite at Baraboo (Greenberg, 1986).

SANDSTONE

A thin cap of sandstone sits atop poorly exposed quartzite along one wall of the quarry. This sandstone, like most other exposures in the area, is correlated with the Upper Cambrian Elk Mound Group. The sands are interpreted as having been deposited on a topographic high of the eroded Precambrian rocks.

Just above the quartzite, the sandstone is very poorly sorted with alternating beds of rubbly conglomerate and finer sand beds (Fig. 7). The rubbly conglomerate contains large angular blocks (to 3 ft; 1 m across) of quartzite. Away from the unconformity, the beds become thinner, with better sorting and flaggy parting.

The Cambrian sediments at Hamilton Mound may have been storm deposits like those which have been described in the Baraboo area by Dalziel and Dott (1970). The Hamilton Mound inlier probably stood above sea level as small islands or stacks during deposition of the flanking sandstone.

REFERENCES CITED

Brown, B. A., 1986, The Baraboo interval in Wisconsin, *in* Greenberg, J. K., and Brown, B. A., eds., Proterozoic Baraboo interval in Wisconsin: Geoscience Wisconsin, v. 10, p. 1–14.

Dalziel, I. W. and Dott, R. H., Jr., 1970, Geology of the Baraboo district, Wisconsin—A description and field guide incorporating structural analysis of the Precambrian rocks and sedimentologic studies of the Paleozoic strata: Wisconsin Geological and Natural History Survey Information Circular 14, 164 p.

Greenberg, J. K., 1986, Magmatism and the Baraboo interval; Breccia, metasomatism, and intrusion, *in* Greenberg, J. K., and Brown, B. A., eds., Proterozoic Baraboo interval in Wisconsin: Geoscience Wisconsin, v. 10, p. 96–112.

Ostrander, A. R., 1931, Geology and structure of Hamilton Mounds, Adams County, Wisconsin [M.S. thesis]: Madison, University of Wisconsin, 27 p.

Taylor, S. M., and Montgomery, C. W., 1986, Petrology, geochemistry, and Rb-Sr systematics of the porphyritic granite at Hamilton Mound, Wisconsin, *in* Greenberg, J. K., and Brown, B. A., eds., Proterozoic Baraboo interval in Wisconsin: Geoscience Wisconsin, v. 10, p. 85–95.

Baraboo Quartzite at Skillett Creek, Wisconsin

B. A. Brown, *Wisconsin Geological and Natural History Survey, 3817 Mineral Point Road, Madison, Wisconsin 53705*

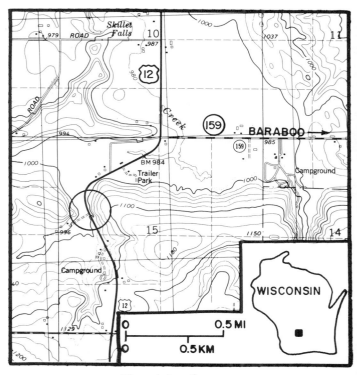

Figure 1. Location map.

Figure 2. Cross-bedding in Baraboo Quartzite, lower part of exposure near road. Lens cap is 2 in (5 cm) in diameter.

Figure 3. Boudinaged and folded beds of quartzite interlayered with phyllite, upper part of exposure, above massive quartzite. Long dimension is approximately 6.5 ft (2 m).

LOCATION

The exposure is on the east side of U.S. 12, 0.3 mi (0.5 km) south of the junction with Wisconsin 159, SW¼,NW¼,Sec.15,T.-11N.,R.6E., North Freedom 7½-minute Quadrangle (Fig. 1). **Caution:** Traffic on U.S. 12 is heavy, and there is a blind curve just north of the outcrop.

SIGNIFICANCE

This outcrop provides an opportunity to examine both the quartzite and phyllite facies of the Baraboo Quartzite. This is a classic exposure that exhibits important sedimentary and tectonic structures typical of the Baraboo interval rocks of Wisconsin (Greenberg and Brown, 1983, 1984; Brown, 1986).

DESCRIPTION

Pink quartzite, dipping 15° north, is exposed at the southern end of the outcrop. Good examples of sedimentary structures typical of the Baraboo Quartzite, including cross-bedding (Fig. 2) and ripple marks, are present at this exposure. Dalziel and Dott (1970) refer to this exposure as an excellent example of the paleocurrent indicators that suggest a southward sediment trans-port direction at Baraboo. Locally, cross-bedding in individual sets of laminae shows contortion, particularly oversteepening, which Dalziel and Dott attributed to synsedimentary deformation.

At the north end of the exposure and on top of the cliff, argillaceous beds up to 6.5 ft (2 m) in thickness occur interbedded with thin (1.5 ft or less; 0.5 m) beds of quartzite, (Fig. 3). The thin quartzite beds within the less competent phyllite provide some spectacular examples of boudinage and parasitic folding. The S_1

cleavage, related to the formation of the Baraboo syncline, is nearly parallel to bedding in the phyllite at this location. Later crenulation cleavages and small-scale conjugate kinks cut the S_1 foliation at high angles. Late veins of white quartz cut the thin quartzite beds at a high angle to bedding. In thin section (Fig. 4), crenulation in the phyllite is quite apparent. Mineralogy is quartz, muscovite, and sometimes pyrophyllite, indicating a maximum of upper greenschist facies metamorphism. Recent road construction has uncovered additional exposures about 300 ft (90 m) to the north, around the curve of U.S. 12. This cut exposes the dip slope of the quartzite and contains some excellent tectonic structures, particularly refracted cleavage, in both the quartzite and phyllite.

This is an exemplary teaching outcrop and field trip stop. Please keep hammering and destructive sampling to a minimum.

REFERENCES CITED

Brown, B. A., 1986, The Baraboo interval in Wisconsin: Geoscience Wisconsin, v. 10, p. 1–18.

Dalziel, I.W.D., and Dott, R. H., Jr., 1970, Geology of the Baraboo district, Wisconsin: Wisconsin Geological and Natural History Survey Information Circular 14, 164 p.

Greenberg, J. K., and Brown, B. A., 1983, Middle Proterozoic to Cambrian rocks in Wisconsin; Anorogenic sedimentary and igneous activity: Wisconsin Geological and Natural History Survey Field Trip Guidebook 8, 50 p.

—— , 1984, Cratonic sedimentation during the Proterozoic; An anorogenic connection in Wisconsin and the upper Midwest: Journal of Geology, v. 92, p. 159–171.

Figure 4. Photomicrograph of crenulated phyllite. Note crenulations at high angle to phyllitic foliation. Field of view is about 8 mm in long dimension.

Middle Ordovician rocks at Potosi Hill, Wisconsin

M. E. Ostrom, Wisconsin Geological and Natural History Survey, 3817 Mineral Point Road, Madison, Wisconsin 53705

LOCATION

Roadcut at east side of U.S. 61 and Wisconsin 35 about 1 mi (1.6 km) northwest of bridge over Platte River and 4 mi (6.4 km) northwest of Dickeyville in the SE¼,NW¼,Sec.7, T.2N.,R.2W., Grant County, Wisconsin in the Potosi 7½-minute Quadrangle (Fig. 1).

SIGNIFICANCE

This is an excellent and easily accessible exposure of the upper few feet of the St. Peter Sandstone and the Glenwood Formation, Platteville Formation, Decorah Formation, and lower part of the Galena Formation (Fig. 2) (Ostrom, 1978). The Platteville, Decorah, and Galena Formations are the principal hosts of zinc and lead mineralization in the southwest Wisconsin zinc-lead mineralized district.

The St. Peter Sandstone was named by Owen (1847) for exposures along the St. Peter River (now the Minnesota River) near St. Paul, Minnesota. The St. Peter consists of very light yellowish gray to white, fine to coarse, subrounded to rounded quartz sand grains. It is typically very friable. It is cross-bedded, thick bedded to thin bedded, and locally massive. In the district it is from 0 to more than 300 ft (0 to 90 m) in thickness and averages about 40 ft (12 m). Variations in thickness are attributed to deposition on an erosion surface. The only fossils noted in the St. Peter are *Skolithos* (vertical straight burrows) and *Corophoides* (U-shaped burrows).

The St. Peter Sandstone is conformably overlain by the Glenwood Formation. The Glenwood was named by Calvin (1906, p. 75) from exposures in Glenwood Township (T.98N.,R.7W.) near Waukon, Iowa. Three members are recognized in the Glenwood Formation in southwest Wisconsin (Templeton and Willman, 1963; Ostrom, 1969). In the base of the Glenwood is the Nokomis Member, which consists principally of sandstone and is transitional with the St. Peter. It is distinguished from the St. Peter Sandstone by a more yellowish and greenish coloration and by a notable change in bedding character from cross-bedded, even-bedded, and uniform-textured sandstone to reworked, burrowed, and poorly sorted sandstone with more or less green clay. It is both silty and argillaceous. The Nokomis ranges from 8 ft (2.4 m) thick near Beetown (16 mi; 26 km northwest of Potosi Hill) to less than 1 ft (0.3 m) thick in the vicinity of New Glarus (about 65 mi; 105 km to the east).

The Nokomis Member is conformably overlain by the Harmony Hill Member, which consists of pale green to greenish gray shale with scattered rounded clear quartz sand grains. It decreases from 3.5 ft (1 m) thick in the western part of the district to zero in the east. The Harmony Hill is conformably overlain by the Hennepin Member. The Hennepin consists of brownish and

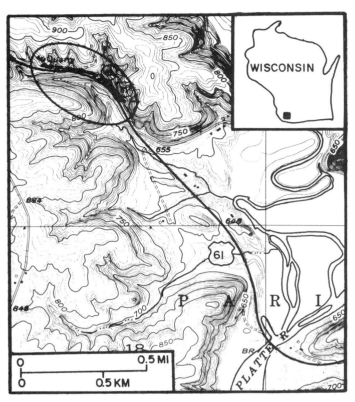

Figure 1. Location of exposure of Middle Ordovician strata in roadcut at Potosi Hill, Wisconsin.

locally calcareous shale with scattered phosphatic nodules and clear rounded quartz sand grains. It thins from 5 ft (1.5 m) thick in the western part of the district to zero in the east.

The Glenwood Formation is conformably overlain by the Platteville Formation, which is subdivided in ascending order into the Pecatonica, McGregor, and Quimbys Mill members. The Pecatonica Dolomite Member was named by Hershey (1894, p. 175) from exposures in the Pecatonica River valley in southwestern Wisconsin near the Illinois border. The Pecatonica is predominantly medium-grained, granular, thick- to thin-bedded dolomite. The lowermost bed, the Chana Member of Templeton and Willman (1963), contains phosphatic pellets and rounded clear quartz sand grains. The Pecatonica ranges from 20 to 25 ft (6 to 7.6 m) in thickness in the district.

The McGregor Limestone Member was named by Kay (1935, p. 286) from an exposure near McGregor, Iowa. It is from 25 to 30 ft (8 to 9 m) thick and consists of irregularly bedded, thin- to medium-bedded, light gray to buff argillaceous dolomite

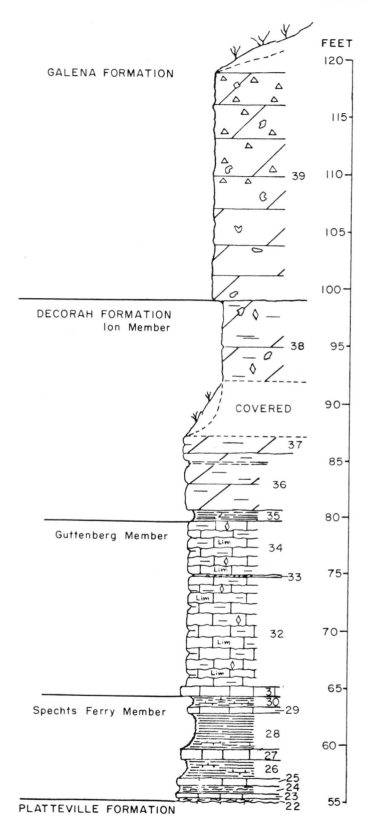

FEET

Galena Formation

Decorah Formation
Ion Member

COVERED

Guttenberg Member

Spechts Ferry Member

Platteville Formation

Ordovician System
Champlainian Series
Sinnipee Group

Galena Dolomite Formation
Cherty Unit
39. Dolomite, yellowish-buff, medium- to coarse-grained, vuggy, abundant white chert in upper 10 ft (3 m).

Decorah Formation
Ion Dolomite Member
(Gray Unit)
38. Dolomite, buff, medium- to coarse-grained, thick- to massive-bedded, vuggy, green shale partings throughout, sparry calcite present. Covered interval.
37. Dolomite, buff, medium-grained, medium-bedded, with green shale partings.
(Blue Unit)
36. Dolomite, purplish-gray, medium-grained, slightly fossiliferous. Green shale present as partings, and as a 0.5 ft (15 cm) bed 0.8 ft (25 cm) below the top of the interval.
35. Shale, green. 0.3 ft (9 cm) green dolomitic shale in middle of interval.
Guttenberg Limestone Member
34. Limestone, purplish-brown, fine-grained to sub-lithographic, fossiliferous, upper 1 ft (30 cm) fine- to medium-grained, brown shale present as partings, calcite and limonite after iron sulfide present in small amounts.
33. Metabentonite, brownish orange, crumbly, sticky when wet.
32. Limestone, purplish-brown, sublithographic, thin wavey bedding, fossiliferous, brown carbonaceous shale present as thin beds and partings, calcite and limonite present. Thin metabentonite bed at base.
31. Limestone brown-gray, fine-grained, thick-bedded.
Spechts Ferry Shale Member
30. Shale, orange-gray, calcareous, and limestone, tan-gray, fine-grained; limestone 0.4 to 0.7 ft (12 to 20 cm) above base.
29. Limestone, gray, fine-grained, thin-bedded.
28. Shale, gray, green, brown, fissle, some beds fossiliferous, limestone present as thin lenses near middle of the interval.
27. Limestone, tan, with iron oxide mottling, fine-grained, thin-bedded.
26. Shale, gray-green-brown. Fissle, with thin lenses of gray fine-grained limestone.
25. Limestone, dark to light gray, thin-bedded fossiliferous.
24. Shale, brown-green-orange-gray, brown carbonaceous shale parting at top, metabentonite near middle.
23. Limestone, purplish-brown, fine-grained, thin-bedded, very fossiliferous, fucoids at base.
22. Metabentonite, orange, sticky when wet, with brown shale partings.

Figure 2 (this and facing page). Section exposed in roadcut at Potosi Hill, Wisconsin.

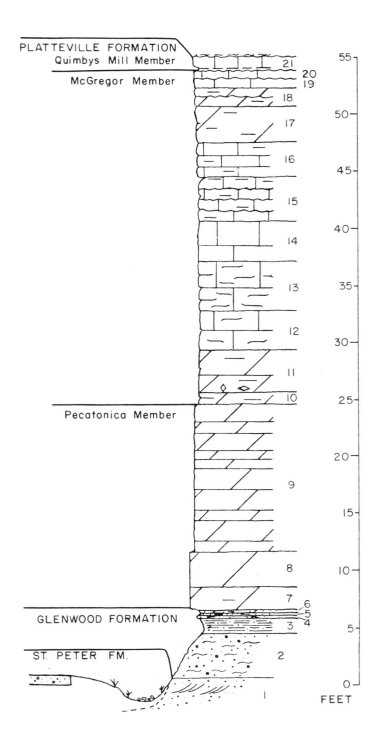

PLATTEVILLE FORMATION
Quimbys Mill Member

McGregor Member

Pecatonica Member

GLENWOOD FORMATION

ST. PETER FM.

FEET

Platteville Formation
Quimbys Mill Member
21. Limestone, purplish-ray-brown, sublithographic, thick-bedded, conchoidal fracture, irregular upper surface, shale at base. Metabentonite in top of shale.
McGregor Limestone Member
20. Limestone, purplish-gray-brown, fine to medium crystalline, thick-bedded; wavy upper surface.
19. Limestone, light gray, very fine crystalline, thin-bedded, fossiliferous; wavy upper surface.
18. Dolomite, yellowish-brown, fine crystalline, sugary texture, argillaceous, thin-bedded; discontinuous beds/nodular appearance.
17. Dolomite, light olive-brown, fine crystalline, sugary texture, argillaceous, thick bedded.
16. Limestone, light brown to greenish-brown, medium to fine crystalline, with some argillaceous partings; some fossil shells and fucoids on bedding planes.
15. Limestone, light brown to light greenish-brown, fine to medium crystalline, thin-bedded and uneven beds with nodular appearance, shale partings. Argillaceous in upper 0.6 ft (18 cm) with abundant fossils.
14. Limestone, light brownish-gray, medium to fine crystalline in medium to thick beds.
13. Limestone, light greenish-gray, fine and medium crystalline, thick-bedded with abundant discontinuous wavy shale partings. Shale partings are greenish-gray, mottled and very fossiliferous.
12. Limestone, same as above but fewer shale partings.
11. Dolomite, light gray, fine crystalline, slightly argillaceous, very fossiliferous, in medium beds, discontinuous faint shale partings. Some clear calcite and dolomite in lower 0.5 ft (15 cm).
10. Dolomite, light gray to light brownish-gray, fine crystalline with thin shale partings up to 1 in (2 cm) thick between beds. Shale is bluish-green and brown. Fossiliferous.
Pecatonica Member
9. Dolomite, brownish-gray, fine to medium crystalline, sugary texture, thin- to medium-bedded, even-bedded, beds 0.1 to 18 in (2 mm to 45 cm) thick. Weathered surface shows distinct but discontinuous thinner beds.
8. Dolomite, brownish-gray, fine crystalline, in single bed, upper 0.5 in (1 cm) stained brown. Scattered dark brown fossil molds and traces. Weathered surface shows wavy horizontal bedding features.
7. Dolomite, brownish-gray, fine crystalline, faint horizontal shale traces, dark brown phosphatic pebbles up to 2 mm, abundant dark brown fossil hash.

Glenwood Formation
Hennepin Member
6. Dolomite, very silty, yellowish-brown, very fine crystalline, abundant phosphatic pellets up to 2 mm, scattered round medium quartz sand grains.
5. Sandstone, brown, very fine- and fine-grained with little medium-grained, abundant grayish-green shale.
4. Sandstone, dark brown, fine- and medium-grained, argillaceous, poorly sorted, iron-oxide cemeted, abundant phosphate pellets up to 2 mm.
Harmony Hill Member
3. Shale, brown and bluish-green in upper 2 in (5 cm) grading downward to bluish-green with some reddish-brown; little rounded medium grained quartz sand.
2. Sandstone, mottled light yellowish-green, light greenish-yellow, and reddish-brown, medium- and fine-grained with some very fine and very coarse grains, poorly sorted, abundant pale green clay in matrix, reworked/bioturbated texture.

St. Peter Formation
Tonti Member
1. Sandstone, light yellowish-gray, very fine- to medium-grained, some light brown stains cross bedded.
Base of exposure in drainage ditch.

and limestone. The overlying Quimbys Mill Member was named by Agnew and Heyl (1946, p. 1585) from a quarry exposure about 5 mi (8 km) west of Shullsburg, Wisconsin. Its thickness varies in the district from less than 1 ft to more than 18 ft (0.3 to 5.5 m) thick and consists of purplish gray-brown, sublithographic, thick-bedded, conchoidally fractured limestone with an uneven upper surface and with shale at its base.

The Platteville Formation is overlain disconformably by the Spechts Ferry Shale Member of the Decorah Formation named from exposures in the city of Decorah, Iowa (Calvin, 1906, p. 61). It thins eastward from 8 ft thick to less than 1 ft (2.4 to 0.3 m). The Spechts Ferry Member consists of fossiliferous, gray-brown limestone with green shale interbeds. At this exposure two thin beds of "metabentonite" occur near its base. The metabentonites are orange to light reddish brown and about 2 in (5 cm) thick. Phosphatic nodules occur locally in the upper 1 ft (0.3 m).

The Spechts Ferry Member is conformably overlain and transitional with the Guttenberg Limestone Member, which consists of hard, fine crystalline, thin-bedded, fossiliferous, light brown limestone with brown petroliferous shale partings and interbeds. The presence of these interbeds has led to the member being referred to as the "Oil Rock." In the district the Guttenberg

Member thins eastward from more than 14 ft to less than 7 ft (4 to 2 m).

The Spechts Ferry is conformably overlain by the Ion Member, which is a gray to blue dolomite, medium-crystalline, and medium to thick bedded with green shale interbeds. The amount of shale decreases to the east. The Ion maintains a thickness of about 20 ft (6 m) across the district.

The Decorah Formation is conformable with and transitional with the overlying Galena Dolomite Formation. The Galena was named (Owen, 1840, p. 19, 24) from exposures in the vicinity of the city of Galena in northwest Illinois. It is a light buff to drab, cherty, thick-bedded, vuggy dolomite with medium to coarse sugary grains. A zone of *Prasopora insularis* Ulrich marks the top of the Ion Member in some areas, but is absent here. In most of the district the Galena is dolomitized and the sparse fossils are poorly preserved. It is from 220 to 230 ft (67 to 70 m) thick throughout the district.

Near the north end of the roadcut, there is a quarry, now occupied by a junkyard, in which, on the southeast wall, can be seen an example of "pitch-and-flat" structure, which is the principal site of zinc and lead mineralization in the district. Here there is no mineralization.

REFERENCES CITED

Agnew, A. F., and Heyl, A. V., Jr., 1946, Quimbys Mill, new member of Platteville formation, Upper Mississippi Valley [U.S.]: American Association of Petroleum Geologists Bulletin, v. 30, p. 1585–1587.

Calvin, S., 1906, Geology of Winneshiek County, Iowa: Iowa Geological Survey Annual Report, v. 16, p. 37–146.

Hershey, O. H., 1894, The Elk Horn Creek area of St. Peter Sandstone in northwestern Illinois: American Geologist, v. 14, p. 169–179.

Kay, G. M., 1935, Ordovician System in the upper Mississippi valley: Kansas Geological Society Guidebook, 9th Annual Field Conference, p. 281–295.

Ostrom, M. E., 1969, Champlainian Series (Middle Ordovician) in Wisconsin: American Association of Petroleum Geologists Bulletin, v. 53, no. 3, p. 672–678.

—— , 1978, Potosi Hill Exposure, *in* Geology of Wisconsin; Outcrop descriptions: Wisconsin Geological and Natural History Survey, Gr-7/2N/2W, 4 p.

Owen, D. D., 1840, Report of a geological exploration of part of Iowa, Wisconsin, and Illinois in 1839; Congressional Documents, U.S. 28th Congress, 1st Session, Senate Executive Document 407, 191 p. (1844); Mineral lands of the United States, Congressional Documents: U.S. 26th Congress, 1st Session, House Executive Document 239, 161 p.

—— , 1847, Preliminary report of progress of geological survey of Wisconsin and Iowa, U.S. General Land Office Report, 1842, Congressional documents: 30th Congress, 1st Session, Senate Executive Document 2, p. 160–173.

Templeton, J. S., and Willman, H. B., 1963, Champlainian Series (Middle Ordovician) in Illinois: Illinois Geological Survey Bulletin 89, 260 p.

Middle Ordovician Platteville Formation, Hoadley Hill, Wisconsin

M. E. Ostrom, Wisconsin Geological and Natural History Survey, 3817 Mineral Point Road, Madison, Wisconsin 53705

LOCATION

Exposure in roadcut at north side of U.S. 151 about 6.5 mi (10.8 km) southwest of Platteville in the W½,NE¼,Sec.12, and the W½,SE¼,Sec.1,T.2N.,R.2W., Grant County, Wisconsin on the Dickeyville 7½-minute Quadrangle (Fig. 1).

SIGNIFICANCE

This is the reference section for the Platteville Formation in the southwest Wisconsin zinc-lead district (Agnew and others, 1956; Ostrom, 1978). The strata exposed here are the upper part of the St. Peter Sandstone, the Glenwood Formation, a complete section of the Platteville Formation, and the lower part of the Decorah Formation (Fig. 2).

The Hoadley Hill exposure shows the interrelationship and lithologic characteristics of the St. Peter, Glenwood, and Platteville Formations in the classic southwest Wisconsin zinc-lead district. The St. Peter Sandstone was named by Owen (1847, p. 170) for exposures in bluffs of the St. Peter River valley (now Minnesota River), near St. Paul, Minnesota. It consists of clear, fine to coarse rounded to subangular quartz grains and is generally poorly cemented. It is white to very light gray and very light buff, thin to thick bedded, locally massive and cross-bedded, and is variable in thickness in the district ranging from 0 to more than 350 ft (0 to 100 m) thick and averaging about 50 ft (15 m) thick. Thickness variations are attributed to deposition of the sand on a deeply dissected erosion surface.

The Glenwood Formation conformably overlies the St. Peter Formation. It was named by Calvin (1906, p. 75) from exposures in Glenwood Township (T.98N.,R.7W.) a short distance northwest of Waukon, Iowa. The classification used here is that of the Illinois Geological Survey (Templeton and Willman, 1963) as modified by Ostrom (1969). Three members are recognized in the Glenwood Formation in southwest Wisconsin. In ascending order these are the Nokomis, Harmony Hill, and Hennepin. The Nokomis Member consists principally of sandstone and is transitional with the underlying St. Peter Sandstone. It is distinguished from the St. Peter Sandstone by light yellowish and greenish coloration and by a notable change in bedding character from cross-bedded, even-bedded and even-textured sandstone to reworked, burrowed, and poorly sorted sandstone with more or less green clay. It is both silty and argillaceous. In the district the Nokomis ranges from 8 ft thick (2.5 m) near Beetown, about 25 mi (40 km) west-northwest of Hoadley Hill, to less than 1 ft (0.3 m) thick in the vicinity of New Glarus, located about 55 mi (88 km) to the east. The Nokomis Member is conformable with the overlying Harmony Hill Member.

The Harmony Hill Member is a pale green to greenish gray shale with scattered rounded clear quartz sand grains. It is up to

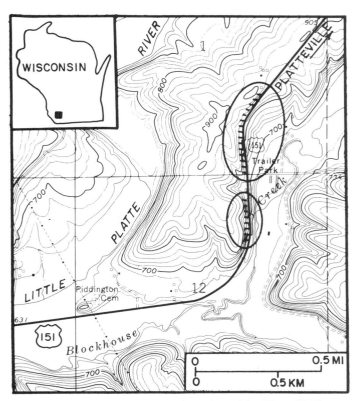

Figure 1. Location of exposure of Middle Ordovician Platteville Formation in roadcut at Hoadley Hill, Wisconsin.

3.5 ft (1 m) thick in the western part of the district and is absent in the east. The Harmony Hill Member is conformable with the overlying Hennepin Member, which consists of brownish and locally calcareous shale with scattered phosphatic nodules and clear rounded quartz sand grains. It is 5 ft (1.5 m) thick in the western part of the district and thins to disappearance in the east.

The Platteville Formation overlies and is conformable with the Glenwood Formation. It is one of the mineralized formations in the southwest Wisconsin zinc-lead mining district and was named by Bain (1905, p. 19) for exposures in the vicinity of Platteville, Wisconsin. It is known throughout the district, and within the Driftless Area, from exposures at the surface and in mines and from drill cuttings. In the district the Platteville Formation ranges in thickness from about 55 ft (17 m) in the west to near 75 ft (23 m) in the east, near Shullsburg.

The Platteville Formation consists of three members, which in ascending order are the Pecatonica, McGregor, and Quimbys Mill. The Pecatonica Dolomite Member was named by Hershey

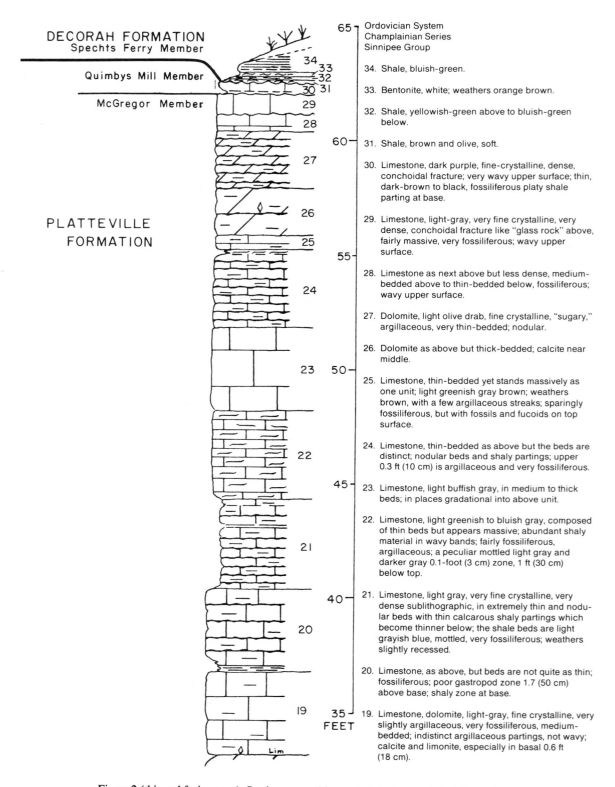

Figure 2 (this and facing page). Section exposed in roadcut at Hoadley Hill, Wisconsin.

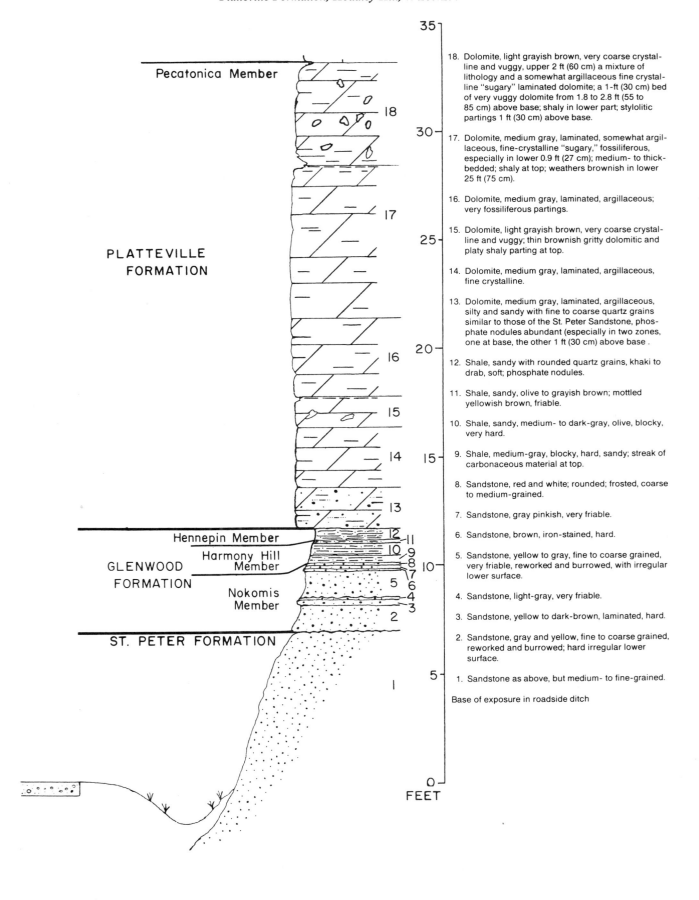

18. Dolomite, light grayish brown, very coarse crystalline and vuggy, upper 2 ft (60 cm) a mixture of lithology and a somewhat argillaceous fine crystalline "sugary" laminated dolomite; a 1-ft (30 cm) bed of very vuggy dolomite from 1.8 to 2.8 ft (55 to 85 cm) above base; shaly in lower part; stylolitic partings 1 ft (30 cm) above base.

17. Dolomite, medium gray, laminated, somewhat argillaceous, fine-crystalline "sugary," fossiliferous, especially in lower 0.9 ft (27 cm); medium- to thick-bedded; shaly at top; weathers brownish in lower 25 ft (75 cm).

16. Dolomite, medium gray, laminated, argillaceous; very fossiliferous partings.

15. Dolomite, light grayish brown, very coarse crystalline and vuggy; thin brownish gritty dolomitic and platy shaly parting at top.

14. Dolomite, medium gray, laminated, argillaceous, fine crystalline.

13. Dolomite, medium gray, laminated, argillaceous, silty and sandy with fine to coarse quartz grains similar to those of the St. Peter Sandstone, phosphate nodules abundant (especially in two zones, one at base, the other 1 ft (30 cm) above base .

12. Shale, sandy with rounded quartz grains, khaki to drab, soft; phosphate nodules.

11. Shale, sandy, olive to grayish brown; mottled yellowish brown, friable.

10. Shale, sandy, medium- to dark-gray, olive, blocky, very hard.

9. Shale, medium-gray, blocky, hard, sandy; streak of carbonaceous material at top.

8. Sandstone, red and white; rounded; frosted, coarse to medium-grained.

7. Sandstone, gray pinkish, very friable.

6. Sandstone, brown, iron-stained, hard.

5. Sandstone, yellow to gray, fine to coarse grained, very friable, reworked and burrowed, with irregular lower surface.

4. Sandstone, light-gray, very friable.

3. Sandstone, yellow to dark-brown, laminated, hard.

2. Sandstone, gray and yellow, fine to coarse grained, reworked and burrowed; hard irregular lower surface.

1. Sandstone as above, but medium- to fine-grained.

Base of exposure in roadside ditch

(1894, p. 175) from exposures in the Pecatonica River valley in southwestern Wisconsin near the Illinois border. It consists predominantly of medium-grained, granular, thick- to thin-bedded dolomite. The lowermost bed, the Chana Member (Templeton and Willman, 1963) contains phosphatic pellets as well as rounded quartz sand grains similar to those in the underlying Glenwood Formation. The Pecatonica is from 20 to 25 ft (6 to 8 m) thick in the district.

The McGregor Limestone Member was named by Kay (1935, p. 286) from an exposure in a ravine 1 mi (1.6 km) west of McGregor, Iowa. The McGregor Member consists of uneven bedded, thin- to medium-bedded, light gray to buff argillaceous dolomite and limestone and is from 25 to 30 ft (8 to 9 m) thick. The McGregor Member contains commercial deposits of zinc-lead ore. The Quimbys Mill Member was named by Agnew and Heyl (1946, p. 1585) from an exposure in a quarry at Quimbys Mill located 5 mi (8 km) west of Shullsburg, Wisconsin. The member ranges from less than 1 ft (0.3 m) in thickness to more than 18 ft (5.5 m). It consists of light brown, thin- to medium-bedded, crystalline sublithographic limestone and finely granular dolomite. This member is locally called the "glass rock" because it breaks with a conchoidal fracture. The Quimbys Mill also contains commercial deposits of zinc-lead ore.

The Platteville Formation is overlain discomformably by the Spechts Ferry Shale Member of the Decorah Formation. The Decorah Formation was named by Calvin (1906, p. 61) from exposures in the city of Decorah, Iowa. The Decorah Formation consists of the Spechts Ferry, Guttenberg, and Ion members. Only the Spechts Ferry is exposed at this outcrop. The Spechts Ferry consists principally of bluish green to brown shale with nodules and discontinuous thin beds of limestone. A thin metabentonite bed, which is believed to be an alteration product of volcanic dust, occurs near its base and can be correlated on a broad regional scale.

REFERENCES CITED

Agnew, A. F., and Heyl, A. V., Jr., 1946, Quimbys Mill, New member of Platteville Formation, upper Mississippi valley: American Association of Petroleum Geologists Bulletin, v. 30, p. 1585–1587.

Agnew, A. F., Heyl, A. V., Behre, C. H., and Lyons, E. J., 1956, Stratigraphy of Middle Ordovician rocks in the zinc-lead district of Wisconsin, Illinois, and Iowa: U.S. Geological Survey Professional Paper 274-K, 312 p.

Bain, H. F., 1905, Zinc and lead deposits in northwestern Illinois: U.S. Geological Survey Bulletin 246, 56 p.

Calvin, S., 1906, Geology of Winneshiek County, Iowa: Iowa Geological Survey Annual Report, v. 16, p. 37–146.

Hershey, O. H., 1894, The Elk Horn Creek area of St. Peter Sandstone in northwestern Illinois: American Geologist, v. 14, p. 169–179.

Kay, G. M., 1935, Ordovician System in the upper Mississippi valley: Kansas Geological Society Guidebook, 9th Annual Field Conference, p. 281–295.

Ostrom, M. E., 1969, Champlainian Series (Middle Ordovician) in Wisconsin: American Association of Petroleum Geologists Bulletin, v. 53, no. 3, p. 672–678.

Ostrom, M. E., 1978, Hoadley Hill Exposure, in Geology of Wisconson; Outcrop descriptions: Wisconsin Geological and Natural History Survey, Gr-1/2N/2W, 4 p.

Owen, D. D., 1847, Preliminary report of progress of geological survey of Wisconsin and Iowa: U.S. General Land Office Report, 1847; Congressional documents: 30th Congress, 1st session, Senate Executive Document 2, p. 160–173.

Templeton, J. S., and Willman, H. B., 1963, Champlainian Series (Middle Ordovician) in Illinois: Illinois Geological Survey Bulletin 89, 260 p.

The Silurian reef at Thornton, Illinois

Donald G. Mikulic, Illinois State Geological Survey, 615 E. Peabody Drive, Champaign, Illinois 61820

LOCATION

The Thornton Reef is exposed in quarries owned by the Material Service Corporation and in roadcuts along I-80 and I-294, about 4.8 mi (8 km) west of the Illinois-Indiana boundary (Figs. 1 and 2), in the W½,Sec.27,N½,Sec.33,T.36N.,R.13E., Harvey and Calumet City 7½-minute Quadrangles, Thornton, Cook County, Illinois.

ACCESSIBILITY

The Thornton Reef may be viewed from the Material Service overlook building. This building may be reached from I-80-294 on Halsted Street (Hwy. 1) south. Proceed 1 mi (1.6 km) to Ridge Road and turn left (east). Follow Ridge Road east and north to Margaret Street and turn right (east). Proceed to Williams Street in the village of Thornton. Turn left (north) onto

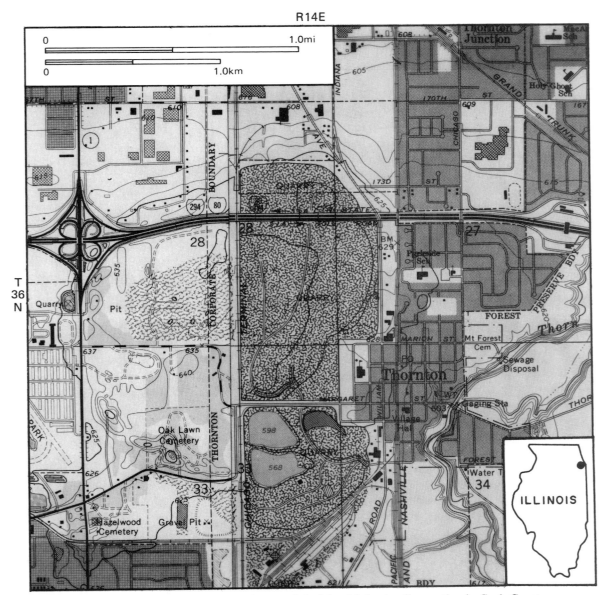

Figure 1. Location of the Thornton Quarry of the Material Service Corporation in Cook County, Illinois.

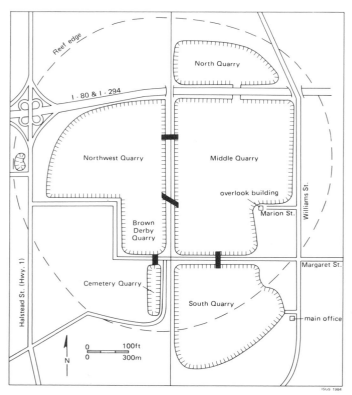

Figure 2. Outline of the Thornton Reef complex and location of the five quarry pits of Material Service Corporation at Thornton.

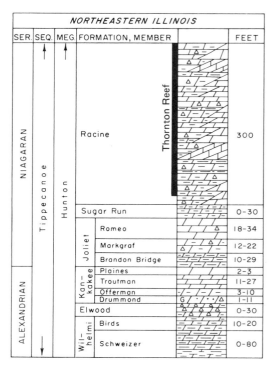

Figure 3. Columnar section of the Silurian System in northeastern Illinois showing the stratigraphic position of the Thornton Reef (from Willman and others, 1975).

Williams and proceed three blocks to Marion (175th) Street. Turn left (west) onto Marion and drive to the end of the road to the overlook parking. If the gate is locked, request permission to enter at the main office of Material Service plant on Williams, three blocks south of Margaret Street. Organized field trips to the reef are conducted by the G.S.A., Tri-State, and other professional groups every few years. Contact Mr. Lyndon Dean, geologist for Material Service, at 300 W. Washington, Chicago, IL 60606, to find out if any trips are being planned.

SIGNIFICANCE

The Thornton Reef probably is the best exposed and most widely known Silurian reef in the world. During the last 50 years it has figured prominently in Silurian reef studies, serving as the focus of several classic papers by Bretz (1939), Lowenstam (1950, 1957), Ingels (1963), and others. Characteristics of this reef have been used to formulate conceptual models for Paleozoic reef development, succession, paleocology, and diagenesis. The reef has considerable economic importance as the largest source of high-quality aggregate in the Chicago metropolitan area.

SITE INFORMATION

The Great Lakes-Upper Mississippi River Valley area is famous for its numerous exposures of Silurian reefs. These structures are some of the most intensively and earliest studied Paleo-

zoic reefs in the world. During the mid to late Silurian this entire region was covered by shallow seas. Probably thousands of reefs developed in the Illinois and Michigan Basins and on the adjacent shelf platform at this time. Large barrier reef complexes formed on the edge of the basins. Thornton Reef is one of the larger reefs to have developed on the broad platform areas surrounding the basins.

Nearly all of Thornton Reef is situated within the Racine Dolomite (Fig. 3), although its development may have begun earlier in one of the underlying rock units. Quarry exposures and subsurface borings indicate that the preserved portion of the reef is about 480 ft (150 m) thick at its center. Originally the reef may have been several hundred meters higher, but the upper portion has been planed off by erosion and Quaternary glaciation. Surrounded by glacial Lake Chicago beach deposits, the reef now appears as a low (10–16 ft; 3–5 m) hill on the Lake Chicago plan.

In plan view, the reef is elliptical, its longest axis has a length of 1.7 mi (2.7 km) along a southwest-northeast trend. The base of the reef has a convex-down bowl-shape, suggesting that it grew upward and outward. Its centrally located core is a small, inverted, cone-shaped structure that appears to expand outward near the bedrock surface. The core is surrounded by radially dipping flank strata, which constitute most of the reef structure. Both the core and flanks consist of high-purity dolomite. Near the core, the flank strata are obscure and dip away at an angle of

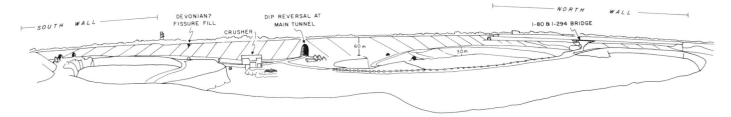

Figure 4. Diagram showing features of the Middle Quarry visible from the overlook building.

about 40°. They become well-defined and level out distally where they interbed and grade into flat-lying, highly argillaceous non-reef strata that surround the entire reef. Large allochthonous blocks and debris flows found in the outer flank areas probably were transported during storms from the higher, now-eroded portions of the reef, suggesting that the reef was subjected to a high-energy environment.

The Thornton Reef contains a high diversity and abundance of fossils (examples are on display in the overlook building). Some lateral and vertical zonation of fossils occurs, but not in as well-defined a pattern as reconstructed by Ingels (1963). Disarticulated pelmatozoan debris is conspicuously abundant in the central core area. The diameter of ossicles is noticeably large in this area, but ossicles decrease in size and abundance towards the outer reef margin. Tabulate corals and stromatoporoids are most common in the upper portion of the reef center surrounding the pelmatozoan-rich area. A higher diversity biota including brachiopods, gastropods, cephalopods, trilobites, bryozoans, and other taxa, occurs in this same area. The diversity of the biota decreases in the outer flanks, but fossils may still be common. Abundant, large, disarticulated, strongly-ribbed pentamerid brachiopods (*Kirkidium*) occur in a conspicuous semicircular zone about 2250 ft (685 m) from the center on the west side of the reef.

Asphalt is found in pores and large vugs throughout the upper 72 ft (22 m) of the reef, indicating that Thornton was once a hydrocarbon reservoir similar to many of the subsurface reefs in the Illinois and Michigan Basins. Erosion of the caprock above the reef allowed the more volatile hydrocarbons to dissipate, leaving behind a thick asphalt residue.

Even though the Thornton Reef has been studied intensively, the conditions under which it developed are still controversial. Two principal models, based on water depth, have been proposed: (1) Most of the exposed portion of the reef developed under shallow-water, high-energy conditions; (2) The exposed portion of the reef developed under low-energy conditions in deep water, although some of the now-eroded portions of the reef may have reached shallow water. There are also other subjects of debate: (1) Was the observed dip original or partly the result of postdepositional compaction? (2) Did an organic reef framework exist or was the reef constructed by inorganic cementation? (3) Were algae ever present? If so, could dolomitization have obliterated evidence of their existence? (4) To what degree did downslope transport contribute to flank strata development?

The overlook building is located near the center of the east wall of the middle quarry, one of six interconnected quarries operated by Material Service Corporation at Thornton. The overlook provides a good view of the west wall of the middle quarry; this wall was not quarried because railroad tracks run along it. The old south quarry and main plant are to the left (south) of Margaret Street. The northwest, Brown Derby, and cemetery quarries are located beyond the tunnels to the west of the middle quarry. The north quarry may be seen to the right (north) just past the I-80 bridge.

The main floor in the middle quarry is 197 ft (60 m) deep, an additional 98 ft (30 m) is exposed in a new lower level. The west wall shows a north-south transect nearly through the center of the reef (Fig. 4). The reef core is located just west of the tunnel in the south half of the wall. To the north of the tunnel, flank strata may be seen dipping to the northeast; to the south they dip to the southeast. Near the tunnel the dip reverses and is less clearly defined.

The I-80 roadcut shows a similar transect through the outer flanks. East of the quarry well-developed flank strata dip towards the northeast; west of the quarry they dip to the north and northwest. The area in which the dip reversed has been quarried away. (Note: this part of I-80 is very heavily traveled and it is illegal and very dangerous to stop along the roadcut. The general features can be seen easily by passengers in a moving vehicle, but drivers should not slow down, change lanes, or stop to observe them.)

To the right (north) of the overlook building large domal structures and other interruption of normal outer flank strata can be seen. These structures originally were thought to be satellite reefs, but may represent debris flows or large allochtonous blocks.

From the overlook, a large, iron-stained fissure is visible south (left) of the main tunnel in the west wall; this clay-filled fissure contains scattered silicified corals and stromatoporoids, possibly of Devonian age.

The extensive quarry operations attest to the economic importance of Thornton Reef. Thornton is one of the largest crushed stone quarries in the country. The high-purity dolomite obtained from the reef is used primarily for aggregate in concrete, but more than 30 products are produced from this rock. Blasted rock is hauled in 55- and 85-ton trucks to the primary crusher near the main tunnel. This crusher may process up to 3500 tons of rock per hour, which is then transported by conveyor belts to the main plant for further processing.

REFERENCES

Bretz, J H., 1939, Geology of the Chicago region. Part I, General: Illinois State Geological Survey Bulletin 65, p. 69–81.

Ingels, J.J.C., 1963, Geometry, paleontology, and petrography of Thornton Reef complex, Silurian of northeastern Illinois: American Association of Petroleum Geologists Bulletin, v. 47, p. 405–440.

Lowenstam, H. A., 1950, Niagaran reefs in the Great Lakes area: Journal of Geology, 58, p. 430–487.

——1957, Niagaran reefs in the Great Lakes area, *in* Ladd, H. S., ed., Treatise on marine ecology and paleocology, volume 2, Paleoecology: Geological Society of America Memoir 67, p. 215–248.

McGovney, J.E.E., 1978, Deposition, porosity evolution, and diagenesis of the Thornton Reef (Silurian), northeastern Illinois: [Ph.D. dissertation]: University of Wisconsin, Madison, 454 p.

Mikulic, D. G., Kluessendorf, J., McGovney, J.E.E., and Pray, L. C., 1983, The classic Silurian reef at Thornton, Illinois, *in* Shaver, R. H. et al., Silurian reef and interreef strata as responses to a cyclical succession of environments, southern Great Lakes area: Geological Society of America Annual Meeting Field Trip No. 12, p. 180–189.

Pray, L. C., 1976, Guidebook for a field trip on the Thornton Reef (Silurian), northeastern Illinois: Kalamazoo, Western Michigan Reserve University and Geological Society of America-North Central Section guidebook, 47 p.

Willman, H. B., 1943, High-purity dolomite in Illinois: Illinois State Geological Survey Report of Investigations No. 90, 87 p.

——1962, The Silurian strata of northeastern Illinois, *in* Silurian rocks of the southern Lake Michigan area: Michigan Basin Geological Society Annual Field Conference Guidebook, p. 62–68.

Willman, H. B., Atherton, E., Buschbach, T. C., Collinson, C., Frye, J. C., Hopkins, M. E., Lineback, J. A., and Simon, J. A., 1975, Handbook of Illinois stratigraphy: Illinois State Geological Survey, Bulletin 95, 261 p.

Wedron Section, Wedron, Illinois: Concepts of Woodfordian glaciation in Illinois

W. Hilton Johnson, Department of Geology, University of Illinois, Urbana, Illinois 61801
Ardith K. Hansel and Leon R. Follmer, Illinois State Geological Survey, 615 E. Peabody Drive, Champaign, Illinois 61820

LOCATION AND ACCESSIBILITY

The Wedron Section is located in the Wedron Silica Quarry, Secs. 8, 9, 10, and 16, T. 34 N., R. 4 E., La Salle County, Illinois, 4 mi (6.4 km) north of I-80 and just west of the Fox River (Fig. 1). From I-80, exit north on Illinois 71 and continue northeast for 3 mi (4.8 km); turn west on Wedron Road and proceed 3 mi (4.8 km) to the entrance to the Wedron Silica Company in Wedron. To enter the quarry, permission must be obtained from the Wedron Silica Company (phone number 815-433-2449). Exposures in the quarry are accessible by vehicle; high axle clearance is usually essential.

SIGNIFICANCE

The Wedron Section is the type section of the Wedron Formation, which consists of glacial diamictons and intercalated stratified deposits of the late Wisconsinan (Woodfordian Subage) glaciation in Illinois, and of the Peddicord Formation. It is one of the thickest and most complete exposures of the Wedron Formation. A complex succession of deposits representing multiple glacial events is exposed in the quarry, and observations and interpretations of the deposits have been important in developing the history of the last glaciation. Sections in the quarry have been studied for more than 70 years, but ongoing studies are developing new interpretations, particularly with respect to the sedimentology of the deposits (Johnson and others, 1985).

SITE INFORMATION

Exposures in the quarry were first described by Sauer (1916), who recognized one main till unit and several units of sand and gravel, and silt and clay. H. B. Willman studied the Wedron Section throughout his long career focusing on the stratigraphy of Illinois. His early work was done in association with Morris M. Leighton; later he worked closely with John C. Frye and H. D. Glass. Concepts developed at Wedron were used in classifying Pleistocene glacial deposits. The early work placed strong emphasis on morphology as a basis for subdividing and interpreting the glacial deposits. End moraines were interpreted as representing still-stands of an otherwise fluctuating ice margin, and it was assumed that each end moraine would have a sheet of till associated with it. The Wedron Section supported this concept because several till units of the last glaciation were exposed, and several end moraines of the last glaciation had been mapped to the west. Thus, the till units were named for the end moraine with which they were assumed to be related; for example, Willman

and Payne (1942) recognized Shelbyville drift, Bloomington drift, Farm Ridge drift, and Marseilles drift at the Wedron Pit.

Frye and Willman (1960), Frye and others (1968), and Willman and Frye (1970) introduced formal lithostratigraphy into the classification of Pleistocene deposits in Illinois, revised the chronostratigraphic classification, and formalized the practice of morphostratigraphic classification. The Wedron Formation, as defined by Frye and others (1968), includes the deposits of glacial till and intercalated outwash and silt of the Woodfordian Sub-stage. Wedron was designated the type section and three till members, Lee Center, Tiskilwa, and Malden, were described in the section. Six morphostratigraphic units (named after end moraines and called drifts) were recognized. Nomenclature used in Willman and Frye (1970), as well as that used earlier (Frye and others, 1968) and in this guide, is summarized in Table 1.

The quarry operation exposes the Starved Rock Sandstone Member of St. Peter Sandstone, a thick, quartz sandstone that is of Champlainian Series (Middle Ordovician) age. The sandstone is medium grained, crossbedded, and friable except for an outer case-hardened surface. More than 100 ft (30 m) of St. Peter are exposed in Pit 1. Because it is almost pure SiO_2, the St. Peter is mined here and elsewhere in the area for silica sand (Kolata, this volume).

Several pits have been worked in the quarry (Fig. 1); Pits 1, 4 and 6 currently are active. Illinoian and pre-Illinoian deposits are exposed in Pits 3 and 4. The following description and discussion focuses on Pits 1 and 6 and is organized by stratigraphic unit. The nongenetic term *diamicton* is used to describe poorly to unsorted deposits. The genetic term *till* is restricted to those diamictons that are interpreted to have been deposited directly from glacier ice with little or no modification after deposition. Other diamictons are interpreted to have been deposited from sediment (mud) flows in the glacial environment.

The bedrock surface contains several valleys that are tributaries of the Ticona Bedrock Valley (Willman and Payne, 1942). These are exposed in Pit 1 (Fig. 2) and are filled with Wisconsinan alluvial and lacustrine deposits. Although the alluvial deposits vary, they are silty for the most part and are included in Robein Silt. The Farmdale Soil, about 4 ft (1.3 m) thick, is developed in the top of Robein Silt (Fig. 2). The A horizon of the Farmdale is dark, cumulic, and contains abundant organic debris and wood. It overlies a weakly developed, gleyed B horizon.

Peddicord Formation

Willman and others (1971) defined the Peddicord Formation to include gray and pink silt that had accumulated in a

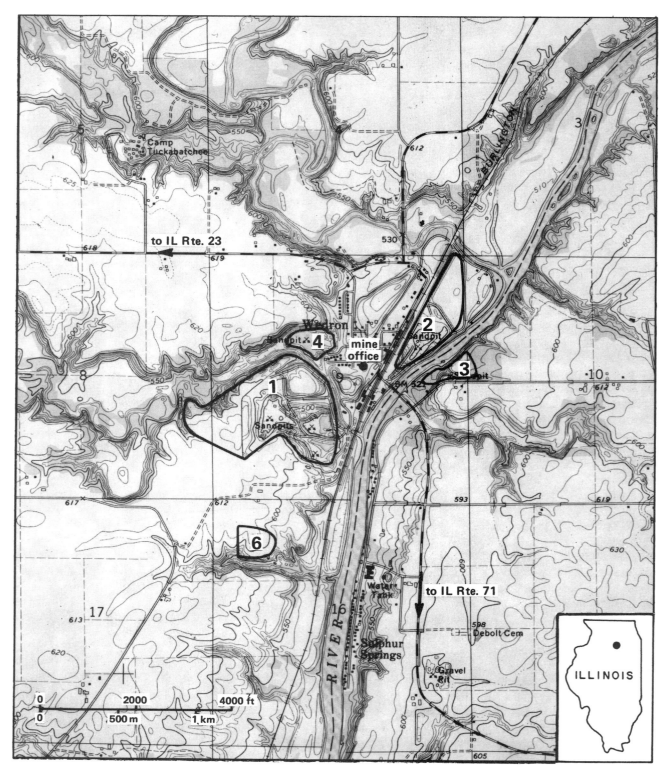

Figure 1. Location of quarry pits at Wedron, LaSalle County, Illinois (Secs. 8, 9, 10, and 16, T. 34 N., R. 4 E., Wedron, Illinois, 7½-minute Quadrangle.)

Table 1 Current and recent nomenclature used for Wisconsinan glacial deposits at Wedron Quarry		
1968 (Frye et al.)	1970 (Willman and Frye)	1986 (this guide)
Richland Loess	Richland Loess	Richland Loess
Wedron Formation Sand and gravel	Henry Formation Batavia Member	Henry Formation Batavia Member
Till (Farm Ridge) Sand and gravel	Wedron Formation Malden Till Member — Farm Ridge Drift (till) Sand and gravel	Wedron Formation Malden Till Member[1] — Upper unit
Till (Cropsey) Sand	Mendota Drift (till) Sand	Middle unit
Till (Cropsey) Silt, some sand	Arlington Drift (till) Silt, some sand	Lower unit
Till (Normal) Sand and silt	Dover Drift (till) Sand and silt	
Till (Bloomington)	Tiskilwa Till Member Bloomington Drift (till)	Tiskilwa Till Mbr. Main unit
Till (Shelbyville)	Lee Center Till Member Atkinson Drift (till)	Lower unit
Sand and gravel (pro-Shelbyville)	Sand and gravel	Peddicord Formation Sand unit
Farmdale Silt	Farmdale Silt	Silt unit
		Robein Silt

[1] Correlations between the Malden units of this guide and drift units of earlier workers are uncertain.

Farmdalian lake confined to valleys of the Ticona drainage system. We recognize these deposits as a silt unit of the Peddicord, and tentatively include overlying sand deposits related to the same drainage system.

The silt unit consists of massive and laminated silt with subordinate beds and laminae of clay and sand. These materials are calcareous and vary in color from gray to reddish brown. Coniferous wood fragments and organic-debris laminae are common, particularly near valley and gully margins. The unit varies in thickness, and up to 43 ft (13 m) have been described at Wedron (Willman and Frye, 1970). It is particularly thick in buried canyons where the silt beds grade to and are interbedded with sandy colluvium derived from St. Peter Sandstone. The unit has the same clay mineral composition as the Tiskilwa Till Member. Radiocarbon dates on detrital wood from this facies are 24,370 ± 310 (ISGS 863), 24,900 ± 200 (ISGS 862), 24,000 ± 700 (W-79), and 26,800 ± 700 (W-871) B.P.

The deposits are interpreted to be typical proglacial lake beds that accumulated in a dammed drainage system during the initial Woodfordian glacial advance in northern Illinois. The color and clay mineral composition suggest the system was dammed by the Tiskilwa ice margin or by outwash from that ice sheet. Lake Peddicord inundated the Farmdale Soil, which had formed in the valleys, and wood and organic debris were washed into the lake from valley sides and adjacent uplands. The lake probably existed in latest Farmdalian and earliest Woodfordian time.

Previous interpretations related the silt deposits either to Lake Kickapoo, interpreted to postdate the initial Woodfordian ice margin advance and hence to be younger (Willman and Payne, 1942), or to Lake Peddicord, considered to predate the earliest Woodfordian glaciation and to be older than the Farmdale Soil (Willman and others, 1971). The latter interpretation of the age of the lake is rejected because of stratigraphic relationships at Wedron. Further regional studies are necessary to determine possible relationships between deposits of Lake Kickapoo and Lake Peddicord.

A sand unit comprising up to 20 ft (6 m) of relatively well-sorted sand and some fine gravel overlies the silt unit of the Peddicord Formation. The sands are calcareous and tan to yellow brown. Beds vary from about 8 to 39 in (0.2 to 1.0 m) thick, and

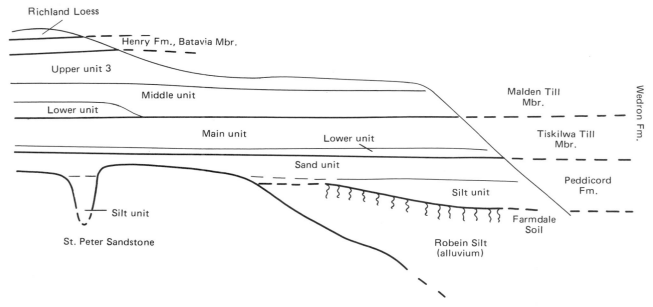

Figure 2. Diagrammatic sketch of southwestern corner, Wedron Quarry, Pit 1; 1984. Not to scale. Units are not exposed continuously.

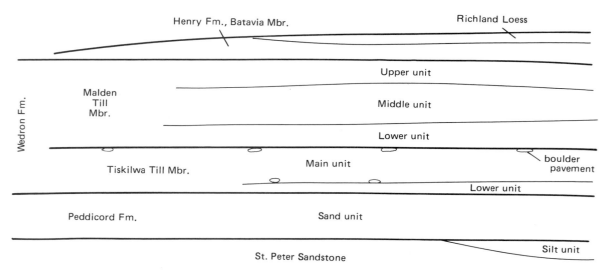

Figure 3. Diagrammatic sketch of west wall exposure, Wedron Quarry,
Pit 6, November, 1984.

are planebedded and trough and planar crossbedded. The sand
unit is more extensive than the silt unit and appears to be continuous across the bedrock surface at Wedron.

This unit is interpreted as representing glaciofluvial sedimentation as the Woodfordian ice margin approached the quarry
area. The proglacial origin agrees with earlier interpretations
(Table 1), except that we relate the unit to the ice sheet that
deposited the Tiskilwa Till Member and not the Lee Center Till
Member or Shelbyville till.

Wedron Formation

The Wedron Formation, which consists of glacial deposits,
overlies the Peddicord Formation. Initially it was subdivided into
three members at Wedron; currently only two members are recognized, although both consist of multiple lithologic units. In
addition, Willman and Frye (1970) recognized six morphostratigraphic units (Table 1). The latter classification is not utilized in
this guide because correlations to end moraines to the west are
uncertain. The Wedron Formation is exposed in all pits, but
currently is best exposed in Pits 1 and 6 (Figs. 2 and 3).

Tiskilwa Till Member

The lower unit of diamicton and intercalated sand and silt is
related here to the Tiskilwa glacial advance, not an earlier advance. The unit is thin, rarely over 3 ft (1 m) thick, and discontinuous. It is highly variable in character and in most places consists
of thin diamicton layers interbedded with stratified sand or silt.
The lenticular beds thicken and thin abruptly. Locally the unit is
uniform pebbly loam diamicton. The diamicton is generally oxidized and has a yellow-brown to pinkish color, similar to pinkish
diamicton in the main Tiskilwa unit. Where unoxidized, it is
distinctly grayer and contains slightly more illite in the clay frac-

tion than the main unit of the Tiskilwa. We interpret the unit to
be till and material that has undergone resedimentation and deformation in the subglacial environment.

This unit was included in the Lee Center Till Member by
Willman and Frye (1970); earlier it had been called Shelbyville
till (Table 1). Subsequent work has demonstrated that Lee Center
Till in the type area is Illinoian, thus the name is inappropriate for
this unit at Wedron. Although the contrasting color and composition suggested to earlier workers that the unit had been deposited
during an earlier glacial event, materials with these characteristics
are not unusual in the lower portion of the Tiskilwa Till Member.
We believe the unit was deposited by the Tiskilwa ice sheet and
that the contrasting characteristics are the result of incorporation
of older drift and local Paleozoic source material.

The lower unit is overlain by 6.5 to 13 ft (2 to 4 m) of
typical Tiskilwa Till, which forms the main unit. The contact is
distinct and locally marked by a concentration of boulders. The
main unit is a massive, relatively uniform loam to clay loam
diamicton that weathers to a distinct pinkish color. The main
body of this unit is interpreted to be till, but locally the upper
portion contains interbedded stratified sands, silts, and diamicton
layers of sediment flow origin. In Illinois, the Tiskilwa is one of
several till units with a reddish-brown hue; its color and composition reflect late Precambrian source materials that occur in the
Lake Superior region north of Lake Michigan. The unit is extensive in northern Illinois and forms several large end moraines
along the western margin of Woodfordian glaciation.

Malden Till Member

At Wedron the Malden Till Member is complex and consists of various lithologic materials. Three main units, numbered
from 1 to 3, are tentatively recognized, and each is variable.

Malden unit 1 is a gray to gray-brown diamicton that oxidizes to a reddish-brown hue and has a variable color and clay mineral composition. The latter characteristics generally are intermediate between those of the main Tiskilwa and Malden unit 2. A discontinuous boulder pavement occurs at the lower contact and azimuths of striae commonly range from 70° to 80°; a boulder pavement is also present within the unit. The diamicton has a pebbly loam texture and the upper part of the diamicton locally is interbedded with stratified deposits. These deposits are overlain by a well sorted silt to fine sand bed that is laminated and continuous in exposures at Wedron. The unit is 3 to 10 ft (1 to 3 m) thick and consists primarily of till and sediment flow and lacustrine deposits.

Malden unit 2, about 10 ft (3 m) thick, consists of a lower silty clay that grades upward to pebbly silty clay diamicton, and an upper pebbly loam diamicton. The lower units range from massive to faintly laminated and are interpreted to be lacustrine in origin. The increased sand and sparse pebble content in the silty clay diamicton likely is ice-rafted material. The overlying gray, pebbly loam diamicton is about 6.5 ft (2 m) thick. It is mostly massive, but locally contains thin streaks of fine sand, block inclusions of older pinkish and clayey diamicton, and interbedded sorted deposits of sand and silt. Although mainly gray, locally the upper part contains diamicton that is pinkish and contains less illite. The diamicton unit is interpreted to consist of basal till and deposits that have undergone resedimentation in the supraglacial environment.

Malden unit 3 consists of a lower sand unit and a discontinuous, overlying fine-grained diamicton unit. The sand unit is stratified, crossbedded, and locally contains multiple coarsening upward sequences. Local lenticular bodies of pea gravel are present at the base or within the unit. The unit is best exposed in Pit 6 where it is up to 6.5 ft (2 m) thick. The overlying diamicton, which contains few pebbles, has a clay texture. The maximum thickness observed in recent exposures is 5 ft (1.5 m). The upper surface of the diamicton has been truncated and locally is marked by a thin pebbly loam lag concentration. The diamicton and subjacent sand and gravel are weathered and locally are part of the solum of the Modern Soil. The regional significance of unit 3

is not known; it may consist of outwash and sediment flow deposits derived from the Yorkville Till Member, which makes up the Marseilles Morainic System located immediately east of the Fox River, or it may have been the result of a younger Malden ice margin advance that extended west of Wedron.

Henry Formation

Several deposits of sand and gravel occur at or near the ground surface and are included in the Henry Formation. They include the sand of Malden unit 3, where it has been exhumed, and thin sand that locally overlies the clay diamicton of Malden unit 3. The deposits, which are interpreted to be outwash, have been weathered and are part of the Modern Soil.

Richland Loess

The uppermost unit at Wedron, Richland Loess, is weathered silt. It is thin, usually about 1.5 ft (0.5 m) thick, but locally approaches 3 ft (1.0 m). This eolian deposit was derived from valley trains and drift surfaces during the middle and latter portions of the Woodfordian Subage. The A and locally the E and/or B horizons of the Modern Soil are developed in the loess. The soils developed in a well-drained position under either forest or grass vegetation and are classified as Udalfs or Udolls depending on the vegetation.

SUMMARY

The Wedron Section exposes a complex succession of deposits typical of many environments occurring in Illinois during the Wisconsinan Age. These include interstadial soils and alluvial deposits, proglacial lacustrine and glaciofluvial deposits, subglacial and supraglacial tills and resedimentation deposits, and eolian deposits. Most accumulated during the early and middle portions of the Woodfordian Subage between about 25,000 and 15,000 B.P.

REFERENCES

Frye, J. C., and Willman, H. B., 1965, [Illinois part of] Guidebook for field conference C—Upper Mississippi Valley (R. F. Black and E. C. Reed [organizers]; C. B. Schultz and H.T.U. Smith [eds.]): International Association Quaternary Research 7th Congress, Nebraska Academy Science, p. 88–90.
——1960, Classification of the Wisconsinan Stage in the Lake Michigan glacial lobe: Illinois Geological Survey Circular 285, 16 p.
Frye, J. C., Willman, H. B., Rubin, Meyer, and Black, R. F., 1968, Definition of Wisconsinan Stage: United States Geological Survey Bulletin 1274-E, p. E16–E17.
Johnson, W. H., Hansel, A. K., Socha, B. J., Follmer, L. R., and Masters, J. M., 1985, Depositional environments and correlation problems of the Wedron Formation (Wisconsinan), northeastern Illinois: Illinois Geological Survey Guidebook 16, 91 p.

Leighton, M. M., and Willman, H. B., 1953, Basis of subdivisions of Wisconsin glacial stage in northeastern Illinois: Guidebook, 4th Biennial State Geologists Field Conference pt. 1 Illinois Geological Survey and Indiana Geological Survey, p. 38, 39.
Sauer, C. O., 1916, Geography of the upper Illinois Valley and history of development: Illinois Geological Survey Bulletin 27, 208 p.
Willman, H. B., and Frye, J. C., 1970, Pleistocene stratigraphy of Illinois: Illinois Geological Survey Bulletin 94, p. 67, 190–191.
Willman, H. B., Leonard, A. B., and Frye, J. C., 1971, Farmdalian lake deposits and faunas in northern Illinois: Illinois Geological Survey Circular 467, 12 p.
Willman, H. B., and Payne, J. N., 1942, Geology and mineral resources of the Marseilles, Ottawa, and Streator Quadrangles: Illinois Geological Survey Bulletin 66, p. 148–149, 307.

Starved Rock, Illinois

Dennis R. Kolata, Illinois State Geological Survey, 615 East Peabody Drive, Champaign, Illinois 61820

LOCATION AND ACCESSIBILITY

Starved Rock is on the south bank of the Illinois River, 1.5 mi (2.4 km) southeast of Utica in La Salle County, Illinois (Sec. 22,T.33N.,R.2E.; Starved Rock 7½-minute Quadrangle). It is part of Starved Rock State Park, which consists of 2,630 acres (1,052 ha) of wooded area along a 7 mi (11.2 km) stretch of the river midway between La Salle–Peru and Ottawa (Fig. 1).

The main entrance to the park is accessible from I-80 or U.S. 51. Leave I-80 at exit 81 and proceed south on Illinois 178 through Utica and across the Illinois River. Turn left at the park entrance 0.25 mi (0.4 km) south of the river. From U.S. 51 turn northeast on Illinois 71 and proceed north through Oglesby. Turn east at Jonesville and continue along Illinois 71 to Illinois 178 and turn left. The park entrance is on the right, 0.75 mi (1.2 km) north of the Illinois 71-178 intersection. Groups of 25 or more people must receive permission from the Site Superintendent, Starved Rock State Park, Box 116, Utica, Illinois 61373 (phone 815-667-4726). Starved Rock can be viewed and photographed on the north side of the Illinois River at the Starved Rock Lock and Dam.

SIGNIFICANCE

Starved Rock is a steep-sided erosional remnant of Champlainian (middle Ordovician) St. Peter Sandstone that rises about 125 ft (37.5 m) above the Illinois River (Fig. 2). Discharge of glacial meltwater from the Chicago area about 10,000 to 14,000 years ago eroded the Illinois River valley to its present configuration, forming Starved Rock and adjacent vertical cliffs.

SITE INFORMATION

According to Indian legend, part of the Illinois tribe was besieged on Starved Rock by the Potawatomi tribe, starved into submission, and then massacred: hence the name Starved Rock. The area has been a source of Indian remains and artifacts for many years.

Starved Rock and the steep bluffs up and down the river from it are composed primarily of St. Peter Sandstone (Ancell Group), which is overlain in places by either the Champlainian Platteville Group or the Pennsylvanian Spoon and Carbondale Formations. The St. Peter is exceptionally pure quartz sand, essentially free of clay, carbonates, and heavy minerals. At nearby Ottawa, the sandstone has been extensively quarried for glass sand. The St. Peter is typically friable and soft; however, in exposures such as Starved Rock it is case-hardened. The lower 25 ft (7.5 m) of Starved Rock consists of the Tonti Sandstone Member (Templeton and Willman, 1963), which is composed of sandstone that is fine grained, white to light buff in color, well sorted,

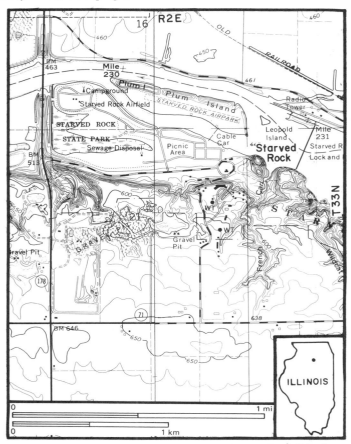

Figure 1. Starved Rock State Park location on Starved Rock 7½-minute Quadrangle.

thick bedded, and cross bedded. About 15 ft (4.5 m) of the Tonti Sandstone lies below, the low-water level of the Illinois River. The Tonti is overlain by about 90 ft (27 m) of sandstone that is medium to coarse grained, white to light buff in color, and massive, which is assigned to the Starved Rock Sandstone Member (Fig. 3). Starved Rock is the type section for both sandstone members of the St. Peter.

In French Canyon, to the south of Starved Rock (Fig. 1), about 10 ft (3 m) of yellowish brown dolomite of the Pecatonica Formation (Platteville Group) overlies the Starved Rock Sandstone Member. In the upper parts of several valleys in the park, the Pennsylvanian Cheltenham Clay Member (Spoon Formation) and overlying Colchester (No. 2) Coal Member of the Carbondale Formation unconformably overlie the St. Peter Sandstone (Fig. 3). The St. Peter unconformably overlies the Canadian (lower Ordovician) Shakopee Dolomite (Prairie du Chien Group). Dolomite of the Shakopee is being quarried 1.4 mi (2.2 km) northwest of Starved Rock on the south side of Utica.

Figure 2. Starved Rock: Ordovician-age St. Peter Sandstone rising above Illinois River.

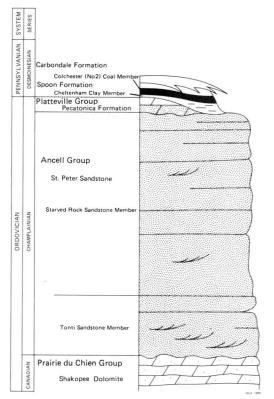

Figure 3. Columnar section of rock strata in Starved Rock area.

Distribution of the bedrock units in the Starved Rock area is largely due to deformation in the La Salle Anticlinal Belt. This complex structure consists of en echelon northwest-southeast–trending folds and troughs that extend through the eastern third of Illinois. The main zone of deformation lies about 2.3 mi (4.5 km) west of Starved Rock. The structure is well exposed along the Vermilion River in nearby Matthiessen State Park (NE¼NE¼NE¼,Sec.31,T.33N.,R.2E.; La Salle 7½-minute Quadrangle). Within this area the La Salle Anticlinal Belt is best described as a westward-dipping monocline. Relatively flat-lying strata of the St. Peter Sandstone seen in the bluffs at Starved Rock dip abruptly into the subsurface on the flank of the monocline. A well drilled in the city of Peru, 3 mi (4.8 km) west of the monocline, encountered the top of the St. Peter Sandstone at a sea-level elevation of –903 ft (–271 m), 1,478 ft (443 m) below the top of Starved Rock. Evidence suggests that the La Salle Anticline Belt underwent minor deformation at least as early as Ordovician time (Cady, 1920; Kolata and Graese, 1983) with major deformation occurring during the late Paleozoic (Clegg, 1965, 1970).

The present Illinois River valley in the area of Starved Rock was formed by two major stages of erosion. The first period occurred during the Valparaiso glaciation about 15,000 years ago when an unusually large volume of meltwater from the general area of Lake Michigan was discharged into the Illinois Valley. The valley was eroded to an elevation of about 550 ft (165 m) m.s.l., just slightly below the top of Starved Rock. The second major period of erosion took place between 14,000 and 11,000 years ago, when overflow from glacial Lake Chicago, an early stage of Lake Michigan, deepened the valley to about 470 ft (141 m) m.s.l. Since that time the river has cut its present inner valley about 30 ft (9 m) deeper into the bedrock.

Starved Rock and Buffalo Rock, which is situated about 3 mi (4.8 km) upstream from Starved Rock, were isolated by erosional deepening of channels developed by glacial meltwater flowing over and around the tops of the hills (Willman and Payne, 1942). The channels may have formed along joints or

zones of poorly cemented sandstone. As floodwaters from Lake Chicago receded, the channel between Buffalo Rock and the valley wall was probably abandoned.

Within Starved Rock State Park the tributaries of the Illinois River are characterized by narrow canyons with vertical overhanging walls cut into the St. Peter Sandstone. Several canyons are 50 ft (15 m) deep and less than 0.5 mi (0.8 km) long. Upstream they end as amphitheater-like enclosures of sheer sandstone walls. The streams drop through a series of falls and steep rapids. The position of many canyons is controlled by fracturing of the sandstone.

REFERENCES CITED

Cady, G. H., 1920, The structure of the La Salle Anticline: Illinois State Geological Survey Bulletin, v. 36, p. 85–179.

Clegg, K. E., 1965, The La Salle Anticlinal Belt and adjacent structures in east-central Illinois: Illinois Academy of Science Transactions, v. 58, no. 2, p. 82–94.

——, 1970, The La Salle Anticlinal Belt in Illinois: Illinois State Geological Survey Guidebook Series No. 8, p. 106–110.

Kolata, D. R., and Graese, A. M., 1983, Lithostratigraphy and depositional environments of the Maquoketa Group (Ordovician) in northern Illinois: Illinois State Geological Survey Circular 528, 49 p.

Templeton, J. S., and Willman, H. B., 1963, Champlainian Series (Middle Ordovician) in Illinois: Illinois State Geological Survey Bulletin, v. 89, 260 p.

Willman, H. B., and Payne, J. N., 1942, Geology and mineral resources of the Marseilles, Ottawa, and Streator Quadrangles: Illinois State Geological Survey Bulletin, v. 66, 388 p.

Cyclothems in the Carbondale Formation (Pennsylvanian: Desmoinesian Series) of La Salle County, Illinois

C. Brian Trask, Illinois State Geological Survey, 615 E. Peabody Drive, Champaign, Illinois 61820

LOCATION

The site is located in Sec.8,T.32N.,R.2E., La Salle County, Illinois (Fig. 1), on the south bank of the Vermilion River, 0.5 mi (0.8 km) west (downstream) of the town of Lowell (Fig. 2). It is in the Margery C. Carlson Nature Preserve owned by the Illinois Department of Conservation. Groups of more than 25 people should check with the site superintendent at Starved Rock State Park before entering the Preserve. For photographic purposes, the site can be viewed from the opposite bank, which is accessible through abandoned clay pits owned by Roy Alleman, Jr. (Village Inn, Tonica, Illinois, 815-442-3535).

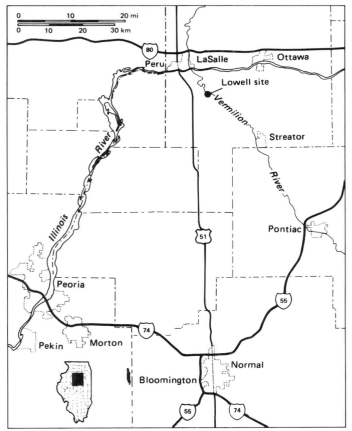

Figure 1. Location of site and major access routes in central Illinois.

ACCESSIBILITY

The site is easily accessible from U.S. 51 or I-80. Take exit 81 from I-80. Proceed south on Illinois 178 through the town of North Utica and across the Illinois River to the intersection with Illinois 71. Continue south from this intersection. The road to the right (west), which is 3.6 mi (5.8 km) south of Illinois 71, leads to the clay pits that provide access to the river bank across from the site. Turn west on La Salle County 14 (N. 2101st Rd) at the end of Illinois 178, 4.4 mi (7.1 km) south of Illinois 71. Turn west on N. 2150th Rd 0.9 mi (1.4 km) west of Illinois 178. Turn north on E. 675th Rd at T intersection 0.5 mi (0.8 km) west of La Salle County 14 and proceed 0.9 mi (1.4 km) to a gravel road on the right side marked by the symbol of the Illinois Nature Preserve System (a white, triangular sign with a red cardinal sitting on an oak twig). From U.S. 51 at Tonica, turn east on La Salle County 14 and drive 2.2 mi (3.5 km) to E. 675th Rd, then north 1.4 mi (2.3 km) to the gravel road.

Park on the shoulder of E. 675th Rd and walk down the gravel road (not passable by vehicles) to a concrete-block building. Go around the left (west) side of the building and walk to your right (east). Follow the path along the top of the bluff to the site. As shown in Figure 2, three exposures offer good potential for an excellent outcrop. Located on a cutbank of the Vermilion River, the outcrops are periodically eroded and exposed by the river.

SIGNIFICANCE

The site is one of the best exposures of cyclothems in the Illinois Basin. Four cyclothems and part of a fifth are present. The Pennsylvanian section forms an angular unconformity with underlying Ordovician rocks. Dolomite of the Ordovician Galena Group can be seen in the riverbed, and the unconformity may be exposed locally.

Bailey Falls lies about a mile to the northwest. It is one of the more prolific localities for collection of natural conodont assemblages (Rhodes, 1952; Collinson and others, 1972).

SITE INFORMATION

Cyclothems have received widespread attention by geologists studying Pennsylvanian strata. A cyclothem is a succession of strata representing a single cycle of deposition. An ideal cyclothem (Fig. 3) consists of 10 units, from a basal sandstone in unconformable contact with an underlying cyclothem, through coal, black shale, and limestone to the uppermost unit, a shale becoming sandy at the top. Figure 4 better illustrates the nature and interrelationships of the various facies present in a cyclothem.

Udden (1912) recognized cyclic repetition of strata near Peoria southwest of this locality; he grouped these units into cycles of deposition beginning with a coal at the base of each cycle. Savage (1927) observed the wide distribution of unconformities at the base of some sandstones and proposed separating

222	*C. B. Trask*

Figure 2. Site location and access (from Tonica and La Salle 7½-minute Quadrangles).

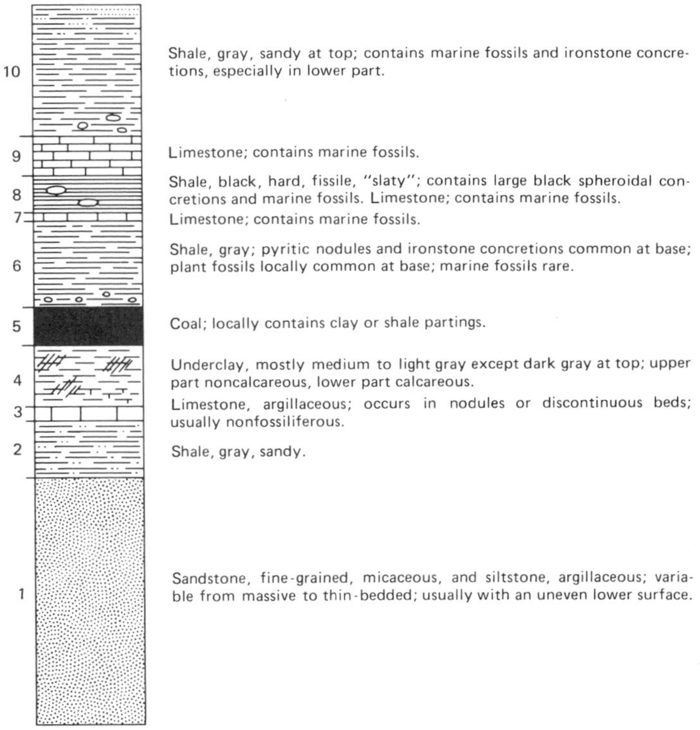
10 Shale, gray, sandy at top; contains marine fossils and ironstone concretions, especially in lower part.
9 Limestone; contains marine fossils.
8 Shale, black, hard, fissile, "slaty"; contains large black spheroidal concretions and marine fossils. Limestone; contains marine fossils.
7 Limestone; contains marine fossils.
6 Shale, gray; pyritic nodules and ironstone concretions common at base; plant fossils locally common at base; marine fossils rare.
5 Coal; locally contains clay or shale partings.
4 Underclay, mostly medium to light gray except dark gray at top; upper part noncalcareous, lower part calcareous.
3 Limestone, argillaceous; occurs in nodules or discontinuous beds; usually nonfossiliferous.
2 Shale, gray, sandy.
1 Sandstone, fine-grained, micaceous, and siltstone, argillaceous; variable from massive to thin-bedded; usually with an uneven lower surface.

Figure 3. Ideal cyclothem (after Willman and Payne, 1942).

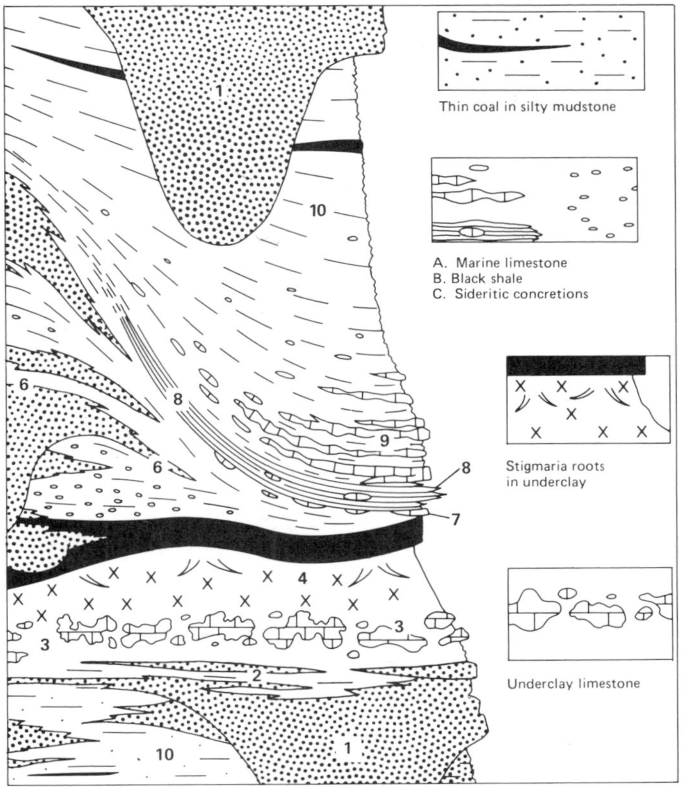
Thin coal in silty mudstone
A. Marine limestone
B. Black shale
C. Sideritic concretions
Stigmaria roots in underclay
Underclay limestone

Figure 4. Facies relationships in a typical cyclothem (from Baird and Shabica, 1980).

the Carbondale Formation from underlying rocks along such an unconformity. Weller (1930) adopted the base of the sandstone as the horizon separating adjacent sedimentary cycles. They have been referred to variously as cycles (Udden, 1912; Weller, 1930), suites (Wanless, 1929), and cyclical formations (Wanless, 1931; Weller, 1931). Wanless and Weller (1932) proposed the term cyclothem "to designate a series of beds deposited during a single sedimentary cycle of the type that prevailed during the Pennsylvanian Period."

The original intent was to use cyclothems for regional stratigraphic correlation (Weller, 1930, 1931; Wanless and Weller, 1932). Wanless and Weller intended the term to have the same rank as formation, as it was used at the time; however, the discontinuous nature of many lithologies and the limited extent of the erosional surfaces proved to be limiting factors in using cyclothems for widespread correlation. Modern usage focuses on depositional environments of Pennsylvanian strata, because an understanding of the origin of cyclothems would help in the study of Pennsylvanian paleoenvironments and paleogeography.

Weller (1964) has summarized several hypotheses to explain cyclic sedimentation. Many investigators feel that cyclothems were formed by repeated advance and abandonment of deltaic lobes, combined with marine transgressions resulting from compaction and subsidence of abandoned lobes. The often-cited model is the Mississippi Delta complex and the numerous delta lobes that have existed throughout Holocene time. Shabica (1979) discusses deltaic models generated under varying sea level conditions for northern Illinois.

The Lowell site is situated on the west flank of the La Salle Anticlinal Belt (essentially a monocline here), which was active during late Paleozoic time. Consequently, 115 ft (35 m) of section at this locality represents the interval occupied by more than 200 ft (60 m) of section in the deep basin. Strata from the Pennsylvanian-Ordovician unconformity to the Vermilionville Sandstone Member are exposed (Figs. 5 and 6). The section comprises five cyclothems: Tonica, Lowell, Summum, St. David, and part of the Brereton. (Hopkins and Simon, 1975, provide a list of cyclothems in Illinois.) The Liverpool Cyclothem is equivalent to the Tonica and Lowell Cyclothems; the former name is used where the Lowell Cyclothem is not well developed. Suggested unit numbers from the ideal cyclothem are shown for each lithology in the explanation accompanying Figure 5.

The Tonica Cyclothem consists of those units above the Ordovician to the base of the siltstone below the Lowell Coal Member. The underclay of the Colchester (No. 2) Coal Member forms an angular unconformity with the underlying dolostone of the Ordovician Galena Group, which can be seen in the riverbed. The claystone between the Colchester Coal and the unconformity includes remnants of several older cyclothems in western Illinois, but here probably represents only the one cyclothem. The Colchester Coal, one of the most widespread coals in the United States, lies as close as 4 ft (1.2 m) to the Galena Group and has been mined out locally.

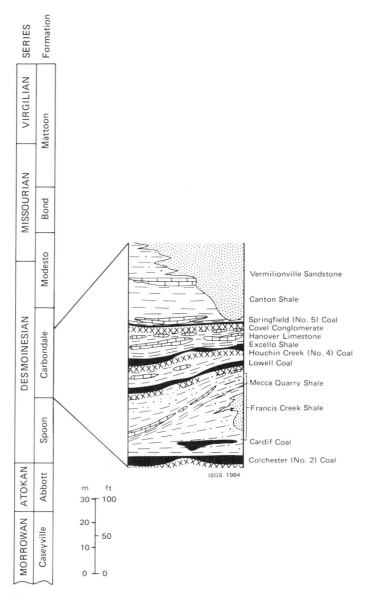

Figure 5. Stratigraphic sequence exposed at Lowell (after Jacobson, 1985).

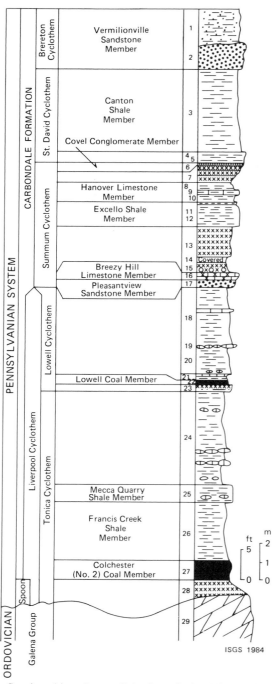

Figure 6. Stratigraphic column of the Lowell site (after Smith, and others, 1970). (Unit numbers in parentheses refer to lithologic units in the ideal cyclothem as shown in Figs. 3 and 4.)

1. Sandstone—brownish gray, thin bedded; interbedded with sandy shale; contains many black carbonaceous partings. (unit 2); 2. Sandstone—brown, fine-grained, poorly sorted, occurs in one massive bed. (unit 1); 3. Shale—gray, lower part fossiliferous (gastropods); contains layers of discoid septarian fossiliferous ironstone concretions; grades into underlying shale. (unit 10); 4. Shale—black, well-bedded, hard, "slaty," contains thin phosphatic lenses and laminae especially in lower part, occasional gray limestone nodule up to (1″) 2.5 cm thick; contains *Aviculopecten* in lower part. (unit 8); 5. Shale—black, very calcitic and fossiliferous; *Marginifera* and crinoid debris, pyritic. (unit 8); 6. Conglomerate—composed of poorly sorted fine-grained limestone particles (<0.5 in; <10 mm) in a

At this locality, the Colchester Coal contains numerous large coal balls as much as 20 in (50 cm) in vertical thickness by 60 in (150 cm) in diameter. The coal balls are stellate in plan and lenticular in cross section. Cursory investigation of the coal ball material reveals poor preservation, with much crystallization and pyritization and with a fair amount of compression, suggesting compaction prior to permineralization. Protuberances radiating from the main body have a central core and resemble cone-in-cone structures in longitudinal section. The material surrounding the central core is continuous with the exterior of the main body. Root-like structures in the underclay are carbonate with large pyrite bodies. These structures also resemble cone-in-cone in longitudinal section, but are crudely radial in cross section. Distally from the coal, these structures are attached to the tops of several spherical to ellipsoidal concretions up to 12 in (30 cm) in diameter.

The Colchester Coal is overlain by a badly slumped Francis Creek Shale Member, which underlies the Mecca Quarry Shale Member. The Mecca Quarry Shale is a black, fissile shale containing phosphatic lenses and nodules and a fauna of vertebrates and coprolites. It is overlain by a succession of shale with a few limestone zones.

The Lowell Cyclothem comprises units from siltstone below the Lowell Coal to the base of the Pleasantview Sandstone Member. This is the type section of both the Lowell Coal and the Lowell Cyclothem. The Lowell Coal is a lenticular coal as much as 5 in (13 cm) thick at this locality. It is probably correlative with the Survant (formerly Shawneetown) Coal Member of southeastern Illinois and southwestern Indiana.

Units from the Pleasantview Sandstone Member to the Covel Conglomerate Member compose the Summum Cyclothem. The Houchin Creek (No. 4) Coal Member (formerly Summum (No. 4) Coal Member) is represented by a very thin coaly zone or is absent. However, its underclay is well developed, consisting of 7 ft (2 m) of gray, calcareous claystone. The Excello Shale Member, which overlies the underclay, is a sili-

ceous, black, hard, fissile shale that forms a prominent ledge. The Hanover Limestone Member, above the Excello Shale, consists of limestone surrounded by shale. The limestone and upper shale are fossiliferous, containing abundant productids and crinoid stems. The Covel Conglomerate consists of pebbles in a sand matrix. Locally clay and silt are present in the matrix. The pebbles and sand are largely limestone. The upper surface of the conglomerate is covered by an algal-like layer (Willman and Payne, 1942). This peculiar lithology and its underlying claystone have not been assigned cyclothem unit numbers in Figure 5.

The St. David Cyclothem, above the Summum Cyclothem, consists of only three units at this locality. The Springfield Coal Member is absent; its position is marked by a shaly, black, fossiliferous limestone band 1 in (2.5 cm) thick, overlain by a black, fissile shale containing thin phosphatic lenses, small limestone concretions, and abundant *Aviculopecten* near its base. The overlying Canton Shale Member is a gray shale, fossiliferous in the lower part and containing discoid septarian ironstone concretions near the base.

The prominent bluff of friable sandstone at the top of the exposure is the Vermilionville Sandstone Member, the only unit of the Brereton Cyclothem present at this locality. It is a brown, fine-grained, poorly sorted, bluff-forming sandstone overlain by a brownish gray, thinly bedded sequence of interbedded sandstone and sandy shale containing black, carbonaceous partings.

---

pyritic matrix, fossiliferous; 7. Claystone—medium dark gray, becomes lighter in color downward with some mottling, reddish in lower 10 in (25 cm) contains irregular calcitic masses up to 1 in (2.5 cm) thick in bottom 1 ft 9 in (.5 m); calcite throughout; 8. Shale—light gray, fossiliferous, as below; contains several lenticular limestone units up to 3 in (7.6 cm) thick. (unit 10); 9. Limestone—light greenish gray, impure, nodular in lower part, fossiliferous, abundant productids and crinoid stems. (unit 9); 10. Shale—medium gray, slightly green. (unit 9); 11. Shale—medium dark gray, mottled with greenish gray; interbedded with medium gray, thinly laminated siltstone beds up to 3 in (7.6 cm) thick. (unit 8); 12. Shale—black, smooth, well laminated, relatively soft, coaly in parts. (unit 8); 13. Claystone—medium olive gray, relatively firm and calcitic especially in lower 4 ft (1.2 m); a few small slickensided surfaces. (unit 4); 15. Claystone—light greenish gray, yellowish cast, silty, noncalcareous, contains sandy limestone nodules up to 1 in (2.54 cm) thick in the lower 8 in (20 cm). (unit 4); 16. Limestone—light greenish gray, sandy, clayey, massive. (unit 3); 17. Sandstone—light greenish gray, fine-grained, calcitic, clayey, thin-bedded. (unit 1); 18. Shale—light greenish gray, fine macaceous, sandy near top, contains small nodules of sandy gray limestone which weather rusty, contains an 8 in (20 cm) mottled, soft red and green shale 1 ft (.3 m) from base, interval mostly covered. (unit 10); 19. Limestone—light gray, weathers reddish in part, septarian, fossiliferous; *Marginifera*, abundant, *Mesolobus, Ambocoelia;* forms a consistent nodular bed. (unit 9); 20. Shale—medium gray, weathers tan, soft, slightly fossiliferous, contains several siderite nodules in lower part, contains a 7 in (18 cm) zone of light olive gray, lithographic septarian limestone nodules 2 ft 4 in (0.7 m) from base, base 14 in (35 cm), poorly bedded. (unit 8?); 21. Shale—dark gray, fossiliferous, *Mesolobus, Marginifera, Neospirifer.* (unit 7?); 22. Coal—contains several dull shaly bands. (unit 5); 23. Siltstones—medium dark gray, sandy, calcitic, micaceous; contains vertical plant impressions and charcoal streaks. (units 1-4?); 24. Shale—dark gray, sandy, micaceous, generally thick bedded, contains two prominent zones of lenticular lithographic septarian limestones up to 1.5 ft (0.5 m) thick and containing a few fossils; several thinner and less persistent nodular limestone zones also present, a few crinoid stem fragments noted near base. (units 9 and 10); 25. Shale—black, hard, "slaty," contains large discoidal concretions of dark gray limestone up to 6 in (15 cm) thick, mostly in lower 1 ft (.3 m). (unit 8); 26. Shale—light gray, soft, thin bedded; contains a few sideritic concretions; generally not exposed. (unit 6); 27. Coal—has been mined out locally. (unit 5); 28. Claystone—gray, noncalcareous; where thicker than 8 ft (2.44 m), commonly consists of three beds: lower gray claystone, thin discontinuous green claystone or shale, and upper gray claystone. (unit 4); 29. Dolostone.

REFERENCES

Baird, G. C., and Shabica, C. W., 1980, The Mazon Creek depositional event: examination of Francis Creek and analogous facies in the Midcontinent region, *in* Langenheim, R. L., and Mann, C. J., eds., Middle and Late Pennsylvanian Strata on Margin of Illinois Basin: Great Lakes Section, Society of Economic Paleontologists and Mineralogists, 10th Annual Field Conference, p. 79–92.

Collinson, C., Avcin, M. J., Norby, R. D., and Merrill, G. K., 1972, Pennsylvanian conodont assemblages from La Salle County, northern Illinois: Illinois State Geological Survey Guidebook 10, 37 p.

Hopkins, M. E., and Simon, J. A., 1975, Pennsylvanian System, *in* Willman, H. B., et al., Handbook of Illinois Stratigraphy: Illinois State Geological Survey Bulletin 95, p. 163–201.

Jacobson, R. J., 1985, Coal resources of Grundy, La Salle, and Livingston Counties: Illinois State Geological Survey Circular 536, 58 p.

Rhodes, F.H.T., 1952, A classification of Pennsylvanian conodont assemblages: Journal of Paleontology, v. 26, no. 6, p. 886–901.

Savage, T. E., 1927, Significant breaks and overlaps in the Pennsylvanian rocks of Illinois: American Journal of Science, v. 14, p. 307–316.

Shabica, C. W., 1979, Pennsylvanian sedimentation in northern Illinois: examination of delta models, *in* Nitecki, M. H., ed., Mazon Creek Fossils: Academic Press, p. 13–40.

Smith, W. H., and others, 1970, Depositional environments in parts of the Carbondale Formation—western and northern Illinois: Illinois State Geological Survey Guidebook 8, 125 p.

Udden, J. A., 1912, Geology and mineral resources of the Peoria Quadrangle, Illinois: U.S. Geological Survey Bulletin 506, 103 p.

Wanless, H. R., 1929, Geology and mineral resources of the Alexis Quadrangle: Illinois State Geological Survey Bulletin 57, 230 p.

—— 1931, Pennsylvanian cycles in western Illinois: Illinois State Geological Survey Bulletin 60, p. 179–193.

Wanless, H. R., and Weller, J. M., 1932, Correlation and extent of Pennsylvanian cyclothems: Geological Society of America Bulletin, v. 43, p. 1003–1016.

Weller, J. M., 1930, Cyclical sedimentation of the Pennsylvanian Period and its significance: Journal of Geology, v. 38, no. 2, p. 97–135.

—— 1931, The conception of cyclical sedimentation during the Pennsylvanian Period: Illinois State Geological Survey Bulletin 60, p. 163–177.

—— 1964, Development of the concept and interpretation of cyclic sedimentation, *in* Merriam, D. F., ed., Symposium on Cyclic Sedimentation: Kansas Geological Survey Bulletin 169, v. II, p. 607–621.

Willman, H. B., and Payne, J. N., 1942, Geology and mineral resources of the Marseilles, Ottawa, and Streator Quadrangles: Illinois State Geological Survey Bulletin 66, 388 p.

Ordovician-Silurian unconformity at Kankakee River State Park, Illinois

Donald G. Mikulic, *Illinois State Geological Survey, 615 E. Peabody Drive, Champaign, Illinois 61820*
Joanne Kluessendorf, *Department of Geology, University of Illinois, Urbana, Illinois 61801*

LOCATION

Upper Ordovician and Silurian rocks are well exposed in a series of natural outcrops along the Kankakee River from Wilmington to Kankakee, Illinois. Some of the best exposures occur in Kankakee River State Park, which extends northwest from the town of Altorf for about 6 mi (9.6 km) (Fig. 1). Three sites are particularly interesting (Fig. 2).

1. Chippewa Campground—a river bluff on the north side of the river just below the campground (NW½NE¼SE¼,Sec. 36,T.32N.,R.10E., Bonfield 7½ minute Quadrangle, Will County).

2. Cowan's Quarry—abandoned stone pits and river bluffs on the north side of the river between the river and State Highway 102 (SE¼SW¼NW¼,Sec.26,T.32N.,R.10E, Bonfield 7½-minute Quadrangle, Will County).

3. Rock Creek—a canyon of a tributary north of the Kankakee River, north and south of the bridge on State Highway 102 (S½SE¼SW¼,Sec.32,T.32N.,R.11E., Bourbonnais 7½-minute Quadrangle, Kankakee County).

All localities except Cowan's Quarry are accessible by passenger car and are a short walk from a highway or parking lot. Cowan's Quarry is reached by an unmarked and unmaintained dirt road leading south to the river from Illinois 102 (Fig. 2). This road is generally in poor condition, but is the only place to park as there is no shoulder on the highway. From the dirt road, walk west about 300 ft (100 m) along the top of the river bluff to the old stone pits.

SIGNIFICANCE

The unconformity at the Ordovician-Silurian boundary is well displayed in this area (Fig. 3). The unusually thin basal Silurian succession in this vicinity demonstrates the depositional control exerted by erosional topographic relief on the underlying Upper Ordovician Maquoketa Group surface. A fairly complete Silurian composite section may be examined at these localities. In addition the only outcrops of the Neda Formation, an unusual oolitic ironstone, occur within this limited area.

SITE INFORMATION

Bedrock is near the surface throughout this region because much of the Quaternary sediment was removed by the floodwaters of the Kankakee Flood during the Wisconsinan glacial period. The Kankakee River and its tributaries have cut into bedrock, producing good bluff and canyon exposures. To the west,

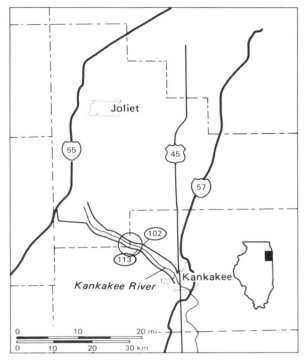

Figure 1. Index map showing highways in the vicinity of Kankakee River State Park (circle).

around Wilmington, these bluffs expose Upper Ordovician rocks. Progressively younger Silurian rocks crop out towards the east because of a slight regional dip in that direction.

In northern Illinois, the upper Ordovician Maquoketa Group consists predominantly of shale with some limestone or dolomite. The lowest Ordovician exposed in this area is the Brainard Formation, a greenish gray shale. The Brainard generally is covered by talus here and only a few inches may be seen beneath the Neda Formation at Chippewa Campground.

Towards the end of the Ordovician, Maquoketa sediments were eroded extensively during a major regressive episode related to glacio-eustatic drop in sea level. The resulting erosional topography displays more than 175 ft (52 m) of relief in portions of northern Illinois, where both the Neda and Brainard Formations have been completely removed and the Silurian rests directly on the Fort Atkinson Formation.

The Neda Formation (uppermost Maquoketa) occurs only where the Maquoketa is thickest. It is uncertain whether the Neda is an erosional remnant of a once more extensive deposit; was deposited only on topographic highs; or is a laterite developed

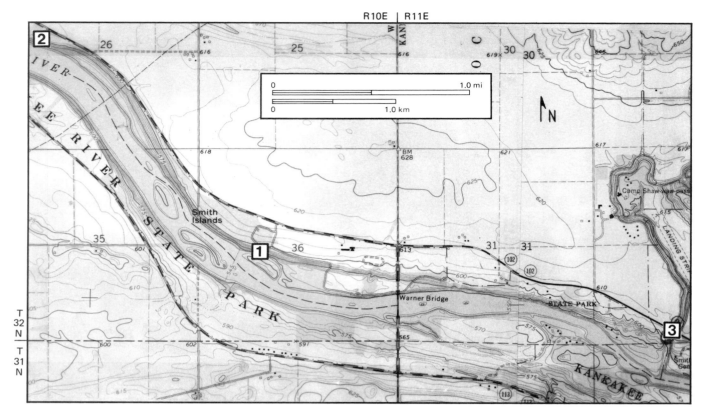

Figure 2. Map showing the location of sites 1, 2, and 3 on the Bonfield and Bourbonnais 7½-minute quadrangles in Will and Kankakee Counties, Illinois.

during emergence of these highs. At its type locality in Dodge County, Wisconsin, the Neda is 35 ft (10.5 m) thick and consists predominantly of ferruginous grain-supported oolite. In Illinois the Neda attains a maximum thickness of only 15 ft (45 m) and is lithologically similar to just the basal Neda in Wisconsin.

The only outcrops of the Neda in Illinois occur in Kankakee River State Park, where it is best exposed below Chippewa Campground. The Neda is distinguished by flattened ferruginous (mostly goethitic) ooids, which are scattered through various reddish brown and purplish red ferruginous sediments, such as dolomicrite, dolosiltite, and shale. In places the Neda is bioturbated and ooids are packed densely into the burrows. Oomoldic porosity, some filled with calcite spar, is common. Phosphatic nodules occur throughout much of the Neda but are most abundant at the top and base of the unit.

Origin of the unusual ferruginous ooids is controversial. They have been attributed variously to iron replacement of normal, marine calcareous ooids; direct formation of ferruginous ooids in iron-rich muds; or lateritic soil development during emergence and weathering of Brainard sediments.

Diagnostic fossils have not been found in the Neda Formation in Illinois and the age of the unit is uncertain. The Neda now is considered late Ordovician, although in the past it was thought to be Silurian.

With the rise of sea level at the end of Ordovician glaciation, early Silurian seas transgressed this area of the midcontinent. The erosional topography of the Maquoketa surface caused significant variations in the thickness and lithologic characteristics of the basal Silurian sediments (Wilhelmi, Elwood, and Kankakee Formations) (Fig. 3). The thicknesses of the Maquoketa and overlying lower Silurian are inversely proportional. Where the Maquoketa has been eroded deeply, the basal Silurian reaches its maximum thickness of about 180 ft (54 m) having filled low areas on the Maquoketa surface. Where the Maquoketa is thickest, the Silurian is extremely thin and the Wilhelmi and Elwood Formations may be absent.

As the seas covered the region, Maquoketa sediments initially were eroded from emergent topographic highs and incorporated into Silurian sediment. This resulted in a transition upward from highly argillaceous basal Silurian sediments, such as the Wilhelmi Formation, to purer carbonates in the Kankakee Formation. Highly argillaceous sediments are pervasive where a thick Silurian succession fills low areas in the Maquoketa; but where a thin Silurian overlies a thick Maquoketa sequence, argillaceous sediments are limited.

These relationships between the Ordovician and Silurian cannot be viewed at any one locality. Thick successions of highly argillaceous lower Silurian overlying the eroded Maquoketa sur-

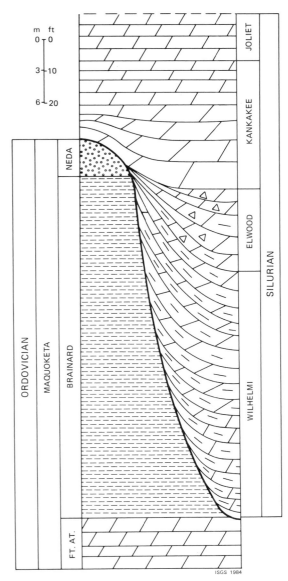

Figure 3. Diagram showing the unconformable stratigraphic relationships between the Ordovician and Silurian rocks in northern Illinois. Ft. At. = Fort Atkinson Formation.

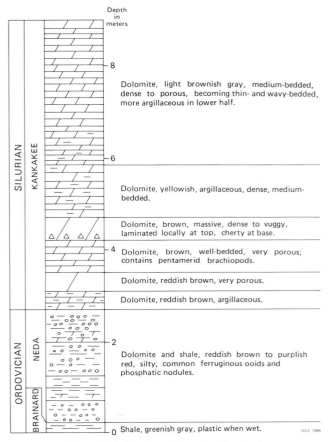

Figure 4(a). Stratigraphic section for site 1.

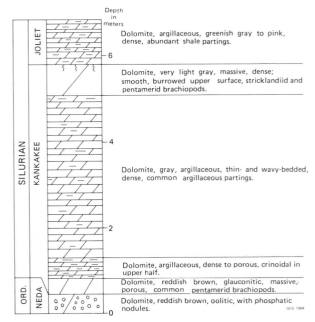

Figure 4(b). Stratigraphic section for site 2.

face are exposed at many places in Illinois, such as along the Des Plaines River between Joliet and Channahon. An extremely thin lower Silurian, however, is exposed only in the Kankakee River State Park area, where Maquoketa topographic highs remained during Silurian transgression.

At Cowan's Quarry and Chippewa Campground the Neda Formation is overlain unconformably by the Silurian Kankakee and Joliet Formations (Fig. 4). The Wilhelmi and Elwood Formations are absent here, with the possible exception of 1 ft (0.3 m) of Wilhelmi lithology. Most of the remaining Silurian is unusually thin, the Kankakee Formation members ranging from one-tenth to one-half their normal thickness.

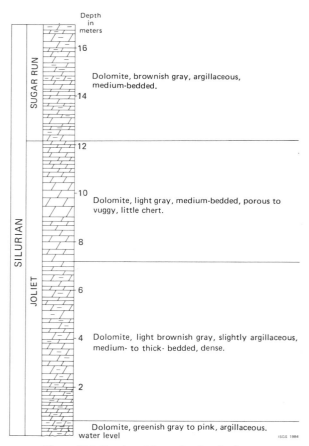

Figure 4(c). Stratigraphic section for site 3.

The contact between the Kankakee and Joliet Formations is interesting and may be observed at Cowan's Quarry. The upper surface of the Kankakee is planar, smooth, and heavily burrowed, suggesting a period of nondeposition and erosion before the Joliet was deposited.

At Rock Creek, a picturesque steep-walled rock canyon exposes the Joliet and Sugar Run Formations (Fig. 4). The section is most complete on the west side of the creek, south of the highway bridge. North of the bridge a fault has caused a downward displacement of about 6 ft (1.8 m). The Joliet Formation here is thinner than normal.

Because of the eastward regional dip, Racine Formation strata are exposed towards the city of Kankakee, particularly at several old quarries within Bird Park in the city itself. Also in the park is a large, old restored lime kiln of historical interest.

REFERENCES

Athy, L. F., 1928, Geology and mineral resources of the Herscher Quadrangle: Illinois State Geological Survey Bulletin 55, 120 p.

Cote, W., Reinertson, D., and Wilson, G., 1967, Geological science field trip—Bourbonnais area: Illinois State Geological Survey Guide Leaflet C, 17 p.

Kolata, D. R., and Graese, A. M., 1983, Lithostratigraphy and depositional environments of the Maquoketa Group (Ordovician) in northern Illinois: Illinois State Geological Survey Circular 528, 49 pp.

Willman, H. B., 1962, The Silurian strata of northeastern Illinois, *in* Silurian rocks of southern Lake Michigan area: Michigan Basin Geological Society Annual Field Conference Guidebook, p. 61–67, 81–96.

——1973, Rock stratigraphy of the Silurian System in northeastern and northwestern Illinois: Illinois State Geological Survey Circular 49, 55 p.

Workman, L. E., 1950, The Neda Formation in northeastern Illinois: Illinois Academy of Science Transactions, v. 43, p. 176–182; Illinois State Geological Survey Circular 170, p. 176–182.

Farm Creek, central Illinois: A notable Pleistocene section

Leon R. Follmer and E. Donald McKay III, Illinois State Geological Survey, 615 E. Peabody Drive, Champaign, Illinois 61820

LOCATION

The Farm Creek Section is located in central Illinois, near the late Wisconsinan glacial margin east of Peoria, Illinois (Fig. 1). The section is a cut bank along Farm Creek in the Farmdale Recreation Area, a public park adjacent to East Peoria (Fig. 2). It is about 5 mi (8.5 km) east of Peoria via I-74 and Illinois 8 to Robein or to Farmdale via Farmdale Road (Fig. 3). Either route provides access to the west entrance of the park. Turn south on Summit Drive 1.5 mi (2.4 km) east of Robein. This road turns into Bittersweet Road and provides a scenic view of the park in the valley on the way to the west entrance. Farmdale Road goes up Farm Creek Valley through Farmdale, a former station on the Toledo, Peoria, and Western Railroad. East of Farmdale, turn north on Summit Road to the west entrance of Farmdale Park.

The east entrance to the park is off School Road about 0.5 mi (0.8 km) east of Summit Drive. During good weather one can drive on a gravel road from one entrance to the other but must cross three concrete-bottom fords. The main geologic section, on the south side of the creek near the middle ford can be seen from the road.

Farm Creek is subject to flash floods that sometimes bury the fords with gravel or cause structural damage. The bottomland road is usually closed for repairs following floods; at such times hiking from the east entrance along the railroad track to the south side of the valley provides the best access to the main section.

The park is open seven days a week in late spring to mid-autumn during daylight hours. When the bottomland road is closed, the campground superintendent will open the gates on request to permit driving on the passable parts of the road. A campground for tents and trailers and a picnic area are located near the west entrance. Hiking and horse trails traverse the 600-acre park. About half the park is heavily wooded; much native flora can be seen in natural areas.

SIGNIFICANCE

The Farm Creek Section is the type section for the Farmdale Soil, Farmdalian Substage, and the Robein Silt. Many workers have used this section as a type or reference section for litho-, chrono-, and pedostratigraphic units. A discontinuous organic paleosol within a sequence of loess that overlies Illinoian till and underlies Wisconsinan till is a principal reference point for Pleistocene researchers in Illinois. The organic soil was first thought to be the Sangamon Soil but later was determined to stratigraphi-

Figure 1. Location of Farm Creek Section in Illinois.

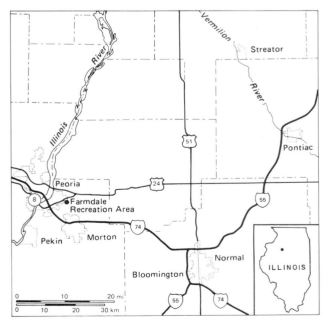

Figure 2. Location of Farmdale Recreation Area.

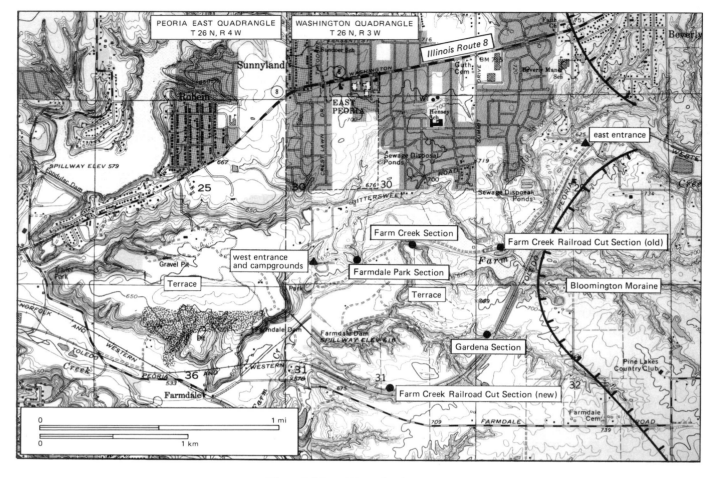

Figure 3. Map of Farm Creek locality.

cally overlie the Sangamon Soil. Excellent exposures in the vicinity have been accessible since Frank Leverett first observed the Farm Creek cut in 1897. The Farm Creek Section and nearby exposures appear to be complete with respect to most stratigraphic elements of the late Pleistocene in central Illinois, and continue to be the best and most accessible exposures in the area.

SITE INFORMATION

Many stratigraphic studies have been completed in the Farm Creek vicinity and much information is available. So many detailed stratigraphic, geologic, and pedologic interpretations have been generated on the features of this site that only a summary of historical points and current interpretations can be made here. Leverett (1899) first described and interpreted the Farm Creek Section in terms of the meaning of the organic soil and weathering zone on the Illinoian till. He related both features to the Sangamon Soil and considered them to be evidence for an interglacial stage (Follmer, 1978).

Leighton (1926) was so impressed with the exposure that he referred to it as "a notable type Pleistocene Section." His general interpretation of the sequence between the overlying Wisconsinan till (Shelbyville) and the Illinoian till below agree with Leverett's (Fig. 4). Leverett's Farm Creek description indicates that he did not resolve the detail that was later found to be present. The terms Iowan and Peorian, first introduced by Leverett, have been confused or misinterpreted and have since been dropped (see McKay in Follmer and others, 1979). The Iowan was thought to be a glacial event between the Illinoian and Wisconsinan, represented here by a calcareous loess; the Peorian was thought to be a loess deposited at the end of a glacial event and weathered during an interglacial event.

Leighton and Leverett agreed that the "Sangamon" (the organic zone) contains coniferous wood and overlies a loesslike silt. The boreal vegetation present caused interpretation problems because the Sangamon was thought to be a time of warmth similar to the present climate. They concluded that the cold-climate indicators reflected either the close of the Sangamon time or the result of the subsequent glaciation. By 1948, Leighton decided that the loesslike silt had been generated by glacial conditions and consequently renamed the unit the Farmdale loess. Leighton and Willman (1950) interpreted this loess as representing the Farmdale substage, the oldest part of the Wisconsin stage. They did not name the organic soil at this time but recognized it as a youthful profile of weathering not sufficient to be designated as an interglacial soil. This interpretation removed the confusion between the organic soil and the profile of weathering on till below the loess, both of which had been called Sangamon. This change brought the basic stratigraphic interpretations into alignment with present concepts, but no agreement on terminology was reached at this time.

In 1960 Frye and Willman proposed a major revision of the Wisconsinan terminology because new data could not be reconciled with the old models. Much new information was developed from their study of the Farm Creek area. Their work culminated with the publication of a comprehensive study of the Pleistocene stratigraphy of Illinois (Willman and Frye, 1970). They correlated and renamed most stratigraphic units present at Farm Creek and designated the section as the type section for the Farmdale Soil, the Farmdalian Substage, and the Robein Silt. A new railroad cut south of the Farm Creek Section (Fig. 3) was designated the type section of the Morton Loess (Frye and Willman, 1960). Most of the changes resulted from the implementation of a system of multiple classification allowing litho-, chrono-, and pedostratigraphic units to be treated independently. In effect, the previous classification system was monotaxonomic, in that all aspects were considered interrelated. This led to confusion of the terms used for materials, time intervals, and paleosols.

The study of Follmer and others (1979), provides the most recent information for the Farm Creek Section. In this study, the classic Farm Creek Section was described using the terminology of Willman and Frye (1970); Figure 5 shows a generalized diagram of the main section. Profiles A and C are in the general area

LEVERETT (1899)		LEIGHTON (1926)	WILLMAN and FRYE (1970)
FARM CREEK	FC RR*	FARM CREEK	FARM CREEK
Shelbyville till	Shelbyville till	Shelbyville till	Delavan Till
Iowan loess	Iowan loess	Peorian loess	Morton loess
	Sangamon peat	Sangamon Soil (Farmdale Loess) (1948)	Robein Silt
	silt		Roxana Silt
Illinoian till	leached Illinoian till	gumbotil	Sangamon Soil in till
	calcareous Illinoian till	calcareous Illinoian till	calcareous Illinoian till

*railroad cut near Farm Creek Section

ISGS 1984

relative thickness (ft): 0, 4, 8, 12, 16, 20, 24

Figure 4. Farm Creek Section stratigraphy (from Follmer and others, 1979).

that had been previously studied. The area where profile B was taken had probably not been studied before, but the materials present there appear to be similar to those which Leverett found in the old railroad cut about 0.5 mi (0.8 km) upstream (east). All profiles show the location of detailed sampling and description. The results are presented in Follmer and others (1979) and are summarized here on Table 1.

General Discussion of the Farm Creek Section

The top of the section is quite irregular and is nearly vertical in places. A late Wisconsinan loess (Richland Loess) is continu-

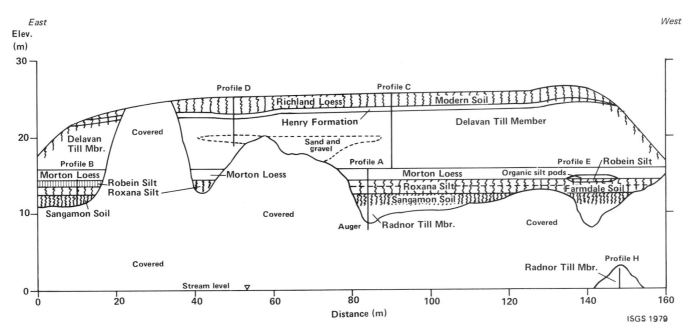

Figure 5. Diagram of the Farm Creek Section (from Follmer and others, 1979).

ous across the top and ranges from 3 to 7 ft (1 to 2 m) thick. In most places a modern soil (Hapludalf) has developed through the loess into the underlying gravel of the Henry Formation. The alteration due to soil formation caused clay enrichment and reddening, particularly in the upper part of the gravel. The gravel is part of the terrace deposits formed by the outwash from the Bloomington Morainic System to the east. Under the gravel is the Delavan Till Member of the Wedron Formation, which is about 25 ft (7 m) thick and contains large lenses of sand and gravel. The Delavan is the current name for basal Woodfordian till in the area and forms the terminal Wisconsinan moraine south of Peoria.

The gray Delavan Till forms a sharp contact with the underlying light-colored Morton Loess that can be easily traced across the outcrop. The Morton is calcareous and appears to be undisturbed except for local and indistinct shear disturbances. At this position Leverett (1899) and Leighton (1926) thought a soil surface or eroded soil surface might be present. No evidence for a soil has been found in recent studies here, but one can be seen at the Gardena Section to the southeast (Fig. 3). The Morton is an early (classic or late Wisconsinan) Woodfordian loess that was overridden by the advancing glacier. Beyond the margin of the late Wisconsinan moraine west of Peoria, the Richland and Morton Loess converge to form the Peoria Loess.

In the main section at Profile A, about 5 ft (1.5 m) of Morton Loess lies over a leached brown silt (the Roxana), which is about 5 ft (1.5 m) thick and contains the fossil remnants of the Farmdale Soil. The main body of the Roxana Silt is eolian and can be traced across the upland of much of Illinois. The Roxana Silt is considered to represent the Altonian Substage or early Wisconsinan in Illinois. At Profile E the Morton-Roxana contact is somewhat masked by organic matter, including wood fragments, carbonized wood, and charcoal. Rounded pods form the organic-rich zone here. Through time, diagenic processes (biogeochemical degradation) eliminate the organic matter, leaving large bleached zones and segregations of secondary iron and manganese minerals. The organic-rich remnants in this occluded form indicate that the Morton was deposited on an organic-rich soil that continued to develop upwards. Therefore, it is likely that most of the Morton and Roxana at this section was initially humic-rich and most of the humic material was removed by diagenic processes leaving most of the Morton and Roxana exposed in the outcrop free of humic material.

At Profile B the environment of burial was sufficient to preserve much of the organic material of the O horizon of the Farmdale Soil. Pollen is present in the richest portion, but it is poorly preserved. The pollen assemblage is dominated by pine and spruce, indicating a cool climate for the formation of the Farmdale Soil. Radiocarbon analysis of samples near the top and bottom of the organic rich zone, designated by Willman and Frye (1970) as Robein Silt, yielded ages of 26,680 ± 380 and 27,700 ± 770 B.P., respectively. The Robein Silt is defined as a resedimented silt derived from the Roxana, but no evidence for resedimentation was found in recent studies reported in Follmer and

others (1979). It appears that organic matter accumulated on the surface of the Roxana Silt during the formation of the Farmdale Soil. This presents a technical problem because this is the type section for the Robein, but no evidence for waterlain stratification can be demonstrated here. Because stratified Robein deposits are present in other localities, a new reference section needs to be designated.

The Roxana Silt was calcareous when deposited, but is now leached of its carbonate minerals in the present exposure. The base of the Roxana is gradational into the top (A horizon?) of the Sangamon Soil developed in Illinoian till. The boundary is hard to identify because it has been blurred by bioturbation or pedogenic processes. The sand content increases downward and the color becomes lighter. Characteristics of fossil A horizons commonly are poorly preserved and organic matter (analogous to soft parts of fossils) is often not preserved. The distinguishing characteristics are the biogenic pores and structures in the probable A and underlying E horizon of the Sangamon Soil profile. The relative high porosity in a bleached (light-colored) matrix serves to identify the "topsoil" of many Sangamon Soil profiles as well as other soils that have developed in a deciduous forest environment. The E horizon overlies a greenish gray Bt horizon that is recognized by the abrupt increase in clay, changes in soil structures from small to large, and the abundant clay skins coating the soil structures. In most places in the outcrop, the Sangamon Soil is poorly drained (gleyed) and was formerly referred to as gumbotil. Because of the gleyed condition, it has been confused with accretion-gley (Follmer, 1978). In recent years the relatively slow bluff erosion has exposed a reddish brown Sangamon Bt horizon near the center of the exposure. This permits

TABLE 1. AVERAGES OF COMMON LITHOLOGIC PARAMETERS

Section Unit Horizon	Grain Size (<2mm)[1]			Carbonate (<74μM)[2]		Clay Minerals (<2μM)[3]		
	Sd (%)	Si (%)	C (%)	Cal (%)	Dol (%)	E (%)	I (%)	K+C (%)
Farm Creek								
Richland Loess	.2	68	30	0	0	58	31	11
Henry (sand & gravel)	65	19	16	19	34	18	68	14
Delavan Till	26	39	35	8	21	12	67	21
Morton Loess	<1	92	8	2	23	37	43	20
"Robein Silt"	2	89	9	0	0	10	60	30
Roxana Silt	2	85	13	0	0	54	26	20
Radnor Till								
E (C/A) horizon	17	65	18	0	0	30	35	35[4]
Bt horizon	16	32	52	0	0	47	34	19
C (oxidized)	24	47	29	5	17	6	80	14
C (unaltered)	27	46	27	5	19	2	75	23
Gardena								
Delavan Till	27	40	33	5	20	10	67	23
Morton Loess	<1	93	6	2	21	30	45	25
"Robein Silt"	5	86	9	1	1	18	50	32[4]
Roxana Silt	<1	84	15	0	0	62	16	22
Farmdale Park								
Radnor Till								
C (oxidized)	26	45	29	5	17	6	82	12
C (unaltered)	30	44	26	5	20	5	71	24
"Vandalia Till"								
C (oxidized)	38	43	19	6	23	7	82	11

[1]Gravel excluded, <4μm clay on tills, <2μm clay on silts and soil horizons.
[2]Weight percent of fine fraction (Chittick method).
[3]Percentages based on sum of three peak heights.
[4]High vermiculite causing K + C value to be relatively high.
Abbreviations: Sd = sand, Si = silt, C = clay, Cal = calcite, Dol = dolomite, E = expandable clay minerals (17Å), I = illite (10Å, K+C = kaolinite and chlorite (7Å).

one to see a Sangamon catena from a poorly drained profile to a moderately-well drained profile.

The B horizon's characteristics of the Sangamon Soil fade with depth into unaltered calcareous, gray Radnor Till of Illinoian age. The top of the Illinoian till must be placed at the top of the Sangamon Soil at this site because no sedimentologic unit can be demonstrated to exist between the Roxana and the till. Lithologic studies (Follmer and others, 1979) support the interpretation that the Sangamon Soil is the highly altered portion of the till. Some problems remain concerning till correlations, which need further study. The relationship of the till exposed at profile A and H is not yet clear. The clay content at H is about 10% higher than average (Table 1), but it contains the amount of illite characteristic of the Radnor. At profile A the texture of the C horizons is characteristic of the Radnor, but the illite content of the unaltered Radnor is about 5% lower than average (Table 1). Two rows of data presented in Follmer and others (1979) are misprinted; all parameters for samples FCA-1 and FCA-3 must be inverted in order to put them into correct stratigraphic order. Lower illite values and higher sand content are regional characteristics of the Vandalia Till Member, which occurs stratigraphically below the Radnor Till in this area and to the south. The till at profile H was at one time assigned to the Hulick, a middle Illinoian unit known in western Illinois (Willman and Frye, 1970). Recent lithologic studies in the Farm Creek area suggest that the best correlation is to the Radnor, the youngest Illinoian till in Illinois. The differences between A and H raise the question of equivalency, but the till at both locations is more like Radnor than Hulick.

Farmdale Park Section

The Farmdale Park Section was studied near the west ford on the north side of the creek (Fig. 3) and presented in Follmer and others (1979). The same stratigraphic units seen at the Farm Creek Section are present here from the top of the section down to the Radnor Till. The Sangamon Soil here is a complete profile of an oxidized, well-drained soil developed in the Radnor Till. The redness of the Sangamon Soil contrasts with the greenish-gray colors that dominate the Sangamon at the main section. This and other physical characteristics indicate that the Sangamon landscape had topographically high (oxidized) and low (wet) landscape positions. This information is used to reconstruct paleo-landscapes.

Two Illinoian tills are visible in the lower part of the exposure. The upper unit is correlated to the Radnor on the basis of stratigraphic position and lithologic characteristics. Most of the Radnor is calcareous and oxidized to a yellowish brown. At the north side of the exposure, the base of the Radnor is unoxidized and gray. Here the gray till forms a distinct boundary with an underlying oxidized, olive brown, calcareous, sandier till. The textural boundary continues across the outcrop, but the color contrast disappears to the south because of oxidization. On the basis of texture and stratigraphic relations, the lower till is correlated to the Vandalia Till Member. But the clay mineral assemblage of the lower unit is typical of oxidized Radnor Till. It is higher in illite than is typical for Vandalia Till to the south. The two units of Illinoian till at the Farmdale Park Section may represent a more complex sequence within the Radnor or the lower unit could be a till older than the Vandalia. Other exposures in the area have not been studied in detail and may contain the stratigraphic information that can resolve the question.

Gardena Section

One can hike across the park or along the new T.P. and W. Railroad track to this rare exposure of a soil developed in the top of the Morton Loess underlying the Delavan Till (Fig. 3) (described in Follmer and others, 1979). A moss layer occurs at the top of the soil, which was dated 19,680 ± 460 B.P. Five species of mosses were found that now range from the northern U.S. to the Arctic. Spruce pollen dominated the pollen from this soil, indicating that this short-lived soil formed during the coldest interval this area experienced during the advance of the Woodfordian (late Wisconsinan) glaciers. A few centimeters of lacustrine clay separate the Delavan Till from the moss layer. The lacustrine clay likely represents the derangement of drainage caused by the advancing glacier to form a lake that was soon overridden.

About 7 ft (2.1 m) of Morton is present here just above the creek level west of the railroad bridge. The lower 2 ft (0.6 m) is dolomitic and black from the decomposed organic matter it contains. The ^{14}C age of wood in the base of this zone is 25,370 ± 310 B.P. Under this is a leached organic zone that yielded an age of 25,960 ± 280 B.P. and is interpreted to be Farmdale Soil developed in Robein Silt. Below water level is the gray (gleyed) horizon of the Farmdale Soil in Robein Silt. The ages of the organic samples and the litho- and pedostratigraphic relations confirm the interpretations drawn from the study of the Farm Creek Section. The analytical data (Table 1) support the correlations between sections and illustrate the value of using lithic parameters in the correlation of glacial deposits. The contrasts within and between sections create problems and questions when attempting litho-, chrono-, and pedostratigraphic correlations. Appropriate features at each section are correlated using stratigraphic and pedologic principles, and lithologic similarities.

REFERENCES CITED

Follmer, L. R., 1978, The Sangamon Soil in its type area—A review, *in* Mananey, W. C., ed., Quaternary Soils: Norwich, England, Geological Abstracts, p. 125–165, (ISGS Reprint 1979-I).

Follmer, L. R., McKay, E. D., Lineback, J. A., and Gross, D. L., 1979, Wisconsinan, Sangamonian, and Illinois Stratigraphy of Central Illinois: Midwest Friends of the Pleistocene Field Conference, Illinois State Geological Survey

Guidebook 13, 134 p.

Frye, J. C., and Willman, H. B., 1960, Classification of the Wisconsinan Stage in the Lake Michigan glacial lobe: Illinois State Geological Survey Circular 285, 16 p.

Leighton, M. M., 1926, A notable type Pleistocene section; the Farm Creek exposure near Peoria, Illinois: Journal of Geology, v. 34, no. 2, p. 167–174.

Leighton, M. M., and Willman, H. B., 1950, Loess formations of the Mississippi Valley: Journal of Geology, v. 58, no. 6, p. 599–623.

Leverett, Frank, 1899, The Illinois glacial lobe: U.S. Geological Survey Monograph 38, 817 p.

Willman, H. B., and Frye, J. C., 1970, Pleistocene stratigraphy of Illinois: Illinois State Geological Survey Bulletin 94, 204 p.

Valmeyer Anticline of Monroe County, Illinois

Rodney D. Norby, Illinois State Geological Survey, 615 E. Peabody Drive, Champaign, Illinois 61820

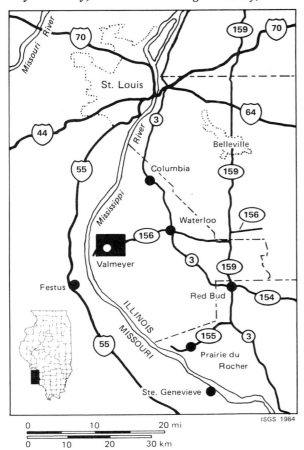

Figure 1. General location of the Valmeyer area approximately (25 km) south of St. Louis (from Illinois Highway Map).

LOCATION AND ACCESSIBILITY

Strata of the Valmeyer Anticline are exposed in road and stream cuts along Illinois 156 (Fig. 1), which passes through Dennis Hollow from the east edge of Valmeyer to 1.5 mi (2.4 km) east of Valmeyer, Monroe County (S½Sec.2,SE¼SE¼Sec.3, and N½Sec.10,T.3S.,R.11W.; Valmeyer 7.5-minute Quadrangle); they are also exposed in the bluff face at Columbia Quarry Plant No. 3, which is 0.9 mi (1.5 km) north of Valmeyer (SE¼NW¼Sec.3,T.3S.,R.11W.). The outcrops along Illinois 156 are easily accessible with no more than a 300 ft (100 m) walk from the highway. Automobile parking is limited to a few wide grassy pullouts adjacent to the south side of the highway and along driveways of property owners. The area north of Illinois 156 has several owners; therefore permission to enter should be obtained from residents nearest a particular outcrop.

SIGNIFICANCE

Following uplift of the Valmeyer Anticline, erosion by the Mississippi River and the stream in Dennis Hollow exposed Ordovician and Mississippian strata. A major unconformity separates the two systems. Several Champlainian (middle Ordovician) and Cincinnatian (upper Ordovician) formations are exposed in the eroded axis of the anticline. The type of the Valmeyeran (middle Mississippian) Series is exposed along Dennis Hollow on the gently dipping northeast flank of the Valmeyer Anticline.

SITE INFORMATION

The Valmeyer Anticline trends northwest-southeast and plunges gently to the southeast (Fig. 2). The anticline has been traced to near Maeystown about 10 mi (16 km) southeast of Valmeyer. To the northwest, the structure intersects the bluffs along the Mississippi River just north of Valmeyer. The Valmeyer Anticline has not been identified on the Missouri side of the river. The southwest limb of this asymmetrical anticline has dips between 15 and 25 degrees, although dip angles as high as 33 degrees have been noted (Odum and others, 1961). The beds on the northeast flank generally dip at angles of 2 degrees or less. The best view of the Valmeyer Anticline can be seen to the southeast from a vantage point 1.0 mi (1.6 km) north of Valmeyer along the bluff-front road to Fountain and Columbia (Fig. 3).

The Valmeyer Anticline is very similar to the Waterloo-Dupo Anticline, which is approximately parallel to it about 10 mi (16 km) to the northeast. The latter structure is also asymmetrical with even steeper dips (up to 40°) on the southwest limb; it is faulted parallel to the axis along part of this limb. Between these two anticlines lies the Columbia Syncline. The Monroe City Syncline just to the southwest of the Valmeyer Anticline completes these structural couplets (Treworgy, 1981).

Minor tectonic activity may have begun as early as the Devonian; however, the major folding and faulting in the area is at least very late Mississippian in age, as Ordovician and Mississippian strata show similar amounts of deformation. Evidence from a small stream cut about 8 mi (12 km) away (NE¼NW¼ NE¼Sec.3,T.2S.,R.10W., Monroe County, Columbia 7.5-minute Quadrangle) suggests the major folding was either late Mississippian or early Pennsylvanian with some activity probably continuing through the Pennsylvanian or even later (Cote and others, 1970). This cut on the west limb of the Waterloo-Dupo Anticline shows shale and underclay of the Desmoinesian Carbondale Formation (Pennsylvanian) overlying early Chesterian (upper Mississippian) limestone and shale with apparent angular unconformity. The Mississippian strata dip westward at an angle of 44 degrees, while the overlying Pennsylvanian beds dip more gently westward at only 9 degrees. The greater dip of the Mississippian strata indicates that uplift occurred on the Waterloo-Dupo Anticline before deposition of the Pennsylvanian strata. The dip of the Pennsylvanian strata also suggests that some movement occurred after Desmoinesian time.

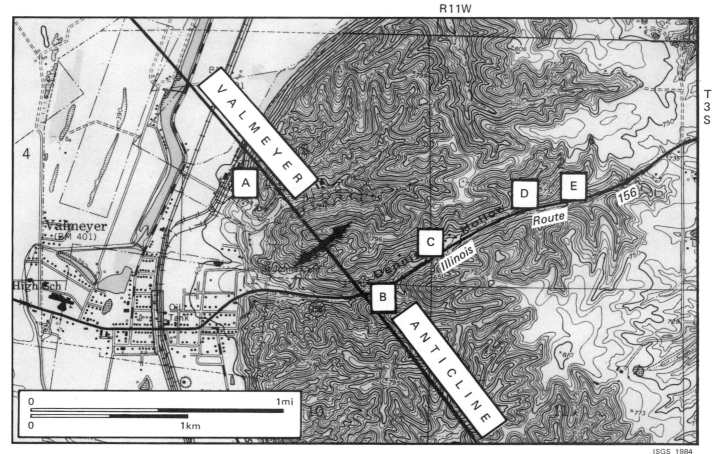

Figure 2. Location of portion of Valmeyer Anticline; circled letters indicate rock exposures mentioned in text (from Valmeyer 7½-minute Quadrangle).

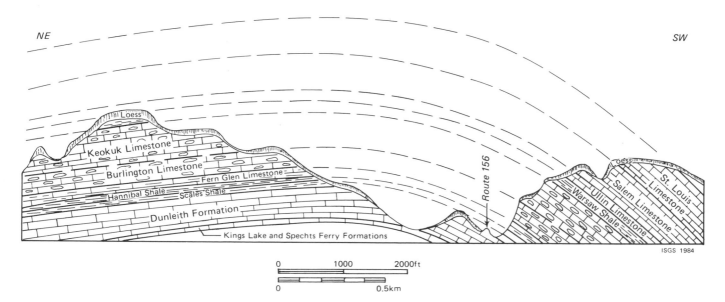

Figure 3. Schematic cross-section of Valmeyer Anticline looking southeast from vantage point 1 mi (1.6 km) north of Valmeyer. The vertical scale is exaggerated 5 times, which increases the apparent dip angles shown on the diagram. The southwest limb is further distorted as this part of the cross section actually is more north-south.

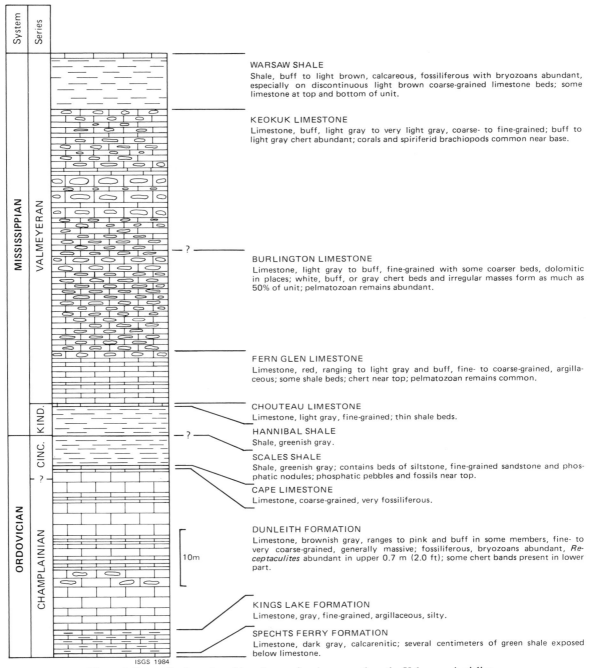

Figure 4. Generalized stratigraphic column of rocks exposed on the Valmeyer Anticline.

Ordovician rocks can be observed in the bluff face at Columbia Quarry Company Plant No. 3 at Valmeyer. Templeton and Willman (1963) described a section here (Fig. 2, exposure A) beginning at the base with less than 3 ft (1 m) of the Spechts Ferry Formation consisting of a dark gray calcarenite and some green shale (Fig. 4). This is overlain by 14 ft (4.2 m) of gray silty limestone belonging to the Kings Lake Formation. The Kings Lake is overlain by 94 ft (28.4 m) of light gray to pinkish gray calcarenite assigned to four members of the Dunleith Formation.

The non-cherty upper part of the Dunleith is the main unit being quarried by underground mining. For many years, some mined-out caverns were used extensively for growing mushrooms. Overlying the Dunleith is the Cape Formation, which consists of 1.5 ft (0.5 m) of coarse-grained limestone. About 12 ft (3.6 m) of greenish gray shale, assigned to the Scales Shale, completes the Ordovician rocks exposed on the bluff face near the crest of the anticline.

Along Dennis Hollow, the first main exposure (Fig. 2, exposure B) occurs about 0.5 mi (0.8 km) east of Valmeyer near the crest of the anticline. All the exposures farther east up Dennis Hollow occur on the gently dipping northeast flank of the anticline. This small outcrop shows about 3 ft (1 m) of Dunleith, 1 ft (0.3 m) of Cape, and 6 ft (2 m) or more of Scales. Workman (1949) indicated that 35 ft (10.6 m) of Maquoketa (Scales Shale) was present at this outcrop and succeeded by limestone of the Fern Glen Formation. Unfortunately this shale unit is not well exposed; however, conodonts (microfossils) indicate that at least the upper 10 ft (3.3 m) of this shale should be assigned to the Hannibal Shale of the Kinderhookian (lower Mississippian) Series rather than to the Scales Shale of the Cincinnatian (upper Ordovician) Series (Lane, 1978). Thus, somewhere in the middle of these 35 ft (10.6 m) of shale is a major unconformity representing all Silurian and Devonian time.

The Mississippian succession continues (Fig. 2, exposure C) with the Chouteau Limestone (uppermost formation of the Kinderhookian Series), which consists of 2 ft (0.7 m) of light gray fine-grained limestone overlying the greenish gray shales of the Hannibal (Fig. 4). The 30-ft (9-m) thick Fern Glen Formation conformably overlies the Chouteau and forms the base of the Valmeyeran Series here in Dennis Hollow. The concept of the middle Mississippian Valmeyeran Series (Osagean and Meramecian of other areas) was developed by Weller and Sutton (in Moore, 1933); the series was named for exposures along Dennis Hollow.

The Fern Glen consists primarily of limestone, much of which is argillaceous and reddish gray with red, green, or gray mottling. Red or gray shale is present with a few prominent chert bands, particularly at the top. The Fern Glen grades vertically into the overlying Burlington Limestone over a distance of 3 ft

(1 m) or more. The Fern Glen can usually be distinguished from the Burlington by being more argillaceous, less cherty, and red.

The Burlington is light gray, fine grained (coarser grained near the top) with olive, gray, and mottled buff chert beds and irregular masses of chert that comprise up to 50 percent of the formation. The Keokuk Limestone, which overlies the Burlington, has been differentiated on the basis of fossils (Collinson and others, 1981), but lithologically it is extremely difficult to subdivide these two highly cherty limestones. Therefore, they are generally referred to as the Burlington-Keokuk Limestone in this area (Lineback, 1981). The Keokuk portion contains more coarse-grained calcarenite than does the Burlington. Approximately 140 ft (42 m) of Burlington-Keokuk are present, if several partial sections are pieced together from exposure C to exposure D (Fig. 2). Driller's logs of two water wells less than 0.8 mi (1.3 km) away indicate that as much as 169 ft (51 m) may be present.

The Warsaw Shale is exposed in the last outcrop (Fig. 2, exposure E) near the top of the hollow 1.3 mi (2.1 km) east of Valmeyer. The Warsaw conformably overlies the Keokuk and consists of buff to light brown fossiliferous calcareous shale with a few limestone strata near the base. Approximately 30 ft (9 m) of Warsaw are present. The upper few feet of poorly exposed limestone and shale, which contain megafossils, probably should be assigned to the Ullin Limestone (Willman and others, 1975). Bryozoans dominate the Warsaw fauna and are particularly abundant near shale-limestone interfaces. Synder (1984) identified 35 genera and 88 species of bryozoans in the Warsaw at this locality. The higher formations of the Valmeyeran are present locally but tend to be covered by Illinoian till or Wisconsinan loess of Pleistocene age.

REFERENCES

Collinson, C., Baxter, J. W., Norby, R. D., Lane, H. R., and Brenckle, P. D., 1981, Mississippian stratotypes: Illinois State Geological Survey Field Guidebook in conjunction with the 15th Annual Meeting of the North-Central Section of the Geological Society of America, 56 p.

Cote, W. E., Reinertsen, D. L., Wilson, G. M., and Killey, M. M., 1970, Millstadt-Dupo area: Illinois State Geological Survey Geological Science Field Trip Guide Leaflet, 38 p.

Lane, H. R., 1978, The Burlington Shelf (Mississippian, north-central United States): Geologica et Palaeontologica, v. 12, p. 165–175.

Lineback, J. A., 1981, The eastern margin of the Burlington-Keokuk (Valmeyeran) carbonate bank in Illinois: Illinois State Geological Survey Circular 474, 23 p.

Moore, R. C., 1933, Historical Geology: New York, McGraw-Hill, 673 p.

Odum, I. E., Wilson, G. M., and Dow, G., 1961, Valmeyer area: Illinois State Geological Survey Geological Science Field Trip Leaflet 1961-F, 15 p.

Reinertsen, D. L., 1981, A guide to the geology of the Waterloo-Valmeyer area:

Illinois State Geological Survey Geological Science Field Trip Guide Leaflet 1981-D, 33 p.

Synder, E. M., 1984, Taxonomy, functional morphology and paleoecology of the Senestellidae and Polyporiidae (Senestelloidea, Bryozoa) of the Warsaw Formation (Valmeyeran, Mississippian) of the Mississippi Valley [Ph.D. dissertation]: University of Illinois at Urbana-Champaign, 802 p.

Templeton, J. S., and Willman, H. B., 1963, Champlainian Series (Middle Ordovician) in Illinois: Illinois State Geological Survey Bulletin 89, 260 p.

Treworgy, J. D., 1981, Structural features in Illinois—A compendium: Illinois State Geological Survey Circular 519, 22 p., 1 plate.

Weller, J. M., 1939, Mississippian System: Kansas Geological Society Guidebook, Thirteenth Annual Field Conference, p. 131–137.

Willman, H. B., Atherton, E., Buschback, T. C., Collinson, C., Frye, J. C., Hopkins, M. E., Lineback, J. A., and Simon, J. A., 1975, Handbook of Illinois Stratigraphy: Illinois State Geological Survey Bulletin 95, 261 p.

Workman, L. E., 1949, Dennis Hollow stratigraphic section, in American Association of Petroleum Geologists Field Conference Guidebook for 34th Annual Convention, p. 26–27.

Horseshoe Quarry, Shawneetown Fault Zone, Illinois

W. John Nelson, Illinois State Geological Survey, 615 East Peabody Drive, Champaign, Illinois 61820

LOCATION AND ACCESSIBILITY

The Horseshoe quarry is located in the SW¼,NW¼,NE¼, Sec.36,T.9S.,R.7E., Saline County; Rudement and Equality 7½-minute Quadrangle Illinois (Figs. 1, 2). From Equality, travel south 2.0 mi (3.2 km) via county road to the foot of Wildcat Hills, then west 2.5 mi (4 km) on gravel road to a point 0.2 mi (0.3 km) past the turnoff to Glen O. Jones Lake. The long-abandoned quarry is in low, wooded hills immediately north of the road, on land administered by the U.S. Forest Service.

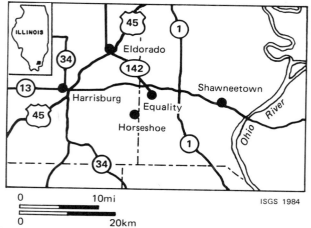

Figure 1. Regional location map for Horseshoe area, southern Illinois.

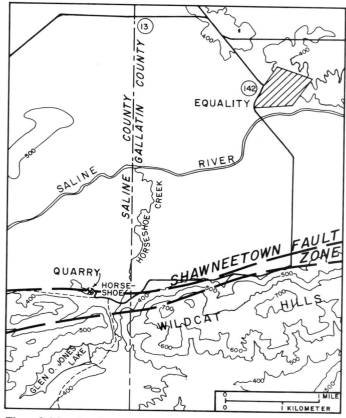

Figure 2. Map showing access route to Horseshoe Quarry from Equality.

SIGNIFICANCE

The unique structural style of one of the greatest fault systems in the midwestern states can be seen in this quarry. Exposed here are steeply tilted and deformed lower Mississippian and uppermost Devonian strata, upthrown approximately 3,500 ft (1,050 m) within the Shawneetown Fault Zone.

SITE INFORMATION

Regional setting. The Shawneetown Fault Zone is the western portion of the Rough Creek–Shawneetown Fault System, which extends 125 mi (200 km) eastward from a point in Pope County, Illinois, 13 mi (21 km) southwest of Horseshoe, to Grayson County in west-central Kentucky (Fig. 3). This fault system is part of the complex series of fractures that surrounds the head of the Mississippi Embayment. Its history is long and complex.

The ancestral Rough Creek–Shawneetown Fault System appeared in late Precambrian or early Cambrian time, forming the northern boundary of the Rough Creek Graben (Schwalb, 1982). The Rough Creek Graben is an east-trending structural trough, more than 100 mi (62 km) long and 30 to 50 mi (60 to 80 km) wide, in southern Illinois and western Kentucky. The western end of the Rough Creek Graben joins the northern end of another buried rift complex, the Reelfoot Rift, which underlies the present Mississippi Embayment. Deep boreholes and seismic data show that the Reelfoot and Rough Creek troughs contain several thousand ft (m) of lower(?) and middle Cambrian sediments, which are absent outside the grabens. Upper Cambrian strata, and to a lesser degree the Canadian (lower Ordovician) beds, are thicker in the rifts than elsewhere; they represent deep-water facies as contrasted with shallow shelf or carbonate-bank facies found outside the troughs. These findings indicate that the Rough Creek Graben and Reelfoot Rift were largely buried in sediment by early Ordovician time.

These ancient rift complexes have been interpreted as aulacogens produced by the massive breakup of continental plates, an occurrence that is believed to have happened near the end of Precambrian time (Braile and others, 1984).

The Rough Creek–Shawneetown Fault System was reactivated periodically throughout the Paleozoic, most notably in late Devonian time (Schwalb, 1982). All exposed bedrock faults, however, developed in post-Pennsylvanian to pre-late Cretaceous

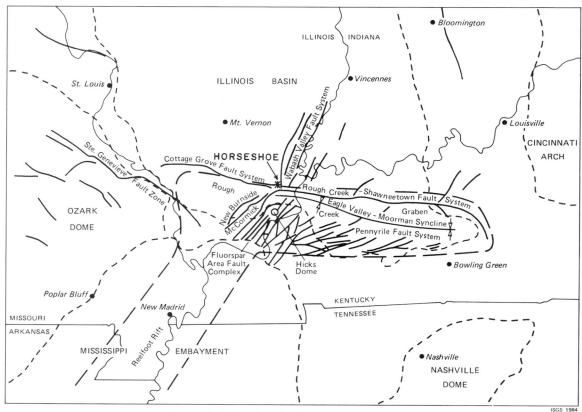

Figure 3. Regional setting of Horseshoe Quarry.

time. It seems likely that these faults developed in Permian time, coincident with the Appalachian orogeny (Nelson and Lumm, 1984).

The complex structure of the Rough Creek–Shawneetown Fault System has inspired several conflicting theories explaining its origin. Some geologists, notably Heyl (1972), advocate strike-slip faulting; Weller (1940); Smith and Palmer (1981); and others favor thrusting caused by horizontal compression from the south. Major wrench faulting can be ruled out by the presence of paleo-channels that cross the fault zone without lateral offset; horizontal compression also seems unlikely because of the high angle of most faults in the system (Nelson and Lumm, 1984).

On the basis of surface mapping, well data, and limited geophysical evidence, I conclude that the major fault movements were vertical. The peculiar structure of the fault zone seen at the Horseshoe Quarry demands at least two separate, very different episodes of dip-slip faulting.

The first post-Pennsylvanian action in the Rough Creek–Shawneetown Fault System was uplift of the southern block (Fig. 3), when the ancestral buried faults of the Rough Creek Graben were reactivated at depth. The fracture propagated upward through the sedimentary cover as a reverse fault steeply inclined to the south. Vertical displacement reached a maximum of at least 3,500 ft (1,067 m) in the vicinity of Horseshoe, diminishing eastward along the fault zone.

Later, as the southern block began to drop back down toward its original position (Fig. 4), the rocks overhanging the main reverse fault became so tightly wedged in place that they could not slip back down the fault surface; this produced very strong drag along the south side of the fault zone. The resulting steep fold is now the northern limb of the Eagle Valley–Moorman Syncline, which parallels the entire length of the Rough Creek–Shawneetown Fault System. As the southern block continued to drop, slices of rock were sheared off and became wedged high in the fault zone. These narrow slivers of old rock trapped between younger rock are a characteristic feature of the Rough Creek–Shawneetown Fault System. The most spectacular example of such a slice is visible in the Horseshoe Quarry.

Features of the Quarry. The rocks exposed in the Horseshoe Quarry are the oldest found at the surface anywhere in the fault zone. The stone that was quarried here belongs to the Fort Payne Formation, of lower Valmeyeran (Mississippian) age (Fig. 5). The Fort Payne is composed of thin-bedded, very silty, highly silicified dark limestone, interbedded with calcareous siltstone and dark gray siliceous shale. Deposited in a relatively deep oceanic basin that extended from Alabama to southern Illinois, it overlaps the Springville Shale and Borden Siltstone, which are bottomset and foreset beds, respectively, of the Borden Delta (Lineback, 1966). The Borden Delta represents the southwestern portion of the great Catskill-Pocono deltaic complex, which

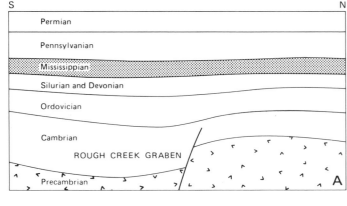

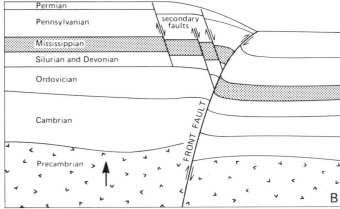

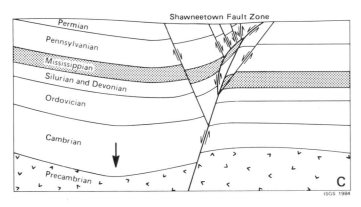

Figure 4. Proposed origin of Rough Creek-Shawneetown Fault System. A, Prior to faulting, in Permian time, the basement is broken by the northern fault of the Rough Creek Graben, which had developed in Cambrian time. B, The southern block is uplifted, reactivating the Cambrian fault as an upthrust. Secondary faults develop under horizontal tension near the top of the upthrown block. The raised block presumably is subjected to erosion. C, The southern block drops back to approximately its original elevation. The overhanging portion of the southern block is wedged in position and held up by friction, so it cannot drop. Flexure and tilting of the sedimentary strata produce the steep northern limb of the Eagle Valley–Moorman Syncline, while narrow slices of rock are sheared off within the fault zone and trapped there. Secondary faults undergo renewed movement, and many additional faults are created.

was derived from erosion of lands uplifted during the Acadian orogeny, and prograded westward in late Devonian and early to middle Mississippian time.

The Springville Shale is reported to have been visible in the quarry in the past (Willman and others, 1975), but cannot be seen today. This shale would be expected to occur along the deep, narrow trenchlike pit that cuts through the hill just north of the Fort Payne exposures. A small amount of vertically dipping New Albany Shale can be observed on the north wall of this trench. The New Albany is part of the highly extensive body of Upper Devonian black shale (including the Chattanooga, Ohio, and Sunbury) found in the eastern and central United States. The lithology seen on the pit wall is typical: black, thinly laminated, fissile, phosphatic, highly organic shale, the product of slow sedimentation in a deep, restricted, anaerobic marine basin (Cluff and others, 1981).

All the rocks in the pit dip steeply. Bedding in the Fort Payne is tilted 42° to 76° south, while New Albany bedding dips 80°S to 80°N (overturned). The strike of the beds, which varies only slightly from an average of N 80°W, is parallel to the large faults in the vicinity. This parallelism of bedding strike with adjacent faults, typical in the Rough Creek–Shawneetown Fault System, is evidence that major movements were dip-slip, not strike-slip. Wrench faulting typically produces folds and tilted upthrown slices that strike obliquely, rather than parallel to, the master fault.

All of the rocks at Horseshoe are pervasively fractured. Three or more sets of fractures commonly are present in a single outcrop. No large faults have been identified, but several small faults or sharp flexures have been observed in the Fort Payne Formation. The best place to look for small structures is among the partially excavated upturned ledges on top of the hill near the east end of the quarry. High-angle reverse faults that have a few inches of displacement and die out upward into tight flexures can be observed there. These features, miniatures of the major faults in the Shawneetown Fault Zone, illustrate its structural style of vertical upthrusting.

Structure adjacent to quarry. The Mississippian and Devonian strata at Horseshoe form an isolated upthrown fault slice, as described previously (Fig. 6). One east-striking fault passes immediately north of the quarry and separates it from the northwestern part of the hill, where the farmhouse stands. Lower Pennsylvanian conglomeratic sandstone crops out on the hillside about 300 ft (91 m) northwest of the house. Another fault separates the sandstone from the middle Pennsylvanian strata that have been revealed by drilling beneath the Saline River bottomlands north of the hill. Relative to the conglomeratic sandstone, the New Albany Shale is upthrown approximately 3,500 ft (1,067 m). This displacement is the greatest known anywhere in the fault system, although an offset of similar magnitude was indicated by the log of a dry oil testhole drilled in 1982 about 0.5 mi (0.8 km) west of Horseshoe. Lower Devonian chert was found in fault contact above lower Pennsylvanian sandstone, shale, and coal in that well.

Figure 5. Steeply dipping Fort Payne Formation in Horseshoe Quarry.

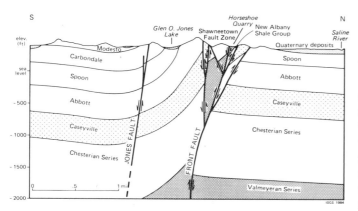

Figure 6. Geologic cross section through Horseshoe Quarry.

The prominent ridge of Cave and Wildcat Hills, south of Horseshoe, is composed of highly resistant lower Pennsylvanian sandstone south of the Shawneetown Fault Zone. The strata in the hills dip southward, forming the north limb of the Eagle Valley Syncline. Dipping beds can be observed alongside the road that passes through Horseshoe Gap between the quarry and Glen O. Jones Lake. A fault cuts the west wall of the gap near its north end. Steeply dipping, shattered conglomeratic sandstone of the Caseyville (lower Pennsylvanian) Formation north of this fault is juxtaposed with gently dipping younger Pennsylvanian sandstone on the south. Yet another fault, of large displacement, is concealed beneath the alluvium between Horseshoe Gap and the Horseshoe Quarry.

To summarize, the structure at Horseshoe testifies to recurrent post-Pennsylvanian movement on the Rough Creek–Shawneetown Fault System. First, the southern block was raised approximately 3,500 ft (2,000 m), placing New Albany Shale of the southern block against middle Pennsylvanian rocks of the northern block. Then the southern block began to drop back down, shearing off a wedge of Fort Payne–New Albany within the fault zone. Other slices of rock were cut off in the same way. The major part of the southern block slid down past the isolated fault slivers, which were jammed in place. At the same time, strong drag on the downward-slipping block rotated the fault slices and formed the north limb of the Eagle Valley Syncline. The southern block eventually returned to nearly its original position; coal beds along the axis of the Eagle Valley Syncline lie at nearly the same elevation as the same coals north of the Shawneetown Fault Zone. The northern block was relatively passive throughout all this activity. Fault slices such as the one at Horseshoe, remain more than 3,000 ft (1,000 m) above their starting points.

REFERENCES CITED

Braile, L. W., Hinze, W. J., Sexton, J. L., Keller, G. R., and Lidiak, E. G., 1984, Tectonic development of the New Madrid Seismic Zone, in Gori, P. L. and Hays, W. W. (editors); Proceedings of the Symposium on The New Madrid Seismic Zone: U.S. Geological Survey Open-File Report 84–770, p. 204–233..

Cluff, R. M., Reinbold, M. L., and Lineback, J. A., 1981, The New Albany Shale Group of Illinois: Illinois State Geological Survey Circular 528, 84 p.

Heyl, A. V., Jr., 1972, The 38th Parallel Lineament and its relationship to ore deposits: Economic Geology, v. 67, p. 879–894.

Lineback, J. A., 1966, Deep-water sediments adjacent to the Borden siltstone (Mississippian) Delta in southern Ilinois: Illinois State Geological Survey Circular 401, 48 p.

Nelson, W. J., and Lumm, D. K., 1984, Structural geology of southeastern Illinois and vicinity: Illinois State Geological Survey Contract/Grant Report 1984-

2, 127 p.

Schwalb, H. R., 1982, Paleozoic geology of the New Madrid area: U.S. Nuclear Regulatory Commission, NUREG CR2909, 61 p.

Smith, A. E., and Palmer, J. E., 1981, Geology and petroleum occurrences in the Rough Creek Fault Zone; Some new ideas, in Luther, M. K., ed., Proceedings of the Technical Sessions, Kentucky Oil and Gas Association, 38th Annual Meeting, June 6-7, 1974: Kentucky Geological Survey, Series IV, Special Publication 4, p. 45–59.

Weller, J. M., 1940, Geology and oil possibilities of extreme southern Illinois, Union, Johnson, Pope, Hardin, Alexander, Pulaski, and Massac Counties: Illinois State Geological Survey Report of Investigations 71, 71 p.

Willman, H. B., and others, 1967, Geologic map of Illinois: Illinois State Geological Survey, scale 1:500,000.

Geologic features near Grand Tower, Illinois: The Devil's Backbone, The Devil's Bake Oven, and Fountain Bluff

George H. Fraunfelter, Department of Geology, Southern Illinois University at Carbondale, Carbondale, Illinois 62901

LOCATION AND ACCESSIBILITY

The Devil's Backbone, the Devil's Bake Oven, and Fountain Bluff are located in extreme southwestern Illinois in Jackson County, adjacent to the Mississippi River (Fig. 1). All three sites are shown along the southeastern edge of the Altenburg 7½-minute Quadrangle; the remainder of Fountain Bluff is shown on the Gorham 7½-minute Quadrangle. Locality 1, the Devil's Backbone, and Locality 2, the Devil's Bake Oven, are situated in Grant Tower City Park on the northwest side of town; Locality 3, Fountain Bluff, is about 1 mi (1.6 km) north of town. All sites are readily accessible via car and short walks.

Locality 1. The Devil's Backbone: abandoned quarry on southeast end of ridge, SE¼SW¼SW¼Sec.24,T.10S.,R.4W., 3rdP.M. (Fig. 2). Locality 2. The Devil's Bake Oven: center of S½NE¼Sec.23,T.10S.,R.4W.,3rdP.M. Locality 3. Draw situated along the southwestern side of Fountain Bluff, SE¼NW¼-Sec.13,T.10S.,R.4W.,3rdP.M. Access to the latter site may be gained by asking the owner, who lives in the white frame house just across the road to the south of the draw, and less than 0.25 mi (0.4 km) northwest of the T-road intersection near the center of the NE¼SW¼Sec.13.

SIGNIFICANCE

Bedrock strata ranging in age from Early Devonian to Early Pennsylvanian are exposed in outcrops and quarries on either side of the Ste. Genevieve Fault Zone, where stratigraphic displacements of as much as 800 ft (250 m) have been recognized. Type sections of the Lower Devonian Backbone Limestone and the Middle Devonian Grand Tower Limestone are in the Devil's Backbone and the Devil's Bake Oven, respectively. Upper Mississippian and lowermost Pennsylvanian rocks crop out in Fountain Bluff. Once part of the western bluffs of the ancient Mississippi River, these erosional remnants were left when the river shifted its course westward because of Pleistocene glaciations and eroded a new channel.

SITE INFORMATION

The Ste. Genevieve Fault Zone—also known as the Rattlesnake Ferry Fault Zone (Treworgy, 1981)—crosses the Mississippi River as two large faults that cut through the southern end of the Devil's Backbone, nearly encircling the Devil's Bake Oven and the Devil's Backbone. The two faults join in the valley east of the Devil's Backbone and cut through Walker Hill to the east (Nelson and Lumm, 1985) (Fig. 2). The fault zone trends generally southeasterly, crossing from the Ozarks in Missouri, through Grand Tower and Alto Pass to Mountain Glen (Cote and others, 1965) (Fig. 1). As a result of this faulting, Devonian rocks in the

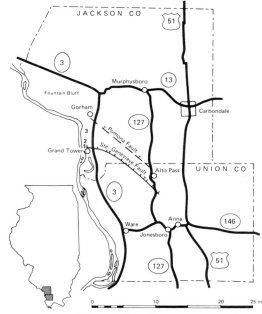

Figure 1. Locality map.

Devil's Bake Oven and in the Devil's Backbone are upthrown and dip about 24° to the northeast into the fault.

Also of interest is Tower Rock, an outlier of Lower Devonian Bailey Limestone Rock located directly west of the village of Grand Tower in the Mississippi River near the Missouri shore. The oldest Devonian unit present in the area, the Bailey, underlies the rock units exposed in the Devil's Backbone (Figs. 2 and 3).

Locality 1 (Fig. 2). The type section of the Lower Devonian Backbone Limestone (Savage, 1920) is in an abandoned quarry at the southeastern end of the Devil's Backbone where only the upper 38 ft (12 m) is exposed. Although the base of the formation is covered here, the upper contact with the Lower Devonian Clear Creek Chert is well exposed.

The Devil's Backbone is about 100 ft (30 m) thick in the outcrop area in Jackson and Union Counties (Fig. 1), but is 200 ft (60 m) thick in some places. It occurs in southern Illinois and Indiana, and is stratigraphically equivalent to the Little Saline Limestone in southeastern Missouri.

The Backbone is a white to light gray, fine- to coarse-grained, largely thick-bedded to massive, somewhat cherty limestone. Lange (1983) identified the following microfacies at the type section:

Depth, ft (m)

Top. Bryozoan, crinoidal wackestone: contains coarse, calcarenitic bioclasts in a very finely crystalline matrix. Minor skeletal

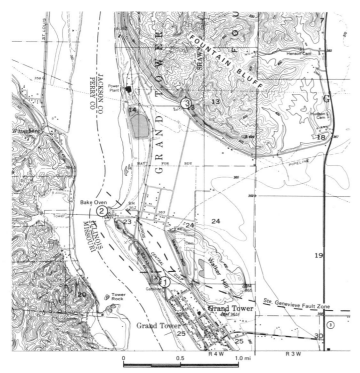

Figure 2. Map showing Ste. Genevieve Fault Zone, Tower Rock, and Localities 1, 2, and 3.

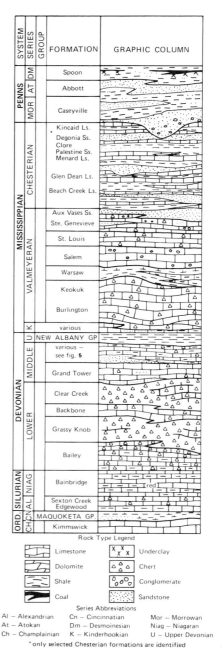

Figure 3. Generalized stratigraphic section along Ste. Genevieve Fault Zone (modified from Nelson and Lumm, 1985).

constituents include brachiopod, trilobite, and *Asphaltina*? fragments.

3.5 (1.1). Cements, primarily intragranular bladed, and fine equant spar. Syntaxial replacement spar, microspar, and dolomite common. Minor authigenic silica.

2.8 (0.8). Crinoidal-Bryozoan packstone: contains coarse calcarenite to fine calcirudite-sized bioclasts. Cement is primarily syntaxial echinoderm rim with fewer coarse- and fine-bladed types. Moderately well sorted with minor microcrystalline calcite and pseudospar. Fibrous skeletal grain recrystallization. Some microstylolites.

15.7 (4.8). Bryozoan, crinoidal wackestone: contains fine calcarenitic to fine calcirudite-size bioclasts in a fine to very finely crystalline micrite matrix. Minor skeletal constituents include brachiopod and *Asphaltina*? fragments. Cement restricted to fine-bladed and granular or intragranular calcite. Neomorphic calcite very common, dolomitic near base, traces of authigenic quartz and chert. Microstylolites are common.

4.5 (1.4). Bryozoan, crinoidal wackestone: contains less than 25 percent total coarse calcarenitic-sized bioclasts in a finely *Asphaltina*? micrite matrix. Brachiopod, trilobite, and crystalline fragments are minor skeletal constituents. Cement is restricted to intragranular areas with some coarse, blocky, second-generation calcite in whole fossil brachiopods. Micrite, commonly neomorphosed. Some fibrous, pseudospar, and syntaxial crinoid replacement rims. Microstylolites and geopetal fabrics present. Minor dolomite, authigenic quartz, and chert.

5.4 (1.6). Bryozoan, crinoidal wackestone: contains fine calcirudite-sized bioclasts in a finely crystalline micrite matrix. Minor trilobite and *Asphaltina*? fragments. Minor intragranular cements. Well-developed microstylolites, dolomitic at bottom. Neomorphic spar is common as matrix rims.

4.9 (1.5). Crinoidal-Bryozoan packstone to grainstone: this zone contains some grainstone and is clearly transitional, going from moderate to well sorted with an increase in calcite cement mainly present as syntaxial echinoderm rims. Bioclasts are slightly larger in the grainstone. Some microstylolites, dolomite, and authigenic silica absent. Recrystallization of some bioclasts appears to be postcementation.

Bottom. Well-preserved, mostly articulate brachiopods are abundant in the lower 4 to 10 ft (1.2 to 3 m) of the formation, whereas echinoderms and bryozoans, mostly worn and broken, dominate the remainder of the formation. The common Backbone fossils are shown in Figure 4.

The composition of the fauna and the accumulations of coarser sediments that punctuate the generally fine-grained sediments suggest that the Backbone Limestone was deposited on an open marine, subtidal shelf (landward of the deeper parts of the Illinois Basin) with shoals at wave base.

The Clear Creek Chert (Worthen, 1866) is exposed in an abandoned quarry at the south end of the Backbone ridge. The Clear Creek is at least 300 ft (91 m) thick in the outcrop area in southern Illinois; it consists mostly of white chert, but also contains medium to light gray, very fine-grained, siliceous limestone. In many places the chert weathers to a rusty red color. For the most part, the Clear Creek conformably overlies the Backbone Limestone; this contact is well exposed at the type section of the Backbone Limestone (Fig. 2). The upper part of the Clear Creek Chert is highly fossiliferous in places. The fauna is dominated by brachiopods (Savage, 1920a). The index fossil for the formation, the brachiopod *Amphigenia* (Fig. 4), has been found in the dark, purplish gray limestones of the uppermost Clear Creek Formation at the base of the Devil's Bake Oven near the bottom of the ferry ramp. Here the Clear Creek is unconformably overlain by the Grand Tower Limestone.

Locality 2 (Fig. 2). The type section of the Middle Devonian Grand Tower Limestone (Keyes, 1894) is in the Devil's Bake Oven, where the entire thickness of the formation, 157 ft (48 m), is exposed. Figure 5 shows a detailed columnar section of the Grant Tower Limestone (also see North, 1969). The only other thick exposure of the Grant Tower Limestone in the area is at Quarry Hill in Ste. Genevieve County, Missouri, near Ozora. The Grand Tower is stratigraphically equivalent to the Jeffersonville Limestone of the Ohio Valley. T. E. Savage (1920a) described some of the abundant fossils found in the Grand Tower (Fig. 4).

In the Wittenberg Trough area in eastern Missouri (Meents and Swann, 1965), carbonate deposition was continuous from the underlying Little Saline Limestone (Lower Devonian) into the Grand Tower. In the region to the east, in the Illinois Basin and adjacent areas, the Grand Tower commonly contains a basal sandstone, the Dutch Creek Member (T. E. Savage, 1920). This member is as much as 30 ft (9 m) thick where it outcrops in southern Jackson, Union, and northern Alexander counties in southern Illinois, but averages about half that thickness (Fig. 5). The Dutch Creek consists for the most part of fine- to medium-grained, white to yellowish or reddish brown, iron-stained, medium- to thick-bedded, quartz sandstone. The abundant fossils, mostly brachiopods and corals, generally are preserved as molds

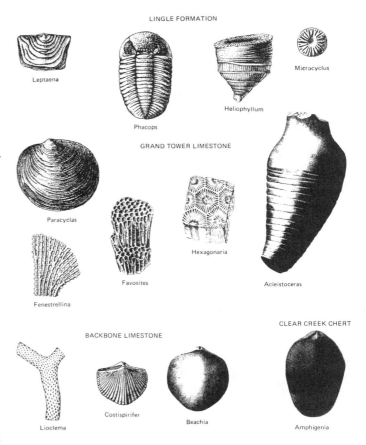

Figure 4. Common Devonian fossils from the Lingle Formation, Grand Tower Limestone, Clear Creek Chert, and Backbone Limestone (from Savage, 1920a).

(Savage, 1910, 1920a). On the basis of its distribution and thickness, and its gradational contact with the overlying carbonates, the Dutch Creek Sandstone Member probably can be interpreted to be sheet sands initially deposited at river mouths and later spread widely by longshore currents, filling lows in the underlying surface.

Above the basal sandstone member, the Grand Tower consists almost exclusively of carbonates and contains only a small amount of clastics. Throughout the region and at various stratigraphic levels in the Grand Tower Limestone, patch reefs or bioherms grew on the edge of the Sparta Shelf and in the shallow parts of the Illinois Basin (Meents and Swann, 1965). The largest constituent of these patch reefs is the colonial coral *Hexagonaria*. The carbonates in the Grand Tower are fine-grained (biomicrite) except at or near the patch reefs, where they are coarser grained (biomicrudite).

The Grand Tower Limestone is unconformably overlain by the Middle Devonian Lingle Formation (Savage, 1920). The Lingle, 50 ft (15 m) thick at its type section, is the approximate equivalent of the St. Laurent Limestone in Missouri. It crops out in Jackson, Union, and Alexander counties in southwestern Illi-

UNIT 19 — Limestone, brownish-gray, argillaceous, thin-bedded. Ranges from biomicrite in lower 20 feet (L1) to biopelmicrite at top (L2, L3). Contains abundant brachiopods, solitary corals, and trilobites.

UNIT 18 — Limestone, brownish-gray, lithographic, massive, lower 4 feet cherty, brachiopod-ostracode-crinoid biomicrite, burrowed.

UNIT 17 — Limestone, brownish-gray, lithographic, massive, upper 9 inches thin-bedded, ostracode-brachiopod-crinoid biomicrite, burrowed.

UNIT 16 — Limestone, bluish-gray, argillaceous, massive, brachiopod-gastropod-ostracode biomicrite.

UNIT 15 — Limestone, pinkish-brown, lithographic, thin-bedded with shaly partings, upper 2.5 feet cherty, brachiopod-ostracode biomicrite with "fish" fragments, some voids filled with sparry calcite, burrowed.

UNIT 14 — Limestone, light gray, very fine-grained, crinoid-brachiopod-ostracode biomicrite, burrowed.

UNIT 13 — Limestone, brownish-gray, crinoid-brachiopod biomicrite in upper part to crinoid-brachiopod biomicrudite with many forams in lower part. Upper part lighter in color than lower part. Ostracodes, *Tentaculites*, gastropods, and corals are also present. Some sparry calcite occurs as crack and pore fillings.

UNIT 12 — Limestone, brownish-gray, crinoid-brachiopod biomicrite, with some sparry calcite in cracks, thin-bedded, some chert at base.

UNIT 11 — Limestone, brownish-gray, cherty, brachiopod-crinoid-ostracode biomicrite.

UNIT 10 — Limestone, brownish-gray, lithographic, brachiopod, crinoid-ostracode biomicrite, conglomeratic, phosphatic, with buff-gray chert at base.

UNIT 9 — Limestone, brownish-gray, very fine-grained, calcarenitic, brachiopod-crinoid-ostracode biomicrite, burrowed.

UNIT 8 — Limestone, dark brownish-gray, lithographic, crinoid-brachiopod biomicrite, thin-bedded, with tan chert at base.

UNIT 7 — Limestone, brownish-gray, thick-bedded, crinoid-brachiopod biomicrite with some bryozoans present. Bioturbate.

UNIT 6 — Limestone, brownish-gray, clayey and silty, crinoid-brachiopod biomicrite with some bryozoans present. Some sparry calcite cement present. Prominent chert band forms reentrant in Bake Oven.

UNIT 5 — Limestone, brownish-gray, massive, crinoid-brachiopod-gastropod biomicrudite with some *Tentaculites* and bryozoans present. Brachiopods are both punctate and impunctate.

UNIT 4 — Limestone, brownish-gray, crinoid-brachiopod biomicrudite with some bryozoans present, mottled. Brachiopods punctate and impunctate.

UNIT 3 — Limestone, crinoid-brachiopod biomicrite in the upper part. Brachiopods are both punctate and impunctate and some bryozoans occur also. The lower part consists of biosparite to biomicrite (crinoid-brachiopod) with some biomicrudite present. Numerous "fucoids," brachiopods, and corals (*Hexagonaria*) are present.

UNIT 2 — Limestone, light gray, medium- to coarse-grained, massive, crinoid-brachiopod biosparite in upper part, crinoid-brachiopod biomicrite with considerable sparry calcite in the middle, and a crinoid-brachiopod biosparudite with some micrite in the lower part. Lower part arenaceous, strongly cross-bedded in lower half, indistinctly crossbedded in upper half, with ripple marks and wavy beds. Silty and clayey at base.

UNIT 1 — Limestone, light gray, medium- to coarse-grained, arenaceous, argillaceous, crinoid-brachiopod biosparudite that is somewhat micritic with interbedded light gray, calcareous sandstones.

Figure 5. Generalized columnar section at the Devil's Bake Oven and Devil's Backbone north of Grand Tower, Illinois (modified from Collison in Meents and Swann, 1965).

nois. Lingle faunas are listed by Fraunfelter (1977); some representative fossils are shown in Figure 4.

Only the lower part of the Lingle Limestone is present near Grand Tower. The lower 20 ft (6 m) of the Lingle (Fig. 5; unit 19, L1), in the Bake Oven and Backbone sections, consists of brownish gray, fine-grained, thin-bedded, argillaceous limestone (brachiopod-crinoid bioplemicrite) that contains *Tentaculites*. Units L2 and L3, which occupy the upper 6 ft (1.8 m) exposed at Locality 2, contain abundant specimens of the brachiopods *Devonochonetes* (Fig. 4) and *Tripidoleptus*, in addition to some solitary corals. The brachiopods present have a restricted size range and therefore were probably winnowed, although they are mostly intact. Norton (1966) and Nance (1968) discuss the depositional environments of the Lingle Formation.

Locality 3 (Fig. 2). Fountain Bluff is an outlier of the Mississippi River bluff cut off by the river during the Pleistocene.

The Chesterian and lower Pennsylvanian rocks, which protect Fountain Bluff from erosion, are well exposed along the south and southwestern side of the bluff. The beds in the bluff show a regional dip of less than 5° to the northeast. The early Pennsylvanian-age Pomona Fault (Fig. 1), which trends northwest-southeast, passes close to the bluff on the north and is downthrown to the northeast. Nelson (personal communication, 1986) described the area, "To the east of Locality 3, the Menard Limestone forms ledges north of the house in the SE¼SE¼Sec.13, where at least 70 ft (21 m) are exposed. Here it dips steeply as much as 26° northwest, but the dip flattens rapidly to the north. The Palestine Sandstone overlies the Menard near the south end of the bluff. It consists of very fine-grained, thin-bedded sandstone, siltstone, and shale, but is poorly exposed here."

The Clore Formation (Weller, 1913), which occurs at the base of the bluff at Locality 3, properly consists of dark gray to greenish gray, calcareous shale and lesser amounts of generally fine-grained, dark bluish gray limestone. The Clore has a thickness of about 80 ft (25 m) in this area; it crops out from south of Chester to Cora in the Mississippi River bluff along Illinois 3 in the type region, as well as in many other places along the southern rim of the Illinois Basin. The section of the Clore exposed at Locality 3 was described by Pickard (1963), but incorrectly assigned to the Menard.

Clore Formation

	Thickness	
	ft	(m)
12. Shale, black, carbonaceous, high-sulphur content, locally coal	0.75	(0.2)
11. Limestone, thin-bedded, brown, slightly fossiliferous, fragmental	1.5	(0.5)
10. Shale, black, fissile	3	(0.9)
9. Limestone, massive, brown, finely crystalline, few fossils	4.5	(1.4)
Covered	10	(3.0)
8. Limestone, massive, gray, finely crystalline, densely fossiliferous, upper few feet shaly, closely jointed	16	(4.9)
Covered	8	(2.4)
7. Shale, gray, fissile	3	(0.9)
6. Limestone, gray, sandy, very fossiliferous	3	(0.9)
5. Shale, gray	1	(0.3)
4. Limestone, thin-bedded, gray, shaly, very fossiliferous	4	(1.2)
3. Shale, greenish gray, fissile	2	(0.6)
2. Shale, red	0.5	(0.1)
1. Shale, black to green, carbonaceous	3	(1.0)

The Clore is highly fossiliferous in many places; it contains such typical species as *Anthracospirifer increbescens, Composita subquadrata,* and *Pentremites sulcatus* (Fig. 6). The Clore was deposited in a shallow marine shelf environment.

The Degonia Sandstone (Weller, 1920), which conformably overlies the Clore Formation, reaches a maximum thickness of 90 ft (27 m) in this area. It is thickest where channel-phases sandstone units are best developed. The Degonia Sandstone is a clastic unit consisting mostly of white to gray, fine- to medium-grained, thin-bedded to massive sandstones, interbedded with dark gray, silty to sandy shales, and dark gray siltstones. The Degonia is well exposed in a number of places along the southwestern and southern rim of the Illinois Basin from the Chester area on the Mississippi River to Hardin County in southeastern Illinois. A section of the Degonia measured at Locality 3 by Pickard (1963) as Palestine Sandstone, follows:

Degonia Sandstone

	Thickness	
	ft	(m)
17. Sandstone, thin-bedded, white to buff, fine-grained, crossbedded	3	(1.0)
16. Sandstone, thin-bedded, white to buff, fine-grained, iron-stained	8	(2.4)
15. Sandstone, massive, fine- to medium-grained, buff, slightly friable	55	(16.8)

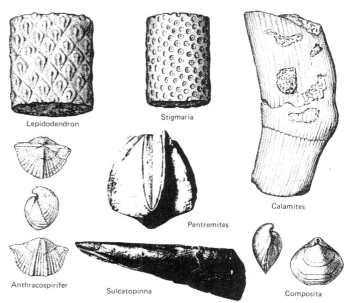

Figure 6. Common Chesterian invertebrate fossils and lower Pennsylvanian plant fossils from the Grand Tower/Fountain Bluff area (from Collinson, 1973; Collinson and Skartvedt, 1960; and Weller, 1920).

	Thickness	
14. Sandstone, thin-bedded, white, fine-grained	2	(0.6)
13. Sandstone, thin-bedded to massive, brown to buff, fine- to medium-grained, clay streaks, iron-stained	11	(3.3)

The Degonia contains numerous plant fossils in many locations. Most commonly preserved are the trunk parts of the fossil tree *Lepidodendron* (Fig. 6). Some beds contain an abundance of trace fossils, mostly horizontal filled burrows. The sediments and sedimentary structures present indicate that the Degonia was deposited under fluvial-deltaic conditions in nonmarine and shallow marine environments.

Nelson (personal communication, 1986) found Kinkaid Limestone (Fig. 3) in Trestle Hollow (NW¼Sec.12) a short distance north, but pre-Pennsylvanian erosion has removed it to the south. Here in Section 13 the basal Pennsylvanian sandstone lies directly on the Degonia Sandstone; the contact is identified by a sudden increase in grain size.

The Caseyville Formation (Owen, 1856) consists mostly of quartzose to slightly micaceous or clayey sandstones and interbedded siltstones and shales. The sandstones, medium- to coarse-grained and medium-bedded to massive, are usually white, weathering to yellowish brown. The bluish gray shales and siltstones occur as thin to thick beds. The unit forms a prominent but intermittent bluff that parallels the rim of the Illinois Basin in southern Illinois.

In Fountain Bluff, the two lowest members of the Caseyville Formation, the Wayside Sandstone and the Battery Rock Sandstone, have been described and measured; however, the rocks here are not typical of the Wayside. The Caseyville is characterized by conglomerate beds, at the base and higher in the formation, that consist of well-rounded, white quartz pebbles, clay pebbles and/or chert pebbles. A basal conglomerate found in the Caseyville but known only at several places on the southwest side of Fountain Bluff is composed of angular chert clasts up to several inches in diameter. Poor (1925) identified Silurian, Devonian, and Mississippian fossils in these chert clasts and suggested that they were eroded from the Ozark Uplift to the southwest. Nelson and Lumm (1985) believe that the southwest side of the Ste. Genevieve Fault Zone was uplifted at the end of Mississippian time, forming a high scarp from which the angular clasts were derived.

In this area the thickness of the Caseyville varies from 50 to 135 ft (15 to 41 m). The Mississippian/Pennsylvanian unconformity is well exposed along the west side of Fountain Bluff. The section above the waterfall at Locality 3 was measured by Pickard in 1963:

Caseyville Formation
Wayside and Battery Rock Sandstone Members

	Thickness	
	ft	(m)
13. Sandstone, massive, coarse- to medium-grained, buff to yellow, friable; shale pebbles, gray upper 6 ft (1.8 m)	33	(10.1)
12. Sandstone, as above; no shale pebbles	15	(4.6)
11. Shale, gray, fissle, silty	2	(0.6)
10. Sandstone, massive, fine- to medium-grained, buff to yellow, friable, shale pebbles	6	(1.8)
9. Conglomerate, massive, medium- to coarse-grained, rounded, quartz and shale pebbles, chert fragments, limonite and hematite cement, sandstone matrix	3	(1.0)
8. Sandstone, massive, medium-grained, brown to dark brown, iron-stained	3	(1.0)
7. Sandstone, massive, fine-grained, buff, friable	21	(6.4)
6. Sandstone, thin-bedded, fine-grained, buff to white, locally shaly, joint trends N70E	5	(1.5)
5. Sandstone, as above	11	(4.3)
4. Sandstone, medium-bedded, medium-grained, yellow, shale pebbles	6	(1.8)
Covered	5	(1.5)
3. Sandstone, thin-bedded, white to buff, calcerous, shale pebbles	3	(1.0)
2. Limestone, thinly and irregularly bedded, white to buff, sandy	2	(0.6)
1. Sandstone, medium- to coarse-grained, buff, shaly	9	(2.7)

Many of the basal channel-fill beds contain plant fossils, especially *Lepidodenron* and *Calamites* (Fig. 6).

Stratified sediments of the mid-Wisconsinan (Pleistocene) valley train are exposed in a borrow pit at the northwest corner of the intersection on Illinois 3 and the road leading west along the south side of Fountain Bluff (Fig. 2). The stratified sediments are overlain by Peorian loess. Here the valley train was 40 ft (12 m) higher than the present adjacent flood plain of the Mississippi River (Stan Harris, personal communication). The exposure, one of the few in this part of the Mississippi River Valley, is on private property, and permission of the owner (who lives in Grand Tower) is required for access.

REFERENCES CITED

Collinson, C. W., 1973, Guide for beginning fossil hunters: Illinois State Geological Survey, Educational Series 4, 37 p., 11 plates.

Collinson, C., and Skartvedt, R., 1960, Field Book; Pennsylvanian plant fossils of Illinois: Illinois State Geological Survey, Educational Series 6, 35 p., 5 plates.

Cote, W. E., Reinertsen, D. L., and Wilson, G. M., 1965, Alto Pass area geological science field trip: Illinois State Geological Survey, Guide Leaflet 1965F, 22 p.

Fraunfelter, G. H., 1977, Middle Devonian stratigraphy of southern Illinois: Field Guide for North Central Section, Geological Society of America, Annual Meeting, v. 2, p. 104–126.

Keyes, C. R., 1894, Paleontology of Missouri, Part I: Missouri Geological Survey, v. 4, p. 30 and 42.

Lange, R. V., 1983, A petrographic study of the Backbone (Little Saline) Limestone (Lower Devonian) in southwestern Illinois and southeastern Missouri [M.S. thesis]: Carbondale, Southern Illinois University, 110 p.

Meents, W. F., and Swann, D. H., 1965, Grand Tower Limestone (Devonian) of southern Illinois: Illinois Geological Survey Circular 389, 34 p.

Nance, R. B., 1968, Petrology of the St. Laurent Limestone (Middle Devonian) of southeastern Missouri [M.S. thesis]: Carbondale, Southern Illinois University, 120 p.

Nelson, W. J., and Lumm, D. K., 1985, Ste. Genevieve Fault Zone, Missouri and Illinois: Illinois State Geological Survey Contact/Grant Report 1985-3, 94 p.

North, W. G., 1969, The Middle Devonian strata of southern Illinois: Illinois Geological Survey Circular 441, 45 p.

Norton, J. A., 1966, Petrology of the Lingle Limestone, Union County, Illinois [M.S. thesis]: Carbondale, Southern Illinois University, 101 p.

Owen, D. D., 1856, Report on the Geological Survey in Kentucky made in the years 1854 and 1855: Kentucky Geological Survey, ser. 1, v. 1, 416 p.

Pickard, F., 1963, Bedrock geology of the Gorham area [M.S. thesis]: Carbondale, Southern Illinois University, 70 p.

Poor, R. S., 1925, The character and significance of the basal conglomerate of the Pennsylvanian System in Southern Illinois: Illinois State Academy of Science Transactions, v. 18, p. 369–375.

Savage, T. E., 1910, The Grand Tower (Onondaga) Formation of Illinois and its relation to the Jeffersonville beds of Indiana: Illinois State Academy of Science Transactions, v. 3, p. 116–132.

—— , 1920, The Devonian formations of Illinois: American Journal of Science, 4th, v. 49, p. 170–171 and 175–177.

—— , 1920a, Geology and economic resources of the Jonesboro Quadrangle, Illinois: Illinois Geological Survey Unpublished Report TES—3, p. 3, 8, 10, 14, 16–31, and 43–48.

Treworgy, J. D., 1981, Structural features in Illinois; A compendium: Illinois State Geological Survey Circular 519, 22 p.

Weller, S., 1913, Stratigraphy of the Chester Group in southwestern Illinois: Illinois Academy of Science Transactions, v. 6, p. 118–129.

—— , 1920, Geology of Hardin County: Illinois State Geological Survey Bulletin 41, 415 p.

Worthen, A. H., 1866, Geology: Geology Survey of Illinois, v. 1, p. 126–129.

Fluvial-deltaic deposits (Caseyville and Abbott Formations) of Early Pennsylvanian age exposed along the Pounds Escarpment, southern Illinois

Russell J. Jacobson, *Illinois State Geological Survey, 615 East Peabody Drive, Champaign, Illinois 61820*

LOCATION

I-24 passes through the lower Pennsylvanian rocks of the Pounds Escarpment, roughly 10 to 20 mi (15 to 30 km) south of Marion, Illinois (Fig. 1). This escarpment is a cuesta formed of resistant basal Pennsylvanian sandstones. Roadcuts along the interstate have made accessible nearly all of the Caseyville Formation and the basal portion of the Abbot Formation. The roadcuts are approximately 20.5 mi (33 km) south of the interchange of Illinois 13 with I-57 at Marion, Illinois (Fig. 1), and 2.7 miles (4.3 km) north of the Vienna interchange of U.S. 45 with I-24.

Figure 2 shows the exact location of the I-24 roadcuts on a portion of the Vienna 7½-minute Quadrangle. The upper roadcut is in the E½NE¼Sec.8,T.12S.,R.3E., Johnson County, Illinois. The lower cut is found in the E½NE¼Sec.17, and the NW¼ Sec.16,T.12S.,R.3E., Johnson County, Illinois.

The Pennsylvanian rocks are well exposed in lower and upper sets of cuts on both sides of the interstate. The shoulders are wide and paved, and provide room for a group of vehicles to pull completely off the road. However, it is still advisable for any large group planning a formal field trip to first check with the state police.

Separate cuts have been made for the northbound and southbound lanes on I-24; thus, four walls of rock are present and separate stops can be made (Fig. 2).

SIGNIFICANCE OF THE I-24 ROADCUTS

These roadcuts provide the best exposure of Lower Pennsylvanian strata in Illinois and contain a great number of primary sedimentary structures typical of fluvial and deltaic deposits. The roadcuts penetrate the scenic Pounds Escarpment, from which the structure of the southern Illinois Basin can be readily visualized. They give a clear picture of the variations in facies that occur in the lower Pennsylvanian of the Illinois Basin. The roadcuts expose units with sedimentary features that indicate deposition in predominantly fluvial, distributary, and marsh environments (Koeninger, 1978; Keoninger and Mansfield, 1979) typical of a river-dominated (high-constructive) delta.

GEOMORPHOLOGY

The resistant, commonly massive sandstone of the basal Pennsylvanian Caseyville Formation (Fig. 3) has produced a prominent south-facing cuesta, the Pounds Escarpment, in southern Illinois. The escarpment curves eastward from southern Jack-

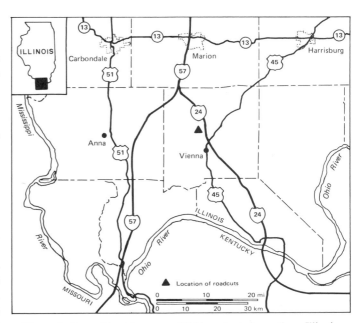

Figure 1. General location map of I-24 roadcuts in southern Illinois.

son County through northern Union, Johnson, and Pope Counties. Farther eastward it is extensively interrupted by faulting. The escarpment is marked by steep slopes that have 200 to 250 ft (60 to 75 m) of relief and by vertical cliffs 100 ft (30 m) or more in height. Headward erosion by south-flowing streams has produced many narrow, scenic salients and reentrants in the scarp. In some areas the cliffs grade into steep slopes that have a succession of rock ledges. This occurs because the massive sandstones grade laterally into thin-bedded sandstones and shales that will not maintain a good cliff face.

South of the Pounds Escarpment, nonresistant shales, limestones, and thin-bedded sandstones of the Chesterian Series (Upper Mississippian) produce a gently rolling topography. Middle Pennsylvanian strata north of the scarp form a moderately rugged landscape, but the Illinoian glacial drift largely masks the influence of the bedrock structure. In fact, the southernmost advance of Pleistocene continental glaciation in North America was approximately 3 mi (5 km) northeast of Goreville.

Roadcuts almost as good as those described here are present on I-57 just south of the Goreville exit; at this exit is one of the best places from which to view the Pounds Escarpment. From this spot, the gentle northward inclination of the bedrock into the Illinois Basin is apparent.

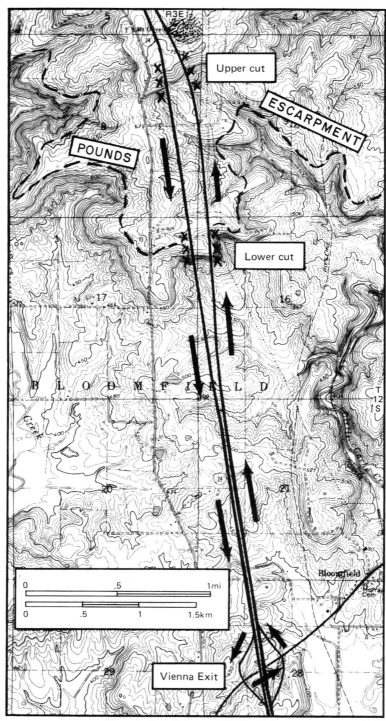

Figure 2. Detailed map of I-24 cuts.

SITE DESCRIPTION

General. The oldest Pennsylvanian rocks in Illinois are assigned to the Caseyville Formation (Fig. 3) of Morrowan Age (Willman and others, 1975). The formation consists dominantly of sandstone, typically massive or cross-bedded; it is commonly conglomeratic, with diagnostic well-rounded granules and small pebbles of white quartz (Simon and Hopkins, 1966). Thin-bedded sandstone, medium- to dark-gray silty shales, discontinuous thin coals and underclays, and, rarely, limestones are also found in the Caseyville. Such lithologies are typical of the basal Pennsylvanian in most of the eastern United States.

In southern Illinois, the Caseyville Formation unconformably overlies the Chesterian Series (upper Mississippian), a cyclic succession of shale and limestone. The surface on which the Caseyville was deposited had a strong relief and numerous valleys cut several hundred ft (m) into a more or less flat erosional surface.

Overlying the Caseyville is the Abbott Formation (Fig. 3) of Atokan age. The Abbott is lithologically similar to the Caseyville, but contains a greater proportion of shale, and the sandstones are less massive and only rarely conglomeratic. Younger Pennsylvanian formations have more shale and contain extensive marine limestones and beds of mineable coal.

In the roadcuts described here, the upper three members of the Caseyville Formation can be seen (Willman and others, 1975; Fig. 3). In ascending order, they are a basal sandstone, the Battery Rock Sandstone Member; a middle shaly interval, the Drury Shale Member; and an upper sandstone, the Pounds Sandstone Member. In some areas of southern Illinois, a fourth member, the Lusk Shale, is recognized below the Battery Rock Sandstone; it is not, however, recognized at this cut. Laterally the Lusk Sandstone grades to sandstone; in such areas the term Wayside Sandstone Member has been applied (Willman and others, 1975).

Description of Roadcuts

The exposures through the escarpment on I-24 consist of lower and upper roadcuts (Figs. 2, 4). These two cuts are separated by a covered interval that represents the Drury Shale Member. Here, strata of the Caseyville and the basal part of the Abbott Formation are exposed.

In the upper cut, about 92 ft (28 m) of sandstone, shale, and coal represent (in descending order) the Grindstaff Sandstone and Reynoldsburg Coal Members of the Abbott Formation, and the Pounds Sandstone Member of the Caseyville Formation. The lower cut exposes about 170 ft (52 m) of the Battery Rock Sandstone Member (Fig. 5), which consists of sandstone that has lesser amounts of conglomerate, siltstone, and shale.

Depositional Environments Seen at the I-24 Roadcuts

Studies by Koeninger (1978) and Koeninger and Mansfield (1979) indicate that river-dominated or high-constructive deltaic sediments compose the Caseyville Formation. Three main facies are distinguished: active channel facies, interdistributary facies, and overbank facies.

Active Channel Facies. Under this facies, Koeninger recognized two basic subfacies: the distributary channel and the point bar (Fig. 4). The distributary channel subfacies commonly cuts into and grades upward into deposits of the interdistributary and overbank facies. Deposits in the distributary channel subfacies are usually medium- and fine-grained sandstone with fewer common lenses of gravelly, coarse-grained sandstone and shale. The main sedimentary structures of this subfacies as seen by

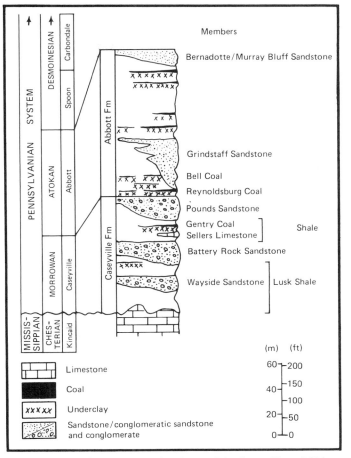

Figure 3. Generalized stratigraphic column of the Caseyville and Abbott Formations in southern Illinois.

Koeninger include planar and trough cross-beds, erosional contacts, casts of plant fragments, and disrupted bedding.

The point-bar subfacies deposits overlie and sometimes cut into deposits of the other facies. This subfacies has rocks that are typically coarser than other channel deposits; the rocks vary from gravelly, coarse-grained sandstone to medium-grained sandstone. Sedimentary structures observed in this subfacies include planar and trough cross-beds, major scour channels, and erosional truncations. The rocks in this subfacies fine upward, with siltstone/mudstone commonly draping each channel fill.

Interdistributary Facies. This facies is subdivided into three subfacies: the interdistributary bay, the abandoned distributary channel, and the deltaic lake (Fig. 4). The interdistributary bay deposits usually overlie distributary channel or abandoned channel subfacies, as well as marsh and swamp subfacies. They may pass upward into marsh deposits (i.e., coal and coaly shales), or be scoured by younger channel subfacies. This subfacies consists mostly of shale, siltstone, and fine-grained sandstone, which are often lenticular and surrounded by shale. These sandstone lenses contain small-scale trough cross-beds and ripple-drift lamination. The shales contain wavy bedding and a few burrows.

UPPER ROADCUT

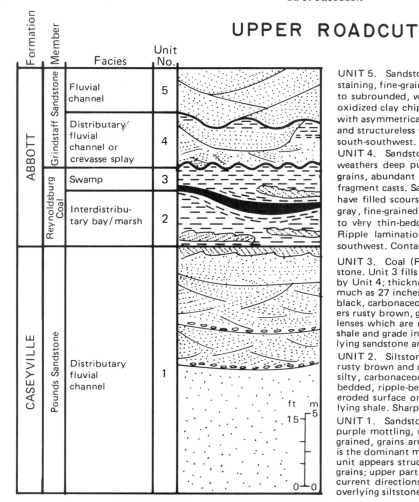

UNIT 5. Sandstone, white, weathers buff with iron-oxide staining, fine-grained to very fine-grained, grains are subangular to subrounded, well-sorted. Quartz is the dominant mineral; oxidized clay chips common. Unit is both planar cross-bedded with asymmetrical, low-amplitude ripples where sorting is poor, and structureless where sorting is good. Paleocurrent directions south-southwest.

UNIT 4. Sandstone and siltstone; sandstone, dark brown, weathers deep purple, fine- to medium-grained, subrounded grains, abundant quartz; some oxidized clay chips and wood fragment casts. Sandstone occurs as thick to thin lenses which have filled scours in siltstone. Siltstone, white, weathers light gray, fine-grained, silty, parts easily along bedding planes; thin- to very thin-bedded; deposits draped over sandstone lenses. Ripple lamination common. Paleocurrent directions west-southwest. Contact with overlying sandstone is sharp.

UNIT 3. Coal (Reynoldsburg Coal Member), shale, and siltstone. Unit 3 fills paleotopographic depressions and is scoured by Unit 4; thickness is variable. The coal is bright banded and as much as 27 inches (70 cm) thick; partly stained yellow. Shale, black, carbonaceous, thin-bedded. Siltstone, dark brown, weathers rusty brown, grains subangular; siltstone occurs in very thin lenses which are rippled and which cut into the underlying shale and grade into the overlying shale. Sharp contact with overlying sandstone and siltstone.

UNIT 2. Siltstone and shale; siltstone, light brown, weathers rusty brown and orange, subangular grains. Shale, gray, slightly silty, carbonaceous, thin-bedded to laminar. Siltstone is thin-bedded, ripple-bedded, and missing locally; fills an irregular eroded surface on underlying shale and grades into the overlying shale. Sharp contact with overlying shale and coal.

UNIT 1. Sandstone; white to pink, weathers rusty tan and buff, purple mottling, with red iron-oxide on bedding planes; fine-grained, grains are subrounded, sorting is fair to good. Quartz is the dominant mineral; contains few clay chips. Lower part of unit appears structureless due to good sorting of constituent grains; upper part contains planar and trough cross-beds. Paleocurrent directions south-southeast. Gradational contact with overlying siltstone and shale.

The abandoned distributary channel subfacies forms deposits that often fine upward into interdistributary bay and marsh deposits, or they may be scoured by younger distributary channels. They are characterized by shale and lenticular bodies of medium- to fine-grained sandstone. Koeninger (1978) believed that some of these deposits were tidally influenced; they are characterized by thin to thick sandstone lenses with small-scale trough cross-bedding. The sandstone lenses are often capped by ripple marks formed by a flow direction that is opposite that suggested by the underlying trough cross-beds. In addition, these tidally influenced abandoned channel subfacies are often characterized by the presence of flaser bedding. Koeninger also believed that abandoned channel subfacies with ripple-marked sandstone that fines upward into normally bedded siltstone and shale lack such tidal influence.

The deltaic lake subfacies conformably overlies the interdistributary bay deposits, but is itself often scoured by younger distributary channel deposits. In this subfacies one can find interbedded black mudstone and a gray, silty mudstone. The units also contain layers of vertical selenite crystals. The bedding is uniformly thin and laterally persistent, resulting in a varvelike appearance.

Overbank Facies. This facies has three subfacies: marsh, floodplain, and swamp (Fig. 4). The marsh subfacies overlies the distributary channel, point-bar, and interdistributary bay subfacies. This subfacies may grade upward into swamp or interdistributary bay subfacies, or it may be cut into by distributary channel subfacies. The marsh deposits also interfinger laterally with the swamp deposits. The marsh subfacies consists mainly of black shales interbedded with thin, fine-grained sandstone lenses. Carbonaceous laminae are common along partings in the shale and bedding in the sandstone.

The floodplain subfacies usually overlies the point-bar subfacies, and is often scoured into by younger distributary channels. Floodplain deposits contain shale that is often color-banded, which probably indicates subaerial exposure. These shales then enclose rhythmically interbedded, uniformly thick, ripple-marked siltstone lenses that represent seasonal floods.

The swamp subfacies of the overbank facies is not well represented in the Caseyville Formation. These deposits consist mainly of coals, and carbonaceous thinly bedded shale with siltstone lenses.

Another facies, not visible in the I-24 cuts, is in the 30-ft (9 m) covered interval between the lower and upper roadcuts. This

LOWER ROADCUT

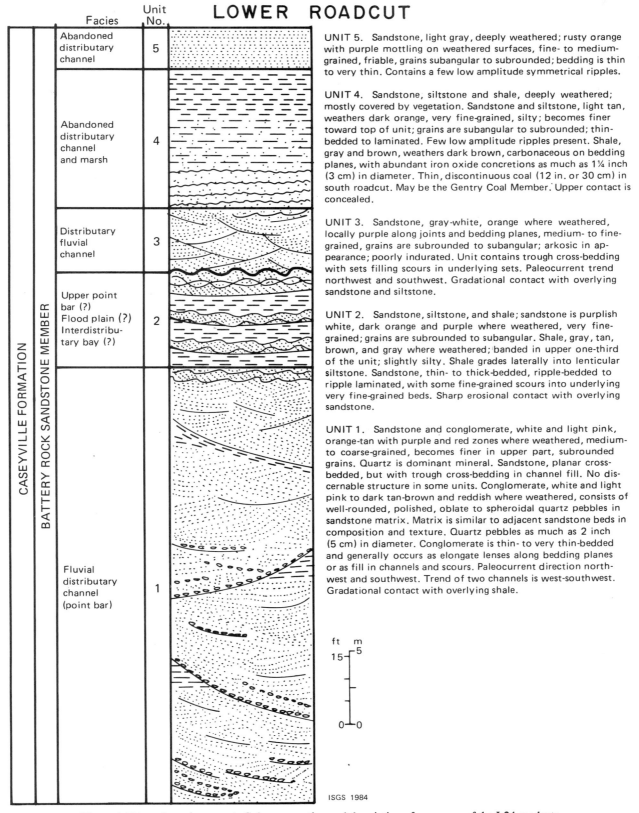

UNIT 5. Sandstone, light gray, deeply weathered; rusty orange with purple mottling on weathered surfaces, fine- to medium-grained, friable, grains subangular to subrounded; bedding is thin to very thin. Contains a few low amplitude symmetrical ripples.

UNIT 4. Sandstone, siltstone and shale, deeply weathered; mostly covered by vegetation. Sandstone and siltstone, light tan, weathers dark orange, very fine-grained, silty; becomes finer toward top of unit; grains are subangular to subrounded; thin-bedded to laminated. Few low amplitude ripples present. Shale, gray and brown, weathers dark brown, carbonaceous on bedding planes, with abundant iron oxide concretions as much as 1¼ inch (3 cm) in diameter. Thin, discontinuous coal (12 in. or 30 cm) in south roadcut. May be the Gentry Coal Member. Upper contact is concealed.

UNIT 3. Sandstone, gray-white, orange where weathered, locally purple along joints and bedding planes, medium- to fine-grained, grains are subrounded to subangular; arkosic in appearance; poorly indurated. Unit contains trough cross-bedding with sets filling scours in underlying sets. Paleocurrent trend northwest and southwest. Gradational contact with overlying sandstone and siltstone.

UNIT 2. Sandstone, siltstone, and shale; sandstone is purplish white, dark orange and purple where weathered, very fine-grained; grains are subrounded to subangular. Shale, gray, tan, brown, and gray where weathered; banded in upper one-third of the unit; slightly silty. Shale grades laterally into lenticular siltstone. Sandstone, thin- to thick-bedded, ripple-bedded to ripple laminated, with some fine-grained scours into underlying very fine-grained beds. Sharp erosional contact with overlying sandstone.

UNIT 1. Sandstone and conglomerate, white and light pink, orange-tan with purple and red zones where weathered, medium- to coarse-grained, becomes finer in upper part, subrounded grains. Quartz is dominant mineral. Sandstone, planar cross-bedded, but with trough cross-bedding in channel fill. No discernable structure in some units. Conglomerate, white and light pink to dark tan-brown and reddish where weathered, consists of well-rounded, polished, oblate to spheroidal quartz pebbles in sandstone matrix. Matrix is similar to adjacent sandstone beds in composition and texture. Quartz pebbles as much as 2 inch (5 cm) in diameter. Conglomerate is thin- to very thin-bedded and generally occurs as elongate lenses along bedding planes or as fill in channels and scours. Paleocurrent direction northwest and southwest. Trend of two channels is west-southwest. Gradational contact with overlying shale.

ISGS 1984

Figure 4 (this and previous page). Columnar section and description of exposures of the I-24 roadcuts (from Palmer and Dutcher, 1979, after Koeninger, 1978; and Koeninger and Mansfield, 1979).

is the interval that is equivalent to the Drury Shale Member. Because it is so shaley and nonresistant, it forms a prominent bench separating the resistant sandstone parts of the escarpment. Locally, it develops into a valley separating the Pounds and Battery Rock Sandstones into two cuestas. This unit appears to be the result of a period of marine transgression between periods of

delta lobe formation represented by the Pounds and Battery Rock Sandstone Members. Koeninger (1978) believes that the Drury Shale probably represents prodeltaic clays. Further evidence supporting this marine transgression is found in the Drury Shale interval in southeastern Illinois, where it contains a thick marine limestone known as the Sellers Limestone Member.

REFERENCES CITED

Koeninger, C. A., 1978, Regional facies of the Caseyville Formation (Lower Pennsylvanian), south-central Illinois [M.S. thesis]: Carbondale, Southern Illinois University, 84 p.

Koeninger, C. A., and Mansfield, C. F., 1979, Earliest Pennsylvanian depositional environments in central-southern Illinois, *in* Palmer, J. E., and Dutcher, R. R., eds., Depositional and structural history of the Pennsylvanian System of the Illinois Basin: Ninth International Congress of Carboniferous Stratigraphy and Geology, Guidebook for field trip 9, part 1, Illinois Geological Survey, p. 76–81.

Simon, J. A., and Hopkins, M. E., 1966, Sedimentary structures and morphology of Late Paleozoic sand bodies in southern Illinois: Illinois State Geological Survey Guidebook Series 7, 67 p.

Willman, H. B., Atherton, E., Buschbach, T. C., Collinson, C., Frye, J. C., Hopkins, M. E., Lineback, J. A., and Simon, J. A., 1975, Handbook of Illinois stratigraphy: Illinois State Geological Survey Bulletin, v. 95, p. 163–183.

The Cache Valley of southern Illinois

John M. Masters and Davis L. Reinertsen, Illinois State Geological Survey, Champaign, Illinois 61820

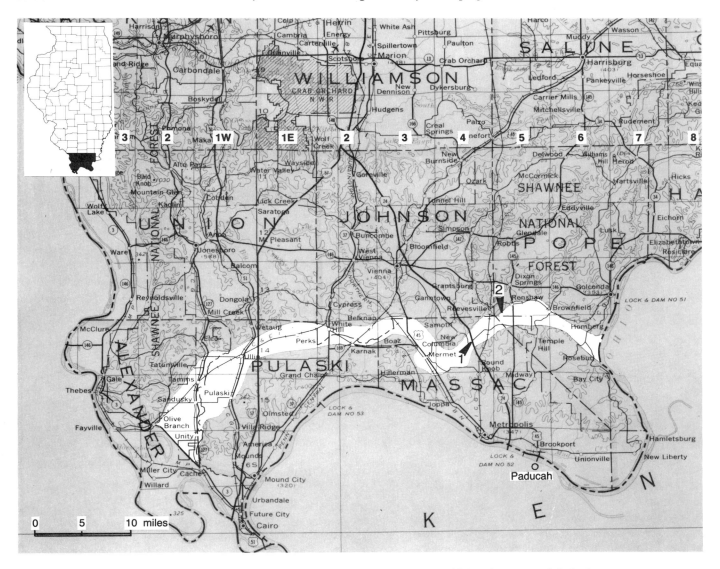

Figure 1. Map of southern Illinois showing location of Cache Valley (white) and recommended viewing sites. Site 1—Big Bay exit from I-24; Site 2—Stafford Bluff, one mile west of Illinois 145 near Renshaw, Illinois.

LOCATION

The Cache Valley in extreme southern Illinois is a major physiographic feature that extends across all or parts of Pope, Massac, Johnson, Union, Pulaski, and Alexander counties, from the Ohio River on the east to the Mississippi River on the west, and cuts across all or parts of the following Third Principal Meridian townships (Fig. 1): T.13S.,R.3,5, and 63; T.14S.,R.1,2,3,4,5, and 6E,R.1 and 2W; T.15S.,R.1,2,3, and 4E.,R.1,2, and 3W; T.16S.,R.1,2,3, and 4W; and T.17S.,R.1 and 2W.

ACCESSIBILITY

I-57 and I-24; U.S. 45 and U.S. 51; Illinois 3, 127, 37, 169, and 145; and a number of county roads cross the Cache Valley. However, forested tracts along the south valley slopes and along the small streams now occupying the valley floor make it difficult even during the winter to comprehend the size and extent of the feature.

The most readily accessible ground location where some notion of the extent of this conspicuous physiographic feature can

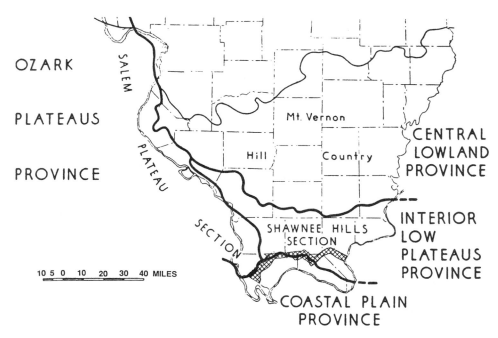

Figure 2. Physiographic divisions of southern Illinois (from Leighton and others, 1948) showing location of Cache Valley highlighted by diamond pattern.

be observed is the overpass of the Big Bay–New Columbia Road at Exit 27 from I-24 (NW¼SE¼Sec. 15,T.14S.,R.4E.,3rd P.M.), Massac County (Mermet 7½-minute Quadrangle). Massive, resistant units of the upper Mississippian (Chesterian) Tar Springs Sandstone form the steep northwest wall of the Cache Valley for several miles. This straight bluff is a fault line scarp eroded along the southwestern extension of the Dixon Springs Graben. The south valley wall, composed of less resistant Cretaceous Gulfian Series strata, has more gentle slopes that generally are obscured by vegetation. Here the valley is about 1.5 mi (2.5 km) wide.

The locality offering the best view of the size and shape of the Cache Valley is situated along Stafford Bluff about 1 mi (1.6 km) west of Illinois 145. To reach this vantage point, take the gravel road (1925N, 185E) west and southwest for 1.3 mi (2.1 km) from its intersection with Illinois 145 at the hamlet of Renshaw, a short distance southwest of Dixon Springs State Park. This view is 0.3 mi (0.5 km) beyond the power substation at an illegal dumping site on the bluff edge along the southeast side of the road (near east edge NE¼,SE¼,NE¼,SE¼,Sec.31,T.13S., R.5E.,3rd P.M.) in Pope County (Reevesville 7½-minute Quadrangle). Occasional burning of the trash at this site has maintained an opening through the tree cover that provides a better-than-average view of the valley to the east-southeast. Although there is considerable vegetation on the valley walls in the distance, the resistant Chesterian rocks that occur on both sides of the valley produce a distinct steep-sided cross-sectional profile. Unfortunately, there is no place that is readily accessible on Staf-

ford Bluff for a comparable view to the west. The Cache Valley is 1.6 (2.6 km) wide at the south end of the bluff.

Descriptions of other stops featuring the Cache Valley and vicinity are available in the following guides to public field trips: Cote and others (1966); Reinertsen and others (1975 and 1981).

SIGNIFICANCE

The Cache Valley is an abandoned segment of the trunk portion of a major drainage system and is one of the best exposed and most widely recognized landforms in Illinois. Physiographically, it forms the northernmost edge of that part of the Coastal Plain Province (Fig. 2) known as the Mississippi Embayment. Here the embayment abuts against the Shawnee Hills Section of the Interior Low Plateaus Province (Fenneman, 1938). Extending nearly 45 mi (72 km) westward from its sharp, angular junction with the Ohio River about 5 mi (8 km) south of Golconda, Illinois, the relatively flat alluvial floor of the Cache Valley ranges from about 1.5 mi (2.4 km) to nearly 4 mi (6.4 km) wide. The valley walls are cut into resistant Paleozoic rocks in the eastern one-quarter and the entire north side of the valley, occasionally following fault zones (Willman and others, 1967; Treworgy, 1981). Where resistant Paleozoic rocks are exposed, the north valley walls are much steeper, 150 to 250 ft (45 to 75 m) high, and better defined than the south side, where all but the eastern one-quarter was eroded in softer, relatively unconsolidated Cretaceous and Tertiary sediments (Fig. 3). The eastern part of the

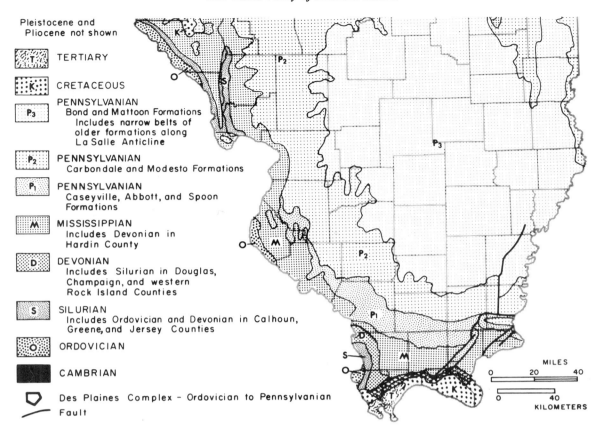

Pleistocene and
Pliocene not shown

T TERTIARY

K CRETACEOUS

P₃ PENNSYLVANIAN
Bond and Mattoon Formations
Includes narrow belts of
older formations along
La Salle Anticline

P₂ PENNSYLVANIAN
Carbondale and Modesto Formations

P₁ PENNSYLVANIAN
Caseyville, Abbott, and Spoon
Formations

M MISSISSIPPIAN
Includes Devonian in
Hardin County

D DEVONIAN
Includes Silurian in Douglas,
Champaign, and western
Rock Island Counties

S SILURIAN
Includes Ordovician and Devonian in Calhoun,
Greene, and Jersey Counties

O ORDOVICIAN

CAMBRIAN

Des Plaines Complex – Ordovician to Pennsylvanian

Fault

Figure 3. Simplified bedrock geologic map of southern Illinois showing Cache Valley along the northern edge of Cretaceous-Tertiary deposits in the Mississippi Embayment.

valley is now occupied by Bay Creek, and the western part by the Cache River, both underfit streams too small to have eroded such an enormous valley. In the subsurface, the deepest part of the valley is incised into Paleozoic strata throughout its extent. The orgin of the Cache Valley and its history have remained matters for research and discussion for more than 45 years. Research projects currently underway along the valley may soon provide definitive answers to some of the problems.

SITE INFORMATION

The earliest fluvial system to occupy the position of the Cache Valley is not known, but it is reasonable to suspect that its beginning dates back to the Paleozoic/Mesozoic erosional unconformity. During part of that time, the Little Bear Soil was developed and Cretaceous Tuscaloosa Gravel was deposited, at least in the vicinity of the Cache Valley (Pryor and Ross, 1962; Ross, 1964). The deposition of the Tuscaloosa chert gravel and the overlying fluvial-deltaic McNairy Formation reflect the northward extension of the Mississippi Embayment into southernmost Illinois.

During the early part of the Late Cretaceous, the northern part of the Mississippi Embayment was on the north flank of an upland known as the Pascola Arch (Fig. 4), with possibly as much as 500 ft (150 m) of relief (Stearns and Marcher, 1962). Streams heading in the upland flowed northeast and east across the embayment into a trunk stream that drained southward around the periphery of the arch. The Tuscaloosa Gravel was deposited in part by this fluvial system. It is not known if any part of the trunk stream occupied the position of the Cache Valley because no remnants of the Tuscaloosa have been identified in the Cache Valley. The strong curvilinear alignment of the Cache Valley with the eastern edge of the Mississippi Embayment and the Cumberland and Tennessee River valleys suggests that its origin may be related to the Cretaceous-age fluvial system, even though during the time of Tuscaloosa deposition, drainage flowed east and south (Marcher and Stearns, 1962), opposite to the modern drainage direction.

In the mid-Tertiary, the northeastern part of the Mississippi Embayment, including most of the area of the Cache Valley, was relatively near sea level in elevation, and the surrounding Paleozoic bedrock regions were also low and of subdued relief (Potter,

Figure 4. Major structural features of the Eastern Interior Basin, U.S. (from Bond and others, 1971).

1955a, 1955b). During the Pliocene, epeirogenic uplift initiated a new erosion cycle. Cert-rich residuum was eroded from the Paleozoic uplands and deposited in the form of alluvial fans in the northeastern Mississippi Embayment by high-velocity braided streams. The ancestral Tennessee and Cumberland were the dominant rivers along the east side of the embayment, building their alluvial fans from the southeast toward the northwest where the Cumberland was joined by the relatively small, preglacial Ohio River near the present upstream end of the Cache Valley.

These low-gradient alluvial fans apparently filled the northeastern part of the low-lying embayment and overlapped onto the east flank of the Paleozoic uplands. The fans were deposited on an erosional surface that is now at an elevation of 450 to 500 ft

(140 to 150 m) above mean sea level (m.s.l.) in the embayment area, and extends upward to 600 to 650 ft (180 to 200 m) on the upland rim areas. Erosion has subsequently dissected the fans and underlying deposits. Northward progradation of these alluvial fans may well have forced the main Cumberland–Ohio, and probably the Tennessee River channels northward into the position of the Cache Valley. Erosion and redistribution of the chert gravels began in the late Pliocene and probably continued into the early Pleistocene (Olive, 1980). Deep entrenchment of the major valleys in the mid-continent may well have begun with the onset of the Pleistocene.

There are few, if any, deposits in the Cache Valley and vicinity to provide evidence of the effect of pre-Illinoian glacial and interglacial stages on the area. Although this area was not glaciated, Illinoian ice did extend southward to within about 25 mi (40 km) of the Cache Valley (Willman and Frye, 1980). Deep weathering and erosion appear to have progressed at variable rates during the early Pleistocene in this area. Early Pleistocene glaciations blocked other drainage systems from Indiana to Pennsylvania and eventually diverted them into the Ohio. This large increase in the size of its drainage basin, coupled with the addition of vast quantities of meltwater, converted the Ohio into a major river. This conversion, accompanied by eustatic lowering of sea level during each pre-Illinoian glaciation (at least two) and probably by epeirogenic uplift, ultimately resulted in the deep entrenchment of the ancestral Ohio River channel into bedrock beneath the Cache Valley. The bottom of this deep valley follows the Ohio, then down the Cache Valley at an elevation of about 120 (35 m) above m.s.l. Drillhole records indicate that the present Ohio valley between Paducah, Kentucky, and Olmsted, Illinois, is not entrenched much below the present river channel, about 220 ft (70 m) above m.s.l. (Olive, 1980). However, a deep valley appears to extend from the mouth of the Tennessee River northeastward within the present Ohio River valley to the upstream end of the Cache Valley. Thus, it seems most plausible that during the early Pleistocene, the Ohio–Cumberland–Tennessee rivers did meet near the sharp bend between Golconda and Bay City, Illinois, to flow westward, carving and entrenching the Cache Valley.

The absence of an equally deep channel in the present Ohio River valley between Paducah and Olmsted is also strong evidence that the pre-Illinoian Ohio River did not flow through its present channel, but rather through the Cache Valley. We do not agree with Alexander and Prior's (1968) interpretation of alluvial deposits along the Ohio River and Cache Valley as indicating that the Ohio followed its present course prior to the existence of the Cache Valley, the latter having been eroded by Cache River and Bay Creek, and that the Ohio used the Cache Valley as a floodwater spillway, widening and deepening it during late Wisconsinan time.

It appears more likely that the present course of the Ohio River formed in late Wisconsinan time. During a period of extremely high meltwater volume and alluviation of the valleys a large slackwater lake formed in the Tennessee River valley, backing up a tributary of the Tennessee toward Metropolis, Illinois. A low divide separating this northeast-flowing drainage from another steep-grascent system flowing westward was eventually topped, and rapidly cut down by the steep-gradient stream. The volume of meltwater was large enough and was sustained long enough to establish a permanent channel southwestward to a juncture with the Mississippi River. Both the Cache Valley channel and the new Ohio River channel were probably used by the Ohio on into the Holocene, especially during times of exceptionally high flooding. However, by studying slackwater lake deposits associated with the Cache Valley, Graham (1985) determined that the diversion of the Ohio was essentially completed by about 8,200 radiocarbon years B.P. Backwaters of the Ohio most recently spilled through the Cache Valley during the record flood of 1937.

Other workers (Weller, 1940; Fisk, 1944; Leighton and others, 1948; Horberg, 1950; Wayne, 1952; Potter, 1955a, 1955b; Walker, 1957; Pryor and Ross, 1962; Ray, 1963; Ross, 1963 and 1964; Willman and others, 1975; Willman and Rye, 1970 and 1980; and Graham, 1985) agree that the Cache Valley was the course of the ancestral Ohio River; that Cache River and Bay Creek are now, and were in the past, too small to erode this enormous feature; and that a divide in the present Ohio valley was breached because of vast increases in meltwater volume and alluviation to change the drainage pattern from the Cache Valley to the present Ohio River valley.

In summary, the geologic history of the formation of the deeply entrenched Cache Valley extends back into the Cretaceous Period and is related to the origin of the northern part of the Mississippi Embayment. A stream flowing east to southeast may have occupied all or a part of the location of the Cache Valley during the erosion of the Pascola Arch. Alternating subsidence and uplift in the embayment resulted in the deposition of shallow marine and fluvial-deltaic sediments in most of the area of the Cache Valley, followed by periods of erosion. Chert gravel deposition in the late Tertiary filled the northeastern portion of the embayment, concentrating the major west- to southwest-draining rivers of the time into the present location of the Cache Valley. The valley was superimposed onto the bedrock by early Pleistocene (pre-Illinoian) deep entrenchment of the major glacial drainage systems. With each glacial cycle the Cache Valley was alternately scoured out and refilled with sediment to varying heights and thicknesses. The shallow sediments underlying the Cache Valley are mostly late Wisconsinan and Holocene in age (12.6 to 8.2 ka), bracketing the time of the diversion of the Ohio, Cumberland, and Tennessee rivers from the Cache Valley into the present Ohio valley.

REFERENCES CITED

Alexander, C. S., and Prior, J. C., 1968, The origin and function of the Cache Valley, southern Illinois, *in* Bergstrom, R. E., ed., The Quaternary of Illinois; A symposium in observance of the centennial of the University of Illinois: Urbana, University of Illinois College of Agriculture Special Publication no. 14, p. 19–26.

Bond, D. C., and 6 others, 1971, Background materials for symposium on future petroleum potential of NPC Region 9 (Illinois Basin, Cincinnati Arch, and northern part of Mississippi Embayment): Illinois State Geological Survey,

Illinois Petroleum, v. 96, 63 p.

Cote, W. E., Reinertsen, D. L., and Wilson, G. M., 1966, Vienna area: Illinois State Geological Survey Guide Leaflet 1966-A, 18 p.

Fenneman, N. M., 1938, Physiography of the eastern United States: New York, McGraw-Hill Book Co., Inc., 714 p.

Fisk, H. N., 1944, Geological investigations of the alluvial valley of the Lower Mississippi River: Vicksburg, Mississippi, Mississippi River Commission, U.S. Army Corps of Engineers, 78 p.

Graham, R. C., 1985, The Quaternary history of the upper Cache River Valley, southern Illinois [M.S. Thesis]: Carbondale, Southern Illinois University, 123 p.

Horberg, L., 1950, Bedrock topography of Illinois: Illinois State Geological Survey Bulletin 73, 111 p.

Leighton, M. M., Ekblaw, G. E., and Horberg, L., 1948, Physiographic divisions of Illinois: Illinois State Geological Survey Report of Investigations 129, 19 p.

Marcher, M. V., and Stearns, R. G., 1962, Tuscaloosa Formation in Tennessee: Geological Society of America Bulletin, v. 73, p. 1365–1386.

Olive, W. W., 1980, Geologic maps of the Jackson Purchase Region, Kentucky: U.S. Geological Survey Miscellaneous Investigations Series, Map I–1217, 11 p., scale 1:250,000.

Potter, P. E., 1955a, The petrology and origin of the Lafayette Gravel; Pt. 1, Mineralogy and petrology: Journal of Geology, v. 63, no. 1, p. 1–38.

—— , 1955b, The petrology and origin of the Lafayette Gravel; Pt. 2, Geomorphic history: Journal of Geology, v. 63, no. 2, p. 115–132.

Pryor, W. A., and Ross, C. A., 1962, Geology of the Illinois parts of the Cairo, La Center, and Thebes Quadrangles: Illinois State Geological Survey Circular 332, 39 p.

Ray, L. L., 1963, Quaternary events along the unglaciated Lower Ohio River Valley: U.S. Geological Survey Professional Paper 475–B, p. B125–B128.

Reinertsen, D. L., Berggren, D. J., and Killey, M. M., 1975, Metropolis area geological science field trip; Massac, Pope, and Pulaski counties: Illinois State Geological Survey Guide Leaflet 1975–A and D, 39 p.

Reinertsen, D. L., Masters, J. M., and Reed, P. C., 1981, A guide to the geology of the Cairo area: Illinois State Geological Survey Field Trip Guide Leaflet 1981–A, 40 p.

Ross, C. A., 1963, Structural framework of southernmost Illinois: Illinois State Geological Survey Circular 351, 28 p.

—— , 1964, Geology of the Paducah and Smithland Quadrangles in Illinois: Illinois State Geological Survey Circular 360, 32 p.

Stearns, R. G., and Marcher, M. V., 1962, Late Cretaceous and subsequent structural development of the northern Mississippi Embayment area: Geological Society of America Bulletin, v. 73, p. 1387–1394.

Treworgy, J. D., 1981, Structural features in Illinois; A compendium: Illinois State Geological Survey Circular 519, 22 p.

Walker, E. H., 1957, The deep channel and alluvial deposits of the Ohio Valley in Kentucky: U.S. Geological Survey Water-Supply Paper 1411, 25 p.

Wayne, W. J., 1952, Pleistocene evolution of the Ohio and Wabash Valleys: Journal of Geology, v. 60, no. 6, p. 575–585.

Weller, J. M., 1940, Geology and oil possibilities of extreme southern Illinois; Union, Johnson, Pope, Hardin, Alexander, Pulaski, and Massac counties: Illinois State Geological Survey Report of Investigations 71, 71 p.

Willman, H. B., and Frye, J. C., 1970, Pleistocene stratigraphy of Illinois: Illinois State Geological Survey Bulletin 94, 204 p.

—— , 1980, The glacial boundary in southern Illinois: Illinois State Geological Survey Circular 511, 23 p.

Willman, H. B., and 8 others, 1967, Geologic map of Illinois: Illinois State Geological Survey Map, scale 1:500,000.

Willman, H. B., and 7 others, 1975, Handbook of Illinois stratigraphy: Illinois State Geological Survey Bulletin 95, 261 p.

Anomalous Paleozoic outliers near Limestone Mountain, Michigan

Randall L. Milstein, Subsurface and Petroleum Geology Unit, Michigan Geological Survey, Lansing, Michigan 48912

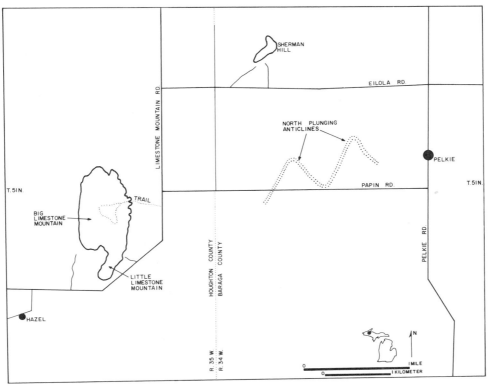

Figure 1. Location of Limestone Mountain, Sherman Hill, and two plunging anticlines near Pelkie, Michigan.

LOCATION

Limestone Mountain, Sec.13,14,23,24,T.51W.,R.35W., Houghton County, Michigan; Pelkie, Michigan, 7½-minute Quadrangle. Limestone Mountain is located 1 mi (1.6 km) northeast of Hazel, Michigan (Fig. 1). Access to Limestone Mountain is by private road or foot trail. The trail head is located on the west side of Limestone Mountain Road, 800 ft (244 m) south of this road's intersection with Papin Road (Fig. 1). Limestone Mountain is located on privately owned wilderness land, and permission should be obtained prior to entry.

SIGNIFICANCE

Limestone Mountain rises more than 300 ft (90 m) above the surrounding regional topography and consists of Ordovician, Silurian, and Devonian carbonates overlying Precambrian sandstone deposits. Considerable deformation is evident at the structure and in smaller, isolated outcrops in the vicinity. Deformation involves major and minor faulting, folding, shearing, missing stratigraphic units, brecciated dolomite, and beds with varied strikes and dips, with some of these beds being pitched near vertical. Structural irregularities also include deformation microstructures in quartz and feldspar grains from the sandstone units.

Limestone Mountain is the westernmost identified outlier of Paleozoic rock in Michigan's Northern Peninsula. The Devonian strata identified at Limestone Mountain are found nowhere else in Michigan's northern peninsula. The nearest outcrops of equivalent-aged rocks are almost 200 mi (320 km) to the southeast.

The Limestone Mountain structure and its surrounding region have been the subject of geologic investigations for nearly 140 years. To date, no positive hypothesis has been presented to account for the preservation of these outliers, or to explain the cause and extent of structural deformation associated with them.

GEOLOGICAL SETTING

The geology of the Keweenaw Peninsula, as shown in Figure 2, has been well defined due to the intense prospecting for

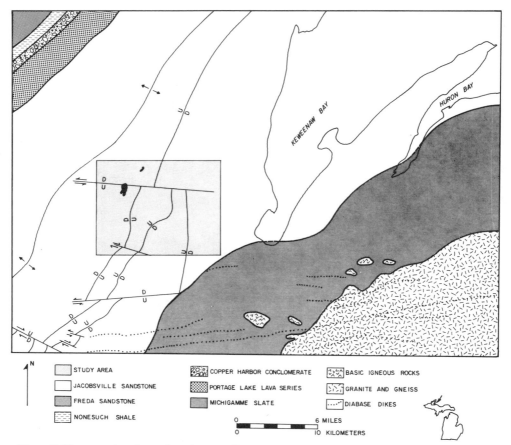

Figure 2. The general geology of southeastern Keweenaw Peninsula, Michigan. Map indicates the location of Limestone Mountain and Sherman Hill. Interpretation of bedrock and structural geology modified from Meshref and Hinze (1970, Plate 2).

copper in the region. A good general outline of the regional geology has been presented by Halls (1966).

The Limestone Mountain region of the Keweenaw Peninsula consists of a fairly flat plain with a few prominent hills rising to a maximum of 1,133 ft (345 m) above mean sea level. The area was glaciated by the southerly movement of the Pleistocene Keweenaw sublobe of the Superior lobe. Pleistocene drift deposits vary in thickness from 0 to 300 ft (90 m).

The Limestone Mountain region is drained northeasterly by the Otter, Silver, and Sturgeon rivers. Hoehl (1981) reports that the Otter River north of Sherman Hill is apparently structurally controlled, whereas there is no evidence of such control influencing the other two rivers. Most regional streams cut sharp-sided valleys into the clayey to silty soil.

The stratigraphy of Limestone Mountain is shown in Figure 3 and is revised from the paleontological work of Case and Robinson (1914).

Outcrops and water-well drilling records establish the Jacobsville Sandstone as bedrock in the area surrounding Limestone Mountain.

Regionally the Jacobsville Sandstone overlies the Powder

Mill Group of lavas (Fig. 3) and in turn is overlain unconformably by the Munising Formation of Upper Cambrian age. In the Limestone Mountain area the Cambrian rocks are overlain by dolomites ranging from Middle Ordovician to Middle Devonian in age.

Hoehl (1981) notes one water-well record from an area southeast of Limestone Mountain identifying limestone as bedrock. Hoehl was uncertain if the limestone was float or bedrock, but seismic tests established a buried high-velocity zone in the area, suggesting the possibility of limestone bedrock.

While Limestone Mountain is the largest and most complex of the Paleozoic outliers in the Pelkie area, it is best to consider all the outliers in the nearby region when considering Limestone Mountain. A summary of the major structural features of the area follows. (1) On the west side of Limestone Mountain, the basal Cambrian sandstone and overlying dolomites are moderately sheared and dip steeply. (2) There is highly disturbed Cambrian sandstone on the southwestern side of Big Limestone Mountain. This may represent soft-sediment deformation during a time of faulting. (3) On the eastern side of Little Limestone Mountain at the top of the Limestone Mountain Quarry, dolomite dips vertically and consists of brecciated dolomite that is healed. (4) At the south end of Little Limestone Mountain, there is an east-west fault along which the strata on the north side have dropped 20 ft (6 m) with reference to those of the south. (5) Sherman Hill northeast of Limestone Mountain is tilted slightly west. (6) Regionally the Jacobsville Sandstone is flat lying and known to be disturbed structurally in very few places. One mile (1.6 km) west of Pelkie a cluster of Jacobsville Sandstone outcrops shows divergent dips but generally northeasterly strikes. The structure resembles that of Limestone Mountain, but lacks dolomite in the vicinity. (7) Geophysical data indicate a thickening of glacial overburden east of Limestone Mountain. (8) Geophysical data suggests a graben or downwarping of near-surface basement rock east of Limestone Mountain. (9) Gravity and magnetic lows form a partial ring encompassing the Limestone Mountain region. (10) Deformation microstructures have been identified in outcrops of the Jacobsville Sandstone at Limestone Mountain. (11) A fault exists at the south end of Big Limestone Mountain that separates it from Little Limestone Mountain. (12) The stratigraphic sequence of Paleozoic rock units found at Limestone Mountain includes a number of unexplained missing units. (13) Two mi (3 km) to the east of Limestone Mountain a pair of compressed anticlines plunging slightly to the east of north are clearly visible on air photographs. The ridge formed by the anticlines is not more than 6 ft (2 m) high and appears in an M-shaped pattern.

DESCRIPTION

Previous work involving the Limestone Mountain region is extensive. The following is a selected listing and summary of studies pertaining to the Limestone Mountain region.

Jackson (1849) first assessed the Limestone Mountain structure and assigned it a Silurian age based on paleontological evidence. The following year, Foster and Whitney (1850) offered the first explanation for the disturbed structures in the Limestone Mountain area. Their conclusion attributes noted structural deformation to core uplifting from some unknown volcanic process (cryptovolcanic).

Rominger (1873) states he could find no positive facts explaining the causes that produced the inclined position of the carbonate beds at Limestone Mountain. Rominger concludes that the structural irregularities are due to underwashing and the subsequent sinking of Paleozoic strata during glaciation, and are unrelated to upheaval at an earlier date.

Wadsworth (1891) reported extensively on the unpublished paleontological and cross-section work done on Limestone Mountain by W. L. Honnold of the Michigan Geological Survey. Based on this earlier work, Wadsworth theorized in 1897 that Limestone Mountain is the remnant of a synclinal or oblong basin-shaped fold.

Lane (1905) reports identifying a four-tiered surface expression on Limestone Mountain and attributed this to wave action from the ancient Great Lakes. Further paleontological investigations of the site were carried out and reported by Lane

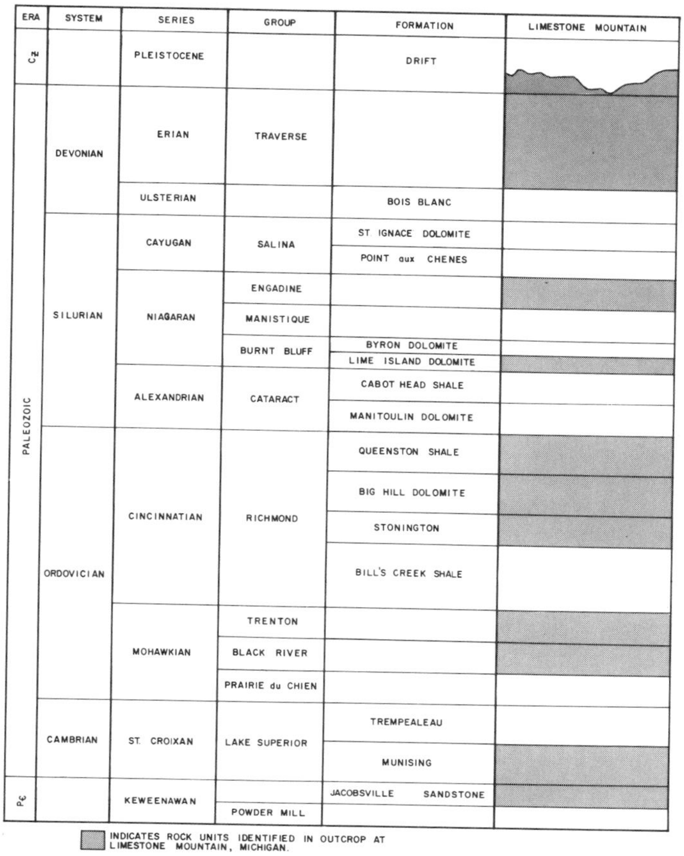

Figure 3. Stratigraphic column indicating rock units both missing and identified in outcrop at Limestone Mountain, Michigan.

(1911). Lane's findings agreed with Jackson's earlier work that Limestone Mountain was of Silurian age.

A major investigation by Case and Robinson (1914) included palentological studies along with structural and stratigraphic interpretations. In this study the authors were unable to propose any hypothesis for the preservation of the Paleozoic outliers. They concluded that the upper layers of Paleozoic rocks showed no peculiar hardness or consistency which would have enabled them to resist erosion and that faulting does not account for their preservation. Case and Robinson speculated that the structural deformation noted is best attributed to landslides or slumps due to undercutting by surface or underground waters in comparatively recent times. Their investigation further suggested the presence of numerous faults associated with the structure at Limestone Mountain. The largest fault is believed to be located between Big and Little Limestone mountains and may be responsible for 80 ft (24 m) of offset in the top of the Jacobsville Sandstone between the two structures (Fig. 4). The major contribution of Case and Robinson's investigation is the exhaustive description and identification of Paleozoic rocks noted at Limestone Mountain. Their findings show that rocks of not only Pre-

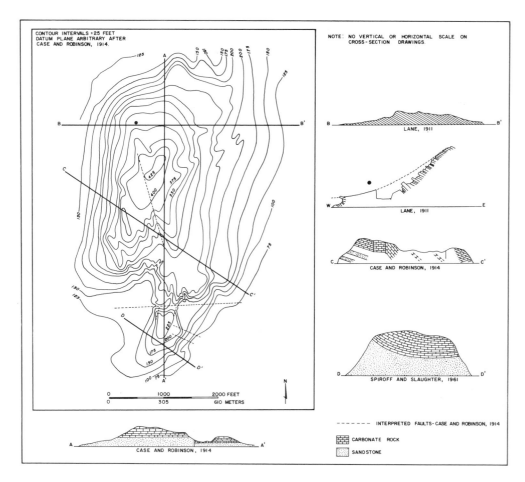

Figure 4. Cross-section interpretations through Limestone Mountain, Michigan, 1911 to 1961.

cambrian age, but those of Cambrian through Middle Devonian age are present and involved in the structural deformation at Limestone Mountain (Fig. 3). This is of great significance as no other rocks of Middle Devonian age appear in outcrop in Michigan's Northern Peninsula. The nearest outcropping of Middle Devonian rocks occur in Michigan's Southern Peninsula west of the Straits of Mackinac on Beaver Island, 175 mi (280 km) to the southeast.

Roberts (1940) concludes that Limestone Mountain is a syncline with a northerly strike and a shallow plunge toward the north. While Roberts defines an easterly striking fault between Big and Little Limestone mountains, with a minimum of 150 ft (46 m) displacement, he is unable to conclude the exact direction of movement. Further, Roberts does not speculate as to whether faulting or folding is the most important factor in creating the structure, though he states that faulting is the most plausible.

Thwaites (1943) speculates that Limestone Mountain may be caused by a major fault on the east side of the structure. As evidence, Thwaites notes vertical bedding and brecciated but healed dolomite on the eastern side of Little Limestone Mountain. Inclined bedding noted at the southeastern corner of Big

Limestone Mountain is attributed to drag along the projected fault.

Spiroff (1952) concludes that Limestone Mountain is the remnant of a graben structure and the Paleozoic rocks found there were preserved by some uncommon glacial movements.

Hamblin (1958) suggests that Limestone Mountain is the remnant of a major regional structure related to compressional forces that caused the Keweenaw fault. He further suggests that Limestone Mountain is a syncline flanked to the west by an anticline of equal or greater magnitude and to the east by an anticline, monocline, or fault.

Based on geophysical data, Aho (1969) suggests Limestone Mountain is the result of basement faults that existed prior to deposition of the Jacobsville Sandstone, and were reactivated during subsequent regional events.

Read (1970) finds that thin sections of quartz and feldspar from the disturbed Jacobsville Sandstone at Limestone Mountain exhibit microstructures similar to those found in quartz and feldspar from generally accepted astroblemes. While not concluding that the Jacobsville Sandstone is "shocked," Read suggests that Limestone Mountain and its accompanying outliers represent the remnants of an encircling graben or downwarp associated with an Ordovician astrobleme.

In a report concerning the potential for diamond-bearing kimberlites in Northern Michigan and Wisconsin, Cannon and Mudrey (1981) speculate that Limestone Mountain and the adjacent Sherman Hill are cryptovolcanic structures formed as the result of collapse over deeper-seated kimberlite pipes. Based on fragments of Early Devonian rocks found at Limestone Mountain, Cannon and Mudrey assign a Devonian or later age to the structures.

In a geophysical investigation of the Limestone Mountain area, Hoehl (1981) concludes that Limestone Mountain is a fault-controlled, north-south striking, synclinal structure flanked on the west by a monocline, with the latest faulting and folding in the area occurring during Late Mississippian time in conjunction with movements in the Michigan Basin. Hoehl's geophysical data indicates that there is no evidence to suggest the existence of a kimberlite pipe or any other high-density, magnetic, near-surface body of igneous rock associated either locally or regionally to the Limestone Mountain–Sherman Hill area.

Additional work by Read (written communication to R. C. Reed, Michigan Geological Survey, 1984) implies the existence of a partial ring of gravity and magnetic lows associated with the Limestone Mountain region (Fig. 5). Read asserts that both anomalies are consistent with a model for a ring graben in which high density and high magnetic basement rock has been dropped down. He maintains, in this instance, the down-dropped rock is basalt of the Powder Mill Group. Read's work further identifies a pair of compressed north-plunging anticlines slightly to the east of north of Limestone Mountain. Read states that the type and pattern of these folded anticlines does not fit with a pattern commonly identified with normal faulting alone. He feels they do, however, compare favorably with the complex combination of faults and folds that are often identified in conjunction with astroblemes found in areas of flay-lying sedimentary rocks.

Milstein (1986), in reviewing structures of a cryptoexplosive nature found in the Michigan Basin, concludes that Limestone Mountain, Sherman Hill, and similar disturbed outcrops of Paleozoic rock found near Pelkie, Michigan, should be classified as cryptoexplosive disturbances. Milstein further states that while no solid evidence implicating impact origin for the structures exists, a large body of interpretive data suggests the possibility that the cryptoexplosive disturbances in the Limestone Mountain area are remnants of an impact structure as defined by Grieve and others (1981).

As can be noted, the Limestone Mountain area has been the subject of numerous investigations. To date, no consensus has been reached to explain these interesting outliers or their accompanying structure. Visitors to Limestone Mountain are reminded that they will be in a rural/wilderness area and that proper precautions and preparations should be taken. Rugged clothing, boots, compass, and insect repellent are essential. Due to the unique nature of these outliers, the taking of samples is discouraged.

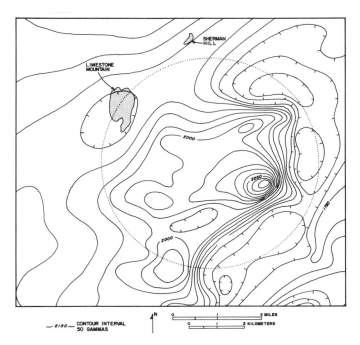

Figure 5. Aeromagnetic map of the Limestone Mountain region. Dotted circle indicating partial ring of magnetic lows (modified from Read, 1984).

SITE DESCRIPTIONS

A series of stops has been outlined to acquaint the visitor with the structure and stratigraphy of the Limestone Mountain outliers.

Stop 1. Sherman Hill. To reach Sherman Hill, proceed north from Pelkie, Michigan, on Pelkie Road 0.5 mi (0.8 km) to the intersection of Eiloca Road. Turn west and continue 1.5 mi (2.4 km). Sherman Hill and a small quarry operation located at the site will be visible on the north side of the road. Sherman Hill is located on private property, and permission to enter the grounds must be obtained from Mr. Reuben Turunen.

While no fault is visible at the surface in the area, at the southwest end of the quarry, fault breccia and fracture surfaces suggest the existence of a major fault striking to the northwest.

Stop 2: Big Limestone Mountain. To reach Big Limestone Mountain from Pelkie, go south on Pelkie Road 0.5 mi (0.8 km) to the intersection of Papin Road. Proceed west on Papin Road 2.5 mi (4 km) to the junction of Limestone Mountain Road. Turn south and continue 0.75 mi (1.2 km). The entrance to Limestone Mountain Quarry is marked by a pair of large gateposts.

This quarry is owned by the Limestone Mountain Company, but there are no local offices. There is no barrier at the entrance to the access road and presumably no objection to visitors. One may drive straight into the quarry.

Active quarry operations have not taken place for more

than 15 years, and exposed dolomite beds appear somewhat discolored by weathering.

This quarry is the only locality among the anomalous Paleozoic outliers where a large mass of rock with steep dips can be seen. Dips throughout most of the quarry are in excess of 45°, but at the quarry's north-end beds dip nearly vertical.

By returning to Limestone Mountain Road and continuing south 0.7 mi (1 km), an access road on the north side will lead to the western flank of Limestone Mountain. On this western flank, large dolomite ledges are exposed. Permission to enter this part of Limestone Mountain must be obtained from Mr. Raymond Clark whose house is located on the access road.

The exposed ledges suggest dolomite reaches a thickness of more than 180 ft (55 m) at this outcrop. The dolomite ledges are continuous all the way to the north end of the mountain, and their continuity is only occasionally broken by beds of buff-colored deformed sandstone.

Stop 3: Little Limestone Mountain. Permission to enter Little Limestone Mountain must be obtained from Mr. Raymond Clark. Little Limestone Mountain is reached by travelling 0.4 mi (0.6 km) on Limestone Mountain Road from the Clark access road. On the north side of the road a north-south section-line fence can be seen. This fence runs to the top of Little Limestone Mountain.

About 500 ft (152 m) in from the road, the ground begins to rise steeply and moss-covered ledges of dolomite can be seen to the northeast. These dolomite beds are similar to those seen in the Big Limestone Mountain Quarry, except the dips are less steep (northwestward).

By following the fence up the slope, the visitor will eventually encounter a small, nearly vertical fault, striking 70° east of north. South of the fault, bedding dips steeply northeast; north of the fault, bedding is nearly vertical.

Eventually the section fence will turn, and at this point the visitor should continue in a north line rather than follow the fence. At a distance of 250 ft (76 m) a small prospect pit can be found. A similar pit can be found about 125 ft (38 m) to the northwest. In both pits, the dip of the bedding is approximately 30° northeast.

Stop 4: Pair of plunging anticlines. To reach the anticlines from Pelkie, travel south 0.25 mi (0.5 km) to the intersection of Papin Road. Turn west on Papin Road and proceed 1 mi (1.6 km) to Marshall Road. At this location the visitor will be at the middle portion of a pair of north-plunging anticlines composed of Jacobsville Sandstone. The anticlines lie on the north side of Papin Road. The folds accompanying these anticlines are greatly eroded and easily escape notice at ground level but are clearly visible in air photos. What is visible at ground level is a low ridge of outcrop whose structure resembles the letter "M" (Fig. 1). It is easier to follow the structure on the eastern anticline. Permission to traverse this property must be obtained from Mr. George Maki, whose house is located on the north side of Papin Road.

REFERENCES CITED

Aho, G. D., 1969, A reflection seismic investigation of thickness and structure of the Jacobsville Sandstone, Keweenaw Peninsula, Michigan [M.S. thesis]: Houghton, Michigan Technological University, 104 p.

Cannon, W. F., and Mudrey, M. G., 1981, The potential for diamond-bearing kimberlite in Northern Michigan and Wisconsin: U.S. Geological Survey Circular 842, p. 7–11.

Case, E. C., and Robinson, W. I., 1914, The geology of Limestone Mountain and Sherman Hill in Houghton County, Michigan: Michigan Geological and Biological Survey Publication 18, Geologic Series 15, p. 167–181.

Foster, J. W., and Whitney, J. D., 1850, Report on the geology and topography of a portion of the Lake Superior land district in the State of Michigan: Thirty-first Congress Executive Document, no. 69, part 1, p. 117–119.

Grieve, R.A.F., Robertson, P. B., and Dence, M. R., 1981, Constraints on the formation of ring impact structures based on terrestrial data, *in* Schultz, P. H., and Merrill, R. B., eds., Multi-ring Basins, Lunar and Planetary Science Conference, 12th, Houston, 1980: New York, Pergamon Press, p. 37–47.

Halls, H. C., 1966, A review of the Keweenawan geology of the Lake Superior region: American Geophysical Union Monograph 10, p. 3–27.

Hamblin, W. K., 1958, The Cambrian sandstone of Northern Michigan: Michigan Geological Survey Publication 51, 121 p.

Hoehl, E. J., 1981, Geophysical investigation of the Limestone Mountain area of Baraga and Houghton counties, Michigan [M.S. thesis]: Houghton, Michigan Technological University, 141 p.

Jackson, C. T., 1849, Message from the President of the United States, to the Two Houses of Congress, at the Commencement of the First Session of the Thirty-First Congress: Thirty-First Congress Executive Document, no. 5, p. 398–401, 452–453.

Lane, A. C., 1905, Correlation across Detroit River: Michigan Geological Survey Annual Report, p. 178–179.

—— , 1911, Globe and challenge explorations, Limestone Mountain: Michigan Geological and Biological Survey Publication 6, Geologic Series 4, v. 2, p. 523–524.

Milstein, R. L., 1986, Impact origin of the Calvin 28 cryptoexplosive disturbance, Cass County, Michigan [M.S. thesis]: Greeley, University of Northern Colorado, 87 p.

Meshref, W. M., and Hinze, W. J., 1970, Geologic interpretation of aeromagnetic data in western Upper Peninsula of Michigan: Michigan Geological Survey Report of Investigation 12, scale 1:250,000, 25 p.

Read, W. F., 1970, Is the Limestone Mountain structure an astrobleme?: Abstracts from the Sixteenth Annual Institute on Lake Superior Geology, Thunder Bay, Ontario, Lakehead University, p. 36.

Roberts, E., 1940, Geology of the Alston district, Houghton and Baraga counties, Michigan [M.S. thesis]: Pasadena, California Institute of Technology, 48 p.

Rominger, C., 1873, Paleozoic rocks: Michigan Geological Survey Bulletin, v. 1, part 3, p. 69–71.

Spiroff, K., 1952, Sandstones near L'Anse, Michigan: Rocks and Minerals, v. 27, nos. 3–4, p. 149–150.

Spiroff, K., and Slaughter, A. E., 1961, Geologic features of parts of Houghton, Keweenaw, Baraga, and Ontonagon counties, Michigan: Michigan Basin Geological Society Annual Field Excursion Guidebook, p. 36.

Thwaites, F. T., 1943, Stratigraphic work in Northern Michigan, 1933–1941: Michigan Academy of Science Paers, v. 28, p. 487–502.

Wadsworth, M. E., 1891, Scientific intelligence: American Journal of Science, v. 42, third series, p. 170–171.

—— , 1897, The origin and mode of occurrence of the Lake Superior copper deposits: American Institute of Metallurgical Engineers Transactions, v. 27, p. 684–685.

Porcupine Mountains Wilderness State Park, Michigan

Robert C. Reed, Michigan Geological Survey, Lansing, Michigan 48909

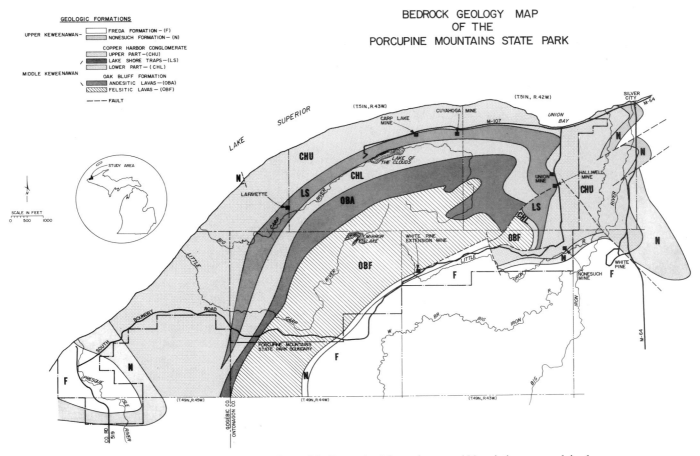

Figure 1. Location and bedrock geology of the Porcupine Mountains area. Abbreviations are explained in the text.

LOCATION

The Porcupine Mountains State Park is located in the northwestern part of Michigan's northern peninsula. The park occupies all of T.50N.,R.45W.; T.51N.,R.43W.; T.51N.,R.44W.; and portions of T.49N.,R.45W.; T.50N.,R.43W.; T.50N.,R.44W.; and T.51N.,R.42W. in Ontonagon and Geogebic Counties (Fig. 1). The area of the park is covered by the Carp River and Tiebel Creek 7½ by 15–minute Quadrangles and the Aldrige Creek, Bergland NE, Government Peak, and White Pine 7½-minute Quadrangles. The principal entrance to the park is via Michigan 107 west of Silver City. Silver City is located 13 mi (21 km) west of Ontonagon on Michigan 64 and 18 mi (29 km) north of Bergland on Michigan 64. Alternate access is provided by County Road 519 16 mi (26 km) north of Wakefield. County Road 519 intersects the park's South Boundary Road which extends north and east to Michigan 107. A visitors' center is located south of the intersection of Michigan 107 and the South Boundary Road. For park visitors, it is extremely advisable not to feed the bears. Further information about Porcupine Mountains State Park may be obtained from Park Manager, Porcupine Mountains State Park, 599 M-107, Ontonagon, Michigan 49953.

SIGNIFICANCE

The Porcupine Mountains Wilderness State Park contains one of the few remaining wilderness areas in the midwest. Also known as the Porkies, it was named by the Chippewa Indians who noted the domal area's resemblance to a crouching porcupine. The Porcupines is the largest of Michigan's state parks, comprising more than 58,000 acres. It contains some 30,000 acres of uncut timber, the largest stand of virgin wilderness/ northern hardwood forest east of the Mississippi River. The park contains four secluded lakes and miles of wild rivers and streams. More than 85 mi (137 km) of foot trails are maintained by the Michigan Department of Natural Resources. The Porkies have some of the highest elevations in the midwest. Summit Peak, in Sec. 11,T.50N.,R.44W., is 1,946 ft. (593 m) above sea level. Summit Peak road and trails leading to the peak are located on the Tiebel Creek Quadrangle.

Among the many attractions the park has to offer are old copper mines. The major mines are located on Figure 1. The geology of the mines has been poorly recorded. Apparently, the LaFayette, Carp Lake, Cuyohoga, Union, and Halliwell mines were explored along the upper part of the Lake Shore Traps and the base of the upper part of the Copper Harbor Conglomerate (Fig. 1). The Nonesuch and White Pine Extension properties were in the Nonesuch Shale. Most of the copper produced in the Porcupine Mountain area, as elsewhere in Michigan's copper range, was in its elemental state and referred to as native copper, although most of the copper from the Nonesuch Shale is from the sulphide chalcocite. At least 53 mining permits were granted in this area. Prospectors were in the Porcupine Mountains as early as 1844, and in 1845 there were two active mining operations, the LaFayette and Union mines. Of those mines located on Figure 1, only two within the park recorded production; the Carp Lake produced 15 tons, and the Nonesuch 180 tons. In the extreme western part of the park, the Nonesuch Shale of the Presque Isle syncline has been explored and reported to contain 95,254,400 tons of ore with an average grade of 1.27 percent copper (Wilband, 1978). A detailed description of the history of early mining in this area is presented by Jamison (1950). The premier mining operation within the Porcupine Mountain area is at White Pine about 2 mi (3 km) east of the park boundary (Fig. 1). The White Pine mine was first opened in 1881, produced 2,104 tons of copper in 1916, and closed in 1921. A new venture began construction in 1951 and from 1954 through 1986 produced 1,451,186 tons of refined copper. The ore-bearing horizon at White Pine, the Nonesuch Shale, contains small amounts of liquid hydrocarbons, commonly referred to as Precambrian oil and, in recent years, has excited some segments of the petroleum industry. This occurrence had previously been reported in a Michigan Geological Survey publication as small quantities of bituminous oil in the Nonesuch beds and as the first sign of life in the Precambrian (Rominger, 1894). The hydrocarbon seeps at White Pine have been investigated by the Geophysical Laboratory, Carnegie Institution of Washington (Hoering and Ableson, 1964). They reported that the petroleum, benzine extract, and kerogen are typical petroleum fluids with a rather high proportion of paraffin hydrocarbons. It was further concluded that the evidence indicates that this formation contains the products of organisms that lived a billion years ago and that it represents the remains of life dating back one-fourth of geologic time (Hoerin, 1965, 1967, 1976). Radiometric age dating of calcites filling late faults which contain petroleum inclusions indicates an age of $1,047 \pm 35$ Ma, which establishes a minimum age for White Pine oil (Kelly and Nishioka, 1985). The Nonesuch Shale within the park boundary has a strike length of approximately 8 mi (13 km) and probably contains hydrocarbons.

DESCRIPTION

The Porcupine Mountains anticline is a large fold on the southern flank of a very large structural syncline referred to as the mid-continent gravity high or rift. Gravity and magnetic surveys trace this feature from eastern Lake Superior west and south to central Kansas. A smaller gravity high extends from eastern Lake Superior to the south through the southern peninsula of Michigan. The Porcupines result from a large domal warp on a rather normal synclinal limb and represent the largest known structural flexure in the exposed Lake Superior syncline. The structural amplitude between the crest of the Porcupine Mountain anticline and the trough of the Iron River syncline to the east is at least 8,000 ft (2,440 m) and may be more than 11,000 ft (3,350 m) (Hubbard, 1975).

Rock formations of the Porcupine Mountains area include one of igneous origin and three sedimentary formations with a combined thickness of about 8,000 ft (2,240 m). They are Middle Proterozoic in age, dated about 1 Ga. Formation descriptions are principally from Hubbard (1975).

The oldest rock unit, the Oak Bluff Formation (formerly Unnamed Formation), has been separated into a series of andesite flows (OBA, Fig. 1) and one or more felsic flows (OBF). The Oak Bluff contains some interflow sediments, principally sandy conglomerate usually containing pebbles and cobbles less than 4 in (10 cm) in diameter, but some boulders as much as 3 ft (1 m) in diameter occur. Most andesite flows are porphyritic and contain oligoclase, pyroxene, and 10 to 30 percent opaque minerals consisting of hematite, magnetite, and ilmenite. Flow tops generally contain sparse vesicles filled with chlorite, epidote, quartz, and calcite. In this area the andesite flows are about 2,000 ft (610 m) thick. The felsite flow or flows form the core of the Porcupine Mountains. This rock contains very sparse quartz phenocrysts in a devitrified cryptocrystalline matrix. It is about 500 ft (155 m) thick.

Above the Oak Bluff lies the Copper Harbor Conglomerate (CH) composed predominantly of sandstone and siltstone with subordinate beds of conglomerate. The composition includes rock fragments, quartz, feldspar, and small amounts of epidote, amphibole, and chlorite. Rock fragments are predominantly intermediate and felsic volcanic rocks, but also include metamorphic rocks. Most of the conglomerate beds are less than 5 ft (1.5 m), although some are as thick as 25 ft (8 m). The pebbles and cobbles are generally less than 3 in (8 cm) but may be as large as 9 in (23 cm). The maximum thickness of the Copper Harbor Conglomerate in the Porcupine Mountains area is about 5,000 ft (1,525 m).

At least six lava flows are interbedded with the Copper Harbor Conglomerate. They have been named the Lake Shore Traps (LS) and form a prominent linear escarpment from which a large area of the park and Lake Superior may be viewed

Figure 2. View of Lake of the Clouds (Fig. 1) from the escarpment of the Oak Bluff Formation.

(Fig. 2). The flows range in thickness from 15 to 70 ft (4 to 21 m). The rocks are fine-grained and generally porphyritic; vesicles are filled with chlorite, quartz, and epidote. Petrographically they resemble flows of the Oak Bluff Formation.

Above the Copper Harbor Conglomerate is the Nonesuch Formation (N). Most of the Nonesuch is a thinly laminated siltstone and very fine to fine-grained sandstone. The basal part is a shale mineralized with chalcacite and with some native copper. The siltstone and sandstone is composed principally of fragments of pre-Keweenawan rocks including schists; mafic, intermediate, and felsic volcanic rocks; and quartz and feldspar. Opaque grains are abundant. Within the Porcupine Mountains area, the Nonesuch Formation is about 600 ft (185 m) thick.

The youngest rocks exposed in the Porcupine Mountains area make up the Freda Formation (F). The Freda is a cross-bedded, moderately well sorted sandstone and coarse siltstone derived from a terrain of metamorphic and plutonic rocks, lower Keweenawan volcanic rocks, and iron-formation. A layer of sandy conglomerate to conglomeratic sandstone 12 to 67 ft (4 to 20 m) thick occurs about 500 ft (150 m) above the base. Along the Montreal River, which separates Michigan from Wisconsin, the Freda Formation is about 12,000 ft (3,660 m) thick.

Many natural features may be observed in the Porcupine Mountains State Park along roads and trails located on the above cited topographic quadrangle maps. Union Spring, near the center of Sec. 20,T.51N.,R.42W., Government Peak Quadrangle, is the second largest natural spring in Michigan. Access is via Union Spring Trail about 2 mi (3.2 km) from the South Boundary Road in Sec.27,T.51N.,R.43W. More than 700 gallons per minute (3,182 L) flow from the spring into the Union River. The Union Spring Trail intersects the Government Peak Trail about 2 mi (3.2 km) west of Union Spring. East of the intersection, where the trail crosses Carp River inlet, outcrops of andesitic lavas of the Oak Bluff Formation may be observed. About 0.5 mi (0.8 km) south and 2 mi (3.2 km) west of the intersection, along Government Peak Trail, is Government Peak in Sec.26,T.51N.,R.43W. This peak, at 1,850 ft (564 m), is the second highest point in the park. The bedrock of Government Peak is felsitic lavas of the Oak Bluff Formation.

Summit Peak, Sec.11,T.50N.,R.44W., Tiebel Creek Quadrangle, at 1,958 ft (597 m) is the highest elevation in the park, and the summit affords an exciting view of the park and of Lake Superior. Summit Peak Road extends 2.5 mi (4 km) north of South Boundary Road from Sec.13, T.50N.,R.44W., and Side Trail continues another 1 mi (1.6 km) to Summit Peak. North of Summit Peak in Sec.2,T.50N.,R.44W., Carp River Quadrangle, along the South Mirror Lake Trail is Mirror Lake, the highest lake in the park at an elevation of 1,532 ft (467 m). Felsitic lavas of the Oak Bluff Formation are well exposed on Summit Peak and surrounding Mirror Lake.

An interesting exposure of Copper Harbor Conglomerate was described by Wright (1905, p. 41). It is located at the mouth of the Carp River, Sec.33,T.51N.,R.44W., Carp River Quadrangle. In addition to rounded pebbles of felsite, melaphyre, and sandstone, jaspilite iron formation pebbles 1 to 6 in (1 to 15 cm) in diameter form part of the conglomerate. The nearest known iron formation is on the Gogebic iron range nearly 30 mi (48 km) to the south. The exposure is along the Lake Superior Trail intersected by the Pinkerton Creek Trail originating from the South Boundary Road in Sec.14,T.50N.,R.45W., Tiebel Creek Quadrangle. For the spirited, undaunted hiker, an infaulted exposure of Nonesuch Shale (Fig.1) is located in Sec.26,T.51N.,R.44W., Carp River Quadrangle, along the Lake Superior Trail approximately 2.5 mi (4 km) northeast of the mouth of the Carp River.

The Lake Shore Traps are well exposed at the end of Michigan 107 in Sec.21,T.51N.,R.43W., Carp River Quadrangle.

Copper exploration and mining in the Nonesuch Shale at the Nonesuch and White Pine Extension mines have been previously mentioned. There are excellent exposures of Nonesuch at and near the mouth of the Presque Isle River in Sec.19 and 30,T.50N.,R.45W., Tiebel Creek Quadrangle. A trail trending north about 0.25 mi (0.4 km) from the termination of County Road 519 leads to a footbridge across the Presque Isle River. Here the Nonesuch, with numerous potholes, may be observed.

South of the Nonesuch Shale in Sec. 30 described above, the Freda Formation is well exposed in and along the Presque Isle River continuing south to the South Boundary Road. Almost continuous outcrops of Freda, Nonesuch, and Copper Harbor are in and along the Presque Isle River in Secs. 4 and 9,T.49N.,R.45W.

An interesting observation was made by Wright (1905) while mapping the Porcupine Mountains area. He noted that the kinds of trees over the various formations differ. The felsite is densely covered with hardwood, while hemlock predominates in other parts. The distribution of tree belts is especially noticeable in this district. In certain parts the boundary lines of various tree belts are so sharply defined that they can be mapped accurately.

REFERENCES CITED

Hoering, T. C., 1965, The extractable organic matter in Precambrian rocks and the problem of contamination: Annual Report of the Director of the Geophysical Laboratory, Carnegie Institution of Washington Yearbook 64, p. 215–218.
——, 1967, Criteria for suitable rocks in Precambrian organic geochemistry: Annual Report of the Director of the Geophysical Laboratory, Carnegie Institution of Washington Yearbook 65, p. 365–372.
——, 1976, Molecular fossils from the Precambrian Nonesuch Shale: Annual Report of the Director of the Geophysical Laboratory, Carnegie Institution of Washington Yearbook 75, p. 806–813.
Hoering, T. C., and Ableson, P. H., 1964, Chemicals from the Nonesuch Shale of Michigan: Annual Report of the Director of the Geophysical Laboratory, Carnegie Institution of Washington, Yearbook 63, p. 262–264.
Hubbard, H. A., 1975, Geology of the Porcupine Mountains in the Carp River and White Pine quadrangles, Michigan: U.S. Geological Survey Journal of Research, v. 3, no. 5, p. 519–528.
Jamison, J. K., 1950, The mining ventures of the Ontonagon Country: Ontonagon, Michigan, Ontanagon Herald Company, 90 p.
Kelly, W. C., and Nishioka, G. K., 1985, Precambrian oil inclusions in late veins and the role of hydrocarbons in copper mineralization at White Pine, Michigan: Geology, v. 13, p. 334–337.
Rominger, C., 1894, Geological report on the upper peninsula of Michigan; Michigan Geological Survey, v. 5, pt. 1, p. 163.
Wilband, J. T., 1978, The copper resources of Northern Michigan: U.S. Bureau of Mines contract J0366067, p. 38.
Wright, F. E., 1905, Report of progress in the Porcupines: Michigan Geological Survey Report 1903, p. 33–44.

Huron River: Precambrian unconformities and alteration at and near Big Eric's Crossing, Michigan

J. Kalliokoski, Department of Geology and Geological Engineering, Michigan Technological University, Houghton, Michigan 49931
Jeffrey S. Lynott, P.O. Box 603, Woodruff, Wisconsin 54568

LOCATION

Big Eric's Crossing (Fig. 1, Site 1) is a county bridge across the Huron River (NW¼NW¼Sec.35,T.52N.,R.30W.), Skanee, Michigan 15-minute Quadrangle, about 12 mi (19 km) northeast of L'Anse, Michigan, along the road that continues north from Main Street. The second locality (Site 2), a small, southeast-facing granite hill, (SE¼SE¼Sec.26) is about 1 mi (1.6 km) northeast of the bridge and lies about 600 ft (200 m) south of the road.

Big Eric's Crossing is a famous fishing spot on the Huron River, and Baraga County maintains a campground at the bridge. The river is high from late September until late May, so the outcrops along the river are best visited during the middle and late summer.

Sites 2 and 3 are on private land. However, so long as visits are brief, and visitors continue to respect the surface and the exposures, property owners in the Upper Peninsula tend not to post their lands.

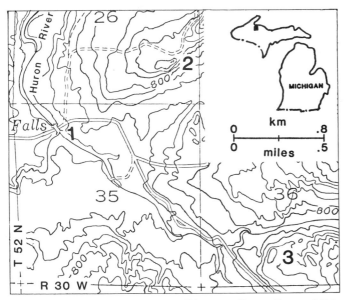

Figure 1. Location map for the Huron River area, Baraga County, Michigan. Site 1, Big Eric's crossing; Site 2, granite bluff.

SIGNIFICANCE

The river outcrop has been known since 1887 for the well-exposed unconformity between the Archean granitic basement and flat-lying Early Proterozoic lean iron-formation, cherts, and calcareous and clastic sedimentary rocks. There is also a patch of Middle Proterozoic red fluvial sandstone. Because of the easy and free access, geology groups have been visiting the unconformity for years, each developing and modifying its observations and interpretations. These notes provide a formal introduction to this area and also add some new ideas regarding the nature of the alteration along the unconformity.

These outcrops show evidence of two contrasting types and ages of alteration (Lynott, in progress). The earlier is restricted to the Archean rocks and is interpreted to be a paleoregolith, a product of subaerial weathering that took place prior to the deposition of the Early Proterozoic sedimentary strata. Subsequent to this deposition, the basement, the paleoregolith, and some of the overlying strata became replaced and cemented extensively by introduced chert and carbonate. This later chert-carbonate alteration is particularly significant because it is identical in mode of occurrence, textures, and structures to that reported from Archean strata below several other Proterozoic iron-formations in North America. We interpret the extensive silicification and carbonatization at the two localities to represent sea-water alteration of those rocks and strata that predate the iron formation, during a marine transgression and subsequent chemical (iron-formation) sedimentation (Lynott, in progress).

ARCHEAN BASEMENT ROCKS AND THEIR WEATHERING

At Site 1, the oldest rocks are a white, coarse-grained, massive granite, cut by one larger and two smaller diabase dikes (Figs. 2 and 3). The dark material can be identified as an intrusive igneous rock by its geometry, a relict chill margin, and relict diabasic texture visible in thin section. Resting on this basement is a 13–16 ft (4–5 m) sequence of Early Proterozoic conglomerate, quartzite, limestone, and lean cherty iron-formation, possibly correlative with the upper part of the Michigamme Formation of the Baraga Group (Cannon and Gair, 1970). These strata are distal from the nearest metamorphic node of James (1955) and consequently show no evidence of Penokean thermal metamorphism. The upper surfaces of some of the Early Proterozoic strata contain joint-controlled depressions, and these are filled by a few centimeters of Middle Proterozoic Jacobsville Sandstone.

At Site 1, in the vicinity of the highest knob in the river bed, the upper surfaces of both the granite and the diabase are altered over a thickness of about 6 ft (2 m) to paleosaprolite in which feldspars are illitized but the granitic and diabasic textures are recognizable, particularly in thin section. This alteration decreases in intensity downward, and on the surface of the largest body of

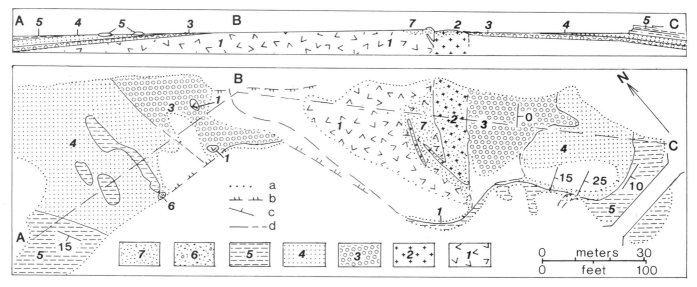

Figure 2. Geology at Site 1, Big Eric's crossing. 1) Archean coarse-grained massive granite; 2) Archean diabase; 3) Lower Proterozoic basal clastics (Fig. 3); 4) laminated chert and clastic material; 5) lean, cherty iron-formation; 6) joint-fillings of Middle Proterozoic Jacobsville Sandstone; 7) area of abundant chert and calcite; (a) limit of outcrop; (b) river bank; (c) inclined bed; (d) contact.

diabase it is associated with spheroidal weathering structures which demonstrably were replaced later by carbonate. Under the overhang on the west bank, the granitic paleosaprolite is overlain by poorly indurated shale that may be either a paleosol or interstratally altered sediment. On both the granitic and diabasic paleosaprolites are small patches of regolith. On granite this is a light colored arkose with rounded grains of illitized feldspar, quartz, and chert, and on the diabase it is a dark calcareous sandstone.

Site 2 (Figs. 1 and 4) is a bluff of garnetiferous, equigranular, gneissic granite with some pegmatitic zones and abundant amphibolite xenoliths. Near the top of the bluff, the feldspars likewise show illite alteration in thin section, and, in places, the illitized gneiss has a faint subhorizontal foliation—almost like a schistosity—interpreted to represent the compaction of weathered granite in early Lower Proterozoic time.

The thickness of the paleosaprolite profiles (up to 3 ft [1 m] at Site 1 and 20 ft [6 m] at Site 2) suggests a shallow groundwater table. Of the two alterations at both localities, the illite alteration is the older in that it has been overprinted by the younger chert-carbonate veining and impregnation.

LOWER PROTEROZOIC SEDIMENTATION

At Site 1 the paleosaprolite and paleoregolith are overlain by a thin (13–16 ft [4–5 m]), laterally continuous sequence of clastic and chemical strata (Fig. 3). The lowest unit is a fairly mature basal conglomerate with well-rounded clasts of quartz, chert, and granite in a highly silicified matrix, overlain by layers of dolomitic sandstone with lenses of cherty limestone and phosphatic material.

The basal clastic unit is overlain by wavy-bedded, laminated chert with interbeds of clastic material, now consisting of quartz grains and flat, angular chert fragments in a chert matrix. Its original lithology is indeterminate. The uppermost unit at Site 1 is a finely laminated, brownish to reddish chert, in part a lean iron-formation, in beds 4–12 in (10–30 cm) thick. Downstream, the iron-formation is overlain along an irregular, jointed surface, by the Middle Proterozoic Jacobsville Sandstone—a fluvial red bed sequence deposited late in the history of the Midcontinent Rift System.

This succession of Lower Proterozoic strata is interpreted as representing initial prograding deposition along a low, sediment starved, gradually submerging shoreline; a period of chert and carbonate deposition with minor clastics represents more stable and quiet conditions. Finally, the deposition of lean iron-formation represents a further decrease in clastic deposition and some fundamental change in the chemistry of the overlying sea water. This succession is much thinner but suggests a sequence of depositional environments somewhat similar to that described by Ojakangas from the Gogebic and Mesabi ranges where sedimentation culminated in the deposition of major iron-formations (Ojakangas, 1983).

SILICIFICATION AND CARBONATIZATION

At Sites 1 and 2, introduced chert and carbonate are widespread in the Archean rocks and, at Site 1, also in strata below the lean iron-formation. Chert and carbonate are most conspicuous as joint fillings, and at both Sites 1 and 2 such veins are more abundant in the upper parts of the granite outcrops. On the northwest side of the granite knob at Site 1 there are several sub-horizontal, symmetrically zoned veins in which thinly banded to colloform gray chert alternates with layers of fine-grained calcite. On the northeast corner of the granite outcrop a taller horizontal cavity is filled with 0.4–1.6 in (1–4 cm) beds of limestone with chert laminae. Chert veinlets also occur in the basal calcareous beds and in some of the higher strata. At Site 2 there are both horizontal and vertical chert veins up to 4 in (10 cm) wide that decrease in abundance downhill. The former are structureless, banded, or colloform, whereas the vertical veins are of chert breccia or chert-granite breccia in a chert matrix. The chert in the vertical veins collapsed, perhaps because of a weak clay surface along the walls. All of these filled structures are interpreted to represent low temperature, open-space deposition from the same seawater from which the overlying chert, lean iron-formation, and carbonate beds were precipitated.

At Site 2 the amphibolite xenoliths show a conspicuous vertical gradation of carbonatization: those xenoliths low on the scarp are fresh, whereas those closer to the top contain progressively more carbonate. At the top of the hill there are masses of dark, impure limestone in highly altered granite, as well as impure limestone blocks resting on the surface.

At both Sites 1 and 2, an additional, inconspicuous silica-carbonate alteration event has taken place. At Site 1, on the highest part of the granite knob, the granite is penetrated by a diffuse network of fine-grained chert-carbonate veinlets (Fig. 2, Unit 7). In thin sections of the granite at Sites 1 and 2, and of the diabase, the previously illitized feldspar can be seen to be almost completely replaced by carbonate. At Site 1 the Lower Proterozoic basal conglomerate on the west bank also contains introduced chert in the thoroughly silicified matrix; chert veinlets

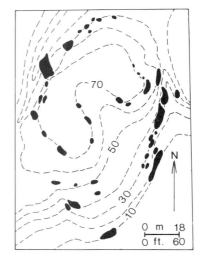

Figure 4. Outcrop map of Site 2; elevations in feet.

also occur in the basal calcareous beds and some of the higher strata. Physically, all of this introduced chert is indistinguishable from that in the Archean basement rocks.

ORIGIN OF THE CHERT-CARBONATE

The widespread occurrence of introduced chert and carbonate in the rocks above and below the Lower Proterozoic unconformity is apparent from field observations. Clues to the origin of these minerals come principally from a consideration of their broader characteristics: 1) The modes of occurrence of chert and carbonate are unlike any described occurrences of modern pedogenic, subaerial caliches or silcretes. 2) The restriction of the introduced chert to a shallow profile at Site 2 suggests that the chert was not introduced from below. 3) Identical carbonatization and silicification of Archean basement rocks below iron-formations has been described from the south margin of the Marquette trough by Gair and Simmons (1969), and from the Mistassini region of Quebec, Canada by Chown and Caty (1983). Silicification below iron formation was recognized by Leith (1925).

Based on these lines of evidence, we propose that the widespread silicification and carbonatization in these outcrops represents the coeval alteration of the basement rocks and preexisting strata by seawater during the general period of iron-formation deposition.

OTHER INTERESTING FEATURES IN THE VICINITY

Microfossils have been reported from the lean iron-formation (Cloud and Morrison, 1979).

Small phosphatic, hummocky features that may be biohermal occur on the west bank, below the iron-formation, downstream from the granite knob (Mancuso and others, 1975). The phosphatic material is radioactive (215 ppm U).

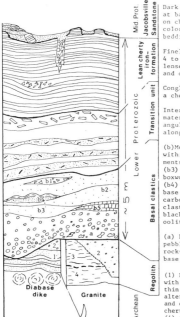

Dark brown, fine to medium grained arkose at base; fills cracks and irregularities on chert surface; overlain by lighter colored medium grained arkose with undulose bedding.

Finely laminated ferruginous chert in beds 4 to 12 inches thick; contains abundant lenses of chert fragments; some beds fractured and oxidized to "soft iron ore".

Conglomerate of silicified angular chips in a cherty matrix.

Interlayered laminated chert and clastic material; latter of quartz grains and flat angular chert fragments; parting occurs along the clastic bedding surfaces.

(b)Medium grained, dark brown quartzite with: (b1) Lenses of angular chert fragments; (b2) Layers of dolomitic sandstone; (b3) Layers of porous sandstone with silica boxwork, give positive test for phosphate; (b4) Radioactive carbonate-apatite, pebbly base overlain by undulating laminae of carbonate and apatite; (b5) Light brown clastic cherty limestone with interlayered black chert, locally steep dips, locally oolitic.

(a) Poorly sorted, rounded quartz and chert pebbles in a silicified, matrix-dominated rock; altered granite clasts abundant at base of unit, above granite.

(1) Black calcareous sandstone; fine grained with very little quartz, interbedded with thin chert layers; (2) Arkose; white rounded feldspars and dark rounded quartz and chert fragments in a black matrix of chert and carbonate; (3) white regolith; (4) red regolith.

Figure 3. Stratigraphic section at Site 1.

At Site 1 the chert veinlets are cut by a radioactive fissure with reddened walls. This may represent Beaverlodge-type uranium mineralization near the sub-Jacobsville unconformity (174 ppm U, 15 ppm Th, [Johnson, 1977]; 1340 ppm U, 150 ppm Cu, [D. Frishman, personal communication]).

On the west bank, downstream from the bridge, two inclined sequences of iron-formation are separated by limey strata, the lithologic repetition suggesting the possible existence of a southeast-dipping thrust fault.

Some parts of the iron-formation have been altered to "soft iron ore." The age of this weathering is unknown.

At Site 1, about 100 ft (30 m) north of the granite knob, some parallel, coarse-grained quartz veins, with galena, cut the granite.

Half-way to Site 2, on the north side of the road, is a low outcrop of chert with a variety of interesting textures, possibly a silicified sedimentary rock.

About 1.3 mi (2.1 km) southeast of Big Eric's Crossing (SE¼SW¼ sec. 36), and 850 ft (260 m) northeast of the gravel road is a small cluster of outcrops on a northwest-facing slope (Fig. 1, Site 3). There is also a shallow prospect shaft. Blocks of carbonatized gneiss and chert are scattered on the surface. In a small pit, a subhorizontal layer of chert rests on granite; some of this chert contains sphalerite. Nearby, pieces of jet-black hydrocarbon can be found.

REFERENCES CITED

Cannon, W. F., and Gair, J. E., 1970, A revision of stratigraphic nomenclature for middle Precambrian rocks in northern Michigan: Geological Society of America Bulletin, v. 81, p. 2843–2846.

Chown, E. H., and Caty, J. L., 1983, Diagenesis of the Mistassini regolith, Quebec: Precambrian Research, v. 19, p. 285–299.

Cloud, P., and Morrison, K., 1979, New microbiotas from the Animikie and Baraga Groups (~2 gyr. old), northern Michigan: Geological Society of America Abstracts with Programs, v. 11, no. 6, p. 227.

Gair, J. C., and Simmons, G. C., 1969, Palmer gneiss—an example of retrograde metamorphism along an unconformity: U.S. Geological Survey Professional Paper 600-D, p. D186–D194.

James, H. L., 1955, Zones or regional metamorphism in the Precambrian of northern Michigan: Geological Society of America Bulletin, v. 66, p. 1455–1488.

Johnson, C. A., 1977, Uranium and thorium occurrences in Precambrian rocks, Upper Peninsula of Michigan [M.S. Thesis]: Houghton, Michigan Technological University, 106 p.

Leith, C. K., 1925, Silicification of erosion surfaces: Economic Geology, v. 20, p. 513–523.

Lynott, J. S., Nature and origin of alteration along overlapping unconformities, Huron River area, Baraga County, Michigan [M.S. Thesis in preparation]: Houghton, Michigan Technological University.

Mancuso, J. J., Lougheed, M. S., and Shaw, R., 1975, Carbonate-apatite in Precambrian cherty iron-formation, Baraga County, Michigan: Economic Geology, v. 70, p. 583–586.

Ojakangas, R. W., 1983, Tidal deposits in the Early Proterozoic basin of the Lake Superior regions—The Palms and Pokegama Formations: Evidence for subtidal-shelf deposition of Superior-type iron-formation, *in* Medaris, L. G., Jr., ed., Early Proterozoic geology of the Great Lakes region: Geological Society of America Memoir 160, p. 49–66.

Pictured Rocks National Lakeshore, northern Michigan

Randall L. Milstein, Subsurface and Petroleum Geology Unit, Michigan Geological Survey, Lansing, Michigan 48912

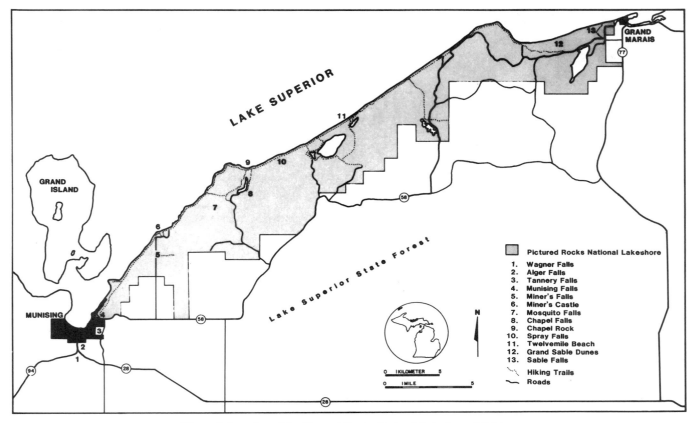

Figure 1. Location of the Pictured Rocks National Lakeshore, Michigan.

LOCATION

Pictured Rocks National Lakeshore, T.48N.,R.19W. and T.49N.,R.13W., Alger County, Michigan; Munising, Wood Island S.E., Grand Portal Point, Trappers Lake, Au Sable Point, Grand Sable Lake, and Grand Marais, Michigan, 7½-minute Quadrangles. The Pictured Rocks National Lakeshore extends from the city of Munising, Michigan, along the Lake Superior shoreline 43 mi (69 km) to Grand Marais, Michigan (Fig. 1). Access to Pictured Rocks by automobile is either from Michigan 28 and Michigan 94 at Munising or Michigan 77 at Grand Marais. Access to the lakeshore can also be gained by hiking trails or boat. Commercial boat tours are available between June 1 and mid-October, weather permitting, and leave the harbor at Munising daily.

SIGNIFICANCE

The geological history of Pictured Rocks National Lakeshore observable to visitors encompasses four distinct systems in geologic time. During the late Precambrian, sediments were deposited in a lacustrine environment over the northern Michigan Basin. During the Cambrian and early Ordovician Periods, sediments were deposited in shallow marine environments that covered the same region. These sediments deposited in the basin's northern portion became the sandstone units that are now exposed within the Pictured Rocks National Lakeshore. Except for their exposure adjacent to Lake Superior, these Cambrian and Ordovician bedrock units are almost completely covered throughout the Michigan Basin by younger sedimentary units or glacial drift.

During the Quaternary Period, ice sheets of all four glacial stages advanced, intermittently, through the Pictured Rocks region. Glacial scouring uncovered the Cambrian and Ordovician units near the shoreline of Glacial Lake Nipissing (ancient Lake Superior). The uncovered bedrock was then exposed to changes in ancient lake levels brought about by fluctuation in water volume and isostatic rebound as the glaciers retreated, freeing the

Figure 2. Cliff faces along the Pictured Rocks stained by surface plant growth and mineral leaching (courtesy Michigan Department of Natural Resources).

Figure 3. Chapel Rocks, Pictured Rocks National Lakeshore, Michigan. (courtesy Michigan Department of Natural Resources).

land of their great weight. Some of the most striking features of the Pictured Rocks National Lakeshore are the well-developed shoreline structures related to the changes in ancient lake levels. These dramatic shore features include sea caves, stacks, arches, and wave-cut cliffs (Fig. 1). Additional glacial features within the national lakeshore include deep karst fractures filled with glacial debris, the outwash deposits of the Kingston Plains and the Grand Sable Banks, numerous kettle lakes, and the magnificent wind-formed Grand Sable Dunes.

In 1966, the Eighty-ninth Congress of the United States created the Pictured Rocks National Lakeshore, the nation's first national lakeshore. The national lakeshore includes a 33,550-acre shoreline zone and a 37,850-acre inland buffer zone. Approximately 420,000 people visit Pictured Rocks National Lakeshore each year.

DESCRIPTION

Cambrian and Ordovician bedrock is best exposed in the western one-third of the Pictured Rocks, where dramatic wave-cut cliffs rise more than 200 ft (60 m) above Lake Superior. These picturesque cliffs extend approximately 17 mi (27 km) from Munising to Beaver Basin. The bedrock remains exposed no more than 350 ft (105 m) inland from the escarpments. Surface plant growth and mineral stains impart a dark streaked appearance upon most of the outcrops and cliff faces (Fig. 2). Elsewhere within the national lakeshore, bedrock is found only in the vicinity of the Grand Sable Banks. Here bedrock forms low bluffs around the north and east sides of Au Sable Point, at the gorge of Sable Creek, and at Sable Falls. These latter outcrops are exposures of the Precambrian Jacobsville Sandstone.

The Jacobsville Sandstone is the oldest formation exposed at Pictured Rocks. It is a feldspar-rich, quartz sandstone deep red in color with white mottlings. Throughout the formation, minor amounts of basalt and iron formation can be found (Hamblin, 1958). Although the Jacobsville has a reported thickness of more than 2,000 ft (610 m; Hamblin, 1958), only the top few feet rise above lake level.

The Middle Cambrian Munising Formation, which lies unconformably above the Jacobsville Sandstone, is divided into two members. Both members were deposited in near-shore marine beach environments. The lower member is the Chapel Rock. Along the Pictured Rocks the Chapel Rock is approximately 50 ft (15 m) thick. The member, however, appears to thin eastward from Grand Marais, and little is known as to the member southward. In the central subsurface Michigan Basin, it is equivalent, at least in part, to the Dresbach, Eau Claire, and Mount Simon Sandstones.

The lower 15 ft (4 m) section of the member is a basal conglomerate. Ninety-five percent of the conglomerate is composed of pebbles of vein quartz, quartzite, and chert. The remaining 5 percent is composed of slate, basalt, granite, iron formation, and sandstone. The upper 10 ft (3 m) of the Chapel Rock is a pink or light buff to brown, well-sorted medium-grained, ortho-quartzitic sandstone characterized by large-scale cross-bedding. Several blue-shale beds also appear in this section (Hamblin, 1958). Mud cracks, ripple marks, clastic dikes, clay pellets, and sand concretions can be found in the Chapel Rock.

The Miner's Castle Member makes up the upper 140 ft (43 m) of the Munising Formation. The member is a poorly sorted, somewhat friable, light-yellow gray, silty-shaley quartz sandstone, which is characteristically cross-bedded (Ostrom and

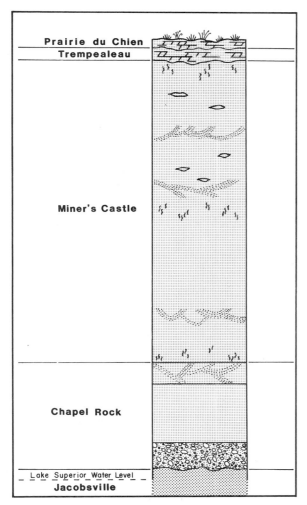

Figure 4. Generalized geologic cross section of lithologies exposed along the Pictured Rocks National Lakeshore.

Figure 5. Sea caves beneath Grand Portal Point, Lakeshore Trail, Pictured Rocks National Lakeshore, Michigan (courtesy Michigan Department of Natural Resources).

Slaughter, 1967). The lower-most beds contain thin lenses of blue shale. Mud cracks, ripple marks, large concretions, and signs of bioturbation are noted. A fossil zone containing trilobite *Prosaukia curvicastata* is present roughly 5 ft (1.5 m) below the contact with the overlying Trempealeau. On the basis of the Prosaukia zone, Hamblin (1958) correlates the Miner's Castle with the Franconia in the central basin. The two members of the Munising Formation are best seen at their respective type localities, Chapel Rock (Fig. 3) and Miner's Castle stack.

Above the Miner's Castle Member is the late Cambrian Trempealeau Formation. The Trempealeau is distinctive in that it forms a cap rock on the weaker underlying Miner's Castle. The Trempealeau Formation is evident in the western half of the national lakeshore, cropping out only along the top edge of the cliffs. The Trempealeau is a hard, buff to light brown, mottled, dolomitic sandstone, containing abundant glauconite and minor amounts of chert (Milstein, 1983).

Above the Trempealeau, and similar in appearance and lithology, is the early Ordovician Prairie du Chien Group. Within the Pictured Rocks National Lakeshore, the Prairie du Chien outcrops only at Miner's Falls. Figure 4 is a generalized cross section of the lithologic units present along the Pictured Rocks National Lakeshore.

During Lake Nipissing time, the Cambro-Ordovician rocks, which make the magnificent Pictured Rock escarpment, were subjected to intense wave action. As the lake level lowered to form modern Lake Superior, different rock units were assaulted by varying degrees of wave intensity. While some units easily resisted the water, others slowly weakened, crumbled, and gave way. The result of this erosive process is a shoreline spotted with sea caves, stacks, chimney promontories, and arches (Fig. 5). The most spectacular of these shoreline features are the stack and complex sea cave called Chapel Rock (Fig. 3) and the Miner's Castle stack.

In addition to the shoreline features, Pictured Rocks National Lakeshore contains the Grand Sable Dunes (Fig. 6). The dunes encompass 4 mi^2 (10 km^2) and rise 380 ft (116 m) above Lake Superior. The Grand Sable Dunes are perched dunes. These perched dunes resulted from wind-blown sand deposited on the tops of glacial moraines and other high glacial features that lay near the water's edge at a time when Great Lakes levels were higher relative to the land elevation.

The Pictured Rocks Natural Lakeshore offers a variety of single- and multi-day hikes for the visitor. A hearty day hike that exposes the visitor to the true wonders and beauty of the lakeshore begins at the Chapel parking area, proceeds to Chapel Falls, then on to Chapel Rock. By turning westward and taking the Lakeshore Trail to Mosquito Campground, the hiker skirts the wave-battered cliffs and can gaze along the shear walls at wave-

Figure 6. The Grand Sable Dunes, Pictured Rocks National Lakeshore, Michigan (courtesy Michigan Department of Natural Resources).

carved pillars, arches, and stacks and hear the air thump and feel the ground shake as waves crash into sea caves (thunder caves) below. From Mosquito Campground the trail circles back to the Chapel parking area. The trip covers 9 mi (14 km).

The Pictured Rock region is also noted for its many picturesque waterfalls (Fig. 1). The Pictured Rocks National Lakeshore is home to more than 100 varieties of wildflowers (Kuenzer, 1972) and to a full range of other vegetation and wildlife native to the northern hardwood-conifer forests that dominate the region. The wildlife includes the North American Black Bear, and campers and hikers should take adequate precautions with food stores. It was the Pictured Rocks area that Longfellow wrote about in his immortal poem, "Song of Hiawatha." The Alger Underwater Preserve extends from Au Train (west of Munising) to Au Sable Point, covering almost the entire length of the national lakeshore. The preserve contains colorful rocks, weedbeds with fish, and more than a dozen shipwrecks. Addi-

tional information about Pictured Rocks National Lakeshore can be obtained by writing the U.S. Department of the Interior, National Park Service, Pictured Rocks National Lakeshore, Munising, Michigan 49862.

REFERENCES CITED

Hamblin, W. K., 1958, The Cambrian sandstones of Northern Michigan: Michigan Geological Survey Publication 51, 141 p.

Kuenzer, D., 1972, Pictured Rocks National Lakeshore wild flowers: U.S. Department of the Interior, p. 1–6.

Milstein, R. L., 1983, Selected studies of Cambro–Ordovician sediments within the Michigan Basin: Michigan Geological Survey Report of Investigation 26, p. 11–12.

Ostrom, M. E., and Slaughter, A. E., 1967, Correlation problems of the Cambrian and Ordovician outcrop areas, Northern Peninsula of Michigan, *in* Ostrom, M. E., and Slaughter, A. E., eds., Michigan Basin Geological Society Annual Field Excursion, 1967: Michigan Basin Geological Society, p. 1–35.

Middle Silurian paleoecology; The Raber Fossil Beds, Chippewa County, Michigan

Randall L. Milstein, Subsurface and Petroleum Geology Unit, Michigan Geological Survey, Lansing, Michigan 48912

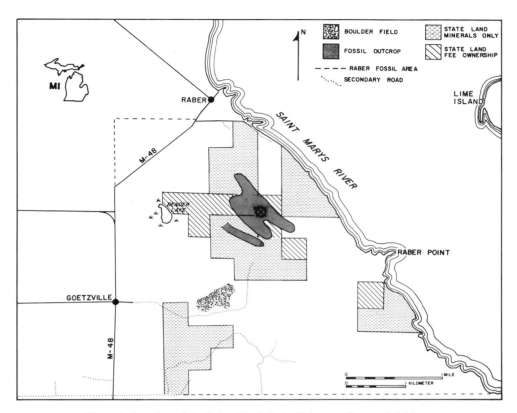

Figure 1. Location of the Raber Fossil Beds, Chippewa County, Michigan.

LOCATION AND SIGNIFICANCE

The Raber Fossil Beds are in Sec.32,33,34,T.43N.,R.3E., and Sec.3–5,8–11,T.42N.,R.3E., Chippewa County, Michigan; Goetzville, Michigan, 7½-minute Quadrangle. The Raber Fossil Beds are located roughly 1 mi (2 km) south and southeast of the village of Raber, on the St. Marys River (Fig. 1).

The area is accessible from two points. A trail from near the Bernard Farm leads in a southeasterly direction 0.5 mi (0.8 km) to the base of a low rock escarpment and follows this escarpment for an additional 0.5 mi (0.8 km) where the trail branches and continues to the top of the ridge. A second point of access is by means of a road extending directly east of Goetzville which is passable for approximately 0.5 (0.8 km) into the area. Other bush trails into the area are not readily passable by automobiles.

The majority of land in which the Raber Fossil Beds exist is owned by the State of Michigan. The Raber Fossil Beds represent a unique natural formation of imposing size. They are significant to geologists and students because of their natural historical value

and to the general public for their unusual scenic appeal. Collecting of fossil specimens on state-owned land is discouraged. Presently, the Raber area is under review as a potential protected Natural Area. Within such a Natural Area the collecting of specimens will be forbidden by law.

Privately-owned land surrounding the state property offers many fine collecting areas. Permission must be gained from land owners prior to entry on their property.

Visitors to the Raber site should prepare themselves adequately for hiking through rough, often uneven wooded terrain. From early spring to late autumn, biting insects can be bothersome.

In addition to the fossil beds, boulder field and bog lake, the Raber area contains the ruins of a structure built of limestone, brick, and lime mortar similar to old fortified buildings on Drummond Island. No historical record of this structure has been uncovered, but its style and location commanding the St. Marys

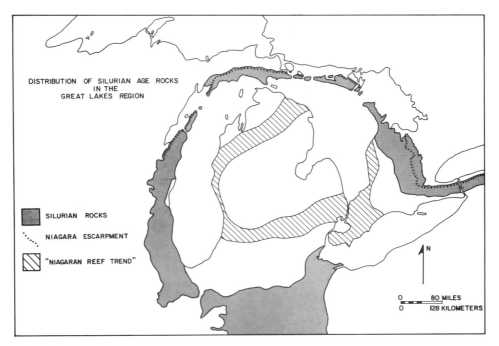

Figure 2. Map shows the distribution of Silurian age rocks in the Great Lakes region, location of the Niagara escarpment and the subsurface "Niagaran Reef Trend."

River channel and St. Joseph Island, and a masonry platform that could well support a cannon, strongly suggest the site was used by the British in connection with their early colonial Fort St. Joseph.

GEOLOGIC SETTING

The topography of the Raber Fossil Area is dominated by an escarpment that runs in a northwest-southeast direction. This escarpment is part of the outer scarp of the Niagara cuesta. The cuesta is a belt of Middle Silurian (Niagaran) limestone and dolomite, which stretches west from the State of New York and makes a great arc around Lakes Huron and Michigan (Fig. 2). In regions west of Lake Ontario, this band of upland slopes gently inward toward the southern peninsula of Michigan and presents a strong scarp on its outer edge. This escarpment is well known where it is crossed by the Niagara River at Lewiston, New York, and forms the northern front of the plateau in which the Niagara gorge has been cut some 250 ft (76 m).

The escarpment forms the peninsula and islands that lie between the Lake Huron Basin and Georgian Bay. This same upland belt continues west to the Garden Peninsula of Michigan where it skirts the east side of Big Bay de Noc and forms the massive Niagara Escarpment at Fayette. From here, the cuesta stretches southward to form a series of islands, the Green Bay Peninsula, and the dominating ridge of eastern Wisconsin. The cuesta forms the high rocky cliffs overlooking Green Bay, as well as the local cliffs near Lake Winnebago.

In the area of the Raber Fossil Beds, Silurian rocks of the

Burnt Bluff and Manistique groups of the Niagaran Series are exposed and form an abrupt north-facing limestone ridge over 90 ft (27 m) high. This ridge slopes gently downward to the south. Along the ridge top on the south slope are extensive exposures of well-preserved fossil colonial corals of unusual size, variety, and quantity. A short distance to the southwest of the main escarpment is a smaller ridge, apparently of glacial origin. This smaller ridge is covered with glacial erratics whose lithology is dominated by fossil remains.

To the west of the smaller ridge is Bender Lake (Fig. 1), a small bog lake of considerable interest because of the variety of vegetation around it. Directly east of Goetzville is a large field of gigantic limestone boulders of unusual scenic and geologic interest (Fig. 1).

During the depositional period of the Silurian rocks exposed at the Raber site, the platform margin of the Michigan Basin was situated in a subtropical environment at about 20° to 25° south of the equator (Ziegler and others, 1977). Accumulating sediments were predominately composed of biogenic carbonate.

Johnson and Campbell (1980) identify three distinct paleocommunities found within Silurian rocks in the Michigan Basin. Each community was adapted to a particular water depth, in which salinity, wave turbulence, and light intensity were controlling factors. The three communities identified were (1) fucoid-ostracode, living in quiet waters close to shore; (2) coral-algal, developing near shore, in shallow active water; and (3) a pentamerid community established in deep off-shore waters.

While all three community types can be identified in out-

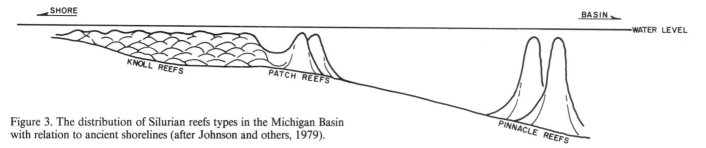

Figure 3. The distribution of Silurian reefs types in the Michigan Basin with relation to ancient shorelines (after Johnson and others, 1979).

crop at the Raber site, the dominant paleocommunity is the coral-algal, and this is best exhibited in the Cordell Dolomite. In the Cordell, reef-forming stromatoporoids, tabulate corals, stromatolites, and other invertebrates established and maintained for a considerable time organic buildups, which formed massive reef colonies.

Johnson and others (1979), find that three types of fossil reefs can be distinguished in the Silurian rocks of the Michigan Basin: pinnacle reefs, patch reefs, and knoll reefs.

Silurian pinnacle reefs in the Michigan Basin are noted for their large size, often covering hundreds of acres and rising vertically over 800 ft (245 m). These pinnacle reefs dominate the heavily drilled "Niagaran Reef Trend" (Fig. 2) and have proven to be the major hydrocarbon producers of the Michigan Basin over the past decade. While having formed in deep water and being of great areal size, the faunal assortment of these pinnacle reefs parallels that of smaller shallow water reefs. Johnson and others (1979) conclude this provides a firm basis for assuming that the pinnacle reefs reached their size as a mark of successful buildup of the colony from generation to generation. Each successive colony would build atop an older colony, keeping pace with basin subsidence and remaining in shallow sunlit and food-rich water (Fig. 3).

Patch reefs developed nearer to the shore than pinnacle reefs (Fig. 3). Because subsidence was less nearer to shore, the patch reefs did not attain the height of pinnacle reefs. A maximum height for a patch reef in the Michigan Basin would be roughly 100 ft (30 m). Patch reefs do not display the large areal size seen in pinnacle reefs. The taller pinnacle reefs were more susceptible to storm erosion and slumping, making them less stable. Lost debris from the pinnacle reef would fall to its base, forming large, ever-expanding rubble piles. The smaller areal size of the patch reef is attributed to the sturdiness and stability of its shorter height, making it less a victim of erosion during its lifetime than the taller pinnacle reef.

Closest to the shoreline of the Silurian Sea, and in very shallow water, were the knoll reefs (Fig. 3). The constant shallowness of the water, due to minimal subsidence, retarded the upward development of the knoll reef and forced growth to expand laterally. Coral colonies were numerous and closely spaced, with many of the colonies becoming intergrown or over-

grown. The extensive exposures of fossils noted in the Cordell Dolomite at the Raber site are from knoll reef colonies.

DESCRIPTION

The lowest stratigraphic unit identifiable at the Raber escarpment is the Hendricks Dolomite of the Burnt Bluff Group. The Hendricks consists of even-bedded dolomites and limestones of a gray to buff color and are slightly argillaceous. The Hendricks Dolomite is very similar to the underlying Byron Formation in color and lithology but is easily distinguished by its abundant fossils. Ehlers (1973) states the most characteristic fossils of the Hendricks are *Clathrodicyon vesiculosum, Favosites, Camarotoechia winiskenses, Rhynchospira lowi, Stokesoceras romingeri, Leperdita fabulina,* and *Isochilina latimagrinata.*

The Schoolcraft Dolomite of the Manistique Group overlies the Hendricks Formation. The Schoolcraft is a massive, coarsely crystalline, buff to brownish gray dolomite. Ranging throughout the brownish dolomite beds are thin, even beds of finely crystalline, blue-gray dolomite. Fossil remains are scarce in the thin, blue-gray beds, while replaced shells and molds of one or more species of the brachiopod *Pentamerus* appear in great abundance in the massive brownish dolomites. Ehlers (1973) finds the *Pentamerus* beds of the Schoolcraft to be a helpful marker horizon of exceptional continuity throughout the region. The top of the Schoolcraft Dolomite contains numerous layers of chert nodules, and these can be used to indicate a proximity to the contact with the overlaying Cordell Dolomite.

The Cordell Dolomite of the Manistique Group consists almost entirely of thin, uneven-bedded, brownish gray to buff colored, siliceous dolomites, interbedded with layers of chert nodules, isolated chert nodules, and silicified fossils. Ehlers (1973) finds the silicified corals of the Cordell to be extremely useful in the recognition of the interval. The most abundant of these silicified corals identified by Ehlers include several species of such genera as *Alveolites, Amplexus, Arachnophyllum (Fig. 4), Favosites, (Fig. 5), Halysites (Fig. 6), Heliolites, Lyellia, Omphysma, Prychophyllum, Streptelasma, Syringopora,* and *Zaphrentis.* In total, Ehlers (1973) identifies and lists the following numbers of invertebrate species from the Cordell Dolomite at the Raber site: hydrozoans, 2; bryozoans, 8; brachiopods, 11; trilo-

Figure 4. *Arachnophyllum striatum* (d'Orbigny), scale in inches (sample courtesy R. T. Segall, Michigan Geological Survey).

Figure 5. *Favosites favosus* (Goldfuss), scale in inches (sample courtesy R. T. Segall, Michigan Geological Survey).

bites, 5; cephalopods, 19; gastropods, 2; pelecypods, 1; and corals, 52.

The fossil invertebrates of the Cordell developed in a warm, shallow near-shore marine environment some 410 m.y. ago during the Middle Silurian. Most of the calcareous material of which the fossils were originally composed has been replaced by silica due to ground water activity. The silica is extremely resistant to weathering and erosion, and the structure of the fossils, especially the corals, is well preserved. As a result of the silica replacement, the corals tend to stand up in bold relief, often 2 in or more (5+ cm) above the carbonate matrix to which they are attached. While the silica replacement makes for striking specimens, from an anatomical standpoint most are too well crystallized for effective microscopic study.

The best exposed outcrops of fossiliferous Cordell Dolomite at Raber are shown on Figure 1. Records of the Michigan Geological Survey indicate large exposures of colonial coral are best viewed in the vicinity of the north quarter–corner of Sec.4,T.-42N.,R3E. This site is marked on Figure 1 by an X.

REFERENCES CITED

Ehlers, G. M., 1973, Stratigraphy of the Niagaran Series of the Northern Peninsula of Michigan: Ann Arbor, University of Michigan, Museum of Paleontology Papers on Paleontology, no. 3, p. 1–200.

Johnson, A. M., Kesling, R. V., Lilienthal, R. T., and Sorenson, H. O., 1979, The Maple Block Knoll Reef in the Bush Bay Dolostone (Silurian, Engadine Group), Northern Peninsula of Michigan: Ann Arbor, University of Michigan, Museum of Paleontology Papers in Paleontology, no. 20, p. 1–33.

Johnson, M. E., and Campbell, G. T., 1980, Recurrent carbonate environments in the Lower Silurian of Northern Michigan and their inter-regional correlation: Journal of Paleontology, v. 54, no. 5, p. 1041–1057.

Ziegler, A. M., Hansen, K. S., Johnson, M. E., Kelly, M. A., Scotese, C. R., and Van Der Voo, R., 1977, Silurian continental distributions, paleogeography, climatology, and biogeography: Tectonophysics, v. 40, p. 13–51.

Figure 6. *Halysites labyrinthicus* (Goldfuss), scale in inches (sample courtesy D. M. Bricker, Michigan Geological Survey).

Mackinac Island State Park, Michigan

Randall L. Milstein, Subsurface and Petroleum Geology Unit, Michigan Geological Survey, Lansing, Michigan 48912

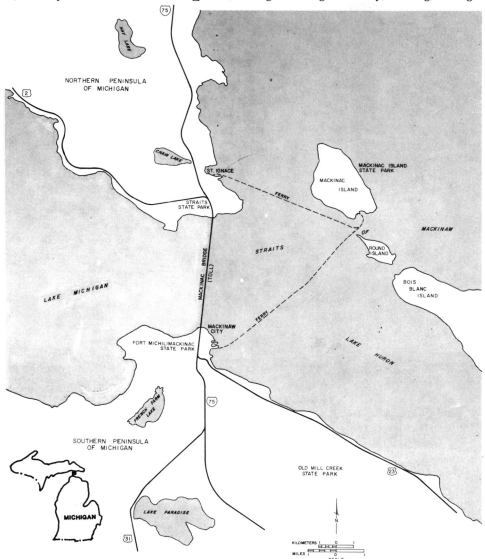

Figure 1. Location of Mackinac Island State Park and landmarks of the Straits of Mackinac Region, Michigan.

LOCATION

Mackinac Island State Park, T.40N.,R.3W., Mackinac County, Michigan; Cheboygan, Michigan, U.S.G.S. 30 × 60 minute series. Mackinac Island is located 3 mi (5 km) east of St. Ignace, Michigan, in the Straits of Mackinac (Fig. 1). Access to the Island from St. Ignace is provided by Arnold Transit or Star Lines from mid-April to late-December. From Mackinaw City, service is provided by Arnold Transit or Shepler's Mackinaw Island Ferry from mid-May to mid-October. Service is at least hourly during the summer months. Air charter service is provided by Great Lakes Charters, St. Ignace, or Michigan Airways, Pellston. From December to mid-April, transportation is by Air Charter Service only from the Mackinac County Airport, St. Ignace. No motorized vehicles are permitted on Mackinac Island. On the Island, transportation is available in the form of rental bicycles, horses and horse drawn carriages.

SIGNIFICANCE

The location of Mackinac Island, and its composition of erosion-resistant limestone breccia, has resulted in the island

LEGEND

1 – Friendship's Alter
2 – Scott's Cave
3 – Sinkhole
4 – Crack-in-the-Island
5 – Chimney Rock
6 – Sunset Rock
7 – Lovers' Leap
8 – Devils' Kitchen
9 – Pontiac's Lookout
10 – Sinkhole
11 – Sugarloaf
12 – Arch Rock
13 – Fairy Arch
14 – Skull Cave

—— Roadway or Trail
≈≈ Shore Cliff and Terrace
░░ Gravel Shoreline Ridge
∿∿ Crevices in Limestone
○   Springs

Figure 2. Location of significant geologic features of Mackinac Island, Michigan.

maintaining a fairly complete and continuous record of historic glacial lake stages of Algonquin and later age. The most striking features of Mackinac Island are the well-developed shoreline structures related to the changes in ancient lake levels (Fig. 2). The most distinctive and dramatic shore features are magnificent limestone structures including a variety of sea caves, stacks, arches, and wave cut bluffs. Additional features include strand lines, bars, terraces, and karst structures.

Though a nearly complete section of Devonian, Mississippian, and Pennsylvanian rocks was deposited within the Michigan Basin, the rocks exposed on Mackinac Island are like no Paleozoic rocks exposed elsewhere within the basin. Probably throughout millions of years the island was above sea level and subjected to intense aerial erosion.

In 1875, the United States government created Mackinac Island National Park. This was the nation's second National Park.

In 1895, the federal government relinquished control of the island to the State of Michigan. The same year, Mackinac Island was designated Michigan's first State Park.

DESCRIPTION

Approaching by boat, Mackinac Island appears to rise in two gigantic steps from the lake shore to the summit of a small, high hill (Fig. 3). The small hill stood as an isolated wave cut island during Lake Alconquin times and is referred to as the "Ancient Island." The lower hill, only about 50 ft (15 m) above present lake level, for the most part is a wave cut cliff related to the Nipissing stage of the modern Great Lakes.

The Ancient Island is slightly more than 2,600 ft (800 m) long and more than 1,300 ft (400 m) wide, with the highest natural point on Mackinac Island (elevation 904 ft; 275 m above mean sea level) recorded at its southeast end.

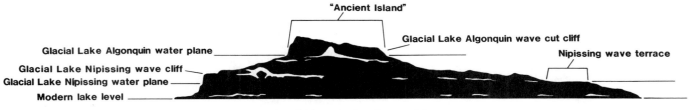

MACKINAC ISLAND

Figure 3. Topographic indicators of Ancient Great Lake levels seen when approaching Mackinac Island, Michigan from the east.

The Ancient Island is almost completely skirted by wave cut cliffs. Only along a portion of the west flank of the Ancient Island are the cliffs absent. Here deposition predominated over wave erosion and the eroded materials were deposited as gravel ridges. At this point the highest shoreline of glacial Lake Algonquin is recorded.

Adjacent to the Ancient Island are two limestone breccia stacks associated with the Algonquin shoreline. These stacks were small rocky islands separated from the coastal cliffs by wave erosion. Sugarloaf stack (Fig. 4), about 300 ft (90 m) east of the Ancient Island, stands 75 ft (23 m) above the surrounding level and is the largest stack on Mackinac Island. Sugarloaf is a magnificent display of Mackinac Breccia (Houghton, 1840: 1841; Landes and others, 1945). The breccia is composed of broken fragments of limestone in varying sizes, cemented into rock masses more solid and resistant than the parent limestone. The most plausible theory for the formation of the breccia is that salt beds deposited during the Silurian Period were dissolved, forming vast caverns into which the overlying Silurian and Devonian limestone beds collapsed and were recemented. Nearly every rock feature of interest on Mackinac Island—stack, cave, or arch—is a result of erosion around a mass of this breccia. A large number of fossils are identifiable within the Mackinac breccia. The most notable fossils being the trilobite, *Anchiopsis anchiops*; the tetracoral, *Acrophyllum oneidaense*; the brachiopod, *Centronella glansfagea*; and the pelecypod, *Conocardium* sp. (Landes and others, 1945).

Skull Cave stack is not as conspicuous as Sugarloaf, but is similarly composed of Mackinac breccia. The stack is approximately 50 ft (15 m) high and about 50 ft (15 m) from the southwest cliffs of the Ancient Island. Skull Cave, the hiding place of Alexander Henry following the massacre at Fort Michilimackinac in 1763, is located in the west side of the stack and is a sea-cave produced by wave action. The entrance to the cave is approximately 3 ft (1 m) in diameter and was awash or slightly submerged at the lake's highest level. The cave's interior is pocketed due to erosion in the softer parts of the breccia. The rounding and smoothing of rock surfaces which characterize wear by waves is not very evident, though this is probably due to subaerial erosion.

Successive shorelines of Lake Algonquin are recorded at several locations about Mackinac Island. Eleven shoreline ridges

are displayed during a 40-ft (12-m) change of elevation on the Short Rifle Range behind Fort Mackinac. Fourteen additional ridges are cut by Custer Road, while British Landing Road passes over 15 ridges. In each case the succession of ridges drops abruptly from the crest of the lowest ridge to flatter ground in front.

The shoreline of Lake Nipissing cannot be traced in a complete circuit about the Island (Stanley, 1945). In some places the distinction between the true Nipissing shore and the shore of a later lake stage is indistinguishable. Terraces evident below the South Sally Port of Fort Mackinac and along bluffs near Scott's Cave represent the true Nipissing shoreline. From the foot of the ramp leading to Fort Mackinac, the Nipissing terrace extends eastward along the high cliffs for more than 2,600 ft (800 m). To the west the terrace encounters a great gravel bar, the result of deposition from storms and currents during Nipissing time.

South of Carver Pond a mass of resistant breccia makes a projection from the cliffs. Similar projections of breccia are at Eagle's Crest, Scott's Cave, and Friendship Altar. Like the stacks of Algonquin age, the Nipissing Friendship Altar is made of Mackinac breccia. The Altar (stack) is about 8 ft (2 m) wide and 13 ft (4 m) high and is separated from the Nipissing bluff by a 10 ft (3 m) gap.

Scott's Cave is an excellent example of Nipissing wave erosion within the Mackinac breccia. The cave is 15 ft (5 m) long and 95 ft (3 m) wide with 9 ft (3 m) ceiling. The ceiling is rough and weathered. The floor and lower wall surfaces show heavy wave scour and rounding.

While looking like stacks, Lover's Leap and Sunset Rock are not true stacks. During the Nipissing stage of the lake they were not islands, but were chimney promontories attached to the cliffs below water level. Regardless, Lover's Leap and Sunset Rock are the result of wave action similar to that responsible for the formation of Friendship Altar. These hard breccia chimneys merely stood up as the softer cliffs wore back around them. A number of smaller detached promontories can be identified elsewhere along the Nipissing bluffs.

Arch Rock (Fig. 5), the most striking geologic feature on Mackinac Island, is of a somewhat similar origin. Though the arch is at a level that would correspond to Lake Algonquin, its proximity to the eroded Nipissing cliffs suggests the later as its time of creation. Arch Rock was more than likely the byproduct

Figure 5. Arch Rock, Mackinac Island, Michigan.

Figure 4. Sugarloaf Stack, Mackinac Island, Michigan. The largest of several stacks on the island, Sugarloaf stands 75 ft (22.5 m) above ground level and is located 300 ft (90 m) east of the Ancient Island.

of wave action undermining and removing softer material and leaving the firmer breccia as a bridge. The arch is approximately 10 ft (3 m) thick with an impressive hollow below. Sanilac Arch, a secondary feature, exists at the base of Arch Rock.

On Mackinac Island, water at the surface and underground has been active in dissolving the bedrock and forming cavities in the rock. While no well-developed sinkholes are on Mackinac Island, numerous small depressions are obviously solution related.

Long fissures in the limestone bedrock are common in many places over the Island. The largest of these fissures is known as "Crack-in-the-Island." The cavities are very likely the result of solution work along major joints in the carbonate bedrock, though some researchers credit Crack-in-the-Island to local earth rifting.

In addition to its importance as a geologic site, Mackinac Island is botanically and historically interesting. Mackinac Island is home to 415 species of wildflowers (Porter, 1984) and has been the site of past botanical studies by such noted botanists as Thomas Nuttall and Henry David Thoreau. At least six terrestrial community types exist on Mackinac Island; these include boreal

forests, marshes, bogs, meadows, beaches, and hardwood forests, each exhibiting rich and diverse plant and animal life. Historically, Mackinac Island was an Indian gathering place, a Jesuit mission, an early fur trading outpost, and a strategic military stronghold, fought over and possessed by the French, British, and American forces during the French and Indian War, the American Revolution, and the War of 1812. Restorations of the Island's historic forts and settlements are major tourist attractions. Mackinac Island affords a spectacular view of the Mackinac Bridge, the world's longest suspension bridge as measured from anchor pier to anchor pier. The bridge spans the Straits of Mackinac and connects Michigan's two peninsulas. Mackinac Island is also home of the Grand Hotel. "The Grand" is the world's largest summer hotel and noted for its traditional glamour and its use as a backdrop for motion pictures. Additional information about Mackinac Island can be obtained by writing The Mackinac Island State Park Commission, Mackinac Island, Michigan 49757.

REFERENCES CITED

Houghton, D., 1840, Third Annual Report of the State Geologist: Accompanying The Journal of the Senate, State of Michigan, vol. 2, no. 27, p. 74–82, 214–222.

——, 1841, Annual Report of the State Geologist: Documents Accompanying The Journal of the Senate, State of Michigan, vol. 1, Senate and House doc. no. 11, p. 481–486.

Landes, K. K., Ehlers, G. M., and Stanley, G. M., 1945, Geology of the Mackinac Straits region: Michigan Geological Survey, Publication 44, Geology Series 37, 198 p.

Porter, P., 1984, The wonder of Mackinac; A guide to the natural history of Mackinac Island: Mackinac Island State Park Commission, 52 p.

Stanley, G. M., 1945, Pre-Historic Mackinac Island: Michigan Geological Survey, publication 43, Geology Series 36, 74 p.

The geology of the Niagara Escarpment, Fayette, Michigan

Timothy M. Dellapenna, Subsurface and Petroleum Geology Unit, Michigan Geological Survey, Lansing, Michigan 48912

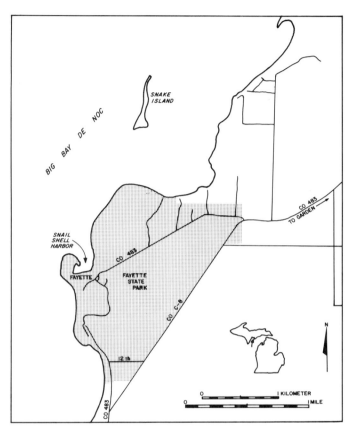

Figure 1. Map showing location of Fayette State Park and site of the restored town of Fayette, Chippewa County, Michigan.

Figure 2. The Niagara Escarpment at Fayette, Michigan (courtesy Michigan Department of Natural Resources).

LOCATION

Fayette State Park, the ghost town of Fayette, and an accessible portion of the Niagara Escarpment are located 8 mi (12.8 km) southwest of the town of Garden, Michigan, on the shore of Big Bay De Noc, Fayette 7½-minute Quadrangle. The sites are reached by U.S. 2 to Garden, then south on Michigan M-183 to Fayette State Park (Fig. 1). The sites may also be reached by boat through entry into Snail Shell Harbor.

SIGNIFICANCE

The Niagara Escarpment at Fayette (Fig. 2) is the largest exposure of the Middle Silurian Burnt Bluff and Manistique groups. These two groups of the Niagaran series are recognized as separate groups in outcrops throughout the Michigan Basin (Fig. 3). The groups are not defined separately in the subsurface of the Michigan Basin, but are collectively identified as the Clinton Group (Harrison, 1985).

The Middle Silurian Engadine Group, not exposed at the Fayette location, but predominant on the eastern shore of the Garden Peninsula, will also be discussed in this paper (Fig. 4). In the subsurface the Engadine Group is identified as the Niagara Group (Harrison, 1985). Biohermal reefs found in the Niagara Group form the rich hydrocarbon producing "Northern Reef Trend" of Michigan's Southern Peninsula (Fig. 5). Recent discoveries of hydrocarbons associated with the Clinton Group are restricted to the central Michigan Basin (Fig. 5).

The town of Fayette was established as an iron-smelting facility in 1867. Dolomite from quarries in the Niagara Escarpment surrounding the town was used as flux in the smelting furnaces. The Burnt Bluff Group and portions of the Manistique Group can easily be viewed in the now-abandoned quarries as well as along the western shore of the Garden Peninsula south of Fayette.

DESCRIPTION

In 1973, Ehlers described the rocks of the Niagaran Series in outcrop in the Northern Peninsula of Michigan. He defined the Burnt Bluff as the succession of limestone and dolomites occupying a position between the top of the Maryville Dolomite and a disconformity at the base of the overlying Manistique Dolomite. The type locality of the Burnt Bluff is on the western shore of the Garden Peninsula (Fig. 4). The largest exposure of the group occurs along the shoreline. Ehlers based the disconformity between the Burnt Bluff and Manistique strata on a faunal break,

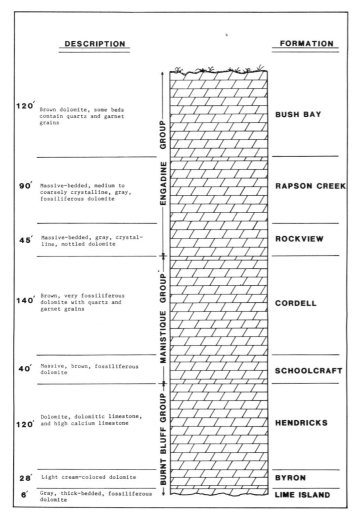

Figure 3. Generalized stratigraphic column of Niagaran Series rocks exposed at Fayette, Michigan.

which is most observable where the uppermost strata of the Burnt Bluff are limestones. According to Ehlers, at these localities, the Burnt Bluff limestones contain a few diagnostic fossils, primarily *Camarotoechia winiskensis, Leperdita fabulina,* and *Isochilina latimarginata,* and are overlain by a thick-bedded, coarsely crystalline dolomite of the Manistique Formation, which contains numerous molds of the brachiopod *Pentamerus.* The disconformity is very poorly defined in most localities because the dolomitization has destroyed the few fossils present. Ehlers (1973) proposed that the Burnt Bluff be a formation of the "Clinton Group," and divided the "Burnt Bluff Formation" into two members, the "Bryon" and the "Hendricks." Ehlers applies the "Byron Member" to the lower member of the group because this member consists of beds "which seems to be continuous with the Byron strata of northeastern Wisconsin."

Ehlers (1973) places the "Hendricks Member" conformably

above the "Byron Member" and below the disconformity at the base of the Manistique.

The Michigan Geological Survey (Ells and others, 1964), recognizes three formational members of the Burnt Bluff Group: the Lime Island Dolomite, Byron Dolomite, and Hendricks Dolomite. The basal member of the Burnt Bluff is the Lime Island, which overlies the Cabot Head Shale of the Cataract Group. At the exposures at Fayette, the Lime Island Dolomite is a buff to buff-gray, thick-bedded dolomite containing molds of the brachiopod *Virginia decussata* (Whiteaves) and fragments of *Favosites* sp., which overlays the Cabot Head Shale of the Catarct Group (Ehlers and Kesling, 1957). The Lime Island dolomite is on the average about 6 ft (2 m) thick at the Fayette exposure.

The Lime Island Dolomite is conformably overlain by the Byron Dolomite, which at Fayette is a light cream-colored dolomite that contains no diagnostic fossils (Ehlers and Kesling, 1957). The Byron Dolomite that is overlain by the Hendricks Dolomite, which consists of beds of dolomite, dolomitic limestone, and high calcium limestone. At Fayette, the Hendricks Dolomite consists of beds of very finely crystalline massive dolomites and coarsely crystalline dolomites, with upper beds containing a few specimens of *Favosites* sp. and stromatoporoids (Ehlers and Kesling, 1957).

The Manistique Group disconformably overlies the Burnt Bluff Group, and is exposed on top of the bluffs on the northwestern side of the Garden Peninsula. The Manistique Group is divided into two formational units: the Schoolcraft Dolomite and the Cordell Dolomite. The Manistique Group is best seen in outcrop at quarries around the town of Manistique, in the Northern Peninsula of Michigan, but the group is also well exposed in the Fayette area.

The basal member of the Manistique Group is the Schoolcraft Dolomite. A disconformity marks the base of the Schoolcraft, the nature of which was previously discussed. The Schoolcraft Dolomite consists of massive, coarsely crystalline, brownish gray to buff dolomites, thin-bedded, brownish gray dolomites, and thin, even-bedded finely crystalline, bluish gray dolomites (Ehlers, 1973). Although fossils are scarce in the thin, even-bedded, bluish gray dolomites, they are numerous in the basal, massive, brownish gray to buff dolomites; the most predominant being replaced shells and molds of the brachiopod *Pentamerus* sp. The top of the Schoolcraft Dolomite is placed at the base of a thin, uneven-bedded, buff to brown dolomite, containing numerous layers of chert nodules, many remains of *Pentamerus* sp., and a few silicified corals and other brachiopods (Ehlers, 1973).

The overlying Cordell Dolomite consists almost entirely of thin, uneven-bedded, brownish gray to buff, siliceous dolomites with interbedded layers of chert nodules, isolated chert nodules, and silicified fossils. The most prominent fossils are corals, which are very abundant, including such genera as *Alveolites, Amplexus, Arachnophyllum, Favosites, Halysites, Heliolites, Lyellia, Omphyma, Ptychophyllum, Streptelasma, Syringopora,* and *Zaphrentis.*

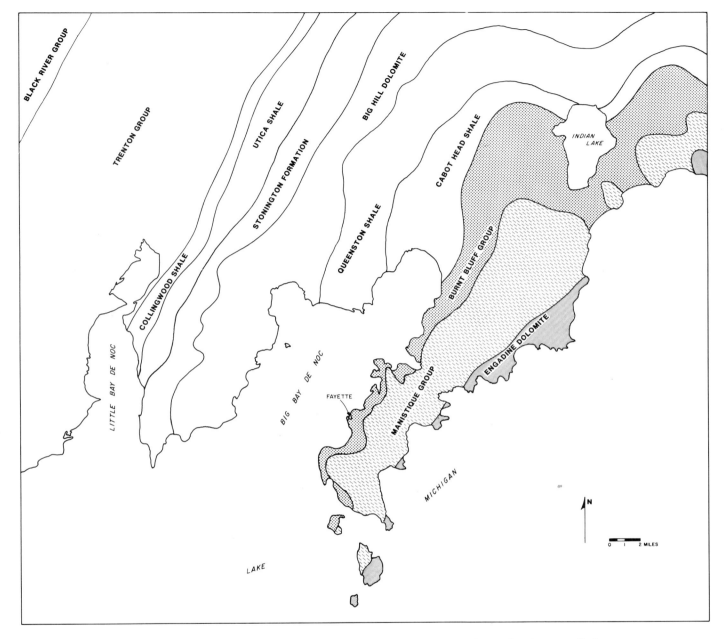

Figure 4. Bedrock geology map of the Garden Peninsula and surrounding region (courtesy R. L. Milstein, Michigan Geological Survey).

The Engadine Group overlies the Manistique Group with a sharp lithologic break between the fossiliferous Cordell Dolomite and the highly crystalline Rockview Dolomite with an apparently conformable contact (Johnson and others, 1979).

The Engadine Dolomite is subdivided into three formational members: the Rockview Dolomite, Rapson Creek Dolomite, and Bush Bay Dolomite (Johnson and others, 1979).

The Rockview Dolomite is the basal formation, which conformably overlies the cherty, fossiliferous Cordell Dolomite. The Rockview Dolomite is characteristically a massive-bedded, gray to bluish gray, medium to coarsely crystalline, mottled dolomite, which weathers to a striking white.

Some fossils have been found mainly in the lower beds of the formation, the most significant being the trilobite *Scutellum laphami,* generally as fragments; the brachiopod *Stricklandia multilirata,* and various algal remnants. However, because the matrix is dolomite, few fossils specimens have good preservation (Johnson and others, 1979).

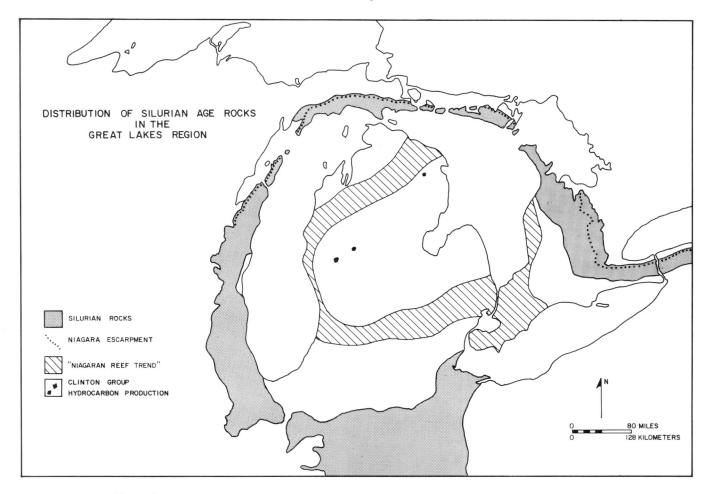

Figure 5. Map showing the distribution of Silurian Age rocks in the Great Lakes region, location of the Niagara Escarpment, the subsurface "Niagaran Reef Trend," and hydrocarbon-producing fields of the Clinton Group (courtesy R. L. Milstein, Michigan Geological Survey).

The Rapson Creek Dolomite overlies the Rockview Dolomite and underlies the Bush Bay Dolomite. The Rapson Creek Dolomite is a massive-bedded, medium to coarsely crystalline, light-gray to buff-gray dolomite, which contains a wide variety of corals and braciopods.

The Bush Bay Dolomite is the uppermost formation of the Engadine Group, which is overlain by the Salina Group. The basal portion of the Bush Bay Dolomite consists of thin to medium-bedded dolomites, containing few fossils and interbedded with white nodular chert layers. The middle portion is massive, but weathered to thin uneven beds, buff-brown, fine-grained to finely crystalline, with local lenses containing quartz and garnet sand grains; fossils are scarce, with a few scattered brachiopods, cephalopods, and corals (Johnson and others, 1979). The upper portion of the Bush Bay Dolomite contains gray colored, medium to thick beds. The beds contain wide irregular dolomite bands, the upper portions composed of vuggy porous interreefal deposits. This portion of the formation is quite rich in fossils, containing a wide variety of stromatoporoids, corals, brachiopods, gastropods, cephalopods, and bivalves.

REFERENCES CITED

Ehlers, G. M., 1973, Stratigraphy of the Niagaran Series of the Northern Peninsula of Michigan: Ann Arbor, University of Michigan Museum of Paleontology Papers on Paleontology, no. 3, 39 p.

Ehlers, G. M., and Kesling, R. V., 1957, Silurian rocks of the Northern Peninsula of Michigan, Michigan Geological Society Annual Geological Excursion: Ann Arbor, Michigan Geological Society, 30 p.

Harrison, W. B., III, 1985, Lithofacies and depositional environments of the Burnt Bluff Group in the Michigan Basin, in Ordovician and Silurian rocks of the Michigan Basin and its margins: Ann Arbor, University of Michigan, Michigan Basin Society Special Paper 4, p. 95–108.

Johnson, J. M., Kesling, R. V., Lilienthal, R. T., and Sorenson, H. O., 1979, The Maple Block Knoll Reef in the Bush Bay Dolostone (Silurian, Engadine Group), Northern Peninsula of Michigan: Ann Arbor, University of Michigan Press, Museum of Paleontology Papers on Paleontology, no. 20, 33 p.

Middle Devonian Transverse Group in Charlevoix and Emmet counties, Michigan

Randall L. Milstein, Subsurface and Petroleum Geology Unit, Michigan Geological Survey, Lansing, Michigan 48912

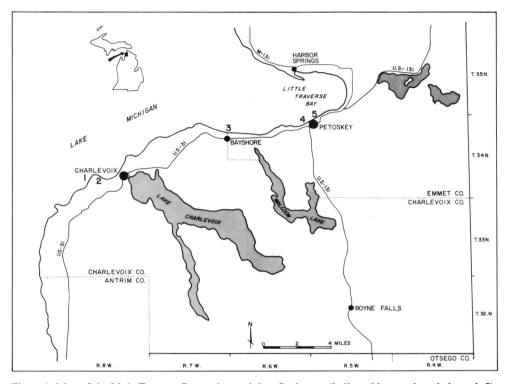

Figure 1. Map of the Little Traverse Bay region and described stops (indicated by numbers 1 through 5).

LOCATION

The Middle Devonian Traverse Group is a sequence of carbonates and shales randomly exposed as shoreline bluffs, roadcuts, river gorges, and quarries in the northern portion of Michigan's Southern Peninsula. Some of the most accessible of these outcrops are in Charlevoix and Emmet counties, where large sections of the group are exposed and are easily viewed by traveling U.S. 31 north along the Lake Michigan shoreline (Fig. 1). The route defined in this chapter can be followed on the Charlevoix, Bayshore, and Petoskey, Michigan 15-minute quadrangles. Most locations described are situated on public lands. Visitors are urged to obtain permission prior to entering sites on private property.

INTRODUCTION AND SIGNIFICANCE

The Traverse Group, exposed in the Little Traverse Bay region of Charlevoix and Emmet counties, presents excellent examples of carbonate and shale sequences resulting from multiple transgressions and regressions that swept a distal muddy sea-floor environment occupying the Michigan Basin during the Middle Devonian. The rate of basinal subsidence, carbonate accumulation, and supply of muds from a distant clastic wedge to the east controlled the rate of Traverse deposition (Gardner, 1974). The Traverse Group is correlative with the Muscatatuck Group of Indiana (Shaver, 1974).

As exposed, the Devonian formations in Charlevoix and Emmet counties appear to be crimped around the Michigan Basin. Kesling and others (1974) state that the "crimps" (folds) observed in the outcrop areas are not confined to the basin periphery; subsurface data indicate the structures continue downdip into the basin at a considerable distance. The observable folds are of various magnitudes, and many appear to be folds within folds. The major folds affect the strike of the beds up to 2 mi (3.2 km) in either limb, whereas the minor folds are on the order of a fraction of a mile. Kesling and others (1974) state that in some locations as many as nine reversals in dip occur within 1 mi (1.6 km). Figure 2 illustrates the major folds.

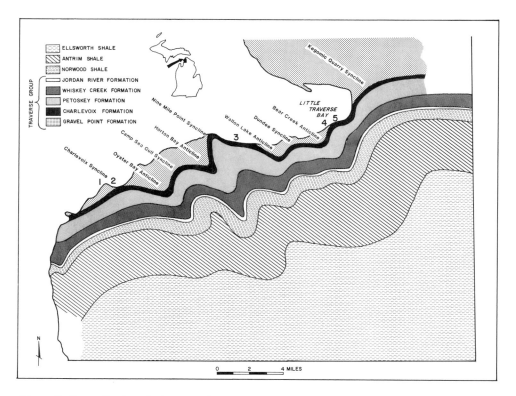

Figure 2. Generalized geologic map of the Little Traverse Bay region with major fold structures identified (modified from Kesling and others, 1974).

SITE DESCRIPTION

Traverse group

The Traverse Group, in the Little Traverse Bay region, consists of five formations, in ascending order: (1) the Gravel Point Formation, (2) the Charlevoix Limestone, (3) the Petoskey Formation, (4) the Whiskey Creek Formation, and (5) the Jordan River Formation (Fig. 3). The outcrops located along the shoreline, or just off U.S. 31, combine to provide excellent exposure of the Gravel Point, Charlevoix, and Petoskey stratigraphic units.

The Gravel Point Formation is a gray to brown, fine to cryptocrystalline, dense, lithographic limestone. The Gravel Point contains shale beds up to 2 ft (0.5 m) thick, as well as zones of chert nodules and petroliferous biohermal masses. The Charlevoix Limestone is a creamy gray to brown, fine to coarsely crystalline, slightly argillaceous, shaly limestone, commonly interbedded with coquina. The Petoskey Formation is a pale buff to grayish brown, fine-grained, arenaceous limestone. The Petoskey Formation contains zones of porous, friable limestone with a strong petroliferous odor.

Recent workers (Runyan, 1976; Gardner, 1974; Kesling and others, 1974) have interpreted the Traverse Group in the Little Traverse Bay region as a product of a shallow, subsiding marine carbonate shelf, similar to the present Bahama Platform. This type of marine environment was conducive to the growth and eventual preservation of the wide variety of fauna now found in the Traverse Group. Notable fauna include trilobites, ostrocods, pelecypods, hydrozoans, bryozoans, crinoids, brachiopods, and corals. Of particular interest is the colonial coral *Hexagonaria,* which occurs throughout the Traverse Group, but is most abundant in the Gravel Point Formation.

Hexagonaria percarinata (Fig. 4) is more commonly known as the "Petoskey Stone." The "Petoskey Stone" is the official stone of the state of Michigan. *Hexagonaria percarinata* was known by many earlier American paleontologists as *Acervularia profunda* and *Cyathophyllum davidsoni.* In the recent past it was usually identified as *Prismatophyllum davidsoni.*

Hexagonaria percarinata existed in massive colonies some 350 m.y. ago. The animal lived anchored to the bottom in deep-water mud flats. Buried by bottom silts, the animals became petrified over geologic time. When the Michigan region was scoured by glaciers during the last ice age, the fossilized coral colonies were exposed and often eroded from their place of origin in the bedrock and redeposited. "Petoskey Stones" are common beach rubble along the shores of Lakes Michigan and Huron, and may also be found in gravel pits, road cuts, and as glacial erratics. While "Petoskey Stones" can be collected throughout Michigan's northern lower peninsula, the most prolific area of collecting is near the city of Petoskey in Emmet County.

	A	B
UPPER DEVONIAN	ELLSWORTH SHALE	
	ANTRIM SHALE	NORWOOD SHALE OR ANTRIM SHALE
	NORWOOD SHALE	
	JORDAN RIVER FORMATION	
		SQUAW BAY LIMESTONE
MIDDLE DEVONIAN — TRAVERSE GROUP	WHISKEY CREEK FORMATION	THUNDER BAY LIMESTONE
	PETOSKEY FORMATION	POTTER FARM FORMATION
		NORWAY POINT FORMATION
	CHARLEVOIX LIMESTONE	FOUR MILE DAM FORMATION
	GRAVEL POINT FORMATION	ALPENA LIMESTONE
	KOEHLER LIMESTONE	NEWTON CREEK LIMESTONE
	GENSHAW FORMATION	GENSHAW FORMATION
	FERRON POINT FORMATION	FERRON POINT FORMATION
	ROCKPORT QUARRY LIMESTONE	ROCKPORT QUARRY LIMESTONE
	BELL SHALE	BELL SHALE
	ROGERS CITY LIMESTONE	ROGERS CITY LIMESTONE

Figure 3. Correlation chart for Middle and Upper Devonian rocks of the north part of the Southern Peninsula of Michigan. A, west side of outcrop area, Antrim, Charlevoix and Emmet counties. B, east side of outcrop area, Cheboygan, Presque Isle and Alpena counties (modified from Pojeta and Renjie, 1986).

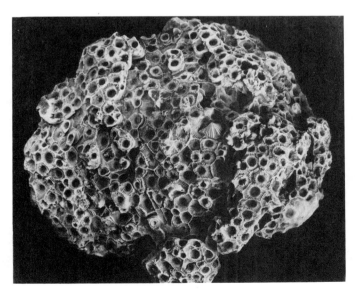

Figure 4. Hexagonaria percarinata, the "Petoskey Stone" (photo courtesy University of Michigan, Museum of Paleontology, specimen UMMP15645).

In outcrop, *Hexagonaria percarinata* may appear in massive colonies, but when found along shorelines, "Petoskey Stones" usually range from pebble size to as big as bowling balls. The most commonly found specimens are about the size of a chicken egg.

The collecting of beach deposited "Petoskey Stones" is a favorite Michigan pastime and visitors are encouraged to search for these prehistoric treasures. Visitors are reminded to seek permission prior to entering private property and that the collecting of "Petoskey Stones" from colonies preserved in outcrop is not encouraged.

Inspection of a *Hexagonaria percarinata* shows the corallites are closely packed, prismatically shaped, often with six sides, and are separated by thin walls. The corallites have been replaced almost entirely by calcite; however, there may be some quartz. The corallites have broad, sloping sides and depressed centers. Radiating septa appear thin, with many tiny cross-bars. The longest septa often meet and intertwine in the center of the calyx. The underside of a colony is covered by dense, smooth, rippled or wrinkled layers. The individual corallites in a specimen will range from a quarter to half an inch (6–13 mm) long and wide.

Stop 1

Stop 1 (Fig. 1) is located at South Point, 1.5 mi (2.5 km) west of the City of Charlevoix. This exposure is the type locality of the Gravel Point Formation. The best exposures appear in the ledges and bluffs along the Lake Michigan shoreline where the limestone is highly crinoidal and contains plentiful pyrite cubes The exposure is nearly on line between Sec. 28 and 29 of Charlevoix Township (see Charlevoix 15-minute quadrangle). The exposure can be reached by traveling south from Charlevoix 2 mi (3.2 km) on U.S. 31 to Bell Bay Road. Turn right (north) and follow Bell Bay Road 2 mi (3.2 km). It is a short walk from the end of the road to the shore. An excellent description of this exposure is given by Kesling and others (1974) under locality number 34-8-28/29.

Stop 2

Stop 2 (Fig. 1) is located at the Medusa Cement Company Quarry at South Point, 1.5 mi (2.5 km) west of the city of Charlevoix. The exposure can be reached by following the same route as to Stop 1, or the visitor may follow Lake Shore Drive 1.5 mi (2.5 km) west from Charlevoix to the quarry. One of the more complete exposures of the Gravel Point Formation is visible at this locality. Complete descriptions of this exposure are given by Segall and Sorenson (1973) and Kesling and others (1974). The local syncline and anticline structure typical of the region can be observed on the quarry faces. Visitors to the quarry must gain permission and sign a waiver prior to entry.

Stop 3

Stop 3 (Fig. 1) is located about 0.5 mi (0.8 km) north of Bay Shore, Emmet County. The exposure is located in the abandoned

Northern Lime Company Quarry and can be reached by traveling north from Charlevoix on U.S. 31 8 mi (13 km) to Bay Shore.
Turn left on Townline Road and proceed about 0.5 mi (0.8 km)
to the end of the road. The Charlevoix Limestone and Petoskey
Formation are exposed at this locality. At present, only the
Charlevoix beds are clearly exposed as slumping has obscured
large portions of the Petoskey exposures. The Charlevoix is represented by a dense sublithographic facies containing stylolites. The
Petoskey is represented by a massive, highly fossilierous, crystal-
line limestone. A complete description of the exposure is given by
Kesling and others (1974).

Stop 4

Stop 4 (Fig. 1) is located at the Dundee Cement Company
Quarry, 2 mi (3.2 km) west of the city of Petoskey off U.S. 31.
The greatest thickness of the Gravel Point, continuous from the
lowest exposed unit to the top of the formation, is exposed in the
quarry. Because of its completeness of exposure the unit has been
exhaustively studied; complete descriptions can be found in
Ehlers (1949), Segall and Sorensen (1973), and Kesling and others (1974). The beds of the Gravel Point are well exposed at this
locality (Fig. 5). The quarry also exposes syncline/anticline structures, rare prehistoric sinkholes, divergent strata, and biohermal
reefs. Visitors must obtain permission to enter the quarry.

Stop 5

Stop 5 (Fig. 1) is located at the abandoned Northern Lime
Company Quarry, bordering Little Traverse Bay in the city of
Petoskey. Both the Charlevoix Limestone and the type locality of
the Petoskey Formation are exposed here. The exposure is north-

Figure 5. South wall, Dundee Cement Quarry, Petoskey, Michigan
(photo courtesy Michigan Geological Survey).

east of the Waterfront Park and Softball Field. It can be reached
from U.S. 31 by turning left (west) on West Lake Street, then
right on Quaintance Street and passing south of the Softball
Field. U.S. 31 traverses the rim of the old quarry.

Excellent exposures of both formations can also be found at
Waterfront Park. Here an intriguing exposure of stromatoporoid
reef can be seen, and the city has built a stairway that climbs up
the side of the outcrop, greatly aiding the visitor. Again, complete
descriptions of these sites can be found in Kesling and others
(1974).

REFERENCES CITED

Elhers, G. M., 1949, The Traverse Group of the northern part of the Southern
    Peninsula of Michigan: The Annual Geological Excursion of the Michigan
    Geological Society Field Guide, p. 17–18.
Gardner, W. C., 1974, Middle Devonian stratigraphy and depositional environ-
    ments in the Michigan Basin: Michigan Basin Geological Society Special
    Papers no. 1, p. 43–47.
Kesling, R. V., Segall, R. T., and Sorensen, H. O., 1974, Devonian strata of
    Emmet and Charlevoix counties, Michigan: University of Michigan Museum
    of Paleontology Papers on Paleontology no. 7, 187 p.
Pojeta, J., Jr., and Renjie, Z., 1986, Devonian rocks and Lower and Middle

Devonian pelecypods of Guangxi, China, and the Traverse Group of Michi-
    gan: U.S. Geological Survey Professional Paper 1394-A-6, p. 55.
Runyon, S. L., 1976, A stratigraphic analysis of the Traverse Group of Michigan
    [M.S. thesis]: East Lansing, Michigan State University, 86 p.
Segall, R. T. and Sorensen, H. O., 1973, Geological significance of the Petoskey-
    Charlevoix area; Its economic and environmental impact, in Geology and
    the environment: Michigan Basin Geological Society Annual Field Confer-
    ence Guide Book, p. 151–181.
Shaver, R. H., 1974, The Mascatatuck Group (new Middle Devonian name) in
    Indiana: Indiana Geological Survey Occasional Paper 3, 7 p.

Devonian shelf-basin, Michigan Basin, Alpena, Michigan

Raymond C. Gutschick, Department of Earth Sciences, University of Notre Dame, Notre Dame, Indiana 46556-1020

INTRODUCTION

This chapter cites two localities, Paxton Quarry and Partridge Point (Fig. 1), and combines field observations from both to illustrate important stratigraphic relations and principles. Focus is on litho- and biostratigraphy of Middle and Late Devonian rocks as they relate to the Michigan Basin (Fig. 2).

Black organic-rich shales deserve our attention due to their energy potential a source rocks and fracture reservoirs for petroleum and natural gas. The Paxton Quarry offers the opportunity to examine an exceptional exposure of black shale (large area in Fig. 3; thick section in Fig. 4) as part of the formational sheet that continues into the Michigan Basin subsurface. These special rocks offer a challenge to observe and test their physical, chemical, and organic composition and structure to decipher origin, paleoenvironments, diagenesis, history, and economic value. A spectacular display of numerous, large calcareous concretions, in situ and free of shale matrix, is present in the quarry. How were they formed?

Partridge Point has an exposed limestone sequence that underlies Antrim black shales beneath Squaw Bay. Applying Walther's Law, one can go up section and down dip from fossiliferous shallow-water oxygenated Middle Devonian carbonate platform rocks at the shelf margin of the Michigan Basin into Upper Devonian transitional deeper water oxygen-deficient pelagic limestones followed by deep-water anaerobic black shales of a euxinic basin.

PAXTON QUARRY

Location and significance. The Paxton shale quarry is located in the N½Sec.30,T.31N.,R.7E., Lachine and Lake Winyah 7½-minute Quadrangles, Alpena County, along the south

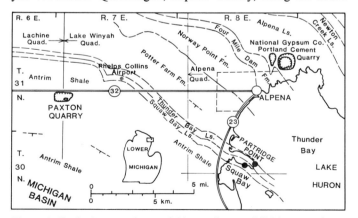

Figure 1. Geologic map of part of Alpena County, Michigan, northeast margin of Michigan Basin, showing the locations of the Paxton Quarry and Partridge Point area in relation to the geography and regional geology. Black dots show type localities. Adapted from Ehlers and Kesling (1970).

Figure 2. Chart to relate the chrono- and biostratigraphy of the rock sequence in the Paxton Quarry and Partridge Point area to standard accepted stratigraphic nomenclature. Chart shows relationship of the Antrim Shale to other widespread Late Devonian black shale units for parts of the United States and Canada. In the conodont zonation column, *Si* stands for *Siphonodella; P., Polygnathus, A., Ancyrognathus;* and *S., Schmidtognathus;* other zones refer to species of *Palmatolepis.*

side of Michigan 32, about 9 mi (14.4 km) west of Alpena (Figs. 1, 3). The quarry is owned by National Gypsum Company, Cement Division, with offices at the limestone quarry in Alpena. One part of black shale from the Paxton Quarry is mixed with four parts of limestone to make cement. Access to the Paxton Quarry can be arranged beforehand with the plant superintendent or directly with the quarry foreman at the site.

The following selected references are recommended for background study of the Late Devonian black shales: sedimentology of shales (Potter and others, 1980); world and United States paleogeography (Heckel and Witzke, 1979; Gutschick and Sandberg, 1983); Euramerican Devonian eustasy (Johnson and others, 1985); euxinic biofacies patterns (Byers, 1977); Antrim and correlative black shale studies (Ells, 1979; Winder, 1966; Russell, 1985; Cluff and others, 1981; Hazenmueller and Woodward, 1981).

The shale section in the Paxton Quarry represents the lower part of the Upper Devonian Antrim Shale (Figs. 2, 4), about 135 ft (41 m) thick. The Antrim Shale in the Alpena area, truncated northern margin of the Michigan Basin, ranges in thickness from an erosional cut-out (zero edge) to about 400 ft (122 m); however, it reaches more than 600 ft (183 m) in the basin depocenter to the southwest (Fisher, 1980). The oldest strata is exposed in the quarry sump (Figs. 3, 4). A greenish-gray shaly calcareous unit forms the bottom 5 ft (1.5 m) and this may be the upper beds of the Squaw Bay Limestone. Conodonts and radiolarians found

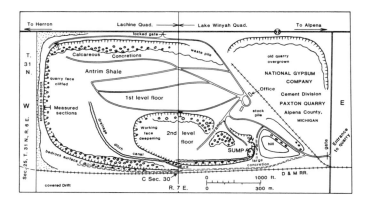

Figure 3. Geologic field sketch map of Paxton shale quarry (lower part of Antrim Shale). The second level is presently being quarried. Topographic quadrangles and oblique aerial views were used for the sketch.

in the sump section identify with the *Polygnathus asymmetricus* Zone. Another unit in the sump exposes at least 12 pairs of thin (few cm) alternating green-gray and black shale layers. The bottom contact of the lighter colored layers is very irregular where organisms burrowed into underlying black muds and backfilled with green-gray mud; whereas the top contact is sharp and planar with the overlying dark muds (Fig. 5, bottom left). Turbidite muds periodically washed into the euxinic basin from the platform margin and may represent distal pro-deltaic tongues of sediment with entrained oxygen.

Another seemingly anomalous interruption in black shale deposition is the light-colored calcareous unit that occurs between 33 and 50 ft (10 and 15 m) above the base of the section (Fig. 4). This unit sharply overlies black concretionary shale below, and in turn is overlain by thin layers of alternating green-gray and black shale (Fig. 5, bottom left). Fossils are present, including conodonts (Lower *gigas* Zone) and leiorhynchoid brachiopods, but this light-colored unit lacks a benthic community. How does one explain this abrupt change from anaerobic black mud conditions to oxygenated calcareous muds and return to anoxic conditions, assuming a deep-water basinal framework with a stratified water column?

Superficially, the black shales in the section look similar from bottom to top, although closer inspection reveals differences. Look for subtle changes of rock color (fresh or weathered), stratification—bedding fissility or the lack of it, sedimentary structures—current scour or cross-stratification, textural changes, jointing, concretionary zones, compaction, fossils (see Fig. 5)—hard and soft-bodied forms, burrows, minerals and crystals, and anything else the eyes can detect. Remember that all clues must fit the "frozen" pattern.

The mineralogy and chemical composition of the Antrim Shale was studied by Ruotsala and others (1981). The Antrim is uniform in composition and contains about 30% Si, 8% Al, 4% Fe, 2% Mg, 3% Ca, 2% S, and 5% organic carbon. Its mineralogy

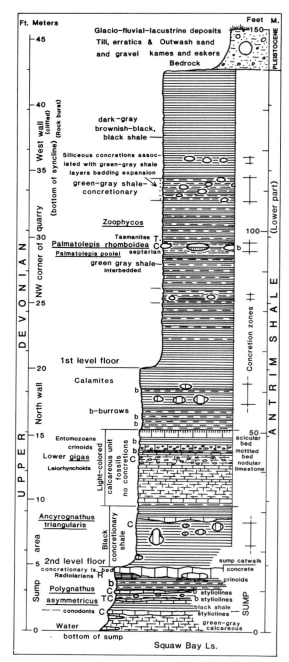

Figure 4. Detailed graphic columnar section of stratigraphy completely exposed in quarry showing current status of biostratigraphy. The rock column represents the lower part of Late Devonian Antrim Shale: C represents conodont occurrences; R, radiolarian concretionary limestone beds; T, *Tasmanites* occurrences; and b, burrowed and/or bioturbated beds.

consists of 50–60% quartz, 20–35% illite, 5–10% kaolinite, 0–5% chlorite, and up to 5% pyrite. When present, calcite and dolomite occur in limestone nodules, lenses, and interbeds as much as 5 ft (1.5 m) thick in the lower half of the Antrim. Organic material, as measured by low-temperature ashing, is present up to 12.8%. Bitumen contents range from 0.2 to 0.8%. Average weights of

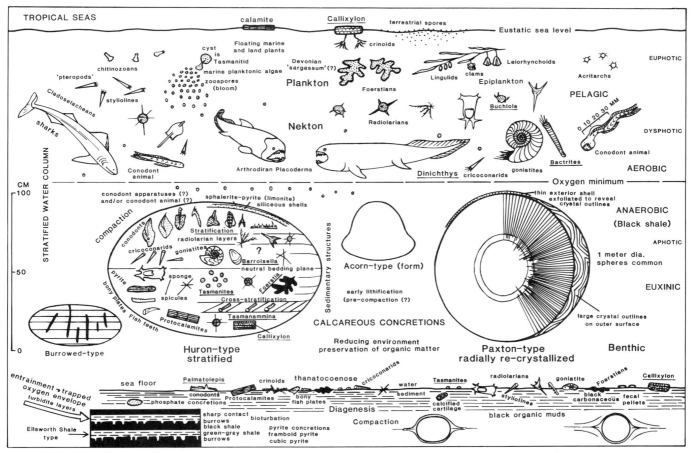

Figure 5. Paleoecological model for biota in Antrim–Kettle Point–Ohio (Huron)–New Albany black concretionary shales of midwestern United States and Canada (Michigan, Ontario, Indiana-Illinois-Kentucky). Size of organisms are not to scale, but scales are given for size of conodont animal and for range of most concretion sizes, except acorn-type, which is greatly reduced. Refer to Sandberg and Gutschick (1984, p. 149, Fig. 13) for Mississippian model.

bitumen components are 360 to 370. Deeper drill holes have higher bitumen contents. Kerogen, which makes up the remainder of organic material, is of low functionality, about 1 functional group per 25 carbon atoms. Hydroxyl (1 group/50 carbon atoms) and alkene bonds (1 group/50 carbon atoms) are the most common groups present.

Concretion biostratigraphy. Concretions may contain many important clues to the age, sedimentology, paleontology, paleoecology, paleogeography, origin, and history of the sediments that contain them (Gutschick and Wuellner, 1983; Sandberg and Gutschick, 1984). Calcareous concretions are present in many Phanerozoic black shales throughout the world. Often they are associated with deep-water, oxygen-deficient, basinal facies where the benthos is very limited or absent, but the pelagic plankton and nekton biota may have flourished (Fig. 5). Fossils, especially microforms, are often present in abundance and well-preserved in the concretions, whereas in matrix shales they may be crushed by compaction and difficult to recover or diagenetically altered beyond recognition.

Large calcareous concretions are common to the Antrim, Kettle Point, Ohio-Huron, New Albany, and other black shale formations. The Kettle Point Formation, Antrim correlative along the eastern shore of Lake Huron, Ontario, is named after the "kettles" (large calcareous concretions) conspicuous at this location. Huron Shale concretions, 13 ft (4 m) maximum diameter at Standardsburg, Ohio (Prosser, 1913), contain Late Devonian radiolarians documented by Foreman (1963), and their origin has been discussed by Clifton (1957). One is overwhelmed by the array of concretions of various sizes and shapes that lie loose on the Paxton Quarry floors and in place in the shale walls. The largest concretion seen in the quarry is 10 ft (3 m) in diameter. Several types—stratified Huron-type, radially crystallized Paxton-type, acorn-form, and burrowed-type (Fig. 5)—are recognized, based on size, shape, internal structure, and host rocks.

Paxton concretions have been studied by Wardlaw and Long (1982), who dealt with their origin. These authors wrote,

"The concretions are composed of ferroan dolomite, contain quartz, and are rimmed by 1-2 cm of pyrite. The weight percent of ferroan dolomite decreases from center to edge (90–50%). In thin section, the dolomite crystals are long, slender, feathery, with curved edges and sweeping extinction. There are no chemical trends from center to edge in the major elements (Ca, Mg, Fe), trace elements (Na, K, Sr, Mn, Zn), total organic carbon, or oxygen isotope values (–9.1‰ PDB). The carbon isotope values range from –10 to –13‰ PDB and increase from center to edge. The data have been interpreted to indicate that the concretions grew below the seawater/sediment interface, within the top 33 ft (10 m) of the sediment surface, and in a system which was open to seawater. Growth occurred before major compaction."

This information apparently applied to Paxton-type concretions.

PARTRIDGE POINT

Location and significance. Partridge Point is a small peninsula of land that is subparallel to the strike of the strata, projecting southeastward into Lake Huron (Figs. 1, 6). The type sections of two formations, the Thunder Bay and Squaw Bay Limestones, are exposed along the shoreline; the former in the W½SE¼Sec.11, and the latter in the SW¼SW¼Sec.11, both in T.30N.,R.8E., 7½-minute Alpena Quadrangle, Alpena County, Michigan. Both exposures are accessible along the beach at the water's edge and can be reached from the unmaintained boat-ramp launch roads (Fig. 6). Otherwise, private home owners should be consulted before crossing their lots to the beach.

The purpose of the Partridge Point study is to examine carefully the Thunder Bay and Squaw Bay Limestones (Fig. 7), reconstruct the sequence with the contrasting overlying Antrim Shale at Paxton Quarry (concealed under Squaw Bay), and interpret the history of the succession in relation to the Michigan Basin. Start at the type Thunder Bay Limestone location and follow the shoreline-beach up section and down dip, around to the Squaw Bay Limestone type section. Section descriptions of formations on Partridge Point with fossil lists and plates are given by Ehlers and Kesling (1970); conodonts are described and discussed in Müller and Clark (1967) and Bultynck (1976). Note that there are covered intervals and that the section is incomplete.

Fossiliferous limestone and shaly beds exposed along the shoreline buttressed by Thunder Bay Limestone are light-colored and replete with coralline and shelly faunas (Fig. 7). Sessile benthos animal skeletons abound with colonial corals, stromatoporoids, bryozoans, crinoids, and blastoids, as well as other mobile benthic types of the reef-like community. This location is close to the shallow-water carbonate platform shelf margin of the Michigan Basin during the Middle Devonian. Note that the medium-gray, fine-grained limestone beds at the top of the section have few fossils. The upper sequence can also be seen south of Partridge Point.

The westernmost projection of land into Squaw Bay (Fig. 6) exposes a remarkably well-jointed limestone pavement covered with *Zoophycos* burrow swirl impressions followed by overlying thin irregular layers of brown-gray crystalline fossiliferous Squaw Bay Limestone. The rock with *Zoophycos* contains a fauna of

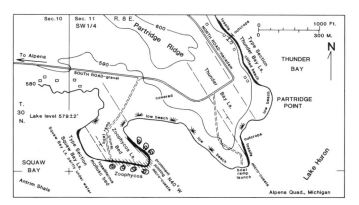

Figure 6. Geologic sketch map for Partridge Point, type localities of Thunder Bay and Squaw Bay Limestones, and for Squaw Bay floored by Antrim Shale bedrock.

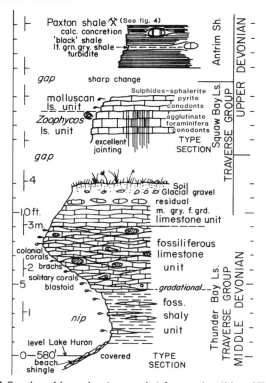

Figure 7. Stratigraphic section (composite) for type localities of Thunder Bay and Squaw Bay Limestones along Lake Huron shoreline, Alpena County and Quadrangle, Michigan. Sequence is interrupted by glacial drift or lacustrine sediment cover and underwater concealed intervals across formational boundaries.

agglutinated saccamminid foraminifera with some tests constructed of glauconite grains, conodonts (possibly pre-*P. asymmetricus*), scolecodonts, and water-worn phosphatic and glauconitic steinkerns of styliolines, ostracods, bryozoans, tiny snails, and pelmatozoans. This unit is discussed separately from the Squaw Bay Limestone, although some include it with the latter.

The Squaw Bay Limestone makes a narrow rock platform along the shoreline at the type section. When the level of Lake Huron is high, or when winds from the southwest pile water onto

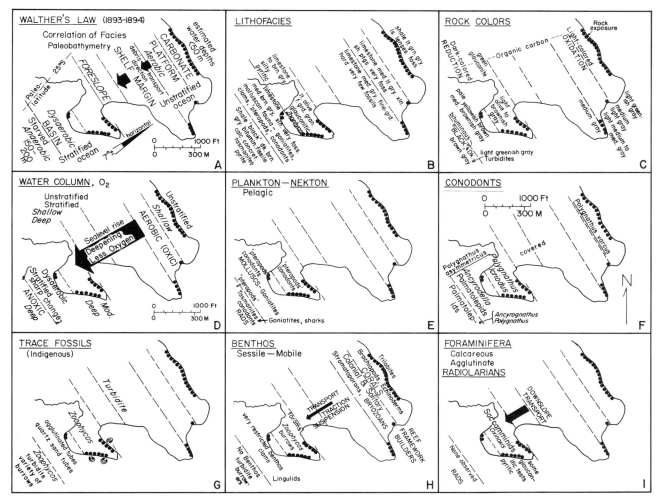

Figure 8. Graphic stratigraphic basin-platform analysis diagrams—framework, rock descriptions, water and biotic realms, important microbiota, and benthic traces (compare with Fig. 6; black dots refer to outcrop exposures). A, Paleoenvironmental framework down-dip across strike of beds from carbonate platform into basin to illustrate Walther's Law. Approximate paleolatitude after Heckel and Witzke (1979). B, Sequence of Thunder Bay and Squaw Bay lithofacies exposed across Partridge Point. C, Distribution of rock colors (standard color chart) exposed across Partridge Point. Rock colors in large part reflect percentage and oxidation-reduction state of organic carbon and iron compounds related to their depositional history (Gutschick and Sandberg, 1983, p. 86, Fig. 7B; Sandberg and Gutschick, 1984, Fig. 15). D, Change in water column with time from shallow oxygenated unstratified conditions on the platform to deep basin stratified conditions of euxinic basin. E, Diagram to emphasize pelagic plankton-nekton aerobic stratification over basin (see Fig. 5). F, Conodont succession reflects zonal and paleoecological changes with time (Sandberg and Dreesen, 1984, Fig. 4, Late Devonian; Sandberg and Gutschick, 1984, Fig. 14, Early Mississippian). G, Map showing distribution of pavement of *Zoophycos* traces on top surface of limestone unit. H, Map illustrates dramatic transition from flourishing sessile benthic communities on platform in aerobic zone to sterile benthos of anoxic waters in euxinic basin. I, Agglutinate foraminiferal fauna of the saccamminid biofacies (Gutschick and Sandberg, 1983, p. 87, Fig. 7E).

the beach, most (if not all) of the Squaw Bay Limestone is barely awash. The best exposures are at times of low water levels of the lake. This pelagic limestone is a coquina of styliolines, cephalopods (goniatites), conodonts, few clams and snails, and poorly preserved wood fragments. Pyrite and sphalerite are also present. Details of the complete section are still forthcoming. Contact with

the Antrim Shale is under water and not exposed at this location, but it may be observed in the quarry sump.

Shelf-basin analysis. There are three formations (litho- and biofacies) in stratigraphic succession—Thunder Bay Limestone, Squaw Bay Limestone, and Antrim Shale—that are superimposed one on top of another in vertical succession.

Walther's Law of Correlation (or Succession) of Facies states that only such facies that occur as contemporaneous deposits laterally adjacent to each other *can succeed one another in vertical sequence* (italics mine), provided there is no break in the succession (Middleton, 1973). The attributes of each formation have been singled out and plotted in a series of diagrams (Fig. 8,A-I); interpretation follows Walther's Law and the models of Gutschick and Sandberg (1983) and Sandberg and Gutschick (1984).

Lithofacies, rock colors, pelagic plankton and nekton faunas and floras, and benthos dictate the paleogeography and paleoecology. Thunder Bay Limestone rock colors are light colored as a result of the oxidation of the organic carbon, and sessile benthic framework builders dominate the biota. This combination represents a shallow-water carbonate platform community habitat. The *Zoophycos* limestone unit has deeper water benthic burrowing organisms and an upper slope saccamminid foraminiferan biofacies, but the presence of water-worn phosphate and glauconite steinkern fragments suggests traction transport downslope into deep water. The Squaw Bay Limestone is dominated by a pelagic molluscan and conodont nekton, the virtual absence of a benthos, and sulfides, which suggest an oxygen-deficient stratified water column of a lower slope or basinal environment. The Antrim Shale is devoid of benthic forms, except for turbiditic green muds with bioturbation, but it has abundant plankton and nekton organisms, especially microforms. The black shale is interpreted as being deposited in deep water at depths of more than 165–330 ft (150–200 m) in an anaerobic basin where there was no oxygen to support bottom life and the organic carbon is in a reduced state.

This chapter provides the reader with orientation, perspective, and focus for sharper resolution of the geology of the two localities described. It directs one's attention toward observational details, sampling for laboratory follow-through, and in-depth library research.

REFERENCES CITED

Bultynck, P. L., 1976, Comparative study of Middle Devonian conodonts from north Michigan (U.S.A.) and the Ardennes (Belgium-France): Geological Association of Canada Special Paper No. 15, p. 119–141.

Byers, C. W., 1977, Biofacies patterns in euxinic basins; A general model: Society of Economic Paleontologists and Mineralogists Special Publication 25, p. 5–17.

Clifton, H. E., 1957, The carbonate concretions of the Ohio Shale: Ohio Journal of Science, v. 57, p. 114–124.

Cluff, R. M., Reinbold, M. J., and Lineback, J. A., 1981, The New Albany Shale Group of Illinois: Illinois State Geological Survey Circular 518, 83 p.

Ehlers, G. M., and Kesling, R. V., 1970, Devonian strata of Alpena and Presque Isle Counties, Michigan: Guidebook prepared for North Central Section Geological Society of America and Michigan Basin Geological Society, p. 1–130.

Ells, G. D., 1979, Stratigraphic cross sections extending from Devonian Antrim Shale to Mississippian Sunbury Shale in the Michigan Basin: Geological Survey Division, Michigan Department of Natural Resources, Report of Investigation 22, 186 p.

Fisher, J. H., 1980, Thickness of Ellsworth Shale, Plate 11, and thickness of Antrim Shale, Plate 12: Dow Chemical Co., U.S. Department of Energy Report No. FE2346-80, scale 1:1,000,000.

Foreman, H. P., 1963, Upper Devonian radiolaria from the Huron Member of the Ohio Shale: Micropaleontology, v. 9, p. 267–304.

Gutschick, R. C., and Sandberg, C. A., 1983, Mississippian continental margins of the conterminous United States, in Stanley, D. J., and Moore, G. T., eds., The shelfbreak; Critical interface on continental margins: Society of Paleontologists and Mineralogists Special Publication 33, p. 79–96.

Gutschick, R. C., and Wuellner, D., 1983, An unusual benthic agglutinated foraminiferan from Late Devonian anoxic basinal black shales of Ohio: Journal of Paleontology, v. 57, p. 308–320.

Hazenmueller, N. R., and Woodward, G. S., eds., 1981, Studies of the New Albany Shale (Devonian-Mississippian) and equivalent strata in Indiana: Indiana Geological Survey, U.S. Department of Energy Report, 100 p.

Heckel, P. H., and Witzke, B. J., 1979, Devonian world paleogeography determined from distribution of carbonates and related lithic paleoclimatic indicators: Palaeontological Association, Special Papers in Palaeontology 23, p. 99–123.

Johnson, J. G., Klapper, G., and Sandberg, C. A., 1985, Devonian eustatic fluctuations in Euramerica: Geological Society of America Bulletin, v. 96, p. 567–587.

Middleton, G. V., 1973, Johannes Walther's Law of Correlation of Facies: Geological Society of America Bulletin, v. 84, p. 979–987.

Müller, K. J., and Clark, D. L., 1967, Early late Devonian conodonts from the Squaw Bay Limestone in Michigan: Journal of Paleontology, v. 41, p. 902–919.

Potter, P. E., Maynard, J. B., and Pryor, W. A., 1980, Sedimentology of Shale; Study Guide and Reference Source: New York, Springer-Verlag, 306 p.

Ruotsala, A. P., Sandell, V. R., and Leddy, D. G., 1981, Mineralogic and chemical composition of Antrim Shale, Michigan [abs.]: American Association of Petroleum Geologists Bulletin, v. 65, p. 983.

Russell, D. J., 1985, Depositional analysis of a black shale by using gamma-ray stratigraphy; The Upper Devonian Kettle Point Formation of Ontario: Canadian Petroleum Geology Bulletin, v. 33, p. 236–253.

Sandberg, C. A., and Dreesen, R., 1984, Late Devonian icriodontid biofacies models and alternate shallow-water conodont zonation, in Clark, D. L., ed., Conodont Biofacies and Provincialism: Geological Society of America Special Paper 196, p. 143–178.

Sandberg, C. A., and Gutschick, R. C., 1984, Distribution, microfauna, and source-rock potential of Mississippian Delle Phosphatic Member of Woodman Formation and equivalents, Utah and adjacent states, in Woodward, J., Meissner, F. F., and Clayton, J. L., eds., Hydrocarbon source rocks of the Greater Rocky Mountain region: Denver, Rocky Mountain Association of Geologists, p. 135–178.

Wardlaw, M. M., and Long, D. T., 1982, Mineralogy, chemistry, and physical setting of carbonate concretions in the Antrim Shale (Devonian, Michigan Basin); Clues to origin: Geological Society of America Abstracts with Programs, v. 14, no. 5, p. 291.

Winder, C. G., 1966, Conodont zones and stratigraphic variability in Upper Devonian rocks, Ontario: Journal of Paleontology, v. 40, p. 1275–1293.

ACKNOWLEDGMENTS

This chapter is dedicated to Richard Shiemke, Paxton Quarry foreman, whose joy in his shale quarry working environment includes visitors who wish to observe, collect, and understand the geology in the quarry. My appreciation is extended to Charles A. Sandberg, U.S. Geological Survey, Denver, Colorado, for identification of the conodonts in the Paxton Quarry; and to former undergraduate students, Dirck Wuellner and Tom Hendrick for field and laboratory help.

Sleeping Bear Dunes National Lakeshore, Michigan

Randall L. Milstein, Subsurface and Petroleum Geology Unit, Michigan Geological Survey, Lansing, Michigan 48912

LOCATION

Sleeping Bear Dunes National Lakeshore, T.27 to 32N.,R.13 to 16W., Benzie and Leelanau counties, Michigan; North Manitou, Maple City, Empire, and Frankfort, Michigan, 15-minute quadrangles. The Sleeping Bear Dunes National Lakeshore begins just north of Point Betsie in Benzie County and extends northward along the Lake Michigan shoreline to Good Harbor Bay in Leelanau County (Fig. 1). The national lakeshore also includes both North and South Manitou islands. Access to the national lakeshore by automobile is by Michigan 22. Access to the Manitou Islands is by private or charter boat service. Charter boat service is from Leland, Michigan, and is provided by Manitou Island Transit from June through December. Service is round trip once daily. The driving of vehicles in Sleeping Bear Dunes National Lakeshore is restricted to established, maintained roads. The driving of offroad recreational vehicles is restricted to designated roadways; offroad travel in sand dunes areas is prohibited. Pets are not allowed on the Manitou Islands.

LEGEND OF THE SLEEPING BEAR

Indian legend states that long before the coming of the white man, Mishe-Mokwa, the mother bear, and her twin cubs were driven from the shores of Wisconsin by fire and famine into the waters of Lake Michigan. Having no other choice, the three set out for the distant shoreline of Michigan. Their hearts were filled with fear, for their journey seemed impossible.

After swimming many hours, the Michigan shoreline came into view. However, the effort had taken its toll, and exhausted by the long swim, the cubs, one at a time, disappeared below the water. Mishe-Mokwa, longing for her cubs, stood upon the shore and waited for them day after day, to no avail. Finally, she laid down to sleep and wait forever.

The Great Spirit Manitou, deeply sensing Mishe-Mokwa's love and longing, caused her two cubs to rise above the waters as islands and named them, in her honor, for himself—South and North Manitou. To mark the resting place from which Mishe-Mokwa would maintain her eternal vigil, the Great Spirit gently covered her with sand, forming the Sleeping Bear Dune.

SIGNIFICANCE

The land-sculpting effect of continental glaciation in northwest Michigan is clearly illustrated in the geologic features of the Sleeping Bear Dunes region. Glacial features such as terminal moraines, kettles, kames, eskers, drumlins, and sand dunes are identifiable. The most spectacular features are the massive perched dunes of the Sleeping Bear Dune complex that rise more than 460 ft (140 m) above Lake Michigan. Additional features

such as wave-cut bluffs, beach terraces, sand bars, ridge and swale formations, and ancient lake plains can be seen and related to the rise and fall of the ancient Great Lakes.

The Sleeping Bear Dunes National Lakeshore was authorized by Congress in October 1970. Including North and South Manitou islands, the national lakeshore comprises approximately 72,000 acres of land and deeded lake surface area.

DESCRIPTION

During the Pleistocene, continental glaciers spread southward from Canada across the Michigan Basin. When the glacial advance halted and then slowly began its retreat, huge quantities of glacially derived sediments were deposited. These deposits formed the numerous glacial landforms now evident in the region of the Sleeping Bear National Lakeshore. In relation to the Sleeping Bear Dunes, the most significant glacial landform is the terminal moraine near Glen Lake, Michigan (Fig. 2). This moraine, called the Manistee moraine, marks the farthest extent of glaciers in the area during the last ice advance and is considered the major glacial landform controlling the development of surficial geology in the Sleeping Bear area.

As ice advanced during the last period of glaciation, immense headlands, formed during previous ice advances along the shores of ancient Lake Michigan, resisted the force of the ice mass and directed the ice flow into existing valleys. The ice lobes scoured debris from the valley floors and walls and deposited it along the sides of the valleys and at the glaciers front. These deposits became the prominent north-south–facing moraines of the Sleeping Bear region (Fig. 2).

As the glaciers began to retreat, many of the valley entrances became blocked by ice in the Lake Michigan basin, causing the valleys to fill with water. The moraine at Glen Lake (Alligator Hill) became an island between the unmelted ice still occupying the Lake Michigan basin and the adjacent moraines. These adjacent moraines are known as Prospect Hill, Miller Ridge, and the Sleeping Bear Plateau.

Meltwaters flowing southward from the receding glacial front formed vast outwash plains in southern Leelanau and northern Benzie counties (Fig. 2). A major southern drainage way sliced through the Manistee moraine forming the extensive outwash area south of the moraine called the Platte Plains. The southward flow of these drainage channels is opposite to the directions in which the area streams run today; when ice blockage in the Lake Michigan basin cleared, the flow reversed and the streams began to run into the Lake Michigan basin (Kelley and Farrand, 1967). The movement of a thinning sheet of ice north of the Manistee moraine formed several vast drumlin fields (Fig. 2).

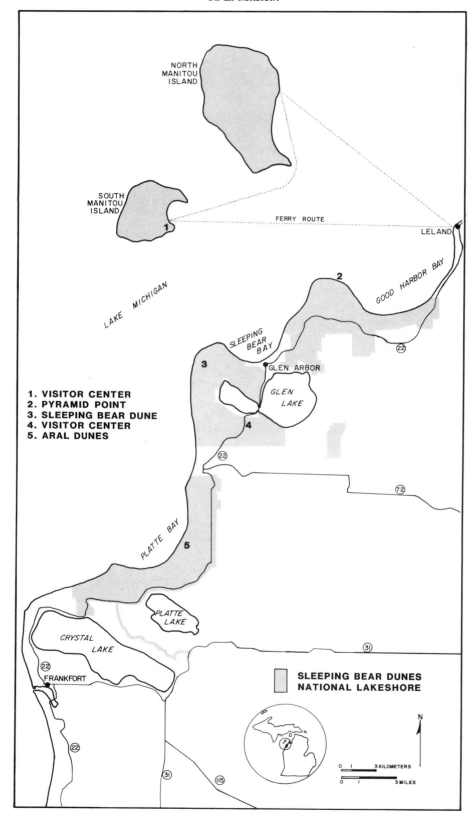

Figure 1. Location of Sleeping Bear Dunes National Lakeshore, Michigan.

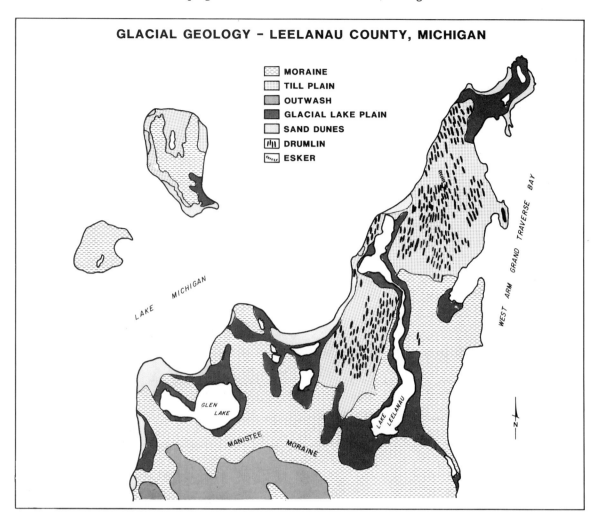

Figure 2. Generalized glacial geology of Leelanau County, Michigan.

The long, rounded drumlin hills are useful in interpreting the Sleeping Bear area because they are oriented in the direction of ice flow. As ice thickness dropped below the level of the Manistee moraine, ice border lakes were formed. Many of these border lakes still exist in modern form (Fig. 2).

As the glacial ice retreated northward, the immense volume of meltwater filled the Lake Michigan basin to form postglacial Lakes Algonquin, Nipissing, Algoma, and Chippewa. The water elevation and extent of these ancient Great Lakes was dependent on the elevation of the lowest outlet available to the sea and blockage by ice still remaining. Evidence of these lakes can be seen in the national lakeshore through features such as wave-cut bluffs, sandbars, beach terraces, ridge and swale formations, and sandy lake plains.

The fluctuating water levels had tremendous impact on landforms in the region. When water receded, current-carried drift was deposited, building sandbars and spits across channels,

damming the waters, and forming many smaller lakes in the embayed areas (Wilson, 1980), for example, Glen Lake (Figs. 1 and 2). When water levels dropped significantly, sand deposits would dry out. Driven before the prevailing westerly winds off ancient and modern Lake Michigan, the blown sand formed the complex variety of dunes in the Sleeping Bear region. The most common dune forms along the national lakeshore are beach dunes and perched dunes (Buckler, 1979).

Beach dunes, or foredunes, are found adjacent and parallel to the lakeshore and are seldom higher than 50 ft (15 m; Kelley, 1971). The Aral Dunes, along Platte Bay's north shore, are good examples of beach dunes. Perched dunes, on the other hand, sit high above the shore on plateaus. In the case of the Sleeping Bear Dune complex, these plateaus are moraines. As the water level in Lake Michigan changed, the dunes migrated. When the shoreline retreated, the dunes stabilized. If the proper conditions were available, new dunes formed along the new shoreline. The older dunes, formed at a higher lake level or atop moraines, remained constant. As the shoreline advanced with a rising lake level, older stable beach and dunes were eroded. The eroded materials were then recycled into the dune building process, creating new dunes or adding to the height of existing perched dunes. The continued inland migration and expansion of the Sleeping Bear Dune complex has forced the move of many buildings in the area as the shifting sand threatens to cover them. The most significant move involved shifting the U.S. Coast Guard Station from Sleeping Bear Point to Glen Haven. Numerous stands of trees have been buried by the migrating sand. As the dunes move on, "ghost forests" of dead trees are exposed. These stark reminders of the dunes' passing are littered throughout the national lakeshore.

The migrating sand dunes of the Sleeping Bear area have effectively buried and preserved numerous fossil vertebrates of the Late Pleistocene and other postglacial ages. Pruitt (1954) and Wilson (1967) described and identified 17 different varieties of mammal and fish skeletons from Sleeping Bear Dune.

The two Manitou Islands are within the authorized boundaries of the national lakeshore. These islands represent the southernmost limit of an island chain that extends north to the Straits of Mackinac. The island chain represents the crest of a high bedrock ridge covered with a blanket of drift.

The west side of South Manitou Island is characterized by high bluffs topped with perched sand dunes. Gull Point on the northeast corner of the island is the nesting ground for a large colony of Herring and Ring-Billed gulls. While the colony is off limits to hikers, surrounding hills afford a good view. The Valley of the Giants, a grove of virgin white cedar trees, is located on the island's southwest corner. The world record white cedar is located within the grove. It measures 17.6 ft (5 m) in circumference and stands more than 90 ft (27 m) tall. A total of 528 growth rings were counted on one of the fallen trees in the grove, dating their existence prior to Columbus. North Manitou Island is undeveloped and will remain a wilderness area. The island is mostly low, sandy open dune country, interfingered with high sand hills and blowout dunes. On the west and northwest sides of the island, 300-ft-high (91 m), deeply gullied bluffs form the shoreline.

Because of its diverse geological character, the Sleeping Bear Dunes region exhibits a wide variety of plant and animal life. At least six terrestrial community types exist within the lakeshore; these include dune, forest, plain, meadow, swamp, and aquatic environments. These communities are home to more than 326 species of birds, 49 species of mammals, 15 species of reptiles, 17 species of amphibians, and 80 species of fish.

Camping spots are plentiful at 12 area campgrounds, and the national lakeshore has numerous hiking trails of varying lengths and difficulty. The Sleeping Bear Point Maritime Museum, the wreck of the merchant ship *Francisco Morazan,* several historic lighthouses, and the dune climb at the Sleeping Bear Dune are all interesting family stops within the National Lakeshore. Additional information about Sleeping Bear Dunes National Lakeshore can be obtained by writing Sleeping Bear Dunes National Lakeshore, 400 Main Street, Frankfort, Michigan 49635.

REFERENCES CITED

Buckler, W. R., 1979, Dune type inventory and barrier dune classification study of Michigan's Lake Michigan shore: Michigan Geological Survey Report of Investigation 23, 20 p.

Kelley, R. W., 1971, Geologic sketch of Michigan sand dunes: Michigan Geological Survey Pamphlet 5, p. 16.

Kelley, R. W., and Farrand, W. R., 1967, The glacial lakes around Michigan: Michigan Geological Survey Bulletin 4, p. 10–17.

Pruitt, W. O., Jr., 1954, Additional animal remains from under Sleeping Bear Dune, Leelanau County, Michigan: Michigan Academy of Science, Arts, and Letters Papers, v. 39, p. 253–256.

Wilson, R. L., 1967, The Pleistocene vertebrates of Michigan: Michigan Academy of Science, Arts, and Letters Papers, v. 52, p. 197–234.

Wilson, S. E., 1980, Michigan's sand dunes: Michigan Geological Survey Pamphlet 7, p. 4.

Mississippian Marshall Formation of the Pointe Aux Barques region, eastern Michigan

Randall L. Milstein, Subsurface and Petroleum Geology Unit, Michigan Geological Survey, Lansing, Michigan 48912

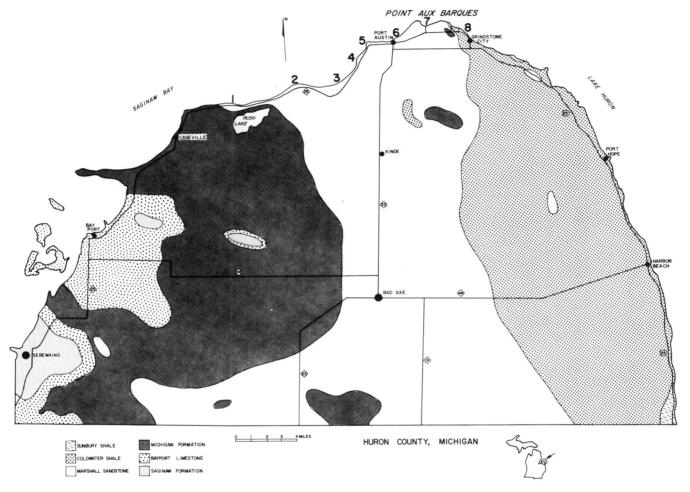

Figure 1. Bedrock geology map of Huron County, Michigan. Numbers indicate described outcrop locations.

LOCATION

Pointe aux Barques Region T.18N.,R.11E. and T.19N., R.13E., Huron County, Michigan; Rush Lake, Port Austin West, Port Austin East, and Kinde West, Michigan, 7½-minute quadrangles. The Pointe aux Barques region of Huron County extends east from Little Oak Point 18 mi (29 km) along the Lake Huron shoreline to Grindstone City (Fig. 1). The Pointe aux Barques area is reached by following Michigan 53 north from Bad Axe to Port Austin. This picturesque shoreline region is also accessible by boat.

INTRODUCTION

The Lower Peninsula of Michigan resembles a man's mitten-covered hand. In such a comparison Saginaw Bay separates the thumb from the rest of the hand. Huron County is located at the very tip of the "thumb." Few counties of the Lower Peninsula have more rock outcrops exposed than Huron County. The majority of Huron County's outcrops are located at its northernmost boundary along the Lake Huron shoreline. The exposed cliffs, ledges, and wave-cut shoreline features make up the thumb's thumbnail.

The name Pointe aux Barques was given to the region by early explorers who noted the resemblance of the shoreline features to the prows of ships (Houghton, 1838). Along the Pointe aux Barques shoreline are numerous examples of sea caves,

Figure 2. The Flagstaff (or Bowsprit) at Pointe aux Barques. This small island was formed by wave action of the ancient and modern Great Lakes (photo courtesy Michigan Department of Natural Resources).

arches, stacks, and wave-cut cliffs (Fig. 2). These dramatic shoreline structures are the result of changes in water level of the ancient and modern Great Lakes.

The outcrops of the Pointe aux Barques region are composed of sandstones of the Mississippian Marshall Formation. Sediments of the Marshall Formation are generally well sorted locally, but lateral gradations in grain size are common. The sandstones and minor siltstones that comprise the Marshall Formation apparently were deposited in a shallow sea frequently disturbed by waves and currents (Monnett, 1948). Cohee (1979) finds the Marshall is time transgressive northward across the Michigan Basin and eventually occurs in the north as lenses of sandstone that interfinger with the overlying Michigan Formation.

The Marshall Formation overlies the Mississippian Coldwater Shale and is about 300 ft (91 m) thick at its most expressive outcrop in southern Michigan. The Marshall Sandstone is comprised of highly angular quartz grains embedded in a softer cement of mica, siderite, and clay. Cross-bedding in the Marshall Formation is subtle, making the rock seem perfectly homogeneous. The Marshall Formation consists of an upper unit, the Napoleon Sandstone Member, and a lower unit, commonly referred to as the lower member. The lower member is a fine-grained, greenish gray sandstone with some siltstone present. The Napoleon Member is a fine- to coarse-grained sandstone, commonly red in color.

From the mid-to-late 1800s the Marshall Formation outcrops of the Pointe aux Barque regions were quarried extensively for the production of grindstones. These grindstones were used for grinding, shaping, and polishing anything from axes and knives to fencing swords. When used for grinding purposes, the softer cement of the Marshall Formation wears away just fast enough to allow the remaining quartz grains to polish. It is the lack of outstanding cross-bedding, evenness of grain, sharpness of grit, and soft-cementing material that gave the stone its peculiar value to the grindstone industry. Users of grindstones from the Pointe aux Barques region considered them to be the finest in the world and they were greatly sought after. Grindstones weighing up to 12,000 pounds (5,443 kg) and as much as 7 ft (2 m) across and 1 ft (0.3 m) thick were produced from the quarries and marketed worldwide. Numerous examples of these grindstones are still on display at Grindstone City, Michigan, or can be found discarded along the Lake Huron shoreline in the region.

The Marshall Formation contains units that are highly fossiliferous. An excellent summary of samples identified in the region can be found in Lane (1900). The most notable unit, a calcareous

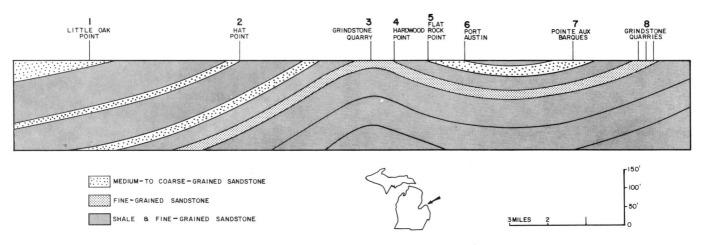

Figure 3. Cross section along shoreline, northern Huron County, Michigan, illustrating regional structure and location of described outcrop locations (modified from Monnett, 1948).

sandstone lying above the quarried grindstone beds, has often yielded remains of the fossil sharks *Orodus* and *Ctenacanthus* (Dorr and Eschman, 1977).

Monnett (1948) defines the major structural features of the Pointe aux Barques region to be an anticlinal fold and its associated syncline (Fig. 3).

SITE DESCRIPTIONS

Site 1. Upper beds of the Mississippian Marshall Sandstone outcrop on the northside of Rush Lake, inland from Little Oak Point. The exposure is in the abandoned Babbit Quarry and can be reached by taking Michigan 25 east from Port Austin 9 mi (14 km) to the intersection of Oak Beach Road. Turn south on Oak Beach Road and proceed 0.75 mi (1.2 km) to Sand Road. Turn west on Sand Road and proceed 2.5 mi (4 km) to the quarry. About 15 ft (4.5 m) of medium- to coarse-grained, very friable sandstone, with abundant limonite throughout, is exposed in the quarry. The top 3 ft (0.9 m) of this exposure is thinly bedded, while the bottom section is massive and cross-bedded. Lane (1900) assigned the strata at this outcrop to the middle part of the Napoleon Member of the Marshall Formation.

Site 2. East of the Little Oak Point/Babbit Quarry exposure, the first outcrop of Mississippian sandstone along the shoreline is at Hat Point. To reach Hat Point, travel west on Michigan 25 from Port Austin 7.5 mi (12 km) to Philip County Park. A short walk west along the beach will bring you to the outcrop. At this exposure, 12 ft (3.5 m) of fine- to coarse-grained, cross-bedded sandstone is exposed. The color ranges from buff to greenish gray. Monnett (1948) describes streaks of small quartz pebbles and flakes of black carbonaceous material occurring throughout the exposure. Lane (1900) assigned this outcrop to the lower Napoleon Member of the Marshall Formation.

Site 3. East of Hat Point is a sandstone outcrop exposed in a

now-abandoned grindstone quarry. The quarry is located roughly 3 mi (5 km) west of Port Austin on Michigan 25 and occupies both sides of the highway. About 9 ft (3 m) of strata are exposed. The upper 7 ft (2 m) are thin-bedded, grayish green, very fine-grained sandstone, but the lower 2 ft (0.6 m) are more massive. Evidence of a more massive but unexposed sandstone unit under the latter is indicated by large discarded quarry slabs lying around the pit. Monnett (1948) describes small pebble zones up to 2 in (5 cm) thick throughout the sandstone; there disappear laterally over short distances. In addition, Monnett (1948) describes a scattering of large individual pebbles throughout the unit. Large fragments of brown, pebbly conglomerate resembling peanut-brittle candy are found on the north side of the quarry. Lane (1900) identifies this and similar conglomerate rock found in the region as "peanut conglomerate."

Site 4. About 4 ft (1 m) of very thin-bedded, fine-grained, greenish to reddish sandstone are exposed at Hardwood Point 2 mi (3 km) west of Port Auston on Michigan 25. The exposure is reached by parking at Jenks County Park and walking north along the shoreline. The sandstone at this outcrop is similar to the sandstone quarried elsewhere within the region to produce grindstones. Winchell (1861) noted the outcrop contained numerous Early Mississippian fossils.

Site 5. The shore of Lake Huron is barren of outcrops from Hardwood Point to Flat Rock Point. From Flat Rock Point to Port Austin, about 25 ft (7.5 m) of medium- to coarse-grained, light-gray sandstone outcrops almost continually. The sandstone is massive, cross-bedded, and friable. Monnett (1948) describes numerous seams of small pebbles, which disappear laterally, throughout the outcrop. Flat Rock Point can be reached by taking Larned Road north from Port Austin to the Lake Huron shoreline and then walking west, or by turning west of Larned Road on any residential road once past the intersection of Michigan 5.

Site 6. One-half mi (0.8 km) west of Port Austin, the outcrops have been broken into large blocks by joints, and the shoreline takes on a very distinct appearance. By following the shoreline east from this point, the massive sandstone is gradually replaced by a thin-bedded, greenish to reddish, fine-grained sandstone. The ability to locate this unit depends entirely on the seasonal water level of Lake Huron. This sandstone unit is capped by an 8-in (20 cm) band of "peanut conglomerate."

Site 7. The sandstone cliffs and picturesque shoreline features of Pointe aux Barques are the first outcrops east of Port Austin (Fig. 2). The outcrops rise about 20 ft (6 m) above Lake Huron and are composed of massive, cross-bedded sandstone, with grain size ranging from fine to coarse. The sandstone is light gray to buff in color, friable, and porous. The Pointe aux Barques scenic area occupies about 0.6 mi (1 km) of the Lake Huron shoreline and can be reached by traveling 2 mi (3 km) east of Pointe aux Barque Road from Port Austin to the Pointe aux

Barques Golf Club Access Road. Turn north on the access road and continue 1 mi (1.6 km).

Site 8. East from Pointe aux Barques, no bedrock is exposed along the Lake Huron shoreline until reaching Eagle Bay. It is in Sections 23, 24, and 25,T.19N.,R.13E. that the sandstone, which was quarried so extensively for grindstone purposes prior to 1900, outcrops intermittently along the shoreline. These "grindstone" beds are a very fine-grained, greenish sandsatone occurring in beds 1 in (2.5 cm) to 1 ft (30 cm) thick. The overall outcrops are roughly 25 ft (7.5 m) thick. Large round quartz pebbles are scattered throughout the beds. The sandstone is moderately hard and argillaceous. Overlying the "grindstone" beds is a layer of "peanut conglomerate."

Visitors are encouraged to stop in Grindstone City 5 mi (8 km) east of Port Austin off Michigan 25. Grindstone City has many parks, displays, and historical markers describing the quarries and grindstone production operations of the last century.

REFERENCES CITED

Cohee, G. V., 1979, Introduction and regional analyses of the Mississippian System, *in* Craig, L. C., and Connor, C. W., eds., Paleotectonic investigations of the Mississippian System in the United States: U.S. Geological Survey Professional Paper 1010, Part I, p. 53–54.

Dorr, J. A., Jr., and Eschman, D. F., 1977, The geology of Michigan: Ann Arbor, The University of Michigan Press, p. 397–398.

Houghton, D., 1838, Report of the State Geologist, Michigan House Documents, 1838, no. 24, p. 149–152.

Lane, A. C., 1900, Geological report on Huron County, Michigan, *in* Lane, A. C., ed., Geological survey of Michigan, Lower Peninsula, 1896–1900: Lansing, Michigan Geological Survey, v. 7, part 2, p. 1–303.

Monnett, V. B., 1948, Mississippian Marshall Formation of Michigan: American Association of Petroleum Geologists Bulletin, v. 32, no. 4, p. 629–688.

Winchell, A., 1861, Observations on the geology, zoology, and botany of the Lower Peninsula; First biennial report of the Geological Survey of Michigan: Lansing, Michigan Geological Survey, part 1, p. 80.

The Ledges of the Grand River, Michigan

Randall L. Milstein, Subsurface and Petroleum Geology Unit, Michigan Geological Survey, Lansing, Michigan 48912

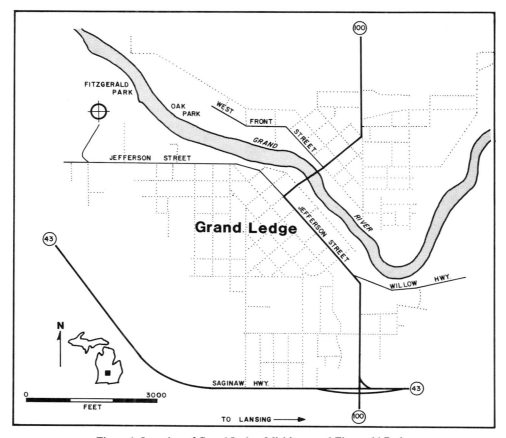

Figure 1. Location of Grand Ledge, Michigan, and Fitzgerald Park.

LOCATION

The Ledges of the Grand River, T.4N.,R.4W., Eaton County, Michigan; Eagle, Michigan, 7½-minute Quadrangle. The Ledges are located in the city of Grand Ledge, Michigan, 11 mi (18 km) west of Lansing, Michigan, on Michigan 43 (Fig. 1). The Ledges occupy both the northeast and southwest banks of the Grand River. Access to the southwest Ledges is by East Jefferson Street to Fitzgerald Park. Access is also possible either by canoe or the River Path Trail that begins at River and Harrison streets in Grand Ledge.

SIGNIFICANCE

The Ledges of the Grand River are the best exposure of Pennsylvanian age rocks in the state of Michigan. Besides their scenic beauty, the Ledges and additional outcrops of the surrounding vicinity provide a unique setting in which to explore and investigate an ancient near-shore marine beach environment. A study of the exposed strata along this portion of the Grand River reveals evidence of cyclic sedimentation (Kelly, 1933). Marine sediments in the section alternate with those deposited under terrestrial conditions.

The sandstone and shale units of the Ledges are fossiliferous and noted for containing numerous varieties of fossilized plants, invertebrates, and trace fossils of ancient vertebrate fish.

DESCRIPTION

The Grand River is one of the most important drainage ways in the southern Peninsula of Michigan. It rises in a series of lakes southwest of Jackson, Michigan, and flows generally north and westward to Grand Haven, Michigan, where it empties into Lake Michigan. The river formed during the final retreat of the last Pleistocene glaciers, developing its course on the newly formed glacially scoured surface and in its initial stage was nourished by water flowing out from the melting ice front. In the relatively short period of its existence, approximately 15,000

311

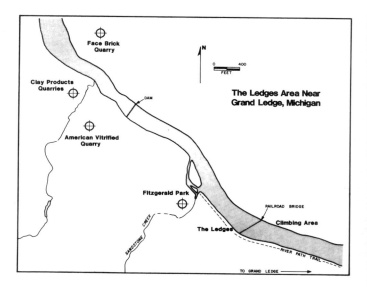

Figure 2. Location of described section, The Ledges area, near Grand Ledge, Michigan.

years, the Grand River has cut the observed gorge through the Pennsylvania strata in the Grand Ledge area. It is probable that the course of the original glacier river in the area of the Ledges developed on joint structures in the bedrock.

The early geologic history of the Ledges is essentially that of any marine beach with an open-water lagoon and swamps behind it (Kelly, 1933). These different yet interrelated environments each produced a distinct set of sediment types. The beach environment, dominated by high-energy waves, produced quartz sand with few fossils. The lagoon environment, in which low-energy currents moved, accumulated black muds, and provided a home for various shellfish. The shores of the lagoon were tidal flats where mud and fine sand settled with the rise and fall of the tides.

Ancient storm waves and winds moved beach sand onto the shore of the lagoon and produced fan-shaped deposits called washover fans. On the landward side of the lagoon, sheltered from the ocean, a marsh area existed where plants grew and muddy sediments accumulated.

The thick, flat-lying, multilayered quartz sandstone beds of the Ledges (Fig. 2), including the climbing area (Fig. 3), are interpreted as an ancient beach environment. The thin, dipping beds occurring in the upper portion of the Ledges outcrop are probably ancient windblown dunes (Fig. 4).

Beds of thin quartz sandstone near the abandoned American Vitrified Quarry (Fig. 2) are similar to the Ledges but also contain layers of coal and shale. The shale was probably deposited in a quiet lagoon sheltered from the ocean by a beach. A coal seam evident in the lower portion of outcrops in the quarry formed from plants that grew near the shores of the lagoon. A layer of

Figure 3. Thick flat-lying multilayered quartz sandstone beds, interpreted as an ancient beach environment, at the climbing area.

graywacke sandstone below the coal seam was deposited by small streams flowing into the lagoon. Numerous fossilized plant roots in the graywacke suggest the lagoon eventually filled up with sediment, changing into a marsh where plants could grow. A sandstone layer above the coal is similar to the sandstone seen at the Ledges, although the grains are smaller. This sandstone was deposited in the otherwise sheltered lagoon from the beach environment by hurricane winds and waves. Each sand layer represents another severe storm or hurricane that breeched the beach and dune complex to invade the lagoon. Muddy layers evident in the outcrop were deposited in quiet waters between these prehistoric storms, forming washover fans.

At the Clay Products Quarry (Fig. 2) the rocks are predominantly shale and siltstone. These rocks formed in the quiet waters of the lagoon. Streams flowing into the lagoon from the land brought in some sandy sediments, while swamps on the landward shore of the lagoon produced coal. The sequence of deposition shows how the lagoon gradually filled up with stream-carried sediments. In the lower part of the north quarry, fine black shale,

Figure 4. Thin, dipping beds of quartz sandstone interpreted as ancient windblown dunes, along the River Path Trail.

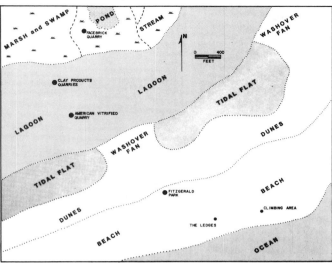

Figure 5. Generalized map showing the approximate location of ancient environments as they may have appeared at one point in geologic time in the area of Grand Ledge, Michigan (after Davis and Bredwell, 1978).

which formed in quiet water in the middle of the lagoon, can be seen. The presence in the shale of the brachiopod *Lingula,* which lived in brackish water, suggests that fresh water from streams and salt water from the ocean mixed together here. The dark color of the shale indicates a large amount of organic decay has taken place. Above the shale is coarse-grained banded-siltstone and sandstone deposited in much the same way as the graywacke at the American Vitrified Quarry, except that the sediment deposition took place farther into the lagoon from the stream mouths. Stream currents would slow as they entered the lagoon, depositing heavier sediments quickly, while still carrying smaller silt-sized sediments farther into the lagoon. Each light-colored siltstone band represents a period of flooding when stream currents moved with greater speed, and could transport sediments farther into the lagoon. Sandstone layers indicate periods of normal deposition. This type of sequential deposition built the lagoon's shoreline outward. After the lagoon was filled in, plants grew in the sediments, and their roots now appear as fossils in the sandstone bands. Above the last sandstone band is a sequence of gray siltstone and coal. The gray siltstone probably formed in marshy areas or large ponds fed by streams. This gray siltstone layer contains a wide variety of well-preserved plant fossils. The excellent state of preservation of these fossils suggests the siltstone was deposited in a quiet water environment; leaves and branches that fell into the water would sink to the bottom without being broken and slowly become covered with sediment. Along the edges of the marshy areas where there was little water movement, the dead plants and leaves accumulated more quickly than they could decay or wash away. This decaying vegetation became peat, but later was buried by other sediments whose weight compressed it and changed it into coal.

Outcrops at the abandoned Face Brick Quarry (Fig. 2) suggest a stream channel, possibly carved by some ancient hurricane or severe storm, cut through older lagoon and swamp deposits. The older deposits of siltstone, coal, and bedded sandstone were carried away by storm-fed streams and replaced by new deposits of sandstone, which filled the newly cut channel. The sandstone contains many broken pieces of shale, coal, and siltstone appearing as rip-up clasts. The sand is apparently reworked beach sediments. One unusual rock type is also present at the Face Brick Quarry: black limestone. This limestone formed when the area was covered either by a large lake or lagoon, which received very little sand or muddy sediment. The dark color of the limestone indicates that a large amount of decay took place, the water was very calm, and the sediments were undisturbed during deposition. The limestone is peculiar because of its proximity to the layers of sandstone and coal deposited in very different yet related environments.

The above descriptions suggest that a number of depositional environments existed in the vicinity of the Ledges, though not necessarily all at the same time. A single paleographic map would not be sufficient to tell the complete history of such a dynamic and rapidly changing ancient environment. Figure 5 is a generalized map showing the approximate location of these ancient environments as they may have appeared at one point in geologic time. Figure 6 is a generalized cross section of the four described sites.

The fossil record is quite extensive at the Ledges, but is dominated by plant remains. The most common animal fossil found is the brachiopod *Lingula.* Occasionally, remains of the

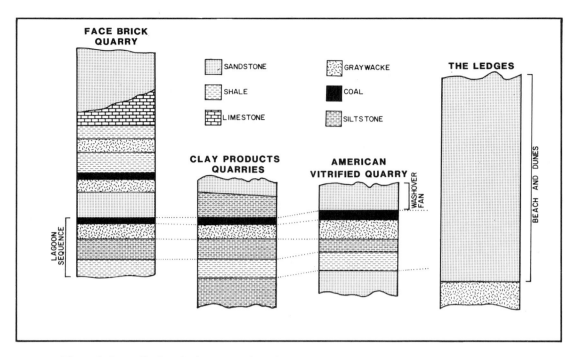

Figure 6. Generalized geologic cross section of the Ledges area. (after Davis and Bredwell, 1978).

fresh-water shark *Pleuracanthus* and the burrows of ancient lung fish can be found (Davis and Bredwell, 1978). Arnold (1949, 1966) identified more than 90 species of fossil plants around the Ledges area. The most common fossil plants include *Lepidodendron, Sigillaria, Calamites, Neuropteris, Sphenophyllum,* and *Cordaites.* It should be noted that collecting of fossils is not encouraged in Fitzgerald Park and that access to the Clay Products Quarry is by permission only.

In addition to the Ledges, Figzgerald Park offers a full range of nature trails, canoe rental, a fish ladder, and is home to the Ledges Playhouse. There is a $1.00 entry fee to the park. Rock climbing at the Ledges is restricted to the marked climbing area on the northeast bank of the Grand River at Oak Park, and all climbers must be top-roped.

REFERENCES CITED

Arnold, G. A., 1949, Fossil flora of the Michigan Coal Basin: Ann Arbor, University of Michigan, Contributions from the Museum of Paleontology, v. 7, no. 9, p. 131–269.
—— , 1966, Fossil plants of Michigan: Michigan Botanist, v. 5, p. 3–13.
Davis, M. W., and Bredwell, H. D., 1978, The nature of Grand Ledge: Grand Ledge Area American Revolution Bicentennial Commission, p. 31–32.
Kelly, W. A., 1933, Pennsylvanian stratigraphy near Grand Ledge, Michigan: Journal of Geology, v. 41, p. 77–88.

The Calvin 28 cryptoexplosive disturbance, Cass County, Michigan

Randall L. Milstein, Subsurface and Petroleum Geology Unit, Michigan Geological Survey, Lansing, Michigan 48912

LOCATION

Geophysical data, geologic mapping, and the extensive drilling of oil and gas test wells have delineated a subsurface structure in Calvin Township, Cass County, Michigan, T.7S.,R.14W.; Adamsville, Michigan, 7½-minute Quadrangle. The center of the structure is approximately 1 mi (1.4 km) southwest of the village of Calvin Center (Fig. 1). The feature maintains a surface topographic expression and a partially encircling drainage pattern, but its major structural complexity is concealed by drift and Paleozoic rock.

The location of the Calvin 28 cryptoexplosive disturbance is shown in Figure 1, which uses an idealized geologic map as a base and to which road numbers have been added. The Calvin 28 cryptoexplosive disturbance can be reached by driving 4 mi (6.4 km) east on Michigan 60 from Cassopolis until encountering the intersection of Calvin Center Road. A right (south) turn on Calvin Center Road will take the visitor toward the structure. After travelling 4.2 mi (6.8 km) on Calvin Center Road, the visitor will cross over the outer encircling rim of the structure. The outer edge of this rim is delineated by a chain of lakes connected by Christiana Creek (Fig. 2). Continuing south on Calvin Center Road 0.5 mi (0.8 km), one encounters the village of Calvin Center. The village is located, subtly, within the annular depression surrounding the central uplift of the structure. Continue south an additional 1 mi (1.6 km) to the junction of Calvin Hill Stree and turn right (west). By traveling 0.5 mi (0.8 km), turning left (south) on an unnamed light-duty road, and following it to its end, the visitor will arrive at the center of the central domal uplift of the structure. By returning to Calvin Center Road and turning right (south) and proceeding for an additional 2 mi (3.2 km) to the intersection of Mason Street, the visitor again enters the annular depression. This time the depressed area is more evident and is seen reflected in the course taken by streams that drain the depression and flow into Lafferty Lake (Fig. 2). By either continuing south on Calvin Center Road or turning right (west) on Mason Street, the visitor will cross the south edge of the encircling anticlinal rim. These high and low areas are evident in the surface topography illustrated in Figure 2. A northwest-southeast geologic section is shown in Figure 3. This section depicts the massive structural upheaval displayed beneath the surface expressions. Although the visible geology at the Calvin 28 cryptoexplosive disturbance is subtle and not particularly exciting, the subsurface geology is of much interest; many researchers consider the structure to be the most anamolous geologic feature in the Michigan Basin.

DESCRIPTION

In 1982, an exploratory well, the Halwell, Incorporated, Hawkes-Adams #1-28, was drilled through the central domal

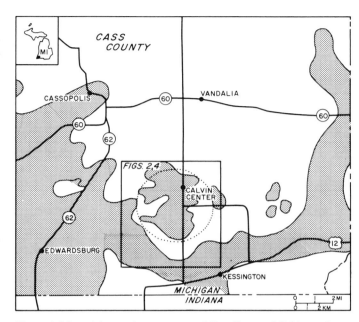

Figure 1. Map of the southeastern part of Cass County, Michigan, showing location of the Calvin 28 cryptoexplosive disturbance (dotted circle). Shaded area, Ellsworth Shale; unshaded area, Coldwater Shale.

structure in Section 28 of Calvin Township, Cass County, Michigan. The well penetrated 1,591 ft (485 m) of regional, although thinned, Paleozoic strata down to the Late Ordovician Cincinnatian Series. Instead of encountering the expected Middle and Lower Ordovician stratigraphic sequence, the well bore entered into a shortened section of the Lower Ordovician Prairie du Chien (Fig. 3). The base of the Prairie du Chien and the top of the Late Cambrian Trempealeau Formation were 1,243 ft (379 m) above normal regional levels. Continued drilling revealed the upper half of the Munising Formation to be absent down to its Eau Claire Member. The Mount Simon Sandstone Member, lying in its proper stratigraphic position below the Eau Claire, showed a vertical displacement of 1,363 ft (415.5 m). Geophysical well logs run on the test well revealed complex faulting in addition to steep and varied dips within the well bore. Continued drilling of more than 100 Devonian test wells in the area has defined the structure as an isolated, nearly circular predominantly subsurface feature with a diameter of 4.5 mi (7.25 km) and consisting of a central dome, an annular depression, and an encircling anticlinal rim. The heavily drilled Devonian horizon reflects the general structural characteristics of the feature, defined by limited deep test wells and seismic profiling (Fig. 4).

Previous work involving the Calvin Township structure is

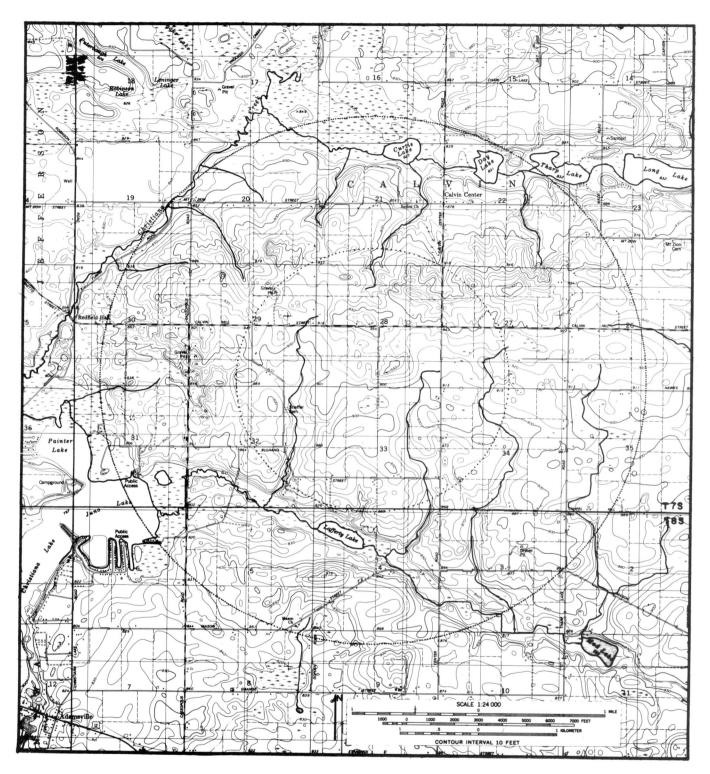

Figure 2. Map of part of the Adamsville 7½-minute Quadrangle showing the surface expression of the Calvin 28 cryptoexplosive disturbance. Circles indicate known subsurface limits of the central uplift and anticlinal rim structures. Structurally related drainage patterns are enhanced.

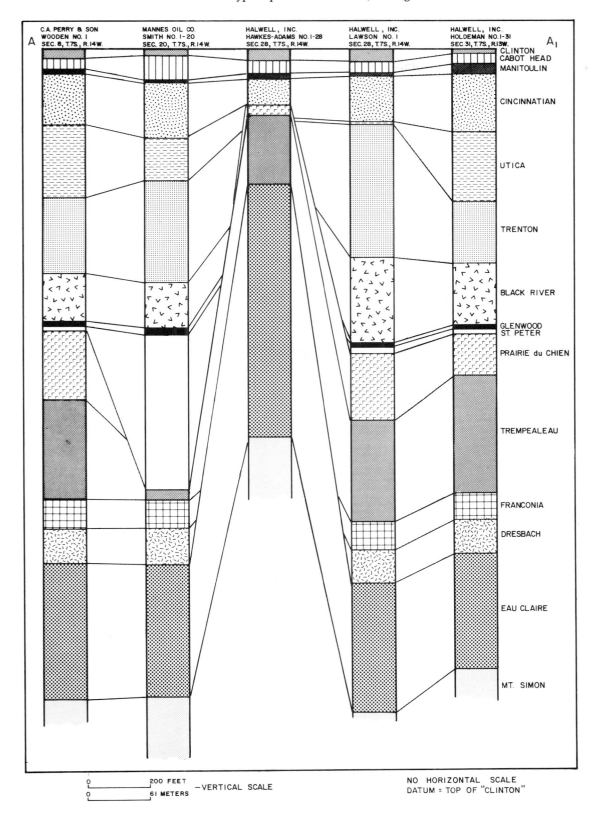

Figure 3. Northwest-southeast cross section indicating stratigraphic uplift and missing lithology associated with the Calvin 28 cryptoexplosive disturbance. Line of section shown on Figure 4.

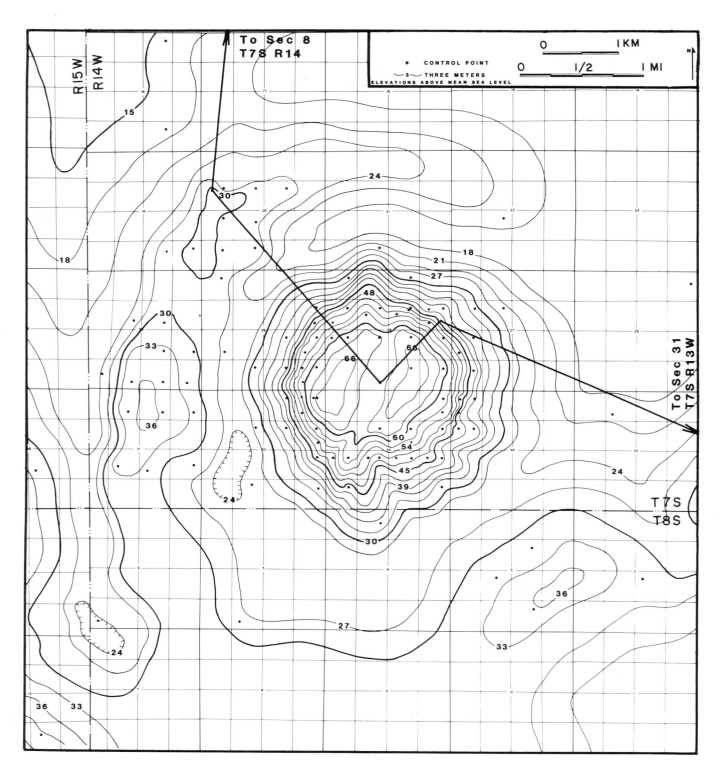

Figure 4. Structure contour map of the Calvin 28 cryptoexplosive disturbance as expressed on the top of the Devonian Traverse Limestone (from Milstein, 1986).

limited. R. J. DeHaas (written communication, 1983), in a presentation to the Michigan Basin Geological Society, provided evidence of dramatic structural uplift associated with a dome beneath the Calvin 28 oil field. He suggested the "cryptoexplosive" nature of the domal feature based on superficial comparisons to other Midwestern and European cryptoexplosive disturbances, and first brought attention to the surface topographic expression of the structure.

Mata and Myles (1985) interpret the feature to be the result of basement faulting in a tectonically "tight" and compressed corner of the Michigan Basin. The southwestern portion of the Michigan Basin, which includes Cass County, is interpreted as a region of compression and stress due to its close proximity to the Illinois Basin. The ensuing room problem caused fault blocks to pop up and fold over.

A geophysical investigation of the structure by Ghatge (1984) indicates no gravity anomalies or magnetic anomalies to 1 milligal associated with the structure. Ghatge concluded that any interpretation of the structure based on its gravity and magnetic surveys would not be able to explain the intense uplift or missing Cambrian and Ordovician sediments. While Ghatge favored origin by impact, he felt his evidence was insufficient to categorize the feature as anything but a "crypto-explosion structure."

In an extensive study, Milstein (1986) defines the complete cryptoexplosive nature of the structure and identifies the feature as the Calvin 28 cryptoexplosive disturbance.

Milstein calculates the age of the event responsible for the formation of Calvin 28 to be prior to Early Silurian time, but after deposition of the Early Cincinnatian Upper Richmond Group. Milstein also suggests that the present surface topographic expression of the feature is due to either slumping and settling of unstable sediments related to complex faulting, the periodic reactivation of faults resulting in the further upward movement of disrupted strata, or continuous rebound action from the initial event forming the structure.

Milstein (1986) finds that the Calvin 28 structure consists of a centrally located uplift 2 mi (3.25 km) in diameter, with structural closure on the Devonian Traverse Limestone of 135 ft (41 m), a surrounding annular depression about 3,280 ft (1 km) wide, and an outer encircling aniclinal feature or rim, roughly 3,280 ft (1 km) wide with a structural relief of 112 ft (34 m) on the Traverse Limestone.

Milstein (1986) concludes that any explanation for the origin of the Calvin 28 structure must accommodate the following observations. (1) The structure is circular, containing a central uplift, surrounding annular depression, and a peripheral anticline. (2) Structural deformation is intense, involving large-scale faulting and upward movements of strata. (3) More than 906 feet (276 m) of strata is missing in portions of the structure, while other locations show highly anomalous thicknesses of units. (4) Deformation wanes with depth beneath, and distance away from, the structure. (5) The structure exists as an isolated feature. (6) No volcanic material is identified in association with the structure. (7) A microbreccia consisting of fractured and unfractured

floating quartz grains in a carbonate matrix is identified in deep well samples. (8) The event responsible for the structure's origin is estimated to have released at least 1×10^{26} ergs of energy, nearly instantaneously, without the development of magma.

Possible origins for such large-scale cryptoexplosive structures as Calvin 28, involve both endogenetic and exogenetic processes. Excellent summaries of both arguments for the origin of cryptoexplosive structures have been given by French (1968) and McCall (1979).

Milstein (1986) compares the Calvin 28 cryptoexplosive disturbance with both endogenetic and exogenetic structures that exhibit similar characteristics. Endogenetic structures that exhibit characteristics consistent with structural patterns identified in the Calvin 28 feature include: maars, diatremes, calderas, kimberlite pipes, igneous intrusives, diapirs, and solution subsidence structures. Exogenetic structures exhibiting similar characteristics are limited to cryptoexplosive/astroblemes and impact craters.

In comparing endogenetic structures to the Calvin 28 feature, Milstein (1986) finds no significant evidence to suggest that volcanic eruption, igneous intrusion, solution subsidence, or a diapiric mass of sedimentary material is responsible for the structure's origin. No igneous material occurs in association with the structure. If igneous material had been present at the structure, even in small amounts, it would be difficult to explain its absence by weathering processes. Diapirism is ruled out by a stratigraphic configuration that would not allow the significant density inversion necessary for flowage. The structural pattern of the feature and lack of soluble strata below the structure rule out the possibility of solution subsidence.

Milstein (1986) identifies eight characteristics of the Calvin 28 cryptoexplosive disturbance that strongly favor origin by impact. (1) Terrestrial surface impact structures with central uplifts (complex craters) show a modelable relationship between stratigraphic displacement in the uplift and the crater form (Grieve and others, 1981). The Calvin 28 structure exhibits this relationship. (2) The waning of structural deformation beneath Calvin 28 is indicated by seismic profiles and dipmeter readings from deep test wells. The lessening of derangement with depth would not be expected from a tectonic or volcanic origin, but would be consistent for structural deformation incurred from a downward-projected shock envelope (Shoemaker, 1960; Lindsay, 1976). (3) The structural pattern of Calvin 28 indicates that a large amount of near-surface energy was involved in the formation of the structure and that the energy was released as a single nearly instantaneous event. (4) No igneous material has been recovered from well cuttings or identified in petrographic studies involving the structure. (5) Calvin 28 is an apparently isolated circular structure of intense deformation in otherwise flay-lying strata. (6) The lack of any magnetic or gravity anomalies associated with the circular structural pattern is a trait commonly identified with impact craters in sedimentary targets. (7) The presence of a microbreccia consisting of fractured and unfractured floating quartz grains in a carbonate matrix is similar to microbreccia associated with impact craters in sedimentary targets (Short and Bunch, 1968).

(8) The energy required to produce the 4.5 mi (7.24 km) diameter structure, the apparent structural relief, the missing strata, and the intense structural deformation is at least 1×10^{26} ergs. While this value exceeds energy estimates for known singular explosive endogenetic events, it would be considered a conservative value for energy released by a hypervelocity impact in a sedimentary target (Shoemaker and Wolfe, 1982).

These eight characteristics would account for all major features listed earlier and considered essential in the evaluation of Calvin 28's origin.

While an endogenetic origin for most cryptoexplosive structures is disputed (French, 1968; McCall, 1979), the possibility exists that a yet unidentified endogenetic process may have formed Calvin 28. It is unlikely though that such an event could generate the tremendous energy required to form Calvin 28, without the presence of magma.

While not ruling out an endogenetic origin, Milstein (1986) concludes that the Calvin 28 cryptoexplosive disturbance is the result of a nearly instantaneous shock event, and that the event can best be attributed to hypervelocity impact.

REFERENCES CITED

French, B. M., 1968, Shock metamorphism as a geological process, *in* French, B. M., and Short, N. M., eds., Shock metamorphism of natural materials: Baltimore, Maryland, Mono Book Corporation, p. 1–17.

Ghatge, S. L., 1984, A geophysical investigation of a possible astrobleme in southwestern Michigan [M.S. thesis]: Kalamazoo, Western Michigan University, 41 p.

Grieve, R.A.F., Robertson, P. B., and Dence, M. R., 1981, Constraints on the formation of ring impact structures, based on terrestrial data, *in* Schultz, P. H., and Merrill, R. B., eds., Multi-ring basins, Lunar and Planetary Science Conference, 12th, Houston, 1980: New York, Pergamon Press, p. 37–57.

Lindsay, J. F., 1976, Lunar stratigraphy and sedimentology; Developments in solar systems and space science, 3: New York, Elsevier Scientific Publishing Company, p. 65–79.

Mata, F., and Myles, J., 1985, Geology, geophysics must be combined for basin exploration: Michigan Oil and Gas News, v. 91, no. 12, p. 65.

McCall, G.J.H., ed., 1979, Astrobleme-cryptoexplosion structures: Stroudsburg, Pennsylvania, Dowden, Hutchinson and Ross, Incorporated, Benchmark papers in geology, v. 50, p. 1–4, p. 4–20.

Milstein, R. L., 1986, Impact origin of the Calvin 28 cryptoexplosive disturbance, Cass County, Michigan [M.S. thesis]: Greeley, University of Northern Colorado, 87 p.

Shoemaker, E. M., 1960, Penetraion mechanics of high velocity meteorites, illustrated by Meteor Crater, Arizona, *in* McCall, G.J.H., ed., Meteorite craters: Stroudsburg, Pennsylvania, Dowden, Hutchinson and Ross, Incorporated, p. 170–186.

Shoemaker, E. M., and Wolfe, R. F., 1982, Cratering time scale for the Galilean satellites, *in* Morrison, D., ed., The satellites of Jupiter: Tuscon, University of Arizona Press, p. 277–339.

The Indiana Dunes area, northwestern Indiana

John R. Hill, Indiana Geological Survey, 611 North Walnut Grove, Bloomington, Indiana 47405

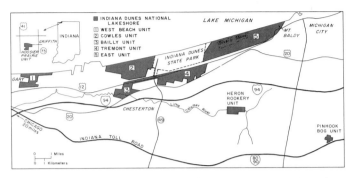

Figure 1. Map of the Indiana Dunes National Lakeshore and the Indiana Dunes State Park and surrounding areas showing park boundaries. Modified from National Park Service Boundary/Unit map for the Indiana Dunes National Lakeshore, 1979.

LOCATION AND SIGNIFICANCE

The area designated as the Indiana Dunes is in Lake, Porter, and LaPorte Counties in northwestern Indiana. Including roughly 45 mi (72 km) of shoreline at the southern tip of Lake Michigan, the dunes area is replete with numerous well-developed geomorphic features that formed in response to the final stages of the Lake Michigan Lobe of the Laurentide Ice Sheet. Attracting such skilled researchers as T. C. Chamberlin, Frank Leverett, William Alden, Frank Taylor, Jack Hough, and J Harlen Bretz, southern Lake Michigan and environs have yielded a wealth of information about the formative stages of the Great Lakes as well as glaciological mechanisms responsible for their genesis. The dunes area of Indiana is of particular interest because of the wide variety of dune forms, well-preserved glacial-lake terraces, and impressive topography of the Valparaiso Moraine that remain relatively unspoiled.

Magnificent examples of blowout dunes, first described by George Babcock Cressey in 1928, can be seen all along the beach of the Indiana Dunes State Park. Access to the park is by Indiana 49 (Fig. 1). Numerous well-marked trails traverse the 2,100-acre (840 ha) park, providing ready access to the foredune complex of the modern beach as well as the interdunal and back-dune areas that formed during the several stages of glacial Lake Chicago.

The Indiana Dunes National Lakeshore affords numerous access points to, perhaps, the greatest variety of geologic features within the Lake Michigan Basin. Interested persons should stop at the visitor center, which is at the intersection of U.S. 12 and Kemil Road. Displays, resource personnel, and many scientific and cultural publications are available there. Access to most of the federal park holdings is by U.S. 12. Ample parking and well-marked trails are provided at West Beach, Cowles, Bailly, Tremont, and East Units (Fig. 1). Only specially guided tours of the Pinhook Bog Unit are permitted, however. Preserved within

the federal parklands are splendid examples of moving foredunes (particularly Mt. Baldy, East Unit, Fig. 1); relict beach, foredune, and lake terraces of glacial Lake Chicago (Indiana 49 from U.S. 20 northward to Lake Michigan); interdunal marshes; varied longitudinal, transverse, and parabolic dune forms; unusual ecosystems developed on lacustrine plain (Hoosier Prairie Unit and Heron Rookery, Fig. 1); and a kettle bog that hosts a unique assemblage of exotic native flora (Pinhook Bog Unit, Fig. 1).

An excellent focal site and vantage point from which to view Lake Michigan, the beach and dune features of glacial Lake Chicago, and the distant crest of the Valparaiso Moraine is Mt. Baldy (NW¼NW¼Sec.31 and SW¼SW¼Sec.30,T.38N.,R.4W., Michigan City Quadrangle; Fig. 2).

GENERAL INFORMATION

The Valparaiso Moraine, which dammed the meltwaters that formed Lake Chicago, attains its maximum width of 17 mi (27.2 km) along the Illinois-Indiana border and its maximum elevation of 950 ft (285 m) north of LaPorte (Fig. 3) in LaPorte County (SW¼Sec.32,T.37N.,R2W., Springville Quadrangle). The moraine narrows to a scant three-fourths of a mile (1.2 km) in width northwest of LaPorte and descends westward

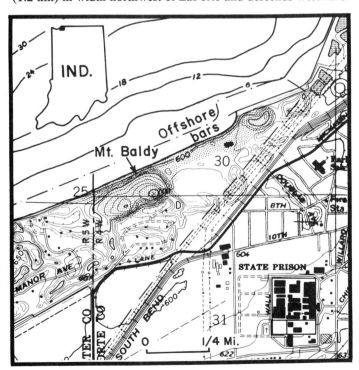

Figure 2. Topographic map of the area due west of Michigan City showing the location of Mt. Baldy. Base modified from Michigan City West 7½-minute Quadrangle, 1980.

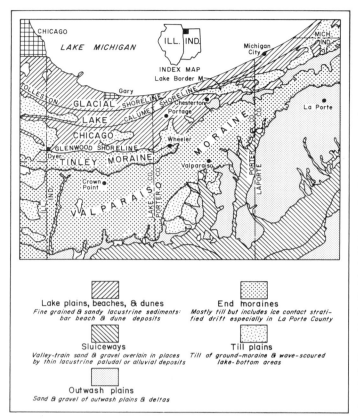

Figure 3. Generalized glacial map of parts of northwestern Indiana and adjoining states. Modified from Schneider (1968).

to an average elevation of about 750 ft (225 m) across much of Porter and Lake Counties. The Tinley Moraine, a prominent feature in Illinois, is distinguished from the Valparaiso Moraine in Indiana by a trough along its distal margin. The Tinley trends east-west along the proximal flank of the Valparaiso Moraine from the Illinois-Indiana state line to an area northeast of Wheeler (Fig. 3) in Porter County, where it loses its identity amid the rolling terrain of the Valparaiso Moraine (Schneider, 1968). The four Lake Border Moraines recorded by Bretz (1955) in Illinois cannot be traced through the Calumet region of Indiana to the southeast side of the lake. However, a single Lake Border Moraine is visible in the Bailly Unit (Figs. 1 and 3) of the National Lakeshore. Lake Border till is exposed along the banks of the Little Calumet River, particularly near U.S. 12 and Indiana 149. The proximal flank of the Lake Border Moraine is quite steep at this locality but has a gentler slope eastward along U.S. 12.

Although the till composing the Valparaiso, Tinley, and Lake Border Moraines is considered to be the result of distinct glacial readvances in northwestern Indiana, it is virtually indistinguishable from one moraine to the other. It is characteristically gray to dark-gray, calcareous, silty clay-loam to silt-loam till of very stiff consistency. The clasts are primarily shale pebbles and granules.

Classic swell-and-swale end-moraine topography typical of

the Valparaiso Moraine is best seen at the intersection of Indiana 49 and U.S. 6. On a clear day, the Valparaiso Moraine can be viewed panoramically from the summit of Mt. Baldy (Figs. 1 and 2).

Final deglaciation of the Lake Michigan Basin, which began during late Woodfordian time (late Wisconsinan), resulted in the genesis of glacial Lake Chicago as the Lake Michigan Lobe retreated from the Lake Border Moraine (Goldthwait, 1908; Leverett and Taylor, 1915; Bretz, 1955; and Hough, 1958 and 1966). The complex interrelationships of shoreline elevations, ice-marginal positions, drainage outlets, and interconnection of the Great Lakes, including the relative and absolute dating of events, are still being worked out. (See Lineback and others, 1979.) Although two(?) stages of the Glenwood level (640 ft [192 m] msl.), two stages of the Calumet level (620 ft [186 m] msl.), repeated occupation of the Toleston (also spelled Tolleston) level (605 ft [181.5 m] msl.), and numerous stages of lower lake levels are recognized (see Hough, 1966), evidence for many of these stages is lacking in the Indiana Dunes area. Kemil Road from U.S. 20 northward to Lake Michigan transects the three major lake levels occupied by Lake Chicago during the past 13,500 years. The Glenwood beach and dunes can be seen just north of the intersection of U.S. 20 and Kemil Road (Fig. 4). Calumet features are evident at the Indiana Dunes National Lakeshore visitor center at Kemil Road and U.S. 12 (Fig. 4). About half a mile (0.08 km) north of U.S. 12, Kemil Road crosses the 605-ft (181.5-km) elevation of the Toleston beach and level (Fig. 4). In the Kemil Road area, the so-called Toleston beach and level is an undifferentiated sequence of dunes from glacial Lakes Algonquin, Nipissing, and Algoma (postglacial Lake Chicago). Few, if any, geomorphic features exist in the Indiana Dunes area to permit distinction of the multiple lake levels at and below 605 ft (181.5 m) msl.

Indirect evidence of the Chippewa low-water stage (postglacial Lake Chicago) was discovered 20 years ago at the base of a lacustrine clay that is exposed along the northern toe of Mt. Baldy. Carbonized wood samples about 5,000 to 6,000 years old were extracted from paludal deposits resting on Lake Border(?) till (Winkler, 1962). The dates correspond to glacial Lake Chippewa (Hough, 1958) and therefore place the shoreline some 50 mi (80 km) north of the wood site at the time of origin.

During the past decade, detailed mapping of glacial-lake levels (Evenson, 1972; Evenson and others, 1976), bottom sampling and fathometer profiling of Lake Michigan (Wickham and others, 1978), and climatologic studies (Saarnisto, 1974) have resulted in the rethinking of some classic interpretations of Great Lakes geology (for example, the Greatlakean Substage as a replacement for the Valderan Substage [Evenson and others, 1976]). Although much has already been done to unravel the late Wisconsinan history of this region, an equal or greater number of questions remains to be answered. Of particular interest is the finite dating of the well-documented Glenwood, Calumet, and Toleston levels of glacial Lake Chicago in Indiana.

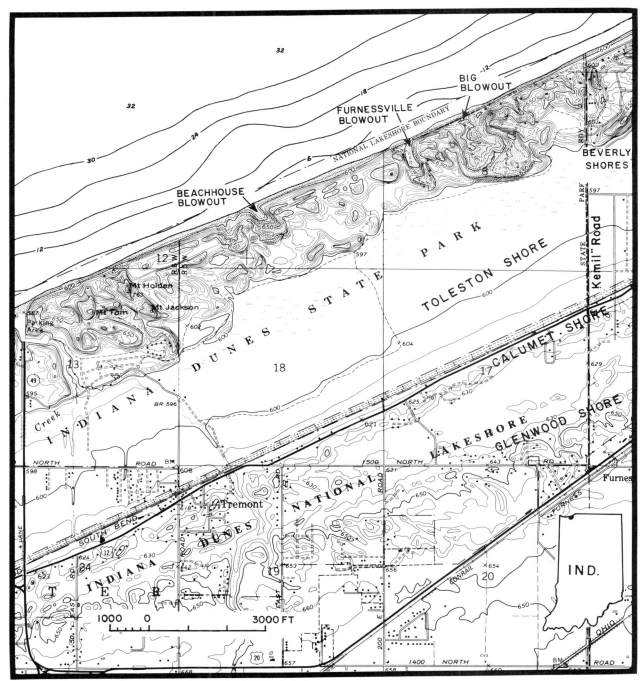

Figure 4. Topographic map of a part of the Indiana Dunes State Park and the Beverly Shores area showing shorelines of glacial Lake Chicago and blowout dunes discussed in the text. Base modified from Dunes Acres 7½-minute Quadrangle, 1953.

Students of aeolian transport will appreciate the many excellent dune forms that can be seen from the beaches and trails of both the Indiana Dunes State Park and the Indiana Dunes National Lakeshore (Fig. 1). Blowout dunes, which originate as a narrow channel on the crest of the foredune where vegetation has been removed by either man or nature, line the shore along most of the Indiana Dunes State Park. Both Mt. Tom and Mt. Holden,

at the west end of the park, rise nearly 200 ft (60 m) above the beach (Fig. 4). Beach House, Furnessville, and Big Blowouts (Fig. 4) are also typical of this dune form. An exhumed buried forest can be seen just south of Big Blowout. The tree graveyard is certainly no older than Toleston time but could date from one of the younger lake levels.

Mobile dunes resulting from increased local sand supply, the sudden loss of vegetative cover, or excessive wind erosion are actively prograding over living forests along the northern flank of Mt. Baldy (Figs. 1 and 2) and in the West Beach area (West Unit, Fig. 1). The rate of dune migration near Mt. Baldy has increased in recent years, primarily in response to lowered lake levels and beach nourishment that increased the sand supply to the foredune ridge. Several buildings are now threatened by the migrating dunes at some localities just north and west of Mt. Baldy.

Various kinds of vegetatively controlled dunes occupy the intradunal and interdunal (zone between the dune complexes of the various lake levels) areas behind the Recent beach. Turret dunes (Cressey, 1928), for example, dot part of the otherwise denuded backdune area at West Beach. Parabolic, longitudinal, and transverse dunes can be observed in the interdunal area between the Calumet and Toleston levels along Kemil Road (Fig. 4). Most of these dunes are best seen in late fall, winter, or early spring when they are not obscured by dense foliage.

From the summit of Mt. Baldy, the observer has a good view of a variety of shoreline processes. Manmade structures that intercept longshore currents, such as the Michigan City Harbor breakwater due east of Mt. Baldy (Fig. 2), create anomalous erosion/deposition patterns. The general littoral-drift flow is from northeast to southwest along this part of the shore. Upcurrent beach accretion has occurred on the east side of the breakwater, and significant beach erosion has resulted on the downcurrent or west side. During the early 1970s, this erosion was greatly aggravated by high lake levels. Excessive loss of beach along the northern toe of Mt. Baldy and in the Beverly Shores area resulted in a beach-nourishment project initiated by the U.S. Army Corps of Engineers. (See Wood, 1976.) More than a quarter of a million cubic yards (190,000 m^3) of sand were placed on the eroding beach in the spring of 1974. Eighty thousand cubic yards (60,800 m^3) of that sand were lost to erosion during the first year of monitoring (Wood, 1976). Lake Front Drive, which traverses the northern perimeter of Beverly Shores, abruptly terminates some 2 mi (3.2 km) west of Michigan City—the result of progressive westward erosion on the downcurrent side of the harbor breakwater. Wave erosion and littoral-drift transport along the foredune at the east end of Lake Front Drive resulted in oversteepening of the dune face (see Hill, 1974, p. 5) in the mid-1970s. Recent lower lake levels permitted the reestablishment of the foredune face at its natural angle of repose of 32°.

Looking north from the top of Mt. Baldy, one can observe on most clear days a series of two or more offshore bars (Fig. 2). The outermost bar, which is about 1,000 ft (300 m) from shore, is a relatively stable feature that according to Wood (1976) is affected only by the intense storms of fall and spring. The inner bar, however, constantly changes configuration in response to the variable intensity of breaking-wave energy. Irregularly distributed ephemeral bars can also be observed in the zone between the shore and the inner bar. (See Wood, 1976; Hawley and Judge, 1969.)

At any given time and throughout the four seasons, the Indiana Dunes area is a fascinating laboratory in which to view both shore- and dune-forming processes. From ephemeral beach features, such as storm-cut berms, to the towering majesty of Mt. Tom or Mt. Baldy, something of interest to nearly every visitor or hiker can be seen within a short walk from readily accessible parking lots or within a few miles' walk through the dunes.

REFERENCES CITED

Bretz, J H., 1955, Geology of the Chicago region—Pt. 2, the Pleistocene: Illinois Geological Survey Bulletin 65, 132 p.

Cressey, G. B., 1928, The Indiana sand dunes and shore lines of the Lake Michigan basin: Geographical Society of Chicago Bulletin 8, 80 p.

Evenson, E. B., 1972, Late Pleistocene shorelines and stratigraphic relationships in the Lake Michigan basin [Ph.D. dissertation]: Ann Arbor, University of Michigan, 88 p.

——, and others, 1976, Greatlakean Substage—A replacement for Valderan Substage in the Lake Michigan basin: Quaternary Research, v. 6, p. 411–424.

Goldthwait, J. W., 1908, A reconstruction of water planes of the extinct glacial lakes in the Lake Michigan basin: Journal of Geology, v. 16, p. 459–476.

Hawley, E. F., and Judge, C. W., 1969, Characteristics of Lake Michigan bottom profiles and sediments from Lakeside, Michigan to Gary, Indiana: International Association of Great Lakes Research, Proceedings of 12th Conference, p. 198–209.

Hill, J. R., 1974, The Indiana dunes—Legacy of sand: Indiana Geological Survey Special Report 8, 9 p.

Hough, J. L., 1958, The geology of the Great Lakes: Urbana, University of Illinois Press, 313 p.

——, 1966, Correlation of glacial lake stages in the Huron-Erie and Michigan Basins: Journal of Geology, v. 74, p. 62–77.

Leverett, Frank, and Taylor, F. G., 1915, The Pleistocene of Indiana and Michigan and the history of the Great Lakes: U.S. Geological Survey Monograph 53, 529 p.

Lineback, J. A., and others, 1979, Glacial and postglacial sediments in Lakes Superior and Michigan: Geological Society of America Bulletin, v. 90, p. 781–791.

Saarnisto, M., 1974, The deglaciation history of the Lake Superior region and its climatic implications: Quaternary Research, v. 4, p. 316–339.

Schneider, A. F., 1968, The Tinley Moraine in Indiana: Indiana Academy of Science Proceedings, v. 77, p. 271–278.

Wickham, J. T., and others, 1978, Late Quaternary sediments of Lake Michigan: Illinois Geological Survey Environmental Geology Notes 84, 26 p.

Winkler, E. M., 1962, Radiocarbon ages of postglacial lake clays near Michigan City, Indiana: Science, v. 137, p. 528–529.

Wood, W. L., 1976, Coastal sedimentation and stability in southern Lake Michigan: Geological Society of America Field Trip Guidebook, 117 p.

Fort Wayne, Indiana: Paleozoic and Quaternary geology

Jack A. Sunderman, Department of Earth and Space Sciences, Indiana University-Purdue University at Fort Wayne, Fort Wayne, Indiana 46805

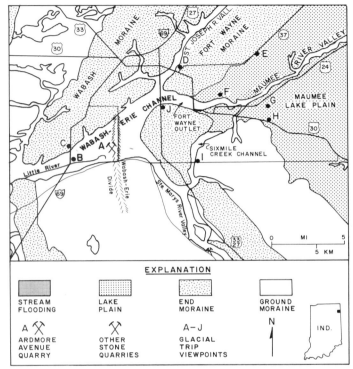

Figure 1. Geomorphic map of the Fort Wayne area, Allen County, Indiana, showing distribution of Pleistocene glacial landforms, location of rock quarries, and viewpoints for Pleistocene geomorphic features. Modified from Bleuer and Moore (1978).

LOCATION AND SIGNIFICANCE

The geology displayed in and around the city of Fort Wayne, Indiana, is interesting and critically important for at least two reasons: (1) a large quarry near the city exposes an unusual variety of Paleozoic strata, some facies of which are not exposed at any other place; and (2) the city is situated at the junction of several Pleistocene glacial features of regional significance (Fig. 1).

The largest and most spectacular quarry in northeastern Indiana is the Ardmore Avenue Quarry of May Stone and Sand, Inc., located along Ardmore Avenue in the NE¼Sec.29,T30N., R.12E. on the southwest edge of Fort Wayne (Figs. 1, 2). This quarry is one of the two largest and deepest crushed-stone quarries in Indiana, measuring more than 2,000 ft (610 m) across in a northeast-southwest direction and about 271 ft (82 m) deep during the fall of 1983 (Fig. 3). The quarry is readily accessible for public viewing from a stand provided by the company on the north side of the quarry. This stand allows such an instructive and spectacular view of the quarry that it has become the focal point for most field trips to the area. Those desiring to enter the quarry

should direct their inquiries to the Ardmore Avenue office of May Stone and Sand, Inc.

The Ardmore Avenue Quarry is also the focal point for the study of Pleistocene features in this area because of the excellent exposures of a complex Pleistocene section at the top of the quarry and the proximity of the quarry to the Pleistocene geomorphic features of the Fort Wayne area. Viewing points for these features are shown on Figure 1 by letter codes keyed to the descriptions below. Starting from the Ardmore Avenue Quarry, a circular trip incorporating these viewpoints can be completed in two to four hours.

SITE 77—PALEOZOIC GEOLOGY

Major Paleozoic bedrock features exposed in the Ardmore Avenue Quarry are the Michigan Basin facies of Silurian rocks, including the northern equivalent of the Louisville Limestone and the impressive reef feature known as the Fort Wayne Bank; the basin facies of Devonian rocks, including the Detroit River and Traverse Formations and an altered volcanic ash bed; and the Silurian-Devonian and Paleozoic-Quaternary unconformities (Fig. 4).

Louisville Limestone Equivalent

The oldest rocks exposed in the Ardmore Avenue Quarry are in the upper part of the northern Indiana equivalent of the Louisville Limestone, of Late Silurian age (Fig. 4). In this area, the Louisville equivalent is part of the Pleasant Mills Formation and probably equates with the A carbonate units of the Salina Group of the Michigan Basin (Droste and Shaver, 1977, 1982). The Louisville equivalent here is gray to brown dolomite that is thick bedded, saccaroidal, and cherty and contains scattered brachiopods, stromatoporoids, *Halysites* and *Favosites* corals, and other fossils. Most of the chert is white and occurs as nodules, lenses, and replacements of fossils. The dense brown matrix of the Louisville equivalent here is characteristic of the A carbonate rocks of the Michigan Basin, which are generally interpreted as having formed in a restricted environment. The fossils of the Louisville equivalent give the rock a somewhat reeflike appearance here, however, and in the upper 10 ft (3 m) of the unit, gray-colored rock and a coquinalike bed of pentamerid *Rhipidium* brachiopods suggest a transition to the more open-marine conditions of the overlying Wabash Formation (Shaver and others, 1983). In the Ardmore Avenue Quarry, rocks of the Louisville equivalent are most accessible near the primary crusher at the bottom of the quarry (Fig. 2).

Wabash Formation and Fort Wayne Bank

Above the Louisville equivalent looms an imposing half-

Figure 2. Wide-angle view of Ardmore Avenue Quarry of May Stone and Sand, Inc., showing exposed Paleozoic and Pleistocene strata. View shows Louisville Limestone equivalent (dark rock) in lower bench wall near crusher, reef facies of Wabash Formation from light band just above Louisville equivalent to top of massive unit in far highwall, well-bedded Devonian strata at top of bedrock section, and about 63 ft (120 m) of Pleistocene drift resting on bedrock. Silurian-Devonian and Paleozoic-Quaternary unconformities are prominent. Water draining into quarry is channeled to main pump in left center of view, then pumped to surface. Photograph by J. A. Sunderman; from Shaver, Sunderman, and others (1983).

mile-long exposure of one of the most impressive geologic features in the midcontinent area—the Fort Wayne Bank (Figs. 2 and 4). During late Niagaran time, this feature may have extended almost continuously around the southern end of the Michigan Basin as a barrier reef (Droste and Shaver, 1982), but its massive, reeflike facies is exposed only in the Ardmore Avenue Quarry. Elsewhere the bank occupies the position of both the Pleasant Mills and Wabash Formations, but here it is developed mainly in the stratigraphic position of the Mississinewa Shale

Member of the Wabash Formation. The bank rock is gray dolomite boundstone that is massive to thick bedded, vuggy, and fossiliferous (Fig. 5). Fossil preservation is poor due to dolomitization, but hemispherical and planar stromatoporoids, which probably acted as the main binding organisms for the reef, are abundant and can be recognized as vaguely laminated bands and laminae-lined vugs. Other recognizable fossils are corals, brachiopods, gastropods, and crinoids. Characteristics of the Fort Wayne Bank rocks can be studied in quarry blocks near the viewing stand and in outcrops along the main ramp road into the quarry (Fig. 3).

Silurian-Devonian Unconformity

The regional Silurian-Devonian unconformity of the Michigan and Illinois Basins (the so-called profound unconformity that separates the Tippecanoe and overlying Kaskaskia Sequences over much of North America) is exposed in both the Ardmore Avenue Quarry and the nearby Allen County Aggregates, Inc., quarry (Fig. 1). A spectacular view of the unconformity is afforded from the viewing platform of the Ardmore Avenue Quarry, and the unconformity can be examined in relative safety along the main ramp road into the quarry (Figs. 2, and 3).

The unconformity here separates Late Silurian strata from overlying Middle Devonian strata, a hiatus representing some 15 million years' time. In spite of its considerable time significance, however, the erosional surface is remarkably smooth, with no more than 1 or 2 ft (0.3 or 0.6 m) of relief along the entire quarry exposure (Figs. 2 and 5), a condition suggestive of marine planation. The unconformity would in fact be difficult to recognize were it not for the contrast in color and bedding characteristics of the Silurian and Devonian rocks and the slight but noticeable dip of the underlying Silurian strata (Fig. 2). The

Figure 3. Pace-and-compass map of Ardmore Avenue Quarry showing distribution of strata exposed at various quarry levels during fall of 1983, location of quarry road, and location of viewing stand along Sand Point Road on north side of quarry. From Shaver, Sunderman, and others (1983).

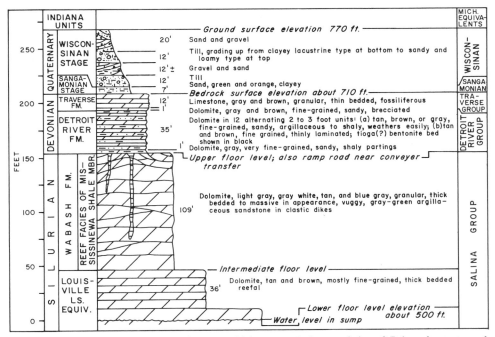

Figure 4. Stratigraphic section showing age, thickness, and characteristics of Paleozoic strata and Pleistocene materials exposed at various levels in the Ardmore Avenue Quarry. Modified from Shaver (1976).

unconformity is a disconformity in most midwestern localities, but it has the characteristics of an angular unconformity in both the Ardmore Avenue Quarry and the Allen County Aggregates Quarry. In these quarries, the underlying Wabash rocks exhibit a local primary dip of 5° or more that is related to the reef structures of the Fort Wayne Bank.

Two unusual features in the upper part of the Wabash rocks exposed in the Ardmore Avenue Quarry give evidence for events that occurred during the time represented by the unconformity. Vugs are especially large and numerous toward the eroded top of the Wabash rocks, and the vugs appear to be concentrated in the upper 20 ft (6 m) of the exposure (Fig. 5). Vug enlargement probably was caused by subaerial and submarine solution of the top of the reef bank during very Late Silurian to Early Devonian time. Sandstone clastic dikes also are found in the upper part of the Wabash rocks (Figs. 4, and 5) and at least one dike extends entirely through the Wabash exposure. The sandstone occupies dilated fractures that probably formed during Early to early Middle Devonian time as a result of local gravitational stress related to reef geometry (Sunderman and Mathews, 1975; Sunderman, 1983). The fractures are filled with gray quartz sandstone that is poorly sorted and argillaceous. The sandstone can be traced directly into a bed of dolomitic quartz sandstone at the base of the overlying Middle Devonian strata (Fig. 5). Similar sandstone dikes also occur in Wabash rocks just beneath the unconformity in the nearby Allen County Aggregates quarry (Fig. 1) and are found in Silurian reefs below the unconformity at several other localities in northern Indiana and Illinois (Sunderman and Mathews, 1975). During the fall of 1983, the best view of the dikes

in the Ardmore Avenue Quarry was along the far western highwall of the quarry.

Detroit River Formation and Tioga(?) Bentonite Bed

At the base of the Detroit River Formation is a ½- to 1-ft (20- to 30-cm) bed of green-gray sandy dolomite and argillaceous quartz sandstone, typical of basal Devonian strata in the north-central United States (Collinson and others, 1967) and probably equivalent to the Pendleton Sandstone Bed of central Indiana (Fig. 5). The sand source for this bed apparently was the sand source for the clasic dikes that occur in this quarry and elsewhere in Indiana and Illinois (Sunderman and Mathews, 1975).

Overlying the sandstone bed is a sequence of 12 or more unfossiliferous beds of alternating massive tan dolomite and finely laminated brown dolomite (Figs. 4 and 5). Farther down dip in both Indiana and Michigan parts of the Michigan Basin, where more saline conditions existed, the Detroit River Formation also contains evaporites. The brown units exposed here contain pellets and crinkled laminae that probably are algal in origin, suggesting that these Detroit River rocks were deposited in shallow water, probably in a restricted intertidal environment. (See Doheny and others, 1975).

About 10 ft (3 m) above the base of the Detroit River Formation is a dark-gray clay bed that contains montmorillonite and heavy mineral grains (Fig. 4). The bed has been interpreted as an altered volcanic ash layer (Droste and Vitaliano, 1973; Sunderman, 1980) that probably was derived from a Devonian volcanic source in central Virginia. (See Dennison and

Figure 5. Close-up view of west highwall of Ardmore Avenue Quarry showing vuggy massive rocks and sandstone clastic dikes of Wabash Formation, Silurian-Devonian unconformity, 1-ft (30-cm) bed of Pendleton Sandstone Bed (?) at base of Detroit River Formation, and part of medium-bedded Detroit River rocks above. Note junction of dikes with Pendleton Sandstone Bed (?). Height of view about 40 ft (12 m).

Textoris, 1970.) This same bed has been correlated in the subsurface with the Tioga Bentonite Bed by Droste and Vitaliano (1973), Droste and Orr (1974), Doheny and others (1975), and others. The Indiana occurrences are unusual, however, in that they lack biotite flakes typical of the Tioga elsewhere. (See discussions in Baltrusaitis, 1974, and Droste and Shaver, 1975.) This bed is about 3 in (7.6 cm) thick here and in three other quarries in Allen County, which suggests that the original volcanic ash layer may have been half a foot or more thick in this area before consolidation.

Traverse Formation

The youngest Paleozoic rocks exposed in the Ardmore Avenue Quarry belong to the Traverse Formation (Middle Devonian) and consist of gray-brown limestone that is thin to medium bedded, wavy bedded, and fossiliferous (Figs. 2 and 4). Fossils found in the Traverse include tentaculites, horn corals, and

large hemispherical masses of *Hexagonaria* corals, all suggestive of a relatively open-marine environment of deposition. *Hexagonaria* corals also occur in the Traverse Group of Michigan, where they have achieved a certain amount of notoriety. Along the upper Lake Michigan shore, near the town of Petosky, the coral masses weather out of their enclosing rock matrix, then become fragmented, rounded, and polished to produce the famous Petosky Stones.

MESOZOIC AND TERTIARY HISTORY

The Missing Record

No strata of Mesozoic or Tertiary age are known in Indiana, but weathering and erosion, possibly during the Mesozoic Era and certainly during the Tertiary Period, have left their imprint on the underlying bedrock.

Paleozoic-Quaternary Unconformity

The Paleozoic-Quaternary unconformity is the most time-significant unconformity in this region, spanning some 375 m.y. in the Fort Wayne area. In the Ardmore Avenue Quarry, the pre-Pleistocene surface exhibits about 10 ft (3 m) of local relief, and some of the Traverse rocks below the unconformity are highly jointed and weathered. About 35 mi (56 km) to the south of Fort Wayne, however, the unconformity surface has 350 ft (105 m) or more of local relief where the buried Teays River valley has been carved into the underlying Paleozoic bedrock (Wayne, 1956). A sweeping view of the unconformity can be obtained from the viewing stand of the Ardmore Avenue Quarry, and details of the surface can be observed in relative safety along the bedrock ledge around the quarry (Figs. 2, 3, and 5).

SITE 78—PLEISTOCENE GEOLOGY

General Features

A complex sequence of glacial tills and outwash is exposed at the top of the Ardmore Avenue Quarry, and the quarry is located within a short driving distance of a large number of interesting Pleistocene geomorphic features. The quarry itself is situated on the crest of the Wabash-Erie Divide and within the large Fort Wayne-to-Huntington sluiceway valley known formally as the Wabash-Erie Channel (Fig. 1). Other nearby geomorphic features include three dune fields, the Fort Wayne Moraine, the southwestern tip of glacial Lake Maumee, the Fort Wayne Outlet and Six Mile Creek Channel through the moraine, and the stream piracy junction of Fort Wayne's "Three Rivers."

Pleistocene Strata

About 63 ft (19 m) of Pleistocene glacial till and outwash

Figure 6. View near viewpoint C of Figure 1 showing appearance of Wabash-Erie Channel during flood of 1982, probably similar to its appearance during Pleistocene time. View along U.S. 24, about 1 mi (1.6 km) west of its intersection with I-69, looking south-southeast.

are exposed in the Ardmore Avenue Quarry (Figs. 2 and 4). At some places at the base of the Pleistocene section is a 7-foot (2.1-m) bed of green sand and gravel that is cherty and argillaceous and that contains weathered igneous and metamorphic pebbles. This bed has been interpreted as a reworked lag deposit of Sangamonian age (Bleuer and Moore, 1972). Overlying the cherty sand and gravel layer is a 2- to 3-ft (0.6- to 0.9-m) bed of calcareous outwash sand and gravel, above which are two tills separated by several feet of outwash sand and gravel that document two separate advances of Wisconsinan ice through the Fort Wayne area (Fig. 4). The lower till probably is equivalent to the Trafalgar Formation, which was deposited by the Erie Lobe during early Woodfordian time. Fossil wood from the base of this unit has been dated at 21,310 ± 350 years B.P. (Bleuer, 1976). The upper till probably is equivalent to the Lagro Formation, which was deposited by the Erie Lobe during late Woodfordian time. The silt and clay components of the upper till probably were derived mainly from lake sediments deposited in the Lake Erie basin during the time between the two ice advances (Bleuer, 1976). Large-scale features of the Pleistocene section can be seen from the viewing stand at the Ardmore Avenue Quarry, and details of the section can be conveniently studied in the changing exposures at the top of the quarry (Fig. 1, viewpoint A; Fig. 2).

Wabash-Erie Divide

The city of Fort Wayne is known as both the Summit City and the City of Three Rivers. The city is situated just east of the low divide called the Wabash-Erie Divide that separates the drainage of the "Three Rivers" (the Maumee and its two tributaries, the St. Joseph and Ste. Marys Rivers) from that of the Wabash River (Fig. 1). Maumee River waters empty into Lake Erie, whereas those of the Wabash River flow to the Gulf of Mexico via the Mississippi River. The near junction of these two drainages connected early travel routes of Indians and traders and helped to establish trading posts and a fort at this site. The Wabash and Erie Canal, which was completed across the divide in

the 1830s, had a similar effect on the early development of Fort Wayne as an industrial and trading center.

In general, the Wabash-Erie Divide is not topographically prominent. It passes from north to south a mile (1.6 km) or so west of the St. Joseph and Ste. Marys Rivers, and just southwest of Fort Wayne it crosses low swampy land that marks the head of the sluiceway valley called the Wabash-Erie Channel. The Ardmore Avenue Quarry is situated almost exactly on the crest of the divide, and the panorama at the viewing stand (Fig. 1, viewpoint A) gives an excellent idea of the nature of the divide where it crosses the sluiceway valley.

St. Joseph and Ste. Marys Rivers and Wabash-Erie Channel

The St. Joseph and Ste. Marys Rivers were established as ice-marginal streams along the western margin of the Fort Wayne Moraine. Meanders, terraces, and other interesting features of these rivers can be conveniently observed in Fort Wayne's several riverside parks (Fig. 1).

These two rivers carried meltwater that eroded a large valley extending from Fort Wayne to Huntington, where the valley joined older segments of the Wabash River valley. Later this same sluiceway valley carried the overflow waters of glacial Lake Maumee, issuing from the Fort Wayne Outlet and the Six Mile Creek Channel through the moraine (Fig. 1). Near Fort Wayne, the valley floor is mainly a stripped surface underlain by till and older stream deposits, but along its northern edge the valley contains a deeply scoured channel now filled with younger stream gravel, sand, silt, and muck (Bleuer and Moore, 1972). Near Huntington, the valley floor is in part a stripped bedrock surface overlain by thin stream sediments.

The segment of the Pleistocene Wabash River drainage between Fort Wayne and Huntington was referred to by Dryer (1889) as the Wabash-Erie Channel (Fig. 6). It is a very prominent topographic feature that is about 3 mi (4.8 km) wide in the vicinity of the Ardmore Avenue Quarry, where the former channels of the St. Joseph and Ste. Marys Rivers merge with the Fort Wayne Outlet to produce the head of the sluiceway. A short distance to the southwest, the channel narrows to a constant width of about 1 mi (1.6 km). In addition to the views of the channel from the quarry viewing stand, excellent views are also available along Lower Huntington Road and U.S. 24, which parallel the valley on the south and north, respectively, and along I-69 where it crosses the valley (Fig. 1, viewpoints A, B, C; Fig. 6).

Pleistocene Dunes

Among the latest Pleistocene events to affect the topography of the fort Wayne area was the formation of sand dunes. Three dune fields are known in Allen County. The most prominent of these is located in the Wabash-Erie Channel about 2 mi (3 km) west of the Ardmore Avenue Quarry in a wooded area called

Figure 7. View at viewpoint F of Figure 1 showing appearance of east end of Fort Wayne Outlet during flood of 1982, probably similar to its appearance during draining of glacial Lake Maumee in late Pleistocene time. View at intersection of Lake Avenue and Maysville Road, looking southwest toward downtown Fort Wayne.

Fox Island (Fig. 1, viewpoint B). These dunes rise 30 ft (9 m) or more above the general level of the sluiceway and rest on a stripped surface of glacial till. A few glacial boulders at the base of the dunes appear to be part of a lag deposit left behind on the sluiceway surface. The dunes are somewhat linear in outline and are elongated west-southwesterly, roughly parallel to the general trend of the sluiceway valley. The source of the sand was stream deposits in the sluiceway, either terraced outwash to the south (Bleuer and Moore, 1978) or stream deposits along the northern edge of the valley. These dunes are the focal point for the Fox Island County Park and Nature Preserve. Nature trails traverse the dunes within the park, and a nature center and other improvements aid the public in interpretation of the Pleistocene history and biology of the dune area.

A second group of low rounded dunes is found on the northern part of the Indiana University–Purdue University Campus just east of the St. Joseph River, which served as the sediment source for the dunes (Fig. 1, viewpoint D). Other small dunes also occur adjacent to the river farther north along St. Joe Road. A third dune field occurs northeast of New Haven, along Doyle Road between U.S. 24 and Indiana 14 (Fig. 1, viewpoint G). These dunes rest on the Maumee Lake plain near low beach ridges or sand bars that served as a source of the sand. Because of the prominence of the dunes and the textural characteristics of their sediments, at least four small cemeteries have been built on dunes in Allen County. One cemetery formerly was located on the IPFW Campus, one is found along St. Joe Road north of the campus, and two are located on the dunes northeast of New Haven.

Fort Wayne Moraine

Another outstanding geomorphic feature of this area is the Fort Wayne Moraine (Fig. 1), one of a set of nested arcuate moraines formed during retreat of the Erie Lobe of the Wiscon-

sinan glacier (Leverett and Taylor, 1915). The origin of the moraine is apparently complex: recent investigations (Bleuer, 1974) suggest that the most recent surface moraine is underlain by an even older Wisconsinan moraine. The surface moraine is only slightly hummocky in the Fort Wayne area and stands only 60 ft (18 m) or so above the adjacent till plains and the glacial Lake Maumee Plain. This, coupled with its gradual rise in height and its merging with the Wabash Moraine to the northwest, makes the moraine difficult to recognize as a topographic feature. Its boundaries are relatively sharp in some places, however, and are easily recognized. They are marked on the west by the valleys of the ice-marginal St. Joseph and Ste. Marys Rivers and on the east by beach ridges of glacial Lake Maumee. Excellent views of the moraine boundaries can be seen on the IPFW Campus and along Indiana 37 northeast of the city (Fig. 1, viewpoints D and E).

Glacial Lake Maumee

Just east of Fort Wayne and New Haven is a remarkably flat lake plain that extends east-northeasterly as far as Lake Erie. This plain was occupied by glacial Lake Maumee (Fig. 1), which was dammed between the Fort Wayne Moraine and the glacial ice front. Outstanding features of the lake site are its flatness, its lack of natural groundwater drainage (much of the area was swampland before being artificially drained), and the prominent beach ridges that mark its western boundary. The most prominent beach ridges are nestled against the Fort Wayne Moraine, one along Indiana 37 northeast of Fort Wayne, and one along U.S. 30 within and east of New Haven (Fig. 1). The lake surface for several miles east of Fort Wayne is underlain by a scoured till surface instead of by the expectable varved lake silts and clays, apparently because the lake water moved rapidly toward the Fort Wayne Outlet during draining of the lake. Excellent views of the lake plain and the beaches along its abrupt western margin can be seen along the highways mentioned above and at viewpoints E and H (Fig. 1).

Fort Wayne Outlet and Six Mile Creek (Trier Ditch) Channel

The overflow waters from glacial Lake Maumee eroded a half-mile (0.8-km)-wide channel, called the Fort Wayne Outlet, through the Fort Wayne Moraine in the site of downtown Fort Wayne (Figs. 1 and 7). The draining of glacial Lake Maumee through the outlet probably took place in three stages, as indicated by the presence of beach ridges at 800-, 780-, and 760-ft (243-, 237-, and 231-m) levels east of Fort Wayne and by the presence of related drainage channels and stream terraces in the Fort Wayne Outlet, Wabash-Erie Channel, and Wabash River valley to the southwest (Bleuer and Moore, 1972). Draining of the lake, once started, may have been catstrophic, eroding deep channels from Fort Wayne to Huntington and producing large-

Figure 8. Panoramic view at viewpoint J of Figure 1 showing appearance of Three Rivers junction in wintertime. St. Joseph River (open water on right) enters view to right of elongate filtration plant building; Ste. Marys River (open water on left) enters view to left of filtration plant building, just to right of Three Rivers apartment building; and Maumee River (ice- and snow-covered) is in immediate foreground. View from Columbia Street bridge in downtown Fort Wayne, looking generally north.

scale erosional and depositional features as far away as the town of Wabash and beyond (Fraser, and others, 1983). Excellent views of the Fort Wayne Outlet can be seen along Lake Avenue and at the Y intersection of Lake Avenue and Maysville Road (Fig. 1, viewpoint F; Fig. 7).

A second drainage channel about a third of a mile wide also was cut through the moraine from New Haven southwest to the Ste. Marys River (Fig. 1), a distance of about 8.5 mi (13.6 km). This channel was named the Six Mile Creek Channel by Dryer in 1889, and this name is now used in most geologic descriptions. The channel is referred to locally as Trier Ditch, however, after the small intermittent streams that flow through it. In the southern segment of the valley, the gradient is to the southwest, and Trier Ditch drains toward the Ste. Marys River, but in the northern segment, the gradient is to the northeast, and Trier Ditch drains northeastward toward the lake plain. The gradient of the valley is very low, so it is likely that the valley served as a second outlet for Lake Maumee and also carried floodwaters of the Ste. Marys River to the lake plain after draining of the lake (Bleuer and Moore, 1972). During the flood of 1913, water flowed from the Ste. Marys River to the Maumee River through the Six Mile Creek Channel. Therefore, it has been suggested that this drainageway could be used to divert modern Ste. Marys River waters directly into the Maumee River to alleviate periodic flooding in the downtown Fort Wayne area. Interesting views of this channel can be seen along Tillman Road where the road crosses the channel (Fig. 1, viewpoint I).

Three Rivers and the Stream-Piracy Event

Several modern drainage features in and around Fort Wayne owe their origin to a major stream-piracy event that occurred some time after the draining of glacial Lake Maumee. The retreat of the Erie Lobe of ice through the Lake Erie Basin uncovered lower drainage outlets to the north and northeast and allowed Lake Maumee to drain to lower levels. The Maumee

River was then established on the lake plain, flowing down the regional slope to the northeast. The St. Joseph and Ste. Marys Rivers, which up to this time had drained into the Wabash-Erie Channel, were then captured by the Maumee River by headward erosion through the Fort Wayne Outlet.

The actual stream piracy event may have occurred as a result of a major flood comparable to those that periodically plague downtown Fort Wayne today. The results of this event include the following major drainage changes: (1) the Wabash-Erie divide was shifted several miles westward, from a position east of the St. Joseph and Ste. Marys Rivers to a position west of them; (2) the junction of the St. Joseph and Ste. Marys Rivers was changed from a normal Y shape, where they flowed into the Wabash-Erie Channel, to the present barbed pattern, where they join the Maumee River (Fig. 8); and (3) with the draining of Lake Maumee and the capture of the St. Joseph and Ste. Marys Rivers, the series of meltwater streams and lake-drainage streams that had occupied the Wabash-Erie Channel were beheaded. The water flow through the channel was then reduced to that of the small stream now known as Little River or Little Wabash River, a grossly underfit stream (Fig. 1). The headwaters of Little River are immediately south of the Ardmore Avenue Quarry, and the small stream can be observed along the south side of Yohne Road just south of Fox Island Park (Fig. 1, viewpoints A, and B). The Three Rivers stream junction can be conveniently viewed from the Columbia Street bridge across the Maumee River in downtown Fort Wayne (Fig. 1, viewpoint J; Fig. 8).

OTHER GUIDES

For previous field guides to parts of the geology of this area, see Shaver and others (1961), Ault and others (1973), Shaver (1974 and 1976), Sunderman and Mathews (1975), and Shaver and others (1983).

REFERENCES CITED

Ault, C. H., and others, 1973, Guidebook to the geology of some Ice Age features and bedrock formations in the Fort Wayne, Indiana, area: Indiana Geological Survey Guidebook, 62 p.

Baltrusaitis, E. J., 1974, Middle Devonian bentonite in Michigan Basin: American Association of Petroleum Geologists Bulletin, v. 58, p. 1323–1330.

Bleuer, N. K., 1974, Buried till ridges in the Fort Wayne area, Indiana, and their regional significance: Geological Society of America Bulletin, v. 85, p. 917–920.

—— 1976, Glacial geology and physiography of field trip area, in Shaver, R. H., Indiana portion of guidebook for a field trip on Silurian reefs, interreef facies, and faunal zones of northern Indiana and northeastern Illinois: Kalamazoo, Michigan, North-Central Section of the Geological Society of America and Western Michigan University Guidebook, p. 29–32.

——, and Moore, M. C., 1972, Glacial stratigraphy of the Fort Wayne area and the draining of glacial Lake Maumee: Indiana Academy of Science Proceedings, v. 81, p. 195–209.

—— 1978, Environmental geology of Allen County, Indiana: Indiana Geological Survey, 72 p.

Collinson, Charles, and others, 1967, Devonian of the north-central region, United States: Calgary, Alberta Society of Petroleum Geologists International Symposium of the Devonian System, v. 1, p. 933–971.

Dennison, J. M., and Textoris, D. A., 1970, Devonian Tioga tuff in northeastern United States: Bulletin Volcanologique, v. 34, no. 1, p. 289–294.

Doheny, E. J., Droste, J. B., and Shaver, R. H., 1975, Stratigraphy of the Detroit River Formation (Middle Devonian) of northern Indiana: Indiana Geological Survey Bulletin 53, 86 p.

Droste, J. B., and Orr, R. W., 1974, Age of the Detroit River Formation in Indiana: Indiana Geological Survey Occasional Paper 5, 5 p.

Droste, J. B., and Shaver, R. H., 1975, Middle Devonian bentonite in Michigan Basin: Discussion: American Association of Petroleum Geologists Bulletin, v. 59, p. 1217–1220.

—— 1977, Synchronization of deposition of reef-bearing rocks on Wabash Platform with cyclic evaporites of Michigan Basin, in Fisher, J. H., editor, Reefs and evaporites—Concepts and depositional models: American Association of Petroleum Geologists Studies in Geology 5, p. 93–109.

—— 1982, The Salina Group (Middle and Upper Silurian) of Indiana: Indiana Geological Survey Special Report 24, 41 p.

Droste, J. B., and Vitaliano, C. J., 1973, Tioga bentonite (Middle Devonian) of Indiana: Clays and Clay Minerals, v. 21, p. 9–13.

Dryer, C. R., 1889, Report upon the geology of Allen County: Indiana Department of Geology and Natural History Annual Report 16, p. 105–130.

Fraser, G. S., Bleuer, N. K., and Smith, N. D., 1983, History of Pleistocene alluviation of the middle and upper Wabash Valley (field trip 13), in Shaver,

R. H., and Sunderman, J., editors, Field trips in midwestern geology: Bloomington, Indiana, Geological Society of America, Indiana Geological Survey, and Indiana University Department of Geology Guidebook, v. 1, p. 197–224.

Leverett, Frank, and Taylor, F. B., 1915, The Pleistocene of Indiana and Michigan and the history of the Great Lakes: U.S. Geological Survey Monograph 53, 527 p.

Shaver, R. H., 1974, Structural evolution of northern Indiana during Silurian time, in Kesling, R. V., editor, Silurian reef-evaporite relationships: Michigan Basin Geological Society Guidebook, p. 55–77, 89–97, 102–111.

—— 1976, Indiana portion of guidebook for a field trip on Silurian reefs, interreef facies, and faunal zones of northern Indiana and northeastern Illinois: Kalamazoo, Michigan, North-Central Section of the Geological Society of America and Western Michigan University Guidebook, 37 p.

Shaver, R. H., and others, 1961, Stratigraphy of the Silurian rocks of northern Indiana: Indiana Geological Survey Guidebook 10, 62 p.

Shaver, R. H., and others, 1983, Silurian reef and interreef strata as responses to a cyclical succession of environments, southern Great Lakes area, in Shaver, R. H., and Sunderman, J. A., editors, Field trips in midwestern geology: Bloomington, Indiana, Geological Society of America, Indiana Geological Survey, and Indiana University Department of Geology Guidebook, v. 1, p. 141–196.

Sunderman, J. A., 1980, Outcrop characteristics and petrology of a Middle Devonian bentonite from northern Indiana [abs.]: Geological Society of America Abstracts with Programs, v. 12, p. 257.

—— 1983, Role of neptunian dikes in structural evolution of reefs [abs.]: American Association of Petroleum Geologists Bulletin, v. 67, p. 554.

——, and Mathews, G. W., 1975, Age and origin of clastic dikes in Silurian reefs of northern Indiana, in Sunderman, J. A., and Mathews, G. W., editors, Silurian reef and interreef environments of northern Indiana: Fort Wayne, Society of Economic Paleontologists and Mineralogists Great Lakes Section and Indiana University–Purdue University at Fort Wayne Department of Earth and Space Sciences Guidebook, p. 72–83.

Wayne, W. J., 1956, Thickness of drift and bedrock physiography of Indiana north of the Wisconsin glacial boundary: Indiana Geological Survey Report of Progress 7, 70 p.

ACKNOWLEDGMENTS

I would like to thank Robert Shaver, Curtis Ault, and Ned Bleuer for critically reading the manuscript and for making valuable suggestions for its improvement.

The Silurian reefs near Wabash, Indiana

Robert H. Shaver, Department of Geology, Indiana University, Bloomington, Indiana 47405, and Indiana Geological Survey, 611 North Walnut Grove, Bloomington, Indiana 47405

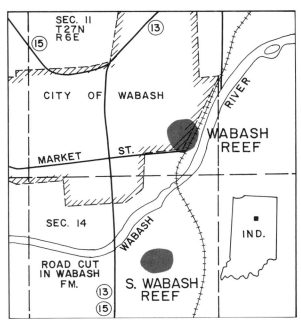

Figure 1. Outline map of Wabash and vicinity, Indiana, showing locations of the Wabash Reef and of a 50-foot (15-m) exposure of the Wabash Formation (Silurian) in a road cut. Map is 1.5 miles (2.4 km) wide. Base is the 7.5-minute U.S. Geological Survey Wabash Quadrangle.

LOCATION AND SIGNIFICANCE

The geology of the area in and near the City of Wabash, located in Wabash County along the upper Wabash River (Fig. 1), has had an important role in the development of modern concepts of Silurian sedimentation and reef growth in the southern Great Lakes area. The geologic significance of this area goes beyond an understanding of Silurian rocks in themselves, as Silurian reef studies have served as models in the development of fossil-reef and attendant sedimentation concepts throughout the geologic column.

One focal site is that of one of the best known fossil reefs in the world, The Wabash Reef, which is exposed on either side of the Wabash Railroad cut that extends more than 1,000 ft (300 m) northeastward from the end of East Market Street and immediately northeast of the site of the former Big Four Railroad Station in Wabash (SE¼SE¼Sec.11,T.27N.,R.6E., Wabash 7½-minute Quadrangle; Figure 1). Generations of geologists on scores of field trips have been inspired by this reef exposure. Noted leaders in the developing study of Silurian reefs who made oft-cited observations here, not all of them correct, include E. M.

Kindle, Edgar R. Cumings, Robert R. Shrock, Marius LeCompte, and Heinz A. Lowenstam.

The exposure is on the property of the Wabash Railroad, a busy rail line. Permission generally is not required, but a local employee of the railroad usually inquires the business of a field-trip group on the site and allows the group to continue its examination after he has advised it of the likelihood of passing trains.

A second especially instructive site at Wabash is along the south bluff of the Wabash Valley where the present route of Indiana 13 and 15 has been cut into a 50-ft (15-m) section of the Wabash Formation (north-central part of sec. 18, T.27N.,R.6E., Wabash Quadrangle; Fig. 1).

GENERAL INFORMATION

The Wabash Reef extends about 750 ft (225 m) along the railroad cut and rises about 40 ft (12 m) vertically above the tracks. Its nearly flat top is an erosion surface, and its bottom is conjectured, with very good inferential evidence, to be as much as 65 ft (20 m) below the track level, there describing in cross section a very broad V shape whose apex is about at the stratigraphic contact between the Pleasant Mills and Wabash Formations (Fig. 2).

Much of the exposed reef mass is made up of gray fine-grained massive dolomite that has few readily identifiable fossils but is characteristic of the so-called cores of Silurian reefs at this stratigraphic level. In the peripheral areas, the dolomite, which is rather pure dolomite in the central area of the reef, is known to increase in clasticity. Such clasticity partakes both of coarsened skeletal debris and of presumably terrigenously derived clay and quartz silt that are abundant constituents of the host rocks, consisting of the Mississinewa Shale Member of the Wabash Formation, with which the wedge-shaped (in cross section) flanking rocks of the reef are interbedded. (See Fig. 5A.) As interpreted by Cumings and Shrock (1928), the flank rocks make up as much as half the volume of the exposed part of the reef, not a large amount in comparison with the amount in larger Silurian reefs.

The Wabash Reef abounds with striking displays of the enigmatic, problematical fossil(?) *Stromatactis* and, in fact, is the type example for the application of this term in North America (by Marius Lecompte in 1938). The identification at Wabash of these calcitic structures as relict fistuliporid bryozoans or other fossils, however, has been denied by some students of fossil reefs. A few algae have been identified here, as well as in South Wabash Reef (Figs. 1 and 3). Altogether, the yield of readily identifiable fossils from the Wabash Reef seems to be surprisingly small to many persons, considering that the term reef has been

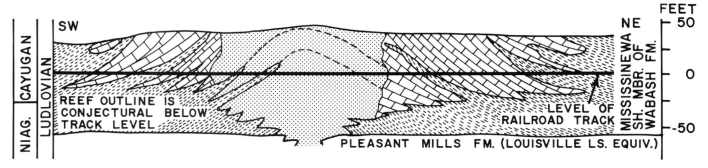

Figure 2. Cross section of the Wabash Reef showing stratigraphy, geometry, and interrelationships of core rocks (crosshatch pattern), flank rocks (brick), and interreef host rocks (laminar). See Figure 1 for location. Note that flank rocks thicken and rise more steeply toward core; also, they are more steeply inclined with higher stratigraphic position. Nearly to scale; reef is about 750 feet (225 m) across. Part of section above track level is from Cumings and Shrock (1928, p. 146); part below is conjectured on basis of observations from several cored and quarried reefs in this area, but level of Pleasant Mills–Wabash contact is not precisely known here.

applied here. This dearth, owed largely to diagenesis, has led some reef students through the years to apply the less committed term carbonate-mud mound and to postulate a deep-water regime inimical to the development of true organic-framework reefs.

The host interreef rocks for not only the Wabash Reef but also for many similar reefs in the upper Wabash Valley area are well exposed in the aforementioned highway cut immediately south of the river at Wabash. Here, about 45 ft (13.5 m) of the upper part of the Mississinewa Shale Member (lower part of the Wabash Formation, Late Silurian) make up most of the bluff. It is a gray dense massive silty dolomite (despite its name) that in

some places in its regionally broad area of distribution contains as much as 50 percent illite clay and quartz silt. Above the Mississinewa are about 8 ft (2.4 m) of upper Wabash rocks, the Liston Creek Limestone Member exhibiting the thin glauconitic Red Bridge Limestone Bed at its base. The Liston Creek characteristically is a light-colored fine-grained well-bedded cherty dolomitic limestone.

Understanding the depositional regime of these Wabash rocks, ranging in thickness on either side of 200 ft (60 m) where uneroded, is critical to understanding the nature of the reefs themselves. For example, and as demonstrable from the Wabash Reef (Figs. 2 and 5A, B) and many other Silurian reefs, the interreef

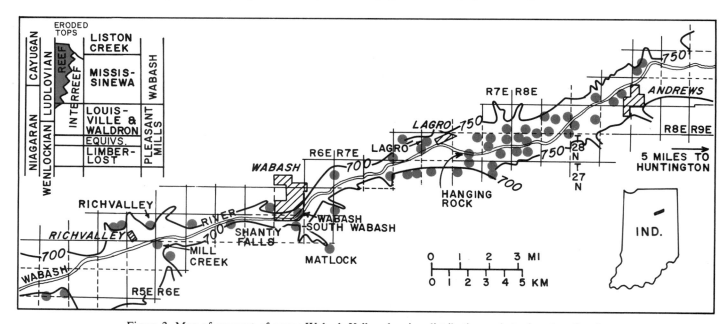

Figure 3. Map of segment of upper Wabash Valley showing distribution and stratigraphy of reefs (stipple) of a Late Silurian generation that have been exhumed, many now standing as klintar, and parts of the Maumee Terrace during late Pleistocene development of the valley. Reefs often mentioned in the literature are labeled with block letters. Topographic contours define positions of valley walls.

rocks, reef-flank rocks, and reef-core rocks are generally contemporaneous with each other. This circumstance, therefore, bears on interpretation of the paleoecologic parameters attendant on both reef and interreef sedimentation. Moreover, only in recent years has it been firmly established that the host rocks, deposited on a marine platform, are near equivalents of the Illinois Basin and Michigan Basin facies made up of the Moccasin Springs Formation and Bailey Limestone (in the former) and of middle and upper parts of the Salina Group (in the latter). These rocks are Late Silurian, also meaning Cayugan, in age, as is the greater volume of salt-bearing rocks of the Michigan and Appalachian Basins. Such correlations explain the ongoing intense debate among geologists over the relationship between the Silurian reef-bearing and salt-bearing rocks.

The Wabash Reef is but one of many classically studied Silurian reefs in the upper Wabash Valley, where they are exposed as klintar within the valley-floor sediments and along the bluffs in especially that stretch of valley between Richvalley, western Wabash County, and just east of Huntington in Huntington County (Fig. 3). Considering their distributional density in the valley, in some places known to be two or three per square mile, many hundreds of such reefs may lie buried beneath the glacial drift in each county in that part of the state. Approximately circular to broadly oval in areal view, they have been called patch reefs.

In the valley they have been alternately exhumed and buried in a complex preglacial and glacial history, when the fledgling Wabash River alternately aggraded and degraded its young valley, including the time during the late glacial episode when the Wabash carried overflow waters from glacial Lake Maumee (precursor of Lake Erie). These reefs, therefore, have erosionally flattened tops (Fig. 5B), and many form the upstream ends of segments of the generally erosively produced Maumee Terrace (reef-defended terrace segments in these examples) that tail both glacial drift and interreef bedrock deposits downstream (Fig. 4). The reef-defended terrace treads are about 50 ft (15 m) above the modern floodplain level in the valley stretch west of Wabash, but upstream near Huntington, the reef tops and the reconstructed terrace level intersect the modern valley floor. This circumstance is yet to be assessed as a clue to the preglacial or early glacial history of this area.

This partially understood history explains the physiographic expressions of the reefs, including that of such craglike exposures as Hanging Rock (Fig. 3), cleaved and oversteepened on their upstream ends, as they were, by the Wabash and its tributaries as these streams assumed their modern regimens. Other klintar, such as that made by the Mill Creek Reef (Fig. 3) before quarrying, appear as small mesas and have rather broad summit areas. Indeed, these reefs are the type example for establishing the American use of the term klint (plural, klintar), which is the term applied earlier in parts of northern Europe to the physiographic expressions of ancient Silurian reefs.

Reef contact is lost west of Richvalley (Fig. 3) through a greatly widened stretch of the valley, because there the modern

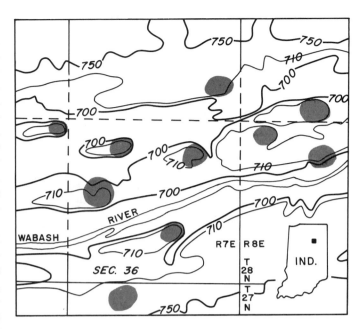

Figure 4. Map of approximately 3 square miles (7.8 km²) of the upper Wabash Valley about 7 miles (11 km) above Wabash, Wabash County, Indiana, showing locations of ten Silurian reefs (gray) and their physiographic effects as klintar and as the upstream bulwarks of rock-defended terraces. Only the 700- and 710-foot (210- and 212-m) contours are shown, except for the 750-foot (225-m) contour, which marks the positions of the valley walls. Base is the 7.5-minute U.S. Geological Survey Lagro Quadrangle.

Wabash crosses over the deeply incised and now-buried bedrock valley of the preglacial Teays River.

As it happens, this upper Wabash Valley segment of the midwestern Silurian reef-bearing section is fairly narrow, stratigraphically considered. Here, the reefs are relatively young as exposed, being about early Ludlovian and early Cayugan (European and North American terms, respectively) in age. Several of these reefs have been cored or quarried out, and all so far recorded bottom out at about the Pleasant Mills–Wabash boundary. This means that they are a late generation apart from many other groups of reefs known in the Great Lakes area. For example, they entirely postdate the buried hydrocarbon-productive reefs of the Michigan Basin, but because of regionally differential erosion, the uppermost remaining parts of these reefs are older than the uppermost remaining parts of the large reefs magnificently exposed in quarries in northwestern Indiana and northeastern Illinois.

Understanding all these relationships—stratigraphic, sedimentational, and physiographic—is necessary to the wise development of the wealth of natural resources contained in the reefs. Such understanding is also necessary to put the Wabash Reef and associated reefs in perspective as only a part of a very long continuum of Silurian reef growth in North America. In the Midwest, this continuum is manifest presently in both immature and mature stages of development. The Wabash Reef, as seen on

Figure 5. Photographs of Silurian reef exposures in the upper Wabash Valley, Indiana. *A,* Northern periphery of the Wabash Reef showing great wedges of bioclastic reef-flank rock interfingering with dense silty argillaceous carbonate rocks of the Mississinewa Shale Member (Wabash Formation); *B,* distal part of a reef at Lagro exhibiting same stratigraphic relations as noted above for the Wabash Reef and a glaciofluvially flattened top that is characteristic of many klintar in the upper Wabash Valley.

present exposure, is rather immature, and its visible geometry hardly suggests its original geometry before erosion.

Given this perspective, the Wabash Reef (as well as many nearby similar structures) should be assigned a narrower role than it once had as a part of the still-developing concept of Silurian reefs and of fossil reefs in general. Its role remains important, however, because it is now more clearly defined by newer understanding of the special geologic circumstances that are outlined here. And, of course, its accessibility and enduring exposure will assure its value to future generations of geologists.

SELECTED REFERENCES

Ault, C. H., and others, 1976, Map of Indiana showing thickness of Silurian rocks and locations of reefs and reef-induced structures: Indiana Geological Survey Miscellaneous Map.

Cumings, E. R., and Shrock, R. R., 1928, The geology of the Silurian rocks of northern Indiana: Indiana Department Conservation Publication 75, 226 p.

Droste, J. B., and Shaver, R. H., 1982, The Salina Group (Middle and Upper Silurian) of Indiana: Indiana Geological Survey Special Report 24, 41 p.

Lowenstam, H. A., 1957, Niagaran reefs in the Great Lakes area, *in* Ladd, H. S., editor, Treatise on marine ecology and paleoecology: Geological Society of America Memoir 67, v. 2, p. 215–248.

Shaver, R. H., 1975, The Silurian reefs of northern Indiana: Reef and interreef macrofaunas: American Association of Petroleum Geologists Bulletin, v. 58, p. 934–956.

—— , and others, 1978, The search for a Silurian reef model: Great Lakes area: Indiana Geological Survey Special Report 15, 36 p.

Shaver, R. H., and Sunderman, J. A., 1980, Silurian reefs at Delphi and Pipe

Creek Jr. Quarry, Indiana, with emphasis on the question of deep vs. shallow water: Purdue University Department of Geosciences and North-Central Section of Geological Society of America Guidebook Field Trip 5, 39 p.

—— , and others, 1983, Silurian reef and interreef strata as responses to a cyclical succession of environments, southern Great Lakes area, *in* Shaver, R. H., and Sunderman, J. A., editors, 1983 GSA field trips in midwestern geology: Geological Society of America, Indiana Geological Survey, and Indiana University Department of Geology, v. 1, p. 141–196.

Shrock, R. R., 1929, The klintar of the upper Wabash Valley in northern Indiana: Journal of Geology, v. 37, p. 17–29.

Sunderman, J. A., and Mathews, G. W., editors, 1975, Silurian reef and interreef environments of northern Indiana: Indiana University-Purdue University at Fort Wayne Department of Earth and Space Sciences and Great Lakes Section Society of Economic Paleontologists and Mineralogists, 94 p.

Textoris, D. A., and Carozzi, A. V., 1964, Petrography and evolution of Niagaran (Silurian) reefs, Indiana: American Association of Petroleum Geologists Bulletin, v. 48, p. 397–426.

The Kentland Dome, Indiana: A structural anomaly

Raymond C. Gutschick, Department of Earth Science, University of Notre Dame, Notre Dame, Indiana 46556

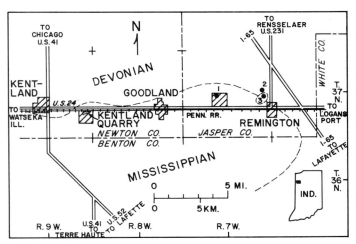

Figure 1. Map of the Kentland area of northwestern Indiana showing access to the Kentland Quarry. Bedrock exposures other than at the quarry: 1, Alter's farmyard—Lower Mississippian; 2, Carpenter's Creek—Upper Devonian black shale; and 3, Fountain Park—Pennsylvanian channel-fill outlier.

SIGNIFICANCE

The Kentland Dome is an enigmatic, thought-provoking, structurally complex anomaly, subsequently covered by glacial drift, in an area surrounded by normal, undisturbed, flay-lying Paleozoic strata. Rocks in the core of the Kentland structure have been uplifted more than 2,000 ft (610 m), folded into a structural dome, and intricately disrupted by faulting. A significant portion of the Ordovician central core is revealed in the spectacular Kentland Quarry. One must see the quarry and its superb rock exposures that challenge the imagination to grasp the magnitude of the anomaly and complexities of its structural pattern.

The oldest quarries at Kentland (McKee and Means) date from more than 100 years ago. Since that time, generations of geologists, applying the scientific method to the exposures in the continually expanding quarry, have built a case history on the evolution of thought to the origin of this unusual structure. The problem has been, and remains, to explain this geometric jigsaw puzzle with respect to its spatial pattern, chronology of disruption, and genesis of deformational mechanics. For example, shatter cones and their orientations in the Kentland Quarry inspired Dietz (1947, 1972) to suggest a meteorite-impact origin simply because the cone apices point upward upon reconstruction of the strata to their normal, flat-lying position. Are Dietz, and other devotees of extraterrestrial origin, correct? Or are there plausible alternative endogenetic explanations?

References to significant developments of the quarries at Kentland and in thought on the origin of the structure can be found in various annual reports of the State Geologist of Indiana (1882–   ) and in Shrock (1937), Shrock and Malott (1929),

Shrock and Raasch (1937), Boyer (1953), Tudor (1971), Laney and Van Schmus (1978), and Gutschick (1961, 1976, 1983). The Kentland Quarry was treated in "Natural Areas in Indiana and Their Preservation" (Lindsey and others, 1969), and the quarry site was given first priority as a geological site in the state even though it is not a naturally exposed site and is in the private domain. The report stressed the importance of access to the quarry by students of geological phenomena.

LOCATION AND ACCESS

The Kentland Quarry of the Newton County Stone Company occupies 100 to 120 acres in northwestern Indiana along the south side of U.S. 24, 3 mi (4.8 km) east of Kentland, CN½Sec.25, T.27N.,R.9W. Kentland 7½-minute Quadrangle, Newton County (Fig. 1). The Newton County Stone Company, Inc. (P.O. Box 147, Kentland, Indiana 47951) is a subsidiary of the Rogers Group, Inc., Bloomington, Indiana. Permission and arrangements to visit the quarry must be made in advance of the visit through the general manager, whose office is at the quarry. Besides individual and group traverses through the quarry, it is possible and well worth the time to arrange for an overview from the fenced-off north and west rims of the quarry.

GENERAL INFORMATON

Regional geology and the Kentland Dome—Kentland is located in the Interior Lowlands physiographic province of the midwestern United States, which is dominated by intracratonic structural basins and intervening arches. The quarry site lies at the extreme northeast margin of the Illinois Basin and, simultaneously, on the southwest flank of the Cincinnati and Kankakee Arches, which separate the Illinois and Michigan Basins. The regional dip near Kentland is off the Kankakee Arch and imperceptibly (to the naked eye, <1°) to the southwest to the Illinois Basin. Paleozoic bedrock is shallow, covered by a thin veneer of Pleistocene glacial deposits, and is made up of Devonian and Mississippian formations and some small Pennsylvanian outliers.

In this regional setting, Paleozoic strata at Kentland have been uplifted to form a structural dome whose south and west flanks are broken by faults (Fig. 2). Erosion subsequent to the deformation and prior to Pleistocene deposition (Fig. 3) has beveled the bedrock surface to produce a concentrically arranged areal pattern of the rock formations. The oldest rocks recognized to date in a central position in the dome belong to the Shakopee Dolomite (Lower Ordovician). Strata dip radially outwards (quaquaversal) from the dome center into the youngest rocks affected by the deformation, those of the Borden Group (Middle Mississippian) and possibly of channel-fill beds of Pennsylvanian age.

The central uplifted Ordovician core is much more complicated than can be shown in Figure 2. Structural complexity is

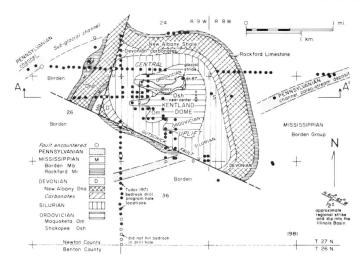

Figure 2. Geologic map of the Kentland structural anomaly based on quarry exposures and subsurface drilling and core information. No bedrock exposures exist outside the Kentland and McKee Quarries for area shown on the map. From Gutschick (1983), but most subsurface information came from Tudor's (1971) study and from cores from holes drilled by the Indiana Geological Survey. All holes to bedrock are shown by black circles. (See Fig. 3 for cross section AA'.)

indicated from quarry exposures and from cores take outside the quarry limits. Comparison of Figures 2 and 5 shows that the quarry is excavated only within the north and northwest flanks of the core; also, only a relatively small portion of the dome has been observed directly in quarry exposures, and the overall gross structure has been determined from subsurface borings and cores. Nevertheless, quarry exposures are very instructive, and quarry expansion continues to reveal stimulating new details of this fascinating structure.

Stratigraphy

Approximately 1,000 ft (305 m) of stratigraphic section are exposed in fault blocks in the quarry, even though the maximum depth of this deep quarry is about 350 ft (107 m). The sequence is continuous from the Shakopee Dolomite (Lower Ordovician) to the Salamonie Dolomite (Middle Silurian) (Fig. 4). In addition, New Albany (Upper Devonian) black and green shales containing *Tasmanites* have been recognized in wadded breccia in a fault wedge. Several key beds among the dominantly carbonate sequence have proved to be useful in identifying the strata and their tops and bottoms within various fault blocks. Rock colors and lithologies are most strikingly displayed after a rain or snowmelt, when rock surfaces are wet and clean; at other times, quarry dust obscures many details of the rock story.

Quarry Structure and Mapping

The central Ordovician area is considered to be the core of the dome, and rocks in the quarry dip steeply off the north and west sides of the dome. Objectives of the quarry operation are the carbonate rocks of the Platteville and Galena Groups and the Joachim Dolomite, all Ordovician in age. As a result, the pit outlines the presence, or former presence, of these units which

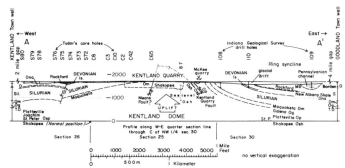

Figure 3. West-east geologic section, AA', of the Kentland Dome showing the faulted asymmetrical domal nature of the structure. (See Fig. 2 for line of section AA'.) Normal, undisturbed stratigraphic sections are shown for deep townsite water wells in Kentland and Goodland. Note 2- and 4-mi 3 and 6 km) gaps in the section adjacent to the town-well sections. The Shakopee Dolomite is the oldest rock exposed in the quarry, so that well-site sections are projected to the top of the Shakopee to provide a datum of reference for structural displacement. From Gutschick (1983).

delineate the structural pattern. One exception is the cross-cut into the Ford W40 Pit that goes through Silurian and Ordovician (Maquoketa) rocks. (See Fig. 4 and the geologic map of Fig. 5.) The west half of the map of the quarry area (Fig. 5) depicts the geology of the bedrock surface shortly after it was stripped of glacial-drift overburden. The east half of the map is based on a combination of bedrock-surface, quarry-walls, and floor exposures, so that contacts are drawn from the bedrock surface to the deep floor of the quarry.

In general, the oldest rocks in the main quarry and in the Ford W40 Pit are to the south, and the youngest rocks are to the north. The beds are right side up, so their dips are to the north. The basic structural style is folding, with a steep NNW-plunging main quarry syncline flanked on both sides by anticlines that are cut transversely by faults on their distal limb. The major Kentland Quarry Fault (KQF) parallels the fold structure and forms the south wall of the main quarry. Faulting (KQF) is along the top of the granular St. Peter Sandstone and the bottom of the brittle Joachim Dolomite.

There is a single sequence of Platteville-Galena carbonate rocks on the west limb of the Kentland Quarry Syncline and in the anticline of the Ford W40 Pit. The eastern half of the main quarry is much more complicated, as is evident from the duplicate pattern of the Platteville-Galena section. This repetition of carbonate rocks accounts for the enlargement of the east half of the main quarry, except for the so-called "narrows."

Folding

Folding is dominant in the main quarry, although faulting seems to be more conspicuous and dramatic than folding. The south quarry wall outlines a continuous steeply plunging anticline-syncline-anticline structural trend. The anticline on the west end is cut off and reappears in the southwest quarry extension, where its orientation is consistent with the fold in the main quarry. Folding is also evident in the eastern part of the main

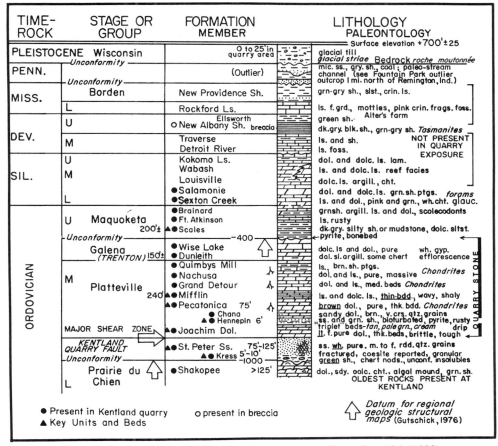

Figure 4. Stratigraphic key for Kentland Quarry geology. From Gutschick (1983).

quarry (Fig. 6). Two small fault blocks, seen in the north wall of the northeast quarry and in the west wall of the main quarry, have strata that are tightly squeezed into steeply plunging anticlines.

Faulting

Important faults identified in the quarry are the Kentland Quarry, McCray, Means, McKee, Crusher, and Ramp Faults (Fig. 5). Direction and movement of these faults still need to be assessed. The Kentland Quarry Fault (KQF) has the largest, most revealing exposure in the quarry. The fault is subparallel to the bedding; consequently, parts of the subjacent formations are cut out by the shearing movement. The white fractured, granulated St. Peter Sandstone forms the footwall, and the hard, brittle Joachim Dolomite and the lowest part of the Platteville constitute the hanging wall of this reverse fault. The KQF zone has sharp planar slickensided polished surfaces, breccias, mullion, quartz lamellae and ductile gouge, and oxidation of iron-sulphide mineralization to mark its presence. The superbly exposed details of the KQF merit close scrutiny. Segments of the KQF have been recognized in several fault blocks, which suggests that it may have been continuous around the Kentland Dome.

The McCray Fault gives the apparent impression that it is a continuation of the KQF; however, the footwall of the McCray Fault is the hanging wall of the KQF, and the structural attitudes across the faults are quite different. The Means Fault, west quarry wall, is a major through-going fault that cuts across the KQF with eastward displacement of the west limb of the north-plunging anticline south of the KQF. The Means Fault has been traced in the subsurface to the southeast and has effected considerable influence on the structure of the west side of the Kentland Dome. The McKee Fault, southeast side of the quarry, cuts transversely across the east limb of the north-plunging anticline of the KQF. Not enough is known to assess the relative importance of the McKee Fault.

The Crusher Fault requires concentrated three-dimensional imagination because it is involved in folding and shearing responsible in part for duplication of the Platteville-Galena sequence. The block that constituted the common hanging wall of the McCray and Crusher Faults was compressed into a syncline anticline fold (Fig. 6B). Parts of this fold are preserved in the north wall of the northeast quarry, except that the limbs of the fold are sheared (Fig. 6A). However, because the crest and trough of the anticline-syncline and common limb are faulted out by the Crusher Fault west of the narrows, only the outer (west) limb of the anticline makes up the hanging wall of the Crusher Fault. Therefore, the east limb of the syncline (KQF hanging wall) and west limb of the anticline (Crusher Fault hanging wall) are sub-

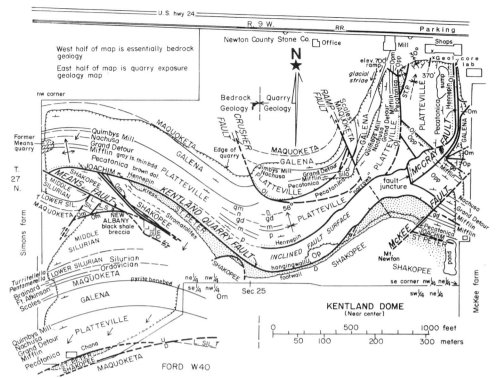

Figure 5. Geologic map of the Kentland Quarry and Ford W40 Pit. West half of map was made by plane table-alidade method on stripped bedrock surface. East half is based on intersection and projection of formations and structures from bedrock to quarry floors. Central part of the dome is directly south of east half of quarry as shown. Glacial striae are shown by arrows. (See Fig. 4 for stratigraphic column.) From Gutschick (1983).

parallel across the Crusher Fault west of the "narrows," and duplication was achieved. The Ramp Fault resulted from accommodation to the confined space between the obliquely trending Crusher Fault and the anticlinal bulge of the KQF at the narrows.

Breccias and Shatter Cones

Rocks in the Kentland structure have undergone considerable stress and deformation, as is manifest in three types of breccia (fault, polymictic, and monomictic) and in shatter cones. Conventional fault breccias can be observed along the Kentland Quarry Fault. Polymictic breccias are characterized by angular clasts of diverse lithologies embedded in a light- to dark-gray, fine-grained, mortarlike groundmass. Clasts include white chert, dark-gray and bright-green shale, brown crystalline dolomite, gray fine limestone, white sandstone, and other rocks found in the Kentland rock sequence. Such breccia with flow lineation is found in fissure- and fracture-filled dikelets most closely associated with faults. Monomictic breccias involve a single lithology from which it appears that the bed or beds have been crushed *in situ* with attendant dilation and then reconstituted by diagenetic recementation. Monomictic breccias occur in most formations in the Kentland structure but are most common in the coarser dolomitic rock layers. The breccias are most conspicuous in carbonate rocks when they are wet and washed clean.

The Kentland Quarry should perhaps be considered as the type shatter cone locality in this country. Shatter cones are unique conical fractures in rock that are characterized by horsetail-type fluting arranged radially outward from the apex on the surface of the cone. Most shatter cones are only a few inches in height; however, very large cones up to several meters in height have been observed. Generally the finer and denser the rock, the better developed are the shatter cones. They occur throughout the Kentland Quarry. Orientations related to the bedding range from near normal in the southeast corner of the quarry to highly oblique in the southwest corner of the quarry. One must be careful to note the difference between cone and the cup counterpart for axes orientations.

Pleistocene Relations

Glacial deposits from the Wisconsinan drift sheet belong to the Cartersburg Till Member of the Trafalgar Formation, which covers the Kentland Quarry area. The thin drift ranges from 1 to 30 ft (0.3 to 9 m) in thickness, being thinnest over the Mifflin beds and thickest over the Maquoketa shale and weathered Galena carbonate rocks. The bouldery-cobblestone-pebbly clay till is weathered to about 8 ft (2.5 m) in depth along a fairly sharp contact between the rust-colored, oxidized till and the gray, unoxidized till. Sand and gravel lenses deposited by glacial melt-

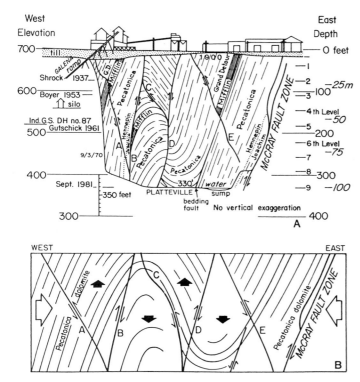

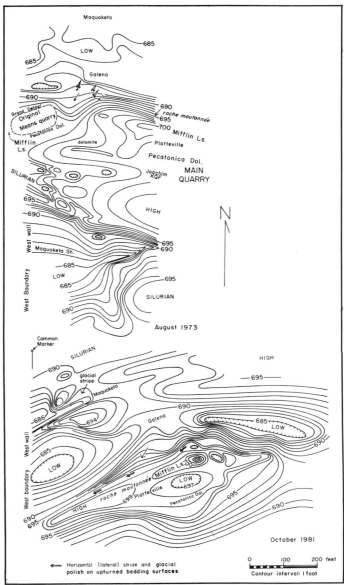

Figure 6. Diagrams showing geologic structure and interpretation of north wall in northeast part of Kentland Quarry. A, West-east cross section of faults and bedding pattern exposed in wall; B, palinspastic structural interpretation of section shown in A. From Gutschick (1983).

water streams are found in the till away from the quarry, and they serve as groundwater aquifers in the area.

The topographic expression of the bedrock surface reflects glacial scour and erosion. Glacial striae are present everywhere the bedrock is exposed, and they generally trend S20° to 30°W (Fig. 7). The glacial vector identifies with the distal margin of the Huron-Saginaw Lobe of the Wisconsinan glacier. Differential ice abrasion produced as much as 15 to 20 ft (5 to 6 m) of relief on the bedrock surface made up of truncated, upturned strata. A direct correlation exists between rock types and glacial erosion, so that the amount of glacial abrasion furnishes an excellent index to the hardness and durability of particular rock units in the quarry.

The Platteville Group was the most resistant to erosion by the ice, and it forms a prominent bedrock platform. Surprisingly, the thin-bedded, micritic limestone of the Mifflin Formation forms a roche moutonnée on the bedrock surface. Upturned bedding surfaces of the Mifflin are polished and striated horizontally where they faced the ice that moved sideways in a U-shaped trough parallel to the strike of the beds before it managed to move up and over. Most bedrock pavement that was once stripped of overburden has been removed by quarry advancement.

SUMMARY AND INTERPRETATION

The following theories have been proposed at various times for the origin of the Kentland anomaly: cryptovolcanism,

Figure 7. Map showing bedrock topography of the Means tract (includes Ford W40 quarry site). Note dramatic effect of differential glacial erosion over steeply dipping strata. From Gutschick (1983).

meteorite or comet impact, fault, diapir, solution collapse of an evaporite dome, and cryptoexplosion (nongenetic).

In attempting to evaluate the origin of the structure, one must account for a major anomalous structural disruption in an otherwise undisturbed area, the geometrical pattern of the structure, and the rock mechanics of the deformation. Structurally, the Kentland Dome is complexly faulted, the Shakopee Dolomite, the oldest bedrock formation, being raised more than 2,000 ft (610 m) vertically from its normal, flat-lying stratigraphic position. The quarry exposes the northwest portion of the Ordovician core of the dome. The dome appears to be faulted on its south and west flanks, and the Shakopee beds in the central part of the dome are all upended.

Folding and faulting are impressive in the central part of the disturbance, as seen in the quarry and as interpreted from subsurface data taken from the area to the south. Fault surfaces are sharp and bold and exhibit slickensided and mullion structure. Intense fracturing is present throughout the St. Peter Sandstone, and granulation gouge, shatter cones, and polymictic and monomictic breccias are distributed throughout the quarry and observed in cores surrounding the quarry. Shatter cones have their apices pointing predominantly stratigraphically upward. Quartz grains have lamellae and cleavages produced by high strain rates of deformation. Asterism is present in the X-ray pattern of the quartz. Coesite has been reported in extremely small size and great dilution (Cohen and others, 1961); however, doubt has been raised about this identification. Polymictic breccias contain clasts that have been derived from rocks that occur throughout the quarry section, and the breccias seem to be very uniform in composition and texture throughout the quarry and in cores outside the confines of the quarry. Middle Ordovician conodonts have been found in the quarry, and the conodonts indicate an anomalously low thermal effect for an impact site (CAI 1½, 50° to 90° range; Votaw, 1980).

The Kentland Dome poses at least two major questions: When? And how? Answers to these questions from convincing evidence are yet to be provided. Some present evidence is equivocal, and so it is used to support different, even diametrically opposed, theories of origin. For example, the breccias can be used to support both extraterrestrial impact and internal diapiric uplift and fracture.

When did it happen? The time window determined from conventional, geologic mapping is very long indeed—Mississippian to Pleistocene. Current field studies of paleomagnetic orientation of framboidal magnetite in carbonate rocks reset at the time of structural deformation show some promise for dating the anomaly. Also, new ideas and advanced technology will find means to recover the date of the structure. Of course, timing will be instantaneous and unique assuming an impact origin, whereas an endogenetic origin could be repetitive and of longer duration.

What convinces advocates of an impact origin that they are correct? Mostly shock metamorphism, but in the example of Kentland, if coesite is ruled out, only relatively low temperature-pressure brittle features are present. Shatter cones and some over-the-threshhold ductile quartz lamellae associated with the faults are present. Are we then to conclude that the Kentland Quarry represents the bottom or lower part of the hemisphere of influence of impact (5 to 10 kb, as suggested by Eugene Shoemaker in field discussion of October 30, 1983)? Geometry and fold-fracture pattern of the anomaly can be equivocal. Whatever the origin, what happened to the topographic cone (crater) or top of the structural dome?

Is an endogenetic origin plausible? Reidel, Koucky, and Stryker (1982) think so for the Serpent Mount (southwestern Ohio), which is similar in many respects to the Kentland structure. These authors provided a discussion and summary for the case against impact.

The Kentland structure provides much opportunity for field study and food for thought, but as yet a complete detailed geologic map of the *entire* structure does not exist, one with such detail as has been produced from available quarry exposures and from which one can evaluate the force fields, fracture pattern, and the nature of the rock mechanics. One can only speculate from the limited available information, and full understanding of the integrated picture is not yet possible.

REFERENCES CITED

Boyer, R. E., 1953, The geology of the structural anomaly near Kentland, Indiana (A. M. thesis): Bloomington, Indiana University, 54 p.

Cohen, A. J., Bunch, T. E., and Reid, A. M., 1961, Coesite discoveries establish cryptovolcanics as fossil meteorite craters: Science, v. 134, p. 1624–1625.

Dietz, R. S., 1947, Meteorite impact suggested by the orientation of shatter cones at the Kentland, Indiana disturbance: Science, v. 105, p. 42

——1972, Shatter cones (shock features) in astroblemes: Montreal, Canada, 24th International Geologic Congress, section 15, Planetology-Terrestrial Explosion Features, p. 112–118.

Gutschick, R. C., 1961, The Kentland structural anomaly, northwestern Indiana: *in* Guidebook for field trips 1961 Cincinnati meeting: Geological Society of America Guidebook, p. 12–17.

——1976, Geology of the structural anomaly, northwestern Indiana: Kalamazoo, Western Michigan University Department of Geology and Geological Society of America North-Central Section Guidebook, 59 p.

——1983, Geology of the Kentland Dome structurally complex anomaly, northwestern Indiana (field trip 5), *in* Shaver, R. H., and Sunderman, J. A., editors, Field trips in midwestern geology: Bloomington, Indiana, Geological Society of America, Indiana Geological Survey and Indiana University Department of Geology Guidebook, v. 1, p. 105–138.

——1959, Arenaceous Foraminifera from the Rockford Limestone of northern Indiana: Journal of Paleontology, v. 33, p. 229–250.

Gutschick, R. C., and Treckman, J. F., 1957, Lower Mississippian cephalopods of the Rockford Limestone of northern Indiana: Journal of Paleontology, v. 31,

p. 1148–1153.

Laney, J. E., and Van Schmus, W. R., 1978, A structural study of the Kentland, Indiana, impact site: 9th Lunar Planetary Science Conference Proceedings, p. 2609–2632.

Lindsey, A. A., Schmelz, D. V., and Nichols, S. A., 1969, Natural areas in Indiana and their preservation, *in* The report of the Indiana Natural Areas Survey: West Lafayette, Indiana, Purdue University Department of Biological Sciences, p. 488–490.

Reidel, S. P., Koucky, F. L., and Stryker, J. R., 1982, The Serpent Mound disturbance, southwestern Ohio: American Journal Science, v. 282, p. 1343–1377.

Shrock, R. R., 1937, Stratigraphy and structure of the area of disturbed Ordovician rocks near Kentland, Indiana: American Midland Naturalist, v. 18, p. 471–531.

Shrock, R. R., and Malott, C. A., 1929, Notes on some northwestern Indiana rock exposures: Indiana Academy Science Proceedings, v. 39, p. 221–227.

Shrock, R. R., and Raasch, G. O., 1937, Paleontology of the disturbed Ordovician rocks near Kentland, Indiana: American Midland Naturalist, v. 18, p. 532–607.

Tudor, D. S., 1971, A geophysical study of the Kentland disturbed area (Ph.D. thesis): Bloomington, Indiana University, 111 p.

Votaw, R. B., 1980, Middle Ordovician conodonts from the Kentland structure, Indiana [abs.]: Geological Society of America Abstracts with Programs, v. 12, p. 259.

The Lovers Leap section and related observations of multiple and cross-cutting glacial drifts in the Great Bend area, Indiana

N. K. Bleuer, Indiana Geological Survey, 611 North Walnut Grove, Bloomington, Indiana 47405
Robert H. Shaver, Department of Geology, Indiana University, Bloomington, Indiana 47401, and Indiana Geological Survey, 611
North Walnut Grove, Bloomington, Indiana 47405

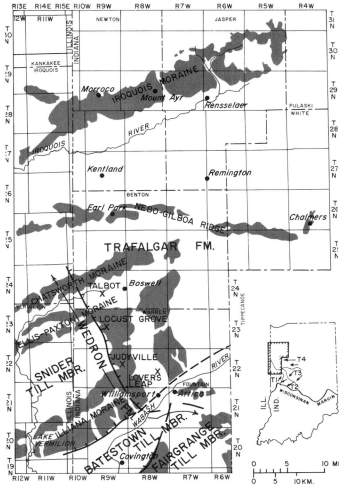

Figure 1. Map of a part of northwestern Indiana and adjacent Illinois showing glacial moraines, stratigraphic-unit boundaries, postulated directions (arrows in index map) of Trafalgar ice movement (Huron-Erie Lobe), and locations of four sites mentioned in the text. Unit boundaries indicated by hachured lines and, where buried, by dashed line. Numbers associated with arrows in index map refer to the numbered Trafalgar tills shown in Figure 4. For details of locations of the four sites, see Figure 2. Partly from Bleuer and others (1983).

LOCATION AND SIGNIFICANCE

The glacial geology of the area north of the Great Bend of the Wabash River in west-central Indiana (Fig. 1) illustrates important early developed principles regarding multiple continental glaciations, the movements of adjoining ice lobes, and the interrelated deposits. Perhaps the first documentation of a lithologically definable till stratigraphy that is basic to interpretation of glacial movement and sedimentation throughout much of the Midwest was made in the Great Bend area. A half century after the first documentation was made, the till stratigraphy here would emerge as the regionally mappable sequence. Further, it would illustrate cross-cutting relationships of morainal ridges and the interbedding of multiple-source deposits. Here, therefore, is a clear demonstration of how deposits first classified on the basis of surface morphology relate to those that are classifiable on a mappable, till-stratigraphy basis.

Four vantage points in the Great Bend area are particularly focal for illustration of the basic concepts that are addressed here. Foremost is the Lovers Leap section (also known historically as the Stone Creek section), which is exposed on the northwest side of Big Pine Creek, a Wabash River tributary, 4 mi (6.4 km) north of Williamsport, Warren County, Indiana, NW¼NW¼ NW¼Sec.23,T.22N.,R.8W., Williamsport 7½-minute Quadrangle (Fig. 2D). Here well exposed are three superimposed tills and other drift materials that a pioneer of continental-glaciation theories, Thomas Chrowder Chamberlin, used to demonstrate his beliefs. In fact, our Figure 3 is taken from Chamberlin's contribution to James Geike's (1894) *The Great Ice Age* and was used as Geike's frontispiece. (See the further history and significance of this site and of Chamberlin's work in Bleuer, 1975.)

Although the Lovers Leap section is on private property, access may be readily obtained by responsible educational and professional groups or individuals by asking permission in advance from the owner, Mr. Deane Braymeyer of Attica, Indiana.

A single exposure such as that at Lovers Leap cannot stand alone as demonstration of the complex, modern interpretation of glacial stratigraphy of the Great Bend area. The locations of three supplementary instructive sites, therefore, are shown in Figures 1 and 2A, B, C, and D and described in the following pages (Talbot, Locust Grove, and Judyville).

The first supplementary site having a key role in our theme is at Judyville, about 5 mi (8 km) west-northwest of Lovers Leap and 2.5 mi (4 km) west of the intersection of U.S. 41 and Indiana 63, Warren County, Indiana, NW¼NE¼NW¼Sec.13, T.22N.,R.9W., West Lebanon 7½-minute Quadrangle (Fig. 2C). The vantage points here are along public roads of free access. The second supplementary site, also of free public-road access, is at Locust Grove about 6 mi (9.6 km) north-northwest of Judyville and 3.8 mi (6 km) west of U.S. 41, center of N-S line between secs. 15 and 16, T.23N.,R.9W., Tab 7½-minute Quadrangle (Fig. 2B). These two sites provide views of morainal ridges left by ice of both northern and eastern sources. Here one may appraise topographic form together with the facts from augering programs and formulate for himself, even though in the footsteps of both

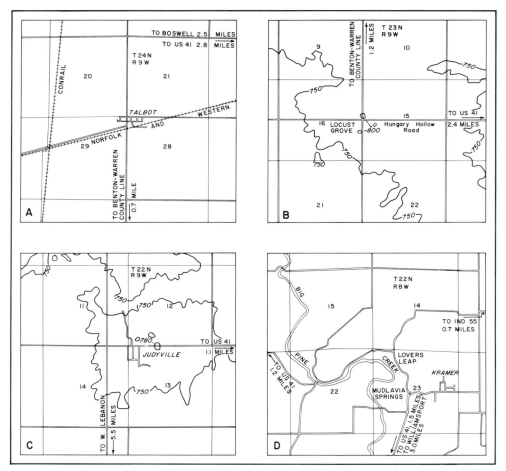

Figure 2. Four maps showing locations of and access to vantage points for field demonstration of glacial geology in the Great Bend area. A, Talbot area; B, Locust Grove area; C, Judyville area; and D, Lovers Leap area. See same locations at a smaller scale in Figure 1. Areas between 750- and 800-ft (225- and 340-m) contours in map B and between 750- and 780-ft (225- and 234-m) contours in map C define the highest parts of the moraine defining the distal Trafalgar margin shown in Figure 1. Bases for the four maps are from U.S. Geological Survey 7½-minute Quadrangle maps, in order from A to D: Boswell and Tab; Tab; West Lebanon, Tab, and Williamsport; and Williamsport.

the pioneers and modern glacial stratigraphers, the answers to questions that have been posed on the cross-cutting nature and continuity of multiple drifts.

A third supplementary site is in the impressive boulder field (glacial erratics) immediately north of the Chatsworth Moraine (Fig. 1) and along the county road immediately south of Talbot, Benton County, Indiana (section line between secs. 28 and 29, T.24N.,R.9W., Baswell 7½-minute Quadrangle (Fig. 2A). Here the erratics are associated with disintegration till, and not on'y were they mapped by Frank Leverett (as recorded by Leverett and Taylor, 1915, pl. 6), but they also served as Chamberlin's (1883) type examples for his concept of englacial drift.

The multiple purpose of this chapter is now evident. To summarize, highlights of the glacial geology of the Great Bend area are to be advanced as especially pertinent to the understanding not only of an important facet of Indiana geology, but also of

the materials and means of continental glaciation in general. At the same time, a recurrent history-of-geology element in our theme is to pay tribute to one of America's great pioneering geologists, Thomas Chrowder Chamberlin (1843-1928), whose higher educational and professional experiences began at Beloit College in Wisconsin and, after many scientific travels and contributions, ended at the University of Chicago. (See Fenton and Fenton, 1956, p. 302–317.)

GENERAL INFORMATION

Lovers Leap Section

Although the section at Lovers Leap on Big Pine Creek (Figs. 1, 2D, and 3) played an important role in Chamberlin's concept of imbricate drift sheets, it was long denied later at-

SECTION OF STONE CREEK, NEAR WILLIAMSPORT, INDIANA:

TO ILLUSTRATE THE IMBRICATION OF THE DRIFT-SERIES. (*See* p. 736).

1. Reddish Till.   2. Old ferruginous gravel.   3. Blue Till.   4. Gravel, fresher than 'old ferruginous gravel' (2).   5. Gray Till.
The Tills differ not only in colour but in constituents.

Figure 3. Photograph (recomposed) showing section at Lovers Leap on Big Pine Creek that was used as the frontispiece of James Geikie's *The Great Ice Age* (1894, 3d ed.). This is T. C. Chamberlin's photograph (labeled as the Stone Creek section), which is numbered 16 in the U.S. Geological Survey Library in Denver, Colorado (I. P. Schultz, written communication, 1973).

tention to its focal role because it was lost under the designation Stone Creek section. As Bleuer's (1975) study showed, no creek by that name ever existed in that area, but comparisons of the early description and photograph (Fig. 3) with the present-day exposure removed all doubt that this section is the one that beautifully supported the principles espoused by Chamberlin in his publications of 1883, 1888, and 1894. This exposure along the northeast bluff of the creek (detailed location given above) is more than 150 ft (45 m) wide, its lowest part being exposed at the far southeast end. It is described in modern stratigraphic terms as follows:

Trafalgar Formation

1. Till, loam; well-developed gray-brown podzolic soil in upper 2-3 ft (0.6-0.9 m); some loess admixture within upper profile, leached in upper 3-3.5 ft (0.9-1.0 m), variably along exposure; rootlets penetrate vertical joints as much as 10 ft (3.0 m); oxidized yellowish-brown (10YR 6/4 to 5/4) to 6 ft (1.8 m), grading to 10YR 5/2 below 10 ft (3 m); variable oxidation depth along exposure about 8-12 ft (2.4-3.6 m) deep; calcareous; coarse fissility developed on outer surface; massive inside; hard and dry till stands as bold vertical cliff, thickest at east end; base of till is

undulatory and has 4-5 ft (1.2-1.5 m) of relief; although uniform in most places, one spur exposes stratified till, sand, and pea gravel in lower 5 ft (1.5 m) and in places the basal few inches are cemented with carbonate; basal 9-10 in (22.5-25 cm) oxidized; unit ranges between 12 and 20 ft (3.6 and 6.0 m) in thickness . . . .
0.0-17.0 ft (0.0-5.1 m)

Wedron Formation, 85 ft (25 m) exposed:

Unnamed at member rank:

2. Gravel, medium to coarse; interbedded sand; scattered carbonate-cemented masses, the upper 5-10 in (13-25 cm) being entirely cemented; thin, highly oxidized gravel at base where immediately above till; unit ranges between 15 and 20 ft (4.5 to 6 m) in thickness. . . .17.0-40.0 ft (5.1-12.1 m)

Snider Till Member

3. Till, clay loam to silty clay loam, grayish-brown (10YR 5/2), calcareous; massive to coarsely angular, blocky jointing; very sticky; abundant shale pebbles; oxidized in upper 3-6 in (7.5-15 cm); marked N-S orientation of elongate pebbles (see Bleuer, 1974, Figure 4); unit ranges between 0 and 11 ft (3.3 m) in thickness. . . .40.0-51.0 ft (12.1-15.5 m)

4. Sand, medium- to coarse-grained, poorly exposed; becomes

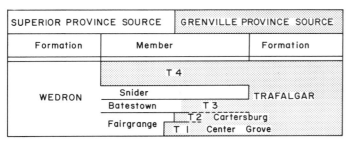

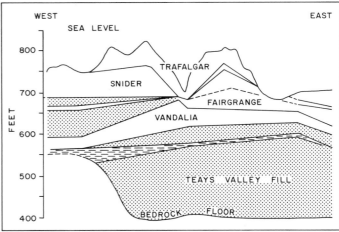

Figure 4. Diagram showing classification of Wisconsinan deposits of the lake Michigan Lobe (source in the Superior Province) and Huron-Erie Lobe (East White Sublobe; source in the Grenville Province). Symbols T1 through T4 represent successively deposited eastern-source tills representing directions of ice movements shown by arrows in the index map of Figure 1.

more gravelly to southeast and merges with upper gravel; basal 3-4 ft (0.9-1.2 m) at southeast end is heavily iron stained, yellowish-brown to olive-brown (10YR to 5YR 4/8); calcareous throughout; unit is 2 to 3 ft (0.6 to 0.9 m) thick at east end and 35 to 40 ft (10.6 to 12.1 m) thick at west end. . . .51.0-62.0 ft (15.5-18.8 m)

Fairgrange Till Member

5. Till, loam; scattered slightly more clayey masses in upper 5 ft (1.5 m); generally pinkish or lavender cast in moist and dry exposures, respectively; top 10 in (25 cm) heavily streaked and oxidized by groundwater; intense brown (7.5YR [to 5YR] 5/6); medium to coarse angular, blocky jointing; few joint coats of 5YR 3/4; below is grayer (5YR 5/1); abundant joints oxidized as much as half an inch on either side and having heavy dark oxide coats; calcareous; base is stream level. . . .62.0-102.0 ft (18.8-31 m)

Although this section is so situated in west-central Indiana that the Batestown Till Member (Wedron Formation) shown in Figs. 1 and 4 is absent, the Big Pine Creek exposure is an excellent example showing that eastern-source till (upper tongue of the Trafalgar Formation) overlaps northern-source tills (Snider and Fairgrange Till Members, Wedron Formation). The latter tills are defined from the nearby Danville, Illinois, area (Johnson and others, 1972). This sequence of an upper brown to gray loam till, a middle clayey till, and a lower pinkish to brown loam till is basic to all of west-central Indiana. Snider till can be recognized by drillers owing to its texture and shale content, and it effects a distinctive signature on gamma-ray logs (Fig. 6). It is, therefore, a key unit for distinguishing the Trafalgar-Wedron relationships, that is, the relationships between tills of eastern and northern sources.

In detailed analyses in sedimentation laboratories, these tills may be further distinguished. A large amount of garnet in relationship to epidote identifies the Trafalgar till as emanating from its eastern source, the Grenville Province, noted in Figure 4, just as the underlying Snider and Fairgrange till mineralogy reveals their northern source as the Superior Province. In addition, high bulk-magnetic susceptibility parallels high garnet: epidote ratios in the eastern-source tills.

Figure 5. West-to-east cross section extending about 24 mi (38.4 km) along the crest of the Ellis-Paxton Moraine from the Illinois state line and extending off the moraine to the eastern Warren County line in Indiana (Fig. 1.) Section shows eastern-source till (Trafalgar Fm.) overlapping buried moraine forms made up of the Snider and Fairgrange Till Members of northern source (Wedron Fm.). The Vandalia Till Member (Glasford Fm.) is the youngest Illinoian till shown. The Teays Valley crosses line of section obliquely from west-northwest to east-southeast. Patterned units consist of stratified sediments.

These stratigraphic relations became fully clear in the late 1970s and early 1980s, and they appear to have settled some standing problems. For example, does the Crawfordsville Moraine (named in Montgomery County, Indiana) extend north of the Wabash River and there continue north to Benton County (the ridge on which Judyville and Locust Grove are shown in Fig. 1) as proposed and mapped by Wayne (1965)? And does this Trafalgar morainal material correlate with the 20,000-year-old Cartersburg Till Member of the Trafalgar Formation at the Cartersburg type locality in Hendricks County, Indiana, there forming the uppermost member of the Trafalgar? If both answers are yes, the entire sequence of central Illinois tills of the Wisconsinan Stage would have been deposited in a span of time of less than 1,000 years. This would have required an uncommonly high rate of advance by Wedron ice, faster than the already known high rate of Trafalgar advance. Unravelling the complexly interfingered eastern and northern till sequence shown in Figures 4 and 6 for west-central Indiana, however, has obviated the unlikely circumstance posed above. The Trafalgar till northwest of the Great Bend of the Wabash must be considered as a younger till than any type Trafalgar till, and the Cartersburg till must be assigned an age approximately that of the Fairgrange. The exposure at Lovers Leap was instrumental in solving this enigma and others, as were intense programs of study by Illinois and Indiana glacial geologists of depositional morphology, of sedimentological parameters, and of sequences investigated through subsurface methods.

Further, these studies, having Lovers Leap as one of the focal points, have fully vindicated Chamberlin's (1888) belief in a "succession of marginal deposits [Wedron Formation] formed by ice movements from the basin of Lake Michigan on the north and [marginal deposits, Trafalgar Formation] from the Erie and Huron basins on the north and northeast" and in a superposition of drifts resulting from the "encroachment of one movement upon the ground of the other."

The highest of the westward-overriding Trafalgar deposits had been mapped by Chamberlin by 1883 as a morainal loop of the so-called Wabash ice lobe (that is, a part of the western Erie, or Maumee, glacier). This morainal loop is the ridge shown at Judyville in Figure 1. The idea of a Wabash ice lobe was also applied northward to account for moraines as far north as the Iroquois Moraine (Fig. 1), and by 1888 the Illinois-Indiana moraines, originating from the Lake Michigan Lobe and truncated along the Trafalgar boundary shown in Figure 1, had been recognized. By 1894, therefore, Chamberlin was able to produce his classic map that first defined an East Wisconsin (now the Wisconsinan) glacial stage.

In the years following exposition of Chamberlin's views, Leverett (1899) reiterated and expanded these views, distinguishing the upper, eastern-source tillsheet as a distinct late Wisconsinan event, but later Leverett and Taylor (1915) proposed a lobe-to-lobe continuity of moraines in and out of reentrant morphologies and all across the Midwest. Glacial stratigraphers beginning with Wayne (1963, 1965) and Wayne and others (1966), however, returned at least in part to Chamberlin's interpretation. The cycle is now full circle, as the later studies of Lineback (1979) and Bleuer and others (1983) have returned completely to Chamberlin's positions.

The Ridges at Judyville and Locust Grove

The ridges at Judyville and Locust Grove (Figs. 1, 2B, and 2C) provide multiple vantage points from which one can appreciate the glacial dynamics that have been outlined. The westward view affords a lasting impression of the vast flatlands of the Wisconsinan till plains of Illinois. Here, the black prairie soils, the mollisols, may be all that Gilbert Imlay would recognize today among what he described in 1792 as "chiefly a morass [that] . . . produces little else, other than hazel, fallow, a species of dwarf poplar, and a very coarse but luxuriant grass; the latter of which covers mostly the whole surface of the earth . . ." Little could he appreciate that the glacial aftermath and the ingenuity of mankind would combine to transform such a morass into general-purpose farmlands that are among the world's very richest, productive, and expensive.

Locust Grove is just south of the area of superposition of young Trafalgar till in the extended Crawfordsville Moraine of Wayne (1965) (that is, the ridge underlying Judyville) on the Snider till of the Ellis-Paxton Moraine (Cropsey Moraine) of Figure 1. The view here of the junction of tills and moraines of the Huron-Erie Lobe (eastern) and of the Lake Michigan Lobe (northern) tends to suggest that the Ellis-Paxton Moraine and till truncate and rise up over the moraine and till on which Locust

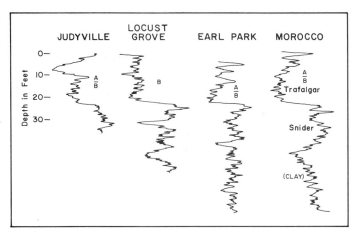

Figure 6. Traces from four gamma-ray logs of shallow holes penetrating Wisconsinan tills north of the Great Bend area, Indiana. Traces show the distinctive Trafalgar and Snider signatures at four locations (shown approximately in Fig. 1) as follows: (1, 2), Judyville and Locust Grove in the area of westward pinchout of the Trafalgar till where it rises over the buried till of the Snider member, the Locust Grove log being for a hole north of Locust Grove and above the crest of the buried Ellis-Paxton Moraine; (3), on the Nebo-Gilboa Ridge at Earl Park; and (4), on the Iroquois Moraine at Morocco. Ablation tills, silts, and sands of the Trafalgar Formation (A) lie above basal Trafalgar till (B) and Snider till.

Grove and Judyville are situated (Fig. 1). Indeed, this was the interpretation of Wayne (1965) and Wayne and others (1966), although, southward, they had proposed the reverse situation: overlap of the Bloomington Moraine, extending from Illinois, by the moraine at Judyville and Locust Grove. This latter interpretation was the beginning of denial of Leverett's (Leverett and Taylor, 1915) assertion of morainal continuity through reentrant lobal junctions.

Nevertheless, the recent studies already noted, which are based on drilling programs, show that the Trafalgar and Ellis-Paxton junction is effected by Trafalgar till lying *atop* Snider till and that the same relationship holds northward to the Kankakee River valley. (See these relationships in Figs. 4 and 5.) Therefore, the eastern ends of all the moraines extending eastward from Illinois (Fig. 1) become buried landforms in western Indiana. Studies of aerial photographs support this conclusion, as the Trafalgar boundary is identified on the basis of a diffuse ice-disintegration landscape developed on the 0- to 20-ft (0- to 6-m) thick Trafalgar till and associated glacio-fluvial deposits. Here again, modern studies have returned completely to the interpretations of T. C. Chamberlin of so long ago.

The Boulder Fields Near Talbot

The ice-disintegration features mentioned immediately above are shown to particular advantage just inside the Trafalgar margin and immediately north of the Chatsworth Moraine (Figs. 1 and 2A), there consisting of ice-walled lake and large circular disintegration deposits. (See Bleuer, 1974, Figure 4.) Large surficial concentrations of boulders are present in this same area, including the area immediately south of Talbot. Fittingly to our historical context, these concentrations are parts of the great boulder belt of Leverett (Leverett and Taylor, 1915, Plate 6) and of Chamberlin's (1890, 1893) type examples of englacial drift. Here, as noted by Chamberlin, the boulder distribution is associated genetically with looping morainal (ice-lobe) margins; the boulders are entirely superficial or partly buried to completely concealed; many were derived from Laurentian granitic and gneissic rocks as well as from Huronian metamorphic rocks as shown, for example, by certain jasper-quartz and slate conglom-

erates (today recognized as the Gowganda Tillite) that had been described in Canada by Sir William Logan; and their angularity and other gross attributes are peculiarly different from the features of boulders found deep within till sheets.

From these and other observations of this suite of erratics, Chamberlin was able to assert forcefully that the boulders were not ice rafted. Also, their gross character showed that they could not have undergone the grinding, abrasive, and otherwise-shaping effects of subglacial transport; they were, indeed, Chamberlin said, englacially and superglacially transported.

REFERENCES CITED

Bleuer, N. K., 1974, Distribution and significance of some ice-disintegration features in west-central Indiana: Indiana Geological Survey Occasional Paper 8, 11 p.

——1975, The Stone Creek section, a historical key to the glacial stratigraphy of west-central Indiana: Indiana Geological Survey Occasional Paper 11, 9 p.

Bleuer, N. K., Melhorn, W. N., and Pavey, R. R., 1983, Interlobate stratigraphy of the Wabash Valley, Indiana: Midwest Friends of the Pleistocene 30th Annual Field Conference, 135 p.

Chamberlin, T. C., 1883, Preliminary paper on the terminal moraine of the Second Glacial Epoch: United States Geological Survey Annual Report 3, p. 291–402.

——1888, The rock scorings of the great ice invasions: United States Geological Survey Annual Report 7, p. 147–248.

——1890, Boulder belts distinguished from boulder trains—their origin and significance [abs.]: Geological Society of America Bulletin, v. 1, p. 27–31.

——1893, Nature of the englacial drift of the Mississippi Basin: Journal of Geology, v. 1, p. 46–60.

——1894, Glacial phenomena of North America, in Geikie, James, The Great Ice Age, 3d edition, New York, D. Appleton & Company, p. 724–774.

Fenton, C. L., and Fenton, M. A., 1956, Giants of geology: Garden City, New York, Doubleday & Company, Inc., 333 p.

Geikie, James, 1894, The Great Ice Age, 3d edition: New York, D. Appleton & Company, 850 p.

Imlay, Gilbert, 1792, A topographic description of the Western Territory of North America, containing a succinct account of its soil, climate, natural history, population, agriculture, manners, and customs: Piccadilly, London, J. Debrett, 598 p.

Johnson, W. H., Follmer, L. R., Gross, D. L., and Jacobs, A. M., 1972, Pleistocene stratigraphy of east-central Illinois: Illinois Geological Survey Guidebook Series 9, 97 p.

Leverett, F., 1899, The Illinois Glacial Lobe: United States Geological Survey Monograph 38, 817 p.

Leverett, F., and Taylor, F. B., 1915, The Pleistocene of Indiana and Michigan and the history of the Great Lakes: United States Geological Survey Monograph 53, 529 p.

Lineback, J. A. [compiler], 1979, Quaternary deposits of Illinois: Illinois State Geological Survey Map.

Wayne, W. J., 1963, Pleistocene formations of Indiana: Indiana Geological Survey Bulletin 25, 85 p.

——1965, The Crawfordsville and Knightstown Moraines in Indiana: Indiana Geological Survey Report of Progress 28, 15 p.

Wayne, W. J., Johnson, G. H., and Keller, S. J., 1966, Geologic map of the 1° × 2° Danville Quadrangle, Indiana and Illinois, showing bedrock and unconsolidated deposits: Indiana Geological Survey Regional Geologic Map 2, scale 1:250,000.

Cataract Lake emergency spillway, southwestern Indiana

Walter A. Hasenmueller and N. K. Bleuer, Indiana Geological Survey, 611 North Walnut Grove, Bloomington, Indiana 47405

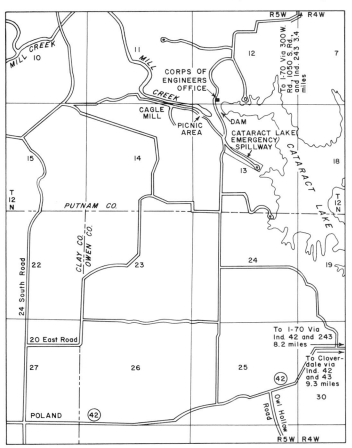

Figure 1. Map showing secondary roads leading to the Cataract Lake emergency spillway area, Putnam County, Indiana. Section lines provide scale at 1 mile (1.6 km) on a side. Base detail from the Poland 7½-minute Quadrangle.

SIGNIFICANCE

The emergency spillway at Cataract Lake provides an unusual opportunity to consider geologic features and problems relevant to Pennsylvanian and Pleistocene stratigraphy in Indiana at an exposure of great extent. Most exposures of Indiana's Pennsylvanian rocks are small because of the extensive blanket of unconsolidated sediments. Surface coal mines create large exposures of Pennsylvanian rocks and Pleistocene sediments; but these are temporary features, and close examination of surface-mine exposures is generally impossible because of safety considerations.

The Cataract Lake emergency spillway is a 60 ft (18 m)-deep trench that is more than 1,000 ft (300 m) long. It slices through a ridge underlain by basal Pennsylvanian rocks. Here geologists have the rare opportunity of walking into a hill and tracing rock units behind and under the kinds of exposures that they usually encounter in fieldwork in Indiana. Economic geologists and sub-surface stratigraphers have the opportunity to ponder how various exploration-drilling patterns might result in miscorrelations of the two coal seams exposed here or to consider how one or two drill holes in this ridge might result in overestimation or underestimation of the coal resources in this area when standard methods of calculating coal resources are applied.

A brief side trip to the base of the dam (see Fig. 1) provides an opportunity to appreciate the hiatus represented by the Pennsylvanian-Mississippian unconformity and the northward overlap of Pennsylvanian rocks onto Mississippian rocks. The Pleistocene exposure at the far east end of the cut offers an opportunity to observe typical Midwestern glacial deposits: tills, lake clays, loess, unconformities in continental deposits, and buried and modern soils. These exposures provide the background for discussion of the past and present classifications of the glacial record.

LOCATION

The Cataract Lake emergency spillway in Putnam County, Indiana, is in the SW¼NW¼Sec.13,T.12N.,R.5W. (Poland Quadrangle). The spillway cut is most easily reached by following the secondary road indicated in Figure 1 north from Indiana 42, a distance of about 2 mi (3.2 km). The secondary road that leads to the spillway cut is identified by a Corps of Engineers sign that points the way to the reservoir dam. The intersection of Indiana 42 with the road to the spillway is about 5 mi (8 km) south of I-70 and is 1.8 mi (3 km) east of Poland, Indiana, and 4.6 mi (7.4 km) west of the intersection of Indiana 243 and Indiana 42 (Fig. 1). The spillway cut is immediately adjacent to the secondary road that runs north from Indiana 42. Because the spillway cut is a restricted area, permission to enter should be obtained at the Corps of Engineers office just north of the dam.

SITE INFORMATION

Pennsylvanian

The rocks exposed here belong to the Mansfield Formation (Early Pennsylvanian), the lowest Pennsylvanian formation in Indiana (Figs. 2, 3). The Mansfield comprises the rocks from the Mississippian-Pennsylvanian unconformity up to the base of the Lower Block Coal Member of the overlying Brazil Formation. The Mansfield and Brazil Formations and the next highest formation, the Staunton Formation, make up the Raccoon Creek Group, which underlies the Carbondale and McLeansboro Groups. These three groups make up all the Pennsylvanian rocks of Indiana and correspond roughly to the subdivision used by nineteenth-century geologists: the Lower Coal Measures, or Millstone Grit, the Middle or Productive Coal Measures, and the

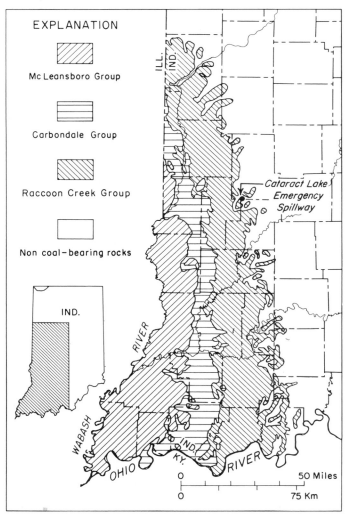

Figure 2. Map of southwestern Indiana showing distribution of the Raccoon Creek, Carbondale, and McLeansboro Groups (Pennsylvanian). From Hasenmueller and Carr (1983, p. 538).

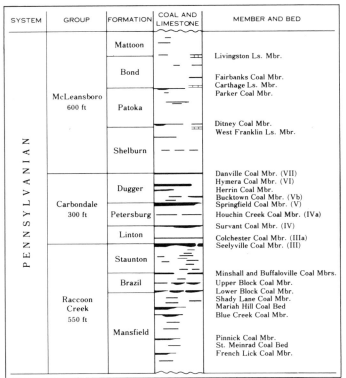

Figure 3. Generalized column showing named coals and important stratigraphic nomenclature of coal-bearing rocks in Indiana. From Hasenmueller and Carr (1983, p. 538).

Upper or Barren Coal Measures. These older terms aptly describe the stratigraphic distribution of Indiana's coal resources. About 85 percent of Indiana's coal production comes from the Carbondale Group (Productive Coal Measures), while the rest of the state's coal production comes from the Raccoon Creek, most of which is production from the Seelyville Coal Member at the top of the group. Only a small amount of Indiana's coal comes from the Mansfield Formation, and this exposure demonstrates the reasons for this. Coals in the Mansfield tend to be irregular and of only local extent. Therefore, minable thicknesses of coal are confined to a few square miles at best and generally cover only a few acres. Mining of the coal resources in the Mansfield is further complicated by the relief that has developed on the Mansfield and underlying Chesterian rocks (Late Mississippian in age). Such resistant rocks as the large sandstone lens exposed here are abundant in the Mansfield, and so a rugged topography of moderately high relief has developed. As a result, coal miners soon after they enter a coal seam at its crop line generally encounter overburden thickness that exceeds economic limits. These conditions, made by irregular coals and irregular overburden, combine to preclude the development of large-scale mining operations in the coals of the Mansfield Formation. The small-mine operators that do work seams in the Mansfield are able to locate and produce coal with exceptionally high quality in many places, however. The sulfur content of the thicker seam exposed in this cut, for example, is 1.5 percent, which is well below the mean sulfur value for most Indiana seams.

A variety of lithotypes occurs in the Mansfield Formation, and these lithotypes are arranged in complex facies patterns that change in a short distance, as the spillway cut dramatically illustrates. The Mansfield includes sandstone, shale, sandy shale, and mudstone with minor amounts of conglomerate, limestone, clay, and coal. Most of these lithotypes can be seen in the spillway cut (Appendix, section 1). Generally, the underclays, coals, black shales, and limestones are the most laterally persistent lithotypes in the Mansfield, and stratigraphers have relied heavily on these lithologies to establish regional lithostratigraphic correlations.

The irregularities of Mansfield coals are well illustrated by the two coals exposed in the spillway cut. The carbonaceous, coaly shale exposed at the west end of the north wall of the

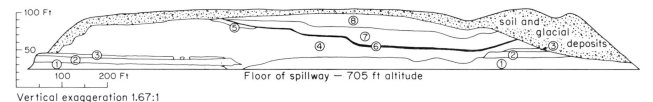

Northwest   Southeast

Vertical exaggeration 1.67:1

Figure 4. Cross section showing lithology of the Mansfield Formation along the north side of emergency spillway of Cataract Lake. Numbers refer to units in section 1 described in Appendix. From Rexroad and others (1983, p. 25).

spillway cut (unit 3 in the measured section) lies at about the same elevation as the thickest part of the coal that makes up unit 6 in the measured section. The nearly continuous exposure in the spillway cut makes it easy to see that two coal seams are present and to hypothesize that the upper seam represents the filling of a small trough, perhaps an oxbow lake, with peat. Compaction of the peat under the load of additional sediment accounts for the draping of the coal seam over the sandstone lens. A geologist working with the natural exposures and road cuts in the area and perhaps with one or two drill holes would probably not detect this relationship, however, Indeed, he or she would very likely miss the coals entirely.

Sedimentary structures abound in the sandstone units exposed in the spillway cut. There is no need to ignore safety restrictions, however, and climb the near vertical walls of the spillway cut to hunt for sedimentary structures. Large talus blocks at the base of the spillway-cut walls provide the best illustrations of small-scale bed forms. Road cuts at the west end of the spillway contain good exposures of small- and large-scale bed forms in place. Other features to be seen include plant fossils, intraformational conglomerates, and scour surfaces.

Although the Mississippian-Pennsylvanian unconformity is not exposed in the spillway cut, its stratigraphic proximity and geologic importance can be appreciated by taking a short side trip to the picnic area at the base of the dam (Fig. 1). The road to the picnic area descends through poorly exposed Mansfield sandstones for a short distance and then drops steeply into a creek bed on the Paoli Limestone (Late Mississippian in age). The unconformity surface is not exposed near the spillway or dam, but its position can be determined within a vertical interval of about 5 ft (1.5 m) in this creek bed. Exposures of the Paoli Limestone and the underlying Ste. Genevieve Limestone (Middle Mississippian) can be seen from the picnic area at the north end of the dam.

Some 500 ft (150 m) of Upper Mississippian rocks known in southern Indiana are absent from this locality because of the northward overlap of Pennsylvanian rocks across Mississippian rocks. In the 200 mi (320 km) from the Ohio River northward to the Kentland area in northwestern Indiana, the unconformity cuts through the entire 1,800 ft (450 m) of Mississippian rocks. In a few scattered outliers at the northern extent of Pennsylvanian

rocks in Indiana, the Mansfield rests directly on the New Albany Shale (Devonian).

Pleistocene

The Cataract spillway cut for many years has been one of Indiana's best examples of the evidence used by glacial geologists in developing the concept of multiple glaciation (Fig. 5). Ample historical background for the interpretations of the stratigraphy of the cut is given by White (1973), Thwaites (1927), and Flint (1965). The site lies outside the Wisconsinan glacial margin on a till plain continuous with the type Illinoian of central Illinois. The weathered zone in the upper till, therefore, presumably equates with the Sangamon Soil (Illinoian Stage) of Illinois. The opening of this cut in the mid-1950s provided the first vivid example of paleosols and till sequences representing possible pre-Illinoian glaciation (Wayne, 1954, 1958).

A brief summary of the stratigraphy of the cut and section description (from Schneider and Wayne, 1967, p. 84–86) follows:

The scientific appeal of this exposure stems from the sequence of glacial and interglacial climatic conditions that are represented and the clarity with which the climatic succession can be recognized. At the top of the cut [Figure 5] is exposed 3 ft [1 m] of Wisconsin (Peoria) loess overlying about 23 ft [7 m] of Illinoian till. The upper 12 ft [3.6 m] of this till shows a well-developed weathering profile, which was formed during the Sangamon Age but which has been partially modified by post-Wisconsin weathering because of the relative thinness of the overlying loess. Beneath the fresh Illinoian till is about 27 ft [8.2 m] of Kansan till and clay, the upper part of which exhibits a Yarmouth profile. The unaltered Kansan till is underlain by 2½ ft [0.75 m] of proglacial lake clay and this by 3 ft [1 m] of fossiliferous loess, which rests on Pennsylvanian rocks.

Until fairly recently, the assignment of ages to continental glacial materials was simply a matter of "counting down" through buried soils, the primary evidence for interglacial climatic conditions. Therefore, the first paleosol found beneath Illinoian deposits was termed the Yarmouthian soil, and the materials in which that soil was developed were termed the Kansan Stage. This was done here, and the section became a type section for the Cloverdale Till Member of Kansan age (Wayne, 1963). Developments of the past several decades have shown that

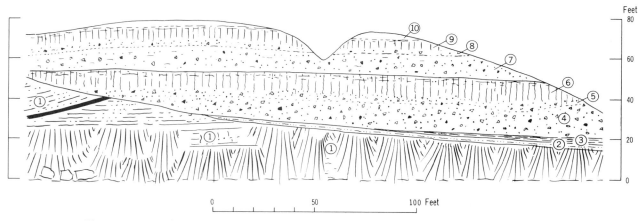

Figure 5. Sketch of the Pleistocene section at emergency spillway of Cataract Lake. Units are numbered in accord with detailed section description in Appendix, section 2. From Schneider and Wayne (1967, p. 84).

the continental record (Kukla, 1970; Boellsdorff, 1978), like the oceanic record (Shackleton and Opdyke, 1976), contains evidence for many more glaciations than the classically defined fourfold model. During this same period of time, rock-stratigraphic concepts have emerged as the basis for studying Midwestern tills, and "counting down" through buried weathered zones is no longer the single basis in classification schemes.

Presently, the tills and soils exposed at the Cataract Lake spillway cannot be related clearly to stratigraphic sequences in central Indiana or Illinois. Wayne's (1958) Cagle Loess Member (Atherton Formation) (unit 2, section 2, Appendix) has normal remanent magnetization (Bleuer, 1976) and presumably postdates the Brunhes-Matuyawa reversal of 700,000 years B.P. Despite its appearance of being an earliest glacial event here, it is predated by

older events recorded farther north in Indiana (Bleuer, 1980; Bleuer and others, 1983).

The principal relationships can be appreciated by viewing the exposure from the center of the spillway and identifying the reddish-brown band that represents the so-called "Yarmouth" soil profile. Although the exposure has deteriorated substantially through the years, numerous rills expose most of the section, although in bits and pieces. Interesting discussion points include: the several unconformities; the sharp, planar, ice-eroded till-paleosol contact (units 7 and 6) and the nature of ice movement and erosion; proglacial environments and landscapes (units 2 and 3); paleontology of the terrestrial gastropod microfauna (unit 2; see Wayne, 1954); and the characteristics of modern, superposed, and buried soils and their classification.

REFERENCES CITED

Bleuer, N. K., 1976, Remnant magnetism of Pleistocene sediments of Indiana: Indiana Academy of Science Proceedings, v. 85, p. 277–294.

——1980, Correlation of pre-Wisconsinan tills of the Lake Michigan Lobe and Huron-Erie Lobe through the Teays Valley fill [abs.]: Geological Society of America Abstracts with Programs, v. 12, p. 219.

Bleuer, N. K., Melhorn, W. N., and Pavey, R. R., 1983, Interlobate stratigraphy of the Wabash Valley, Indiana: West Lafayette, Indiana, Purdue University, Midwest Friends of the Pleistocene, 30th Annual Field Conference, Field Trip Guidebook, 135 p.

Boellsdorff, John, 1978, Chronology of some late Cenozoic deposits from the central United States and the ice ages: Nebraska Academy of Sciences Transactions, v. 6, p. 35–49.

Flint, R. F., 1965, Introduction: Historical perspectives, in Wright, H. E., and Frey, D. G., editors, The Quaternary of the United States: Princeton, New Jersey, Princeton University Press, 922 p.

Hasenmueller, W. A., and Carr, D. D., 1983, Indiana, in Nielson, G. F., editor, Keystone coal industry manual: New York, McGraw-Hill, Inc., 1461 p.

Kukla, J., 1970, Correlations between loesses and deep-sea sediments: Geologiska Foereningen i Stockholm Föerhandlingar, v. 92, p. 148–180.

Rexroad, C. B., Gray, H. H., and Noland, A. V., 1983, The Paleozoic systemic boundaries of the southern Indiana–adjacent Kentucky area and their relations to depositional and erosional patterns (Field Trip 1), in Shaver, R. H. and Sunderman, J. A., editors, Field trips in midwestern geology: Blooming-

ton, Indiana, Geological Society of America, Indiana Geological Survey, and Indiana University Department of Geology, v. 1, p. 1–35.

Schneider, A. F., and Wayne, W. J., 1967, Pleistocene stratigraphy of west-central Indiana (Field Trip No. 3), in Schneider, A. F., editor, Geologic tales along Hoosier trails: Bloomington, Indiana, Field Trip Guidebook, Geological Society of America, North-Central Section and Indiana University, 103 p.

Shackleton, N. J., and Opdyke, N. D., 1976, Oxygen-isotope and paleomagnetic stratigraphy of Pacific Core V28-239 late Pliocene to latest Pleistocene, in Cline, R. M., and Harp, J. D., editors, Investigation of late Quaternary paleooceanography and paleoclimatology: Geological Society of America Memoir 145, p. 449–464.

Thwaites, F. T., 1927, The development of the theory of multiple glaciation in North America: Wisconsin Academy of Science Transactions, v. 23, p. 41–164.

Wayne, W. J., 1954, Kansan till and a pro-Kansan faunule from Indiana [abs.]: Geological Society of America Bulletin, v. 65, p. 1320.

——1958, Early Pleistocene sediments in Indiana: Journal of Geology, v. 66, p. 8–15.

——1963, Pleistocene formations in Indiana: Indiana Geological Survey Bulletin 25, 85 p.

White, G. W., 1973, History of investigation and classification of Wisconsinan drift in north-central United States, in Black, R. F., Goldthwait, R. P., and Willman, H. G., editors, The Wisconsinan Stage: Geological Society of America Memoir 136, p. 3–34.

APPENDIX

Section 1: Description of units illustrated in Figure 4 from Rexroad and others (1983, p. 24–25).

Pennsylvanian System
Mansfield Formation

8. Sandstone, very light gray to light-yellow-brown; thin uneven stratification; very fine grained; sorting poor, clay chips locally abundant; numerous trails, cast flow markings, and many large *Stigmaria;* basal contact sharp and uneven; 20.0 ft [6.0 m].

7. Sandstone, medium-gray, with much thinly interstratified dary-gray shale; very fine grained; sorting poor; much relief on lower surface, but no evidence of scour; 0–23.0 ft [7.0 m].

6. Coal, canneloid, black; rests on an evidently depositional surface with at least 45 ft (14 m) of present relief; 0.5–3.0 ft [0.15–1.9 m].

5. Clay, locally present and poorly exposed; 0–3.0 ft [0.9 m].

4. Sandstone, light-yellow-brown; irregular to even and thin to thick stratification; some tangential cross stratification in lower part and some thinly interstratified shale in upper part; fine grained; sorting fair; plant impressions common; basal contact sharp and uneven and showing much evidence of scour and at least 12 ft [3.6 m] of relief; 6.0–53.0 ft [2.0–16.0 m].

3. Shale, carbonaceous, partly coaly, dark-gray; very thin even stratification; grades laterally into sandstone that is silty, dark gray, nonstratified, and intensely disturbed by root action and burrowing; 3.0–4.0 ft [0.9–1.2 m].

2. Clay, silty, medium-gray, partly shaly; lower 2 ft contans lenticular aggregates of ferruginous oolites; 7.0–9.0 ft [2.1–2.75 m].

1. Shale, dark-gray; indistinct thin even stratification; many siderite nodules as much as 1.5 ft [0.5 m] in diameter and 0.3 ft [0.4 m] thick; scattered lenses of ripple-stratified sandstone; lower part seen in gullies west of spillway; 20.0 ft [6.0 m].

Section 2: Description of units illustrated in Figure 5. From Schneider and Wayne (1967, p. 84–86).

	Thickness (ft) [m]
10. Silt: clayey, yellowish-brown, noncalcareous	3.0 [0.9 m]
9. Till: brown, noncalcareous; secondary limonite deposits along joints	12.0 [3.64 m]
8. Till: clayey, light-brown, calcareous	5.3 [1.6 m]
7. Till: clayey, dark-gray, calcareous	6.0 [1.8 m]
6. Clay: silty, sandy, brown to greenish-gray, noncalcareous	3.5 [1.06 m]
5. Till: silty, sandy, brown, noncalcareous	8.5 [2.6 m]
4. Till: silty, sandy, reddish brown in upper part, brownish gray below, calcareous; contains wood fragments in basal few feet	15.0 [4.5 m]
3. Clay: silty, brownish-gray, laminated, highly calcareous; contains scattered wood fragments throughout; unit is lenticular and pinches out toward west in exposure	2.3 [0.7 m]
2. Silt: grayish-brown, calcareous, fossiliferous, contains wood, peat, and humus along upper contact; unit is lenticular and pinches out toward west in exposure	3.0 [0.9 m]
1. Bedrock overlain by as much as 12 ft [3.6 m] of coluvial debris	——

The Indiana building-limestone district, south-central Indiana

John B. Patton and Donald D. Carr, Indiana Geological Survey, 611 North Walnut Grove, Bloomington, Indiana 47405

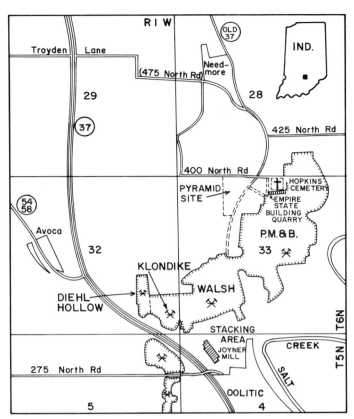

Figure 1. Outline map of the Pyramid Site and vicinity in south-central Indiana showing access routes and active and abandoned dimension-stone quarries. Section lines (1 mi (1.6 km) on a side) provide scale.

LOCATION

The best public access to quarries from which the noted Indiana Limestone (trade name) is produced is at a park and demonstration center in the NE¼NW¼Sec.33,T.6N.,R.1W., Lawrence County, Indiana (Oolitic 7½-minute Quadrangle), and lies 1 mi (1.6 km) north of Oolitic, which is, in turn, some 3 mi (4.8 km) north of Bedford, the county seat of Lawrence County (Fig. 1).

The site may be reached by passenger car or bus. Visitors approaching from the south by 4-lane Indiana 37 should pass the turnout to the north part of Oolitic and proceed 2.3 mi (3.7 km) to Trogden Lane, turning right (east) there and passing through Needmore to the old 2-lane route of Indiana 37 (1.2 mi [1.9 km]). Turn right (southeast) and proceed half a mile (0.8 km) to the entry for the Indiana Limestone Tourist Park and Demonstration Center.

Visitors approaching from the north by Indiana 37 (4-lane) will pass a stoplight intersection with Indiana 45 at the southwest edge of Bloomington and proceed 7.7 mi (12.3 km) to the place

where the road descends from the upland in a deep bedrock cut through the lower part of the Salem Limestone, the entire Harrodsburg Limestone and Ramp Creek Formation, and the upper few feet of the Borden Group (Middle Mississippian). The highway crosses Clear Creek about 0.3 mi (0.5 km) south of the cut, and 2.2 mi (3.7 km) south of the bridge the new highway intersects old Indiana 37 at a blinker light. Turn left (southeast) and proceed 3.8 mi (6 km), passing Needmore on the right, to the entrance of the Center.

At the time of this writing, it is planned that the Center will be open (entry fee will be charged when fully in operation) from late spring until early fall. At other times, special arrangements must be made through Mr. Alvin Walker, Bedford Chamber of Commerce, 2999 West 16th Street, Bedford 47421 (phone: 812/275-4493). Even during the summer, it would be well for visitors to check in advance concerning open days and hours.

Just south of the Center, beyond the south end of the former highway, the Indiana Limestone Company quarries dimension stone during the season (approximately April to November), and to cross onto company property, visitors must obtain advance permission from Indiana Limestone Company, Inc., 405 I Street, Bedford 47421 (phone: 812/275-3341).

SIGNIFICANCE

Material quarried from the Salem Limestone has dominated the building-limestone market in the United States for nearly a century. In early years, the Salem was used principally for local markets, but with the coming of the railroads in the 1850s, the market for the Salem greatly increased. By the 1870s, a substantial volume of stone was reaching Indianapolis, Chicago, New York City, and Washington, D.C., points encompassing an area that currently is the principal market for Indiana Limestone. The names of buildings using Indiana Limestone are legend, but a few well-known examples are the Pentagon, Washington Cathedral, Empire State Building, Rockefeller Center, Chicago Museum of Science and Industry, Indiana state capitol, and American United Life Insurance Company building in Indianapolis, the state's tallest.

Although the outcrop belt of the formation extends for more than 100 mi (160 km) within southern Indiana and quarrying formerly took place at widely dispersed localities, the industry has for some 70 years been localized within a region that extends from a short distance south of Bedford to some 15 mi (24 km) northwest of Bloomington. The Center site is in the most intensively exploited part of the district. Visitors can view vast areas of quarried land, study the bedrock stratigraphic section and the character of the rock quarried and the overburden, and observe the methods of removal. To the south, active quarrying may be seen (by access permission) on property of the Indiana Limestone Company.

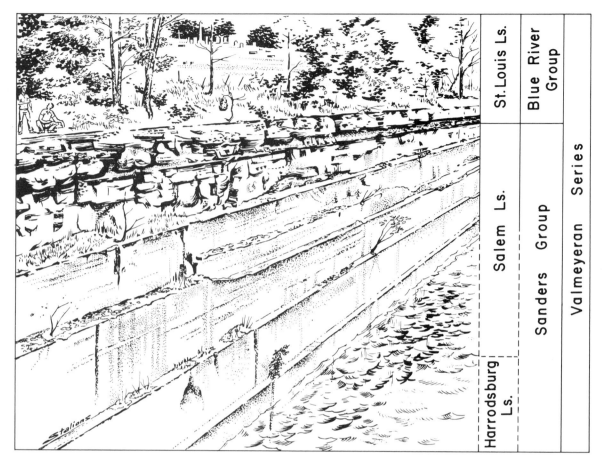

Figure 2. Sketch, as seen from the west side, of the north face of the Empire State Building Quarry showing general stratigraphy of the Indiana building-limestone district. This quarry is part of the large complex of active and abandoned quarries of the Indiana Limestone Company near the Pyramid Site as shown by Figure 1.

SITE INFORMATION

The Center is on a tract of land donated by the Indiana Limestone Company and adjacent to a complex of quarries covering more than 300 acres (Fig. 1).

Directly east of the Center (across a former highway and accessible by foot along a road surfaced with crushed stone) is the quarry from which stone for the Empire State Building was taken (Fig. 2). The stratigraphic section in the north wall exposes, beneath thin residual soil, a variable thickness of the St. Louis Limestone, identifiable from the observation point by its visibly bedded units ranging from a few inches to a foot (30 cm) or more in thickness. Beneath the St. Louis is the massive Salem Limestone, from which the building material is taken, but the uppermost 10 ft (3 m) of the massive section is the "bastard stone" of local terminology—an impure silty dolomitic limestone. Both it and the overlying limestone in the lower part of the St. Louis formation were and are discarded as overburden where present. Directly below the bastard stone is an 8-ft (2.4 m) cut, called "second ledge," meaning that the stone is inferior grade but that

parts of it are usable for some purposes. Below the second ledge is about 60 ft (18 m) of usable building stone, but at the time of writing in October 1983, a part of this stone was below water level. The lowermost exposed limestone is the Salem, giving a total thickness of about 80 ft (25 m), but the contact with the Harrodsburg Limestone is not far below. In places, the uppermost beds of the Harrodsburg so resemble the Salem that building-stone quarriers have dropped their lowest ledges below the contact. In general, however, the upper Harrodsburg, even where it resembles the Salem building stone, has coarser texture, is less well sorted, and tends to lose the massive character that permits the removal of sound blocks.

At lower positions relative to the topographic surface, where Salem stone of dimension quality directly underlies the soil, deep, vertical, clay-filled solution joints called "grikes" are likely to be present and cause some loss of recoverable blocks.

The part of the formation that came to be developed as a premium building material of national and international renown is a freestone—one that has virtually the same workability in all directions and shows little preferential direction of splitting. It is a relatively pure, massive, evenly granular, fine-grained, moderately porous limestone, crossbedded in places, that consists of small fossils and fossil fragments cemented by calcite. The original color ranges from light gray to bluish-gray, but the rock is oxidized to shades of buff to varying depths below the surface, in some places massively and in others along bedding planes and joints. This oxidation gives rise to the three principal color categories—gray, buff, and variegated (mixed gray and buff).

Although quarrying may not be taking place, at least within sight at the time of a visit, even viewing the inactive quarries enables spectators to understand the geologic situation and the practices and methods of the industry.

Overburden may be: (1) residual soil generally red immediately above the stone; (2) bedrock not suitable for building stone, either the lower part of the overlying St. Louis Limestone or ledges of Salem that are impure or too hard for satisfactory milling; and (3) surficial layers that have the lithology of the building stone but that are too weathered, irregular, or jointed to yield sound blocks of usable size. Unconsolidated overburden is removed by bulldozer and scraper, and clay from irregularities is removed by some hand labor. Rock overburden may be loosened by drilling and blasting before removal, but explosives must be used with wariness to prevent damage to the underlying building stone. The lowermost part of the rock overburden, the second ledge, whether stone or unsuitable lithology or weathered and irregular building stone, is generally removed by the same methods used for producing mill blocks. In some places, blocks can be salvaged for housing veneer or for uses other than cut stone.

Waste stone so quarried must be placed somewhere, and in opening a new quarry, operators may need to stack waste blocks adjacent to the quarry opening. Therefore, the ground needed for later quarry expansion may be covered, and the waste blocks may have to be handled a second time. Besides the overburden removed as block, some of the stone quarried from the working ledge must generally be set aside because of irregular shapes that do not justify gang sawing or because of weathering, color, texture, or flaws that cause its rejection. Some of the "grout" piles, as they are called, are stacked in orderly fashion, but others are tumbled masses of mixed large blocks and unshaped fragments.

Once a quarry opening has been developed to its full intended depth, block overburden, nonblock overburden, and rejected material from adjacent working ledges may be placed in the abandoned opening. Many such former quarry holes are no longer recognizable, as they have been graded over and even paved or vegetated.

The principal methods for separating stone from the ledge involve the use of channeling machines or of wire saws. Channelers are mobile devices that travel on narrow-gauge tracks, moving slowly back and forth over the prescribed length of the particular cut that they are making. They deepen a channel on one side of the tracks, or two channels straddling the tracks, a fraction of an inch to several inches at each pass through the pounding of a set of bits at the bottom of sheaved, chisel-ended rods actuated by a cam or directly by a piston. The depth of the channel cut may range from as little as 5 or 6 ft (1.5 to 1.8 m) to 13 ft (4 m) or more and may be determined by a change in the character of the stone at some depth that would make it impractical to add another foot or two of a different lithology to the ultimate block. In other places the lithology remains unchanged from the ledge being channeled to the underlying ledge, and in such places the depth of the channel cut will be the maximum that equipment can cut or transport effectively or that the mill can use efficiently. For the most part, the successive ledges in a deep quarry are not bottomed at bedding planes or other necessary limitations, and the ledges are commonly of fairly consistent height. A channel cut cannot be made flush with the quarry walls, and therefore each successive downward ledge steps inward about a foot (.3 m).

Wire saws are now considered to be more economical to operate than channeling machines. A power-driven endless helical wire is drawn, initially over the surface of the ledge and later in its own slot, carrying a mixture of water and quartz sand as abrasive. The wire is under tension, exerting constant pressure on the bottom of the slot and feeding over sheaves that slowly descend as the slot deepens. The resulting cut is slightly convex but may be flattened at the end of the run by continued operation under tension. Typically a single length of wire, actuated by a power unit that may be out of sight over the quarry rim, is so rigged over an elaborate assembly of pulleys and posts that it is performing many cuts at different places in the quarry simultaneously.

At the lateral extremities of the floor to be quarried, other cuts must be made perpendicular to the parallel channels or wire-saw cuts thus described; and unless the entire width of the ledge is to be handled in the next step as a single unit, intervening crosscuts are made at intervals of 30 to 60 ft (9 to 18 m) and to the same depths. Once a mass of stone, perhaps 4 ft (1.2 m) wide by 10 ft (3 m) high by 30 (9 m) or 40 ft (12 m) long and weighing more than 100 tons, has one long exposed face and vertical cuts at the back and ends, it must be freed from the bedrock below. Jackhammer holes, spaced less than a foot apart, are drilled as near the base and as nearly horizontal as possible. Into these, wedges are driven, forcing apart metal half sleeves called "slips" (to transmit the force back into the stone and prevent it from simply shattering around the hole). Broken stone less than a cubic foot in size is stacked on the new lower floor in a row of piles called "pillows," and the underdrilled and wedged-free cut is turned down, generally by using wire rope attached through sheaves to a power source, onto the pillows, which are intended to prevent breakage of the cut as it topples. Once on its side, the cut is sectioned into quarry blocks of the size desired by using wedges and slips in pneumatic-drill holes.

In this part of the American Midwest, clastic sedimentation was dominant during both the beginning and ending of Mississippian time, but carbonate deposition prevailed during Middle Mississippian time, accumulating to more than 400 ft (120 m) of stratigraphic section in this area. Although this thick carbonate section was marked by episodes of clastic and evaporitic deposition, the part of the Salem that is used for building stone is thought to have been deposited in warm, clear, shallow-marine water—warm enough to sustain an abundant invertebrate fauna; clear enough to have yielded almost no clay or other inorganic sediments; and sufficiently shallow to permit wave action and marine currents to macerate the larger units of shell material, to winnow the organic detritus to remarkably uniform grain sizes, and to distribute calcarenite sand consisting of small fossils and well-graded fragments of larger organisms over thousands of square miles of Mississippian sea floor.

Bimodal orientation of crossbedding suggests that these marine sands moved back and forth on a southwestward-dipping paleoslope in response to ebb-and-flood tidal currents (Sedimentation Seminar, 1966, p. 112–113, Figure 6), but how far eastward the building-stone lithology extended from the belt of present exposure is unknown. Rock of similar lithology is not present eastward from the Cincinnati Arch, but no evidence is found in three-dimensional study of the building stone to suggest that the present outcrop belt approaches an eastward terminus. A landmass, one that contributed terrigenous debris, must have been tens to hundreds of miles away, because its contribution to the Salem building-stone facies was slight. Discontinuous beds of silty and less calcareous materials were deposited locally, which interrupted the sequence of building-stone deposition in some places and capped it in others. Westward into the subsurface, the building-stone lithology of the Salem extends to the Illinois-Indiana line (Pinsak, 1957, p. 25–26, 34–36, plate 1) and beyond the center of the Illinois Basin, a distance of 150 to 200 mi (240 to 320 km) from the Indiana outcrop.

Some of the material apparently underwent little distance of transport, as mats consisting largely of delicate fronds of fenestelloid bryozoans are found in places—broken, it is true, to fragments only a few millimeters in long dimension, but still probably too fragile to have withstood any major distance of transport. The greater part of the building-stone bank is heterogeneous in organic composition, the only common character being particle size, and we must conclude that the greater part of the building stone is composed of winnowed material that was not derived mainly from invertebrate fauna of the immediate vicinity. A considerable vogue existed in the early part of the century for considering the Salem to contain a so-called "dwarf fauna" (Cumings and Beede, 1906, p. 1189–1200; Smith, 1906, p. 1220, 1237–1242), and speculations were advanced to explain environmental conditions that could have caused stunting of various species. The Salem fauna rather than being dwarfed, however, consists of minute species, of infantile forms of larger species, and of fragments from organisms of normal size.

The Salem has become a remarkable building stone not only because geologic events leading to its discovery were just as they should be, but also because its location and suitability for exploitation allowed a product to be produced that people wanted and could afford. The market for the Salem has varied through the years, but performance of the stone has held up for well over a century in buildings and monuments of all sorts, distributed through a wide diversity of climates and environments. Aesthetic appeal, reasonable cost, and permanance are properties of the stone that have brought the Indiana Limestone industry to its prominence today; these properties ensure its success in the future.

REFERENCES CITED

Batchelor, J. A., 1944, An economic history of the Indiana Oolitic Limestone industry: Bloomington, Indiana University School of Business, Indiana Business Studies 27, 382 p.

Cumings, E. R., and Beede, J. W., 1906, Fauna of the Salem Limestone of Indiana: Indiana Department of Geology and Natural Resources Annual Report 30, p. 1189–1218.

Hopkins, T. C., and Siebenthal, C. E., 1897, The Bedford Oolitic Limestone: Indiana Department of Geology and Natural Resources Annual Report 21, p. 289–427.

Patton, J. B., and Carr, D. D., 1982, The Salem Limestone in the Indiana building-stone district: Indiana Geological Survey Occasional Paper 38, 31 p.

Pinsak, A. P., 1957, Subsurface stratigraphy of the Salem Limestone and associated formations in Indiana: Indiana Geological Survey Bulletin 11, 62 p.

Rooney, L. F., 1970, Dimension limestone resources of Indiana: Indiana Geological Survey Bulletin 42-C, 29 p.

Sedimentation Seminar, 1966, Cross-bedding in the Salem Limestone of central Indiana: Sedimentology, v. 6, p. 95–114.

Smith, E. A., 1906, Development and variation of *Pentremites conoideus:* Indiana Department of Geology and Natural Resources Annual Report 30, p. 1219–1242.

The Borden Group: Ancient and modern perspectives from the Knobs Overlook, Indiana

Robert H. Shaver, *Department of Geology, Indiana University, Bloomington, Indiana 47405, and Indiana Geological Survey, 611 North Walnut Grove, Bloomington, Indiana 47405*

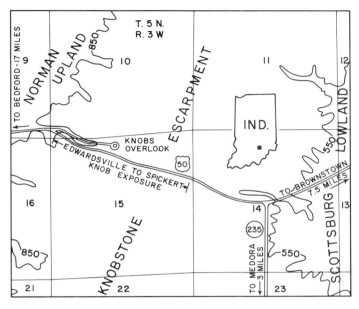

Figure 1. Map of the Knobs Overlook area near Medora, Indiana, showing access and relationships to major physiographic features and to exposures of the Borden Group ranging from the Edwardsville Formation down through much of the Spickert Knob Formation. Contour lines (550 and 850 ft) represent approximate boundaries between physiographic features. Base detail from the Kurtz and Medora 7½-minute Quadrangles. Boundary lines of section 14 are 1 mi [1.6 km] long.

LOCATION AND SIGNIFICANCE

One of many remarkable exposures of the Borden Group (lower Middle Mississippian) is on either side of a nearly continuous 1.5-mi (2.4-km) road cut along U.S. 50 leading down through the Knobstone Escarpment from the Knobs Overlook (a convenient parking area for field-trip groups) in the northern third of section 10 and central western part of section 14,T.5N., R.3E. (Medora and Kurtz 7½-minute Quadrangles), Jackson County, Indiana (Fig. 1). From the time of its earliest scientific note by one of America's great pioneer geologists (David Dale Owen in 1838), the wedge of Borden rocks, as much as 700 ft (210 m) thick in an area extending from central Illinois eastward and southeastward into north-central Tennessee, has been intensively studied by scores of geologists.

Perhaps the two most significant of the earlier, pioneering Indiana studies were conducted by Borden (1874) and Stockdale (1931). These and other studies of the time were preoccupied with descriptive explorations and with establishing ages of the

Borden divisions within the Subcarboniferous Group, a term that Owen had correctly applied about 150 years ago. Only during a modern period of intensive study in the broader area noted above did geologists come to realize that the Borden Group is a superb example of a vast ancient delta system complete with bottomset, foreset, and topset beds together with many sedimentologic facies and ecologic niches afforded by prodelta, delta-slope, delta-platform, distributary-channel, and interdistributary-flat environments. (See, for example: Swann and others, 1965; Lineback, 1966; Suttner and Hattin, 1973; Kepferle, 1977; Gray, 1979; Sable, 1979; Ausich and others, 1979; Ausich and Lane, 1980; and Kammer and others, 1983.)

No one section or area can demonstrate more than a fraction of the great array of rock types, sedimentary structures, and fossil communities associated with the complex Borden delta system, nor can it be wholly convincing of the tremendous importance, both scientific and economic, that deltas and their interpretations have come to assume in modern geologic study. The Knobs Overlook section, however, is exemplary in itself and becomes even more so when considered in the context of other Borden sections, regionally considered. For more reason than this, however, the Knobs Overlook site permits further appreciation of the essence of Indiana geology: a view of striking landscape features in the classic area of physiographic study of partly unglaciated bedrock terrains in southern Indiana (parts of the Interior Low Plateau Province and the Central Lowland Province of Fenneman, 1938) (Figures 2 and 3).

SITE INFORMATION

Borden Delta

The graphic section of the Borden Group shown in Figure 4 is a composite one. The upper and middle parts from the Edwardsville Formation down through much of the Spickert Knob Formation extend continuously from the top of the Knobstone Escarpment eastward along U.S. 50 nearly to the east line of section 15 (Fig. 1). The lower part of the Borden Group (New Providence Shale) is not well exposed in this area but is present in the subsurface.

This section of the Borden Group needs to be interpreted in the light of synthesis of the ancient Borden delta. Early Mississippian (Kinderhookian) and early Middle (Osagean) Mississippian sedimentation in the Midwest reflects the lingering phases of the Acadian Orogeny during which large quantities of clastic sediments were shed from eastern and northeastern highland areas

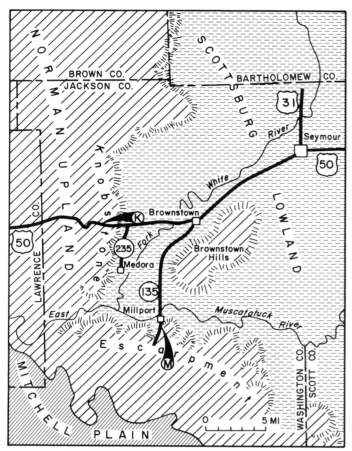

Figure 2. Map of the Medora-Brownstown area of western Jackson County, Indiana, showing physiographic and other natural features, place names, and sites of the Knobs Overlook section (K) and Millport section (M). See index map of Figure 1 for location within Indiana, and compare with block diagram and cross section of Figure 3. Modified from Burger and others (1966, Figure 5).

and carried westward and southward through two major drainage basins of what usually are portrayed as the Ontario River and Michigan River systems (Fig. 5A). During the course of a few million years and southwestward from a southwestward-prograding and northwest-southeastward-trending shoreline, the shifting distributaries of these systems produced an interleaved complex of deltaic lobes (Fig. 5A). The front, or delta slope, of this complex formed the shoreward scalloped margin of a deeper water (few to several hundred feet) embayment, or basin, centered in the tristate area of Illinois, Indiana, and Kentucky (Fig. 5B). The cross-sectional perspective through this delta front and the facing basin (Fig. 5C) illustrates the ideal case for deltaic sedimentation fitted to the formations exposed in the Knobs Overlook vicinity.

In such a case, Walther's law (on the relationship between lateral facies and vertical succession in an integrated sedimentational province) is satisfied, and the Knobs Overlook section and many others should reveal coarsening-upward sequences representing fine-grained prodelta, bottomset deposits (New Provi-

dence); fine-grained, but coarser delta-slope, foreset deposits (Spickert Knob); and still coarser grained, delta-platform, topset deposits (Edwardsville). Further, individual formations and the entire Borden suite should thin southwestward (downdip and into the subsurface in present structural circumstances), and they do. In the Knobs Overlook area, the Borden totals about 600 ft (180 m) in thickness, but in an 85-mi (136-km) distance to the modern Wabash River area, it thins to about 100 ft (53 m) (Pinsak, 1957; Ausich and others, 1979; Kammer and others, 1983). The prodelta part itself (New Providence) thins from about 250 ft to about 50 ft (76 to 15 m) in the same distance.

Moreover, correlation charts (based on time) should show the diachroneity of the different named units of single lobes making up the Borden delta proper, diachroneity among different delta lobes, and some contemporaneity of nondeltaic carbonate rocks that in part prograded eastward from the modern Mississippi River area and that in some areas were later encroached on by younger Borden delta deposits (Fig. 6). (See Ausich and Lane, 1980; and Lineback, 1966, 1969, 1981.) In these assumptions, the correlation chart shows that the Warsaw Shale and Borden Siltstone, representing the Borden delta lobe of southern Illinois (Fig. 5A), are younger than the Borden Group of southern Indiana, representing one or more older delta lobes in that area.

Some paleontological support of these principles has been presented, for example, through the conodont biostratigraphy of Nicoll and Rexroad (1975). Also, the section along U.S. 50 descending from the upland exposes a calcareous bed containing large spiriferoid brachiopods called *Syringothyris texta,* one of the means by which clastic Borden delta sediments have been correlated with the carbonate-bank section (Burlington and Keokuk Limestones) that accumulated on the west side of the basin into which the ancient Michigan and Ontario Rivers debouched. Further, a crinoid fauna found in the Borden delta-platform facies of eastern Kentucky shows that these deposits are late Burlington-depositional time in age and that these deposits are older than the Edwardsville delta-platform facies in Indiana (Lane and DuBar, 1983). Such a finding is significant, because it not only helps to establish age relations between deltaic and nondeltaic deposits, but also shows progradation within (that is, time lapse between the different parts of) the one delta facies.

Much additional proof is needed, however. It will not be easy because of the intercomplexities of the diachronously developed delta lobes (Fig. 5A). Therefore, just as single-named formations may have partly different chronologies if distributed in more than one lobe (see above), so may the delta fossil communities have repeated themselves at successive times. Difficulties arise, therefore, in attempts to distinguish between age correlations that are based on true age differences and those correlations that are based (incorrectly and unknowingly) only on similarity between environments. (See discussions in Ausich and Lane, 1980, p. 39–42; and Sable, 1979, p. E80-E82.

From all these considerations, one may appreciate how imperfect the Knobs Overlook section, or any other section, is to

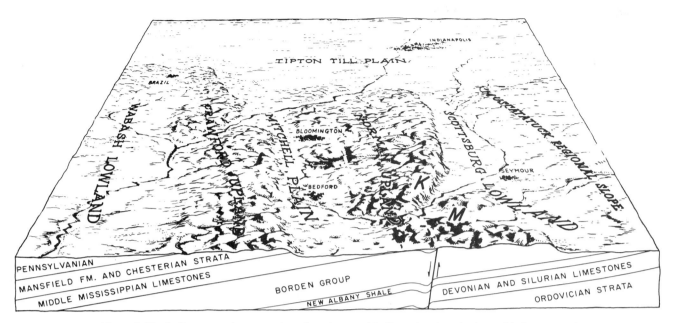

Figure 3. Block diagram and cross section of south-central Indiana showing physiographic features and their relationship to regional stratigraphy and structure. View is northward. K and M, locations of Knobs Overlook and Millport sections. Cross section is about 115 mi (185 km) long. Modified from Burger and others (1966, back cover).

represent such a fascinating episode in geologic history. For example, the clastic part of the Edwardsville in the Knobs Overlook section seems to be represented more by interdistributary deposits than by distributary-channel deposits that may be seen in other sections; also, turbidite sandstones seem to be lacking from the Spickert Knob interval. Therefore, in-depth field observations must include other exposures. One such exposure of easy access and near the Knobs Overlook is where Indiana 135 ascends the Knobstone Escarpment just south of Millport and the Muscatatuck River (Fig. 2). (See measured sections presented by Burger and others, 1966, p. 18; and Sunderman, 1968, p. 81.) Whether one's interests extend especially to delta-toe sandstone turbidites, delta-platform distributary-channel deposits, delta-platform carbonate banks and bioherms and their communities, many kinds of small-scale sedimentary structures, or to other delta facets, modern field guides provide necessary illustrations, including some for the Knobs Overlook area. Among them are those by Suttner and Hattin (1973), Ausich and Lane (1980), and Kammer and others (1983). Moreover, most of Stockdale's (1931) beautifully detailed measured sections remain easily accessible, although they need to be interpreted in light of the deltaic theory advanced above and in the cited sources of information.

Physiography

The Borden rocks have had a profound effect on the physio-

graphic development of southern Indiana, which is demonstrable from the Knobs Overlook (Figs. 2 and 3). Relatively resistant to weathering and erosion in the midwestern humid climate, the middle and upper Borden rocks are those exposed in the eastward-facing Knobstone Escarpment and those that hold up the maturely dissected upland called the Norman Upland. These two features together make up the most conspicuous physiographic province of southern Indiana, which is actually one of a series of cuestas and eastward-facing scarps that extend for nearly half the length of Indiana from the great glacial till plain (Tipton Till Plain) south by southeast to the Ohio River and beyond. At the Knobs Overlook, the relief between the Knobstone summits and the adjacent Scottsburg Lowland to the east is more than 300 ft (90 m), which is about average for the scarp in Indiana. South of Indianopolis, the scarp becomes buried in glacial sediments, but it extends far north as a subdrift feature; relief in the Ohio River area reaches more than 600 ft (182 m).

From the Knobs Overlook vantage point, the Scottsburg Lowland can be seen to extend northward and southward. This feature, too, is an effect of the Borden delta in part of its detail because the relatively nonresistant New Providence shaly pro-delta sediments, together with the New Albany Shale (Devonian and Mississippian), are the foundation of this lowland. Distal lobes of successive pre-Wisconsinan continental glaciers were directed southward through the lowland, along the Knobstone, and to the present Ohio River and beyond, but the Knobstone was nearly everywhere too much of an impediment for the ice to override in this area and the area southward.

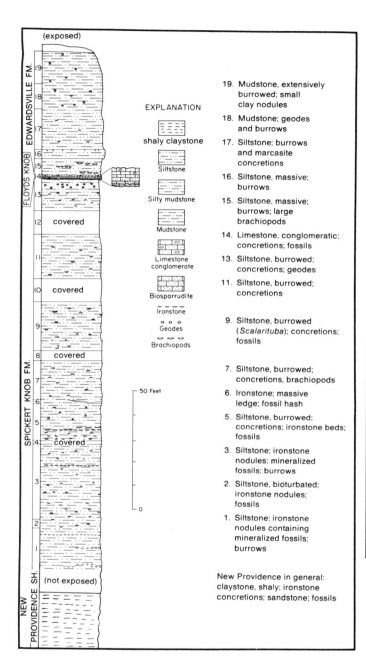

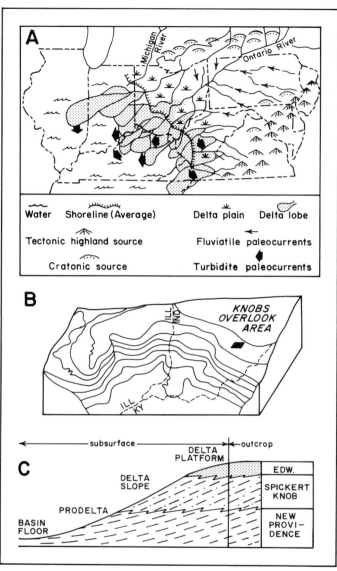

Figure 4. Graphic and descriptive section of the Borden Group along U.S. 50 at the Knobs Overlook, western Jackson County, Indiana. See Figure 1 for location detail and Figures 5 and 6 for correlation and interpretation. The Floyds Knob Limestone Member is part of the Edwardsville Formation. Mostly adapted from Suttner and Hattin (1973, Figure 28 and p. 105–106).

Figure 5. Drawings showing paleogeographic interpretation of the Borden delta of southern Indiana and its stratigraphic relations. A, Paleogeographic map of a segment of the eastern United States showing development of Early to Middle Mississippian drainage systems and the Borden delta; the Knobs Overlook section located at position of paleocurrent arrow in Indiana; modified from Kepferle (1977); paleocurrents added from Lineback (1968), Suttner and Hattin (1973), and University of Cincinnati Sedimentation Seminar (1981). B, Block diagram of the tristate area of Illinois, Indiana, and Kentucky showing interpretive Borden delta front and adjacent basin area during early Middle Mississippian time; solid lines are form lines; from Ausich and Lane (1980). C, Idealized cross section normal to the Borden delta front in the Knobs Overlook area showing interrelations of delta environments, structure, sediments, and named rock units; EDW., Edwardsville; modified from Ausich, Kammer, and Lane (1979).

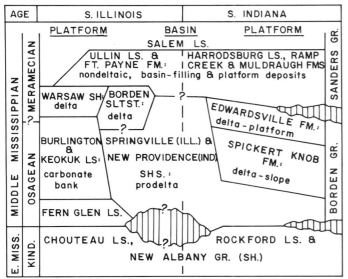

Figure 6. Correlation chart for Early Mississippian (Kinderhookian) and Middle Mississippian (Osagean and Meramecian) rock units of Illinois and Indiana showing chronologic relations of different Borden delta units, including prodelta units, and only partly contemporaneous carbonate-platform and mostly younger, deeper water, basin-filling, nondeltaic units. Adapted from the discussions and portrayals of especially Kepferle (1972), Lineback (1966, 1968, 1969), Nicoll and Rexroad (1975), Ausich and Lane (1980), and Shaver and others (1985).

Other physiographic features that can be seen from the overlook include the Brownstone Hills, a large outlier of the Norman Upland that became separated probably during late Tertiary (and early Quaternary?) adjustments of major drainage lines. Also, on a clear day, the dip slope of the broad cuesta called the Muscatatuck Regional Slope, which is held up by Middle Silurian and Middle Devonian carbonate rocks, can be seen. Other landscape features and questions abound in this area and are pertinent to Quaternary geomorphic development. They include terraced valley-train deposits and sand-dune tracts along the trunk streams that can be viewed along U.S. 50 between Seymour and the Knobs Overlook. Also, a few miles south of the overlook is a large noteworthy reentrant in the Knobstone through which the East Fork White River flows. Does this oversized (for the present river) reentrant and, indeed, the Muscatatuck Lowland itself hold clues to the course of major pre-Pleistocene drainage in this area, that is, to a precursor of the modern Ohio River? (See Malott, 1922, p. 59–256; Burger and others, 1966, p. 13–17; Schneider, 1966, p. 43–47; and Suttner and Hattin, 1973, p. 91–92.)

REFERENCES CITED

Ausich, W. I., Kammer, T. W., and Lane, N. G., 1979, Fossil communities of the Borden (Mississippian) delta in Indiana and northern Kentucky: Journal of Paleontology, v. 53, p. 1182–1196.

Ausich, W. I., and Lane, N. G., 1980, Field trip 2, Platform communities and rocks of the Borden siltstone delta (Mississippian) along the south shore of Monroe Reservoir, Monroe County, Indiana, in Shaver, R. H., editor, Field trips 1980 from the Indiana University campus, Bloomington: Bloomington, Indiana University Department of Geology and Geological Society America North-Central Section, p. 36–37.

Borden, W. W., 1874, Report of a geological survey of Clark and Floyd Counties, Indiana: Indiana Geological Survey Annual Report 5, p. 133–189.

Burger, A. M., Rexroad, C. B., Schneider, A. F., and Shaver, R. H., 1966, Excursions in Indiana geology: Indiana Geological Survey Guidebook 12, 71 p.

Fenneman, N. M., 1938, Physiography of the United States: New York, McGraw-Hill Book Co., Inc., 714 p.

Gray, H. H., 1979, The Mississippian and Pennsylvanian (Carboniferous) Systems in the United States—Indiana: U.S. Geological Survey Professional Paper 1110-K, 20 p.

Kammer, T. W., Ausich, W. I., and Lane, N. G., 1983, Paleontology and stratigraphy of the Borden delta of southern Indiana and northern Kentucky (field trip 2), in Shaver, R. H., and Sunderman, J. A., editors, Field trips in midwestern geology: Bloomington, Indiana, Geological Society of America, Indiana Geological Survey, and Indiana University Department of Geology, p. 37–71.

Kepferle, R. C., 1972, Stratigraphy, petrology, and depositional environment of the Kenwood Siltstone Member, Borden Formation (Mississippian), Kentucky and Indiana [Ph.D. thesis]: Cincinnati, University of Cincinnati, 158 p.

—— 1977, Stratigraphy, petrology, and depositional environment of the Kenwood Siltstone Member, Borden Formation (Mississippian), Kentucky

and Indiana: U.S. Geological Survey Professional Paper 1007, 49 p.

Lane, N. G., and DuBar, J. R., 1983, Progradation of the Borden delta: new evidence from crinoids: Journal of Paleontology, v. 57, p. 112–123.

Lineback, J. A., 1966, Deep-water sediments adjacent to the Borden Siltstone (Mississippian) delta in southern Illinois: Illinois Geological Survey Circular 401, 48 p.

—— 1968, Turbidites and other sandstone bodies in the Borden Siltstone (Mississippian) in Illinois: Illinois Geological Survey Circular 425, 29 p.

—— 1969, Illinois Basin—sediment-starved during Mississippian: American Association of Petroleum Geologists Bulletin, v. 53, p. 112–126.

—— 1981, The eastern margin of the Burlington-Keokuk (Valmeyeran) carbonate bank in Illinois: Illinois Geological Survey Circular 520, 24 p.

Malott, C. A., 1922, The physiography of Indiana, in Logan, W. N., and others, Handbook of Indiana geology: Indiana Department of Conservation Publication 21, p. 59–256.

Nicoll, R. S., and Rexroad, C. B., 1975, Stratigraphy and conodont paleontology of the Sanders Group (Mississippian) in Indiana and adjacent Kentucky: Indiana Geological Survey Bulletin 51, 36 p.

Owen, D. D., 1838, Report of a geological reconnoissance of the State of Indiana, made in the year 1837, in conformity to an order of the legislature: Indianpolis, J. P. Chapman, 37 p.

Pinsak, A. P., 1957, Subsurface stratigraphy of the Salem Limestone and associated formations in Indiana: Indiana Geological Survey Bulletin 11, 62 p.

Sable, E. G., 1979, Eastern Interior Basin region, in Craig, L. C., and Connor, C. W., coordinators, Paleotectonic investigations of the Mississippian System in the United States. Part I. Introduction and regional analyses of the Mississippian System: U.S. Geological Survey Professional Paper 1010, p. E59–E106.

Schneider, A. F., 1966, Physiography, in Natural features of Indiana: Indiana Academy of Science, The Indiana Sesquicentennial Volume, p. 40–56.

Shaver, R. H., and others, 1985, Midwestern arches and basins region: American

Association of Petroleum Geologists Correlation of Stratigraphic Units of North America chart MBA.

Stockdale, P. B., 1931, The Borden (Knobstone) rocks of southern Indiana: Indiana Department of Conservation Publication 98, 330 p.

Sunderman, J. A., 1968, The geology and mineral resources of Washington County, Indiana: Indiana Geological Survey Bulletin 39, 90 p.

Suttner, L. J., and Hattin, D. E., editors, 1973, Field conference on Borden Group and overlying limestone units, south-central Indiana: Bloomington, Indiana University Department of Geology, 113 p.

Swann, D. H., Lineback, J. A., and Frund, Eugene, 1965, The Borden Siltstone (Mississippian) delta in southwestern Illinois: Illinois Geological Survey Circular 386, 20 p.

University of Cincinnati Sedimentation Seminar, 1981, Mississippian Pennsylvanian section on Interstate 75 south of Jellico, Campbell County, Tennessee: Tennessee Division of Geology Report of Investigations 38, 42 p.

Madison, Indiana: Geomorphology, and Paleozoic and Quaternary geology

Stanley M. Totten, Department of Geology, Hanover College, Hanover, Indiana 47243
Helen B. Hay, Department of Geology, Earlham College, Richmond, Indiana 47374

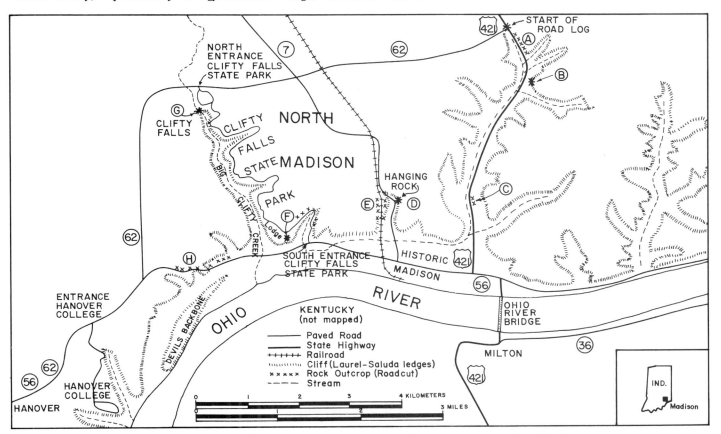

Figure 1. Outline map of Madison and vicinity, Indiana, showing roadlog and physiographic features. Base is the Canaan, Clifty Falls, Madison East, and Madison West 7½-minute Quadrangles.

LOCATION AND SIGNIFICANCE

The Madison area in southeastern Indiana has long been acclaimed for its geomorphic features, Upper Ordovician fossils, and scenic beauty. Madison is located on the Ohio River (Figure 1) almost midway between Cincinnati, Ohio, and Louisville, Kentucky. It is easily accessible by several state highways, and the U.S. 421 bridge over the Ohio River at Madison provides access from Kentucky. Clifty Falls State Park offers outstanding recreational and camping facilities. Madison actually consists of two towns separated from each other by a 300 ft (90 m)-high rocky bluff: so-called Historical Madison located on the terraces and floodplain and the newer North Madison located on the upland.

Upper Ordovician and Silurian strata, well exposed in the Madison area along numerous road cuts and tributaries of the Ohio River, dip gently westward along the western flank of the Cincinnati Arch and the eastern margin of the Illinois Basin. The best and most easily accessible outcrops are along U.S. 421 just north of Madison where the highway rises from the Ohio River valley to the uplands.

The fossiliferous Ordovician outcrops at Madison are among the classic Cincinnatian fossil-collecting localities of southeastern Indiana, southwestern Ohio, and northern Kentucky. In addition, the distinctive dolomitic and sparsely fossiliferous Saluda Formation near the top of the Ordovician is well exposed and reaches its maximum thickness in the Madison area.

A roadlog at the end of this article and keyed to Figure 1 describes access to the many points of geologic interest in the Madison area. Important vantage points are shown by letters in Figure 1.

GEOMORPHOLOGY AND QUATERNARY GEOLOGY

The Madison area lies along the eastern edge of the Muscatatuck Regional Slope, a subprovince of the glaciated Central Lowlands physiographic province. The Ohio River is considered as the southern boundary of this province, though at least one glacier advanced into Kentucky south of the river. The Muscatatuck Regional Slope is a relatively featureless westward-sloping plain that is developed on relatively resistant westward-dipping carbonate rocks of Silurian and Devonian age. The eastern edge of the slope, which extends northward from Madison, represents a major drainage divide that attains elevations in excess of 900 ft (274 m). Prior to Pleistocene glaciation, there was no Ohio River in this area, and this divide separated the Kentucky River-Teays drainage basin to the east from the ancestral Ohio River headwaters to the west (Malott, 1922; Fowke, 1925; Wayne, 1952). The advance of an early Pleistocene ice sheet (Nebraskan?) into southwestern Ohio and southeastern Indiana ponded the northeastward-flowing drainage and ultimately forced a reversal of flow across the divide at Madison into the southwestward-flowing stream, giving birth to the modern Ohio River.

The rapid entrenching of the Ohio Valley during the early Pleistocene left the short tributary streams hanging at their original preglacial level, which was above the resistant Saluda Formation. Consequently, all of the drainage near the Ohio on the Indiana side in the Madison area flows over an estimated 80 waterfalls, the most famous of which is Clifty Falls. Clifty Falls, which has a vertical drop of nearly 70 ft (21 m), has retreated 2.04 mi (3.4 km) from the bluffs of the Ohio at a rate estimated at 0.24 in (6 mm) per year.

The Ohio River is considered to be the southern margin of Illinoian glaciation in the Madison area, as evidenced by the presence of very thin Illinoian till on the bluff edge and the absence of Illinoian till in Kentucky. Illinoian till, typically less than 10 ft (3 m) thick, attains a thickness of 16.5 ft (5 m) in North Madison and is leached 10 to 11.5 ft (3 to 3.5 m). The till is overlain by loess 2 to 3 ft (0.6 to 0.9 m) thick.

A thin, considerably older (Kansan?) and much weathered till underlies the Illinoian till in many places. Kansan(?) till occurs on the uplands in Kentucky for about 15.6 mi (24 km) south of the Ohio River at Madison, and significant patches of Kansan(?) till occur within the Ohio gorge and its tributaries. A brick-red regolith of probable Tertiary age and consisting of large, angular chert blocks, clay, and pisolitic goethite occurs in many places on the upland beneath Kansan(?) till. The Kansan(?) till is virtually indistinguishable from the underlying regolith.

The Ohio River floodplain is narrow, and three terraces, at elevations of 495, 470, and 460 ft (150, 143, and 140 m) are preserved in the valley bottom. The 495-ft (150-m) terrace, on which much of Historic Madison is built, is the most extensive and is considered to be constructional in origin. The terraces are remnants of Pleistocene valley trains and are composed of gravel

of probable Wisconsinan age. The lower terraces were cut into gravels of the 495-ft (150-m) terrace most likely during the late Wisconsinan. The terrace gravels attain a maximum thickness of 99 to 148.5 ft (30 to 45 m) and provide water for Madison and Hanover. An overburden of silty alluvium, as much as 10 ft (3 m) thick, overlies the terrace gravels.

ORDOVICIAN ROCK—STRATIGRAPHIC CLASSIFICATION AND LITHOFACIES

Cincinnatian rock-stratigraphic classification for southeastern Indiana and southwestern Ohio has undergone substantial revision during the past 30 years. Figure 2 outlines these changes and gives the nomenclature in use in Kentucky. The formations and members used by Caster and others (1955) were, at least in part, faunal divisions. Workers in southwestern Ohio (Weiss and Sweet, 1964; Ford, 1967) defined on lithologic criteria four formations in the lower part of the Cincinnatian: the Kope and Fairview Formations, Miamitown Shale, and Bellevue Limestone. Brown and Lineback (1966) and Gray (1972a) grouped all strata between the Kope and the Whitewater Formation or Saluda Formation into the Dillsboro Formation. Hay (1981) and Hay and others (1981) extended the Fairview, Miamitown, and Bellevue Formation names throughout the outcrop area of Indiana and Ohio, and named a new unit, the Brookville Formation, for strata between the top of the Bellevue and the base of the Saluda or Whitewater Formation. The Brookville Formation, equivalent to the upper part of the Dillsboro, is divided into five members, three of which occur in the Madison area.

All of the Cincinnatian Series of Ohio and Indiana, with the exception of the Saluda, consists of interbedded thin- to medium-bedded fossiliferous limestones and thin to thick calcareous shales. The limestones include a variety of types (Martin, 1975), but the most common is biomicrosparrudite. Lithofacies are distinguished by percentages of shale, bedding characteristics, textures and structures of the limestones, and character of the shales, including the presence or absence of conspicuous lenses and/or nodules of limestone in the shales (referred to as limy shales).

Six lithofacies types occur in the Madison road cuts: 1a, 2a, 2b, 3a, 3b, and 3c (Fig. 3). The numbers 1, 2, and 3 of the first part of the facies code refer to shale percentage: 1, greater than 70 percent of shale; 2, 55 to 70 percent; and 3, less than 55 percent. The second part of the facies code refers to other stratigraphic and lithologic aspects: the letter "a" indicates rather even-bedded limestones and shales without limestone lenses and/or nodules; "b," well-bedded limestones interbedded with limy shales, although some of the limestone beds of these facies may be argillaceous and rubbly; and "c," poorly bedded, rubbly argillaceous limestones and very limy shales such that a given bed may be called either argillaceous limestone or limy shale.

ORDOVICIAN STRATIGRAPHY AND PALEONTOLOGY OF MADISON U.S. 421 ROAD CUTS

Cincinnatian strata are exposed in each of the three road

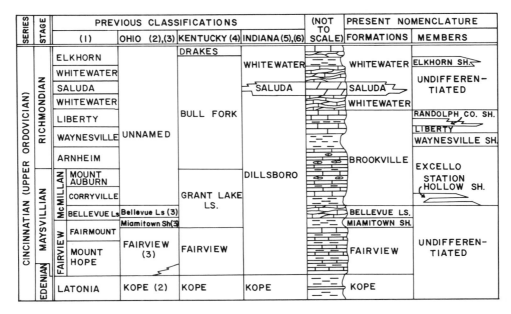

Figure 2. Chart showing earlier rock-stratigraphic classifications and the classification used here: 1, Caster, Dalvé, and Pope, 1955; 2, Weiss and Sweet, 1964; 3, Ford, 1967; 4, Peck, 1966; 5, Brown and Lineback, 1966; and 6, Gray, 1972a. After Hay (1981) and Hay and others (1981).

cuts along U.S. 421 north of Madison (Fig. 1A, B, C; Fig. 3). The southernmost section, 0.5 mi (0.8 km) north of Madison, includes the upper 7.6 ft (2.3 m) of the Bellevue Limestone and about 99 ft (30 m) of the Excello Member of the Brookville Formation. Toward the north is a small road cut which exposes 16 ft (4.9 m) of the Excello Member. This section overlaps the top of the south road cut and is close to the top of the Excello Member. The Waynesville Shale Member lies in a covered stratigraphic interval of about 60 ft (18.3 m) between the top of the middle road cut and the base of the long road cut just to the north. The third and northernmost road cut exposes the Liberty Member of the Brookville Formation, the Saluda and Whitewater Formations, and the Silurian formations.

The Cincinnatian fauna includes abundant and diverse representatives of many phyla. The major rock formers are bryozoans, brachiopods, echinoderms, and mollusks. Trilobites are common, and corals occur in the upper part of the section. Ostracods are abundant in parts of the Saluda. An excellent and inexpensive reference for identification of these and many other Ordovician fossils is available from the Cincinnati Museum of Natural History (Davis, 1981).

Bellevue Limestone

The Bellevue Limestone is approximately 40 ft (12 m) thick in the Madison area, but only the top 7.6 ft (2.3 m) are exposed in the south road cut, where it is assigned to rubbly facies 3c (Fig. 3). Notable fossils include the brachiopods *Platystrophia ponderosa, P. laticosta, Hebertella sinuata, Plectorthis,* and *Rafinesquina,* which occurs throughout the Cincinnatian. Massive and encrusting bryozoans and *Hallopora,* a dendritic genus, are major faunal components.

The Bellevue was probably deposited in very shallow, slightly agitated normal-marine water. The fine-grained terrigenous clastics were not winnowed out, and the environment, therefore, was not likely to have been one of extremely high energy.

Excello Member of the Brookville Formation

The Excello Member constitutes the remainder of the south road cut and the middle road cut. As Figure 3 indicates, the Excello is composed of intervals of facies 1a, 2a, and 3a. Most of the outcrop by far is classified as facies 2a but could be more finely subdivided into smaller intervals of 3a alternating with 1a. Compared with the Bellevue, the limestones are generally thicker and more evenly bedded, and the shales are fissile or blocky, not limy. The south road cut has four levels separated by three benches. *Platystrophia ponderosa* and *P. laticosta* are found as high as the second level but not higher. Other brachiopods are *Rafinesquina, Zygospira,* and *Plectorthis; Leptaena* occurs in level three. The Excello has abundant and diverse mollusks. *Ctenodonta,* a small clan, is particularly abundant at the top of the middle road cut. Large, coarse-ribbed clams, gastropods including *Cyclonema inflatum,* and cephalopods also occur. Echinoderms, bryozoans, and trilobites occur in the Excello.

The Excello sediments indicate a transgressive phase with sediments deposited in slightly deeper water than the Bellevue, probably below normal wave base but subject to periodic reworking by storm waves and strong currents that accounted for the interbedding of limestones and shales. Martin (1975) and Harris and Martin (1979) believed that the limestones are *in situ* biogenic accretions representing progressive community development after colonization of the muddy substrate.

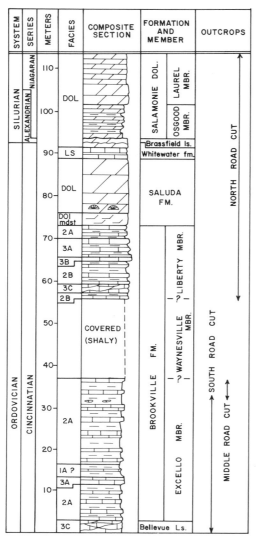

Figure 3. Composite section representing road cuts along U.S. 421 north of Madison, Indiana.

Liberty Member of the Brookville Formation

The lower half of the Liberty Member (Fig. 3), exposed at the base of the north road cut, consists of several facies intervals (2b, 3c, 2b, 3b upward), which vary in shale percentage but share the characteristic of limy shales. Except for a generally higher percentage of shale, these intervals are similar to the facies of the Bellevue and probably represent a similar depositional environment. The upper half of the Liberty Member has more limestone than the lower half and has nonlimy shales. These upper beds were probably deposited in a higher energy environment close to normal wave base. The Liberty-Saluda sequence indicates shoaling of the sea, a regressive phase. Prominent Liberty brachiopods include *Thaerodonta, Plaesiomys, Hebertella, Rafinesquina, Catazyga, Strophomena,* and *Hiscobeccus* (= *Lepidocyclus*). The horn coral *Streptelasma* is present.

Saluda Formation

The Saluda contains a number of facies but is distinguished from the other formations by the preponderance of fine-grained dolomite and dolomitic mudstones, having lesser amounts of dolomitic limestone in the upper and lower parts. The base consists of 7.5 ft (2.3 m) of slabby dolomitic mudstone that appears quite massive in fresh cuts. This is overlain by interbedded limestone, dolomite, and mudstone containing the large colonial corals *Favistella* and *Tetradium*. The thickest unit of the Saluda is mud-cracked and ripple-marked argillaceous and silty dolomite that exhibits color mottling in its lower part. This laminated, rather homogeneous rock appears massive in the fresh road cut. The total thickness of the Saluda is 51 ft (15.5 m).

The main body of the Saluda is nearly barren of fossils, but some beds and bedding planes bear a normal-marine fauna that perhaps was washed into the area or that may indicate times of normal-marine salinity. Ostracods and burrows are the dominant fossils in the lower and upper dolomitic and calcareous zones. Hatfield (1968) interpreted the paleogeographic setting of the Saluda as a hypersaline lagoon rimmed by colonial coral reefs. The mud-cracked sediment was intermittently exposed to the atmosphere. Regionally, it is a lens-shaped body of rock that pinches out northward between Brookville and Richmond, Indiana (Hay, 1981).

Whitewater Formation

The Whitewater Formation is 12.5 ft (3.8 m) thick at Madison if, as it is here, the base is taken to be at the change from Saluda dolomite to dolomitic limestone. If the *Lophospira hammeli* bed is taken as the base of the Whitewater (Foerste, 1903), the thickness of the Saluda is 59.4 ft (18 m) and the Whitewater is 4.6 ft (1.4 m) thick at Madison. The Whitewater lithology farther north in Indiana is typically that of rubbly facies 3c, but at the road cut at Madison it consists, in descending order, of the following six units (modified from Conklin, 1977) that do not fit the facies classification used for other Ordovician strata: (1) fine-grained mottled dolomite (1.8 ft [0.56 m]), (2) black shale (2.56 m [6.4 cm]), (3) fossiliferous micritic limestone with the gastropod *Lophospira hammeli* (2.9 ft [0.87 m]), (4) black shales (0.2 in [0.4 cm]), (5) silty dolomitic biomicrosparite with ostracods (9 in [22.9 cm]), and (6) ostracod-bearing-coarse-grained, greenish-gray burrowed dolomitic limestone 8 ft [2.44 m]) placed by some authors at the top of the Saluda rather than in the Whitewater.

The Whitewater represents a return to normal-marine conditions, although the origin of the fine-grained dolomite at the top is uncertain.

SILURIAN STRATIGRAPHY

The Silurian formations exposed along U.S. 421 are the Brassfield Limestone and the Osgood and Laurel Members of the Salamonie Dolomite (Fig. 3). Useful references include Pinsak and Shaver (1964), Rexroad (1967), Nicoll and Rexroad (1968), and the articles on the Salamonie Dolomite and Brassfield Limestone in Shaver and others (1970).

Brassfield Limestone

At Madison, the Brassfield is 2.3 feet (0.70 m) thick and

overlies disconformably the nearly barren, fine-grained dolomite at the top of the Whitewater Formation (Ordovician). These two units, however, make a rather massive, resistant 5-ft (1.5 m) bed exhibiting the sharp systemic contact near its middle. The Brassfield is coarsely crystalline, grayish-orange dolomite bearing poikilitic-appearing calcite as void fillings. It contains clasts and reworked fossils derived from the Whitewater below.

Salamonie Dolomite—Osgood Member

Both the Osgood (26 feet [7.8 m]) and the Laurel (40 ft [12.1 m]) Members are well exposed at Madison. The Osgood Member consists of fine-grained, very sparsely fossiliferous dolomite interbedded with shale. The shales range in thickness from thin seams to a bed 3 feet (0.9 m) thick in the upper third of the member. Dolomite occurs in thin-bedded intervals alternating irregularly with beds approximately 1.6 ft (0.5 m) thick.

Salamonie Dolomite—Laurel Member

The Laurel Member consists of massive dolomite in the lower half and medium- to thick-bedded dolomite separated by shale seams in the upper half. There are scattered chert nodules in the upper half and one conspicuous chert bed located about 4 ft (1.3 m) below the top of the outcrop. Fossil molds of a normal-marine fauna attest to the replacement origin of the dolomite.

It seems likely that the Osgood sediments formed in a near-shore transgressive setting, one possibly favorable to the formation of syngenetic dolomite. In contrast, the Laurel sediments were deposited under more stable conditions on a shallow open-marine shelf where the supply of clastics was very low. The Laurel was dolomitized some time after deposition of the calcareous fossiliferous sediments.

ROADLOG

The road log (Fig. 1) begins at the intersection of U.S. 421 and Indiana 62, SE¼Sec.14,T.4N.,R.10E., Canaan 7½-minute Quadrangle, near the northeast corner of North Madison, about 4 mi (6 km) north of the Ohio River. Three quadrangles are traversed on this field trip: Canaan, Clifty Falls, and Madison West. The area also is covered by the geologic map of the Louisville 1° × 2° Quadrangle (Gray, 1972b). Important vantage points are shown by letters in the roadlog and in Figure 1.

Miles (Kilometers)

0 (0) Intersection of U.S. 421 and Indiana 62. Proceed south on U.S. 421 toward Madison. Road begins descent into the Ohio Valley following the valley of Crooked Creek.

0.2 (0.3) Upper cut begins (point A, Fig. 1). Exposures of the Laurel Member and the underlying Osgood Member of the Salamonie Dolomite.

0.4 (0.6) Cut interrupted by small tributary valley on left. Note small waterfall. Larger waterfall is visible on right, looking back (upstream) along Crooked Creek. The Ordovician-Silurian contact occurs in the middle of the ledge below the shale.

0.5 (0.8) Laminated beds of Saluda dolomite are well exposed on both sides of road. Large heads of the coral *Favistella* occur below the massive dolomite ledge.

0.6 (1.0) Fossiliferous interbedded limestone and shale of the Liberty Member of the Brookville Formation.

1.2 (1.9) Small exposure of the fossiliferous Excello Member of the Brookville Formation on left (B).

2.0 (4.7) Lower cut begins (C). Fossiliferous Bellevue rocks at base overlain by fossiliferous Excello beds of the Brookville Formation.

3.6 (5.8) High terrace. Most of downtown Madison is built on the highest of three terraces that occur in the Ohio Valley.

3.9 (6.3) Intersection of U.S. 421 and Indiana 56. The Ohio River is visible straight ahead. Turn right (west) on Indiana 56 and proceed west along Main Street. Indiana 56 to left (east) follows the course of the Ohio River upstream. Numerous scars and hummocky topography of debris slides and flows are visible on the steep valley sides adjacent to the highway east of Madison.

4.7 (7.6) Intersection of Indiana 56 and Indiana 7. Proceed straight ahead on Indiana 56. Indiana 7 to the right traverses the steep valley wall known as Hanging Rock Hill. A ledge of Saluda dolomite is called "Hanging Rock" (D), which juts out over the road at a sharp curve.

4.9 (7.9) Railroad bridge. The railroad leading up the steep slope to the right is the steepest noncog railroad grade in the world. Good fossil collecting is available for those who walk up the tracks to the cuts (E) seen in the distance.

5.9 (9.5) South entrance to Clifty Falls State Park. Turn right, go past entrance gate (fee collected during summer season), and proceed up steep hill to the top of the valley wall. Good exposures of Brookville, Saluda, and Laurel along road.

6.7 (10.8) Clifty Falls Lodge on left (F). Turn left into parking lot, park, and walk to the bluff edge just beyond the lodge. Panoramic view of Ohio River valley: Historic Madison to the left looking upvalley, Clifty Creek Power Plant and settling basins to the right, and Devils Backbone in the distance to the right beyond and adjacent to the settling basins. The Devils Backbone is a bedrock ridge that probably originated when an ice advance blocked the earlier channel to the north of the backbone and diverted the Ohio River a short distance south to its present valley. The Ohio River bends southward at this point. Note the accordant summit levels directly across the valley in Kentucky. The valley is about 1 mi (1.6 km) wide and 363 ft (110 m) deep at this point.

6.8 (11.0) Leave lodge parking lot, turn left, and follow main park road to Clifty Falls.

10.1 (16.3) Lookout Point. Good view of Clifty Falls.

10.6 (17.1) Turn left on road to Clifty Falls parking lot.

10.8 (17.4) Parking on the right. Park and walk across road to trail leading to Clifty Falls (G). The upper part of the trail goes past weathered outcrops of Laurel dolomite and Osgood shale. A series of wooden steps leads to the bottom of Clifty Falls. The falls' cap rock is the ledge of Saluda dolomite, and the vertical distance of fall is about 69 ft (21 m). Total walking time to the falls and back is about 30 minutes. Several scenic hiking trails are maintained, including one along the bottom of the gorge.

10.8 (17.4) Clifty Falls parking lot. Road to right from parking lot leads to north entrance and Indiana 62. Turn left from parking lot and retrace route to south entrance.

15.6 (25.2) South Entrance. Turn right (west) on Indiana 56.

16.4 (26.5) Beginning of a series of road cuts in the Brookville, Saluda, Whitewater, Brassfield, and Salamonie Formations along Indiana 56 (H). Parking ahead on left at road leading to the Devils Backbone. Some of the best fossil collecting in the Madison area occurs in the Brookville Formation exposed in these cuts.

17.2 (27.7) Massive laminated Saluda dolomite.

17.4 (28.1) Ordovician-Silurian unconformity just below the two driveways on the right. Stromatoporids of the Whitewater Formation have been truncated and are disconformably overlain by the Brassfield Limestone. Parking available on the left.

17.8 (28.7) Blinker light; intersection with Indiana 62. To return to starting point of field trip, turn right on Indiana 62 and proceed 4.8 miles (7.7 km) to North Madison.

REFERENCES CITED

Brown, G. D., Jr., and Lineback, J. A., 1966, Lithostratigraphy of Cincinnatian Series (Upper Ordovician) in southeastern Indiana: American Association of Petroleum Geologists Bulletin, v. 50, p. 1018–1032.

Caster, K. E., Dalvé, E. A., and Pope, J. K., 1955, Elementary guide to the fossils and strata of the Ordovician in the vicinity of Cincinnati, Ohio: Cincinnati Museum of Natural History, 47 p.

Conklin, B. J., 1977, Stratigraphy and petrography of the Madison, Indiana U.S. 421 road cut, Upper Ordovician through Middle Silurian [bachelor's thesis]: Richmond, Indiana, Earlham College, 75 p.

Davis, R. A., 1981, Cincinnati fossils: Cincinnati Museum of Natural History, 58 p.

Foerste, A. F., 1903, The Richmond Group along the western side of the Cincinnati Anticline in Indiana and Kentucky: American Geologist, v. 31, p. 333–361.

Ford, J. P., 1967, Cincinnatian geology in southwest Hamilton County, Ohio: American Association of Petroleum Geologists Bulletin, v. 51, p. 918–936.

Fowke, Gerard, 1925, The genesis of the Ohio River: Indiana Academy Science Proceedings, v. 34, p. 81–102.

Gray, H. H., 1972a, Lithostratigraphy of the Maquoketa Group (Ordovician) in Indiana: Indiana Geological Survey Special Report 7, 61 p.

——1972b, Geologic map of the 1° × 2° Louisville Quadrangle, Indiana, showing bedrock and unconsolidated deposits: Indiana Geological Survey Regional Geologic Map 6, scale 1:250,000.

Harris, F. W., and Martin, W. D., 1979, Benthic community development in limestone beds of the Waynesville (upper Dillsboro) Formation (Cincinnatian Series, Upper Ordovician) of southeastern Indiana: Journal of Sedimentary Petrology, v. 49, p. 1295–1306.

Hatfield, C. B., 1968, Stratigraphy and paleoecology of the Saluda Formation (Cincinnatian) in Indiana, Ohio, and Kentucky: Geological Society of America Special Paper 95, 33 p.

Hay, H. B., 1981, Lithofacies and formations of the Cincinnatian Series (Upper Ordovician), southeastern Indiana and southwestern Ohio [Ph.D. thesis]: Oxford, Ohio, Miami University, 236 p.

Hay, H. B., Pope, J. K., and Frey, R. C., 1981, Lithostratigraphy, cyclic sedimentation, and paleoecology of the Cincinnatian Series in southwestern Ohio and southeastern Indiana, in Roberts, T. G., editor, GSA Cincinnati '81 field trip guidebooks: Washington, D.C., American Geological Institute, v. 1, Stratigraphy, sedimentology, p. 73–86.

Malott, C. A., 1922, The physiography of Indiana, in Handbook of Indiana geology: Indiana Department of Conservation Publication 21, p. 59–256.

Martin, W. D., 1975, The petrology of a composite vertical section of Cincinnatian Series limestones (Upper Ordovician) of southwestern Ohio, southeastern Indiana, and northern Kentucky: Journal of Sedimentary Petrology, v. 45, p. 907–925.

Nicoll, R. S., and Rexroad, C. B., 1968, Stratigraphy and conodont paleontology of the Salamonie Dolomite and Lee Creek Member of the Brassfield Limestone (Silurian) in southeastern Indiana and adjacent Kentucky: Indiana Geological Survey Bulletin 40, 73 p.

Peck, J. H., 1966, Upper Ordovician formations in the Maysville area, Kentucky: U.S. Geological Survey Bulletin 1244-B, p. B1-B30.

Pinsak, A. P., and Shaver, R. H., 1964, The Silurian formations of northern Indiana: Indiana Geological Survey Bulletin 32, 87 p.

Rexroad, C. B., 1967, Stratigraphy and conodont paleontology of the Brassfield (Silurian) in the Cincinnati Arch area: Indiana Geological Survey Bulletin 36, 64 p.

Shaver, R. H., and others, 1970, Compendium of rock-unit stratigraphy in Indiana: Indiana Geological Survey Bulletin 43, 229 p.

Wayne, W. J., 1952, Pleistocene evolution of the Ohio and Wabash Valleys [Indiana]: Journal of Geology, v. 60, p. 575–585.

Weiss, M. P., and Sweet, W. C., 1964, Kope Formation (Upper Ordovician): Ohio and Kentucky: Science, v. 145, p. 1296–1302.

Sub-Pennsylvanian disconformity near Shoals, Indiana

Henry H. Gray, Indiana Geological Survey, 611 North Walnut Grove, Bloomington, Indiana 47405

LOCATION AND SIGNIFICANCE

This site is along a spur railroad cut in the W½NE¼Sec.28, T.3N.,R.3W., Martin County, Indiana (Fig. 1). It is shown on the Huron 7½-minute Quadrangle and is accessible by foot from a county road that crosses the railroad just south of the cut. The cut is on the property of the National Gypsum Company, Shoals Plant, and permission to visit should be requested, particularly for groups.

This is the best and most revealing exposure of the sub-Pennsylvanian disconformity in southern Indiana and is one of the best in the entire Illinois Basin. A small pre-Pennsylvanian valley with about 50 ft (15 m) relief is well shown, and many features of the rocks above and below the contact are also illustrated. This is an excellent introduction to the stratigraphy of much of southwestern Indiana.

GENERAL INFORMATION

Throughout most of the Illinois Basin, rocks of the Pennsylvanian System are separated from those beneath by a prominent regional unconformity, a surface that represents uplift and erosion of the underlying rocks before deposition of the overlying sediments. Northward across the basin, this unconformity truncates the entire Mississippian System, and in parts of north-central Illinois, where minor folding of the pre-Pennsylvanian strata occurred, basal Pennsylvanian rocks lie on Devonian, Silurian, and, in limited areas, on Ordovician rocks as old as the St. Peter Sandstone.

In Indiana this surface is a disconformity because strata above and below the surface are essentially parallel. The main outcrop trace of this disconformity extends from the Ohio River near Cannelton, where Pennsylvanian rocks rest on uppermost Chesterian strata, north-northwestward to northern Warren County, where Pennsylvanian rocks overlie rocks of the Borden Group (Figs. 2, 3). In outliers well north of the main outcrop, Pennsylvanian rocks rest on the New Albany Shale. Though this transection is impressive, it represents a loss of only 1,800 ft (550 m) of strata in a lateral distance of 190 mi (300 km), or less than 6 ft (2 m) per km, and thus it is substantially unnoticeable except from a regional perspective.

Because these rocks dip only gently westward, at the rate of about 45 ft/mi (7 m/km), the trace of the disconformity is much dissected by the valley pattern and there are many outliers; and although the difference in dip between Pennsylvanian and Mississippian strata is slight, the northward transection of Mississippian rocks demonstrates that the strike of the older rocks is somewhat more to the northwest than is the strike of the Pennsylvanian strata (Fig. 3). This too is only regionally perceptible.

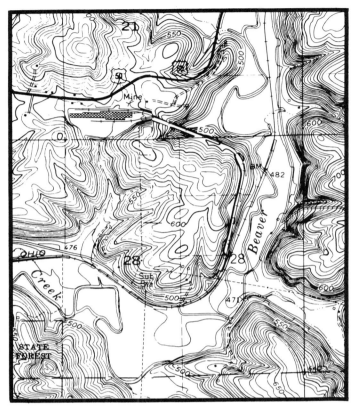

Figure 1. Map showing location of railroad cut at National Gypsum Company plant near Shoals, Indiana (bracketed); Shoals is about 2 mi (3 km) west on U.S. 50; Bedford is about 20 mi (35 km) east. From Shoals and Huron 7½-minute Quadrangles. For location in Indiana, see Figure 3.

The surface over which Pennsylvanian sediments transgressed may be visualized as an eroded plain, one possibly similar to the inner margin of the present coastal plain. Local relief on this surface in most areas is 100 to 150 ft (30 to 50 m), but in some places major valleys more than 350 ft (100 m) deep are known. Resistant strata among the older rocks form subtle northeastward-facing cuestas.

In most places where it is well exposed, the sub-Pennsylvanian disconformity is a prominent feature. It is not always easy, however, to distinguish Pennsylvanian from Mississippian rocks, and local disconformities occur within both systems that mimic the major one. The best and most revealing exposure of the systemic disconformity in Indiana is in this cut, which exposes the contact for a linear distance of a few hundred meters (Fig. 4).

In the area surrounding this cut, the basal Pennsylvanian Mansfield Formation commonly rests on the Haney Limestone (H, Fig. 2) or on rocks just above or below the Haney position.

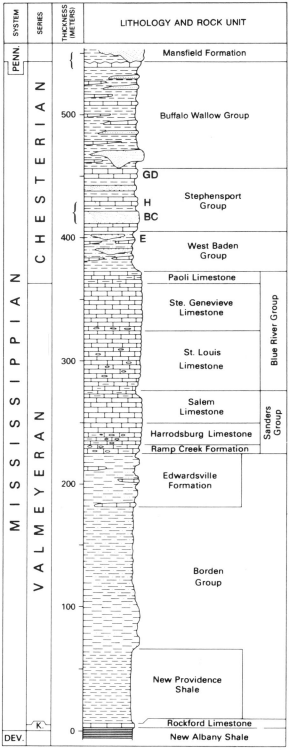

Figure 2. Columnar section showing exposed Mississippian rocks in Indiana. K. = Kinderhookian; other letters identify formations referred to in text. Sub-Pennsylvanian disconformity transects this entire section from south to north along the Indiana outcrop. Brackets to left of column identify range of stratigraphic units seen at National Gypsum Company cut. From Gray (1979).

Local relief on the disconformity, however, carries the contact as high as the Glen Dean Limestone (GD) and as low as the Elwren Formation (E, Fig. 2). Thus, about 350 ft (100 m) of Mississippian strata that are represented farther south in Indiana are missing here.

Exposures in the cut show a small valley eroded through the Big Clifty Formation (BC, Fig. 2) and filled with strikingly different sandstone of the Mansfield Formation. The three sections below and the diagram (Fig. 4) were prepared in 1957, when the cut was new. Some features seen then are no longer as clear as they once were.

Section A: 400 ft (125 m) from south end of cut.
PENNSYLVANIAN SYSTEM
Mansfield Formation
4. Sandstone, yellow-brown, interstratified with minor amounts of gray shale; fucoidal markings common; basal contact transitional .... 13+ ft (4.0+ m)
3. Sandstone and shale interstratified, gray and yellow-brown; wavy to uneven stratification; fucoidal markings common; basal contact sharp and uneven with about 5 ft (1.5 m) relief .... 2.0–7.7 ft (0.6–2.4 m)
2. Sandstone, yellow-brown; indistinct contorted stratification; lenticular; basal contact transitional but showing about 5 ft (1.5 m) relief .... 0–6.9 ft (0–2.1 m)
MISSISSIPPIAN SYSTEM
Big Clifty Formation
1. Sandstone, yellow-brown; medium horizontal and cross stratification; base of exposure slightly below track level .... 6+ ft (1.8+ m)

Section B: 1,000 ft (300 m) from south end of cut.
PENNSYLVANIAN SYSTEM
Mansfield Formation
4. Sandstone, yellow-brown; uneven medium stratification; locally cross stratified and contorted; basal contact transitional .... 42+ ft (12.8+ m)
3. Ironstone, sandy, red-brown; indistinct cross and contorted stratification; abundant ironstone and shale pebbles in a quartz sand matrix with pervasive red and yellow-brown iron oxide cement; lenticular; basal contact sharp and disconformable with about 40 ft (12 m) relief .... 0–10 ft (0–3.0 m)
MISSISSIPPIAN SYSTEM
Big Clifty Formation
2. Siltstone, mudstone, and marlstone, gray and olive-gray; basal contact sharp and undulatory with relief of about 3 ft (1 m) and wavelength of about 33 ft (10 m) .... 0–3 ft (0–0.9 m)
1. Sandstone, yellow-brown; medium horizontal and cross stratification; upper 3 ft (1 m) locally carbonate cemented; base of exposure slightly below track level .... 20+ ft (6.1+ m)

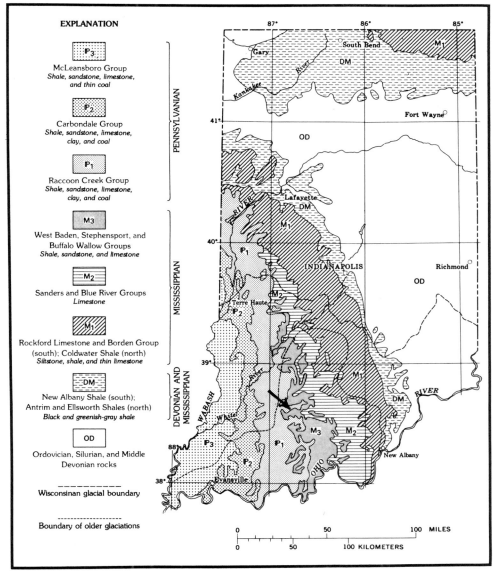

Figure 3. Map of Indiana showing distribution of Mississippian and Pennsylvanian rocks. Arrow shows approximate location of National Gypsum Company cut. From Gray (1979).

Section C: 1600 ft (500 m) from south end of cut.

MISSISSIPPIAN SYSTEM

Haney Limestone

4. Limestone, biomicrite, yellow-brown, cross stratified; abundant fossils and fossil fragments .... 2.6+ ft (0.8+ m)

Big Clifty Formation

3. Shale and mudstone, olive-gray and red-brown in lower third, gray in upper two thirds; bryozoans, brachiopods, and conulariids abundant in upper part .... 17 ft (5.2 m)

2. Siltstone, mudstone, and marlstone, as unit 2, section B .... 1.6–5 ft (0.5–1.5 m)

1. Sandstone, as unit 1, section B; base of exposure 5.6 ft (1.7 m) below track level .... 13+ ft (4.0+ m)

Many features of this exposure call for some discussion, although all questions cannot be definitively answered. Fucoids in units 3 and 4 of section A possibly suggest a marine origin, yet these units grade laterally into the very different sandstone of unit 4, section B, which conventionally is interpreted as a fluvial channel deposit. Unit 2, section A is transitional into the unit below, but its upper contact is sharp. It is assigned to the Mansfield Formation on the premise that it is a residuum derived from the underlying sandstone and is likely to be Pennsylvanian in age.

Unit 3, section B resembles bog iron ore in appearance, but its distribution suggests, instead, some postdepositional effect controlled by the disconformity. Perhaps circulating groundwater precipitated the iron in response to scattered plant remains in the basal part of the sandstone shortly after deposition of the sandstone. A hundred years ago, similar iron ore was mined about 5

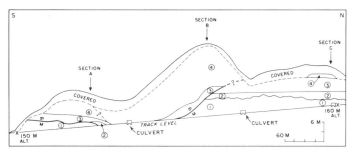

Figure 4. Profile of west side of railroad cut at National Gypsum Company plant near Shoals, Indiana. See Figure 1 for location. Approximately to scale; vertical exaggeration 5 times. Letters and numbers refer to measured sections and units described in text. From Gray and others (1957).

mi (8 km) north of this point and was smelted in a small furnace near Shoals. Although the ore was low grade, it was blended with ore from Missouri, and the iron produced was said to have been of excellent quality.

One of the important characteristics useful for distinguishing Mississippian from Pennsylvanian sandstones in this area is displayed here, namely, the relative uniformity of the Mississippian (in this case, Big Clifty) sandstone as compared with the greater variability of the commonly somewhat coarser, less mature sandstone of the Mansfield. This is not to say, however, that there are no sandstones of similar appearance; many Mississippian sandstones resemble those of the Mansfield, probably because they too were deposited in some kind of limited deltaic environment.

Sandstone of the Big Clifty Formation (unit 1, all sections), however, probably represents some kind of longshore beach and bar complex. For about 90 mi (150 km) along its outcrop in Indiana, it exhibits remarkable uniformity in character and thickness, but its continuity is abruptly lost downdip. In the subsurface of southwestern Indiana, where this formation is a minor producer of oil, sandstone in this stratigraphic position occurs mainly in discontinuous lenses.

The undulatory upper surface of the Big Clifty sandstone (unit 1, sections B and C) resembles megaripples. This feature has not been observed elsewhere, mostly for lack of good exposures, and indeed it is not as well exposed here as it was when this cut was new. Internal stratification does not support the megaripple hypothesis, and the origin of the undulations remains obscure. The overlying siltstone, mudstone, and marlstone appear to be lagoonal and are transitional upward into fossiliferous shale that contains problematic conulariids in association with a fully marine fauna; this is overlain by the open-water Haney Limestone.

Section C thus exposes a nearly complete sedimentary cycle of the type that characterizes lower and middle Chesterian strata in southwestern Indiana. Upper Chesterian cycles are subtly different, but by Mansfield time, the locus of cyclic sedimentation had shifted to a deltaic environment from which marine deposits were almost excluded.

This exposure and others nearby are described in Gray and others (1957) and in Rexroad, Gray, and Noland (1983). Basic geology of the surrounding area is outlined in Gray, Jenkins, and Weidman (1960). Other relevant references, many having a regional context, are listed below.

REFERENCES CITED

Atherton, Elwood, and others, 1960, Differentiation of Caseyville (Pennsylvanian) and Chester (Mississippian) sediments in the Illinois Basin: Illinois Geological Survey Circular 306, 36 p.

Bristol, H. M., and Howard, R. H., 1971, Paleogeologic map of the sub-Pennsylvanian Chesterian (Upper Mississippian) surface in the Illinois Basin: Illinois Geological Survey Circular 458, 14 p.

Gray, H. H., 1962, Outcrop features of the Mansfield Formation in southwestern Indiana: Indiana Geological Survey Report of Progress 26, 40 p.

——1979, Mississippian and Pennsylvanian (Carboniferous) Systems in the United States—Indiana: U.S. Geological Survey Professional Paper 1110-K, 20 p.

——Jenkins, R. D., and Weidman, R. M., 1960, Geology of the Huron area, south-central Indiana: Indiana Geological Survey Bulletin 20, 78 p.

Gray, H. H., and others, 1957, Rocks associated with the Mississippian-Pennsylvanian unconformity in southwestern Indiana: Indiana Geological Survey Field Conference Guidebook 9, 42 p.

Malott, C. A., 1931, Geologic structure in the Indian and Trinity Springs locality, Martin County, Indiana: Indiana Academy of Science Proceedings, v. 40, p. 217–231.

——1946, The Buddha outlier of the Mansfield Sandstone, Lawrence County, Indiana: Indiana Academy of Science Proceedings, v. 55, p. 96–101.

Rexroad, C. B., Gray, H. H., and Noland, A. V., 1983, The Paleozoic systemic boundaries of the southern Indiana–adjacent Kentucky area and their relations to depositional and erosional patterns (Field Trip 1), in Shaver, R. H., and Sunderman, J. A., editors, Field trips in midwestern geology: Bloomington, Indiana, Geological Society of America, Indiana Geological Survey, and Indiana University Department of Geology Guidebook, v. 1, p. 1–35.

Shaver, R. H., and others, 1986, Compendium of Paleozoic rock-unit stratigraphy in Indiana; A revision: Indiana Geological Survey Bulletin 59, 203 p.

Siever, Raymond, 1951, The Mississippian-Pennsylvanian unconformity in southern Illinois: American Association of Petroleum Geologists Bulletin, v. 35, p. 542–581.

Swann, D. H., 1964, Late Mississippian rhythmic sediments of Mississippi Valley: American Association of Petroleum Geologists Bulletin, v. 48, p. 637–658.

Vincent, J. W., 1975, Lithofacies and biofacies of the Haney Limestone (Mississippian), Illinois, Indiana, and Kentucky: Kentucky Geological Survey Thesis Series 4, 64 p.

The Orangeville Rise and Lost River, Indiana

Richard L. Powell, Geosciences Research Associates, Inc., Bloomington, Indiana 47405

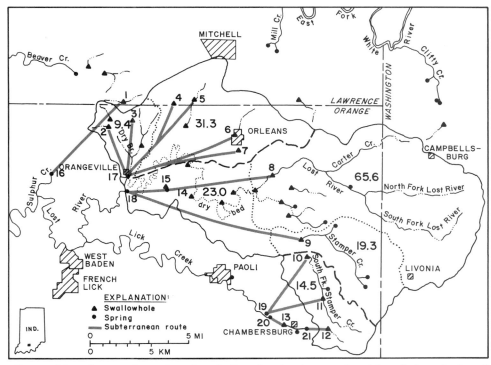

Figure 1. Map of a part of central southern Indiana showing the upper Lost River drainage basin and results of dye traces (gray lines). Numbered localities: 1, Wadsworth Hollow Sink; 2, Showfarm Cave; 3, Black Hollow Cave; 4, Maple Grove Cave; 5, Sink of Mt. Horeb Drain; 6, Flood Creek Sink; 7, Orleans Sewage Plant; 8, Sink of Lost River; 9, Sink of Stampers Creek; 10, Matity Sinks; 11, Wolfe Sink; 12, Clements Sink; 13, Sinks of Lick Creek; 14, Tolliver Swallowhole; 15, Wesley Chapel Gulf; 16, Rise of Sulphur Creek; 17, Orangeville Rise; 18, Rise of Lost River; 19, Spring Mill on Lick Creek; 20, Half Moon Spring; and 21, Dillard Cave Spring. Modified from Murdock and Powell (1968).

LOCATION AND SIGNIFICANCE

The Orangeville Rise is the spring head of Lost River in the town of Orangeville in northwestern Orange County. The site is about 6 mi (9.6 km) northeast of French Lick and 9 mi (14.4 km) southwest of Mitchell. (See location in Fig. 1; also, the inset map in Fig. 5 shows locations of other features mentioned here.) The Orangeville Rise is owned by the Nature Conservancy and is a National Natural Landmark registered by the U.S. National Park Service. The spring is the most accessible of several sites that are commonly visited on field trips and are considered to be focal points that depict significant aspects of subterranean drainage in the Lost River basin. In addition to the Orangeville Rise, the other nearby sites singled out here are the Rise of Lost River, Tolliver Swallow Hole, and Wesley Chapel Gulf.

The Lost River drainage basin heads on the Mitchell Plain, a low limestone plateau developed on the westward-sloping vale of an east-facing cuesta. The dip of the strata is only slightly greater than the general slope of the plain. To the west, the Crawford

Upland is a rugged hilly area separated from the Mitchell Plain by the east-facing Chester Escarpment. Similar physiographic units to the south in Kentucky are the Mammoth Cave Plateau and the Pennyroyal Plateau separated by the Dripping Springs Escarpment. A few major streams flow in entrenched sinuous valleys westward across the Mitchell Plain and the Crawford Upland.

Most of the eastern part of the Mitchell Plain is mantled with clay that is in places as much as 30 to 35 ft (900 to 1050 cm) thick. The western part and areas adjacent to the entrenched drainage are characterized by numerous sinkholes and thinner soils. The two longest caverns in Indiana, each with 20 mi (32 km) of mapped passages, drain parts of the sinkhole plain and discharge into the entrenched streams.

The eastern part of the Crawford Upland, where the limestones that are the surface bedrock on the western part of the Mitchell Plain are exposed in the lower valley walls, is in places characterized by dry and perched karst valleys and sinkhole tracts

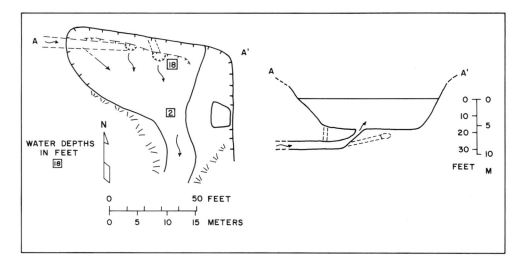

Figure 2. Sketch map and cross section of the Orangeville Rise, Lost River drainage basin, by S. D. Maegerlein (circa 1979). See Figure 1 for location.

on bedrock terraces. Many long caverns underlie the valley walls and ridges of the Crawford Upland. Numerous caverns extend from headwaters in sinkholes along the western margin of the Mitchell Plain to discharge at springs along valleys in the eastern margin of the Crawford Upland. The Lost River drainage system is an example of a large drainage basin within this setting.

The subterranean drainage features of Lost River are known to karst researchers and students around the world owing to the detailed studies by Malott (1922, 1932, 1948, and 1952) as well as to some works by previous and subsequent authors. One of the earliest geological notes on the unusual drainage of the Lost River area was by Owen (1862, p. 154–155), but about 40 articles and guidebooks have subsequently been published. Numerous newspaper articles have also been written about the Lost River area, particularly the Orangeville Rise, indicating to some extent the layman's interest in the area. Scarcely a weekend passes without a visitor to the Orangeville Rise "where a river flows out of the ground."

Most of the landowners in the Lost River watershed are aware of the unusual karst drainage in the area, and many are accustomed to visitors. Nearly all the landowners will grant permission to visit particular features if asked courteously and at a reasonable time. The area is covered by the Georgia, French Lick, Mitchell, and Paoli 7½-minute Quadrangles.

SITE INFORMATION

Orangeville Rise

Orangeville Rise is the second largest spring in Indiana and is the resurgence or upwelling of underground drainage from water-filled cavernous passages. The spring outlet is a pool about 18 ft (5.4 m) deep at the base of a limestone bluff that is about 12 ft (3.6 m) high above normal pool level and 110 ft (33.4

m) long (Fig. 2). The pool shallows about 50 ft (15.2 m) from the bluff to where the spring run flows southward within a steep-walled alluvial channel. A recent low-flow measurement of stream flow was about 9 cfs (.25 m³/sec), and a measurement at

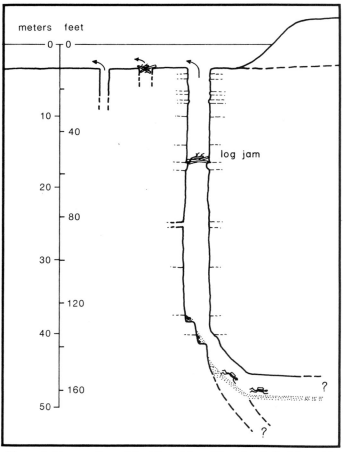

Figure 3. Cross section of the Rise of Lost River, by S. D. Maegerlein (circa 1978). See Figure 1 for location.

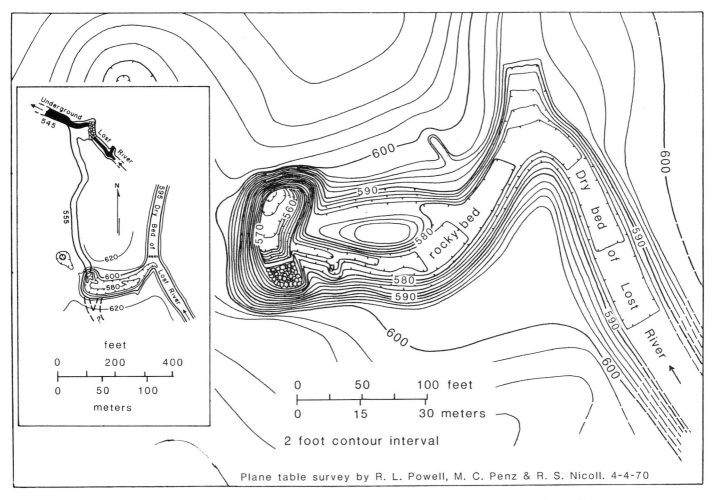

Figure 4. Map of Tolliver Swallowhole, Lost River drainage basin. See Figure 5, inset map, for location. Modified from Malott (1952).

bank-full stage amounted to 185 cfs (5.2 m³/sec) (Bassett, 1974, Fig. 12). Highest flood flow stage is estimated to be about 200 cfs (5.6 m³/sec), at which time the limestone bluff and some of the alluvial lands adjacent to the spring are inundated by the muddy, roiling storm waters. Peak flows are affected by leakage into other storm-water rises and back flooding from the lower end of the dry-bed channel. The water issuing from the Orangeville Rise surges upward from openings beneath a ledge of the Ste. Genevieve Limestone (upper Middle Mississippian). One of these openings has been explored by divers for an estimated distance of about 60 ft (18.2 m) at a measured depth of 25 ft (7.6 m). Other openings are partially blocked to divers by large blocks of rock.

The subterranean drainage system that discharges at the Orangeville Rise heads in the area to the north and northeast of the spring (Childs, 1940) (Fig. 1). Several dye-trace tests and outlines of surface-drainage basins characterized by sinking streams indicate that the drainage basin should encompass about 41 mi² (105 km²) (Bassett, 1976, p. 80; and Murdock and

Powell, 1968, p. 251). Much of the drainage basin lies within the eastern margin of the Crawford Upland. It is characterized, therefore, by sandstone-capped ridges and valleys with numerous large and small sinking streams. Several of the valleys are karsted valleys, that is, they contain numerous sinkholes in the relatively flat valley bottoms. Part of the drainage basin extends east of the Chester Escarpment onto the Mitchell Plain, where sinkholes are more numerous, but the plain is also in part drained by a few large sinking stream catchments. One square mile (2.5 km²) (a section) just southwest of Orleans was mapped and found to contain 1,022 sinkholes (Malott, 1945, p. 12).

Dissolved solids in the water that discharges from the Orangeville Rise at moderate to high flow stages are dominantly $CaCO_3$; most other karst springs in the region share this characteristic (Bassett, 1976). At low flow rates, however, appreciable amounts of Mg and SO_4 and a higher equilibrium CO_2 partial pressure are detected. Waters are closer to saturation with respect to $CaCO_3$ at low-flow stages. Concentrations of the major chemical species are inversely proportional to discharge. The molar

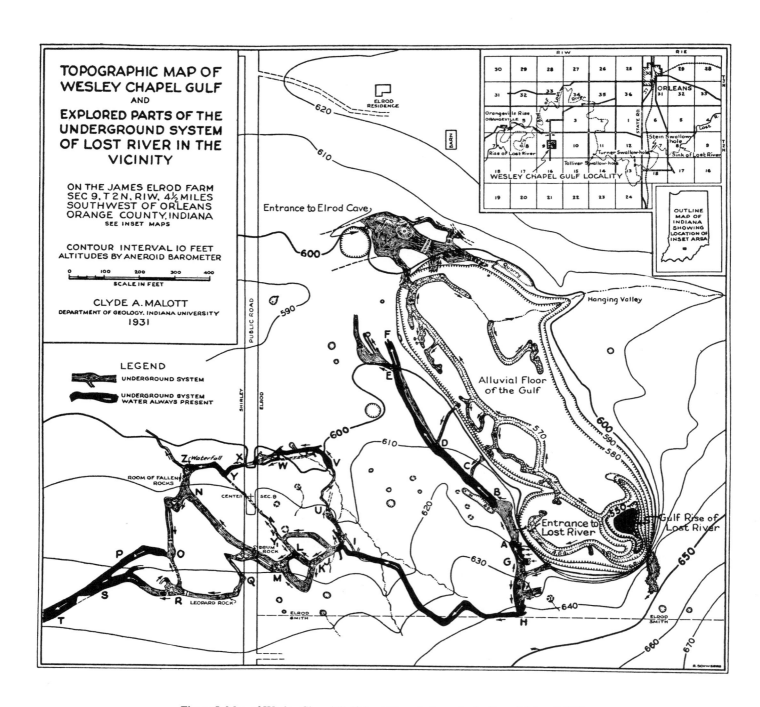

Figure 5. Map of Wesley Chapel Gulf, Lost River drainage basin. From Malott (1932).

ratios, Ca/Mg and HCO_3/SO_4, relate directly to discharge, but the Ca/Mg molar ratio is inversely related to SO_4 concentration. Saturation with respect to calcite in the spring waters is controlled both by discharge and by CO_2 partial pressure. The CO_2 pressures are high in the summer and late fall. Sulfur isotopic studies show a mean $\delta 34S_{(SO_4)}$ of $11.51^0/_{00}$, indicating a shallow-flow input and low residence time of the water (Noel Krothe, 1983, personal communication). The SO_4 component may be derived from seepage from an underlying sulfur-water bearing zone.

Rise of Lost River

The Orangeville Rise is the spring head of Lost River but not the actual resurgence of underground Lost River. The Rise of Lost River is located about a mile (1.6 km) to the south of and downstream from the Orangeville Rise (Fig. 5, inset map). The Rise of Lost River appears to be a simple water-filled channel or cul-de-sac to the east of Lost River that is about 100 ft (30.4 m) long, 30 ft (9 m) wide, and 11 ft (3.3 m) deep. This unimpressive alluvial channel is actually the third largest spring in Indiana, as well as the deepest. There are several small openings in the bottom of the channel, but one, a nearly vertical slot, has allowed divers to descend to a depth of 165 ft (50 m) (Fig. 3). Waters rise from a large unmeasured void at the bottom that may be a horizontal conduit.

Two dye traces have shown that sinking streams at the first principal sinks of Lost River and Stampers Creek are tributary to the Rise of Lost River via cavernous underground routes that are at least 8 to 9 mi (12.9 to 14 km) long (Fig. 1). Thus, surface drainage from about 85 mi^2 (217 km^2) at low-flow to low flood-flow stages resurges at the Rise of Lost River.

A sinuous dry bed extends downstream from the first principal sinks of Lost River (5, Fig. 1) a distance of about 22 mi (35 km) to near the Orangeville Rise. The dry-bed channel is deeply entrenched into bedrock within the sinkhole plain and through the Chester Escarpment. Heavy storm waters flow the entire length of the channel a few times each year. There are several sets of swallowholes along the dry bed that divert various amounts of storm waters into subterranean conduits. Each swallowhole diverts the storm runoff underground until the surface flow becomes greater than the capacity of the swallowhole, at which point the surplus flow continues downstream to the next swallowhole, perhaps to overflow the entire length of the dry bed.

Tolliver Swallowhole

Tolliver Swallowhole (14, Fig. 1), a National Natural Landmark, is one of several major swallowholes along the dry bed of Lost River. Northward-flowing floodwaters in the dry bed initially flow westward into the steep rocky channel leading to the swallowhole and into a mapped portion of underground Lost River (Fig. 4). The swallowhole becomes ponded when storm waters fill the underground conduits and, depending upon the size of the storm, may backflood to a depth of about 38 ft (11 m) into the dry bed so that the storm waters spill over a threshold and flow on downstream in the dry bed. Storm waters that enter Tolliver Swallowhole apparently flow westward to Wesley Chapel Gulf.

Wesley Chapel Gulf

Wesley Chapel Gulf (15, Fig. 1), a National Natural Landmark, is a collapse sinkhole with an area of 8.3 acres (3.35 ha). It is about 1,000 ft (300 m) long, 350 ft (106 m) wide, and ranges from about 25 ft (7.6 m) to 95 ft (29 m) deep (Fig. 5). Malott (1932, p. 287–288) defined a gulf as a sinkhole with steep walls that characteristically has an alluviated floor and that always contains a stream that rises and sinks within it. He suggested that gulfs enlarge from smaller collapse sinkholes or from karst windows overlying a network of cavern passages.

The alluvial floor of the gulf, 6.1 acres (2.5 ha), is gently sloping from southeast to northwest. A rise pool about 14 ft (4.2 m) deep during normal stage, with a surface water level 22 ft (6.7 m) below the alluvial floor of the gulf, occurs in the southeast corner of the gulf. Water from the rise pool flows a few feet (meters) to the south into several small openings during low to normal flow stages but during flood-flow stages roils up and surges into a large overflow channel around the southeast end of the alluvial floor of the gulf to where it reenters the underground system. High-flood flows overtop the rise pit and the deep channel to fill a shallow channel with as many as 100 swallowholes along its west side next to the wall of the gulf. A floodwater depth of about 5 ft (1.5 m) over the entire floor of the gulf has been recorded. The swallowholes in the gulf divert water back into the cavern network west of the gulf. The subterranean waters from Wesley Chapel Gulf resurface at the Rise of Lost River.

The rise pit overlies a water-filled opening about 3 ft (0.9 m) in diameter that slopes downward to where it intersects a larger passage about 160 ft (50 m) from the rise pit and about 45 ft (14 m) below normal pool level. The large passage is 10 ft (3 m) high, 30 ft (9 m) wide, and extends northeastward at least 200 ft (61 m) (S. D. Maegerlein, 1983, personal communication).

The numerous sinkholes and swallowholes, thedry bed, storm-water rises, and other drainage features of the Lost River area have been described in detail by Malott (1952). The physical appearances of these features are best observed during a dry period or low-flow stage, but the dynamic hydrologic aspects of the subterranean drainage can only be surmised or pondered by observation of the system at various stages of storm-water discharge.

REFERENCES CITED

Bassett, J. L., 1974, Hydrology and geochemistry of karst terrain, upper Lost River drainage basin, Indiana [A.M. thesis]: Bloomington, Indiana University, 102 p.

—— 1976, Hydrology and geochemistry of the upper Lost River drainage basin, Indiana: National Speleological Society Bulletin 38, p. 79–87.

Childs, Lewis, 1940, A study of a karst area in Orange and Lawrence Counties, Indiana [A. M. thesis]: Bloomington, Indiana University, 111 p.

Malott, C. A., 1922, The physiography of Indiana, in Handbook of Indiana geology: Indiana Department of Conservation Publication 21, p. 59–256.

—— 1932, Lost River at Wesley Chapel Gulf, Orange County, Indiana: Indiana Academy of Science Proceedings, v. 41, p. 285–316.

—— 1945, Significant features of the Indiana karst: Indiana Academy of Science Proceedings, v. 54, p. 8–24.

—— 1949, Hudelson Cavern, a stormwater route of underground Lost River, Orange Co., Indiana: Indiana Academy of Science Proceedings, v. 58, p. 236–243.

—— 1952, The swallowholes of Lost River, Orange County, Indiana: Indiana Academy of Science Proceedings, v. 61, p. 187–231.

Murdock, S. H., and Powell, R. L., 1968, Subterranean drainage of Lost River, Orange County, Indiana: Indiana Academy of Science Proceedings, v. 77, p. 250–255.

Owen, Richard, 1862, Report of a geological reconnoissance of Indiana made during the years 1859 and 1860 under the direction of the late David Dale Owen, M. D., State Geologist: Indianopolis, H. H. Dodd and Co., 368 p.

The Falls of the Ohio River, Indiana and Kentucky

Carl B. Rexroad, *Indiana Geological Survey, 611 North Walnut Grove, Bloomington, Indiana 47405*
Richard L. Powell, *Geoscience Research Associates, Inc., Bloomington, Indiana 47401*

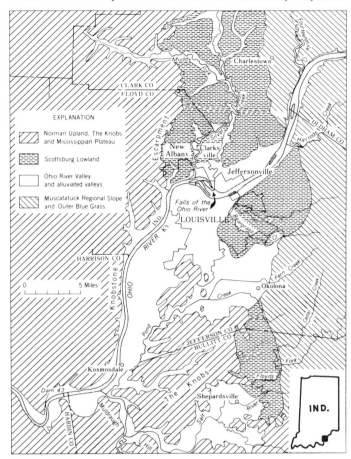

Figure 1. Map showing location of The Falls of the Ohio River and the nearby physiographic units. Slightly modified from Powell (1970).

LOCATION AND SIGNIFICANCE

The Falls of the Ohio River is internationally renowned for the profusion of corals and other fossils found in the Devonian beds at this site. Indeed, the earliest paper on its corals was published in Belgium by Rafinesque and Clifford in 1820. Subsequently, about 75 papers have described some 600 fossil species from the falls. Other studies have addressed the geomorphology of the area, including in particular the origin and history of the Ohio River and of the falls. The Falls of the Ohio is also important in understanding the Devonian history of southern Indiana and adjacent Kentucky.

In reality a series of rapids, the Falls of the Ohio encompassed a stretch of the Ohio River about 2 mi (3.2 km) long between Louisville, Kentucky, on the south bank of the river, and New Albany, Clarksville, and Jeffersonville, Indiana, on the northern bank (Fig. 1). "In these rapids, the river has a descent of twenty-two and a half feet, . . . but in no case, has it a perpendicular fall of more than three. At high water, an acceleration of current, not usual to other parts of the river, is all that is perceived; but at low water, it cannot be passed by loaded boats, without great risk and danger" (Lapham, 1828). Therefore, the Falls Cities owe their initial growth and development to the fact that during times of low water, boats following this major artery of westward expansion either had to off-load and portage or had to wait for high water.

In May 1778, Lieutenant Colonel George Rogers Clark reached the falls and established a military base on Corn Island en route to the capture of Kaskaskia (Illinois) and control of other French settlements. Part of his force returned to the falls during the winter of 1778–79 and established Fort Nelson on the mainland just a few blocks west of the present Jefferson County courthouse. This gave rise to Louisville, and the Falls Cities were platted during the next several decades. Population increased rapidly as many migrants coming down the Ohio River or through the Cumberland Gap elected to settle the area. By 1825, Louisville was a city measuring four blocks by twelve blocks and was a gateway to the west.

Presently, river traffic bypasses the falls by using the Louisville and Portland Canal. Construction of the original Louisville and Portland (or Shippingsport) Canal was authorized by the Kentucky legislature in 1825 (Fig. 2). The present huge canal created Shippingsport Island, where the defunct town of Shippingsport was located. McAlpine Dam further modified the river at the falls so that much of the original rapids is now inundated. Although most of the area of the Falls of the Ohio is in Kentucky because the state boundary is along the north bank of the Ohio River, access to the falls, a National Natural Landmark, is almost entirely through Clarksville. The area, which includes parking for cars and buses, is shown on the New Albany, Indiana-Kentucky 7½-minute Quadrangle.

GEOMORPHIC SETTING AND ORIGIN OF THE FALLS OF THE OHIO

The Ohio River valley, itself a distinct land type, crosses parts of three regional geomorphic units within a short distance of the falls (Figure 1). As one looks to the west from the area of the falls, the Knobstone Escarpment, the most prominent single geomorphic feature in Indiana, looms nearly 600 ft (180 m) above the Ohio River. The escarpment continues across the river in Kentucky where it is called Muldraugh Hill. Slightly east of Muldraugh Hill, a series of hills called The Knobs has been isolated from the upland to the west (Fig. 1). The Ohio River is one of the few streams to have cut through the escarpment.

The Knobstone Escarpment marks the boundary between the Scottsburg Lowland to the east and the Norman Upland to

381

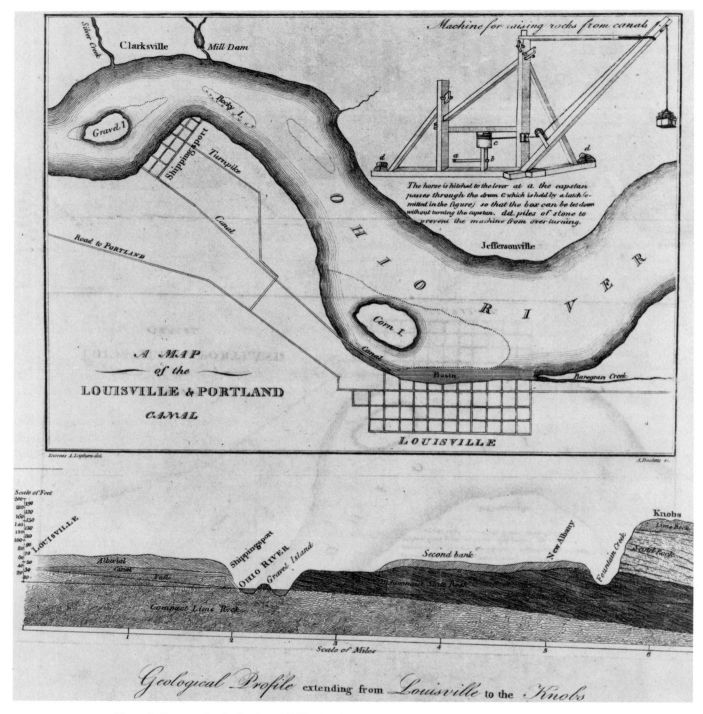

Figure 2. Cross section by Lapham (1828) showing the area of The Falls of the Ohio River at that time and his stratigraphic unit.

the west. The Norman Upland is upheld by deltaic sandstones, shales, and siltstone of the Borden Group and by the lower part of the overlying Middle Mississippian carbonate sequence. Because the area of the Falls of the Ohio River is near a primary margin of the Borden delta, clastic sediments are less prominent southwest of the falls in Kentucky. There the corresponding upland area is

called the Mississippian Plateau. For many miles downstream from the escarpment, the Ohio flows through the uplands in a narrow, youthful valley.

The Scottsburg Lowland developed on the relatively easily eroded New Albany and New Providence Shales. The thin Rockford Limestone, as much as 3 ft (1 m) thick in this area, gen-

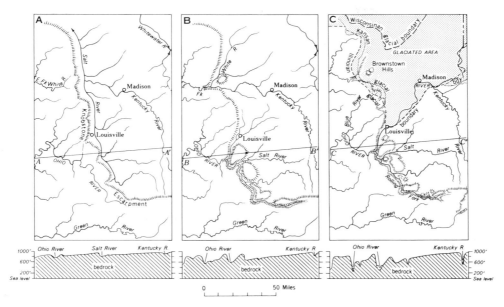

Figure 3. Maps showing the drainage routes near The Falls of the Ohio River during mid-Tertiary time (A), late Tertiary or early Pleistocene time (B), and mid-Pleistocene time (C). From Powell (1970).

erally intervenes between the shales but has little topographic effect. Although essentially an Indiana feature, the Scottsburg Lowland can be recognized for a few miles south of Louisville. The Falls of the Ohio is within this lowland area, where the Ohio River valley is as much as 5 mi (8 km) wide.

Eastward, the Scottsburg Lowland merges inperceptibly with the Muscatatuck Regional Slope in Indiana or the Outer Blue Grass in Kentucky. Beyond the southern limit of the Scottsburg Lowland in Kentucky, the Outer Blue Grass is juxtaposed with The Knobs or with the Mississippian Plateau. The Muscatatuck Regional Slope and Outer Blue Grass formed on the dip slopes of Middle Silurian and Middle Devonian carbonate rocks. The regional slope of the land is only slightly less than the dip of bedrock. Like the Scottsburg Lowland, relief is low except along the meandering, incised streams. For about 12 mi (19.2 km) above Harrods Creek, the Ohio River flows through a narrow gorge averaging about three-quarters of a mile wide and 250 ft (75 m) deep. Much of the Scottsburg Lowland and the Muscatatuck Regional Slope is veneered with a thin mantle of glacial drift.

Except for drainage derangement and the forming of the modern Ohio River, the glacial epochs wrought only minor changes in the topography of southern Indiana. The major geomorphic units as described above originated mostly during mid-Tertiary time by erosion of the Lexington Peneplain during a series of uplifts. Powell (1970) suggested that during the dissection of the peneplain, the Salt River developed as the major northward-flowing stream east of the emerging Knobstone Escarpment (Fig. 3). Then, during late Tertiary or early Pleistocene time, capture by the East Fork White River to the north and the Ohio River to the south diverted much of the former Salt

River drainage. The final integration into the modern pattern resulted from glaciation.

The major geomorphic features greatly influenced the course of the ice. The farthest advance of ice was during the Illinoian Age, but an older drift, presumably Kansan, reaches nearly as far south. Generally, the ice was deflected around the southern parts of the Norman Upland and the uplands immediately west of it—the Mitchell Plain and the Crawford Upland. Even so, east of the Norman Upland, the ice reached a southern limit approximating the present position of the Ohio River. Illinoian drift is common north of the Ohio River, but only scattered patches have been found in Kentucky.

The advance of the early ice sheet, a presumed Kansan advance, blocked the route of northward-flowing streams. These streams were ponded against the ice margin; the waters rose, overtopped the divides, and overflowed to the west as ice-marginal streams. Large amounts of melt water liberally laced with clastics and flowing with a relatively high gradient quickly cut troughs in the old divides and ultimately cut a bedrock channel to a level below present stream level. This early drainage was integrated into what is essentially the modern Ohio River. During subsequent advances and retreats of glacial ice, meltwater scouring alternated with outwash alluviation and formed a complex of overlapping, interbedded, and truncated channel deposits of clays, sands, gravels, and cobbles. At times outwash deposits blocked the tributary streams, ponding them and resulting in lake deposits that are seen as large, flat areas in and around Louisville and around New Albany. During one stage, the Ohio River remained near the south side of the valley, and the resulting bedrock channel, subsequently filled, runs beneath the city of Louisville.

Although the Wisconsinan ice margin was about 60 mi (96 km) north of Louisville, outwash deposits filled the valley to an elevation about 50 ft (15 m) above the bedrock now exposed at the Falls of the Ohio. With the waning of the meltwater, a relatively small stream flowed on this surface, but the retreat of the ice greatly increased the drainage basin to the north and thereby increased the flow in the Ohio. This, in turn, resulted in erosion of the glacial deposits. By chance the river shifted to a position above bedrock before major downward erosion. Within about the last 10,000 years it has cut nearly 70 ft (21 m) below the adjacent areas of outwash and about 10 to 20 ft (3 to 6 m) into bedrock. The erosion at this position resulted in The Falls of the Ohio. This is the only spot between Pittsburgh and the mouth of the Ohio where bedrock is exposed entirely across the river.

STRATIGRAPHY AND PALEONTOLOGY OF THE FALLS OF THE OHIO

In 1828, Lapham subdivided the bedrock units present from the Falls of the Ohio to the top of The Knobs into four units in ascending order: Compact Lime Rock, Bituminous Slate Rock, Sand Rock, and Lime Rock (Fig. 2). Near the top of the lowest of these he distinguished a "water limerock" and above that a coarse-grained limestone "probably oolite or roestone."

Eventually, Lapham's water limerock was named the Silver Creek Hydraulic Limestone by Siebenthal (1901), and the "roestone," which in reality is a crinoidal limestone, was named the Beechwood Limestone by Butts (1915). An additional unit immediately below the Silver Creek and only a foot (30 cm) or so thick just north of the falls was named the Speed Member by Sutton and Sutton (1937). In Indiana, these three units are now considered to be members of the North Vernon Limestone (named by Borden in 1876) (Fig. 4). The Speed and Silver Creek are in a complex facies relationship with each other and are separated from the Beechwood by an unconformity.

On the basis of the presence of *Catenipora [Halysites] escharoides,* Hall (1842) correlated the basal rocks at the falls with the Niagara Limestone. These rocks were generally called the *Catenipora* beds until Foerste (1897) named the Louisville Limestone to which this unit was then referred. A series of abandoned quarries along Middle Fork of Beargrass Creek in Louisville on I-64 constitutes part of the type section of the Louisville. These quarries yielded much of the stone for Louisville's oldest buildings.

In 1899, the limestone between the *Catenipora* beds below and the Sellersburg Limestone above (that is, the North Vernon) was named the Jeffersonville Limestone by Kindle, who referred to the excellent exposures at the Falls of the Ohio. These exposures are considered to be the type section of the formation. It is in the Jeffersonville that corals are so profuse and that many other kinds of fossils are abundant. At present, only the Jeffersonville is readily accessible at the Falls of the Ohio (Fig. 5).

Perkins (1963) refined earlier subdivisions of the Jeffersonville and recognized five zones at the falls, each zone with charac-

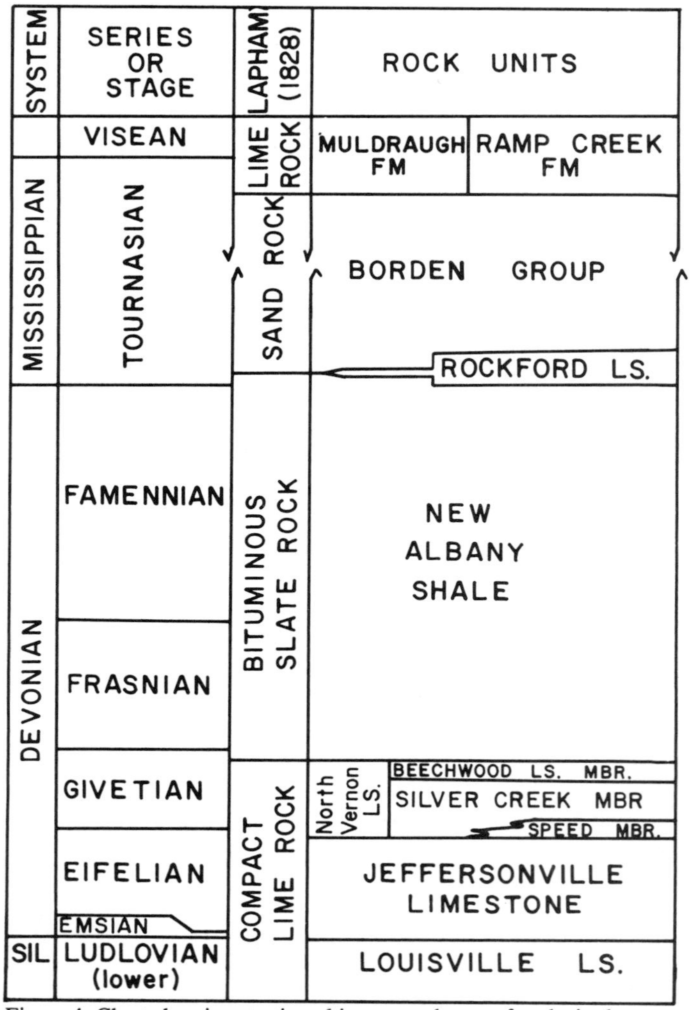

Figure 4. Chart showing stratigraphic nomenclature of rocks in the area of The Falls of the Ohio River.

teristic fossils and associated rock types. The zones in ascending order are: Coral Zone, *Amphipora* Zone, *Brevispirifer gregarius* Zone, Fenestrate Bryozoan-Brachiopod Zone, and *Paraspirifer acuminatus* Zone. The Jeffersonville is truncated by erosion a short distance southward in Kentucky, and the above zones merge northward with different facies and become unrecognizable.

The lower lithologic units of the Coral Zone, grain-supported biomicrite and above that biosparite, contain many upright branching corals, very large corals, and stromatoporoids. The upper unit, a biomicrudite, contains abundant solitary corals and branching coral fragments. A detailed paleoecologic analysis by Kissling and Lineback (1967) of a single, very broadly exposed bedding surface within the Coral Zone showed that the beds accumulated as a coral bank below effective wave base in a manner somewhat analogous to nonreef corals on the south Florida carbonate bank. The linear orientation of the fossils suggests action of a gentle tidal current. About 10 ft (3 m) of the Coral Zone is exposed at average water level.

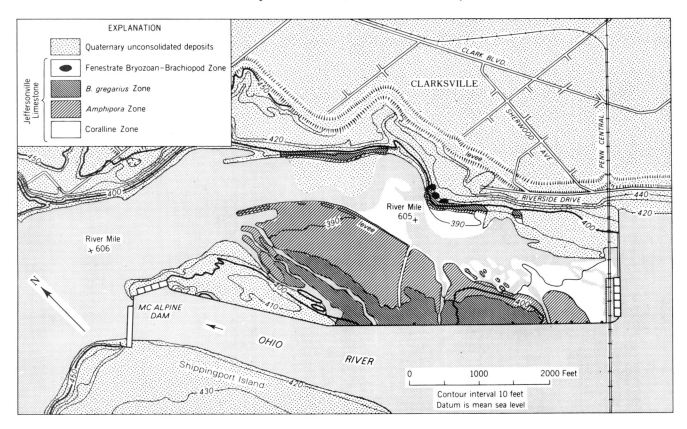

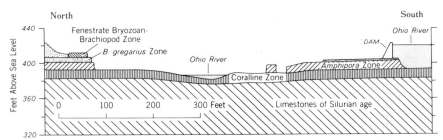

Figure 5. Geologic map of The Falls of the Ohio River showing the bedrock exposures of strata of Devonian age and areas covered with Pleistocene and Recent unconsolidated deposits, above, and general cross section across The Falls of the Ohio River showing the relationship of zones within the Jeffersonville Limestone, below. From Powell (1970).

The *Amphipora* Zone is about 9 ft (2.7 m) thick at the falls, is of variable lithology, and represents an interval of shoaling. Fragments of *Amphipora* and matlike stromatoporoids are the major faunal elements. In the next zone up, *Brevispirifer gregarius,* the name giver for the zone, and echinoderms dominate the 5-ft (1.5-m)-thick zone, which is mostly grain-supported biomicrite. Continued regression of the sea is suggested. A second Jeffersonville transgression began with the Fenestrate Bryozoan-Brachiopod Zone, which is about 6 ft (2 m) thick. During that time, corals and stromatoporoids returned as important parts of the fauna. The final depositional unit, the *Paraspirifer acuminatus* Zone, consists of about 6 ft (2 m) of grain-supported biomicrite and biosparite and is the most widespread of the five zones because of maximum marine transgression during this latter interval. It is not exposed at the falls, however.

Above the Devonian limestones, Lapham's Bituminous Slate Rock was named the New Albany Slate by Borden in 1874. It is part of the extensive regional black shales that include the Ohio, Chattanooga, and Antrim Shales, which are undergoing investigation now because of their potential as oil shales. The New Albany is Late Devonian and Early Mississippian (early Kinderhookian) in age, but the Mississippian part is progressively removed by erosion southward and is not represented a few miles south of Louisville (Rexroad and others, 1983).

Lapham did not include the very thin Rockford Limestone

(of Owen and Norwood, 1847) in his section, but this is not surprising, as it pinches out along an erosional margin near the Ohio River. The Rockford is of Kinderhookian age in the falls area but to the north includes early Valmeyeran beds (Rexroad and Scott, 1964).

The Sand Rock of Lapham apparently coincides with the Borden Group of present usage, a name replacing the older non-geographic name Knobstone (Cumings, 1922). The Borden encompasses the rocks between the Rockford Limestone or the New Albany Shale below and the Muldraugh or Ramp Creek Formations above and is Valmeyeran in age. Although called the Sand Rock by Lapham, considerable siltstone and shale, including the dominantly gray plastic lower shale, the New Providence Shale named by Borden in 1874, are also represented. In the Louisville area of Kentucky, the Borden Formation includes the Kenwood Silstone Member (Butts, 1915), which interfingers with the New Providence. Above these two units, the Borden includes the Nancy Member (Weir and others, 1966) and the Holtsclaw Siltstone member (Butts, 1915). In Indiana, subdivision above the New Providence is difficult (Stockdale, 1931, 1939).

For a more detailed overview of the "Geology of the Falls of the Ohio River," Powell's 1970 report by that title is excellent.

REFERENCES CITED

Borden, W. W., 1874, Report of a geologic survey of Clark and Floyd Counties, Indiana: Indiana Geological Survey Annual Report 5, p. 133–189.

——1876, Jennings County: Indiana Geological Survey Annual Report 7, p. 146–180.

Butts, Charles, 1915, Geology and mineral resources of Jefferson County, Kentucky: Kentucky Geological Survey, series 4, v. 3, pt. 2, 270 p.

Cumings, E. R., 1922, Nomenclature and description of the geological formations of Indiana, in Logan, W. N., Handbook of Indiana geology: Indiana Department of Conservation Publication 21, pt. 4, p. 403–570.

Foerste, A. F., 1897, A report on the geology of the Middle and Upper Silurian rocks of Clark, Jefferson, Ripley, Jennings, and southern Decatur Counties, Indiana: Indiana Department of Geology and Natural Resources Annual Report 21, p. 213–288.

Kindle, E. M., 1899, The Devonian and Lower Carboniferous faunas of southern Indiana and central Kentucky: Bulletins of American Paleontology, v. 3, no. 12, p. 131–239.

Kissling, D. L., and Lineback, J. A., 1967, Paleoecological analysis of corals and stromatoporoids in a Devonian biostrome, Falls of the Ohio, Kentucky-Indiana: Geological Society of America Bulletin, v. 78, p. 157–174.

Lapham, I. A., 1828, Notice of the Louisville and Shippingsport Canal, and of the geology in the vicinity: American Journal of Science, v. 14, p. 65–69.

Owen, D. D., and Norwood, J. G., 1847, Researches among the Protozoic and Carboniferous rocks of central Kentucky made during the summer of 1846: St. Louis, Keemle and Fields, 12 p.

Perkins, R. D., 1963, Petrology of the Jeffersonville Limestone (Middle Devonian) of southeastern Indiana: Geological Society of America Bulletin, v. 74, p. 1335–1354.

Powell, R. L., 1970, Geology of the Falls of the Ohio River: Indiana Geological Survey Circular 10, 45 p.

Rafinesque, C. S., and Clifford, J. D., 1820, Prodome d'une monographie des Turbinolies fossiles du Kentucky (dans l'Amerique Septentrionale): Bruxelles, Annales Generales des Sciences Physiques, v. 5, p. 231–235.

Rexroad, C. B., Gray, H. H., and Noland, A. V., 1983, The Paleozoic systemic boundaries of the southern Indiana-adjacent Kentucky area and their relations to depositional and erosional patterns (Field Trip 1): in Shaver, R. H., and Sunderman, J. A., editors, Field trips in midwestern geology: Bloomington, Indiana, Geological Society of America, Indiana Geological Survey, and Indiana University Department of Geology, v. 1, p. 1–35.

Rexroad, C. B., and Scott, A. J., 1964, Conodont zones in the Rockford Limestone and the lower part of the New Providence Shale (Mississippian) in Indiana: Indiana Geological Survey Bulletin 30, 54 p.

Siebenthal, C. E., 1901, The Silver Creek Hydraulic Limestone of southeastern Indiana: Indiana Department of Geology and Natural Resources Annual Report 25, p. 331–389.

Stockdale, P. B., 1931, The Borden (Knobstone) rocks of southern Indiana: Indiana Department of Conservation Publication 98, 330 p.

——1939, Lower Mississippian rocks of the east-central interior (U.S.): Geological Society of America Special Paper 22, 248 p.

Sutton, D. G., and Sutton, A. H., 1937, Middle Devonian of southern Indiana: Journal of Geology, v. 45, p. 320–331.

Weir, G. W., Gualtieri, J. L., and Schlanger, S. O., 1966, Borden Formation (Mississippian) in south- and southeast-central Kentucky: U.S. Geological Survey Bulletin 1224-F, p. F1-F38.

The Chesterian section near Sulphur, Indiana

Alan Stanley Horowitz and Stuart M. Kelly, Department of Geology, Indiana University, Bloomington, Indiana 47405

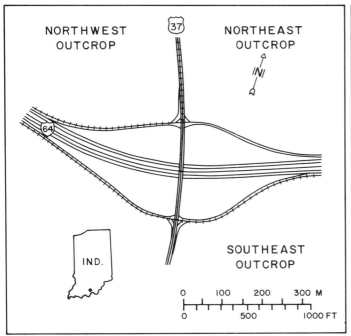

Figure 1. Map showing location of junction of I-64 and Indiana 37 in southern Indiana, 1 mi (1.5 km) north of Sulphur, Crawford County, Indiana. Modified from Visher (1980b).

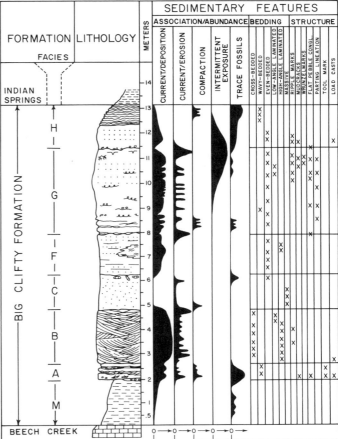

Figure 2. Section and table showing association of sedimentary features in the facies of the Big Clifty Formation along I-64 near Sulphur, Indiana. From Visher (1980b).

LOCATION AND ACCESS

A remarkable exposure of early and middle Chesterian (Late Mississippian) rocks is found at the junction of I-64 and Indiana 37 approximately 1 mi (1.5 km) north of Sulphur, Crawford County, Indiana (N½S½Sec.24,T.3S.,R.1W., Beechwood 7½-minute Quadrangle; Fig. 1). Some of the geological boundaries in the area are shown on the geologic maps prepared by Perry and Smith (1958, plate 2) and Gray and others (1970).

This site can be reached by exiting I-64 at its junction with Indiana 37 north from Tell City or south from Indianapolis, Bloomington, Bedford, or Paoli. Safe offroad vehicle parking on blacktop is present along Indiana 37 at its junction with all four interstate ramps. Examination of the upper part of the section requires the climbing of exposures that are revealed to best advantage in the northwestern road cuts.

SIGNIFICANCE

This locality at the eastern edge of the Illinois Basin in south-central Indiana contains an almost continuous exposure (rare in the Illinois Basin) of the lower to middle Chesterian section ranging upward from the Reelsville Limestone through the Haney Limestone. Exposures display textures, structures, and

fossils generally used for interpreting Chesterian sedimentary environments. The road cuts are especially notable for the extent of exposure of the Beech Creek Limestone, Big Clifty Formation, and Haney Limestone. The richly fossiliferous interbedded limestones and shales of the Indian Springs Shale below the Haney have been intensively collected by amateurs and professionals.

The Big Clifty at this locality has been the subject of a guidebook article (Suttner and Visher, 1979) and an unpublished masters thesis (Visher, 1980a, 1980b). The overlying fossiliferous Indian Springs Shale is presently being studied paleoecologically by S. M. Kelly. The Chesterian rocks, especially the sandstones in the tristate area of Indiana, Kentucky, and Illinois, long have been targets for petroleum exploration in the Illinois Basin. Sections such as the sandstones in the Big Clifty Formation at Sulphur provide geologic insights not available through subsurface studies alone. The Big Clifty Formation and Indian Springs Shale are singled out for discussion below. The remaining formations in the exposure await future studies.

Facies	Structure	Process
H	wavy bedding	alternating bedload and suspension deposition tidal or storm-induced
H	horizontal tracks, trails, bioturbation	biogenic activity during periods of relatively slow deposition
H	size grading, decreasing thickness	primarily suspension settling under conditions of waning flow
G	horizontal lamination, parting lineation	upper flow regime deposition
G	transition from lunate to current ripples	shoaling transition accompanying decrease in H_2O depth
G	wrinkle marks	near-exposure wind-induced rippling
G	interference ripples, mudchip-filled scour depressions	near-exposure shoaling, local opposition of flow, intermittent scour
G	laterally extensive high-relief erosional surfaces	widespread scour accompanying major storm event
G	mudcracks	intermittent emergence, desiccation
G	vertical burrows: Ophiomorpha Skolithos (?) Monocraterion (?)	construction of nearshore (agitated water) dwelling structures
F	horizontal lamination	low flow regime suspension settling
F	clay drapes, clay-filled scour depressions	fluctuating current strength, alternation of erosion and suspension deposition
F	thin bifurcating wavy-flaser intervals	fluctuating current strength, tidal or storm deposition
F	local climbing-ripple to micro-X bd. strata	rapid deposition
E	high-angle wedge-planar cross beds (single sets = unimodal, set pairs = bimodal bipolar)	low velocity traction and saltation deposition, sand wave migration
D	low-angle parallel laminations	
C upper	parallel laminae burrows (horiz + vert) algal structures	gradual decrease in sedimentation rate; transition from vertical "escape" burrows to horizontal forms; subaqueous growth of algal mats
C lower	erosional lower boundary; "massive" texture grades upward into parallel laminae	rapid deposition following extensive scour; rate of sediment accumulation decreases towards upper facies boundary
B upper	multi-directional low-angle wedge planar cross-beds, set thickness decreases upwards within facies	bedload transport (accretion*) on slopes of megaripples or complex dunes wherein set thickness decreases with increasing flow velocity *where dip avg. = 22°
B lower	multi-directional high-angle cross-beds bi-directional trough and current ripples; abundant truncations; sharp set boundaries herringbone X-bedding	tidal current bedload transport (avalanching*) of megaripples or dunes (as above); reversals and non-uniformity of flow, and general instability of bedforms *where dip≥ 28°
A	wavy-bedding; even and interbedded sand/mud ripples & clay drapes	alternating traction and suspension deposition tidal or storm induced
A	horizontal trace fossils (Cruziana ichnofacies); bioturbation of some beds	abundant biogenic activity during periods of slow or halted deposition (at subtidal depths)
A	load casts syneresis cracks	differential loading, compaction, dewatering, and shrinkage during rapid sedimentation

Figure 3. Summary chart showing characteristic sedimentary structures and bedforms of each Big Clifty facies and one or more interpretations for the probable mechanism or process responsible for each group of features. Modified from Visher (1980b).

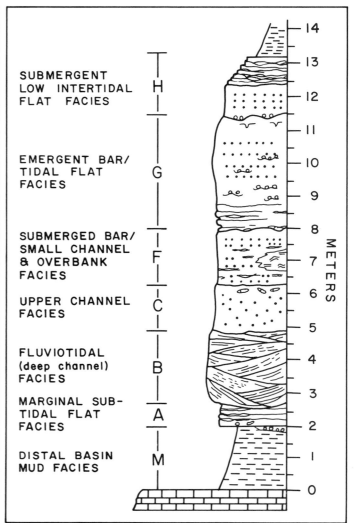

Figure 4. Section showing environmental interpretations proposed for each depositional facies present within a generalized section of the Big Clifty Formation near Sulphupr, Indiana. Modified from Visher (1980b).

SITE INFORMATION

Regionally, the site is within the unglaciated Interior Low Plateaus physiographic province of Fenneman (1938, p. 411) and near the boundary between his Highland Rim and Shawnee Hills Sections of the province. In Indiana, Malott (1922) assigned this area to the Crawford Upland, an unglaciated and dissected terrain underlain by the alternating limestones and clastic rocks of the Chesterian Series (Late Mississippian) and by the basal Pennsylvanian clastic rocks of the Mansfield Formation. The beds dip gently, approximately 25 to 50 ft/mi (5 to 10 m/km) to the west into the Illinois Basin. Horowitz (1979) has reviewed previous geologic studies, including a number of unpublished Indiana University theses, pertaining to this area. Since that review, Visher (1980a, b) has completed a thesis on the Big Clifty Sandstone that elaborates on the paper by Suttner and Visher (1979).

Table 1 lists the stratigraphic units exposed along the interstate road cuts. The uppermost clastic beds of the Sample Formation and the sandy beds of the Reelsville Limestone are exposed only along the southeastern ramp to I-64, where the most complete exposures of the Elwren Formation also are present. The Beech Creek Limestone, Indian Springs Shale, and Haney Limestone occur at all four corners of the interstate junction, and the lower beds of the Hardinsburg Formation are exposed high in the northwest exposures.

The most intensively studied unit is the Big Clifty Sandstone, which Visher (1980b) has interpreted as a tidally influenced delta

TABLE 1. SEQUENCE OF STRATIGRAPHIC UNITS IN THE
SULPHUR SECTION, CRAWFORD COUNTY, INDIANA

Series	Group	Formation
Chesterian	Stephensport	Hardinsburg Sandstone Haney Limestone Indian Springs Shale Big Clifty Formation Beech Creek Limestone
	West Baden	Elwren Formation Reelsville Limestone Sample Formation

lying along the eastern margin of the Late Mississippian epicontinental (cratonic) sea. Figure 2 illustrates the sedimentary features associated with the seven facies recognized by Visher in the Big Clifty Sandstone. Figure 3 summarizes the structures, bedforms, and probable processes responsible for the facies, and Figure 4 is a summary interpretation of the clastic facies of the Big Clifty Formation. Visher (1980b, p. 70) concluded that grain-size distributions, sorting, and textural maturity of the sandstones in the Big Clifty were consistent with the ranges of these factors reported from delta-front or tidal-flat deposits. Paleocurrent data was found to be consistently northeast-southwest (perpendicular to the inferred shoreline) in several facies but showed a wide variation in other facies so that a uniform trend was not present throughout the Big Clifty at this site.

Regionally, the sediment source was from the east-northeast from both the present area of the Canadian Shield and from rising highlands (Appalachia of preplate tectonic geologic literature) related to the collision of the North African and North American plates during Late Paleozoic time.

Malott and Thompson (1920, p. 521) did not designate a type section for their Indian Springs Shale, and the unit is not presently recognized by the Indiana Geological Survey. We use the Indian Springs Shale for the commonly fossiliferous interbedded shales and limestones lying between the Big Clifty Sandstone and the massive limestone units of the Haney Formation. Units of the published section of Gray and others (1957, p. 16, units 2 and 3; also, Rexroad and others, 1983, p. 23) can be used as a reference section until the publication of additional stratigraphic studies on the Indian Springs in Indiana.

At this site, the Indian Springs Shale consists, at its base, of a lower unit composed of poorly fossiliferous purplish-gray to gray shales, which grade upward into green shales. Local lenses of carbonate nodules or limestone contain a low-diversity association of ostracods and mollusks. The contact with the upper unit of the Indian Springs Shale is marked by a sharp color change from a green shale to a dark-gray (almost black) shale. The contact exhibits semiconical structures of interpenetrating green and dark gray shales that are probably the result of fluid escape from the lower shale during compaction. The lower part of the Indian Springs Shale, which is conformable with the submergent low-intertidal flat facies (facies H of Visher, 1980b), probably repre-

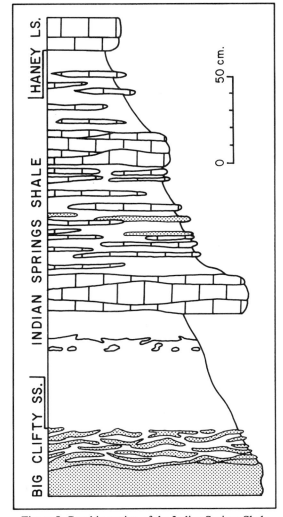

Figure 5. Graphic section of the Indian Springs Shale.

sents the high-intertidal flat facies grading into a tidal estuary at the top. The ostracod/mollusk limestone bed probably represents a tidal channel deposit within the estuary.

The upper unit of the Indian Springs Shale consists of alternating gray shales and discontinuous lenticular limestones. The lowest shale of the upper unit contains black corroded pebbles of phosphatized limestone and a phosphatized and corroded fauna of low density and diversity. These features are consistent with condensed sequences described from younger geologic rocks. Sedimentation in the remainder of the upper unit was more continuous.

Although the lenticular limestones probably were current generated, the composition of the fauna within the limestones does not differ from that of the shales. Consequently, the limestones are considered in situ accumulations of fossil debris. Many of the fossils are disarticulated or fragmented, and sorting is poor to moderate, but the fossil grains do not exhibit evidence of extensive abrasion. The diverse fauna of the upper unit are almost

TABLE 2. SUMMARY OF NICHE PARTITIONS IN THE FAUNA OF THE INDIAN SPRINGS
SHALE NEAR SULPHUR, INDIANA

Feeding Strategy	Suspension feeders		Detritus feeders (browsers and scavengers)	Predators
	Microphagous	Macrophagous		
Low (on or near bottom)	brachiopods, bryozoans, Sanguinolites conularid, Spirorbis, Tubulelloides	cup corals	?scolecodonts, ?ostracods, trilobites, non-platycerid gastropods, Phestia	?scolecodonts, ?ostracods, cephalopods, vertebrates
Medium	blastoids, Archimedes, and fenestrate fronds, short-stemmed crinoids	small Taxocrinus		cephalopods, vertebrates
High	long-stemmed crinoids	large Taxocrinus and Onychocrinus		cephalopods, vertebrates

totally marine and minimally consist of foraminifers (2 genera), coelenterates (2 genera), bryozoans (17 genera), brachiopods (16 genera), annelid (?) tubes (2 genera), scolecodonts, pelecypods (4 general), gastropods (7 genera), cephalopods (2 genera), ostracods (10 genera), trilobites (1 genus), echinoderms (26 genera), vertebrates (20 form genera of fish teeth), and *incertae sedis* (3 genera). The diversity is typical of Late Paleozoic filter-feeding communities dominated by bryozoans, brachiopods, and echinoderms. Table 2 indicates some of the niches occupied by the fauna.

REFERENCES CITED

Fenneman, N. M., 1938, Physiography of eastern United States: New York, McGraw-Hill Book Company, Inc., xiii + 714 p.

Gray, H. H., and others, 1957, Rocks associated with the Mississippian-Pennsylvanian unconformity in southwestern Indiana: Indiana Geological Survey Guidebook 9, 42 p.

Gray, H. H., Wayne, W. J., and C. E. Wier, 1970, Geologic map of the 1 × 2 Vincennes Quadrangle and parts of adjoining quadrangles, Indiana and Illinois, showing bedrock and unconsolidated deposits: Indiana Geological Survey Regional Geologic Map 3, Vincennes Sheet.

Horowitz, A. S., 1979, Notes on the geology of the Interstate 64 and Indiana Highway 37 exposures (stop 3), *in* Horowitz, A. S., editor, Mississippian rocks: New Albany to Indianapolis, Indiana, via Interstate 64 and Indiana Highway 37: Bloomington, Indiana, Guidebook prepared for Ninth International Congress of Carboniferous Stratigraphy and Geology Field Trip 7—Day 4, 20 May 1979, p. 3–18.

Malott, C. A., 1922, The physiography of Indiana, *in* Logan, W. N., editor, Handbook of Indiana Geology, part II: Indiana Department of Conservation Publication 21, p. 59–256.

——, and Thompson, J. D., 1920, The stratigraphy of the Chester Series of southern Indiana: Science, new series, v. 51, p. 521–522.

Perry, T. G., and Smith, N. M., 1958, The Meramec-Chester and intra-Chester boundaries and associated strata in Indiana: Indiana Geological Survey Bulletin 12, 110 p.

Rexroad, C. B., Gray, H. H., and Noland, A. V., 1983, The Paleozoic systemic boundaries of the southern Indiana–adjacent Kentucky area and their relations to depositional and erosional patterns (Field Trip 1), *in* Shaver, R. H., and Sunderman, J. A., editor, Field trips in midwestern geology: Bloomington, Indiana, Geological Society of America, Indiana Geological Survey, and Indiana University Department of Geology, v. 1, p. 1–35.

Suttner, L. J., and Visher, P. M., 1979, Guide to field study of the Interstate 64 outcrop of the sandstone of the Big Clifty Formation (Mississippian), *in* Horowitz, A. S., editor, Mississippian rocks: New Albany to Indianapolis, Indiana via Interstate 64 and Indiana Highway 37: Bloomington, Indiana, Guidebook prepared for Ninth International Congress of Carboniferous Stratigraphy and Geology Field Trip 7—Day 4, 20 May 1979, p. 19–34.

Visher, P. M., 1980a, Anatomy of an ancient tidally influenced delta: the Big Clifty Formation (Mississippian) of southern Indiana. Geological Society of America Abstracts with Programs, v. 12, p. 259.

——1980b, Sedimentology and three dimensional facies within a tidally-influenced Carboniferous delta: the Big Clifty Formation, Sulphur, Indiana [A.M. thesis]: Bloomington, Indiana University, × + 152 p.

Lake Erie: Deposition, erosion, and the effect of harbor structures near Fairport Harbor, Ohio

Charles H. Carter, Department of Geology, University of Akron, Akron, Ohio 44325

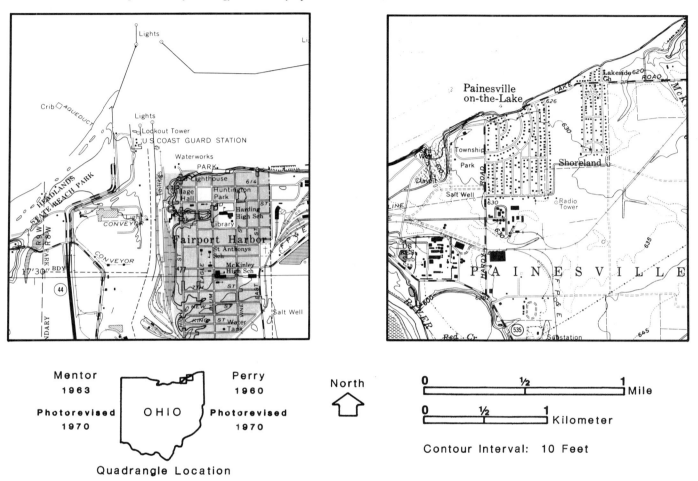

Figure 1. Location maps of Headlands Beach State Park (Mentor, Ohio 7½-minute quadrangle) and Painesville Township Park (Perry, Ohio 7½-minute quadrangle).

LOCATION AND ACCESSIBILITY

Headlands Beach State Park is located adjacent to the mouth of the Grand River, and Painesville Township Park is located about 2 mi (3.2 km) east of the mouth of the Grand River; the sites are near the middle of the south shore of Lake Erie (Fig. 1). Each site can be reached from roads running north of Ohio 2 (a major, four-lane, divided highway): Headlands Beach from Ohio 44 that runs directly into the park at a distance of about 2.4 mi (3.8 km) from Ohio 2 and Painesville Township Park from Ohio 535 to Hardy Road at a distance of about 2.1 mi (3.4 km) from Ohio 2. One can also follow a circuitous route—mostly Ohio 535 and 283—between the two sites (Fig. 1).

Buses can be driven to both sites. At Headlands Beach it is easiest to park in the easternmost parking lot, then walk northwest through the dunes to the beach and shoreline, then northeast to the jetty, and then return south along the jetty. At Painesville Township Park one can look at the observation point just beyond the east side of the park where the road (John Bailey Drive) has been cut off by erosion and then take a path at the western end of the park to reach the beach.

SIGNIFICANCE OF SITES

Headlands Beach, a depositional stretch that has advanced

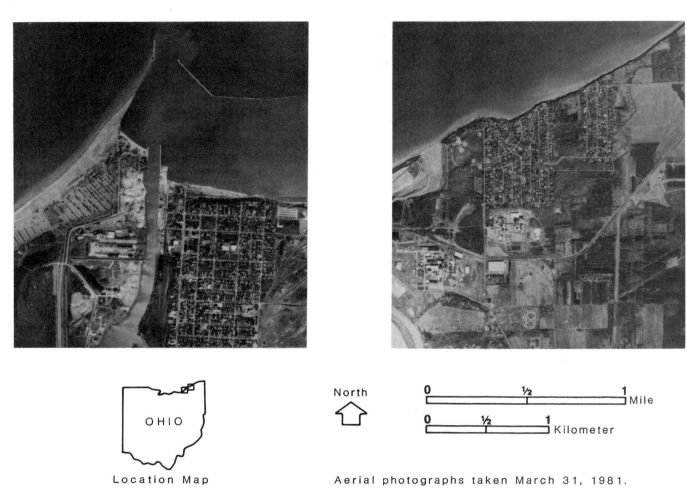

Figure 2. Vertical aerial photographs of Headlands Beach State Park and Painesville Township Park. Note west to east movement of suspended sediment along Headlands Beach and the headland-embayment couple at Painesville Township Park.

lakeward as much as 2,000 ft (600 m) since the mid-1820s, and Painesville Township Park, an erosional stretch that has retreated landward as much as 500 ft (150 m) since the mid-1870s, illustrate well the marked effect of major harbor structures on modern coastal processes, particularly the longshore transport of sand (Fig. 2). The results of these processes are evident in depositional features such as coastal dunes, a wide backshore, and multiple nearshore bars and erosional features such as the mass wasting of Pleistocene deposits, a narrow beach, and a nearshore devoid of sand.

HEADLANDS BEACH STATE PARK

The beach is the direct result of the Fairport Harbor structures. The first structures were jetties, constructed in the mid-1820s. These structures were lengthened so that by 1876 the west jetty was over 1,970 ft (600 m) long. Breakwaters were then constructed in the early 1900s; like the jetties, they were subsequently lengthened so that the present west breakwater has a length of about 0.75 mi (1.2 km). These structures were modified to keep pace with the tremendous buildup of sand on the west side of the structures from the net west to east longshore transport system (the prevailing westerly winds more than offset the effect of the infrequent, but intense, northeasters). Sand deposition on the west side has been calculated to be about 3,885,000 ft^3/yr (110,000 m^3/yr) (Bajorunas, 1961), a sizable rate especially for the Great Lakes. This buildup of sand has occurred despite the dredging of several hundred thousand cubic meters of sand from the outer harbor area (Hartley, 1964). Aside from the tremendous buildup of sand, a comparison of beach widths in 1876 and 1968 shows that beach widths have increased for about 0.9 mi (1.4 km) to the west of the structures, and the orientation of the shoreline has changed from nearly east-west in the mid-1820s to northeast-southwest in 1975 (Fig. 3).

The beach is characterized by a narrow (no more than 3 to 6.5 ft [1 to 2 m]) foreshore, a wide (tens of meters) backshore, and then a zone of incipient to fully developed dunes. The beach sand is largely derived from erosion of the shore, as the tributary streams transport little sand-size sediment to the lake. The sand is compositionally immature, with abundant shale clasts derived from the till bluffs to the west, and texturally not well sorted, as rapid deposition has led to less than normal abrasion and poorer sorting. The dunes naturally are made up of finer, better sorted sand. Beach plants such as switchgrass and beach grass are important in the development of the dunes, as they act as baffles to trap the windblown sand; willows and cottonwoods grow on the stable dunes farther landward.

PAINESVILLE TOWNSHIP PARK

This site, which lies about 2 mi (3.2 km) east of the Grand River, has also been profoundly affected by the Fairport Harbor structures, but in an erosional rather than a depositional way. Beach widths have decreased from 1876 to 1968 for at least 4 mi (6 km) to the east of the Grand River. By almost completely cutting off the longshore transport of sand, the harbor structures have caused the decrease in beach widths, a decrease that in turn has led to greater wave energy reaching the shore and thus accelerated erosion. A wide beach is the best form of shore protection, because wave energy is greatly attenuated by the interaction of the waves with the beach. The accelerated erosion does contribute more than a normal quantity of sand to the system, but erosion must take place for a stretch of several kilometers east of the structures to offset the enormous loss of sand to the Fairport Harbor structures to the west.

The overlook is characterized by a 60-ft (18-m)-high till bluff and an eroded highway that offers mute testimony to the overall erosional nature of the Lake Erie shore. The bluff is fronted by a narrow beach and, at its western end, by rotational slumps and manmade debris. Farther to the east, block falls are the dominant process; here a nearly homogeneous till and/or more rapid wave erosion has led to a different mode of slope failure.

The stretch fronting the park is characterized by a 60-ft (18-ft)-high hummocky till slope fronted by a slightly wider beach and by 85 to 121-ft (26 to 37-m)-long groins. The hummocky slope is the result of rotational slumps that likely have slipped along zones of stratified drift. The groins, by trapping some sand, have helped protect the slumped slope. This stretch, which forms a headland, and the stretch east of the overlook, which forms an embayment, illustrate well the tremendous effect

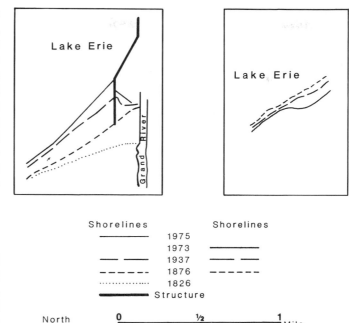

Figure 3. Headlands Beach State Park stretch with harbor structures and historic shorelines, and Painesville Township Park with historic recession (bluff) lines.

of moderately long groins on a longshore transport system nearly lacking sand. Recession rates prior to construction of the groins along this reach of shore were about 2 ft (6 m)/yr. After construction of the groins in the early 1940s, the recession rates west (updrift) of the structures were about 0.7 ft (0.2 m)/yr, whereas the rates east (downdrift) of them were about 7 ft (2.2 m)/yr (Fig. 3).

The till exposure, just beyond the west side of the park, consists of about 3 ft (1 m) of a gray, pebbly (10–15% gravel-size clasts) basal till, overlain by 1.5 ft (0.5 m) of a stratified pebbly clay and intercalated silt and clay, in turn capped by about 10 ft (3 m) of a brown flow till with less than 5% gravel-size clasts. The gray till is probably the "coastal" till, and the brown till is probably the Ashtabula till as defined by White (1980). In contrast to the beach at Headlands, the beach here is much narrower (usually 33 to 49 ft [10 to 15 m] wide) and the sand is coarser and more poorly sorted. The beach reflects the relatively small sand supply in the longshore transport system as well as proximity to the source materials (the till just to the west).

REFERENCES CITED

Bajorunas, L., 1961, Littoral transport in the Great Lakes: Proceedings 7th Coastal Engineering Conference, The Hague, Netherlands, 1960, p. 326–341.

Carter, C. H., 1976, Lake Erie shore erosion, Lake County, Ohio: Setting, processes, and recession rates from 1876 to 1973: Ohio Geological Survey Report of Investigation 99, 105 p.

Carter, C. H., Guy, D. E., Jr., and Fuller, J. A., 1981, Coastal geomorphology and geology of the Ohio shore of Lake Erie: Geological Society of America Field Trip Guidebook, v. 3, p. 433–456.

Feldmann, R. M., Coogan, A. H., and Heimlich, R. A., 1977, Southern Great Lakes geology field guide: Kendall/Hunt, Dubuque, Iowa, 241 p.

Hartley, R. P., 1964, Effects of large structures on the Ohio shore of Lake Erie: Ohio Geological Survey Report of Investigation 53, 30 p.

U.S. Army Corps of Engineers, 1952, Appendixes, III, VII, XII, Ohio shore line of Lake Erie between Fairport and Ashtabula, beach erosion control study: U.S. 82nd Congress House Document 351, 46 p.

White, G. W., 1980, Glacial geology of Lake County, Ohio: Ohio Geological Survey Report of Investigation 117, 20 p.

Kelleys Island: Giant glacial grooves and Devonian shelf carbonates in north-central Ohio

Rodney M. Feldmann, Department of Geology, Kent State University, Kent, Ohio 44242
Thomas W. Bjerstedt, Department of Geology and Geography, West Virginia University, Morgantown, West Virginia 26506

LOCATION

Kelleys Island is the easternmost of the Lake Erie Islands group, located north of Marblehead Peninsula and Sandusky, Ohio, (Fig. 1) on the Kelleys Island 7½-minute Quadrangle. Access to the Island is provided by the Newman Boat Lines which operates a ferry from Marblehead (April through November) and from Sandusky (June through Labor Day). Throughout the year, the Island is also served by Island Arilines from Port Clinton Municipal Airport. On the Island, transportation is available in the form of rental bicycles, golf carts, or private cars ferried from the mainland.

SIGNIFICANCE

The Island is particularly well-known for its giant glacial grooves and striated bedrock produced by the intense glacial scouring to which it was subjected during the Pleistocene. Although most of the Island is mantled by till, several excellent exposures of these features are available for examination, the

most famous of which is enclosed within Glacial Grooves State Memorial.

The bedrock consists of Middle Devonian (Eifelian) carbonate rocks, the Lucas Dolostone, and the Columbus Limestone. The two units are excellent examples of carbonate platform sedimentation.

GLACIAL FEATURES

The largest single exposure of glacial grooves on Kelleys Island is preserved as Glacial Grooves State Memorial, near the north end of the Island. Within the Memorial, an outcrop nearly 400 ft (120 m) long exposes giant glacial grooves in a depression about 30 ft (9 m) wide and 15 ft (4.5 m) deep (Fig. 2). Some of the grooves, notably near the east edge of the exposure, are relatively straight, smooth, and show striated surfaces. Farther to the west, however, the path of the individual grooves becomes far more tortuous, and striations are not as much in evidence. It appears that the primary sculpting of the grooves was produced by subglacial tools and that further modification, at least of these

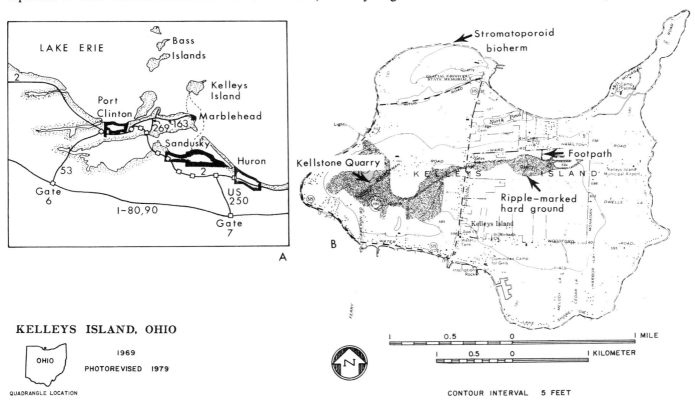

Figure 1. Location map for Kelleys Island (A) and several key geological features on the Island (B).

Figure 2. Glacial grooves preserved in Glacial Grooves State Memorial near the northwestern corner of Kelleys Island (A). Termination of the grooves at the edge of an abandoned quarry (B).

grooves, was the result of solution and polishing by stream activity, perhaps subglacial. This exposure was once far more extensive than it is currently. Quarrying in the early 1900s removed many spectacular grooves. At the west edge of the exposure, the grooves terminate abruptly at the wall of an abandoned quarry (Fig. 2).

Although these are the most spectacular of the grooves, other glacial grooves, striations, and crescentic marks can be seen on various surfaces around the Island. Additional excellent exposures occur along the north shore of the Island (Fig. 3), near the stromatoporoid bioherm (Fig. 1), where large grooves, well-developed striations, and crescentic gouges can be observed. Along the eastern shore and offshore, glacially striated pavement on the thin-bedded Venice Member of the Columbus Limestone extends below lake level at a gentle dip. Vigorous wave activity tears up thin slabs of this pavement and deposits them on a rubbly beach forming a gravel step of glacially striated slabs. Most beach-face property, however, is private.

MIDDLE DEVONIAN
CARBONATE SEDIMENTATION

The bedrock on Kelleys Island was formed on a shallow-water carbonate platform along the eastern margin of the Wabash Platform (Droste and others, 1975). A summary of several lithologic components, typical of the units exposed, is given in Figure 4. The Lucas Dolostone was deposited in a shallow-water, restricted, marginal marine environment which was progressively overstepped westward by Columbus facies during Eifelian time. The timing of this event has been addressed by Sparling (1983) using conodont zonation. The Lucas is intensely dolomitized and, with the exception of localized algal mounds and stromatoporoids (Fig. 5), is sparsely fossiliferous. Although several species of stromatoporoids in four genera have been described from Kelleys Island (Fagerstrom, 1982; Bjerstedt and

Figure 3. Glacially pitted and striated surface located on the north shore of Kelleys Island just east of the stromatoporoid bioherm (Fig. 1).

Feldmann, 1984) the most common forms are species of *Syringostroma*.

Progressive flooding of the Wabash Platform throughout early Middle Devonian time resulted in the onset of Columbus Limestone deposition characterized by more "normal marine" conditions and a much more abundant and diverse fauna. Columbus environments developed in open ramp settings eastward and basinward from the platform margin. Small, localized stromatoporoid mounds and biostromes accumulated in Lucas facies near the margin and were subsequently transgressed by Columbus facies. The contact between the Lucas Dolostone and the Columbus Limestone in the deeper Kellstone Quarry (Fig. 1) is distinct on Kelleys Island, and it would appear that sedimentation was continuous.

The Bellepoint Member, lowermost unit of the Columbus Limestone, consists of a gray wackestone-packstone facies marking the advent of sedimentation in higher energy regimes. The

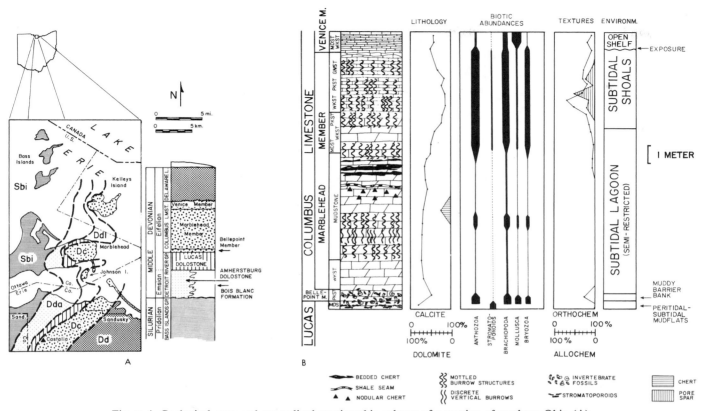

Figure 4. Geological map and generalized stratigraphic column of a portion of northern Ohio (A) showing the distribution of Silurian and Devonian rocks cropping out in the vicinity of Kelleys Island as well as details on the stratigraphy of the Island (B). The units depicted in Figure 4B are exposed at sites described in the text. Summaries of lithologic, biotic, textural, and environmental variation are taken from Bjerstedt and Feldmann (1984) and the generalized geologic map is modified from Forsyth (1971).

Bellepoint contains a large number of rugose corals, gastropods, rostroconchs, and brachiopods, as well as fragments of stromatoporoids torn up from stromatoporoid accumulations in the underlying Lucas Dolostone.

Variations in the lithologies observed within the Columbus Limestone are related primarily to changes in water depth and aeration with attendant variations in wave energy. The gray, thick-bedded, cherty, lower Marblehead Member is a dolomitic, heavily-burrowed mudstone (Fig. 6) deposited in lagoonal conditions below mean wave base. At certain horizons, the sea floor may have been stirred up, resulting in the deposition of allochthonous wackestone or these more fossiliferous accumulations may have resulted from more aerated subenvironments within this facies. Higher in the Marblehead Member, bedding becomes thinner as shoaling facies displaced faunally sparse environments. These were associated with a decrease in water depth to the approximate level of mean wave base and a proportionate increase in wave turbulence. The sediments become coarser and cleaner, and are either packstones or grainstones. The upper surface of the Marblehead Member is marked by a rippled, hardground surface representing subaerial erosion or maintenance of shallow water depth for an interval of time followed by subsidence of the platform below wave base and deposition of the mudstones and wackestones of the uppermost Venice Member. A noteworthy example of this hardground surface can be seen on the uppermost bench of an old quarry (Fig. 1).

Throughout the section there are excellent examples of a variety of biota. In the uppermost Lucas Dolostone, for example, stromatolitic algal heads and a stromatoporoid bioherm (Fig. 1) are well preserved. The bioherm varies vertically from crust-like mats to hemispherical and irregular forms, thought to represent colonizing and stabilizing adaptations respectively. Digitate or fasciculate forms, characteristic of a somewhat firmer substrate, occur in the highest horizon of the bioherm (Bjerstedt and Feldmann, 1984). Burrowed textures are common throughout most of the Columbus interval and, where exposed by differential weathering along the lake shore, they form spectacular surfaces on bedding planes. Finally, the Columbus yields an abundant Onondaga fauna of tabulate and rugose corals, articulate brachiopods, gastropods, pelecypods, cephalopods, rostroconchs, trilobites, ramose and fenestrate bryozoans, and pelmatozoan echinoderms. These can be observed and collected in old abandoned quarries in the Columbus Limestone (Fig. 1). The most richly fossiliferous horizons are in the packstones and grainstones of the upper Marblehead Member (Fig. 4) which is well exposed in the older abandoned quarries on the Island.

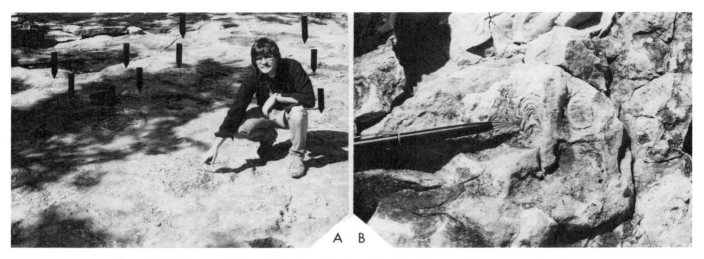

A B

Figure 5. (A) Uppermost Lucas Dolostone showing a broad surface exposing numerous low domes (arrows) probably formed over stromatolitic algal mounds, and (B) erect stromatoporoids exposed in a loose block of Lucas Dolostone littering the domed surface. Photos taken in the area indicated on Figure 1 as the stromatoporoid bioherm.

ACCESS TO EXPOSURES

Excellent exposures of the Lucas and Columbus sequence are available in natural exposures, mostly on public land along the north shore of Kelleys Island as well as in abandoned quarries, including the presently inactive Kellstone Quarry (Fig. 1). The Lucas Dolostone is exposed only in the lower level of the Kellstone Quarry, now flooded to the top of the Bellepoint Member, and along the northern edge of the Island. This weathered exposure, stretching several hundred meters along the shoreline, provides a good view of the relationship between the Lucas and the Bellepoint members of the Columbus Limestone as well as the lower part of the Marblehead Member. Exposures in the Kellstone Quarry, just northeast of Bookerman Road, provide nearly complete exposures of the Marblehead Member. The ripple-marked hard ground surface and the Venice Member, a tan, fossiliferous unit truncated by glacial erosion, can be seen only in the highwalls of the older, abandoned quarry on the eastern part of the Island which furnishes relatively fresh exposures of the upper Marblehead and Venice members. Access to the quarry is provided by a footpath south from Ward Road (Fig. 1).

Figure 6. Spindley, pitted pattern caused by lake front weathering of an extensively burrowed flat near the base of the Columbus Limestone. Photo taken in the area indicated on Figure 1 as the stromatoporpoid bioherm.

REFERENCES CITED

Bjerstedt, T. W. and Feldmann, R. M., 1984, Stromatoporoid paleosynecology in the Lucas Dolostone (Middle Devonian) on Kelleys Island, Ohio: Journal of Paleontology, v. 59, p. 1033–1061.

Droste, J. B., Shaver, R. H., and Lazor, J. D., 1975, Middle Devonian paleogeography of the Wabash platform, Indiana, Illinois, and Ohio: Geology, v. 3, p. 269–272.

Fagerstrom, J. A., 1982, Stromatoporoids of the Detroit River Group and adjacent rocks (Devonian) in the vicinity of the Michigan Basin: Geological Survey of Canada Bulletin 339, 81 p.

Feldmann, R. M., Coogan, A. H., and Heimlich, R. A., 1977, Field Guide; Southern Great Lakes: Dubuque, Iowa, Kendall/Hunt Publishing Company, 241 p.

Forsyth, J. L., 1971, Geology of the Lake Erie islands and adjacent areas: Michigan Basin Geological Society, Annual Field Excursion, 27 p.

Sparling, D. R., 1983, Conodont biostratigraphy and biofacies of lower Middle Devonian limestones, north-central Ohio: Journal of Paleontology, v. 57, p. 825–864.

Garfield Heights: Quaternary stratigraphy of northeastern Ohio

Barry B. Miller, *Department of Geology, Kent State University, Kent, Ohio 44242*
John P. Szabo, *Department of Geology, University of Akron, Akron, Ohio 44325*

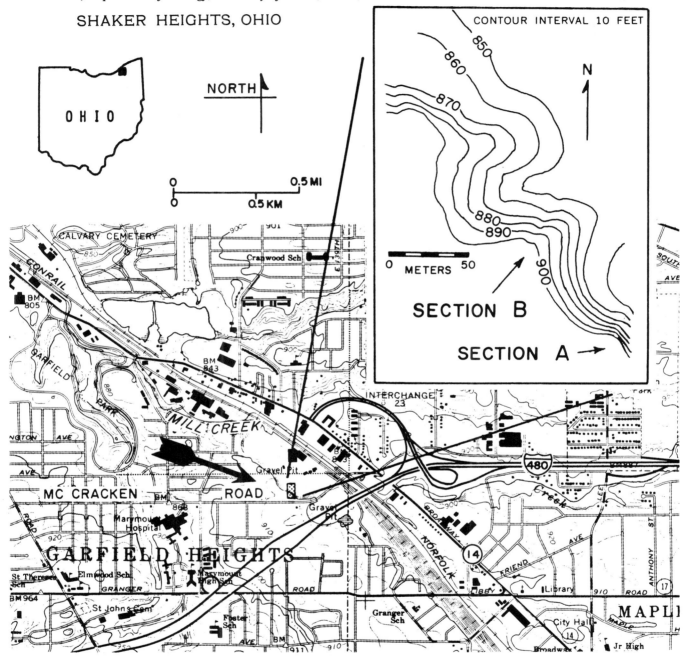

Figure 1. Index map showing location of Garfield Heights site.

LOCATION

Complex Quaternary deposits are well exposed along the valley of Mill Creek, in Garfield Heights (Shaker Heights 7½-minute Quadrangle), Cuyahoga County, Ohio (Fig. 1). Details of the stratigraphy can be examined best at two sections exposed along the steep-wall of an abandoned sand and gravel pit north-west of McCracken Road. The pit has been used as a landfill through the years and is now part of an industrial park. A major interstate highway, I-480 (Fig. 1), passes through and partially covers some of the units that were previously exposed in the steep-wall just southeast of McCracken Road. Access to the northwest pit may be made through an entrance road to the industrial park, which joins the northwestern side of McCracken Road about 100 ft (30 m) southwest of the intersection of the Conrail right-of-way and McCracken Road. Vehicles should be parked close to the steep-wall of the pit and as far as possible from buildings in the industrial park.

SIGNIFICANCE OF SITE

The site is of significance because of the diversity of Quaternary events recorded here, the variety of studies that have been published on the deposits, and the ongoing discovery of new geological facets of the site. The stratigraphy of the steep-wall northwest of McCracken Road, as first described by White (1953), included a basal Illinoian sand and gravel, overlain by a Sangamon paleosol and two fossiliferous Wisconsinan loesses (Leonard, 1953), "varved" clay and silt, and a till. Subsequent studies of the steep-walls exposed northwest and southeast of McCracken Road have resulted in the recognition of: 1) three Wisconsinan tills (Kent, Lavery and Hiram) above the "varved" clay and silt (White, 1965; 1968; Fullerton and Groenwold, 1974); 2) at least one till (Garfield Heights Till) between the Sangamon paleosol and the lower loess (Berti, 1971; 1975; Dreimanis, 1969); 3) plant remains, insects, and molluscs from the basal portion of the "varved"clay and silt (Coope, 1968; Miller and Wittine, 1972; Morgan and others, 1982; Berti, 1971; 1975); 4) two Illinoian tills (Fullerton and Groenwold, 1974); and 5) cryoturbation involutions and frost-cracks in the upper and lower loesses (Dreimanis, 1969; Morgan and others, 1982). The lower loess unit, which contains the fossil gastropod *Omalodiscus pattersoni,* represents the stratigraphically youngest and the geographically easternmost record of this extinct species. There is probably no other area of comparable size in northeastern Ohio that offers documentation for such a variety of glacial, interglacial, and interstadial events.

STRATIGRAPHY

At the site are two sections, one of which exhibits Late Woodfordian deposits overlying Farmdalian loess and two paleosols (Fig. 2, Section A). The other exposes Early Wisconsinan Altonian loess underlying Farmdalian loess and a much thicker exposure of the upper paleosol and an Altonian till (Fig. 2, Section B).

The youngest unit in Section A (Fig. 2) is a dark brown, friable to firm, blocky, calcareous till containing black shale fragments. This till is correlative with the Woodfordian Hiram and/or Lavery Till that are commonly difficult to distinguish unless they are separated by a stone line or loess such as exists in the gravel pit south of McCracken Road at this site (Fullerton and Groenwold, 1974). Another problem with the upper unit here is that it has been disturbed repeatedly by humans, as evidenced by a buried cast iron pipe near the section.

A sharp contact separates the Woodfordian till from the underlying unit, which is predominantly light olive-brown, friable, calcareous silt (Fig. 2). The silt, which resembles loess, contains 3 to 4 m (7 to 10 cm)-thick interbeds of dark-brown to dark-gray, calcareous clay silt and yellowish-brown, very fine sand.

Approximately 8 ft (2.5 m) of light brownish-gray to gray, calcareous, laminated lacustrine silt and clay underlie the silt and represent the "varves" of White (1968). This unit fines downward and contains organic matter that has been dated from 23,000 to 25,000 B.P. (White, 1968). It overlies about 20 in (50 cm) of light brownish-gray, friable to firm, calcareous, clay silt interbedded with yellowish-brown sand that has been interpreted as colluvium by Miller and Wittine (1972).

The colluvial zone (Fig. 2) grades into the upper part of the underlying Farmdalian loess, which has been dated at about 28,000 B.P. (White, 1968). This light olive-brown, friable, calcareous loess is nearly 3 ft (1 m) thick, and it contains a terrestrial snail fauna (Leonard, 1953; Miller and Wittine, 1972). The loess dips to the northwest and overlies 20 in (50 cm) of diamicton, which White (1968) interpreted as a Sangamonian accretion gley. The upper 10 in (25 cm) of this gley is brown, very firm, platy, and noncalcareous and contains yellowish-brown laminae and wavy organic bands. The lower half is very dark grayish-brown.

In sharp contact with this unit are 3 to 10 ft (1 to 3 m) of an underlying sand and gravel. Clay and iron coat the pebbles and penetrate about 2 ft (0.6 m) into the gravel. This zone has been interpreted as a B3 horizon of a Sangamonian paleosol (White, 1968). The sand and gravel beneath the truncated paleosol are calcareous.

The Farmdalian loess and the accretion gley form a weakly developed scarp that can be traced from Section A about 330 ft (100 m) northwest to Section B (Fig. 2). Here the accretion gley thickens, and an older, Altonian loess separates it from the Farmdalian loess.

At the top of Section B, approximately 4.3 ft (1.3 m) of Woodfordian laminated silt and clay are exposed beneath material disturbed by man. Wood and organic matter in the yellowish-brown, more sandy layers lie parallel to the contact with the underlying Farmdalian loess. The light grayish-brown, friable, calcareous loess contains snails and overlies a 1.6-in (4-cm)-thick, very dark gray, wavy, organic band in the top of the older Altonian age loess, which may be the involutions described by Berti (1975). This brown to yellowish-brown, slightly fossiliferous, friable loess apparently is leached of calcite. Laboratory analyses have shown, however, that the loess is composed of up to 25 percent dolomite.

The two loesses overlie about 7 ft (2 m) of greenish-gray to light olive-brown, firm accretion gley. Fractures in the upper 10 in (25 cm) of the gley contain yellowish-red iron and black manganese stains. The next 2 to 4 ft (0.6 to 1.3 m) of very firm, grayish-brown gley, which is mottled yellowish-brown, is cut by manganese-coated fractures. Highly weathered pebbles are scattered throughout this zone, which is underlain by 20 in (50

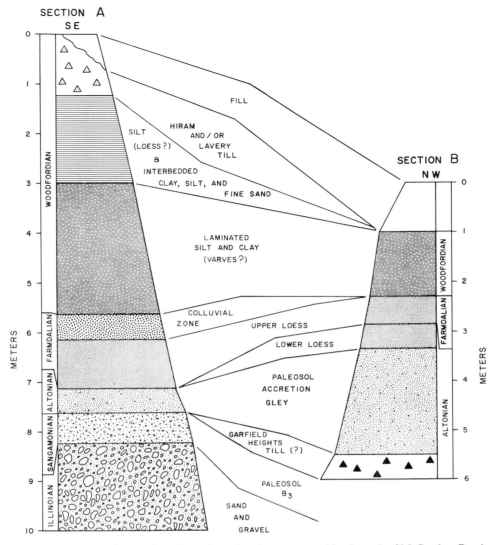

Figure 2. Representative stratigraphic sections of sediments exposed in pit north of McCracken Road.

cm) of light olive-brown to gray, platy gley containing gravel pockets and pebble bands. The lowermost 20 in (50 cm) of the section is yellowish-brown, firm, pebbly, iron- and manganese-stained diamicton, which may be the Garfield Heights Till cited in Dreimanis and Goldthwait (1973) and Berti (1971). This changes the interpreted age of the accretion-gley from Sangamonian to Altonian.

INTERPRETATION OF THE FOSSILS

A variety of plant and animal remains have been reported from the Garfield Heights sediment. These include beetles, ants, fever flies, a centipede, a spider, mites, mastodon bone fragments (Coope, 1968; Morgan and others, 1982), molluscs (Leonard, 1953; Miller and Wittine, 1972), spruce needles (*Picea*), the leaves of birch (*Betula*) and *Dryas,* and other plant macrofossils (Morgan and others, 1982; Berti, 1971; 1975).

Leonard (1953) described 16 taxa of gastropods from the lower loess unit (Table 1). These include five aquatic forms,

which he considered to be "indicative of a pond environment, on the surface of which the loess was being deposited." The associated terrestrial snails are mostly small-sized forms that require an environment with a moist organic litter on a calcareous substrate. Berti (1971) reported some pine, spruce, and oak pollen and spores from this unit, which he interpreted as indicative of ". . . boreal or even more severe conditions . . ." (p. 84).

The molluscan fauna from the upper loess is composed entirely of terrestrial gastropods, some of which are characteristic of wooded areas. Wood and insects from near the top of this unit indicate an environment with scattered trees (Morgan and others, 1982). Pollen suggests the presence of pine and spruce (Berti, 1971).

The laminated clay and silt unit has been interpreted as " . . . pro-Kent, deposited in a lake formed by damming of the drainage by the advance of the Kent ice" (White, 1968, p. 751). The molluscan and insect faunas from this unit (Miller and Wittine, 1972; Coope, 1968; Morgan and others, 1982) pose a problem with this interpretation. With the one possible exception of

TABLE 1. MOLLUSCS FROM THE GARFIELD HEIGHTS SECTION

Species	Lower Loess	Upper Loess	"Varved" Clay and Silt	Terrestrial (T) or Aquatic (A)
Carychium exile canadense	x[1]	-	-	T
Cionella lubrica	x[1]	x[1]	x[2]	T
Columella alticola	-	x[1]	x[2]	T
Deroceras laeve	-	x[2]	-	T
Discus cronkhitei	-	x[2]	-	T
Discus macclintocki	x[1]	-	-	T
Discus patulus	-	x[1]	-	T
Euconulus fulvus	x[1]	x[1]	x[2]	T
Fossaria dalli	x[1]	-	-	A
Gastrocopta armifera	-	x[2]	x[2]	T
Gyraulus parvus	x[1]	-	-	A
Hawaiia minuscula	x[1]	-	-	T
Helicodiscus parallelus	-	x[2]	x[2]	T
Helisoma anceps	x[1]	-	-	A
Hendersonia occulta	x[1]	x[1]	x[2]	T
Menetus sp.	x[1]	-	-	A
Nesoviterea electrina	-	x[2]	x[2]	T
Omalodiscus pattersoni	x[1]	-	-	A
Punctum minutissimum	-	x[2]	x[2]	T
Pupilla muscorum	x[1]	x[1]	-	T
Stenotrema leai	-	x[1]	x[2]	T
Strobilops labyrinthica	x[1]	-	-	T
Strobilops sp.	-	x[2]	-	T
Succineids (two forms)	x[1]	x[1]	-	T
Triodopsis multilineata	-	x[2]	-	T
Vertigo alpestris oughtoni	x[1]	x[1]	x[2]	T
Vertigo elatior	-	x[2]	x[2]	T
Vertigo gouldi hannai	-	x[2]	x[2]	T
Vertigo pygmaea	x[1]	x[1]	-	T

x = taxon present; - taxon not present; [1] = reported by Leonard, 1953; [2] = reported by Miller and Wittine, 1972.

the beetle, *Helophorus,* which frequents the banks of water bodies, but which is such an active flier that it is frequently found far removed from water, all of the mollusca and insects are terrestrial species. Coope (1968, p. 755) has suggested that the fossils may not have been deposited in a permanent lake, "... but were the products of successive floods, giving rise to shallow bodies of water that dried up in summer when water beetles would otherwise have been likely to colonize them." The plant material identified from this unit and the insects imply forest-tundra conditions (Berti, 1975).

REFERENCES CITED

Berti, A. A., 1971, Palynology and stratigraphy of the Mid-Wisconsinan in the eastern Great Lakes region, North America [Ph.D. thesis]: London, Ontario, University of Western Ontario, 160 p.

—— 1975, Paleobotany of Wisconsin interstadials, eastern Great Lakes region, North America: Quaternary Research, v. 5, p. 591–619.

Coope, G. R., 1968, Insect remains from silts below till at Garfield Heights, Ohio: Geological Society of America Bulletin, v. 79, p. 753–755.

Dreimanis, A., 1969, The last ice age in the eastern Great Lakes region of North America: VIII International Association Quaternary Research, p. 69–75.

Dreimanis, A., and Goldthwait, R. P., 1973, Wisconsinan glaciation in the Huron, Erie and Ontario Lakes *in* Black, R. F., Goldthwait, R. P., and Willman, H. B., eds., The Wisconsinan Stage: Geological Society of America Memoir 136, p. 71–106.

Fullerton, D. S., and Groenwold, G. H., 1974, Quaternary stratigraphy at Garfield Heights (Cleveland) Ohio; Additional observations: Geological Society of America Abstracts with Programs, v. 6, p. 509–510.

Leonard, A. B., 1953, Molluscan faunules in Wisconsinan loess at Cleveland, Ohio: American Journal of Science, v. 251, p. 369–376.

Miller, B. B., and Wittine, A. H., 1972, The origin of Late Pleistocene deposits at Garfield Heights, Cuyahoga County, Ohio: Ohio Journal of Science, v. 76, p. 305–313.

Morgan, A. V., Morgan, A., and Miller, R. F., 1982, Late Farmdalian and Early Woodfordian insect assemblages from Garfield Heights, Ohio: Geological Society of America, Abstracts with Programs, v. 14, p. 267.

White, G. W., 1953, Sangamon soil and Early Wisconsinan loess at Cleveland, Ohio: American Journal of Science, v. 251, p. 369–376.

—— 1965, Northeast Ohio *in* Schultz, C. B., and Smith, H.T.V., eds., Guidebook for Field Conference G, VII Congress International Association of Quaternary Research: Nebraska Academy of Science, Lincoln, Nebraska, p. 82–90.

—— 1968, Age and correlation of Pleistocene deposits at Garfield Heights (Cleveland), Ohio: Geological Society of America Bulletin, v. 79, p. 749–752.

The Cuyahoga Valley National Recreation Area, Ohio: Devonian and Carboniferous clastic rocks

Joseph T. Hannibal, *Cleveland Museum of Natural History, Wade Oval, University Circle, Cleveland, Ohio 44106*
Rodney M. Feldmann, *Department of Geology, Kent State University, Kent, Ohio 44242*

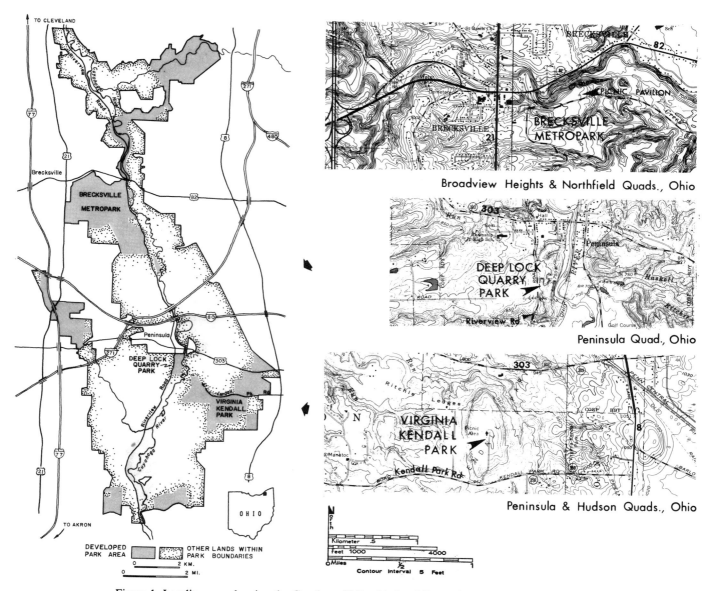

Figure 1. Locality map showing the Cuyahoga Valley National Recreation Area and the three sites discussed.

LOCATION

The Cuyahoga Valley National Recreation Area is located along the Cuyahoga River Valley between Cleveland and Akron, Ohio (Fig. 1). Major interstate highways, including I-77, I-80, and I-271, as well as several Ohio state routes, pass near, or through, the National Recreation Area. The sites discussed are located within three parks that are part of the National Recrea-

tion Area. Although one can drive to each of the three sites, final access requires some hiking, primarily along park trails.

SIGNIFICANCE OF SITES

The Cuyahoga Valley National Recreation Area contains excellent exposures of several of northeastern Ohio's classic and well-studied Paleozoic rock units. In Brecksville Metropark rocks

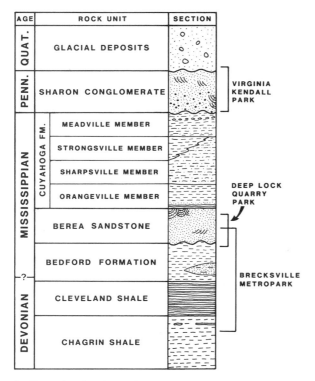

Figure 2. Chart showing the major rock units exposed in northeastern Ohio. Rock units found at the sites discussed are indicated by brackets.

ranging in age from Devonian to Mississippian, the Chagrin Shale through Berea Sandstone, can be examined (Fig. 2). Deep Lock Quarry Park and Virginia Kendall Park provide superior views of the Mississippian Berea Sandstone and the Lower Pennsylvanian Sharon Conglomerate, respectively (Fig. 2). These rock units have been the subject of study by numerous geologists during the last 100 years. Their mode of deposition, source area, and, in some cases, geologic age have been ongoing topics of debate. The rock units of the Cuyahoga Formation (Fig. 2), which lie between the Berea Sandstone and the Sharon Conglomerate, are also exposed in and near the National Recreation Area, as at Akron Gorge Park to the southeast.

BRECKSVILLE METROPARK

Outcrops of northeastern Ohio's oldest rocks can be examined along more than 0.6 m (1 km) of Chippewa Creek, within the Brecksville Metropark, part of the Cleveland Metropolitan Park System. The most convenient parking areas are at the park entrance, off Ohio 82 in Brecksville, and at a picnic pavilion along the park road (Fig. 1) that runs south of, and parallel to, Chippewa Creek. In outcrops along Chippewa Creek the Chagrin Shale, Cleveland Shale, Bedford Formation (including the Euclid Member), and the Berea Sandstone are exposed (Fig. 3).

The Chagrin Shale is predominantly a silty, soft gray shale. At this locality the unit contains several thin siltstone beds usually less than 2.8 in (7 cm) thick and commonly filled with small,

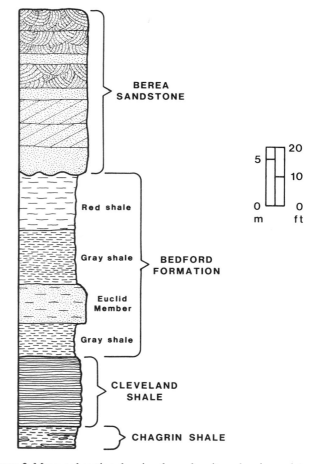

Figure 3. Measured section showing the rock units and major rock types exposed along Chippewa Creek within the Brecksville Metropark. (Adapted from Feldmann and others, 1977, Fig. 2.52.)

branching burrows. Pyrite layers and small concretions are also present. The Chagrin Shale is uncharacteristically fossiliferous here, containing an assemblage of body fossils dominated by articulate brachiopods. Inarticulate brachiopods, cephalopods, and other types of fossils are also present. The fossils can be studied best on the south side of the stream in and along the stream bed. The Chagrin Shale is usually considered to be Upper Devonian, specifically Famennian, in age.

Overlying the Chagrin Shale is a thin, discontinuous, pyritic layer containing carbonized plant fragments and water-worn fish bones. This bed marks the base of the Cleveland Shale. The remainder of the unit is predominantly black, fissile shale with some gray shale and siltstone beds. In outcrop, the black shales have a blocky appearance, in part due to jointing. The shale at this site contains small pyrite concretions and some plant material, the latter preserved as carbonaceous films. The Cleveland Shale has yielded numerous remains of large placoderms and sharks at nearby localities in the Cleveland area. The unit can be examined best on the north side of the stream.

The Bedford Formation, which overlies the Cleveland Shale, consists of gray shale, a massive to flaggy sandstone and

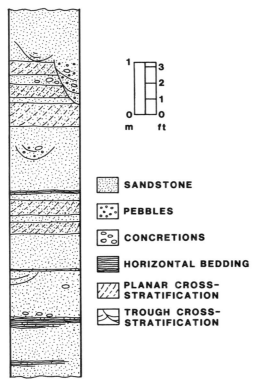

Figure 4. Measured section of Berea Sandstone at Deep Lock Quarry Park.

siltstone unit known as the Euclid Member, and red shale. Fossils are not common in the Bedford at this locality, but sedimentary structures are abundant. The Euclid Member contains interference ripple marks, load casts, and ball-and-pillow structures (Coogan and others, 1981) that can be examined readily on blocks of float in, and alongside, the stream. In situ Bedford rocks can be examined near the point at which the Ohio 82 bridge crosses the stream. The age of the Bedford is problematical. According to one review of the biostratigraphic evidence (DeWitt, 1970), the lowermost Bedford is either Upper Devonian or Lower Mississippian in age and the remainder of the unit is Mississippian.

Cropping out above the Bedford, the Berea Sandstone is a predominantly planar to trough cross-stratified sandstone, composed primarily of fine- to medium-grained quartz sand. These features can be observed easily within large blocks of the Berea on the south side of the stream valley. Most consider the Berea Sandstone to be Mississippian, specifically Kinderhookian, in age.

Rocks of the Cuyahoga Formation, including the Orangeville and Sharpsville members, are exposed in and along Chippewa Creek west of the Ohio 82 bridge, outside of the Brecksville Metropark. These rock units are discussed in Coogan and others (1981).

DEEP LOCK QUARRY PARK

Deep Lock Quarry Park is part of the Akron Metropolitan Park System. The park entrance is on Riverview Road, about 0.8 mi (1.3 km) south of the intersection of Riverview Road and Ohio 303 (Fig. 1). A trail from the parking lot leads north to the quarry. Deep Lock Quarry is one of the few inactive Berea Sandstone quarries in which good exposures of the formation can be examined readily. Sandstone was quarried here for hulling stones (used for preparing grain) and for building stone. Berea Sandstone was probably also used in construction of the Ohio and Erie Canal lock preserved in the park.

As seen in the old quarry floor and walls, the Berea Sandstone is a clean, fine- to medium-grained sandstone composed primarily of quartz. Dark brownish-red concretions, usually less than 2 in (5 cm) in length, and small quartz pebbles are common in some intervals. Both planar and trough cross-beds, as well as both massive and horizontal bedding, can be seen readily along the quarry walls (Fig. 4). There is also some shale of the Bedford Formation exposed beneath the level of the quarry in this Park.

VIRGINIA KENDALL PARK

Virginia Kendall Park is located on Ledges Road, which can be reached via Ohio 8 and Ohio 359 (Fig. 1). Containing one of the largest and most varied exposures of the Sharon Conglomerate, the park can be reached by following the trails (Fig. 5) north and east from Ledges Parking Lot or south from the park information center located off Ohio 303. At this locality and elsewhere the Sharon Conglomerate consists predominantly of sandstone and conglomerate. The sandstone at Virginia Kendall Park is typically medium-grained, relatively clean, and friable. Conglomerate layers and lenses are common. Most of the pebbles within the conglomerate layers and in the more sandy layers are milky white quartz, although some are quartzite, jasper, or limestone. Trough cross-stratified cut-and-fill structures and high angle, planar cross-beds are abundant here as are several layers displaying overturned cross-beds (Fig. 6). Near Ice Box Cave (Fig. 5), a conglomerate-filled channel is well exposed. Jointing is an obvious feature in the Sharon, and the cave is formed by movement along joints and slippage on the upper surface of the Meadville Member of the Cuyahoga Formation, exposed on the floor of the cave. Additionally, some of the vertical joint surfaces show well-developed honeycomb weathering of the conglomerate. Based upon the rare occurrence of certain fossil plants in the unit, the Sharon Conglomerate has been determined to be Lower Pennsylvanian in age.

INTERPRETATION

The Chagrin Shale through Berea Sandstone sequence in northeastern Ohio has been interpreted in several ways over the years. Early workers tended to consider each unit as representing a disjunct depositional event, separated from the adjacent units by significant unconformities. However, more recently these rocks have been interpreted as representing integrated depositional sequences (see, for example, Kohout and Malcuit, 1969; Coogan and others, 1981).

Although the Chagrin Shale is typically devoid of body fossils, the fauna of the unit preserved at Brecksville Metropark is indicative of a well-oxygenated, shallow marine paleoenviron-

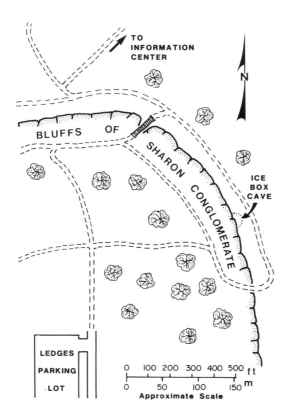

Figure 5. Map showing outcrops of Sharon Conglomerate, trails leading from the Ledges Parking Lot, and the southern portion of trails leading to the Information Center at Virginia Kendall Park. (Adapted from Feldmann and others, 1977, Fig. 2.66).

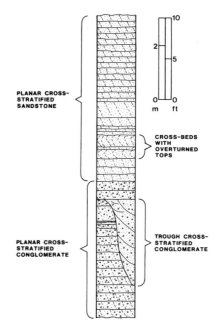

Figure 6. Measured section of Sharon Conglomerate at Virginia Kendall Park. (Adapted from Feldmann and others, 1977, Fig. 2.67a).

ment. The Cleveland Shale, due to its high organic content and the extreme scarcity of fossils of bottom-dwelling organisms, is usually considered to have been deposited in an anaerobic marine environment. However, there are widely divergent views on the depth of water in which this shale was laid down. Details of the sedimentology, stratigraphy, and depositional environment of the Chagrin and Cleveland shales have recently been summarized by Broadhead and others (1982).

In 1954, Pepper, DeWitt, and Demarest published a classic

study interpreting the Bedford and Berea sediments as a delta fan deposit. Coogan and others (1981) have presented an alternative scenario based upon the rock units as exposed in Brecksville Metropark and other localities. In their view, the gray shales of the Bedford were deposited in a bay, or lagoonal and offshore marine, environment and the red shales of the Bedford were deposited in a tidal flat environment. They interpret the Berea Sandstone, as exposed in Brecksville Metropark and in Deep Lock Quarry Park, as channel deposits, accounting for the lithology and bedding structures at these localities and the erosional contact of the Berea Sandstone with the Bedford Formation. Recent studies (Mrakovich and Coogan, 1974; Coogan and others, 1974) have found the sequence of cross-bedding types within the Sharon Conglomerate at Virginia Kendall and other localities to be indicative of braided stream deposits.

REFERENCES CITED

Broadhead, R. F., Kepferle, R. C., and Potter, P. E., 1982, Stratigraphic and sedimentologic controls of gas in shale—Example from Upper Devonian of northern Ohio: American Association of Petroleum Geologists Bulletin, v. 66, p. 10–27.

Coogan, A. H., Feldmann, R. M., Szmuc, E. J., and Mrakovich, J. V., 1974, Sedimentary environments of the Lower Pennsylvanian Sharon Conglomerate near Akron, Ohio, p. 19–41, *in* Heimlich, R. A., and Feldmann, R. M., eds., Selected Field Trips in Northeastern Ohio: Ohio Division of Geological Survey Guidebook, No. 2.

Coogan, A. H., Heimlich, R. A., Malcuit, R. J., Bork, K. B., and Lewis, T. L., 1981, Early Mississippian deltaic sedimentation in central and northeastern Ohio, p. 113–152 *in* Roberts, T. G., ed., Stratigraphy, Sedimentology (Geological Society of America, Cincinnati 1981 Field Trip Guidebooks): American Geological Institute.

DeWitt, W., 1970, Age of the Bedford Shale, Berea Sandstone, and Sunbury Shale in the Appalachian and Michigan Basins, Pennsylvania, Ohio, and Michigan: U.S. Geological Survey Bulletin, 1294-G, 11 p.

Feldmann, R. M., Coogan, A. H., and Heimlich, R. A., 1977, Field Guide: Southern Great Lakes: Dubuque, Iowa, Kendall/Hunt Publishing Company, 241 p.

Kohout, D. L., and Malcuit, R. J., 1969, Environmental analysis of the Bedford Formation and associated strata in the vicinity of Cleveland, Ohio: Compass, v. 46, p. 192–206.

Mrakovich, J. V., and Coogan, A. H., 1974, Depositional environment of the Sharon Conglomerate Member of the Pottsville Formation in northeastern Ohio: Journal of Sedimentary Petrology, v. 44, p. 1186–1199.

Pepper, J. F., DeWitt, W., and Demarest, D. F., 1954, Geology of the Bedford Shale and Berea Sandstone in the Appalachian Basin: U.S. Geological Survey Professional Paper 259, 106 p.

Wellsville Hill: Exposures in the Pennsylvanian Allegheny and Conemaugh Groups of eastern Ohio

Robert G. Wiese, Jr., Department of Geology, Mount Union College, Alliance, Ohio 44601

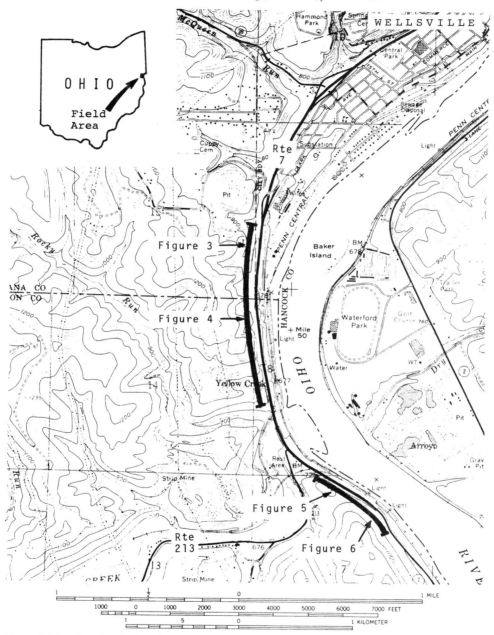

Figure 1. Map showing the location of the northern and southern sections of the Wellsville Hill area. Wellsville, Ohio–West Virginia 7½-minute Quadrangle (1978).

LOCATION

The Wellsville Hill section consists of two large roadcuts located along Ohio 7 just south of Wellsville, Ohio, on the Wellsville Ohio–West Virginia 7½-minute Quadrangle (Fig. 1).

Ohio 7 is a divided highway with wide shoulders adequate for parking at the exposures. Access to the area from western Pennsylvania and West Virginia is by bridges across the Ohio River at East Liverpool, Ohio, to the north and at Steubenville, Ohio, to the south of the field area.

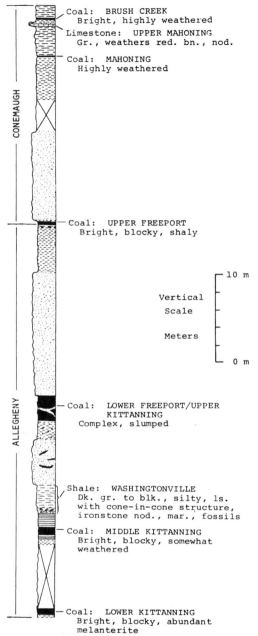

Figure 2. Wellsville Hill section, along Ohio 7 just south of Wellsville. (Modified after Denton and others, 1961).

SIGNIFICANCE OF THE SITE

The Pennsylvanian Allegheny and Conemaugh Groups exposed at this locality contain a number of the important coal beds of eastern Ohio. Although major coal seams are not well represented in the Wellsville Hill section, stratigraphic evidence of the fluvial environment associated with the coals is particularly well displayed. Extreme lateral and vertical variations in the sediments are evident, leading one to question the validity of the time-

honored concept of cyclothems. Because of their location along a sloping highway, the exposures present a fine opportunity for examination of about 180 ft (55 m) of complex Pennsylvanian stratigraphy. Further, the exposures afford an excellent illustration of the problems of correlation, even over short distances, within the coal-bearing sequence.

INTRODUCTION

The general geology of this part of Ohio has been described by Stout and Lamborn (1924) and by Lamborn (1930). General discussions of the Pennsylvanian sequence of eastern Ohio are found in Collins (1979), Frye (1979), and Kovach (1979). The Wellsville Hill section has been described by Denton et al. (1961) and by Ferm and Cavaroc (1969). For the purpose of this guide the Wellsville Hill exposures are divided into a northern section consisting of about 1 mi (1.6 km) of roadcuts extending south from Wellsville to Yellow Creek and a southern section consisting of approximately 0.5 mi (0.8 km) of roadcuts beginning 0.2 mi (0.3 km) south of the intersection of Ohio 7 with Ohio 213 (Fig. 1).

A number of suggestions by P. K. Spencer of Mount Union College and R. A. Heimlich of Kent State University have been most helpful in the preparation of this field guide.

NORTHERN SECTION

The general stratigraphic section of the northern series of roadcuts in the Wellsville Hill area is given in Figure 2. A detailed section is included in the guidebook by Denton and others, (1961, p. 188), and an interpretive stratigraphic section is given by Ferm and Cavaroc (1969, Fig. 8). The most easily identified stratigraphic marker is the Upper Freeport (No. 7) coal zone exposed high in the north part of the first big roadcut just south of Wellsville (Fig. 3). The massive, faintly reddish sandstone above the coal is the lower part of the Conemaugh Group. The Upper Freeport coal is nearly at road level just south of the Columbiana-Jefferson County line (Fig. 4). Here the coal is overlain, and partly scoured and cut out, by a massive sandstone unit. The Mahoning coal overlies the massive sandstone at this location (Fig. 4).

The Middle Kittanning (No. 6) coal is the lowest coal exposed in the Wellsville Hill section. It crops out just below the level of the railroad tracks on the north side of the mouth of Yellow Creek. Ferm and Cavaroc (1969) suggest that this may be the Lower Kittanning (No. 5) coal and place the Middle Kittanning coal above the tracks at this locality. Correlation problems of this kind are typical in this part of the Pennsylvanian System. The Middle Kittanning coal is below the level of the new highway at the north end of the big roadcut.

The sedimentary sequence of the Upper Allegheny Group exposed along the highway consists of irregularly interbedded sandstones and shales. The section exhibits lateral and vertical facies variations and contains many coal streaks and erratic and

Figure 3. Roadcut just south of Wellsville. The Upper Freeport coal zone is at the top of interbedded sandstones and shales of the Allegheny Group and is overlain by a basal sandstone of the Conemaugh Group.

Figure 4. Exposure just south of the Columbiana County–Jefferson County line. The Upper Freeport coal is overlain by sandstone. The Mahoning coal is present near the top of photograph.

Figure 5. Roadcut at the north end of the southern section showing discontinuous coal seams in the sandstones of the Freeport Formation.

Figure 6. Roadcut in the southern part of the southern section. The irregularly interbedded sandstones, siltstones, and shales are considered to be in the Freeport Formation of the upper part of the Allegheny Group.

discontinuous coal seams. The truncation of the Upper Kittanning coal by the Lower Freeport coal reported by Denton et al. (1961) is well exposed approximately 0.2 mi (0.3 km) north of the county line. Ferm and Cavaroc (1969) interpret this occurrence to be the result of slumping of the Upper Kittanning coal.

SOUTHERN SECTION

The southern series of roadcuts (Fig. 1) exposes the same general stratigraphic interval between the Middle Kittanning coal and the Upper Freeport coal. Here, however, the Upper Freeport coal is not easily visible from the road. The Middle Kittanning coal is exposed in the railroad cut just south of the rest area at the intersection of Ohio 213 with Ohio 7. The sandy clay below the coal contains the remains of roots and stumps. Ferm and Cavaroc (1969) interpret this exposure to be the lower part of a bar deposit.

The irregular coal seams occurring upsection in the north part of this series of roadcuts most likely are correlative with the Upper Kittanning coal or with the Lower Freeport coal (Fig. 5) and are channeled and cut out by the associated sandstones.

In the south part of the southern section, the stratigraphic succession consists of irregularly interbedded sandstones, siltstones, and shales, which display a variety of sedimentary structures including large- and small-scale cross bedding, discontinuous sand lenses, and cut-and-fill structures (Fig. 6). The sequence clearly represents an extensively channelized deltaic floodplain environment, possibly that of a filled interdistributary bay as described by Ferm and Cavaroc (1969).

OTHER NEARBY SITES

The Middle Kittanning shales are well exposed along the

south side of Yellow Creek on Ohio 213 west of the intersection with Ohio 7. Marine fossils, including brachiopods, pelecypods, and a few cephalopods, and some plant fossils have been found in the shales. The Middle Kittanning (No. 6) coal, the roof shale, and the Washingtonville Shale, including a zone with cone-in-

cone textures, are exposed across Ohio 213 from the post office in Hammondsville a few miles farther west. Exposures downsection in the lower part of the Allegheny Group and the upper part of the Pottsville Group of the Pennsylvanian System occur along Ohio 7 north of Wellsville.

REFERENCES CITED

Collins, H. R., 1979, The Mississippian and Pennsylvanian (Carboniferous) Systems in the United States—Ohio: U.S. Geological Survey Professional Paper 1110-E, 26 p.

Denton, G. H., and others, 1961, Pennsylvanian geology of eastern Ohio. Field trip 4 *in* Geological Society of America Guidebook for field trips. Cincinnati meeting, 1961: New York, Geological Society of America, p. 131–205.

Ferm, J. C., and Cavaroc, V. V., Jr., 1969, A field guide to Allegheny deltaic deposits in the upper Ohio Valley with a commentary on deltaic aspects of Carboniferous rocks in the northern Appalachian Plateau [1969 spring field trip]: Ohio Geological Society and Pittsburgh Geological Society, 19 p.

Frye, C. I., 1979, Generalized description of Carboniferous rocks of eastern Ohio *in* Ettensohn, F. R., and Dever, G. R., Jr., eds., Carboniferous geology from the Appalachian Basin to the Illinois Basin through eastern Ohio and Ken-

tucky: University of Kentucky, Lexington, p. 11–18.

Kovach, J., 1979, Introduction to Upper Carboniferous stratigraphy and depositional environments of eastern Ohio *in* Ettensohn, F. R., and Dever, G. R., Jr., eds., Carboniferous geology from the Appalachian Basin to the Illinois Basin through eastern Ohio and Kentucky: University of Kentucky, Lexington, p. 19–29.

Lamborn, R. E., 1930, Geology of Jefferson County: Ohio Geological Survey Bulletin 35, 304 p.

Stout, W. B., 1943, Generalized section of rocks in Ohio: Ohio Geological Survey Circular No. 4

Stout, W. B., and Lamborn, R. E., 1924, Geology of Columbiana County: Ohio Geological Survey Bulletin 28, 408 p.

Black Hand Gorge State Nature Preserve:
Lower Mississippian deltaic deposits in east-central Ohio

Robert J. Malcuit and Kennard B. Bork, *Department of Geology and Geography, Denison University, Granville, Ohio 43023*

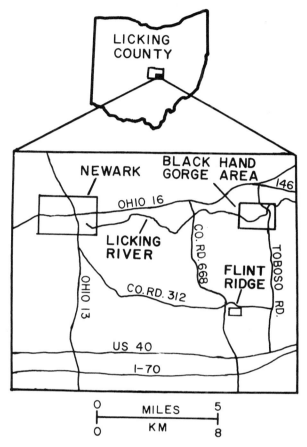

Figure 1. Index map showing location of Black Hand Gorge as well as other features (e.g., Flint Ridge) of southeastern Licking County.

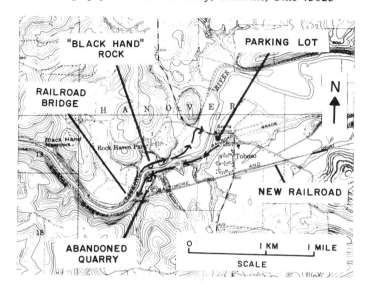

Figure 2. West-central portion of Toboso 7½-minute Quadrangle showing location of Black Hand Gorge and other features of geological and/or historical interest as well as a suggested field trip route through the eastern end of the gorge.

LOCATION AND ACCESSIBILITY

Black Hand Gorge State Nature Preserve is located about 9 mi (15 km) east of Newark, Ohio. It is readily accessible from Ohio 16 and 146, as shown on Figure 1. The area became a state nature preserve in 1975 and is accessible on a year-round basis during daylight hours. Canoeing in the river and bicycling along a recently constructed bicycle trail are possible when weather and water level permit. The gorge is located on the west-central area of the Toboso (Ohio) 7½-minute Quadrangle (Fig. 2).

SIGNIFICANCE OF SITE

Black Hand Gorge State Nature Preserve is the type locality of the Black Hand Sandstone Member of the Cuyahoga Formation. In addition to a superb exposure of about 65 ft (20 m) of

Black Hand Sandstone, the lower two members of the superjacent Logan Formation, the Berne and Byer, are well exposed here. The best rock exposures occur at the eastern end of the gorge. Here, one can view the units on naturally weathered outcrops, in railroad cuts, and in an abandoned sand quarry, as well as in a tunnel used in earlier times for an electric interurban streetcar track.

The gorge owes its scenic beauty to Pleistocene glacial activity. At least one Pre-Wisconsinan ice sheet advanced from a westerly direction to the vicinity of the gorge (Mather, 1909; White, 1939; Forsyth, 1966). Processes associated with the ice front, such as morainal and ice damming, created a meltwater lake whose water subsequently helped to cut the gorge.

The gorge area is also of interest for its archaeological and cultural history. The Black Hand, for which the gorge is named, was an ancient, large, dark, hand-shaped petroglyph that was carved into the sandstone by native American inhabitants of the area. In 1828, construction workers blasted away the petroglyph to make way for a towpath for the Ohio and Erie Canal. Later, this natural gorge area was host to a railroad line on the south side of the river and a streetcar track on the northern side. Many vestiges of these historic works are preserved in the gorge and, indeed, some of the best rock outcrops are exposed in cuts and excavations that were originally made for commercial/industrial purposes, such as the large quarry developed in the nineteenth century by the Everett Glass Company.

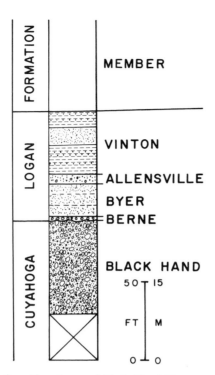

Figure 3. Stratigraphic column of Black Hand Gorge area (modified from Coogan and others, 1981). Lithologies are as follows: Black Hand Member—conglomeratic sandstone; Berne Member—sandstone conglomerate; Byer Member—fine-grained sandstone; Allensville Member—conglomeratic sandstone, siltstone, and shale; Vinton Member—sandstone, siltstone, and shale.

LOWER MISSISSIPPIAN CLASTIC SEDIMENTATION

A stratigraphic section for the area of Black Hand Gorge is shown in Figure 3. The Black Hand Sandstone is the only unit of the Cuyahoga Formation exposed in the gorge. To the west, along Ohio 16, the Black Hand thins, by way of a facies change with the subjacent Raccoon Shale, to extinction just west of Granville, Ohio. To the east of the gorge area, the Black Hand dips into the subsurface. A well drilled at the Everett Glass Sand Quarry (Fig. 2, "abandoned quarry") penetrated over 325 ft (100 m) of Black Hand Sandstone before reaching the underlying Raccoon Shale (Ver Steeg, 1947).

Two units of the Logan Formation, the Berne Conglomerate and the Byer Sandstone, are well exposed in the gorge area. The upper two members, the Allensville and Vinton Members, are well exposed in the Bowerston Shale Quarry, which is located about 1.25 mi (2 km) north of the western end of the gorge. (See Coogan and others, 1981, for outcrop description.)

The most striking features of the Black Hand Sandstone in the gorge area are the conglomeratic nature of the sandstone and the abundance of cross-bedding within it. Quartz and chert are the main pebble lithologies and trough cross-bedding is predominant. No fossils have been found in the Black Hand Sandstone

Figure 4. Erosional surface between Black Hand Sandstone (below) and Berne Conglomerate (above). Hammer handle is 15 in (38 cm) long.

within the gorge. However, some marine fossils have been found in the uppermost unit of the Black Hand Sandstone in outcrops along Ohio 16 just northwest of the gorge. Trace fossils can occasionally be found on iron-oxide-rich bedding planes in the gorge area.

The Black Hand Sandstone in the preserve is overlain by 0 to 5 ft (0 to 1.5 m) of Berne Conglomerate of the Logan Formation (type locality is Logan, Hocking County, Ohio). The upper surface of the Black Hand is an obvious erosional surface, as evidenced by the presence of channels and truncation of sets of crossbeds (Fig. 4). Berne pebbles have petrologic characteristics and size range that are very similar to Black Hand sediments. Many investigators have concluded that the Berne Conglomerate was derived by reworking material of Black Hand lithology.

The fine-grained Byer Sandstone Member of the Logan Formation has very few characteristics in common with the subjacent Berne Conglomerate. However, the contact between the two units is conformable, ranging from sharply conformable to

Figure 5. Typical outcrop of Black Hand Sandstone along bicycle trail in eastern end of Black Hand Gorge showing characteristic cross-bedding and honeycomb weathering. Geologic pick is 11 in (28 cm) long.

gradational over a vertical distance of 6 to 8 in (15 to 20 cm) in places. Marine fossils can be found in the area, occasionally in the Berne and frequently within the Byer Member. For more details on the stratigraphy and paleontology of the region, see Bork and Malcuit (1979) and Coogan and others (1981). The classic work of Hyde (1915, 1953) offers numerous details, of descriptive and interpretive nature, about the Lower Mississippian strata and paleontology of central and southern Ohio.

Bork and Malcuit (1979) use a sequence of block diagrams to present a depositional model for the Cuyahoga and Logan Formations. In their model, the source region for the Black Hand Sandstone is to the east—"Poconoland" of Pelletier (1958). In time the deltaic complexes prograded into the area of Black Hand Gorge. During their formation, these deltaic complexes underwent strong marine influence and some of the sediments were worked into a barrier bar complex trending in a north-northwesterly direction. Eventually, coarse clastic sedimentation ceased, and the top part of the Black Hand Sandstone was reworked by marine processes. After some sea-level rise or basin subsidence, the widespread marine sheet sands of the Logan Formation were deposited over the east-central Ohio area.

SUGGESTED FIELD TRIP ROUTE

The reader is referred to Figure 2 for a tour of the eastern part of the gorge. Starting at the main parking lot of the nature preserve, one can proceed westward along a bicycle trail located on an abandoned railroad bed on the southern bank of the Licking River. In a few hundred meters there are very good naturally weathered outcrops of Black Hand Sandstone on the left of the trail. This series of outcrops shows both the typical complex trough cross-bedding (Fig. 5) and the characteristic honey-

comb weathering of this unit. At about the 0.6 mi (1 km) point there is a rock cut that was excavated for railroad construction during the late nineteenth century. Just beyond the rock cut is an abandoned quarry where sand was mined for use in the nineteenth century glass industry in Newark. The contacts between the Black Hand Sandstone, the Berne Conglomerate, and the Byer Sandstone can be examined along the quarry walls.

The more recent railroad grade is located a few tens of meters beyond the abandoned quarry. Since there is a wooden-floored pathway on the high-level railroad bridge, it is possible to cross over the river. However, the tracks are currently IN USE and extreme CAUTION is suggested. (An alternate route to the northern side of the gorge is to return to the parking lot, cross the bridge over the Licking River, and proceed westward on a footpath on the northern side of the river.)

As one travels down the footpath that passes under the bridge, on the northern bank of the river, one can examine several outcrops of Black Hand Sandstone. A few hundred meters eastward on this path (which is the grade for an abandoned interurban system) is "Black Hand" Rock. This mass of Black Hand Sandstone is an erosional remnant from the excavation of the gorge by glacial meltwater. A portion of the towpath for the Ohio and Erie Canal is located on the river side of "Black Hand" Rock and it was for this towpath that the Black Hand petroglyph was destroyed. After returning to the path and proceeding a few tens of meters eastward, one comes to the entrance of a tunnel cut in the Black Hand Sandstone. There is an excellent display of cross-bedding on the western end of the tunnel. A few hundred meters beyond the eastern end of the tunnel are remnants of a canal lock constructed from blocks of Black Hand Sandstone. A wooded embankment running from the lock southeastward to the river leads one to the site of a former timber-rock dam constructed for the Ohio and Erie Canal. When the canal was operating, the Licking River was dammed up at the eastern end so that canal boats could be towed along a placid Black Hand Lake. From the canal lock, the footpath continues eastward to the Toboso Road. One can then turn right (south) and walk across the bridge to the main parking lot of the nature preserve.

GLACIAL HISTORY

The following rendition of the glacial age history of the Black Hand Gorge area is a modified version of Mather's (1909) analysis. County-wide and regional views of the glacial geology are found in White (1939), Dove (1960), and Forsyth (1966). As stated previously, the excavation of Black Hand Gorge was caused by processes associated with drainage of a pre-Wisconsinan ice-dammed lake (Fig. 6). The Pre-Pleistocene drainage of the area was to the west, by way of a major tributary of the ancient Teays River system (the Pre-Pleistocene Cambridge River) that occupied the broad valley (Hanover Valley) just north of the gorge area along which Ohio 16 is located (Figure 6a and d). Currently Ohio 16 follows the ancient Cambridge River valley from Newark to Coshocton, Ohio. The

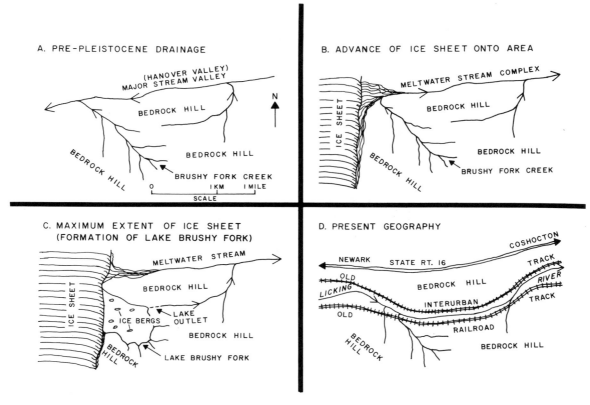

Figure 6. Diagrammatic sketches showing three stages in a model for the evolution of Black Hand Gorge and the present geography of the gorge area. Description of evolutionary sequence is in text.

direction of drainage in Hanover Valley was changed at some time prior to the excavation of the gorge (Dove, 1960).

As the ice sheet approached the western end of the gorge (Fig. 6b), the meltwater flowed from the ice front and passed eastward out through the broad valley north of the gorge. Eventually the apron of outwash debris at the ice sheet terminus in Hanover Valley became high enough to cause a lake to form in Brushy Fork Valley to the south and west of the gorge (Fig. 6c). As still more outwash accumulated in Hanover Valley, the level

of the debris eventually became higher than the outlet of Lake Brushy Fork. The topographic low of the periphery of the lake then became a spillway for large quantities of glacial meltwater. Vertical downcutting predominated as the lake outlet was lowered by erosion, a process that ultimately resulted in the steep walls of the Black Hand Gorge area. Since that time the flow of the Licking River has been eastward, through the steep-walled gorge, rather than through the much broader ancient valley to the north of the gorge.

REFERENCES CITED

Bork, K. B., and Malcuit, R. J., 1979, Paleoenvironments of the Cuyahoga and Logan Formations (Mississippian) of central Ohio: Geological Society of America Bulletin, v. 90, Pt. I, p. 1091–1094; Pt. II, p. 1782–1838.

Coogan, A. H., Heimlich, R. A., Malcuit, R. J., Bork, K. B., and Lewis, T. L., 1981, Early Mississippian deltaic sedimentation in central and northeastern Ohio in Roberts, T. G., ed., GSA Cincinnati '81 Field Trip Guidebooks, v. 1, Stratigraphy, Sedimentology: American Geological Institute, p. 113–152.

Dove, G. D., 1960, Water resources of Licking County, Ohio: Ohio Department of Natural Resources Division of Water, Bulletin 36, 96 p.

Forsyth, J. L., 1966, Glacial map of Licking County, Ohio: Ohio Geological Survey, Report of Investigations No. 59.

Hyde, J. E., 1915, Stratigraphy of the Waverly Formations of central and south-

ern Ohio: Journal of Geology, v. 23, p. 655–682.

——1953, The Mississippian formations of central and southern Ohio: Ohio Geological Survey Bulletin 51, 355 p.

Mather, K. F., 1909, Age of the Licking Narrows: Bulletin of the Scientific Laboratories of Denison University, v. 14, p. 175–187.

Pelletier, B. R., 1958, Pocono paleocurrents in Pennsylvania and Maryland: Geological Society of Americ Bulletin, v. 69, p. 1033–1064.

Ver Steeg, Karl, 1947, Black Hand Sandstone and Conglomerate in Ohio: Geological Society of America Bulletin, v. 58, p. 703–728.

White, G. W., 1939, Illinoian drift of eastern Ohio: American Journal of Science, v. 237, p. 161–174.

Flint Ridge, Ohio: Flint facies of the Pennsylvanian Vanport Limestone

Ernest H. Carlson, Department of Geology, Kent State University, Kent, Ohio 44242

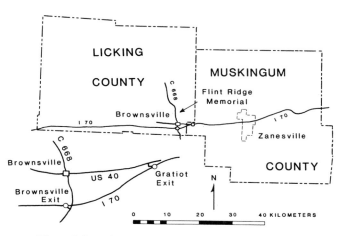

Figure 1. Location map showing the route to Flint Ridge.

LOCATION AND ACCESSIBILITY

Flint Ridge is located in southeastern Licking and western Muskingum counties in east-central Ohio (Fig. 1). Geological interest centers on ⌐ ↳₁Ridge State Memorial and a roadcut exposure nearby (Fig. 2). This area can be reached easily from Brownsville by taking I-70 to the Brownsville exit, if traveling east, or to the Gratiot exit and continuing west on U.S. 40 for 1.9 mi (3 km), if traveling west (Figs. 1, 2). The memorial is situated 3 mi (5 km) north of Brownsville where Licking County Road 668 intersects Flint Ridge Road (Licking County Road 312). Access to the backbone of the ridge is provided by Flint Ridge Road, which runs east and west from the memorial.

The memorial includes a museum (built over a prehistoric Indian quarry), a trail that winds through old flint workings, picnic facilities, and a parking area. It is open from April through October, but closed Mondays. The roadcut exposure is reached by walking 0.2 mi (0.3 km) north of the memorial along Licking County Road 668. No parking is available here, and due to the sharp curve and traffic along the road, caution must be observed at all times. Exposures of flint can also be observed on two tracts of private land if appropriate arrangements are made. The owners are Clayton Mason, in Licking County, 0.9 mi (1.5 km) east of the memorial on Flint Ridge Road, and John Nethers, in western Muskingum County on Muskingum County Road 8 just northwest of its intersection with Hopewell Township Road 292 (Fig. 3).

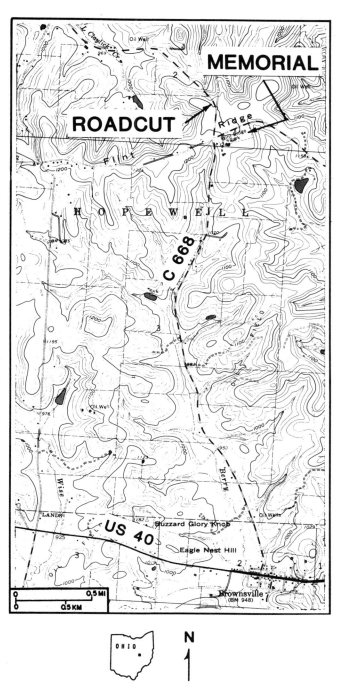

Figure 2. Part of the Glenford, Ohio, 7½-minute Quadrangle (photorevised 1972), showing the locations of Flint Ridge Memorial and the roadcut exposure.

E. H. Carlson

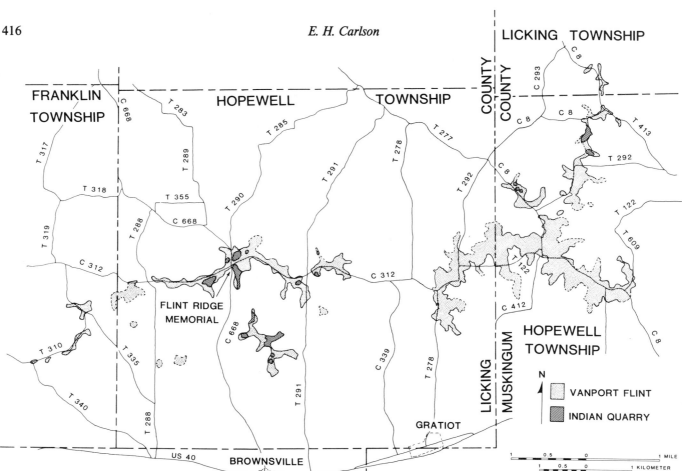

Figure 3. Map of Flint Ridge showing the lower contact of the flint facies and the location of the prehistoric Indian quarries. The contact of the flint is modified from DeLong (1972) and the quarry locations are modified from Mills (1921).

SIGNIFICANCE OF LOCALITY

Flint Ridge is upheld by a sheetlike body of massive flint that is a locally developed facies of the Pennsylvanian (Lower Allegheny) Vanport Limestone. The Vanport flint and several similar Pottsville and Lower Allegheny flint beds in the Appalachian Plateau collectively form an important type of siliceous deposit in which structures and paleogeographic setting differ from those of both the true bedded cherts (e.g., the Monterey chert) and the nodular cherts in limestones. The Flint Ridge deposit is a famous locality for this kind of material, which is associated with marine sediments that were marginal to Pennsylvanian deltas.

VANPORT LIMESTONE IN THE FLINT RIDGE AREA

The Vanport Limestone is a widespread marine horizon in the Lower Allegheny Group (Pennsylvanian) of eastern Ohio and western Pennsylvania (Lamborn, 1951). Along Flint Ridge, the upper part of the Vanport horizon is occupied by a flint facies that

shows patchy development in areas adjacent to the ridge (Stout, 1918; Stout and Schoenlaub, 1945; DeLong, 1972; Fig. 4 of this report). Because of its resistance to weathering, the flint forms the cap rock of the ridge. Historically, the rock in this region has been termed flint (although typically light colored), and that usage is continued here. The flint overlies gray limestones and interbedded dark gray shales that are more persistent laterally. According to Stout (1918), the contact of the flint with the shaly limestone below is sharp and conformable; the contact is usually covered, however. The Vanport Limestone is distinguished by a fusulinid zone that consists of *Fusulina carmani* and *Wedekindellina euthysepta* (Smyth, 1957). These fusulinids dominate the flint portion of the formation along with a few brachiopods, gastropods, pelecypods (Stout, 1918), and sponge spicules (Mills, 1921). The limestone and shale units, on the other hand, carry an abundant megafauna (Stout, 1918). The Vanport Limestone is underlain by the Clarion Shale and Sandstone and overlain by the Kittanning Shale and Sandstone (Fig. 4).

CHARACTER OF THE FLINT

The flint forms a massive, light-colored layer that lacks

prominent bedding planes. The sheetlike body is 8 mi (13 km) long, and it reaches a maximum width of 3 mi (5 km) at the eastern end of the ridge (Fig. 3). It is continuous except for a small area just east of the ridge center, where the flint is absent (Stout and Schoenlaub, 1945; DeLong, 1972). Thickness averages 4 ft (1.2 m), with a maximum of 12 ft (3.7 m) near the southeastern corner of the memorial.

The purity of the flint varies across and along the ridge. Prehistoric Indian quarries mark locations where quality is highest (Fig. 3); the Indians could chip and shape only pure materials (Mills, 1921). The quarries are located in two main regions: 1) along the ridge near the memorial and the spur extending southeast from the memorial, and 2) along the northeastern section of the ridge. The impure flint is found in broad areas of the western and eastern parts of the ridge; pioneers utilized this material and fashioned it into grinding stones.

The chemical composition of two samples, one from Indian quarries along an 5 mi (8 km) length of the ridge and the other from a newer quarry at the southeastern corner of the memorial, was reported by Stout (1918) and Stout and Schoenlaub (1945), respectively. The weight percentage of SiO_2 (96.4% in the former sample and 98.9% in the latter) indicates that the purity of the Flint Ridge material ranks among the highest in the world (Frondel, 1962). This pure flint is recognized by its conchoidal fracture and waxy luster, while impure material shows an uneven to even fracture and a dull surface.

The writer (Carlson, unpublished manuscript) recognizes four types of flint in the area on the basis of color, purity, and structure. Previous workers have not described the textures and structures, which are an important feature of these rocks, and, except for Mills (1921), the presence and significance of sponge spicules have not been recognized. The four varieties include: 1) massive, white flint, 2) bluish-gray flint, 3) ribbon flint, and 4) porous, light-brown flint. The four types are composed chiefly of equidimensional grains of microcrystalline quartz that average less than five microns in diameter. The first two kinds of flint can be observed at the roadcut exposure. The white flint is an impure, massive, opaque-appearing material with an uneven to conchoidal fracture. The bluish-gray flint is pure and is translucent with a conchoidal fracture. Both types of flint contain silicified fusulinids but lack sponge spicules. In some places, the two types occur together as a breccia composed of angular clasts of white flint enclosed by the gray variety. The ribbon flint, which can be observed on the Nethers property, is massive and has an uneven to conchoidal fracture. This flint displays alternating light- and dark-colored laminations that vary from 0.05 to 1 mm in thickness, averaging 0.5 mm. The colored layers are commonly brown, red, or gray and contain minute grains of brown and red iron oxides. Variation in concentration of the iron oxides accounts for the darkness of the layers. Although the layers are contorted, individual laminae are uniform in thickness and can be traced laterally; they appear to represent a type of rhythmic bedding. No fusulinids or spicules have been observed in the ribbon flint.

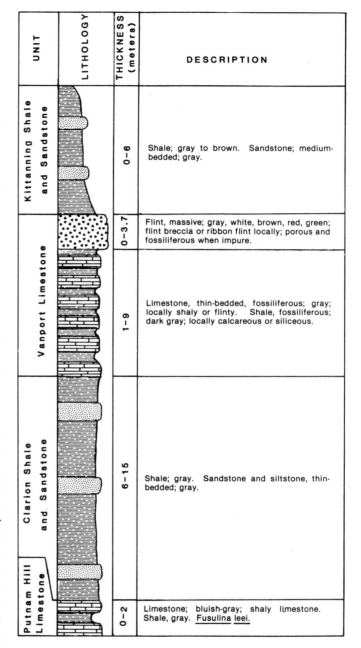

Figure 4. Generalized stratigraphic section for Flint Ridge.

The porous flint occurs along the western and southeastern portions of the ridge. This impure material has an uneven fracture, is typically light brown to white, and is riddled with molds of fusulinids and irregularly shaped cavities. Because of its dense mat of sponge spicules, the porous material can be termed a "spiculite." Although not well exposed, loose fragments of it can be observed in ditches and gullies. It also occurs in some areas as breccias composed of clasts of the porous brown material imbedded in a massive gray flint that is locally translucent. The gray flint, in contrast to the porous material, contains silicified fusulinids and fewer spicules. The four flint lithologies are cut by an irregularly spaced network of uneven fractures and vugs, the latter being fracture controlled. The fractures and vugs are filled with fibrous chalcedony and megaquartz.

ORIGIN OF THE DEPOSIT

The paleogeography, fauna, textures, and structures of the flint provide clues that bear on its origin. There is evidence that silica was precipitated both during and after sedimentation, but before diagenesis. The Vanport Limestone host was deposited during a widespread marine transgression from the west into eastern Ohio and western Pennsylvania (Ferm, 1970; Wanless and others, 1970). The Vanport fusulinids indicate shallow marine water of normal salinity (Dunbar, 1957). Apparently, the Flint Ridge area was located near the northern border of the Vanport sea, since north of the ridge the marine sediments of that horizon are replaced by Clarion beds (Stout, 1918; Lamborn, 1954; Cavaroc and Ferm, 1968; DeLong, 1972). Both the Clarion and overlying Kittanning sediments have been interpreted as parts of Lower Allegheny deltaic sequences (Ferm, 1970; Wanless and others, 1970).

The occurrence of siliceous sponge spicules in impure phases of the Vanport flint shows that the deposit is partially biogenic. The purity of the flint points to a depositional site that was isolated from a source of detrital sediments. The light color and the presence of trace amounts of iron oxides, rather than pyrite, indicate that deposition of the silica occurred in an oxidizing environment. Evidence of replacement is provided by the purer types of flint that contain fusulinids and other fossils that have been silicified. The paucity of spicules in these latter rocks, combined with replacement textures, suggests that the spicules may have been dissolved completely, and the released silica was then reprecipitated as microgranular quartz. The ultimate source of the silica for the sponge populations appears to be land-derived, silica-bearing waters, based on proximity of the depositional site to the Vanport shore.

Two characteristics of the breccias suggest that they are sedimentary rather than tectonic and that the matrix of the breccias was a siliceous sediment. These features are: 1) clasts that consist of a single lithology, and 2) a breccia matrix composed of microgranular quartz rather than megaquartz and fibrous chalcedony. The megaquartz and chalcedony are localized exclusively in later cross-cutting fractures and vugs.

REFERENCES CITED

Carlson, E. H., Minerals of Ohio: Unpublished manuscript submitted to the Ohio Geological Survey.

Cavaroc, V. V., Jr., and Ferm, J. C., 1968, Siliceous spiculites as shoreline indicators in deltaic sequences: Geological Society of America Bulletin, v. 79, p. 263–271.

DeLong, R. M., 1972, Bedrock geology of the Flint Ridge area, Licking and Muskingum Counties, Ohio: Ohio Geological Survey Report of Investigation 84, one sheet with text.

Dunbar, C. O., 1957, Fusuline foraminifera in Treatise on marine ecology and paleontology: Geological Society of America Memoir 67, v. 2, p. 753–754.

Ferm, J. C., 1970, Allegheny deltaic deposits in Deltaic sedimentation, modern and ancient: Society of Economic Paleontologists and Mineralogists Special Publication 15, p. 246–255.

Frondel, C., 1962, The system of mineralogy, v. 3, silica minerals: New York, Wiley and Sons, 334 p.

Lamborn, R. E., 1951, Limestones of eastern Ohio: Ohio Geological Survey Bulletin 49, 377 p.

——1954, Geology of Coshocton County: Ohio Geological Survey Bulletin 53, 245 p.

Mills, W. C., 1921, Flint Ridge: Ohio State Archaeological and Historical Quarterly, v. 30, p. 90–161.

Smyth, P., 1957, Fusulinids from the Pennsylvanian rocks of Ohio: Ohio Journal of Science, v. 57, p. 257–283.

Stout, W., 1918, Geology of Muskingum County: Ohio Geological Survey Bulletin 21, 351 p.

Stout, W., and Schoenlaub, R. A., 1945, The occurrence of flint in Ohio: Ohio Geological Survey Bulletin 46, 110 p.

Wanless, H. R., Baroffio, J. R., Gamble, J. C., Horne, J. C., Orlopp, D. R., Rocha-Campos, A., Souter, J. E., Trescott, P. C., Vail, R. S., and Wright, C. R., 1970, Late Paleozoic deltas in the central and eastern United States in Deltaic sedimentation, modern and ancient: Society of Economic Paleontologists and Mineralogists Special Publication 15, p. 215–245.

John Bryan State Park, Ohio: Silurian stratigraphy

William I. Ausich, *Department of Geology and Mineralogy, The Ohio State University, Columbus, Ohio 43210*

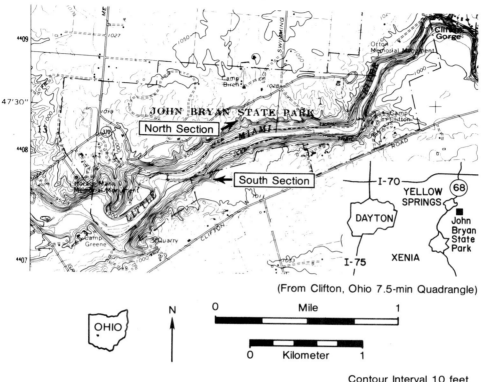

Figure 1. Locality map for the North Section and South Section at John Bryan State Park, Greene County, Ohio.

LOCATION

John Bryan State Park is located 3 mi (4.8 km) southeast of Yellow Springs, Ohio. Outcrops are highwall exposures of a gorge cut by the Little Miami River (Fig. 1). Access to the two sections in John Bryan State Park is by foot trails on either side of the Little Miami River. The trails originate from the lower picnic area parking lot. A foot bridge spans the Little Miami 0.25 mi (0.4 km) east of the lower picnic area parking lot. The North Section is beneath a waterfall and in a stream bed 0.5 mi (0.8 km) east of the bridge along the north river trail, and the South Section is beneath a waterfall and in a stream bed 0.25 mi (0.4 km) east of the bridge along the south river trail. Locations are as follows: North Section, NW¼SW¼NW¼Sec.1,T.4,R.8; South Section, SE¼NW¼SE¼Sec.1,T.4,R.9; both sites on Clifton U.S.G.S. 7½-minute Quadrangle, Greene County, Ohio (Fig. 1).

SIGNIFICANCE OF SITE

Exposed at John Bryan State Park are parts of the entire Silurian section of this part of Ohio. Included at the sites in ascending stratigraphic order are the upper 24.3 ft (7.4 m) of the Brassfield Formation, Dayton Formation, Osgood Shale, Laurel Limestone, Massie Shale, Euphemia Dolomite, Springfield Dolomite and the lower 17.7 ft (5.4 m) of the Cedarville Dolomite (Fig. 2). The Early Llandoverian through Middle Wenlockian is represented with only four time breaks. Each formation is accessible, and each is displayed in its typical character in the composite north and south section. The type sections for the Dayton Formation, Massie Shale, Springfield Dolomite and Cedarville Dolomite are all within a 15 mi (24 km) radius of John Bryan State Park.

STRATIGRAPHIC NOMENCLATURE

Nomenclature of the Silurian System of Ohio reflects the varied, complex facies relationships of these rocks. The Greene County area was situated on the Cincinnati Platform during the Silurian. Facies and the resulting rocks varied north along the

Cincinnati platform and into the Michigan Basin, east into the Appalachian Basin and southeast obliquely into the Appalachian Basin. Rexroad and others, (1965), Horvath and Sparling (1967), Horvath (1967, 1969), Janssens (1977), Berry and Boucot (1970) and Rexroad and Nicoll (1972) have established and discussed Ohio Silurian stratigraphy; however, no comprehensive work has been completed on the entire Silurian section of the Greene County area since Foerste (1935). Stout, VerSteeg and Lamb (1943), Norris (1950) and Horvath and Sparling (1967) each discussed the Silurian stratigraphy of Greene County, yet the nomenclature used by each is different. The nomenclature of Horvath and Sparling (1967) is used herein, because this represents the more commonly followed scheme of local geologists. Nomenclature of Horvath and Sparling (1967) was followed basically by Berry and Boucot (1970). The Silurian stratigraphy of Highland and Adams counties has received more attention than that of the Greene County area (Rexroad and others, 1965). Correlation of Silurian outcrop stratigraphy between Greene County and the Highland-Adams County area is given in Figure 3.

LOWER AND MIDDLE SILURIAN GEOLOGY

Brassfield Formation

The upper 24.3 ft (7.4 m) of the Brassfield is exposed in the South Section (Fig. 2). The entire thickness of the Brassfield is approximately 26 ft (8 m) in this area, and the Belfast Member of the Brassfield is present in the subsurface (Stith and Stieflitz, 1979). The Brassfield is a white to grayish-orange to pink crinoid-bryozoan grainstone. Bedding is lenticular with bedding thickness ranging from 1 to 12 in (2 to 30 cm) and averaging approximately 4 in (10 cm). Crinoidal debris, bryozoans, trilobites, brachiopods, corals and stromatoporoids are the dominant fossils present. Brassfield conodonts (Rexroad, 1967) and corals (Laub, 1979) have been studied recently.

Dayton Formation

The Dayton Formation is a grayish-orange to white, saccharoidal, coarsely crystalline dolomite. It is 12.5 ft (3.8 m) thick and even bedded. Dolomite crystals are euhedral to subhedral, irregular zones of limonite staining are present, and remnant echinoderm fragments are present in thin sections. Fossils are not common in the Dayton Formation at John Bryan State Park. In the Dayton area, Foerste (1935) reported a dominantly brachiopod fauna in the Dayton Formation; but corals, cephalopods, bryozoans and crinoids are also present.

Osgood Shale

At John Bryan State Park, 20.3 to 26.6 ft (6.2 to 8.1 m) of the Osgood Shale are present. The Osgood is characterized by interbedded blue-gray silty shale and mottled dark gray to green-gray very sparsely fossiliferous, recrystallized micrite. In the lower

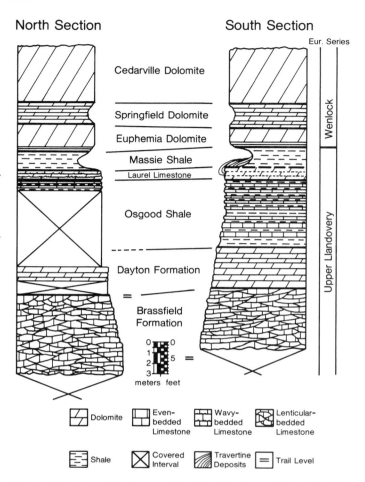

Figure 2. Stratigraphic columns of the North Section and South Section at John Bryan State Park. The entire section is Lower Silurian. Note the trail level for each section.

three-quarters of the Osgood, shale predominates over limestone; however, approximately 6.6 ft (2 m) above the base of the formation, two prominent limestone beds (5.9 in [15 cm] thick) are present. The upper quarter of the Osgood is composed of approximately equal proportions of limestone and shale. In the upper quarter, limestone and shale beds average approximately 3.9 in (10 cm) in thickness.

The Osgood contains considerable burrow mottling. In thin section the limestones contain echinoderm fragments and (?)ostracods. In outcrop fossil abundance is relatively low; but crinoids, bryozoans, brachiopods, gastropods, trilobites and corals have been collected. Foerste (1935) reported a coral and brachiopods from the Osgood.

Laurel Limestone

The Laurel Limestone is 2.9 ft (0.9 m) of recrystallized and dolomitized crinoid packstone. It is dark gray in color with common euhedral to subhedral dolomite rhombs. Average bed thickness is approximately 3.9 in (10 cm), and irregular contacts are

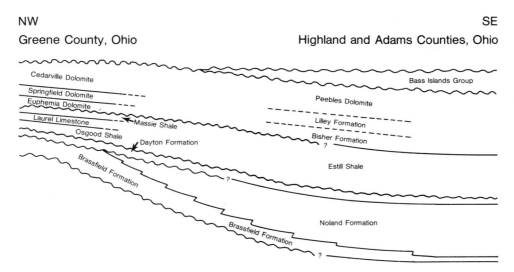

Figure 3. Correlation of Silurian outcrop stratigraphy from Greene County to Highland and Adams counties, Ohio. Note that the thicknesses decrease and that unconformities are more common in the Greene County area, which was situated on the middle of the Cincinnati Platform (modified from Horvath and Sparling, 1967).

present between beds of otherwise uniform thickness. Crinoids, brachiopods and corals are present.

Massie Shale

The Massie Shale is a dark, blue-gray shale 5.6 ft (1.7 m) thick. Silty layers are rare, and rare fossils include crinoids, bryozoans and brachiopods. Foerste (1935) reported a fauna of a crinoid, brachiopods, a gastropod and a trilobite from the Massie. The Massie may weather to form small caves with the Euphemia as a roof.

Euphemia Dolomite

The Euphemia Dolomite is a mottled, light-gray dolomitized echinoderm packstone with limonite staining. Some limonite staining occurs as botryoidal infilling, and some dolomite occurs as euhedral rhombs. The Euphemia varies from 5.2 to 8.5 ft (1.6 to 2.6 m) in thickness at John Bryan State Park. Four 3.9 in (10 cm) thick beds are present at the base of the Euphemia. The remainder of the formation is a massive, vuggy-weathering dolomite.

The Euphemia Dolomite is the lowest of three units that form the highwall of the gorge. Normal marine fossils are present, and a discontinuous zone of *Pentamerus* brachiopods is present near the contact between the Euphemia and Springfield dolomites.

Springfield Dolomite

The Springfield Dolomite varies in thickness from 6.2 to 7.2 ft (1.9 to 2.2 m) in the State Park area. It is a grayish-orange

dolomicrite or a finely crystalline dolomie. Irregular patches with larger euhedral dolomite rhombs are present. On outcrop the Springfield weathers as a recess between the Euphemia and Cedarville dolomites. The Springfield is even bedded with bed thickness averaging approximately 3.9 in (10 cm). Rare ostracods and *Pentamerus* brachiopods are present, and an insoluble residue of quartz silt is high compared with the super- and subjacent formations (Stout, 1941; Elliott, 1984).

Cedarville Dolomite

As between the Euphemia and Springfield, a discontinuous zone of *Pentamerus* brachiopods is present near the contact of the Springfield and Cedarville dolomites. The Cedarville is a light gray to grayish-orange, mottled, very coarsely crystalline dolomite. It was probably an echinoderm grainstone originally. The lower 17.7 ft (5.4 m) of the Cedarville is exposed. Total thickness of the Cedarville in Ohio was estimated to have been as much as 100 ft (30 m) (Foerste, 1935). It is a massive, vuggy-weathering unit. Biomoldic porosity is well developed. A normal marine fauna is present in the Cedarville. Molds of brachiopods, cystoids, crinoids and (?)corals are present. Lithologically and faunally, the Cedarville was probably quite similar to the Euphemia.

GEOLOGIC HISTORY

Deposition on the Cincinnati Platform during the Silurian was greatly influenced by fluctuations in sea level. The Brassfield Formation was deposited as the initial Silurian transgression reached this area. Brassfield deposition began during the late early or early middle Llandoverian (Rexroad, 1967). The Belfast Member is a discontinuous unit at the base of the Brassfield. This lower member is an argillaceous, fine-grained carbonate disconformably separated from the remainder of the Brassfield. Apparently a minor regression caused the break within the Brassfield. Renewed deposition of the Brassfield was that of a very shallow-water carbonate environment with crinoidal grainstones predominating. A diverse fauna was supported in this environment, and small patch reefs were locally present.

Brassfield deposition was terminated by a regression. Evidence for this regression is from missing conodont zones (Rexroad, 1967) and from local truncation of Brassfield bedding. Again deposition of the Dayton Formation began as the seas transgressed across the Platform, and deposition ended with a regression. Normal marine conditions are indicated for the Dayton Formation.

The Osgood Shale was the result of a renewed transgression in the late Llandoverian. Deposition was continuous from the Osgood through the Cedarville Dolomite. Osgood Shale through Massie Shale deposition fluctuated between carbonates and fine clastics. Above the Massie, carbonate deposition prevailed. The fauna of the Osgood through Massie is sparse but indicates that normal marine conditions prevailed.

The Euphemia and Cedarville dolomites are lithologically very similar and are indicative of a normal marine offshore environment free from clastic influence (Elliott, 1984). These dolomites are separated by the Springfield Dolomite, which lacks a normal marine fauna. The Springfield is a dolomicrite or finely crystalline dolomite with scattered pentamerid brachiopods and ostracods. At both the upper and lower contacts of the Springfield are discontinuous zones of pentamerid brachiopods in life position. It is suggested that deposition of the Springfield was the result of a minor regression which, aided by the pentamerid brachiopod banks, resulted in restricted conditions (Elliott, 1984). Normal marine conditions returned with deposition of the Cedarville Dolomite.

ACKNOWLEDGMENT

Steven R. Hiller, Park Manager, John Bryan State Park, has allowed continued access to study rocks in the park. Without his cooperation, this report would not have been possible. Drafting was completed at Wright State University, and Janet Bishop typed the manuscript.

REFERENCES CITED

Berry, W.B.N., and Boucot, A. J., 1970, Correlation of the North American Silurian Rocks: Geological Society of America Special Paper 102, 289 p.

Elliott, M. D., 1984, Petrology of the Middle Silurian dolomites of Clark and Greene Counties, Ohio. [M.S. thesis]: Dayton, Ohio, Wright State University, 84 p.

Foerste, A. F., 1935, Correlation of Silurian Formations in southwestern Ohio, southeastern Indiana, Kentucky, and western Tennessee: Journal of the Scientific Laboratories of Denison University v. 30, p. 119–205.

Horvath, A. L., 1967, Relationships of Lower Silurian strata in Ohio, West Virginia, and northern Kentucky: Ohio Journal of Science, v. 67, 341–359.

—— 1969, Relationships of Middle Silurian strata in Ohio and West Virginia: Ohio Journal of Science, v. 69, p. 321–342.

Horvath, A. L., and Sparling, D., 1967, Silurian geology of western Ohio: Guide to the 42nd Annual Field Conference of the Section of Geology of the Ohio Academy of Science, University of Dayton, Dayton, Ohio, 25 p.

Janssens, A., 1977, Silurian rocks in the subsurface of northwestern Ohio: Ohio Geological Survey Report of Investigations 100, 96 p.

Laub, R. S., 1979, The corals of the Brassfield Formation (Mid-Llandovery; Lower Silurian) in the Cincinnati Arch region: Bulletins of American Paleontology v. 75, no. 305, 457 p.

Norris, S. E., 1950, The water resources of Greene County, Ohio: Ohio Department of Natural Resources, Division of Water Bulletin 19, 52 p.

Rexroad, C. B., 1967. Stratigraphy and conodont paleontology of the Brassfield (Silurian) in the Cincinnati Arch Area: Indiana Geological Survey Bulletin 36, 64 p.

Rexroad, C. B., Branson, E. R., Smith, M. O., Summerson, C., and Boucot, A. J. 1965, The Silurian formations of east-central Kentucky and adjacent Ohio: Kentucky Geological Survey, Series X, Bulletin 2, 34 p.

Rexroad, C. B., and Nicoll, R. S., 1972, Conodonts from the Estill Shale (Silurian, Kentucky and Ohio) and their bearing on multielement taxonomy: Geologica et Palaeontologica Sonderbuch v. 1, p. 57–74.

Stith, D. A., and Stieflitz, R. D., 1979, An evaluation of "Newberry" analysis data on the Brassfield Formation (Silurian), southwestern Ohio: Ohio Geological Survey Report of Investigations 108, 25 p.

Stout, W., 1941, Dolomites and limestones of western Ohio: Ohio Geological Survey Bulletin 42, 468 p.

Stout, W., VerSteeg, K., and Lamb, G. F., 1943, Geology of water in Ohio: Geological Survey of Ohio, Fourth Series, Bulletin 44, 694 p.

Hueston Woods State Park: Wisconsinan glacial stratigraphy in southwestern Ohio

David P. Stewart, *Department of Geology, Miami University, Oxford, Ohio 45056*
Barry B. Miller, *Department of Geology, Kent State University, Kent Ohio 44242*

Figure 1. Index map showing the location of the Oxford-Eaton region of southwestern Ohio.

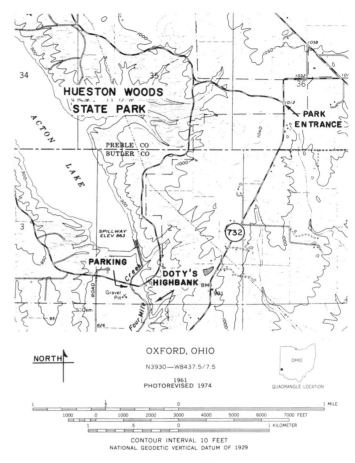

Figure 2. Portion of the Oxford (Ohio) quadrangle showing the location of Doty's Highbank.

LOCATION AND ACCESSIBILITY

The exposure described in this report is located in Hueston Woods State Park, in Butler County, southwestern Ohio. It is southwest of Eaton, which is 7 mi (12 km) south of Interstate 70 on U.S. 127, and it is 25 mi (40 km) north of Cincinnati and I-71 and I-75 (Fig. 1). The site (Doty's Highbank) is located on the east side of Four Mile Creek, 3 mi (5 km) north of Oxford (Oxford 7½-minute Quadrangle).

Access to the site is by Ohio 732 (Fig. 2). Turn west on Park Road and continue for about 0.5 mi (0.8 km) to a junction, turn left (south), and follow this road until it crosses Four Mile Creek. Park southwest of the bridge, walk across the bridge, and follow the east side of the creek downstream 750 ft (230 m) to a high stream-cut exposure.

SIGNIFICANCE OF SITE

Although the terminus of Wisconsinan glaciation in southwestern Ohio is 10 mi (16 km) south of Oxford, three surface tills and three subsurface tills of Wisconsinan age have been identified in the Oxford-Eaton region (Fig. 3). The sequence at Doty's Highbank shows multiple tills separated by interstadial deposits and includes three of the Wisconsinan tills. This classic exposure is probably the most studied of any in southwestern Ohio, and it is still the most confusing (Goldthwait and others, 1981). The site is one of the best dated (6 radiocarbon-dates) Pleistocene sequences in the state and represents one of the few localities where fossiliferous deposits representing the New Paris, Sidney, and Connersville interstages can be viewed in one section.

DESCRIPTION

Doty's Highbank is a stream-cut exposure approximately 75

WISCONSINAN	Camden	Knightstown Phase
		Crawfordsville Interphase
		Crawfordsville Phase
GLACIAL	Stade	Shelbyville Interphase
		Shelbyville Phase
	Connersville Interstade	
STAGE	Fayette Stade	
	Sidney Interstade	
	Fairhaven Stade	
	New Paris Interstade	
	Whitewater Stade	
SANGAMON INTERGLACIAL STAGE	WEATHERING	
ILLINOIAN GLACIAL STAGE	Richmond Stade	
	Abington Interstade	
	Centerville Stade	

Figure 3. Subdivisions of the Wisconsinan and Illinoian stages in southwestern Ohio.

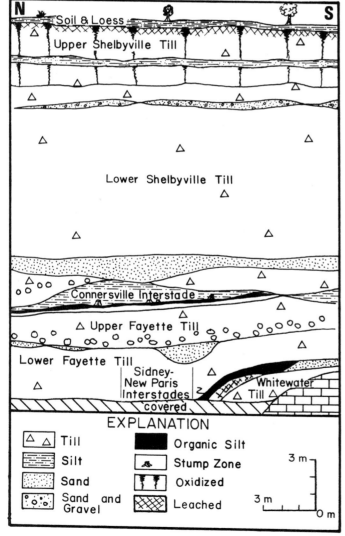

Figure 4. Field sketch of Doty's Highbank.

ft (23 m) high. The lowest part of the section is immediately upstream from limestone bedrock exposed at stream level (Fig. 4). The glacial sequence includes the following units:

Unit	Thickness in meters	Description (See Figure 4)
7	.6	Soil and loess, reddish brown, leached, and oxidized.
6	1.1- 2.1	Till, tan to brown, upper .6-.9 m (24-35 in), leached oxidized. Loosely compact, sandy texture. Gray mottling in large joints. N.E. fabric.
	.3	Silt and sand, buff, calcareous.
	1.5- 1.8	Till, brown, calcareous, oxidized, sand lenses. N.E. fabric. (Upper Shelbyville Till)
5	1.5	Till, gray-blue, calcareous, compact, clay rich. N.W. fabric.
	.3- .6	Coarse sand and gravel, rust, calcareous, poorly sorted.
	10.7-12.2	Till, gray-blue, calcareous, many sand lenses. N.W. fabric.
	0.0- 1.5	Clay, sand, and gravel, thin clay layer at the top. Buff sand and gravel below.
	1.8- 2.1	Till, brown to gray, calcareous, blocky structure. Wood from the base of the unit dated 20,500 ± 420 B.P. (I-10, 184). N.W. fabric. (Lower Shelbyville Till)
4	.9- 1.5	Sand, silt, and gravel, variable thickness and composition. One section contains calcareous lake sediments and flow till. Unit contains the stump zone that has been dated at 20,210 ± 210 B.P. (ISGS-761). Snails occur in organic layers. (Connersville Interstadial deposits)
3	2.1	Till, greenish-gray, calcareous, sandy matrix. Well jointed. Boulder concentration of flaggy limestone occurs at the bottom. N.W. fabric.
	3.0	Till, gray, calcareous, exposed surface weathers to reddish-brown. Bouldery at the base. Two wood samples from the contact with interstadial deposits below dated 21,070 ± 100 B.P. (ISGS-604) and 21,350 ± 60 B.P. (QL-1373). Wood collected 46 cm (18 in) above base dated 21,250 ± 440 B.P. (I-10, 185). N.E. fabric. (Fayette Till)
2	.6- .9	Organic silts, gray to black, calcareous, grades into sand toward the south end of the exposure. Contains molluscs, wood, and fine organics. Wood from the top of the unit dated 21,500 ± 60 B.P. (QL-1372).
	.3	Layer containing till lenses, leached.
	.3- .6	Silt and clay, gray to black, partially leached near top. (Sidney[?] and New Paris Interstadial deposits)
1	.9- 1.5	Till, blue-gray, compact, calcareous. Well-developed joints. N.W. fabric. (Whitewater Till)

When Gooding (1963) identified and named the Shelbyville Till, he noted that the ice that deposited it came from both the northeast and northwest. He believed that the ice invasions were contemporaneous and that the two tills therefore never occurred one above the other. Goldstein (1968), however, found the two tills occurring one above the other in the same exposure and designated them the Upper and Lower Shelbyville. The two tills are exposed at this location. The great thickness of the Lower Shelbyville is believed to be the result of an actively fluctuating ice margin and the termination of the ice against the valley wall.

TABLE 1. MOLLUSCUS FROM DOTY'S HIGHBANK

Species	Connersville	Sidney-New Paris	Habitat
Carychium exile*	3[++]	1[++]	1
Cionella lubrica*	2	1	1
Columella alticola[+]	–	4	1
Discus cronkhitei*	1	6	1
Euconulus fulvus*	–	1	1
Fossaria parva*	2	2	1
Gastrocopta pentadon*	1	2	1
Hendersonia occulta[+]	1	5	1,2
Steotrema leai*	1	1	2
Succineids*	15	15	?
Vallonia "albula"[+]	1	–	2
Vertigo modesta[+]	5	3	1
Vertigo elatior*	2	2	1
Vertigo alpestris oughtoni[+]	2	1	1
Vertigo gouldi paradoxa[**]	–	2	?
snail egg	1	–	?

Note: Numerals in Habitat column refer to: (1) swamp, damp lowland close to water; moist areas available all year, under twigs, bark, rocks; (2) moderately drained, moist to seasonally dry sites beneath bark, twigs, and logs.
* = now living in this area of the state.
[+] = not now living in the state.
[**] = extinct.
[++] = number of individuals.

The Fayette Till at this site was identified on the basis of Carbon[14] dates above and below it. Part of the difficulty with past correlations (and there have been many) was the occurrence of an upper Fayette till unit with northwest fabric and a boulder concentration at its base. The unit is unique to this particular locality. It was formerly believed that the lower Fayette till, and therefore the units below it, were Illinoian because of its bright reddish color. Recent studies have shown that the bright color was a surface weathering phenomenon that has subsequently been eroded (Oldfield, 1977; Stewart and Oldfield, 1978).

The interstadial deposits between the Whitewater and Fayette tills are believed to represent a long interval of deposition and weathering that spans both the New Paris and Sidney interstades. A weathered zone near the base of the unit shows weathering intensity equal to, or greater than, the Sidney profile at Sidney, Ohio (Pritchard, 1980; Forsyth, 1965). The intervening Fairhaven Till is not present inasmuch as the terminus of that ice was 10 to 12 mi (16 to 19 km) to the north.

The New Paris–Sidney interstadial deposits contain water and, as a result, the Fayette Till invariably slumps over the units below. It is, therefore, usually necessary to dig off the slump to expose the New Paris interstadial deposits and the Whitewater Till.

The wood from the stump zone (Connersville Interstadial), which has been identified as *Picea,* is of particular interest in that it frequently occurs *in situ* as rooted stumps that extend upward into the overlying till. The base of this till has zones of fragmented wood that extend in the down-ice direction from the stumps, implying that they were broken off by the advancing ice (Goldthwait and others, 1961).

Two small assemblages of predominantly terrestrial gastropods (Table 1) have been identified from screen-washed matrix recovered from units 2 and 4. As is often the case with Wisconsinan interstadial molluscan assemblages collected from this region (Forsyth, 1965; LaRocque, 1966–70; Wayne, 1963; Nave, 1969), the two assemblages include an association of species (e.g., *Vertigo alpestris oughtoni* and *Stenotrema leai*) for which there are no known modern analogs.

The two assemblages are quite similar and include several "cool climate" species that now reach the southern limits of their range in the Great Lakes region near the north shore of Lake Superior. These taxa occur together with species that now have distributions that include southern Ohio and that now approach the southern range limits of the "cool climate" species group in the area around the north shore of Lake Huron. The southern range limits of the "cool climate" species appear to be limited by the summer temperature, whereas the species whose distributions now include the study area have their northern range limited by the length and severity of the winter. Based on these modern distribution patterns the two assemblages appear to have lived at a time that combined summer-month temperatures (June-July-August) of about 59°F (15°C) and a summer-month precipitation of about 9 in (232 mm), similar to what now occurs along the north shore of Lake Superior (NOAA, 1979). The patterns also imply a longer frost-free growing season of about 130 days, similar to that which now occurs along the north shore of Lake Huron (Fremlin, 1974, p. 51–52).

REFERENCES CITED

Forsyth, J. L., 1965, Age of the buried soil in the Sidney, Ohio area: Geological Society of America Bulletin, v. 68, p. 1728.

Franzi, D. A., 1980, The glacial geology of the Richmond–New Paris region of Indiana and Ohio (M.S. thesis): Miami University, Oxford, Ohio, 134 p.

Franzi, D. A., and Stewart, D. P., 1980, The Sidney interstade and Early-Middle Wisconsinan history of the Miami Sublobe (abstract): Geological Society of America, Abstracts with Programs, v. 12, no. 5, p. 226.

Fremlin, G. (ed.), 1974, The National Atlas of Canada: 267 p., MacMillan, Toronto.

Goldstein, F. R., 1968, The Pleistocene geology of a portion of Butler County, southwestern Ohio (M.S. thesis): Miami University, Oxford, Ohio, 102 p.

Goldthwait, R. P., Durrell, R. H., Forsyth, J. L., Gooding, A. M., and Wayne, W. J., 1961, Pleistocene geology of the Cincinnati region (Ohio, Kentucky, and Indiana): Guidebook for field trips, Cincinnati meeting, Geological Society of America, p. 58–98.

Goldthwait, R. P., Stewart, D. P., Franzi, D. A., and Quinn, M. J., 1981, Quaternary deposits of southwestern Ohio: Geological Society of America, Guidebook for field trips, Cincinnati meeting, v. 3, p. 409–432.

Gooding, A. M., 1963, Illinoian and Wisconsinan glaciations in the Whitewater basin, southeastern Indiana and adjacent areas: Journal of Geology, v. 71, p. 665–682.

——1975, The Sidney Interstadial and Lake Wisconsinan history in Ohio and Indiana: American Journal of Science, v. 275, p. 993–1011.

Larocque, A., 1966–70, Pleistocene mollusca of Ohio: Ohio Geological Survey, Bulletin 62, Parts 1-4, 800 p.

Nave, F. R., 1969, Pleistocene mollusca of southwestern Ohio and southeastern Indiana: [Ph.D. thesis]: Ohio State University, 159 p.

NOAA, 1979, World weather records, 1961-1970, v. 1, North America, 290 p., National Climatic Center, Asheville.

Oldfield, J. H., 1977, A study of the Sidney interstadial weathering profiles and Middle Wisconsinan till in portions of Preble and Butler counties, southwestern Ohio (M.S. thesis): Miami University, Oxford, Ohio.

Pritchard, G. F., 1980, Trace element studies of glacial deposits in southwestern Ohio (M.S. thesis): Miami University, Oxford, Ohio.

Stewart, D. P., and Oldfield, J. H., 1978, The Middle Wisconsinan of southwestern Ohio (abstract): Geological Society of America, Abstracts with Programs, v. 10, no. 6, p. 285.

Wayne, W. J., 1963, Pleistocene formations in Indiana: Indiana Geological Survey Bulletin 25, 85 p.

Copperas Mountain: Black shale facies of the Upper Devonian Ohio Shale in south-central Ohio

Ernest H. Carlson, Department of Geology, Kent State University, Kent, Ohio 44242

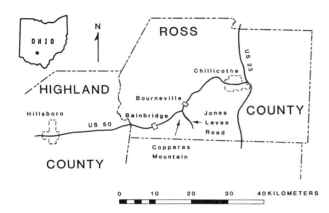

Figure 1. Location map showing the route to the Copperas Mountain area, Ross County, Ohio.

LOCATION AND ACCESSIBILITY

Copperas Mountain is located on the south side of Paint Creek Valley in southwestern Ross County, south-central Ohio (Fig. 1). Erosion by Paint Creek has produced a steep face along the northwestern side of this northerly trending ridge that forms one of the most remarkable exposures in the state (Fig. 2). The site is reached by traveling southwest on U.S. 50 from Bourneville for 1.0 mi (1.6 km), turning south on Jones Levee Road for 0.9 mi (1.5 km), continuing west on Spargursville Road for 1.5 mi (2.2 km), turning west and following Orms Road and then Storm Station Road for 0.7 mi (1.2 km), and finally bearing sharply west onto a narrow township road for 0.7 mi (1.2 km) (Figs. 1, 2). Vehicles should be parked at the side of the road just north of the cliff, as the road at the foot of the cliff is narrow and treacherous. The cliff area is particularly hazardous in the spring and after periods of rain, due to slick rock and rock falls. The road provides access to the lower portions of the section, while trails lead to the upper parts of the face.

SIGNIFICANCE OF LOCALITY

Erosion by Paint Creek has produced a spectacular cliff exposure of the Upper Devonian black shale facies at Copperas Mountain. The Ohio Shale here is equivalent to the Chattanooga and New Albany Shales, which were deposited over widespread portions of the Appalachian and Illinois Basins during the Devonian. These black shales are characterized by abundant organic matter and are of economic interest as a potential source of hydrocarbons.

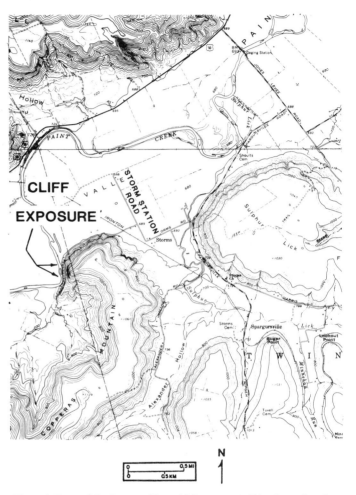

Figure 2. Parts of the Bourneville and Morgantown 7½-minute Quadrangles (1961) showing the location of Copperas Mountain.

STRATIGRAPHY AND LITHOFACIES

The exposure at Copperas Mountain includes the Ohio Shale, Bedford Shale, and Berea Sandstone. With the base of the formation lying below Paint Creek, the Ohio Shale comprises the vertical part of the cliff and three-fourths of the section (Fig. 3). It consists of two major units, the Huron (lower) Member and the Cleveland (upper) Shale Member. At Copperas Mountain these members are separated by the Three Lick Bed, a thin unit consisting of three beds of green mudstone and interlayered black shale

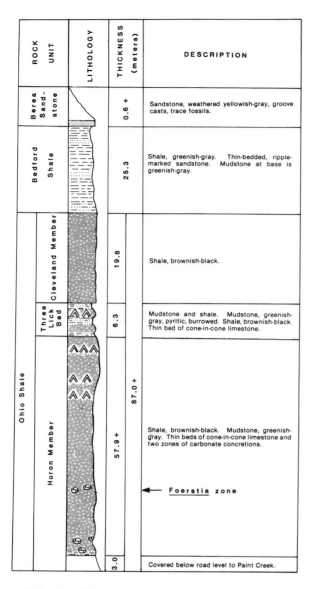

Figure 3. Stratigraphic section for Copperas Mountain at Paint Creek. The section is modified from Kepferle and Roen (1981).

(Provo and others, 1977; Kepferle and Roen, 1981). The Three Lick Bed, which can be spotted easily along the cliff from U.S. 50, thickens to the northeast where it correlates with the Chagrin Shale.

Lithologies of the Ohio Shale are dominated by black shale, with minor green mudstone present locally. The shale is brownish-black to black, finely laminated, and fissile and weathers locally into a series of thin ledges and recesses. The mudstone is greenish-gray and massive and weathers easily, resulting in recesses between ledges of black shale in the cliff exposure. Three thin beds of cone-in-cone limestone occur in the Huron Member and one is present in the Three Lick Bed. Two zones of dolomitic concretions occur within the Huron Member, the most prominent

zone appearing at road level. Compaction of the shale is notable around the concretions, which vary from 1 to 6 ft (0.3 to 2 m) in diameter. They are nearly spherical in shape, but slightly flattened in the plane of bedding. The larger concretions are characterized by a resistant shell and a soft, septarian core that weathers readily, leaving a hollow interior.

The black shales of the Appalachian Basin typically contain 15–20 percent organic matter (by weight), 25–50 percent clay (chiefly illite), 10–15 percent pyrite, and 20–25 percent quartz. The green mudstones, on the other hand, average about 5 percent organic matter, 80 percent clay, and 5–10 percent quartz (Conant and Swanson, 1961; Ettensohn and Barron, 1981; Maynard, 1981; Roen, 1981). The organic matter, which has both marine and terrestrial components, is composed chiefly of spores, algae, woody material, opaque macerals, and organic films that are disseminated through the rock. *Foerstia*, a fossil alga that has a restricted stratigraphic range, is found in loose rock from the Huron Member near the base of the cliff (Provo and others, 1977).

The organic carbon content, which determines the amount of hydrocarbons that can be extracted from the rocks, increases as the color of the shale darkens. Although no commercial gas production has been reported in the vicinity (Potter and others, 1982), notable yields of oil have been obtained experimentally by retorting black shales from Copperas Mountain (Krumin, 1951). Assays of 30 samples ranged from 1.2–8.3 gallons of oil per ton and averaged 3.6 gallons per ton, with highest yields obtained from the upper 20 ft (6 m) of the Cleveland Member. The black shales also show relatively high concentrations of radioactive elements and heavy metals. The greatest enrichments of these elements occur in shales having the highest organic content due to the ability of organic matter to scavenge metals.

The Huron Member also contains considerable pyrite, which occurs as silt-size grains and as flattened nodules up to 6 in (15 cm) in diameter. Copperas Mountain received its name from the prominent accumulations of melanterite (copperas) and other efflorescences, like halotrichite-pickeringite, which result from the weathering of this pyrite. Veinlets of coarsely crystalline barite, calcite, dolomite, ferroan dolomite, and quartz occur as septa fillings in the large concretions (Carlson, unpublished manuscript).

The Bedford Shale, which overlies the Cleveland Member, can be observed near the top of the cliff. The contact, well-exposed in a slump scar, is sharp but conformable. The Bedford section consists of poorly exposed greenish-gray shale and mudstone as well as thin-bedded siltstone and sandstone. Typical sandstone structures include ripple marks and sole marks. The cliff is capped by the Berea Sandstone, which conformably overlies the Bedford Shale, weathers yellowish gray, is thin bedded, and displays sole markings.

PALEOENVIRONMENT OF THE BLACK SHALES

The paleogeographic conditions that existed during the Upper Devonian were unique for the development of the black

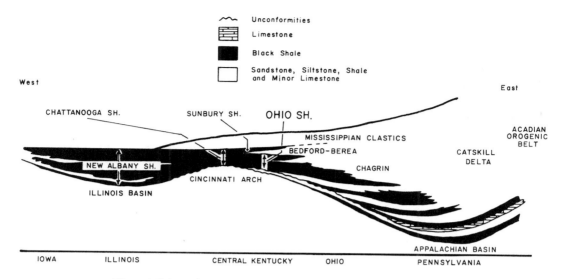

Figure 4. Schematic cross section showing the distribution of black shales and related facies from Pennsylvania to Iowa. Not drawn to scale. Modified from Ettensohn and Barron (1981).

shales (Fig. 4). A regional unconformity of low relief separates Middle Devonian and older rocks from those of younger age in the east-central United States. Beginning in the late Middle Devonian, subsidence on the cratonic side of the uplifted Acadian belt produced the Appalachian Basin, which extended from New York to Tennessee (Gutschick and Moreman, 1967; Ettensohn and Barron, 1981). During the Upper Devonian, the Ohio Shale was deposited in the west-central part of this basin. The Catskill Delta, which received its detritus from the Acadian uplift, bordered the basin on the east, and the Cincinnati Arch bordered it on the west. The center of the basin, as measured by the greatest thickness of strata, was located in Pennsylvania and West Virginia. Isotopic studies of carbon show that the terrestrial organic matter in the shales originated from the east side of the basin and suggest that the Cincinnati Arch was submerged and not a source of sediment (Maynard, 1981). Thus, the nature of the sediments represented by the Ohio Shale was determined in large part by the volume of clay and coarser clastics supplied to the Catskill Delta. The basinal (black shale) facies was deposited when the volume of detritus was low, allowing the organic matter to accumulate slowly with minimal dilution from clay. The deltaic marine (green shale) facies was formed when the amount of clay increased and the rate of sedimentation increased correspond-

ingly. The latter facies accumulated as the delta prograded westward. Thus, the black shale facies, which is characteristic of the western part of the Appalachian Basin, thins eastward, where it is replaced by much thicker deposits of green shale and coarser clastics. Similarly, the amount of organic carbon and, therefore, the potential for hydrocarbon production are greatest from shales in the western portion of the basin (Maynard, 1981; Potter and others, 1982).

The black shales were deposited in a basin where circulation of oxygen was restricted, and the green shale accumulated where circulation was free (Ettensohn and Barron, 1981). Also, warm (equatorial) seas would favor a high organic productivity. These basinal conditions are indicated by the nature of the fauna and the relative abundance of organic matter and pyrite in the shales. The black shales are characterized by numerous planktonic forms and few benthic fossils, denoting restricted conditions. Conversely, the fauna of the green shale is characterized by abundant and diverse benthic forms, attesting to an open environment. Large quantities of organic matter, which characterize the black shales, can be preserved only in a restricted environment. Similarly, significant accumulations of sedimentary pyrite, which distinguishes the black shales as well, could occur only if reducing conditions existed at the site of deposition.

REFERENCES CITED

Carlson, E. H., Minerals of Ohio: Unpublished manuscript submitted to the Ohio Geological Survey.

Conant, L. C., and Swanson, V. E., 1961, Chattanooga Shale and related rocks of central Tennessee and nearby areas: U.S. Geological Survey Professional Paper 357, 91 p.

Ettensohn, F. R., and Barron, L. S., 1981, Depositional model for the Devonian-Mississippian black shales of North America: A paleoclimatic-paleogeo-

graphic approach *in* Geological Society of America, Cincinnati 1981 field trip guidebooks, v. 2, p. 344–357.

Gutschick, R. C., and Moreman, W. L., 1967, Devonian-Mississippian boundary relations along the cratonic margin of the United States *in* International symposium on the Devonian system, Alberta Society of Petroleum Geologists, v. 2, p. 1009–1023.

Kepferle, R. C., and Roen, J. B., 1981, Chattanooga and Ohio Shales of the southern Appalachian Basin *in* Geological Society of America, Cincinnati 1981 field trip guidebooks, v. 2, p. 259–323.

Krumin, P. O., 1951, Some studies of Ohio coals and oil shales, pt. 1 of Further studies of Ohio coals and oil shales, Ohio State University Engineering Experiment Station Bulletin 143, p. 1–29.

Maynard, J. B., 1981, Some geochemical properties of the Devonian-Mississippian shale sequence *in* Geological Society of America, Cincinnati 1981 field trip guidebooks, v. 2, p. 336–343.

Potter, P. E., Maynard, J. B., and Pryor, W. A., 1982, Appalachian gas bearing Devonian shales: statements and discussions: Oil and Gas Journal, v. 80, p. 290–318.

Provo, L. J., Kepferle, R. C., and Potter, P. E., 1977, Three Lick Bed; useful stratigraphic marker in Upper Devonian shale in eastern Kentucky and adjacent areas of Ohio, West Virginia and Tennessee: U.S. Energy Research and Development Administration, Morgantown Energy and Research Center, Report MERC/CR-77-2, 56 p.

Roen, J. B., 1981, Regional stratigraphy of the Upper Devonian black shales in the Appalachian Basin *in* Geological Society of America, Cincinnati 1981 field trip guidebooks, v. 2, p. 324–330.

The Serpent Mound disturbance, south-central Ohio

Frank L. Koucky, Department of Geology, College of Wooster, Wooster, Ohio 44691
Stephen P. Reidel, Geosciences Group, Rockwell International, Richland, Washington 99352

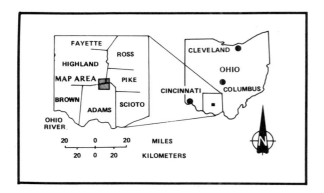

Figure 1. Index map of the Serpent Mound disturbance and vicinity. From Reidel and Koucky (1981), American Geological Institute, with permission.

LOCATION

The Serpent Mound disturbance is located at the common boundaries of Adams, Pike, and Highland counties in south-central Ohio (Fig. 1). The locality lies on the eastern edge of the Interior Low Plateau province and just west of the Appalachian Plateau. Serpent Mound State Memorial, a prehistoric Indian site of the Adena culture, is situated on the west side of the disturbance above Ohio Brush Creek.

Two paved highways serve the area; all other roads within the disturbance are gravel. Ohio 73 crosses the southwest part of the disturbance, providing access to the Serpent Mound State Memorial, and Ohio 41 passes through the east side of the disturbance (Fig. 2). Pertinent 7½-minute Quadrangles are the Byington and Sinking Spring sheets. Access to most geologic sites within the area is by foot. All land within the disturbance is privately owned, and permission must be gained to visit it. Landowners, however, understand the scientific significance of the area and have granted access graciously.

SIGNIFICANCE OF SITE

In 1838, John Locke of the Ohio Geological Survey noted that near the join of Adams, Highland, and Pike counties more geologic formations could be seen in a small disturbed area than at any other locality in Ohio. The uniqueness of the region was also noted by Walter Bucher (1933, 1936), who mapped and described the area in a classical geologic study. He ascribed the origin of the disturbance to endogenic processes and labeled it a cryptovolcanic structure after the Steinheim Basin. This study sparked a debate (Boon and Albritton, 1936) as to the possibility of an exogenic origin (astrobleme), a debate that continues today. Many similar structures occur throughout the world (Freeberg,

1966, 1969) and thus to remove genetic implications from the name the neutral descriptive terms disturbance and cryptoexplosion structure were introduced (Dietz, 1946).

The Serpent Mound disturbance provides a natural laboratory for scientific investigations into this type of feature. The accessibility, level of erosion, and exposure of the site and the presence of a detailed stratigraphy provide the tools necessary to examine the structure and interpret the geologic history (Bucher, 1933; Reidel, 1975; Reidel and Koucky, 1981; Reidel and others, 1982). In addition to the geologic significance, the Serpent Mound disturbance played an important role in Ohio's prehistoric Indian history, and it contains areas of original undisturbed prairie vegetation.

STRATIGRAPHY AND STRUCTURE

Upper Ordovician through Lower Mississippian bedrock is exposed in the disturbance with Illinoian glacial till occurring just north of the locality. Major streams within the disturbance also contain glacial outwash. The stratigraphy is summarized in Figure 3, and descriptions of the units can be found in Reidel and Koucky (1981) and Reidel and others (1982). An examination of the stratigraphy outside the disturbed area is advisable before attempting to recognize the disturbed rock.

The Serpent Mound disturbance is nearly circular in plan (Fig. 4) with a slight north-south elongation and a diameter between 4.5 and 5 mi (7.3–8.0 km). It covers approximately 16 sq mi (41 sq km) and formed between the Late Mississippian and the Pleistocene.

The broad three-fold structural subdivision of Serpent Mound—consisting of a central uplifted core, outer ring-graben, and intermediate transitional area—is an intricate part of other disturbances as well and was originally recognized here by Bucher (1933, 1936). The uplifted central core occupies about 9% of the surface area and has been folded and faulted into seven anticlines that radiate from the center of the disturbance. The anticlines are asymmetrical, with the south and east limbs having the steepest dips. Six grabens separate the anticlines on the south and east sides, with only faults on the west and north sides. One thousand ft (300 m) of uplift is estimated to have occurred here.

The intermediate transition area consists of faulted and folded rock having both concentric and radial orientations. This zone occupies about 19% of the surface area and is characterized by rock that has been neither greatly uplifted nor downdropped.

Beyond the transition area occurs a series of doubly plunging, generally faulted synclines and basins termed the outer ring-graben. This zone occupies approximately 71% of the surface area and is structurally lower than rock outside the disturbance.

The style of folding within the disturbance is primarily de-

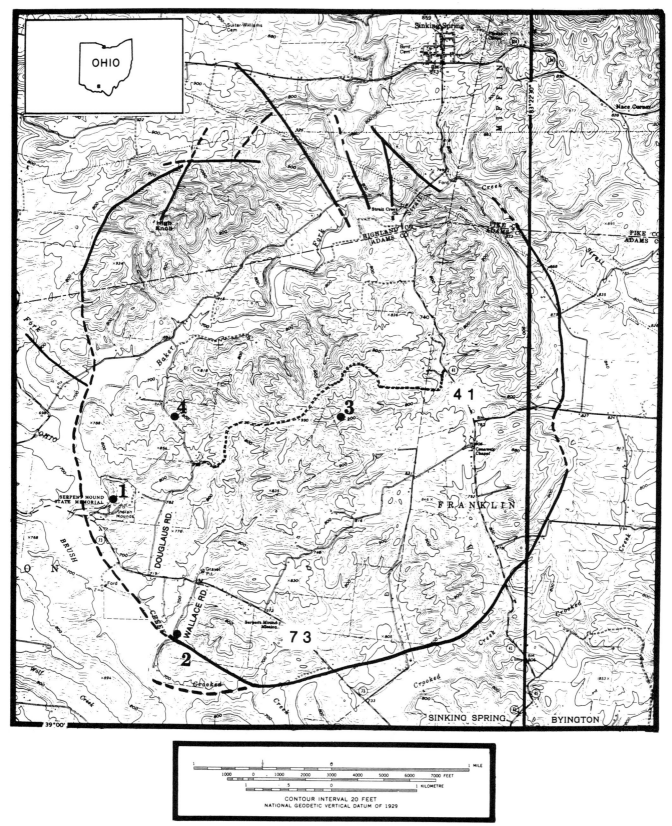

Figure 2. Detailed location map for the Serpent Mound disturbance.

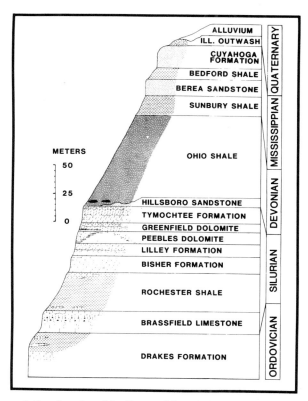

Figure 3. Stratigraphy of the Serpent Mound disturbance. From Reidel and Koucky (1981), American Geological Institute, with permission.

turbance and as a graben in the Plum Run quarry. The NW fault zone coincides with gravity and aeromagnetic anomalies that Janssens (1973) interpreted as the Grenville boundary. It is now recognized that the Grenville front occurs to the west (Lidiak and others, 1983), paralleling the Cincinnati Arch. A recent aeromagnetic anomaly map of southern Ohio (USGS, 1983) shows that the Serpent Mound disturbance lies on a NW-trending anomaly (Fig. 5) that farther north has been interpreted by Burke (1980) and Lidiak and others (1983) as a relic of a late Proterozoic rift complex. Although the exact nature of this zone is uncertain, its coincidence with the fault zone suggests an inherent zone of crustal weakness.

Any hypothesis for the origin of the disturbance must account for the structural characteristics, timing of geologic events, and regional geologic setting. Although no single explanation can satisfactorily account for all these features, an endogenic process rather than an exogenic process can best explain the Serpent Mound disturbance.

SITES

Four sites within the locality are described as an introduction to the geologic nature of this disturbance. These sites are readily accessible and demonstrate typical geologic characteristics of this feature. More complete descriptions of these sites can be obtained from Reidel and Koucky (1981).

Outer Ring-graben

Site 1 (Figs. 2 and 4): Serpent Mound State Memorial. Serpent Mound is the largest and finest Serpent effigy in the United States. It is an embankment of earth nearly a half kilometer long and represents a serpent in the act of uncoiling. The park is operated by the Ohio Historical Society, which maintains an excellent museum with both archaeological and geological displays.

The park is situated on dolomite of the Peebles, Greenfield, and Tymochtee formations that have been folded into a syncline. The syncline can be observed by walking along the nature trail that follows Ohio Brush Creek. The limbs of the syncline dip between 10 and 17 degrees. Zones of faulting and breccia are exposed locally, and some sphalerite can be found on the park grounds north of the museum. This area is not typical of the outer ring-graben because the amount of displacement is only about 240 ft (75 m), much less than elsewhere.

Site 2: Wallace Road Section. Wallace Road, 1 mi (1.6 km) southeast of Serpent Mound State Memorial along Ohio 73, is a gravel road that provides one of the best cross sections of the outer ring-graben. About 0.4 mi (0.6 km) south of Ohio 73, a dirt track leads from Wallace road up a dip slope (30 degrees N) of the Tymochtee Dolomite. The dolomite is shattered, and a set of fractures generally parallel to the strike shows offsets of less than 1 cm normal to the bedding. Brecciation increases to the south where the boundary fault juxtaposes

pendent upon the lithologies. The tectonically competent carbonate and sandstone deform by brittle deformation. Simple shear and cataclastic flow, the principal mechanisms, typically produce parallel and similar type folds. The tectonically incompetent shales behave as ductile masses producing a wide variety of folds.

Faulting is the principal deformational mechanism in the disturbance. Faults are typically both high-angle normal and reverse, with normal faults dominating. Fault contacts are either sharp or are marked by zones of breccia and gouge, many of which are cemented by calcite and less commonly by sphalerite. Faulting within the disturbance has a preferred NW-SE trend.

Shock features are rare in the Serpent Mound disturbance. Shatter cones are the only recognized feature, and they occur only in the central uplift.

A minor amount of sphalerite occurs within the disturbance, typically filling fractures, cementing breccia fragments, and replacing dolomites. The sphalerite was deposited in faults formed by the initial deformation and then brecciated by later deformations.

The Serpent Mound disturbance occurs within a complex regional geologic setting (Fig. 5). The disturbance lies near an inflection in the Precambrian basement and at the intersection of a NE-trending monocline and a NW-trending fault zone. The monocline has been interpreted to extend south to the West Hickman fault zone and north to connect with a basement fault. The NW-trending fault zone is best exposed adjacent to the dis-

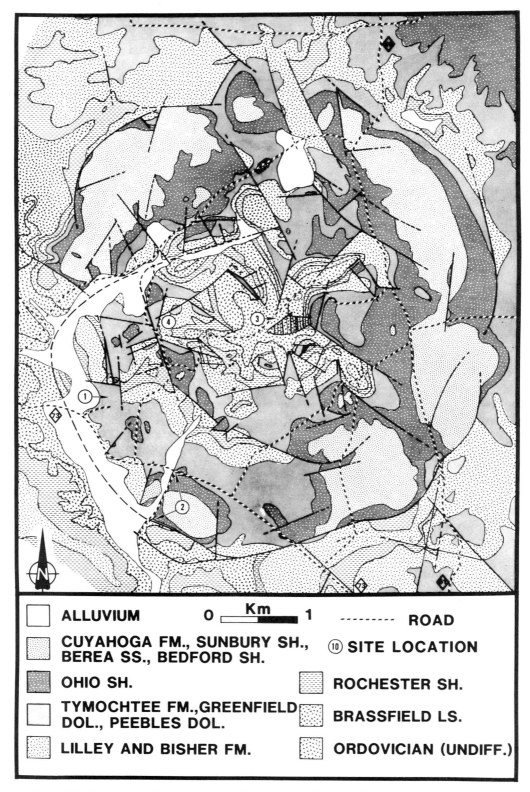

Figure 4. Geologic map of the Serpent Mound disturbance. Known or inferred faults marked by solid and dashed lines, respectively. From Reidel and others (1982), American Journal of Science, with permission.

nearly horizontal Brassfield Limestone. The fault surface is covered with clay, and an erosional gully to the east marks the trace of the fault.

Proceeding north along Wallace road, the Ohio Shale is observed overlying the Tymochtee Dolomite. Mississippian sandstone overlies the Ohio Shale in an abandoned quarry that also marks the axial trace of the ring-graben. The contact between the Ohio Shale and Tymochtee Dolomite is exposed farther north at the junction of Wallace road and Ohio 73. About 0.12 mi (0.2 km) north of Ohio 73 along a gravel road, the Peebles, Greenfield, and Tymochtee dolomites are exposed in a quarry. Lying along the quarry floor are blocks of fractured dolomite with small offsets similar to that near the boundary fault. Minor faults can also be seen in the quarry wall.

Central Uplift

The best access to the central uplift (Site 3) is by a gravel road that joins Douglaus road and Ohio 41. Easiest access is from Douglaus road, which is 0.4 mi (0.6 km) west of Site 2 on Ohio 73. The gravel road begins on the right about 1.4 mi (2.2 km) north of Ohio 73 (mileage 57.4 of Reidel and Koucky, 1981). Approximately 1.3 mi (2.1 km) from Douglaus road is the center of the disturbance. (Should this road be in poor repair, the road also joins Ohio 41 approximately 0.2 m (0.3 km) north of Dutch Thomas road. Access may also be obtained from the Bernard Brown ranch, Stop 5 of Reidel and Koucky, 1981.)

At the center of the disturbance, Upper Ordovician rock is poorly exposed. Shatter cones can be found in the Ordovician limestones and Brassfield Limestone. By walking from the center of the disturbance due east into a ravine toward a fence, one can observe, east of the fence, vertically standing Ordovician and Silurian rocks in one of the grabens that occurs between the radiating anticlines. The broad structure of the rock in the graben is that of a faulted anticline. A complete Silurian and Devonian section is exposed and locally brecciated near the junction with the next drainage to the north. At the junction of the two creeks, by crossing the fence to the north and walking up the drainage, one moves outside the graben and onto the southeast flank of one of the radiating anticlines. The creek leads down section into Silurian rocks. Where the creek divides, a small anticline of Brassfield Limestone (Fig. 7, Reidel and others, 1982) is exposed. This

Figure 5. Structural features of south-central Ohio. Modified from Reidel and Koucky (1981), American Geological Institute, with permission.

anticline appears to have behaved in a ductile fashion, but closer examination reveals that it formed by a combination of simple shear and cataclastic flow. Following the creek on the left leads back to the road.

Transition Area

By returning to Douglaus road and traveling about 0.4 mi (0.6 km) to the right, one encounters a small quarry in the Peebles, Greenfield, and Tymochtee dolomites (Site 4). It is this locality where Heyl and Brock (1962) first found sphalerite and documented evidence for several deformations in the area. The rock is part of a wide breccia zone along the principal fault that parallels the road here. Near the base of the quarry is a coarse, angular breccia, but mylonite to protomylonite comprise the upper part of the quarry. Away from this fault zone, the rock is gently folded.

REFERENCES CITED

Boon, J. D., and Albritton, C. C., Jr., 1936, Meteorite craters and their possible relationship to "cryptovolcanic structures": Field and Laboratory, v. 5, no. 1, p. 1–9.
Bucher, W. H., 1933, Ueber ein typische cryptovulkanische Storung im sudlichen Ohio: Geologisches Rundschau, v. 23a (Solomon-Calvi Festchr.), p. 65–80.
——1936, Cryptovolcanic structures in the United States: 16th International Geological Congress Report, v. 2, p. 1060–1064.
Burke, K., 1980, Intracontinental rifts and aulacogens, *in* Continental tectonics: Washington, D.C., National Academy of Sciences, p. 42–49.
Dietz, R. S., 1946, Geological structures possibly related to lunar craters: Popular

Astronomy, v. 54, no. 9, p. 465–467.
Freeberg, J., 1966, Terrestrial impact structures—A bibliography: U.S. Geological Survey Bulletin 1220, 91 p.
——1969, Terrestrial impact structures—A bibliography: U.S. Geological Survey Bulletin 1320, 39 p.
Heyl, A. V., and Brock, M. R., 1962, Zinc occurrence in the Serpent Mound structure of southern Ohio: U.S. Geological Survey Professional Paper 450-D, p. 95–97.
Janssens, A., 1973, Stratigraphy of the Cambrian and lower Ordovician rocks in Ohio: Ohio Geological Survey Bulletin 64, 197 p.

Lidiak, E. G., Ceci, V. M., Hinze, W. J., and McPhee, J. P., 1983, Tectonic framework of basement rocks in the eastern Midcontinent: Geological Society of America Abstracts with Programs, v. 15, no. 6, p. 627.

Locke, J., 1838, Report, *in* Second annual report on the geological survey of the State of Ohio: Geological Survey of Ohio, p. 203–274.

Reidel, S. P., 1975, Bedrock geology of the Serpent Mound cryptoexplosion structure, Adams, Highland, and Pike Counties: Ohio Geological Survey Report of Investigations 95.

Reidel, S. P., and Koucky, F. L., 1981, The Serpent Mound cryptoexplosion structure, southwestern Ohio, *in* Roberts, T. G., ed., Geological Society of America, Cincinnati, 1981, Field Trip Guide Books: American Geological Institute, v. 8, p. 391–403.

Reidel, S. P., Koucky, F. L., and Stryker, J. R., 1982, The Serpent Mound disturbance, southwestern Ohio: American Journal of Science, v. 282, p. 1343–1377.

U.S. Geological Survey, 1983, Aeromagnetic map of southern Ohio: U.S. Geological Survey Open-file Rept. 83-59, scale 1:250,000.

Index

[Italic page numbers indicate major references]

Typeset by WESType Publishing Services, Inc., Boulder, Colorado
Printed in U.S.A. by Malloy Lithographing, Inc., Ann Arbor, Michigan

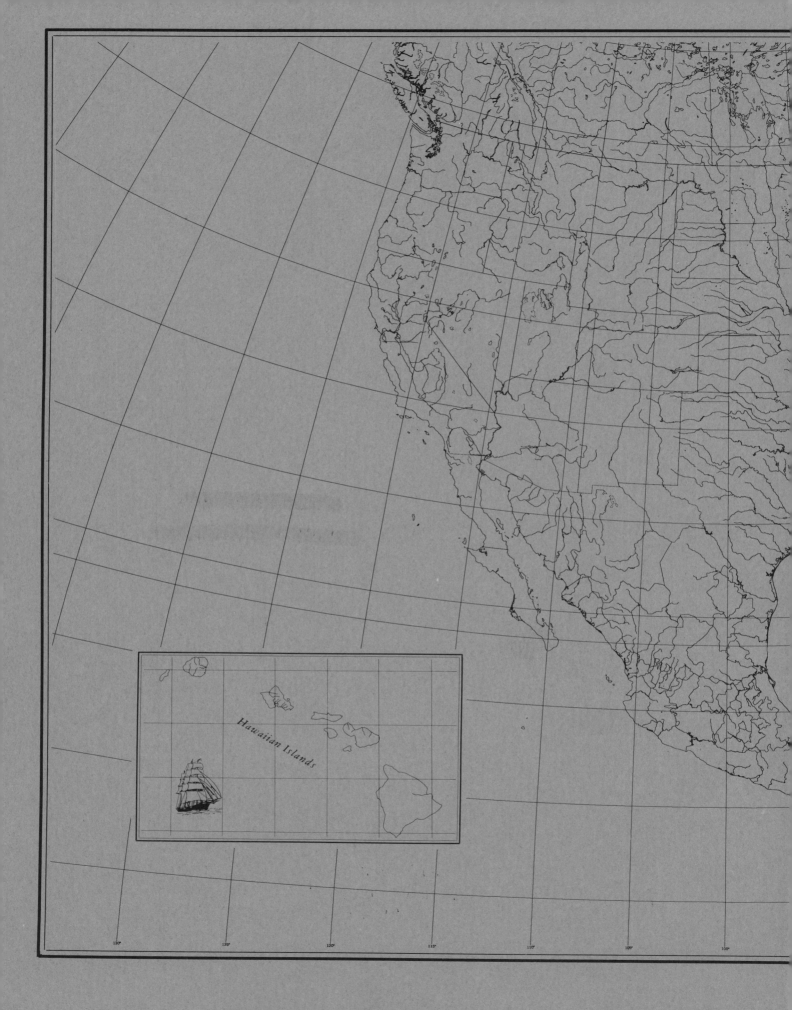

Hawaiian Islands